15th International Conference on Optical Methods of Flow Investigation (OMFI-2019)

Journal of Physics: Conference Series Volume 1421

Moscow, Russia
24 - 28 June 2019

ISBN: 978-1-7138-1326-2
ISSN: 1742-6588

Printed from e-media with permission by:

Curran Associates, Inc.
57 Morehouse Lane
Red Hook, NY 12571

Institute of Physics
Dirac House, Temple Back
Bristol BS1 6BE UK

Phone: 44 1 17 929 7481
Fax: 44 1 17 920 0979

techtracking@iop.org

Additional copies of this publication are available from:

Curran Associates, Inc.
57 Morehouse Lane
Red Hook, NY 12571 USA
Phone: 845-758-0400
Fax: 845-758-2633
Email: curran@proceedings.com
Web: www.proceedings.com

TABLE OF CONTENTS

TIME-RESOLVED INTERFERENCE MICROSCOPY FOR STUDYING NONIDEAL PLASMA FORMED BY HIGH-POWER FEMTOSECOND LASER PULSES 1
D S Sitnikov

APPLICATION OF HYPERSPECTRAL IMAGING FOR ANALYSIS OF EMBRYONIC DEVELOPMENT OF LOWER VERTEBRATES 10
A B Burlakov, S V Shirokov, D D Khokhlov, V I Kuzmin, A F Gadzaov

NUMERICAL SIMULATION OF SUPERSONIC FLOW AROUND A CONE IN A THREE-DIMENSIONAL FORMULATION IN THE SOFTWARE PACKAGE OPENFOAM 16
A E Kuvshinnikov

THE STUDY OF VORTEXES BASED ON THE MULTI COLORED PARTICLE IMAGE VELOCIMETRY TECHNIQUE 21
S S Usmanova, N M Skornyakova, A V Anikeev, O A Evtikhieva

APPLICATION OF COMPUTER VISION AND DEEP LEARNING FOR FLAME MONITORING AND COMBUSTION ANOMALY DETECTION 27
S Abdurakipov, E Butakov

EXPERIMENTAL OBSERVATION OF TURBULENT EXCHANGE BETWEEN HETEROGENEOUS VORTEX RING AND SURROUNDING MEDIUM BY THE SHADOWGRAPH METHOD 38
V V Nikulin, R A Panenko

INVESTIGATION OF TEMPERATURE FIELD OF WEAKLY IONIZED GAS JET BY THE BACKGROUND ORIENTED SCHLIEREN METHOD 43
I I Korotkikh, Yu I Malakhov, A V Kuchmenko

THE EXPERIMENTAL INVESTIGATION OF WATER SPRAY DISPERSE PARAMETERS BEHIND THE WATER INJECTING SYSTEM AT THE AIRCRAFT ENGINE INLET BY APPLYING PSV 51
A A Loginova, O G Chelebyan, B I Mineev, A Yu Vasilyev

RESEARCHES OF TWO-PHASE STREAM BY METHODS OF REGISTRATION OF FLUORESCENCE OF DROPS OF LIQUID AND SHADOWGRAPH 58
O G Chelebyan, A Y Vasilyev, A A Sviridenkov, A A Loginova

OPTICAL STUDY OF THE CONCENTRATION AND PARTICLE SIZE DISTRIBUTION OF OCTADECYLAMINE IN WATER 65
N M Skornyakova, A V Anikeev, V A Parshin

DIAGNOSIS OF LOW-SPEED FLOWS BY PARTICLE SHADOW VELOCIMETRY METHOD 72
P T Khant, V Yu Barinov

THE CHOICE OF AN IMAGE PROCESSING ALGORITHM FOR INCREASING SENSITIVITY OF THE SURFACE PLASMON RESONANCE METHOD 76
I N Pavlov, A V Vedyashkina, G M Yanina

IN-PLANE INTERFEROMETRY OF TERAHERTZ SURFACE PLASMON POLARITONS 82
A K Nikitin, O V Khitrov, V V Gerasimov, I Sh Khasanov, T A Ryzhova

LONG-FOCUS LOW-COST LENS DEIGNED FOR BIOMEDICAL HYPERSPECTRAL CAMERA 93
V I Batshev, A E Gavlina

APPLYING PIV TO STUDY A FLUID FLOW IN THE VICINITY OF A CIRCULAR STREAMLINED CYLINDER 97
K G Dobroselsky

DEVELOPMENT OF A CYLINDRICAL ACOUSTIC LENS TO CONTROL THE AMPLITUDE AND PHASE CHARACTERISTICS OF OPTICAL RADIATION 103
A N Bykhanov

PROSPECTS OF APPLYING THE ELASTIC LIGHT SCATTERING EFFECT FOR THE DEVELOPMENT OF AN OPTICAL METHOD TO THE COLLOIDAL SOLUTIONS DIAGNOSITIC 109
M V Sapronov, N M Skornyakova

APPLICATION OF COMPUTER VISION AND NEURAL NETWORK ANALYSIS TO STUDY THE STRUCTURE AND DYNAMICS OF TURBULENT JETS 118
S Abdurakipov, M Tokarev, E Butakov, V Dulin

PECULIARITIES OF THE ACOUSTO-OPTICAL INTERACTIONS OF BESSEL LIGHT BEAMS IN CRYSTALS 127
V N Belyi, G V Kulak, G V Krokh, P I Ropot, O V Shakin

MODELING PICTURES OF THE FLOW PAST ROUGH AND SMOOTH SURFACES IN THE PROGRAM COMSOL MULTIPHYSICS 134
Win Thu, D V Ilin

COMPACT INTERFEROMETER FOR PRECISE SHAPE TESTING OF LARGE-SIZE CONVEXASPHERICAL MIRRORS 143
A E Gavlina, D A Novikov, M V Sergeeva

APPLICATION OF ACOUSTIC EMISSION AND LASER OPTOACOUSTICS AT VARIOUS STAGES OF DEFECT FORMATION DURING FRICTION STIR WELDING 147
K Stepanova, I Kinzhagulov, Y Yakovlev

OPTICAL-ACOUSTIC DIAGNOSTICS OF CRACKS ON THE SURFACE OF SOLID BODIES 157
N S Kazak, G V Kulak, T V Nikilaenko, A G Matveeva, O V Shakin

SEPARATE BOS-VISUALIZATION OF TEMPERATURE AND PRESSURE DISTRIBUTIONS IN A SOLID MEDIUM 161
D G Sychev

DEVELOPMENT OF THE SYSTEM FOR QUANTITATIVE GAS-LIQUID FLOW VISUALIZATION 168
A V Kuchmenko, D A Koshelev, N M Skornyakova, P P Shcherbakov

MICRO-PTV TECHNIQUE APPLICATION TO VELOCITY FIELD MEASUREMENTS IN IMMISCIBLE LIQUID-LIQUID PLUG FLOW IN MICROCHANNELS 175
A V Kovalev, A A Yagodnitsyna, A V Bilsky

ELIMINATION OF DISTORTIONS ON IMAGES IN THE EXISTENCE OF TURBULENCE 180
S Yu Belov

LASER DIAGNOSTICS OF THE INSTABILITY ONSET IN THE CONFINED VORTEX FLOWS 188
I V Naumov, M A Tsoy

HIGH SPEED VISUALIZATION OF SELF-OSCILLATION OF AXISYMMETRIC CAVITATING CAVITY IN A SQUARE VORTEX CHAMBER 195
S G Skripkin, M A Tsoy, P A Kuibin, Zhigang Zuo, Shuhong Liu

VISUALIZATION OF PLASMA FLUXES BY A MEASUREMENT OF OWN ROTATION OF A DUST PARTICLE 202
E S Dzlieva, V Yu Karasev, I R Krylov, I Ch Mashek, L A Novikov, S I Pavlov

COMPENSATION OF SPECTRAL IMAGE SHIFT IN AOTF-BASED SYSTEM 208
A A Bykov, P V Zinin, K M Bulatov, D D Khokhlov, I B Kutuza

A REVIEW OF NON-POLAR LIQUIDS AS MATERIALS FOR BULK ACOUSTO-OPTIC DEVICES OPERATING WITH TERAHERTZ RADIATION 212
P Nikitin

SPECIAL FEATURES OF LASER DIAGNOSTICS OF GAS FLOWS IN A THICK-WALLED GLASS PIPE 219
K M Lapitskiy, D A Poyatsyka, I A Lapitskaya, A Yu Poroykov, N S Ilina

DEVELOPMENT OF A SERIES OF INTERCHANGEABLE OPTICAL TIPS FOR A VIDEO ENDOSCOPE 225
V I Batshev, A S Machikhin, A V Gorevoy, D D Khokhlov

MEASUREMENT OF HIGH-SPEED GAS-DROPLET FLOW CHARACTERISTICS BY PULSE PHOTOGRAPHY METHOD 229
N I Bokuchava, F M Inochkin

LED ILLUMINATION MODULE FOR FLUORESCENCE AND SPECTROSCOPIC STUDIES IN MICROSCOPY 239
V I Batshev, A A Blokhin, A S Belov

APPLICATION OF THE PARTICLE TRAJECTORY IMAGING FOR MODELLING DUSTY PLASMA LEVITATION ON THE MOON 243
A Yu Poroykov, S A Bednyakov, A V Zaharov, G G Dolnikov, A N Lyash, I A Shashkova, I A Kuznetsov

CALCULATION OF THE OPTICAL VARIOSYSTEMS WITH TUNABLE OPTICAL POWER LENSES 249
P A Nosov, D E Piskunov, M A Vinogradov, V O Tigaev, A A Yablokova

PHOTOTHERMOCAPILLARY DETECTION OF CONDUCTIVE TRACK RUPTURES ON A PRINTED CIRCUIT BOARD COATED WITH A PROTECTIVE FILM 257
A Yu Zykov, N A Ivanova

EFFECT OF THE NANOPARTICLE COMPOSITION ON THE FLUORESCENCE OF CARBON NITRIDE COATINGS 261
A B Bogomolov, P V Zinin, S A Kulakov, V A Kutvitsky, M F Bulatov, I B Kutuza

COMPUTER MODELING OF ACOUSTIC STANDING WAVES IN THE COOLANT OF NUCLEAR POWER PLANTS 266
K N Proskuryakov, A V Anikeev, E Afshar
DETERMINATION OF PARAMETERS OF THE COMPELLED CAPILLARY DISINTEGRATION OF LIQUID JETS 275
A V Boukharov, S I Kukanov
VISUALIZATION AND PIV-STUDY OF THE FORMATION OF A COOLING FILM WHEN FLOWING AROUND THE LEADING EDGE OF THE GAS TURBINE NOZZLE BLADE 283
S V Veretennikov, O A Evdokimov
ESTIMATING THE QUALITY OF STEREOSCOPIC ENDOSCOPIC SYSTEMS 290
A A Naumov, A V Gorevoy, A S Machikhin, V I Batshev, V E Pozha
ON COMPUTER MODELLING OF OPERATIONS, GEOMETRIC CHARACTERISTICS AND METHODS OF PULSED OPTICAL TOMOGRAPHY OF NUCLEAR FUEL MICROOBJECTS 302
E P Domoratskiy, T N Baybikova
HYPERSPECTRAL CAMERA APPLICATION FOR REMOTE GAS ANALYSIS 309
S M Neverov
STUDY OF CONVECTIVE STRUCTURES AND PHASE TRANSITION INDUCED IN A WATER LAYER BY NON-STATIONARY BOUNDARY CONDITIONS 315
Yu N Dubnishchev, V A Arbuzov, E V Arbuzov, V S Berdnikov, O S Zolotukhina, S A Kislitsin
RESEARCH OF THE DYNAMICS OF TWO-PHASE FLOWS BY OPTICAL METHODS PIV/IPI 323
A I Guryanov, S V Veretennikov, O A Evdokimov, M M Guryanova, K L Kalinina
INFRARED FIBER BASED ON AGCL-AGBR AND AGBR-TLI CRYSTALS TO TRANSFER THERMAL RADIATION IN PULSED 329
A S Shmygalev, A S Korsakov, B P Zhilkin, A Turabi
STUDY OF EFFECTS OF ELECTROMAGNETIC FACTORS ON THE EARLY DEVELOPMENT OF BIOLOGICAL ORGANISMS USING COMPREHENSIVE MULTISPECTRAL OPTICAL AND SCANNING ACOUSTIC MICROSCOPY APPROACH 334
A B Burlakov, D D Khokhlov, V L Domanskiy, S A Titov
APPLICATION OF THE STRUCTURED LASER ILLUMINATION TO PLANAR OPTICAL DIAGNOSTICS OF THE FLOWS WITH NONUNIFORM DENSITY 340
D Sharaborin, R Tolstoguzov, E Frolova
EXPERIMENTAL STUDY OF THE MULTIPHASE FLOW IN A PORE DOUBLET MODEL 347
Yu S Zamula, E S Batyrshin, R R Latypova, O A Abramova, Yu A Pityuk
DETAILED INVESTIGATION OF BAG-BREAKUP PROCESS IN THE CONTEXT OF SPRAY OF DROPLETS GENERATION WITHIN WIND-WAVE INTERACTION 356
A A Kandaurov, D A Sergeev, Yu I Troitskaya
THE METHOD OF COMPENSATION FOR LOCAL DISPLACEMENTS OF IMAGES OF CAPILLARIES IN THE EVALUATION OF CAPILLARY BLOOD FLOW PARAMETERS 363
M V Volkov, N B Margaryants, A V Potemkin, I P Gurov
MULTI-SPECTRAL PHASE IMAGING BASED ON ACOUSTO-OPTIC SELECTION OF LIGHT IN COMMON-PATH INTERFEROMETER 370
L Burmak, L Zykova, A Machikhin, B Chichigin
SPECTRAL OIL ANALYZER 375
A A Lashova, V S Korsakov, E A Korsakova, I A Kashuba, L V Zhukova
STUDIES BY PIV-METHODS OF SELF OSCILLATING HYDRODYNAMIC PROCESSES IN THE ELEMENTS OF PIPELINE 383
D A Sergeev, A A Kandaurov
TECHNICAL IMPLEMENTATION OF ACOUSTO-OPTICAL INSTRUMENTS: BASIC TYPES 389
V E Pozhar, M F Bulatov, A S Machikhin, V A Shakhnov
INSTALLATION FOR INVESTIGATION OF OPTICAL AND MECHANICAL SYSTEM OF FIBER OPTICAL SENSORS OF PARAMETERS OF LIQUID OR AIR FLOWS 397
E Badeeva, E Shachneva, T Murashkina, A Kukushkin
ANALYSIS OF THE STATISTICAL ERRORS OF THE EMISSIVITY MEASUREMENT OF A LASER HEATED SURFACE BY ACOUSTO-OPTICAL TUNABLE FILTER 405
Yu V Mantrova, P V Zinin, A A Bykov, K M Bulatov
THE SPECIFICS OF SINGLE-MODE LASER MODULES APPLICATION IN OPTICAL-ELECTRONIC COMPLEXES BASED ON LIGHT SCATTERING 412
V Parshin, V Bliznyuk, O Evtikhieva
MEDICAL CONTROL OF A PULSE WAVE FIBER OPTIC SENSORS 420
M Gerashchenko, E Badeeva, T Murashkina, S Gerashchenko

OPTICAL METHODS FOR DROPLETS EVAPORATION ON GRAPHENE NANOCOMPOSITE SURFACES 428
D D Babenko, A S Dmitriev
THE COMBINATION OF PIV AND HEAT FLUX MEASUREMENT IN STUDY OF FLOW AND HEAT TRANSFER NEAR A CIRCULAR FINNED CYLINDER 434
S Z Sapozhnikov, V Yu Mityakov, V V Seroshtanov, A A Gusakov
DESIGN FEATURES OF THE GLOSSMETER SYSTEM 444
A S Beliaeva, G E Romanova
IMMERSION LASER-INDUCED ULTRASOUND IMAGING OF SOLIDS WITH COMPLEX MACROSCOPIC GEOMETRY OF SURFACE 453
V Zarubin, A Bychkov, V Simonova, E Cherepetskaya, A Karabutov
ACOUSTIC PROPERTIES OF LNM CRYSTAL 458
D Yu Velikovskii, A I Kokshaiskii, M G Milkov, V A Sukharev
APPLICATION OF A LASER-DOPPLER ANEMOMETRY FOR THE STUDY OF UNSTEADY FLOW STRUCTURE IN A MODEL MICRO-HYDRO TURBINE 465
S I Shtork, I V Litvinov, E Yu Gorelikov, D G Mukhin, S V Dremov, D A Suslov
SELECTIVE SCHEME OF GAS CONCENTRATION MEASUREMENT BASED ON ACOUSTO-OPTIC SPECTRAL FILTERS WITH MULTI-FREQUENCY CONTROL SIGNAL 472
Yu V Pisarevsky, S A Kolesnikov, V A Lomonov, A V Vinogradov
LASING IN SLURRY-LIKE ACTIVE MEDIUM ON $LIF:F_2$ MICROPARTICLES 476
O A Burdukova, V A Konyushkin, V A Petukhov, Yu V Senatsky, P G Zverev
ESTIMATION OF THE IMAGE QUALITY CORRECTION EFFICIENCY IN OPTICAL MEASUREMENT SYSTEMS UNDER SCHEIMPFLUG CONDITION 484
O V Pechinskaya
SPHERICAL SURFACE CONTROL BY BACKGROUND ORIENTED SCHLIEREN METHOD 491
M M Nyein
Author Index

Journal of Physics: Conference Series **1421** (2019) 011001 doi:10.1088/1742-6596/1421/1/011001

Preface

The conference "Optical Methods of Flow Investigation" (OMFI) has been regularly held since 1989 at intervals of 2 years. The All-Union Seminar was held in 1989, I and II All-Union Conferences (founder – Institute of Thermal Physics of the Siberian Branch of the Russian Academy of Sciences) were held in 1991 and 1993. Since 1995, conferences have been held at the Moscow Power Engineering Institute (MPEI) under the auspices of the Siberian Branch of the Russian Academy of Sciences and the Ministry of Education of the Russian Federation, the co-founders were MPEI and the Institute of Thermophysics of the Siberian Branch of the Russian Academy of Sciences. Since 2019 Scientific and Technological Center of Unique Instrumentation of the Russian Academy of Sciences has also become a co-organizer of the conference.

This issue of the Journal is devoted to research and studies presented at the XV International Scientific and Technical Conference "Optical Methods of Flow Investigation" (OMFI-2019). The conference was held in Moscow Power Engineering Institute, Moscow, Russia, on 24–28 June 2019. The National Research University “Moscow Power Engineering Institute” is one of the leading technical universities in Russia having about 15,000 students from more than 60 countries. MPEI is one of the largest technical universities of Russia. It provides the specialist’s training and scientific research in the field of energetic, electrical engineering, radio electronics, computer engineering.

The conference brought together experts from different organizations that had a great opportunity to share knowledge during oral and poster presentations, to initiate discussions on topics that are of interest to the conference attendees. The conference papers presented in the following areas: laser anemometry, flow visualization, development of quantum-optical devices, computer methods of signal and image processing, laser and optical interferometry, application of optical methods, acousto-optic and optoacoustics, optical methods in biomedicine and ecology, also included papers of the Youth workshop «Modern methods of flow investigation – 2019».

Conference Committees

Conference co-chairs:

D.M. Markovich - Correspondent member of Russian Academy of Sciences

A.V. Klimenko - Academician of Russian Academy of Sciences

Program Committee Co-Chairs:

Professor Yu.N. Dubnishchev

Professor B.S. Rinkevichyus

Deputy Chairman of the Organizing Committee:

Ph.D., Associate Professor N.M. Skornyakova

Scientific Secretary:

Ph.D. A.Yu. Poroykov

Members of the organising committee

T.V. Bazhenova (JIHT RAS, Moscow)
A.E. Bondarev (KIAM RAS, Moscow)
A.Yu. Varaksin (BMSTU, Moscow)
G.N. Vishnyakov (VNIIOFI, Moscow)
V.M. Gordienko (MSU, Moscow)
V.A. Gourashvili (TRINITI, Troitsk)
O.A. Evtikhieva (MPEI, Moscow)
G.M. Zharkov (ITAM SB RAS, Novosibirsk)
P.G. Zverev (GPI RAS, Moscow)
V.M. Zemlyansky (Kiev)
I.A. Znamenskaya (MSU, Moscow)
S.I. Inshakov (TsAGI, Zhukovsky)

Journal of Physics: Conference Series **1421** (2019) 011002 doi:10.1088/1742-6596/1421/1/011002

Members of the working committee

Peer review statement

All papers published in this volume of *Journal of Physics: Conference Series* have been peer reviewed through processes administered by the proceedings Editors. Reviews were conducted by expert referees to the professional and scientific standards expected of a proceedings journal published by IOP Publishing.

Journal of Physics: Conference Series **1421** (2019) 012001 doi:10.1088/1742-6596/1421/1/012001

Time-resolved interference microscopy for studying nonideal plasma formed by high-power femtosecond laser pulses

D S Sitnikov

Joint Institute for High Temperatures of the Russian Academy of Sciences, Izhorskaya 13 Bldg 2, Moscow 125412, Russia

E-mail: sitnik.ds@gmail.com

Abstract. Measurement results of characteristic size of plasma density inhomogeneity on the surface of bulk iron target formed under action of femtosecond laser pulses with intensity of 10^{16} W/cm^2 are presented. Investigations are performed using time-resolved interference microscopy technique. Femtosecond laser pulses with high time contrast of 10^7 generated by chromium:forsterite laser system are applied. Efficacy of the chosen technique is demonstrated; the size of plasma inhomogeneity is shown to be less than 30 nm.

1. Introduction

Rapid development of physics and technologies has made possible laser systems with unique characteristics to emergence-tremendous intensity ($I > 10^{21}$ W/cm^2), high contrast (up to 10^{14}) [1], and short duration (several tens of femtoseconds) [2-6]. Investigation of interaction of laser radiation with such intensities with matter is not only of fundamental, but also of practical importance, ranging from generation of X-ray and gamma radiation to ion acceleration in cancer therapy [7-9], etc.

To generate efficiently characteristic X-ray radiation, electrons with appropriate energies are required. To accelerate electrons, not only structured targets (for example, nanowires [10] or grating targets for generation of propagating surface plasmons on their surface [11]) are used, but also parameters of laser pulse (chirp [12] and polarization [13]) are manipulated. Methods for modeling laser-plasma interaction processes are also constantly being improved; about a dozen of theoretical papers on this topic have been published in 2018 only (for example, see [14, 15]).

At the same time, correct simulation of the considered processes requires experimental data to validate itself, since plasma properties on the target surface depend not only on characteristics of the latter (structure, composition, thickness and density), but also on such parameters of laser pulse as duration, energy, power density, and polarization. As intensity of laser pulse increases, its time profile comes first, since the presence of a pre-pulse can lead to formation of a pre-plasma on the target surface, which affects the efficiency of fast (or hot) electron generation [16, 17]. In these papers, the effect of pre-plasma on both the generation processes of fast electrons (spectrum, spatial distribution, and angular divergence) and their temperature was studied. At intensities of a laser pulse, for which the condition $I\lambda^2 \leq 10^{18}$ W/cm^2 μm^2 is satisfied, generation of hot electrons can be caused mainly by two mechanisms: the vacuum heating and the resonant absorption. Numerical simulation by means of particle-in-cell (PIC) method [18] has shown that vacuum heating can be more efficient than resonant absorption in many cases.

To assess the effect of a particular mechanism, X-ray characteristic radiation generated in plasma due to impact of a *p*-polarized femtosecond pulse is used as a diagnostic tool. Earlier in [19], to determine the prevailing mechanism of hot electrons generation, experimental dependences were obtained of intensity of K_α radiation on the angle of incidence of laser pulse on the target surface. However, the obtained results have demonstrated that generation of hot electrons in the whole range of angles can be due to both of the above mechanisms; the decisive role is played by the characteristic size of the plasma density inhomogeneity in the vicinity of the critical density.

In this paper a method of time-resolved interferometric microscopy is proposed for the first time to study the size of pre-plasma on the target surface when exposed to a femtosecond laser pulse.

2. Experimental setup

Similar to previous study [19], an infrared terawatt femtosecond laser system based on the chromium:forsterite active element is used to generate X-ray characteristic radiation. It provides laser pulses with an energy of up to 90 mJ and a repetition rate of 10 Hz, ~ 80 ÷ 20 fs in duration at a wavelength of 1240 nm and a spectral width of 26 nm (FWHM) [4].

Time-resolved interference microscopy technique [20] is used to study the dynamics of preplasma formation and expansion as a result of the impact of femtosecond pulse with an intensity of 10^{16} W/cm^2. The applied approach allows to obtain information on the changes of amplitude r_{ind} and phase Ψ_{ind} of the complex reflection coefficient. Previously, this technique has been used to study such processes as the formation of electron-hole plasma in semiconductors [21], phase transitions in solids [22], laser ablation of the surface under the action of ultrashort laser pulses [23], and determining the optical and transport properties of plasma at an laser intensity of 10^{15} W/cm^2 [24], as well as to determine the strength of metals in the liquid phase [25, 26]. It is based on the pump-probe scheme, in which laser radiation is divided into two beams – the pump (more powerful) and the probe ones. To change the energy of laser pulses, a polarization attenuator (a half-wave plate and a prism polarizer) is installed in each arm, and the energy is controlled using a calibrated germanium photodiode. The probe pulse (λ = 620 nm, frequency is doubled in DKDP crystal) is used to illuminate the sample area under study. Standa 8MT160-300 delay-line used in the scheme allows changing the time delay t_{delay} between the pump and the probe pulses with 8 fs steps. The uniformity of illumination of the sample area is reached using a spatial filter assembled in the probe beam; it is a Kepler 1:1 telescope with a diaphragm 30 μm in diameter installed in the common focal plane of the lenses.

Michelson interferometer is assembled in a vacuum chamber. The chamber has four flanges, including optical windows, electrical connectors, and inlet valves. Laser radiation comes through the input window with antireflective coating for both the fundamental and the second harmonic frequencies of laser radiation. The evacuation of the chamber is performed up to a pressure of ~ 10^{-3} Torr.

The image transfer of the target surface to the CCD plane (1024 × 1024 pixels) with magnification M ~ 30 is performed using a micro-lens 9×, NA = 0.2. The second arm of the interferometer is formed by a micro-lens with similar parameters and a reference mirror. To align the intensity in the interferometer arms, a set of neutral optical filters installed between the reference mirror and the micro-lens is used. The probe beam reflected from the target surface (called "object" beam) interferes with the "reference" beam in the CCD plane of the camera placed outside the vacuum chamber. Application of CCD camera (SensiCam QE, PCO) with 12-bit depth allows interferograms to be recorded with a smaller quantization step of intensity than common 8-bit cameras that increases the sensitivity of the experimental setup. The intrinsic plasma emission is cut off by a set of optical filters consisting of a narrow-band interference filter with transmission at a wavelength of the probe pulse λ = 620 ± 10 nm, as well as SZS-23, KS-11 and OC-4 ones. A couple of interferograms of the target surface are recorded in the experiment – prior to laser impact and at the moment of laser irradiation (at a certain t_{delay} value). Processing of these interferograms [20] results in spatial distribution of the amplitude and phase changes of the complex reflection coefficient. Further, only the phase change

$\Psi_{ind}(x,y)$ of the reflection coefficient will be of interest. Its accuracy in interference microscopy setup is up to $\Delta\Psi_{ind} \sim 0.01$ rad.

Similar to [23], the target is a bulk cylindrical sample of iron (30 mm in diameter and 35 mm in height) with a side surface roughness not worse than $R_z = 10$. The target is set on a 3D node consisting of three motorized linear translators (Standa 8MT-173-20) and a rotation one (Standa 8MR-174-11). The target is oriented so that the probe laser pulse falls normally the cylindrical surface and the pump one has 45° angle of incidence. The pulp pulse is focused by an off-axis parabolic mirror with a focal length of 2" (Thorlabs MPD229-M01) 2" in diameter.

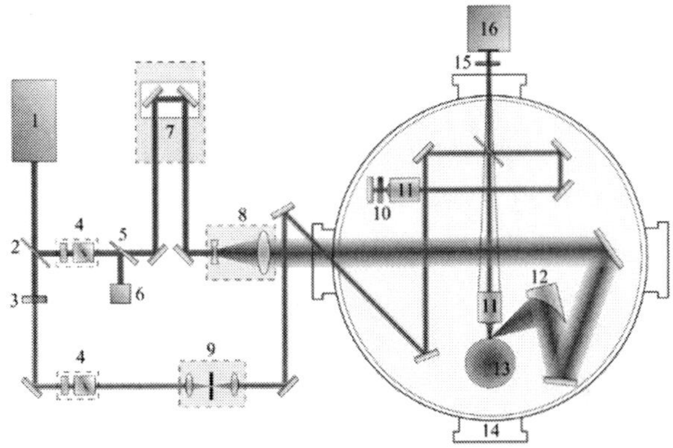

Figure 1. Time-resolved interference microscopy scheme. 1 – terawatt femtosecond laser system, 2 – beam splitter, 3 – SHG, 4 – polarizing attenuator, 5 – glass wedge, 6 – photodiode, 7 – delay line, 8 – telescope, 9 – spatial filter, 10 – set of neutral light filters, 11 – micro-lens, 12 – parabolic mirror, 13 – sample, 14 – vacuum chamber, 15 – interference filters, 16 – CCD camera.

3. Determination of parameters of laser pump pulse

Figure 2 demonstrates temporal profile of femtosecond laser pulse in picosecond time range obtained by third-order cross-correlator with dynamic range of 10^9. Temporal profile of laser pulse as well as temporal contrast are the key features of high-power femtosecond laser pulse. Temporal contrast is the ratio of the peak power to the nanosecond "pedestal" caused by amplified spontaneous emission (ASE) in active medium of laser. As is seen from the figure, the pulse contrast at a wavelength of 1240 nm is 10^7 in the range of ± 2 ps and in nanosecond range as well; while at the time of ± 1 ps laser pulse intensity equals ~ 10^{-4} of the maximum; pulse duration is 100 fs FWHM.

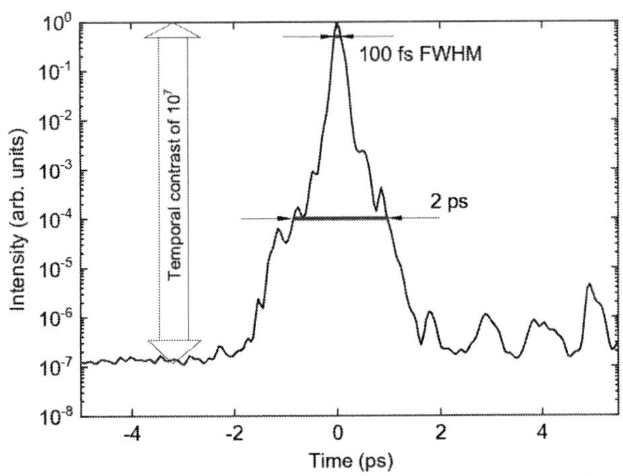

Figure 2. Third-order cross-correlation function of femtosecond laser pulse.

Using an off-axis paraboloid for laser radiation focusing imposes strict requirements on the accuracy of alignment of this element, since field aberrations of an aspheric optics exceed greatly aberrations of common spherical one. To estimate spatial distribution of energy density in the beam waist, a technique described in [27] is used. According to it, a square of radius of the crater formed on the target surface as a result of ablation is linearly related to the logarithm of the laser pulse energy applied. For his purpose, iron target has been temporarily replaced with a sample of polished monocrystalline silicon. When the threshold value F_{abl} ~ 200 mJ/cm^2 is exceeded, craters on the surface become clearly distinguishable. Figure 3 (a) demonstrates experimental dependences of the square of the minor and major axes of the ablation crater on the energy of laser pump pulse. The dependences obtained are well approximated by a family of two straight lines with different angles of inclination. This means that spatial distribution of intensity in the focal plane of the off-axis parabolic mirror is described by a combination of two Gaussian curves with parameters r_{01} and r_{02} in figure 3 (b) for central and peripheral parts respectively. Due to oblique incidence of laser beam and elliptic shape of craters, we have different values for X and Y axes.

This spatial distribution may result from aberrations of laser beam wavefront either due to an error in the manufacturing accuracy of the surface of the focusing element (less than $\lambda/2$ @ λ = 633 nm, according to the manufacturer), or due to distortions in the active elements of laser source itself. In practice, this leads to the fact that the central tightly focused part of the laser spot 24×32 μm at $1/e^2$ level contains 95% of the entire pulse energy. This in turn means that energy of E = 3.1 mJ is required to achieve laser intensity $I_{max} = 10^{16}$ W/cm^2 on target surface.

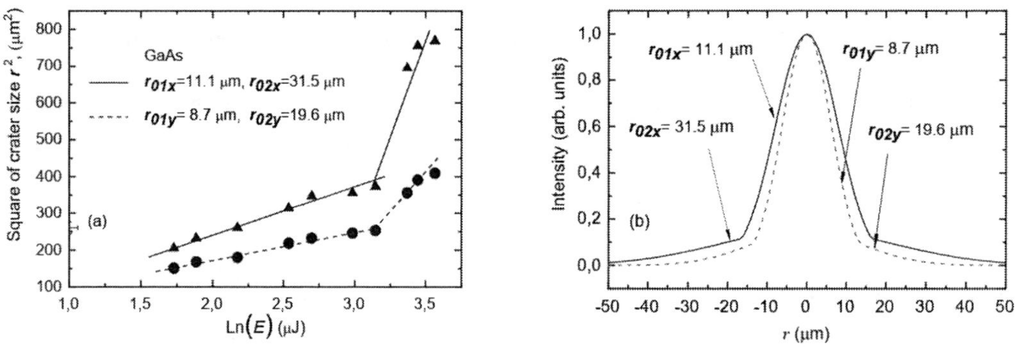

(a) (b)

Figure 3. (a) Dependence of the square of crater size on the incident energy of pump pulse for GaAs target. Triangles and circles – experimental values for major and minor axes respectively; lines – linear approximations. (b) Spatial distribution cross-sections of the intensity of laser radiation on the target surface.

4. Results and Discussion

The change in the phase of the reflected wave Ψ_{ind} of the probe pulse depends on the processes running at the target surface and the intensity of the pump pulse as well; in general terms, it can be represented as a sum of several components:

$$\Psi_{ind} = \Psi_{oc} + \Psi_{shift} + \Psi_{pl}, \quad (1)$$

where Ψ_{oc} is the phase change due to the change in the optical constants of the material due to melting process, Ψ_{shift} is the phase change caused by surface movement due to expansion of the target, and Ψ_{pl} is the phase change due to propagation of probe radiation through plasma layer of subcritical density. The change of Ψ_{oc} due to melting of target material may have various signs: in semiconductors, for example, Si and GaAs, it is negative, while in metals it can have both positive (gold) and negative (tungsten) values.

Figure 4 (a) demonstrates an example of spatial distributions of phase changes Ψ_{ind} for Si and Au targets after laser exposure with laser fluence five times above the ablation threshold value for delay times t_{delay} of 200 fs and 5 ps. In experimental studies [28], it was shown that information on optical constants could be obtained on the basis of changes in amplitude and phase of the reflected wave (Fresnel equations) for time delay of 200 fs only if pump pulse intensity does not exceed $2{\cdot}10^{13}$ W/cm^2. At intensities above the specified value, phase change caused by layer movement of critical density, from which probe pulse is reflected, exceeds the phase change caused by alternation of optical constants of target material. The change in Ψ_{oc} due to the change in optical constants upon laser action on Fe target has a positive sign [29], and due to high intensity of the pump pulse $I_{\text{max}} = 10^{16}$ W/cm^2, we can assume $\Psi_{oc} \ll \Psi_{shift}$.

The physical meaning of Ψ_{pl} term becomes clear from figure 4 (b), where two cases of reflection of the probe pulse are presented; before the pump pulse impact (at the top) and when the plasma formed at the target surface expands as a result of laser exposure (at the bottom). In the first case, electron density profile has a stepped form corresponding to the target-vacuum interface. In the second case, probe radiation reflected from the layer of critical density propagates back through plasma layer, acquiring a positive phase shift $\Psi_{pl} > 0$. Thus, all three terms in equation (1) are assumed to have non-negative values.

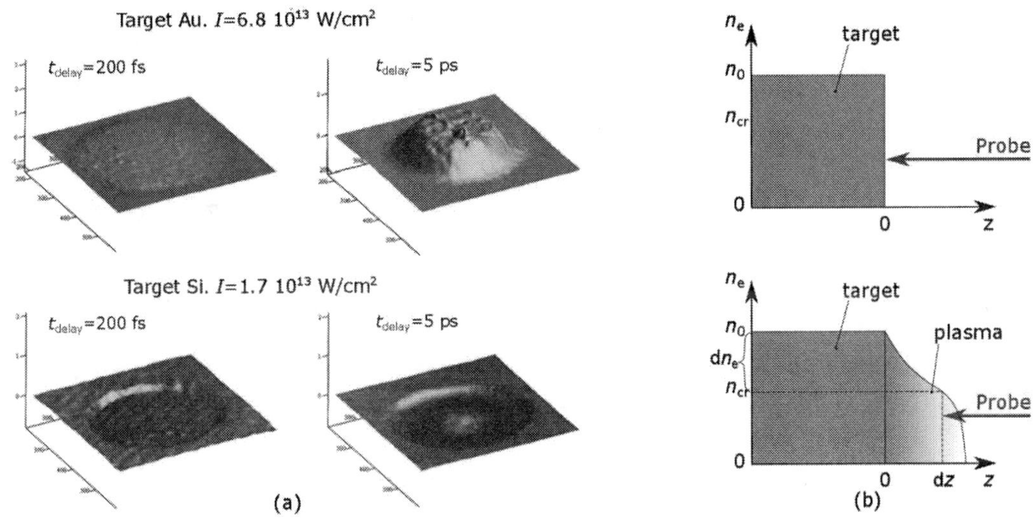

Figure 4. (a) Spatial distribution of phase change $\Psi_{ind}(x,y)$ of the reflected probe radiation when gold (at the top) and silicon (at the bottom) targets are exposed to femtosecond laser pulse. (b) Reflection of the probe pulse from the target surface: before (at the top) and after (at the bottom) the impact of the pump pulse.

As it has been shown previously [19], a key issue in determining the mechanism for creating fast electrons is the value of the hydrodynamic expansion of pre-plasma prior to the impact of the maximum of intensity of the laser pulse. The pre-plasma mentioned above can be formed either as a result of ionization of vapor cloud caused by evaporation of the target exposed to ASE of nanosecond duration at the front edge of laser pulse, or as a result of the impact of a prepulse with sufficient energy. To study the effect of both the nanosecond ASE pedestal and the prepulse on target surface, measurements of phase changes Ψ_{ind} have been carried out in the range of negative delay values of the probe pulse relative to the pump pulse. The moment of $t_{\text{delay}} = 0$ designates the coincidence in time of the maxima of intensity profiles of pump and probe pulses. For time delay $t_{\text{delay}} = -2$ ps, pump pulse intensity on target surface is $\sim 10^{-7}{\cdot}I_{max}$ that is below evaporation threshold of target material.

Figure 5 (a) demonstrates time profile of intensity of the pump pulse on a logarithmic scale over a time range of ± 3 ps relative to the intensity maximum. Figure 5 (b) shows the dependence of the phase change of the complex reflection coefficient (left axis, Ψ_{ind}) on time delay between pump and probe pulses in the range of – 1.7 ps < t_{delay} < 1 ps. From the obtained experimental data it can be seen that for t_{delay} = 0 phase change is $\Psi_{ind} \sim 0.58 \pm 0.12$ rad. The error in determining phase change is caused only by the accuracy of the Fourier method of interferogram processing, but also by the error of setting a zero delay, which is Δt_{delay} ± 60 fs (determined by the duration of the laser pulse), and surface roughness. Changes of Ψ_{ind} in the region of negative delays ($t_{delay} \geq -300$ fs) correspond to laser pulse intensity $I \geq 4 \cdot 10^{13}$ W/cm^2 that exceeds ablation threshold of target material.

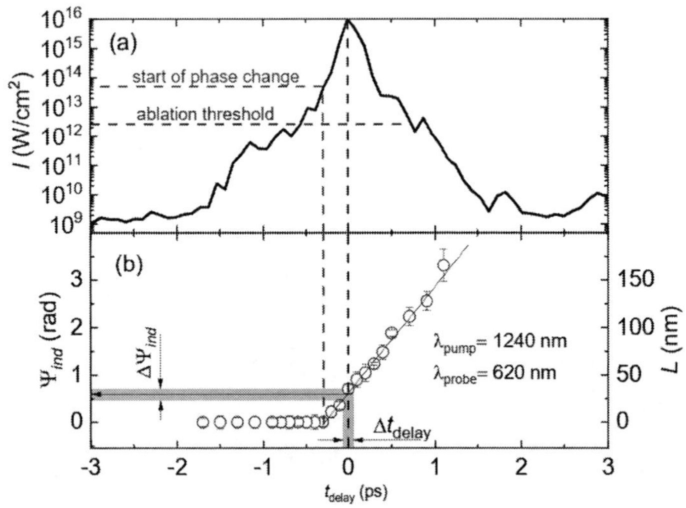

Figure 5. (a) Time profile of laser pump pulse. (b) Dynamics of phase change Ψ_{ind} (left axis) of the complex reflection coefficient of plasma and the magnitude of the characteristic scale inhomogeneity L of plasma density (right axis).

Considering the assumptions made above, the experimentally obtained phase changes Ψ_{ind} of the complex reflection coefficient consist mainly of two components: 1) phase change Ψ_{pl} associated with the propagation of laser probe pulse in plasma with inhomogeneous dielectric constant before and after reflection from the layer with critical density and 2) phase shift Ψ_{shift} due to the displacement of plasma layer, from which the probe radiation is reflected:

$$\Psi_{ind} = \Psi_{shift} + \Psi_{pl} = 0.58 \pm 0.12 \text{ rad.} \tag{2}$$

Since each of these summands is non-negative, it can be argued that the phase change caused by pre-plasma expansion does not exceed the value of Ψ_{ind} i. e. $\Psi_{shift} < 0.58$ rad. In this case, displacement of the plasma layer with critical density of $n_{cr} = 2.9 \cdot 10^{21}$ cm^{-3} for probe radiation at a wavelength of λ = 620 nm does not exceed $dz = \lambda_{probe}\Psi_{shift}/4\pi < 29$ nm, and the expansion velocity of the layer with critical density is about 10^5 m/s.

The size of characteristic inhomogeneity of plasma density $L = [(1/n_e)(dn_e/dz)]^{-1}$ [30] calculated in linear approximation and corresponding to the registered phase change of the complex reflection coefficient is shown as vertical axis on the right in figure 5 (b). As the magnitude of the change in electron density, we took the difference between the initial and critical densities (for probe pulse wavelength of λ = 620 nm) $dn_e = n_0 - n_{cr}$ shown in figure 4 (b). Thus, experimental studies have shown that at the moment of Fe target exposure to the maximum intensity of laser pulse, plasma with a characteristic size of inhomogeneity L < 30 nm is formed on the surface.

As noted in previous study [19], optimal conditions for creating electrons with energies of about 10 keV for generating X-Ray characteristic radiation, accelerated through the vacuum heating mechanism, are observed in the case when the size of plasma inhomogeneity L is small compared to the amplitude of oscillations of electrons in the electric field component of laser radiation normal to the target. According to [18], the criterion for the action of the vacuum heating mechanism is the fulfillment of the condition $L/\lambda < 0.1$, which corresponds to the characteristic plasma inhomogeneity size $L \sim 120$ nm for an infrared laser pulse.

5. Conclusions

The results of experimental studies of plasma formation and expansion when bulk iron target is exposed to radiation of femtosecond chromium-forsterite laser system at a wavelength of 1240 nm and pulse intensity of $I_{max} = 10^{16}$ W/cm^2 with a high temporal contrast of 10^7 are presented. Time-resolved method of interference microscopy was applied to study the characteristic size of inhomogeneity of pre-plasma for the first time. Assuming that the recorded changes in the phase of the probe pulse reflected from the pre-plasma layer with a critical density are determined primarily by plasma expansion, the characteristic size of plasma inhomogeneity at the moment of impact of the maximum intensity of laser pulse is $L < 30$ nm. Obtained results suggest that vacuum heating may be the prevailing electron acceleration mechanism in this case.

All experimental studies were carried out at the UNU "Laser Terawatt Femtosecond Complex", which is a part of the center for collective usage "Laser Femtosecond Complex" of JIHT RAS.

References

[1] Braenzel J, Andreev A A, Platonov K, Klingsporn M, Ehrentraut L, Sandner W and Schnürer M 2015 Coulomb-Driven Energy Boost of Heavy Ions for Laser-Plasma Acceleration *Phys. Rev. Lett.* **114** 124801

[2] Yang X, Xu Z, Leng Y, Lu H, Lin L, Zhang Z, Li R, Zhang W, Yin D and Tang B 2002 Multiterawatt laser system based on optical parametric chirped pulse amplification *Opt. Lett.* **27** 1135

[3] Waxer L J, Maywar D N, Kelly J H, Kessler T J, Kruschwitz B E, Loucks S J, McCrory R L, Meyerhofer D D, Morse S F B, Stoeckl C and Zuegel J D 2005 High-Energy Petawatt Capability for the Omega Laser *Opt. Photonics News* **16** 30

[4] Agranat M B, Ashitkov S I, Ivanov A A, Konyashchenko A V, Ovchinnikov A V and Fortov V E 2004 Terawatt femtosecond Cr : forsterite laser system *Quantum Electron.* **34** 506–8

[5] Lozhkarev V V, Freidman G I, Ginzburg V N, Katin E V, Khazanov E A, Kirsanov A V, Luchinin G A, Mal'shakov A N, Martyanov M A, Palashov O V, Poteomkin A K, Sergeev A M, Shaykin A A, Yakovlev I V, Garanin S G, Sukharev S A, Rukavishnikov N N, Charukhchev A V, Gerke R R and Yashin V E 2006 200 TW 45 fs laser based on optical parametric chirped pulse amplification *Opt. Express* **14** 446

[6] Rus B, Bakule P, Kramer D, Naylon J, Thoma J, Green J T, Antipenkov R, Fibrich M, Novák J, Batysta F, Mazanec T, Drouin M A, Kasl K, Baše R, Peceli D, Koubíková L, Trojek P, Boge R, Lagron J C, Vyhlídka Š, Weiss J, Cupal J, Hřebíček J, Hříbek P, Durák M, Polan J, Košelja M, Korn G, Horáček M, Horáček J, Himmel B, Havlíček T, Honsa A, Korouš P, Laub M, Haefner C, Bayramian A, Spinka T, Marshall C, Johnson G, Telford S, Horner J, Deri B, Metzger T, Schultze M, Mason P, Ertel K, Lintern A, Greenhalgh J, Edwards C, Hernandez-Gomez C, Collier J, Ditmire T, Gaul E, Martinez M, Frederickson C, Hammond D, Malato C, White W and Houžvička J 2015 ELI-Beamlines: development of next generation short-pulse laser systems *Proc. SPIE 9515, Research Using Extreme Light: Entering New Frontiers with Petawatt-Class Lasers II* ed G Korn and L O Silva p 95150F

[7] Bulanov S V, Esirkepov T Z, Khoroshkov V S, Kuznetsov A V and Pegoraro F 2002 Oncological hadrontherapy with laser ion accelerators *Phys. Lett. A* **299** 240–7

[8] Malka V, Fritzler S, Lefebvre E, D'Humières E, Ferrand R, Grillon G, Albaret C, Meyroneinc S, Chambaret J-P, Antonetti A and Hulin D 2004 Practicability of protontherapy using compact laser systems *Med. Phys.* **31** 1587–92

[9] Fourkal E, Velchev I, Fan J, Luo W and Ma C-M 2007 Energy optimization procedure for

treatment planning with laser-accelerated protons *Med. Phys.* **34** 577–84
[10] Jiang S, Krygier A G, Schumacher D W, Akli K U and Freeman R R 2014 Effects of front-surface target structures on properties of relativistic laser-plasma electrons *Phys. Rev. E* **89** 013106
[11] Macchi A 2018 Surface plasmons in superintense laser-solid interactions *Phys. Plasmas* **25** 031906
[12] Salamin Y I and Carbajo S 2019 A Simple Model for the Fields of a Chirped Laser Pulse With Application to Electron Laser Acceleration *Front. Phys.* **7** 1–6
[13] Wen M, Salamin Y I, Keitel C H, En M E N G W, Alamin Y O I S and Eitel C H H K 2019 Electron acceleration by a radially-polarized laser pulse in a plasma micro-channel *Opt. Express* **27** 557–66
[14] Cantono G, Sgattoni A, Fedeli L, Garzella D, Réau F, Riconda C, Macchi A, Ceccotti T and Riconda C 2018 Extensive study of electron acceleration by relativistic surface plasmons *Phys. Plasmas* **25** 031907
[15] Svedung Wettervik B, Gonoskov A and Marklund M 2018 Prospects and limitations of wakefield acceleration in solids *Phys. Plasmas* **25** 013107
[16] Yang X H, Ma Y Y, Xu H, Shao F Q, Yu M Y, Yin Y, Zhuo H B and Borghesi M 2013 Generation of hemispherical fast electron waves in the presence of preplasma in ultraintense laser-matter interaction *Laser Part. Beams* **31** 379–86
[17] Lin X X, Li Y T, Liu B C, Liu F, Du F, Wang S J, Chen L M, Zhang L, Liu X L, Liu X L, Wang Z H, Ma J L, Lu X, Dong Q L, Wang W M, Sheng Z M, Wei Z Y and Zhang J 2012 Directional transport of fast electrons at the front target surface irradiated by intense femtosecond laser pulses with preformed plasma *Laser Part. Beams* **30** 39–43
[18] Gibbon P and Bell A R 1992 Collisionless absorption in sharp-edged plasmas *Phys. Rev. Lett.* **68** 1535–8
[19] Agranat M B, Andreev N E, Ashitkov S I, Ovchinnikov A V, Sitnikov D S, Fortov V E and Shevel'ko A P 2006 Generation of characteristic x rays by a terawatt femtosecond chromium-forsterite laser *JETP Lett.* **83** 72–4
[20] Sitnikov D S, Komarov P S, Ovchinnikov A V and Ashitkov S I 2009 Femtosecond Fourier interferometry of a nonideal plasma *Tech. Phys.*
[21] Ashitkov S I, Ovchinnikov A V. and Agranat M B 2004 Recombination of an electron-hole plasma in silicon under the action of femtosecond laser pulses *J. Exp. Theor. Phys. Lett.* **79** 529–31
[22] Inogamov N A, Petrov Y V, Khokhlov V A, Anisimov S I, Zhakhovskiĭ V V, Ashitkov S I, Komarov P S, Agranat M B, Fortov V E, Migdal K P, Il'nitskiĭ D K and Émirov Y N 2014 The effect of an ultrashort laser pulse on metals: Two-temperature relaxation, foaming of the melt, and freezing of the disintegrating nanofoam *J. Opt. Technol.* **81** 233
[23] Ashitkov S I, Komarov P S, Ovchinnikov A V, Struleva E V, Zhakhovskii V V, Inogamov N A and Agranat M B 2014 Ablation and nanostructuring of metals by femtosecond laser pulses *Quantum Electron.* **44** 535–9
[24] Komarov P S, Ashitkov S I, Ovchinnikov A V, Sitnikov D S, Veysman M E, Levashov P R, Povarnitsyn M E, Agranat M B, Andreev N E, Khishchenko K V and Fortov V E 2009 Experimental and theoretical study of Al plasma under femtosecond laser pulses
[25] Ashitkov S I, Inogamov N A, Komarov P S, Zhakhovsky V V, Oleynik I I, Agranat M B, Kanel G I and Fortov V E 2012 Strength of metals in liquid and solid states at extremely high tension produced by femtosecond laser heating *AIP Conference Proceedings* pp 120–5
[26] Ashitkov S I, Komarov P S, Ovchinnikov A V, Struleva E V and Agranat M B 2016 Strength of liquid tin at extremely high strain rates under a femtosecond laser action *JETP Lett.* **103** 544–8
[27] Liu J M 1982 Simple technique for measurements of pulsed Gaussian-beam spot sizes *Opt. Lett.* **7** 196

[28] Morikami H, Yoneda H, Ueda K I and More R M 2004 Detection of hydrodynamic expansion in ultrashort pulse laser ellipsometric pump-probe experiments *Phys. Rev. E - Stat. Nonlinear, Soft Matter Phys.* **70** 1–3

[29] Struleva E V, Ashitkov S I, Komarov P S, Khishchenko K V and Agranat M B 2016 Strength of iron melt at high extension rate during femtosecond laser ablation *J. Phys. Conf. Ser.* **774** 012098

[30] Brunel F 1987 Not-so-resonant, resonant absorption *Phys. Rev. Lett.* **59** 52–5

Journal of Physics: Conference Series **1421** (2019) 012002 doi:10.1088/1742-6596/1421/1/012002

Application of hyperspectral imaging for analysis of embryonic development of lower vertebrates

A B Burlakov[1], S V Shirokov[2], D D Khokhlov[2,3], V I Kuzmin[4,5] and A F Gadzaov[4,5]

[1] Department of ichthyology, Faculty of Biology, Moscow Lomonosov State University, Leninskie Gory GSP-1, Moscow, 119991, Russian Federation
[2] Laboratory of acousto-optic spectroscopy, Scientific and Technological Center of Unique Instrumentation of Russian Academy of Sciences, 15 Butlerova, Moscow, 117342, Russian Federation
[3] Department of Electrical Engineering and Introscopy, Institute of Automatics and Computer Engineering, National Research University "Moscow Power Engineering University", 14, Krasnokazarmennaya street, Moscow, 111250, Russian Federation
[4] Department of Applied Mathematics, Russian Technological University (MIREA), 78 Vernadskiy avenue, Moscow, 119454, Russian Federation
[5] Moscow Technological University, 38a Leninskii avenue, Moscow, 119334, Russian Federation

E-mail: khokhlov.dd@ntcup.ru

Abstract. The task of analyzing the characteristics of the embryonic development of lower vertebrates using non-invasive optical methods is considered. We propose to apply hyperspectral imaging for segmentation of embryo zones with different physical and chemical properties. This approach allows one to obtain information on the spatial distribution of optical spectral properties of the observed object. An experimental setup based on a microscope coupled with a spectral imaging add-on module utilizing a tunable acousto-optic filter has been created to implement this approach. Hyperspectral data arrays are formed as a series of spectral narrowband images obtained by successive tuning of the acousto-optic filter within the working spectral range. A series of experiments was conducted in which the development of embryos of the Misgurnus fossilis loach during their transition from the 32nd to the 33rd developmental stage was observed using the described experimental setup. Characteristic spatial regions corresponding to certain parts of the loach embryo were identified. The time dependences characterizing the change in the spectral properties of the selected regions were obtained. It is shown that joint processing of a series of spectral images obtained in different spectral bands within the working spectral range of an acousto-optical spectral imaging module allows one to effectively identify differences in the kinetics of the optical transmission spectral density for various regions of the axial mesoderm.

1. Introduction

The use of the embryonic development model in experimental research is promising since it gives an ability to trace the influence of various factors on living systems in a period of rapid complication, in which even minor disturbing impact can lead to disruptions in the development program and the appearance of visible anomalies. According to this approach lower vertebrate embryos are objects of special interest as their development occurs in the surrounding water and is available for direct observation. Bony fish eggs are polylecithal. Their yolk is formed during the gametogenesis to support the embryo development at the early stages. At further stages of development, yolk transforms into the yolk sac, which is retained until the end of embryogenesis. The main function of the yolk is to provide the metabolic processes of the embryo. It contains energetic substrates, vitamins and different proteins such as vitellogenin, lipovitellin, phosphitin etc. Recent studies show that the mentioned yolk components take part in congenital immune defences as proteins act like polyvalent recognition receptors [1-4]. Vitellogenin and phosphitin provide antioxidant properties and protect the embryo from the oxidative stress [2, 4]. During embryogenesis, certain specific fish egg yolk regions perform functions of organs being developed further, such as blood formation, osmoregulatory and excretory functions [5-7] etc. All these processes are connected with the embryo development stages.

The conditions and the dynamics of yolk components local changes must be traced accurately with respect to the embryo morphogenesis features and the changes of physical and chemical properties of the yolk itself as the extraembryonic systems are being formed. This task requires the use of intravital non-invasive research methods. Hyperspectral imaging is a kind of non-invasive methods successfully tested for studying the structure of animal organs and tissues and their vital cycles [8-12]. This method allows quantitative characterizing the spatial distribution of spectral properties. In this paper, we analyzed the functional conditions of different regions of Missgurnus fossilis loach embryo yolk before the beginning of spontaneous motor activity using acousto-optic (AO) spectral imaging.

2. Experimental setup

Different scientific instruments and methods can be used to obtain spectral images, but the method based on tunable optical filters (liquid crystal tunable filters, acousto-optic tunable filters, etc.) is considered to be one of the most versatile. AO tunable filters (AOTFs) provide high optical throughput, the ability to vary the transmission function, rather high spectral and spatial resolution and small tuning time, while being completely PC-controlled devices with no need of mechanical adjustment. AOTF operation principle is based on anisotropic Bragg diffraction of wideband optical radiation on the ultrasound wave [13]. AO cells installed in the AOTF used in the described setup are made of a uniaxial birefringent crystal. The ultrasonic wave excited by the piezotransducer in the AO cell induces the Bragg grating due to elasto-optic effect. The transmission wavelength of the AOTF is defined by the period of the grating which can be controlled by tuning the frequency of the ultrasonic wave. Imaging AOTFs conserve the spatial structure of the image for each narrow spectral band. In this case, one can electronically adjust the frequency of the ultrasonic wave and scan the working wavelength range of the AOTF to implement narrow-band imaging (NBI) and hyperspectral imaging (HSI) techniques.

Hyperspectral imaging data arrays are formed from the series of spectral images obtained by the acousto-optic spectral imaging add-on module [14]. The schematic drawing of the used module is presented in figure 1. The experimental setup is created on the basis of a conventional bright-field microscope with a wide-band Koehler illumination system which consists of wide-band light source (1) (halogen bulb), collector lens (2), field diaphragm (3), mirror (4), condenser aperture diaphragm (5) and condenser (6). Koehler illumination system is commonly used in modern scientific light microscopy to provide even illumination of the observed object. The specimen (8) (loach embryo) is installed on the 2-axis linear translation stage (7) to allow position alignment and located in the front focal plane of the infinity-corrected microscopic objective (9). Light emerging from this objective forms a parallel beam. Beamsplitter cube (10) splits the input light power into two optical trains. Beamsplitting ratio is 90:10 for optical trains (11-18) and (19-20), respectively. Elements (19) and (20) form a wide-band channel used for specimen orientation and positioning. Projection lens (19) focuses light from an objective (9) and forms a color image captured by the RGB camera (20). Spectral channel of the microscopic acousto-optic imager consists from elements (11-18). The key element of the spectral channel is the AOTF (12-16). To provide high image quality, AOTF operates in parallel beams scheme [15]. To reduce spectral drift two identical AO cells (13) and (15) are

Journal of Physics: Conference Series **1421** (2019) 012002 doi:10.1088/1742-6596/1421/1/012002

installed and rotated at 180° in the diffraction plane. As the AO cells implement anisotropic diffraction, the appropriate polarization is selected by the polarizer (12). Polarizers (14) and (16) cut off the undiffracted light. The coupling system 11 conserves the structure of the parallel beam and allows correct optical matching of the microscopic objective (9) and the AOTF. Projection lens (17) focuses the narrow-band light from the AOTF and forms an image captured by the monochrome camera (18). AOTF controller unit (21), monochrome camera (18) and RGB camera (20) are connected with a personal computer (PC) (22).

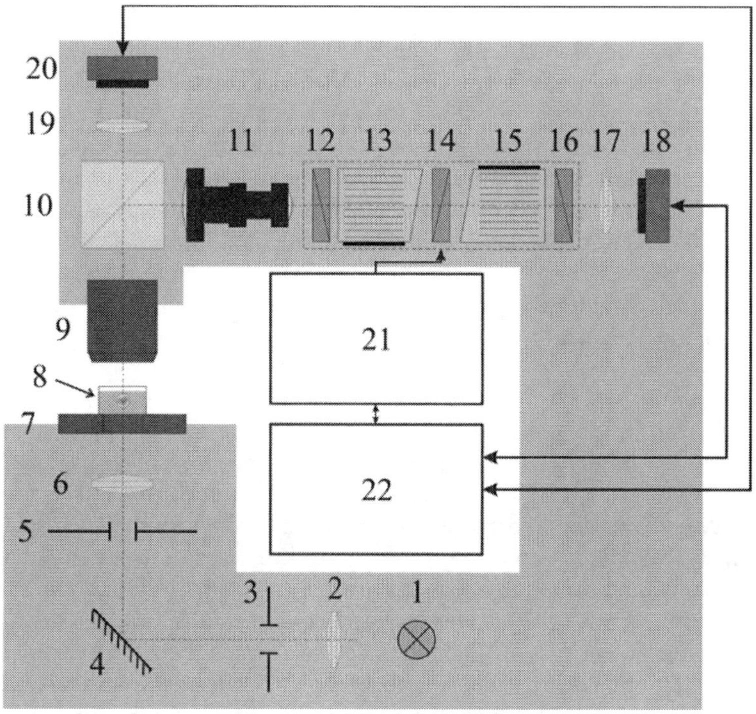

Figure 1. Microscope-based setup with AO spectral imaging module. 1 – wide band light source; 2 – collector lens; 3 – collector lens field diaphragm; 4 – mirror; 5 – field diaphragm; 6 – condenser lens; 7 – 2-axis linear translation stage; 8 – specimen; 9 – microscopic objective; 10 – beamsplitter cube; 11 – coupling system; 12, 14, 16 – crossed polarizers; 13, 15 – identical AO cells; 17, 19 – projection lenses; 18 – monochrome camera; 20 – RGB camera; 21 – AOTF controller unit, 22 – PC.

The embryo development between 32nd and 33rd stages of the loach Misgurnus fossilis which is considered to be a typical object of developmental biology was observed in the experiment. Females brought from natural inhabitant were kept in a refrigerator at the temperature of 4-5 °C. Accelerated maturation of females was performed by hormonal stimulation of chorionic gonadotropin at room temperature. The artificial insemination was performed according to the standard method, described by Kostomarova in [16]. The fertilized eggs were washed with two portions of fresh water. After that, some of the embryos (50 pcs) were placed in isolated storage with a stabilized temperature of 17 °C. Developmental stages were determined according to the data tables of normal development of the loach [17, 18]. The embryo of a desired developmental stage was washed with two portions of fresh water and placed in the object plane of a spectral imager. Spectral image series of the developing embryo were registered every 2 minutes for approximately an hour. We analyzed the yolk optical transmission spectra in four regions of interest shown in figure 2. The regions of interest (ROIs) were located near the head of the embryo (1), near the tail (2), near the differentiated myotomes (3) and in the ventral part of the egg (4).

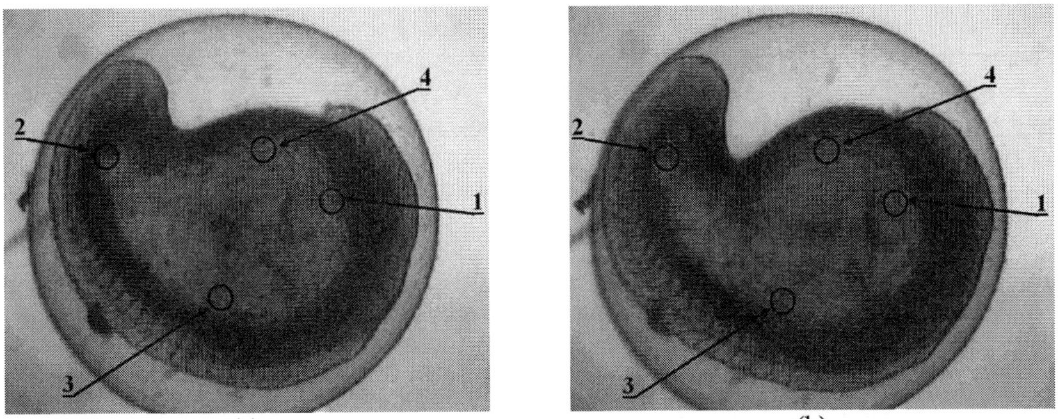

(a) (b)

Figure 2. Spectral images of the loach embryo ($\lambda = 635$ nm) at the beginning (32nd stage – a) and in the end (33rd stage – b) of the experiment with highlighted ROIs analyzed in the yolk: 1 – near the head; 2 – near the tail; 3 – near the differentiated myotomes; 4 – in the ventral part of the egg.

3. Results and discussion

Having analyzed the obtained optical transmission spectra for the described ROIs, we noticed the different character in spectral distributions during the transition of the embryo from the 32nd to the 33rd stage of development. The differences first appeared at the certain moment corresponding to the growth of the number of body somites. Several transformations took place through the experiment, are described below. The embryo length increased and the tail started differentiating. As the yolk continued transforming and its shape became more "pear-like", the intensive movement of the yolk granules occurred. The eye formation began from the creation of the lens placodes. The motor activity of the embryo began at the end of the experiment (at the 33rd stage of development).

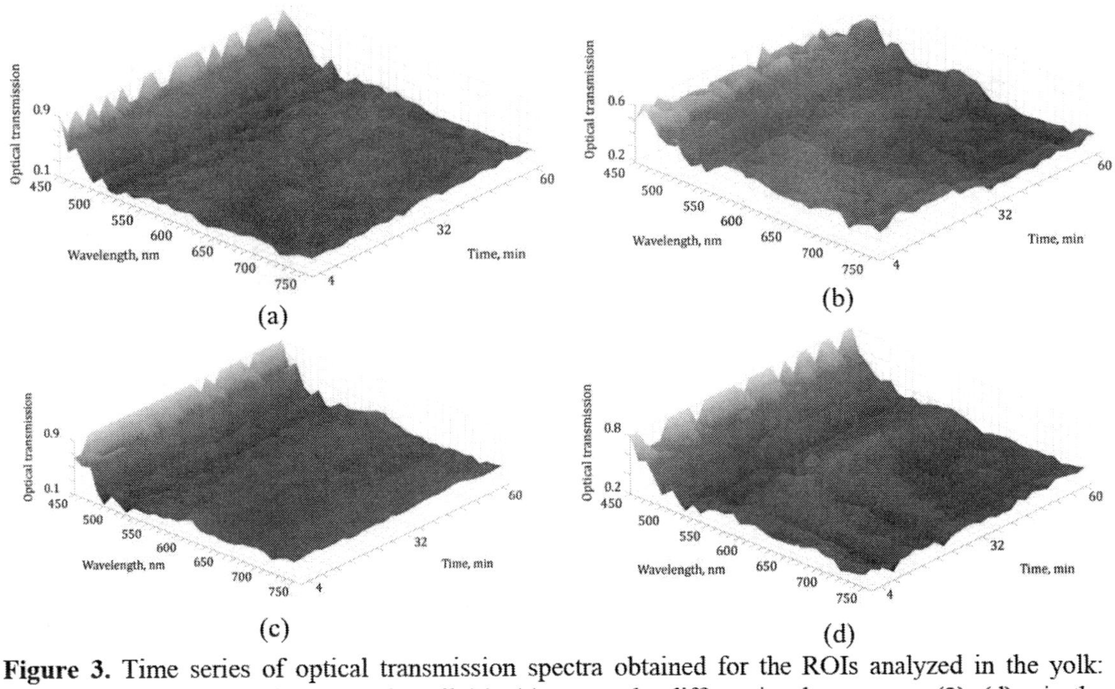

(a) (b) (c) (d)

Figure 3. Time series of optical transmission spectra obtained for the ROIs analyzed in the yolk: (a) – near the head (1); (b) – near the tail (2); (c) – near the differentiated myotomes (3); (d) – in the ventral part of the egg (4).

Journal of Physics: Conference Series **1421** (2019) 012002 doi:10.1088/1742-6596/1421/1/012002

The entire time series of optical transmission spectra obtained for the regions of interest are demonstrated in figure 3. A change in the character of the transmission spectra that occurs at the 30th - 32nd minute of the experiment is common for all of the ROIs. This change is conditioned by the regulative features of the central nervous system. Time series demonstrate partially matching local extrema of optical transmission for different ROIs. Figure 4 illustrates the time series cross-sections corresponding to the chosen wavelengths. The local minima common for regions near the head and in the ventral part of the egg indicate the synchronization of the transformation processes occurring in the yolk. Though, the additional local extrema for the ROIs near the tail and near the differentiated myotomes and the difference between the extrema positions indicate a specific heterogenic time delay of biochemical reactions in the yolk. Such heterogeneity may occur due to forming a hierarchy of connections in the central nervous system of a developing embryo.

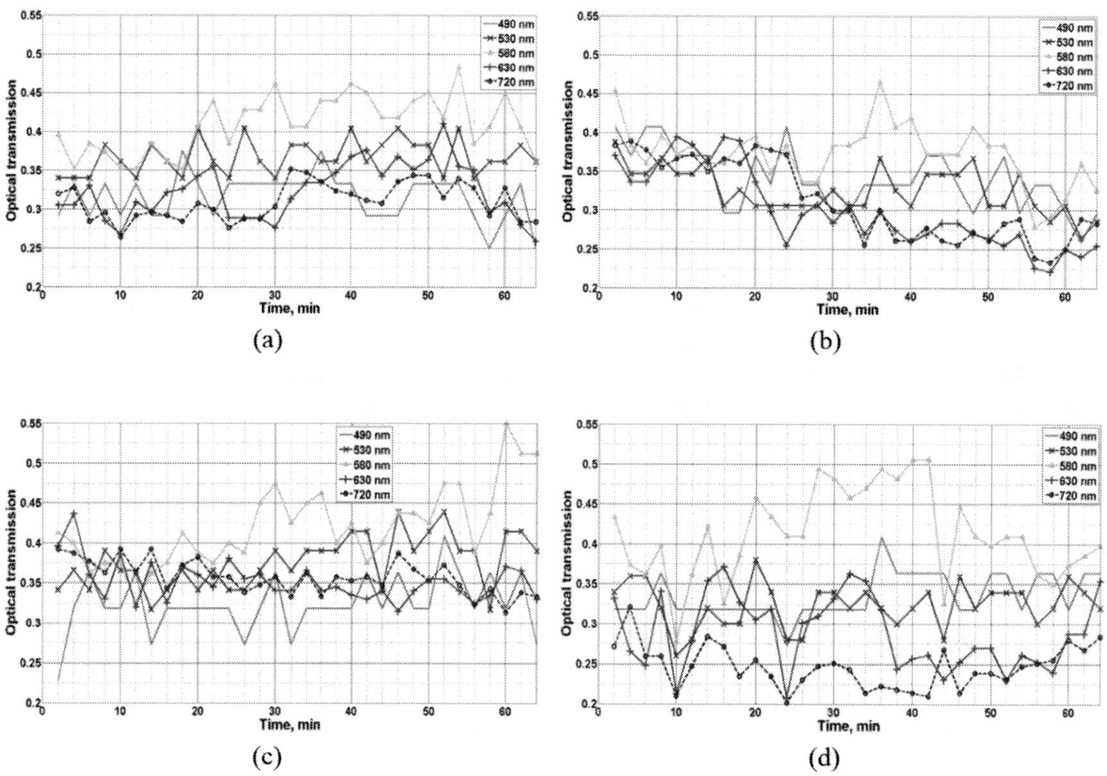

Figure 4. Cross-sections of the optical transmission spectra time series (see figure 3) corresponding to the wavelengths of 490 nm, 530 nm, 580 nm, 630 nm and 720 nm for the ROIs analyzed in the yolk: (a) – near the head (1); (b) – near the tail (2); (c) – near the differentiated myotomes (3); (d) – in the ventral part of the egg (4).

4. Conclusion

Analysis of the experimental results shows that spectral imaging of a developing biosystem is a promising technique able to provide data characterizing the features of specific morphological structures. The observation of the developing embryo indicated synchronization and the local differences of the biochemical processes in different areas the yolk. The implementation of the described technique can lead to a better understanding of the processes inside the biological system during its development.

References

[1] Wang S, Wang Y, Ma J, Ding Y and Zhang S 2011 Phosvitin plays a critical role in the immunity of zebrafish embryos via acting as a pattern recognition receptor and an antimicrobial effector *J Biol Chem* **286(25)** pp 22653-64
[2] Sun C and Zhang S 2015 Immune-Relevant and Antioxidant Activities of Vitellogenin and Yolk Proteins in Fish *Nutrients* **7(10)** pp 8818-29
[3] Zhang S, Dong Y and Cui P 2015 Vitellogenin is an immunocompetent molecule for mother and offspring in fish *Fish Shellfish Immunol* **46(2)** pp 710-5
[4] Li H and Zhang S 2017 Functions of Vitellogenin in Eggs *Results Probl Cell Differ* **63** pp 389-401
[5] Hyodo S, Kakumura K, Takagi W, Hasegawa K and Yamaguchi Y 2014 Morphological and functional characteristics of the kidney of cartilaginous fishes: with special reference to urea reabsorption *Am J Physiol Regul Integr Comp Physiol* **307(12)** pp R1381-95
[6] Takagi W et al 2014 Urea-based osmoregulation in the developing embryo of oviparous cartilaginous fish (Callorhinchus milii): contribution of the extraembryonic yolk sac during the early developmental period *J Exp Biol* **217(8)** pp 1353-62
[7] Takagi W, Kajimura M, Tanaka H, Hasegawa K, Ogawa S and Hyodo S 2017 Distributional shift of urea production site from the extraembryonic yolk sac membrane to the embryonic liver during the development of cloudy catshark (Scyliorhinus torazame) *Comp Biochem Physiol A Mol Integr Physiol* **211** pp 7-16
[8] Hernandez O, Pietrajtis K, Mathieu B and Dieudonné S 2018 Optogenetic stimulation of complex spatio-temporal activity patterns by acousto-optic light steering probes cerebellar granular layer integrative properties *Sci Rep* **8(1)** p 13768
[9] Ronzitti E et al 2017 Submillisecond Optogenetic Control of Neuronal Firing with Two-Photon Holographic Photoactivation of Chronos *J Neurosci* **37(44)** pp 10679-10689
[10] Pégard N C, Mardinly A R, Oldenburg I A, Sridharan S, Waller L and Adesnik H 2017 Three-dimensional scanless holographic optogenetics with temporal focusing (3D-SHOT) *Nat Commun* **8(1)** p 1228
[11] Venet C, Bocoum M, Laudereau J B, Chaneliere T, Ramaz F and Louchet-Chauvet A 2018 Ultrasound-modulated optical tomography in scattering media: flux filtering based on persistent spectral hole burning in the optical diagnosis window *Opt Lett* **43(16)** pp 3993-96
[12] Schwarz M et al 2016 Acousto-Optic Cerebral Blood Flow Monitoring During Induction of Anesthesia in Humans *Neurocrit Care* **24(3)** pp 436-41
[13] Goutzoulis A P and Rape D R 2004 *Design and Fabrication of Acousto-Optic Devices* (Boca Raton(FL): CRC Press)
[14] Polschikova O, Machikhin A, Batshev V, Ramazanova A, Belov A and Pozhar V 2017 AOTF based optical system of a microscope module for multispectral imaging techniques *Proc. SPIE*, **10592** 105920H
[15] Machikhin A, Batshev V and Pozhar V 2017 Aberration analysis of AOTF-based spectral imaging systems *J. Opt. Soc. Am.* A **34(7)** pp 1109-13
[16] Kostomarova A A 1975 *Objects of developmental biology* (Moscow: Nauka) pp 309–23
[17] Jirmunsky A V and Kuzmin V I 1990 *Critical levels in the development of natural systems* (Leningrad: Nauka)
[18] Fujimoto T, Kataoka T, Sakao S, Saito T, Yamaha E and Arai K 2006 Developmental stages and germ cell lineage of the loach (Misgurnus anguillicaudatus) *Zoological Science* **23(11)** pp 977-89

Journal of Physics: Conference Series **1421** (2019) 012003 doi:10.1088/1742-6596/1421/1/012003

Numerical simulation of supersonic flow around a cone in a three-dimensional formulation in the software package OpenFOAM

A E Kuvshinnikov

Keldysh Institute of Applied Mathematics RAS, Miusskaya sq., 4, Moscow, 125047, Russia

E-mail: kuvsh90@yandex.ru

Abstract. Three solvers of the OpenFOAM package are chosen to numerically solve the problem of supersonic flow around a cone. The cone half-angle and the angle of attack varied. The obtained results are compared with the well-known numerical solution of this problem. It has been established that the rhoCentralFoam solver provides minimal error for the pressure field and pisoCentralFoam solver for velocity field.

1. Introduction

This work is part of a series of articles [1, 2], aimed at creating a new method for calculating the flow around elongated bodies of rotation (EBR). Such developments are important due to the fact that currently there are many software packages that solve such problems. Thus, the researcher has a question, which package is better suited for such calculations. The authors have proposed to recreate at the level of modern high-performance computing tools computing technology developed earlier in the Keldysh Institute of Applied Mathematics A.E. Bondarev and V.A. Cherkashin under the direction of A.V. Zabrodin. The essence of this technology is that the aerodynamic drag coefficient C_x is considered as a sum of three other coefficients: C_p – coefficient for inviscid flow, C_f – coefficient for viscous friction and C_d – coefficient for near wake pressure. Such approach was widely used in the problems of mass industrial analysis of the aerodynamic properties of an EBR and proved to be very effective [3].

To calculate the aerodynamic characteristics of the non-viscous flow around EBR, the OpenFOAM [4] software package was used. It is a freely distributed software product created for solving problems of hydro and gas dynamics. It is used in many areas of science and technology, both in commercial and in academic organizations. OpenFOAM contains a number of solvers with different computational properties.

It is necessary to clarify that some solvers were previously created by developers of the OpenFOAM package [5-7], but users can create their own solvers [8].

Tabular solutions from [9] were used as in the previous works of the authors. These tables were obtained using finite-difference methods in a wide range of Mach numbers and cone half-angles with variation of the angle of attack.

In past works [1, 2], the case of a flow around EBR at a zero angle of attack was investigated. Now our main goal is to calculate the problem when the angle of attack varies.

It should be noted that similar comparisons of solvers were carried out in [10-12]. However, these comparisons were made on other examples and do not give clear recommendations on the choice of a solver for the class of problems in question.

2. Formulation of the problem

The statement of the problem is presented in full accordance with the work [9], where the results of inviscid flow around cones with different cone half-angles and angles of attack at different Mach numbers are considered.

We study the flow of an EBR placed in uniform supersonic flow of an ideal gas at an angle of attack $\alpha = 0°, 5°, 10°$ with a Mach number $M = 3$. The investigated body is a cone with a half-angle $\beta = 20°$. The incoming flow conditions at the input are denoted by the index "∞", and at the output by the index ξ, since the solution is self-similar and depends on a dimensionless variable. The flow pattern is shown in figure 1.

For calculation, the Euler system of equations is used. The system is supplemented by the ideal gas equation of state.

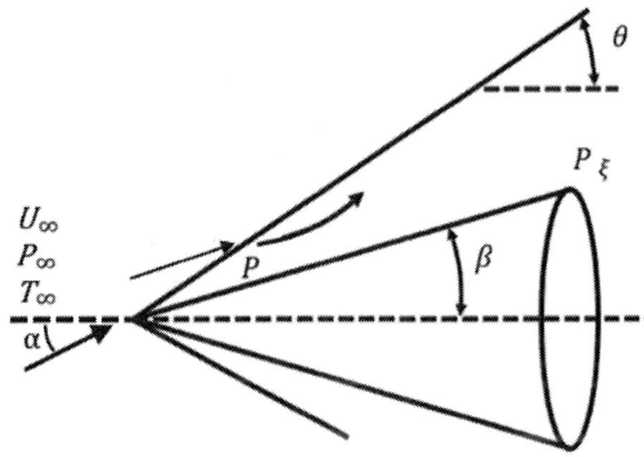

Figure 1. Flow scheme.

3. OpenFOAM solvers

For comparison, 3 solvers were selected from the OpenFOAM software package:

- rhoCentralFoam – is based on a central-upwind scheme, which is a combination of central differential and upwind schemes [5, 6]. The main idea of central-upwind schemes is a special choice of a control volume containing two types of domains: the first type around the boundary points, the second type around the center point. The advantage of these schemes is that, using the appropriate technique of reducing the numerical viscosity, it is possible to achieve good solvability for discontinuous solutions (shock waves in gas dynamics) and for solutions where viscous phenomena play a major role.
- sonicFoam is based on the PISO (Pressure Implicit with Splitting of Operator) algorithm [7]. The basic idea of the PISO method is that two differential equations are used to calculate the pressure to correct the pressure field obtained from discrete analogs of the moment and continuity equations. This approach is due to the fact that the velocities corrected by the first correction may not satisfy the continuity equation; therefore, a second corrector is introduced that allows one to calculate velocities and pressures that satisfy the linearized equations of momentum and continuity.
- pisoCentralFoam is a combination of a central-upwind scheme with the PISO algorithm [8].

Calculations for all solvers were carried out using the OpenFOAM version 2.3.0.

4. Organization of calculations

4.1. Grid, initial and boundary conditions
Figure 2 shows the calculated area. At the boundaries indicated in the figure 2 as “top”, “bottom”, “front”, “back” and “outlet” a zero gradient condition is set for gas-dynamic functions. On the left border, denoted by "inlet", the parameters of the incoming flow are set: pressure P = 101325 Pa, temperature T = 300 K, speed U = 1041.75 m/s. On the boundary of the cone “cone”, the pressure and temperature conditions are set to zero gradient, and the speed conditions are set to “slip”, which corresponds to the non-flow condition for the Euler equations. The number of grid cells is 336000.

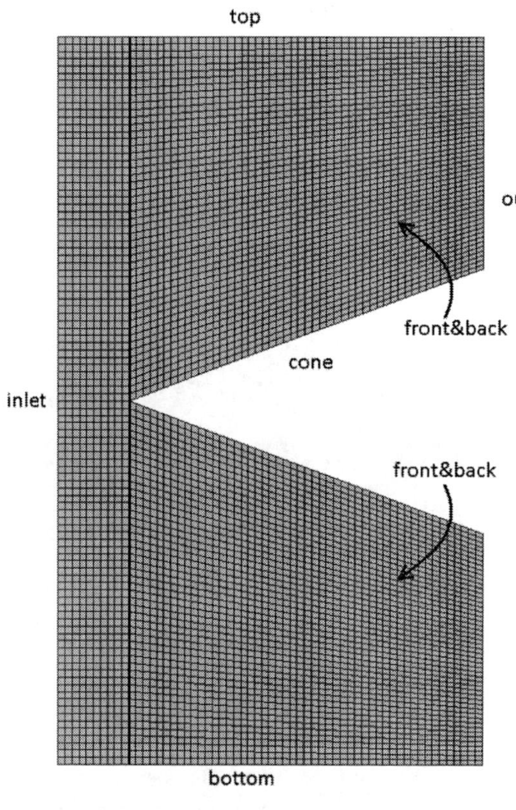

Figure 2. Computational domain.

4.2. Solvers parameters
In the OpenFOAM package, there are two options for changing the approximating of differential operators: directly in the solver code or using the fvSchemes and fvSolution configuration files. In order for the comparison to be correct, we used the same parameters where it was possible.

In the file fvSchemes: ddtSchemes — Euler, gradSchemes — Gauss linear, divSchemes — Gauss linear, laplacianSchemes — Gauss linear corrected, interpolationSchemes — vanLeer. In the file fvSolution: solver — smoothSolver, smoother — symGaussSeidel, tolerance — 10^{-9}, nCorrectors — 2, nNonOrthogonalCorrectors — 1.

5. Flow calculation

Figure 3 shows the steady-state flow field for pressure using the interpolation of the tabular solution from [9].

Tables 1–3 show the results of calculations in the form of an analog of the L_2 norm:

Journal of Physics: Conference Series **1421** (2019) 012003 doi:10.1088/1742-6596/1421/1/012003

$$\sqrt{\sum_m \left|y_m - y_m^{exact}\right|^2 V_m} \Big/ \sqrt{\sum_m \left|y_m^{exact}\right|^2 V_m},$$

where y_m are the velocity component U_x, pressure p and density ρ in the cell, V_m is the cell volume of the cone half-angles $\beta = 10–20°$. Mach number $M = 3$, angle of attack $\alpha = 0°, 5°, 10°$. The minimum values are highlighted in bold. Here, y_m^{exact} values are obtained by interpolating tabular values from [9] onto grid cells. It should be noted that the authors of the tables [9] indicate the admissibility of interpolation for all parameters and table values.

Further we will use abbreviations for solvers: rCF (rhoCentralFoam), pCF (pisoCentralFoam), sF (sonicFoam).

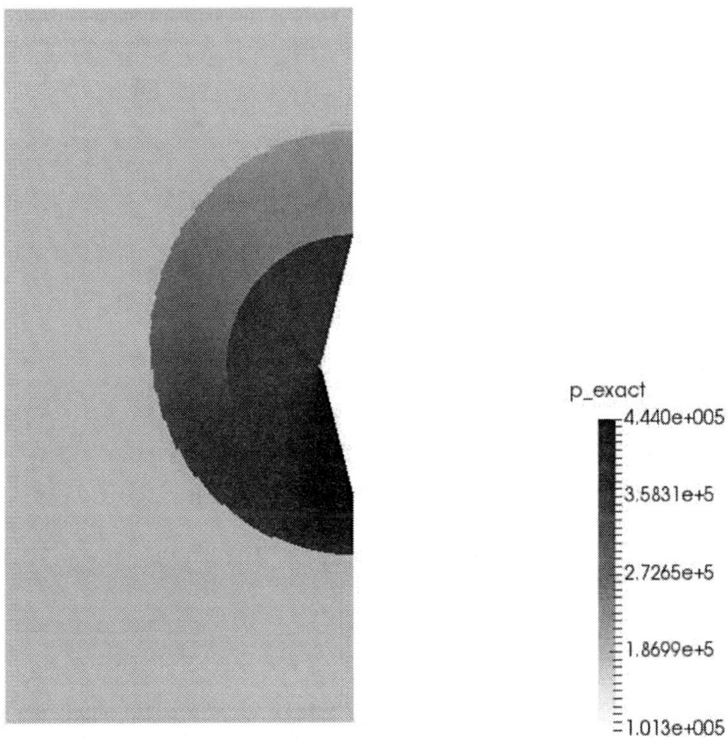

Figure 3. Pressure field.

Table 1. Deviation from the exact solution, $\beta = 10°$.

Angle of attack	Pressure p rCF	Pressure p pCF	Pressure p sF	Velocity component U_x rCF	Velocity component U_x pCF	Velocity component U_x sF
0	**0.026174**	0.032426	0.046394	0.005242	**0.004086**	0.004455
5	**0.030636**	0.037702	0.067424	0.005573	**0.004241**	0.005796

Table 2. Deviation from the exact solution, $\beta = 15°$.

Angle of attack	Pressure p rCF	Pressure p pCF	Pressure p sF	Velocity component U_x rCF	Velocity component U_x pCF	Velocity component U_x sF
0	**0.046490**	0.058198	0.091404	0.010903	**0.009031**	0.012235
5	**0.050298**	0.060187	0.116237	0.011135	**0.008931**	0.014591
10	**0.060519**	0.069622	0.145829	0.011158	**0.009040**	0.016186

Table 3. Deviation from the exact solution, $\beta = 20°$.

Angle of attack	Pressure p			Velocity component U_x		
	rCF	pCF	sF	rCF	pCF	sF
0	**0.060614**	0.069128	0.123338	0.017077	**0.014436**	0.022867
5	**0.065373**	0.075543	0.149785	0.017155	**0.015589**	0.026898
10	**0.072673**	0.081022	0.172609	**0.020015**	0.020474	0.031979

It is clearly seen that the error increases both with increasing cone half-angle and with increasing angle of attack.

6. Conclusion

Obtained results show that the rhoCentralFoam solver has a minimum field error rate for the pressure field and the pisoCentralFoam for the velocity filed. The only drawback of rhoCentralFoam is the appearance of oscillations near the surface at the head of the cone. And solver pisoCentralFoam does not have such oscillations. The results are qualitatively similar to the axisymmetric case.

Solver sonicFoam has oscillations at the shock wave front. This adversely affects the error rate. Thus, it can be argued that the rhoCentralFoam and pisoCentrlFoam solvers provide better accuracy for the class of problems and can be used in the construction of the computational flow calculation technology for the elongated bodies of rotation.

Acknowledgments

This work was supported by RFBR grant № 18-31-00320.

References

[1] Bondarev A E and Kuvshinnikov A E 2018 Analysis of the accuracy of OpenFOAM solvers for the problem of supersonic flow around a cone *Computational Science – ICCS 2018* 221–30
[2] Bondarev A E and Kuvshinnikov A E 2017 *Keldysh Institute Preprints* **12** 1–16
[3] Krasil'shchikov A P and Gur'yashkin L P 2007 *Eksperimental'nye Issledovaniya Tel Vrashcheniya v Giperzvukovykh Potokakh* (Moscow: FIZMATLIT) [In Russian]
[4] OpenFOAM Available at: http://www.openfoam.org (Accessed 25 April 2019)
[5] Kurganov A and Tadmor E 2000 *J. Comp. Phys.* **160** 241–82
[6] Greenshields C, Weller H, Gasparini L and Reese J 2009 *Int. J. Num. Meth. Fluids* **63** 1–21
[7] Issa R 1986 *J. Comp. Phys.* **62** 40–65
[8] Kraposhin M, Bovtrikova A and Strijhak S 2015 *Proc. Comp. Science* **66** 43–52
[9] Babenko K I, Voskresenskii G P, Lyubimov A N and Rusanov V V 1984 *Prostranstvennoe Obtekanie Gladkikh Tel Ideal'nym Gazom* (Moscow: Nauka) [In Russian]
[10] Karvatskii A Ya, Pulinets I V, Lazarev T V and Pedchenko A Yu 2015 *Kosm. Nauka Tehnol.* **21** 47–52 [In Russian]
[11] Gutierrez L F, Tamagno J P and Elaskar S A 2012 *Mecanica Computacional* **XXXI** 2939–59
[12] Lorenzon D and Elaskar S A 2015 *Rev. Fac. Cienc. Exactas Fis. Nat.* **2** 65–76 [In Spanish]

The study of vortexes based on the multi colored particle image velocimetry technique

S S Usmanova, N M Skornyakova, A V Anikeev and O A Evtikhieva

National Research University «Moscow Power Engineering Institute», Krasnokazarmennaya 17, Moscow, 111250, Russia

E-mail: shirinusmanova25@mail.ru

Abstract. The paper is concerned with a prospective method of qualitative and qualitative diagnostic technique – multi colored particle image velocimetry. The method proposed relies on illuminating of the studied flow by three laser planes with different wavelengths. The advantages of applying the technique in practice to investigate various flows have described. An efficiency verification of multi colored particle image velocimetry method to study vortex structures has been conducted. An experimental setup has assembled and an algorithm of the recorde images processing procedure has been developed. To assess the quality of MPIV measurements the experimental images have been compared with the computer model.

1. Introduction

Spatial velocity distribution is the main parameter that allows determining flow behavior in a gas or liquid media. The existing techniques for velocity flow measurements, based on the applying of laser light, belong to the remote, non-invasive methods allowing registering the instantaneous velocity fields in the measurement plane. Various remote sensing airs flows devices in the surface layer of the atmosphere to determine the direction and speed of the wind at the time of emergency toxic and radioactive substances emissions into the atmosphere, with the view to subsequently predict the consequences and infestation outbreaks are in high demand in the chemical and nuclear industry In turn information about the full flow vector velocity field is necessary to solve a number of current problems in aero hydrodynamics. For instance, timely and accurate information on the air flow structure in the surface atmosphere layer will increase the safety of various aircrafts and reduce the percentage of accidents and other destructive disasters.

Accordingly the development of new non-invasive flow diagnostics methods, allowing measuring the velocity vector field and the subsequent three-dimensional flow structure visualization, took on particular relevance. Currently particle image velocimetry has become a widely used flow diagnostics and visualization methods in gas and liquid medium. The method allows the two components of flow velocity vectors to be measured simultaneously in a non-invasive manner [1].

The operation method principle is based on the laser light decomposition into a sheet in order to illuminate the investigated plane in the flow volume. Further, images are recorded due to digital camera and processed by a specialized program [2].

Recent advances in the PIV technique have been directed toward obtaining all three components of flow velocity vectors in a plane or in a volume simultaneously to allow wider applications of the PIV [3-6].

Existing advanced modifications of the PIV technique to study compex flow, such as stereo or tomographic PIV method in comparison with the planar PIV technique, allow obtaining the most complete information about flow structure. The main advantages of these methods are the capability to measure the flow velocity distribution and three-dimensional visualization. Nevertheless, the methods application in practice causes a number of difficulties involving the necessity of labor-intensive preparatory training and expensive equipment [7-11].

This paper discusses a new a prospective method of diagnostic and visualization flow technique – multi colored particle image velocimetry.

2. Multi colored particle image velocimetry

The multi colored particle image velocimetry (MPIV) technique mainly differs from the known modifications of PIV method. The proposed method allows fixing the vector velocity field in several planes simultaneously.

It relies on seeding flow with tiny tracer particles and fixing their position at short time intervals. The flow velocity is determined basing on the presumption that tracer particles follow the flow without introducing any disturbances into the flow. In order this is within reach if the particles have a small volume and mass in order not to fall under the gravity, but sufficient not to fly out of the flow. For MPIV measurements, the region of interest is illuminated by three RGB laser sheets, than the positions of particles are recorded by a color digital camera.

Each image recorded will have with different intensity three main colors. As a result, by applying in the program Pivview the cross correlation to the processed images pairs for each color channel, it is possible to obtain information about the flow vector velocities field distribution in three planes. On the basis of the data obtained, distributions of the vertical or horizontal projection of particle velocities over the selected cross sections of the studied flow could be constructed. Further, by constructing in the region of the vertical and horizontal projection particle velocity distributions for the channels of red, blue and green color for each of the sections and applying the approximation procedure of the obtained planes it is capable to visualize the studied flow three-dimensional velocity field.

Figure 1 shows the schematic of the optical-electronic setup to achieve MPIV measurements. The measuring complex consists of a generating with different wavelengths three parallel laser sheets unit; receiving optical system involving a color CMOS matrix and a computer with specialized software. The laser unit includes a set of three semiconductor lasers of different wavelengths and an optical system. To obtain more accurate MPIV measurement results is necessary to ensure high power density, energy characteristics uniformity and small laser sheet thickness in the region of interest as a requirement for the formation optics.

When the experimental image is decomposed into three color channels (red, green and blue) , the scattered light each of the sheets shall conform to one of the colors, ideally, i.e. in the green color channel there should not be a signal from the other two colors. Herein lays the MPIV measurements technical basis.

3. Experimental setup and measurement technique

During MPIV work capacity testing, an experimental setup to investigate vortexes was developed. A vortex structure formed by a chemical stirrer model RW 16 basic in liquid pre-seeded polyamid particles 20 micrometers in diameter was the study object.

The Chroma 10 series krypton-argon laser of the Spectra-Physics company was a light source in experimental testing setup. The laser consists of a metal-ceramic discharge tube, which on a mixture of argon and krypton generates laser beam in the visible spectrum range from ultraviolet to near IR.

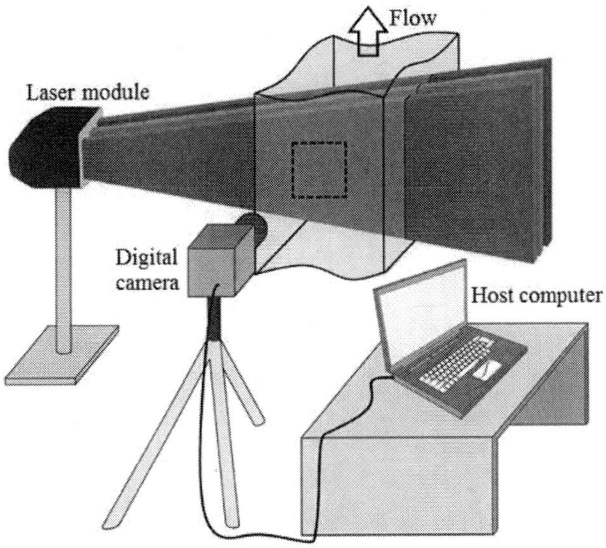

Figure 1. The optical-electronic setup schematic.

Figure 2 shows the experimental setup light sheets optics. The optics were included a dispersion prism, a set of four mirrors and a cylindrical lens. The need to use a spectral prism has been associated with some light source features and was to decompose the laser beam into a spectrum and select only three spectral components, which would account for the intensity maxima. As a result, the intensity peaks were observed in blue, green and red beams.

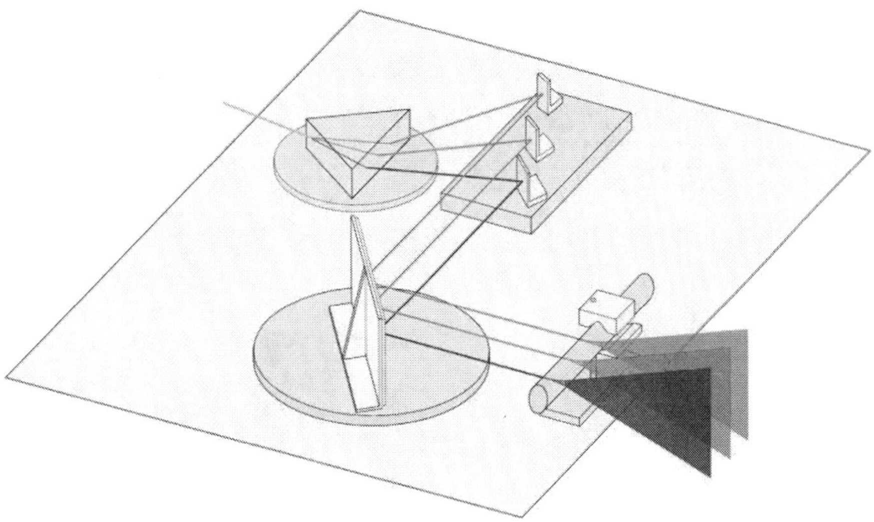

Figure 2. The testing experimental setup light sheets optics.

As a MPIV measurements result, scattered laser light of RGB sheets have been recorded. In turn, each image recorded was contained information about the flow velocity vector field distribution in three sheets located in 2 mm from each other. The recording on a digital camera Nikon 1J5 the video possibility with a 1200 frames/s frequency and 144×400 pixels resolution was made. Each image recorded had three primary colors with different intensity.

Journal of Physics: Conference Series **1421** (2019) 012004 doi:10.1088/1742-6596/1421/1/012004

4. The experimental measurements processing results

The experimental results were processed as follows. At first, a video has been broken into frames. Then, over each pair of recorded images, pre-processing was carried out, which was lied in the image decomposition into three main color channels. Then the cross-correlation in the program Pivview to images pair of each color channel was applied.

In the course of the technique efficiency testing a series of experiments at chemical stirrer different modes was performed. The processing results of images pair obtained at 400 rpm stirrer speed are presented below. For each color channel there is a speed change of trace particles displacement, as shown in the figures.

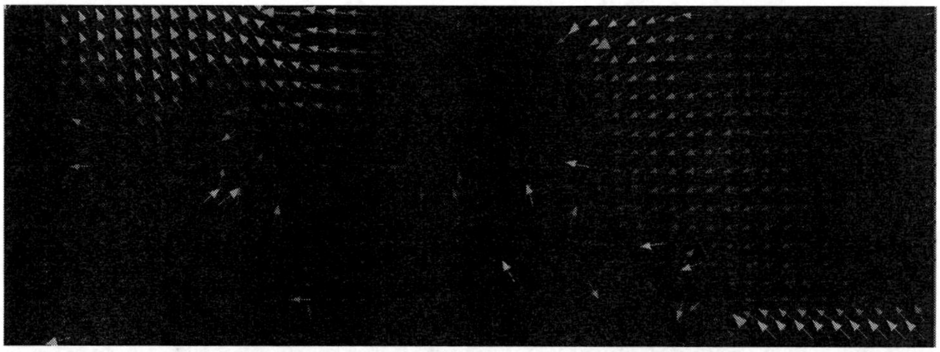

Figure 3. A vector field in the blue color channel.

Figure 4. A vector field in the green color channel.

Figure 5. A vector field in the red color channel.

5. The experimental results comparison with simulated ones

To assess the results quality, a chemical stirrer computer model developed in the Flow Vision software package has been considered. As shown in the figures below, in the chemical stirrer plane, a vortex in the upper right corner is formed, as well as currents in the blade plane are observed. It should be noted that the obtained results coincide qualitatively with the vector field directions of the three-dimensional model, comparing the experimental results with the simulated ones. But, nevertheless, the technique requires significant improvement to up its accuracy. The MPIV measurements in turn require the development of a lighting module and an obtaining vector fields program.

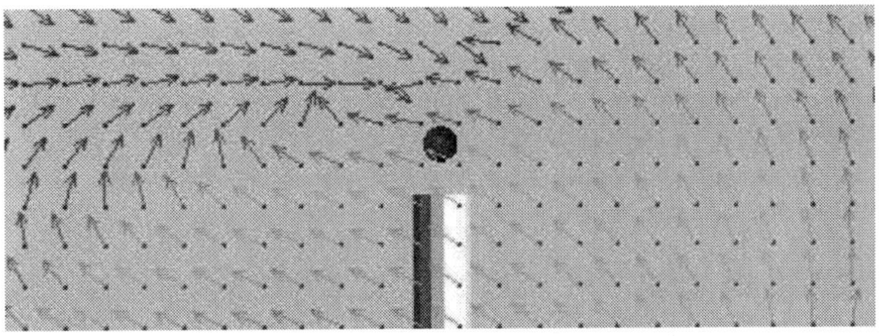

Figure 6. A vector field in the blue color channel.

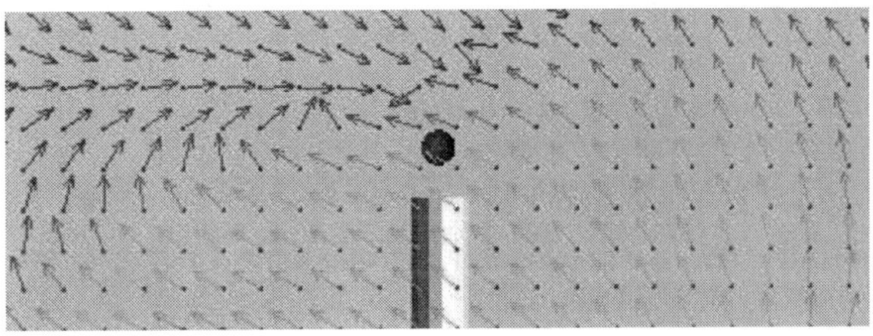

Figure 7. A vector field in the green color channel.

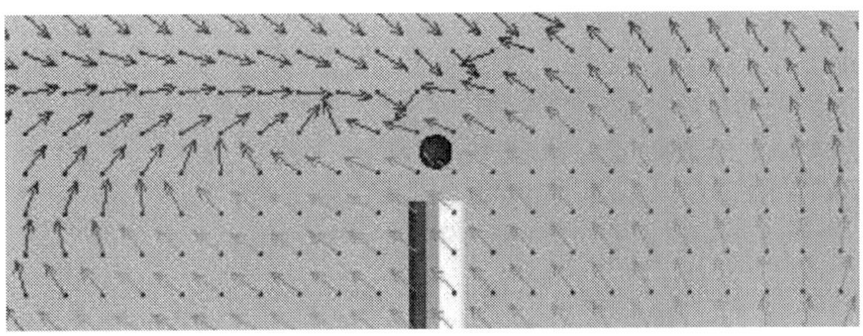

Figure 8. A vector field in the red color channel.

6. Conclusion

The major laser measuring instruments advantages in comparison with other measuring instruments are noninvasive, remote, high temporal and spatial resolution, as well as the full automation possibility of the measuring process.

The work results indicate that the MPIV method allows obtaining a flow velocities vector field in three parallel sheets simultaneously. Therein lays the main advantage over the classic PIV technique, which gives a distribution in only one selected sheet.

A qualitative experimental data comparison with those modeled by the Flow Vision program have been showed the results validity. Further MPIV measurement improvement requires the development of program for approximating the experimental results to obtain 3D images.

Acknowledgments
This work was funded by Russian Foundation for Basic Research (project 19-07-00921A).

References

[1] Raffel M, Willert C E and Kompenhans J 1998 Particle image velocimetry. A Practical Guide Corrected 3rd printing
[2] Brossard C, Monier J C, Barricau P, Vandernoot F X, Le Sant Y, Champagnat F, Le Besnerais G 2009. Principles and Applications of Particle Image Velocimetry Optical Diagnostics of Flows AL01-03.
[3] Elsinga G E, Scarano F, Wieneke B and van Oudheusden B W 2006 Tomographic particle image velocimetry *Experiments in fluids* **41** 933–47
[4] Prasad A K 2014 Stereoscopic particle image velocimetry *Experiments in Fluids* **29** 103 – 16
[5] Pan G 2002 Digital Holographic PIV for 3D Flow *Measurement ASME International Mechanical Engineering Congress & Exposition : Proceedings of IMECE'02*
[6] Hinsch K D 2014 Holographic particle image velocimetry means Sci. technol. Vol. 13 61–72
[7] Barnhart D H, Adrian R J and Papen G C 1994 Phase-conjugate holographic system for high-resolution particle image velocimetry *Applied optics* **33** 7159–70
[8] Zhang J, Tao B and Katz J 1997 Turbulent flow measurement in a square duct with hybrid holographic PIV *Experiments in fluids* **23** 373–81
[9] Willert C 1997 Stereoscopic digital particle image velocimetry for application in wind tunnel flows *Measurement science and technology* **8** 1465–79
[10] McCormick D C and Bennett J C 1994 Vortical and turbulent structure of a lobed mixer free shear layer *AIAA Journal* **32 (9)** 1852–59
[11] Katz J and Sheng J 2010 Applications of holography in fluid mechanics and particle dynamics *Annual Review Fluid Mechanics* **42** 531–55

Application of computer vision and deep learning for flame monitoring and combustion anomaly detection

S Abdurakipov and E Butakov*

Kutateladze Institute of Thermophysics SB RAS, Prospekt Akademika Lavrent'eva 1, Novosibirsk, 630090, Russia

*E-mail: e_butakov@mail.ru

Abstract. This paper is devoted to study combustion of pulverized coal in a 5 MW thermal furnace with tangential scroll supply of coal-air suspension and cylindrical reaction chamber. Deep learning approaches were used to monitor the combustion of a coal flame in a furnace and to determine combustion anomalies from flame images. We have developed a deep neural network autoencoder, which is a combination of convolutional layers, fully-connected layers and upsampling layers. The autoencoder was trained to reconstruct the combustion regimes that corresponded to high values of the coefficient of excess air. The trained autoencoder was then used to identify abnormal burning regimes with a lower excess air ratio, in which there is an increase in the amount of unburned coal dust. The best classification model had AUC ROC quality metric value on a test image sample AUC ROC = 0.8. The average precision of the model was 77% and the average recall was 66%. The metrics obtained is limited by the quality of the image labeling due to poorly controlled experimental conditions. The paper concluded that the use of upsampling with convolutional layers show themselves better than the deconvolutional layers, and the combination of convolutional layers with fully-connected layers constitutes the optimal architecture of the model.

1. Introduction

Energy is one of the main life-supporting industries. At present, about 40% of the world's electricity is generated by coal-fired power plants [1]. The use of ordinary coal in power, especially in "small" energy, is associated with its inefficient combustion: low efficiency of existing boiler equipment and increased losses, unstable combustion, slagging of heating surfaces, higher harmful emissions into the atmosphere. Continuously growing energy consumption and environmental safety issues require maximum optimization of the process of coal combustion.

The development of modern machine learning algorithms has provided researchers and engineers with new effective tools for data mining of large complex multivariate databases and generalizing contained valuable information by building mathematical model of process. The application of such intellectual data driven approaches is relevant for improving the performance of technical devices, preventing failures and unplanned stops, because trouble-free operation of complex technical systems is one of the priorities of any production process [2-4]. Neural network autoencoders are widespread unsupervised machine learning algorithms whose primary goal is to find low-dimensional latent feature space. Autoencoders are widely used for a number of tasks, for example, noise reduction, image compression, dimension reduction, construction of a probability distribution model, etc. [5-7].

This work was aimed at studying the applicability of various neural network architectures of autoencoders to solve the problem of detecting anomaly combustion regimes of a coal flame in a combustion chamber. Abnormal burning regimes correspond to low values of the coefficient of excess air, in which there is an increase in the amount of unburned coal dust.

2. The experiment

The burning process were tested in a 5 MW pilot-scale combustion rig with dry (cyclone) slag removal (figure 1). Such a relatively large-scale set-up was chosen in order to study the possibility of using mechanically activated coal in industrial boiler and this experiment is the first step in the investigation of different conditions on the ignition and combustion stability. The rig consists of two stages. The first stage (denoted in the figure 1 (a) as burner) is a cylindrical swirl chamber of a diameter of 315 mm and length of 1515 mm into which micronized coal premixed with the primary air by an ejector was fed tangentially into the burner through a vaneless spiral. The second stage (denoted in figure 1 (a) as afterburner chamber or combustor) is a coaxial cylindrical duct with interior diameter of 1000 mm and length of 2820 mm, also equipped with a tangential entry for secondary air (which also reinforces the swirling motion). It is envisaged that additional ordinary pulverized coal can be fed together with the secondary air, but this option was not considered in the present experiment. The burner spiral has an aperture for a gas torch used to ignite the coal-air mixture. The main purpose of the first stage (burner) is the thermochemical preparation, ignition and sustenance of combustion of micronized coal so that it can ensure continued combustion in the furnace (chamber), but also the ignition and sustenance of additional non-activated pulverized coal that can be supplied to the afterburner chamber in real-scale installations. The main purpose of the secondary air was to ensure the complete burning of coal and its products (CO, CH) in the combustion chamber. Thus, its flow rate, while being monitored and adjusted to fulfill the above mentioned requirement, was not precisely controlled, as the main focus of the experiment was on the burner. The rig has adjustable systems for coal feeding, for primary and secondary air supplies, and for ash recovery. It is also equipped with a pressure controller in the combustion chamber.

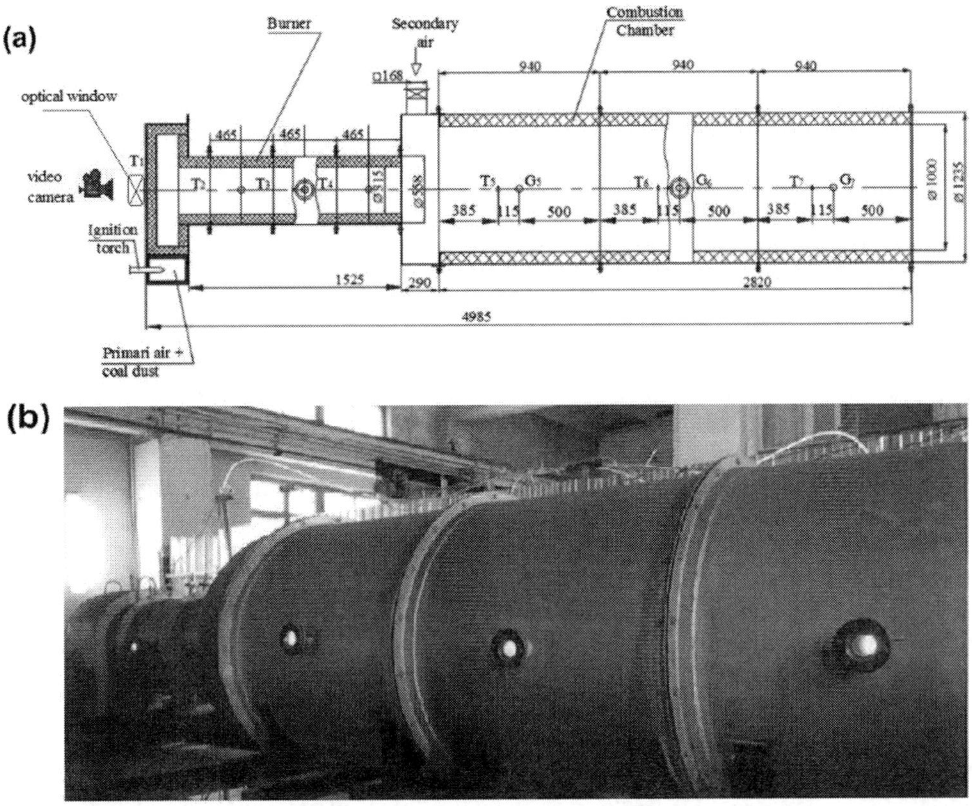

Figure 1. Experimental rig of 5 MW power. (a) Schematics; T1–T8 denote thermocouples and G5 – G7 the gas sampling apertures. (b) View at the furnace (burner and combustion chamber) in autothermal operation. Inspection windows show stable and uniform flame along the whole combustor length.

For combustion monitoring, the rig was also equipped with a number of apertures along the burner and combustion chamber for thermocouples and gas sampling probes. The temperature was measured by platinum-rhodium thermocouples of type B (TPR/1-0679, 'Vakuummash'', Izhevsk, Russia), with 30% and 6% of rhodium in one and the other conductor respectively. The B type thermocouples have proven to have superior thermoelectric and tensile properties, and to be exceptionally stable and resistant to oxidizing atmosphere at high temperatures. The wires of 0.3 mm thickness are placed in a ceramic sheath of 4 mm in diameter. The tip of the ceramic sheath is recessed with a 0.5×3 mm slot so that the thermocouple junction is directly exposed to the measuring medium. (The external ceramic sheath of 10 mm diameter, provided by the manufacturer for measurements in molten metals, was removed). The calibration data, provided by the manufacturer and approved by the Russian National Standards (GOST), complying with the international standard (IPTS-68) guarantee the nominal temperature range of 600-1650 °C for long continuous operation (up to 6000 h) with accuracy of $\pm 0.004t$, where this the measured temperature in °C. The measuring range can be extended to 1800 °C for a shorter use. It is noted that none of the thermocouples was used longer than 10 h in total, with a typical measurements campaign lasting no longer than 15 min.

The gas samples probes are connected to a multi-component (O_2, CO_2, CO, CH NO, NO_2) gas analyzer (Boner-VT) with real-time computer processing of the measured data.

To study the burning process, 5 MW thermal furnace with tangential scroll supply of coal-air suspension and cylindrical reaction chamber was used. Brown coal (one of the main coal deposits of Siberia) was used in experiments. The technical composition of coal is as follows (mass %, dry):

Journal of Physics: Conference Series **1421** (2019) 012005 doi:10.1088/1742-6596/1421/1/012005

moisture W_{rl} – 30.8, ash A_r – 11.1, volatile content – 49.1, sulfur S_{dl} – 0.29, high heat value Q_{daf} – 6900 kcal/kg. Coal was fed to the high-energy mills and then was directed by an ejector with transport air to the scroll burner.

3. Introduction to algorithms

Neural autoencoder – unsupervised machine learning algorithm. The main idea of the autoencoder is to reduce the dimension by mapping the original feature space into the latent feature space of a smaller dimension (encoding step), and then in reconstructing the input data based on the latent representation (decoding step) (see figure 2). By reducing the dimension of feature space, we thereby train the model to learn only the most important information from which initial data can be reconstructed [3]. Autoencoders consist of two parts: encoder g and decoder f. The encoder transforms the input signal into its representation (code): $h = g(x)$, and the decoder reconstructs the signal by its code: $x = f(h)$. The autoencoder, varying f and g, tends to learn the identity function $x = f(g(x))$, minimizing some error functional $L(x, f(g(x))$. Property of dimension reduction is achieved by forming the encoder structure in the form of a bottleneck, significantly reducing the dimension of the last hidden layer. The goal of the autoencoder is to minimize the reconstruction error between the input and output data. To reduce the reconstruction error, during training process we "propagate" in the opposite direction along the neural network and update the weights of the neural network. Weights of neurons are updated depending on how much they contribute to the reconstruction error. Autoencoder can be considered as a data compression algorithm [5]. At the same time, the families of functions of the encoder g and decoder f are somehow limited, so that the autoencoder is forced to select the most important properties of the signal. The critical limitation of autoencoders is that the data should belong to the general population on which the network was trained. Having trained the autoencoder on digit numbers, it cannot be used to encode something else (for example, human faces). Moreover, some of their developments, such as the variational autoencoder (VAE), as well as its combination with generative adversarial networks (GAN), give very interesting results and are now at state-of-the-art of generative models.

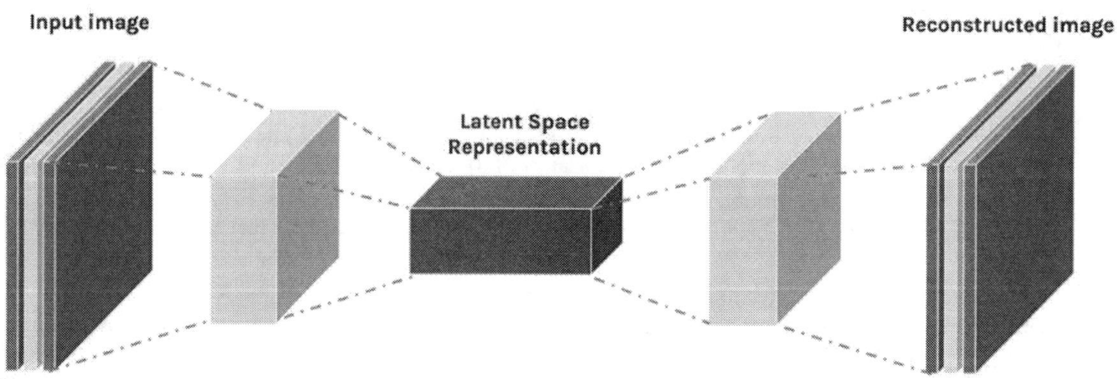

Figure 2. Scheme of autoencoder [5].

The neural autoencoder captures the non-linear features of the data, and therefore has an advantage, for example, over the linear principal component analysis approach (PCA). To build an autoencoder, it is necessary: 1) to define a coding method, 2) a decoding method, 3) a loss function for comparing the output of the autoencoder with the target result. In this paper, the root-mean-square error was used as a loss function. The simplest version of a neural network autoencoder is an encoder-decoder structure consisting of fully-connected layers, without feedback, most similar to a perceptron. A more complex option, more suitable when working with images is to replace full-connected layers with convolutional layers. A convolutional layer is a set of maps; each map has a convolution operator or filter. The filter moves through the image, multiplies the filter values by the original pixel values of the image (elementwise multiplication). All these multiplications are summed, and as a result one number is obtained-the result of convolution. The number of feature maps is determined by the requirements of the task; if you take a large number of feature maps, the recognition quality will increase, but the computational complexity will increase too. Based on the analysis of scientific articles, in most cases it is proposed to take the ratio of one to two, that is, each feature map of the previous layer (for example, the first convolutional layer, the previous one is the input one) is associated with two maps of the convolutional layer.

One of the important implementations of this paper is the use of upsampling layers in autoencoder architecture. Inverting a convolutional layer can be quite complicated due to overlapping of the kernels. Most deep learning architectures use deconvolutional layers (sometimes referred to as transposed convolution layers), which are designed to invert the convolutional layer. Even though the transposed convolution layers are intuitive for the reconstruction of the input data, they have a few well-known drawbacks, such as the generating checkboard artifacts. To overcome this limitation, the transposed convolution layers were replaced with oversampling layers and simple convolutional layers. We also tested this solution on a handwritten number dataset.

4. Model description

During the study, we worked on the following neural network architectures:

1) Simple autoencoder with five hidden fully-connected layers (basic model);

2) Convolution autoencoder, which consists only of convolutional layers in the encoder and transposed convolutional layers (or deconvolutional layers) in the decoder;

3) Another convolutional model that uses convolution and subsampling layers in the encoder, and upsampling with convolutional layers in the decoder;

4) Combination of convolutional and fully-connected layers.

To simplify the implementation of the neural network, we used Python development using Keras framework with Tensorflow backend. A simple fully-connected autoencoder had the architecture: Input (128 × 128) – Flatten (16384) – Dense (1024) – Dense (1024) – Dense (512) – Dense (1024) – Dense (16834) – Reshape (128 × 128), where Input is the layer of input parameters, Flatten is a straightening operator, Dense is a fully-connected layer, Reshape is a shape changing operator. The second model is a convolutional autoencoder, which consists only of convolutional and deconvolutional layers. In the encoder, the input data passes through 12 convolutional layers (Conv2D) with 3 × 3 cores and filter sizes starting from 8 and increasing to 32. The decoder reconstructs image using transposed convolutional layers (Conv2DTranspose). Using only convolutional layers may seem unjustified, but in this case the goal is to compare methods instead of achieving the highest possible accuracy. The latter architecture, which was analyzed, is a convolutional autoencoder with convolutional layers, pooling layers, and upsampling layers (see figure 2.). The encoder is the following sequence of layers: Conv2D – Con2D – MaxPooling2D – Conv2D – Conv2D – MaxPooling2D – Flatten – Dense. Thus, alternating convolutional (Conv2D) and pooling layers (MaxPolling2D), ending in a fully-connected layer (Dense). At the exit from the last convolutional layer of the decoder, a reconstructed image with an initial resolution of 128 × 128 is obtained.

Data preparation. Most of the original image containing the shaded area has been cropped. The original data resolution has been reduced to 128 × 128 pixels. The data were splitted into a training, test and validation sample. In the training sample got 2000 images, in the test 200 images and 200 in the validation. In the training sample, there were only combustion regimes with a large excess fuel ratio, which are characterized by a high burning intensity of the coal torch.

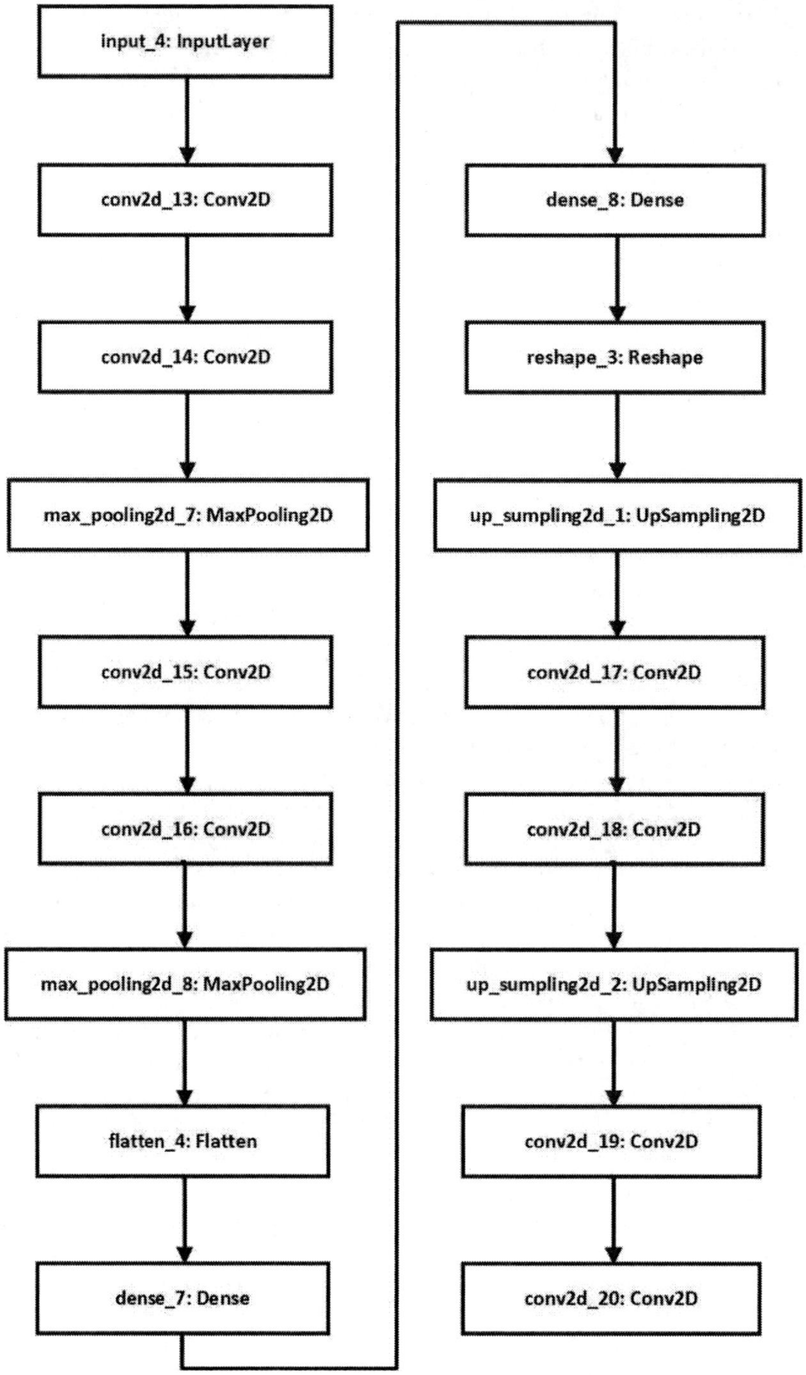

Figure 3. The architecture of the combined model of autoencoder (a) encoder and (b) decoder.

Ideally, we want to train the neural network so that it remains resistant to image distortions and noise in the data, but our model can be trained only on the basis of those samples that we provided to it, despite the fact that it performs some kind of statistical analysis of the training set and extrapolates him. In this paper, we applied a simple but effective solution: to artificially expand the training data with distorted versions of images during training. This means the following: before setting an example for the model input, we apply all the transformations we deem necessary, and then let the network directly observe what effect it has on applying to the data and training it to behave well on these examples. For example, here are some examples of shifted, scaled, deformed, rotated images (see figure 4). To implement this procedure, we used Keras framework, which provides a valuable interface for expanding the training set – ImageDataGenerator class. We initialize data generator and choose, what types of transformation we want to apply to the images, and then we run the training data through the generator, calling the fit method, and then the flow method, getting a continuously expanding iterator for the batch types we replenish. There is even a special method model. fit-generator that will teach our model using this iterator, which greatly simplifies the code. ImageDataGenerator also provides us with the ability to perform horizontal vertical shifts, random turns, scaling, warping, and specular reflection.

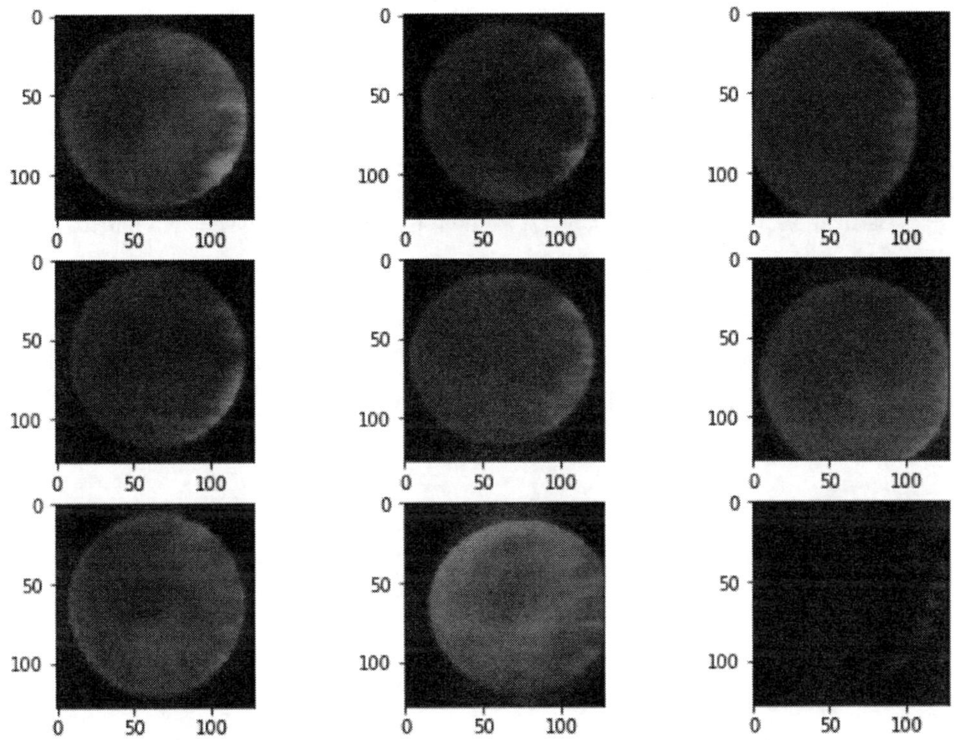

Figure 4. Examples of augmented images in training set. Shifts, small angle rotations, blur, zoom were applied.

Training. Neural network optimization method, hyperparameters, batch size, number of learning epochs, etc. We used the ModelCheckpoint procedures to monitor and preserve the neural network in training process and EarlyStopping procedure for loss function on a validation sample with a patience *stop* = 10 (training was stopped if the loss function did not decrease within 10 epoch). Number of *epochs* = 500, *batch_size* = 8. The mean square error (MSE) was used as a loss function. Adam's adaptive inertia method was used as an optimization method, which is currently the standard option for most tasks.

Metrics. After calculating the root-mean-square reconstruction error, it was normalized to one for its interpretation as the probability of belonging to the class of anomalous images. The higher the probability, the higher the confidence of the classification model. Then, the cut-off was selected according to the probability at which the smallest number of false positives of the model was observed and the quality metrics were calculated. Precision, recall and areas under ROC and PRC curves were used as quality metrics. The ROC curve is a very useful metric for assessing the quality of binary classifiers, and the area under it is the most popular feature of the quality of binary classifiers [7]. AUC ROC is equal to proportion of image pairs (image of class 1 – anomaly, image of class 0 – normal), which the algorithm correctly ordered, i.e. the first object goes in the ordered list before. The ROC curve reflects the dependence of true positive rate versus false positive rate. The advantage of ROC curve is that the area under the curve does not depend on the probability cutoff. We want the blue curve to be as close as possible to the upper left corner, then the area under the curve will tend to unity. However, due to the imbalance of classes, ROC curve may not be indicative. Therefore, we also use precision and recall metrics, as well as the area under of precision-recall curve (PRC). Precision is the proportion of images that belong to this class relative to all images that the model has assigned to this class. The recall is the proportion of images found by the classifier that belong to the class relative to all images of this class in the test set. The area of the precision-recall curve may be more indicative than AUC ROC in the case of strong class imbalance, but this quality metric is not interpretable, unlike AUC ROC.

5. Results

Our model seems to catch a lot of abnormal burning situations, when a large amount of coal dust is present when burning. However, the number of ordinary situations classified as an anomaly is indeed large, but the sensitivity of the detector can be adjusted by varying the probability threshold. The average classification precision was 77% and average recall was 66%. The area under precision-recall curve was AUC PRC = 0.54.

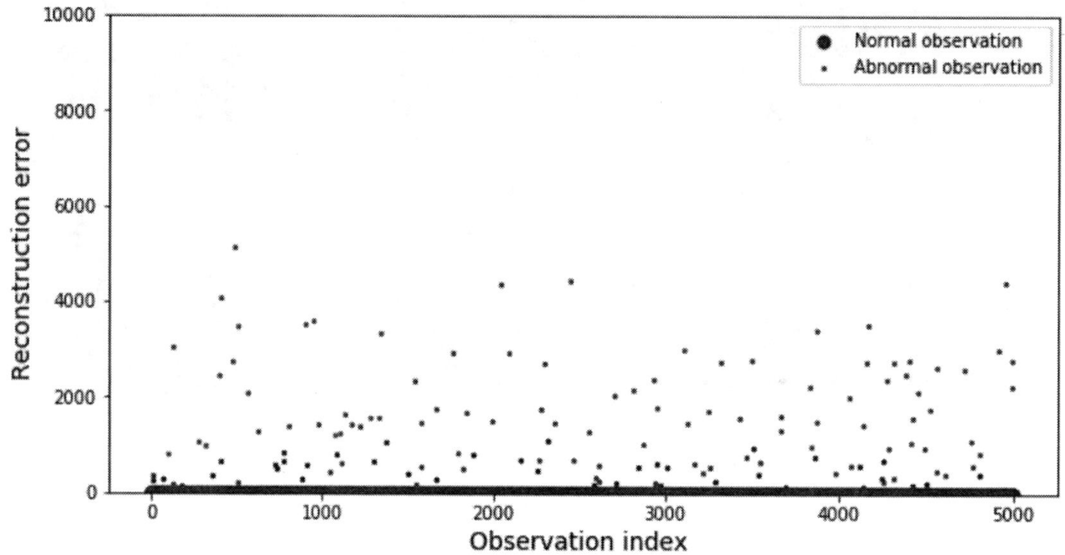

Figure 5. Reconstruction error on test sample. The blue dots show the reconstruction error of normal images and red ones – anomaly images. The classification probability is proportional to the reconstruction error.

Figure 5 shows the error of reconstruction of images of an autoencoder trained by using only normal images. As a result, it can be seen that the error in reconstructing images with high values of the coefficient of excess air is much lower compared with the regimes at low values of the coefficient. Thus, based on the analysis of the reconstruction error, it is possible to classify the normal and abnormal regimes of coal combustion. If the reconstruction error is scaled to unit interval, then it can be interpreted as the probability or degree of confidence of the classifier when the image is assigned to one or another class. The cutoff for the root-mean-square error equal to 200 is selected. Figure 6 shows the ROC curve and classification error matrix. The area under the ROC curve, demonstrating the quality of the binary classification is 0.8. According to the analysis of the model's confusion matrix, it can be seen that the model has a sufficiently large number of false negatives, that is, we do not detect a significant part of the abnormal images. Figure 7 shows the output of a fully-connected layer of the encoder, demonstrating how the model separates normal and anomalous observations. Basically, the model relies on the size of the region of intense luminescence on the periphery. The resulting number of errors is probably due to errors when labeling images due to insufficient control of parameters during the experiment. In Figure 7 image (a) model correctly classifies to normal. The image (c) is correctly assigned to the abnormal state, which is probably associated with a small area of intense luminescence from combustion, which is especially noticeable at the periphery. This seems logical because the intensity and size of the burning area is related to the amount of unburned coal dust. Image (b) is anomalous, but the model falsely classifies it as a normal combustion regime.

Tuning. Usually the process of developing a neural network begins with the development of a simple network (baseline model), either directly applying those architectures that have already been successfully used in literature to solve similar problems or using those hyperparameters that have previously produced good results. Ultimately, we hope, we will reach a level of performance that will serve as a good starting point, after which we can try to change all the fixed parameters and extract the maximum performance from the network. This process is commonly referred to as setting up hyperparameters, because it involves changing the network components that must be determined before starting the training. To improve the quality of models, the following approaches will be applied in the next step:

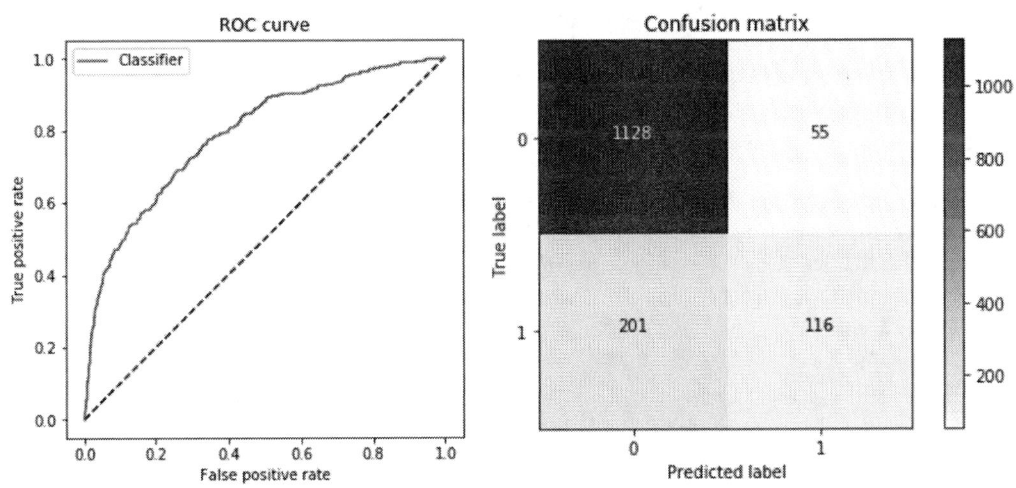

(a) (b)

Figure 6. ROC curve and confusion matrix.

1) *Regularization.* One of the main problems of machine learning is overfitting, when the model, in pursuit of minimizing the cost of training, loses the ability to generalize. Regularization makes it possible to reduce the complexity of the model. One of the effective ways to regularize a neural network is Dropout approach. A more direct approach to regularization, this is a penalty for a higher weights rate (L2 regularization) [8].
2) *Weight initialization.* When calculating initial weights of neural network, one can rely on a probability distribution (uniform or normal) with a dispersion equal to $D = 2 / (n_{in} + n_{out})$, where n_{in} and n_{out} are the number of neurons in the previous and next layer, respectively. The

He initialization method is more suitable for the ReLu activation function, compensating for the fact that this function returns zero for half the definition domain ($D = 2 / n_{in}$) [9].

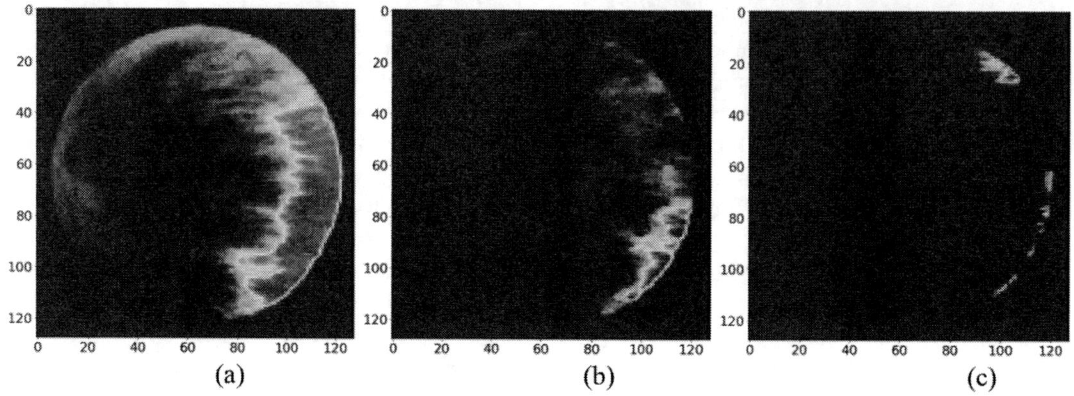

Figure 7. Visualization of output of fully-connected layer for neural network autoencoder. Examples of flame images in the combustion chamber as the excess fuel ratio decreases (from left to right). Image (a) is a true classified normal observation. Image (c) is a true classified anomaly. Image (b) is a false negative error.

3) *Batch normalization.* One of the popular methods for accelerating deep learning. The method solves the following problem, which impedes the effective training of neural networks: as the signal propagates through the network, even if we normalize it at the input, passing through the inner layers, it can be greatly distorted both by expectation and dispersion, which is fraught with serious inconsistencies between gradients at various levels. Therefore, we must use stronger regularizers, thereby slowing down the pace of learning. Batch normalization offers a very simple solution to this problem: to normalize the input data in such a way as to get a zero expectation and a unit variance. Normalization is performed before entering each layer.
4) *Cyclic learning rate.* The learning rate is one of the most important hyperparameters during training the neural network. A recent idea is called the cyclical learning rate, which is to increase the learning rate from time to time. This procedure allows to find another local minimum if the current minimum is not sustainable, and makes the model better generalized with new data [10].

6. Conclusions

In this work, a study was conducted of the possibility of using computer vision approaches, in particular, deep neural networks to detect abnormal burning regimes of a coal dust in the combustion chamber, corresponding to low values of the air excess coefficient, according to the analyzed gas composition. We have developed a deep neural network autoencoder, which is a combination of convolutional layers, fully-connected layers and upsampling layers. The autoencoder was trained to reconstruct the combustion regimes that corresponded to high values of the coefficient of excess air. The trained autoencoder was then used to identify abnormal burning regimes with a lower excess air ratio probably associated with an increase in the amount of unburned coal dust. The best classification model had AUC ROC quality metric value on a test image sample AUC ROC = 0.8. The average precision of the model was 77% and the average recall was 66%. Despite the large number of false negatives, the model can be useful for detecting anomalies in the combustion chamber. The precision and recall of the model can be significantly enhanced by collecting more detailed data with controlled combustion parameters, which will make the image labeling more accurate. The paper concluded that the use of upsampling with convolutional layers show themselves better than the deconvolutional layers, and the combination of convolutional layers with fully-connected layers constitutes the optimal architecture of the model. In addition, complementary regularization methods could help generalization, but this seems unnecessary, since a convolutional autoencoder with layers of high sampling could achieve almost as good results with a network ten times smaller. The combination of these two types of layers in the end ensured the best performance with a reasonable architecture.

Acknowledgements
This work was carried out within the framework of the state assignment for the IT SB RAS.

References

[1] Chernetskiy M, Burdukov A, Butakov E, Anufriev I and Strizhak P 2016 Using ignition of coal dust produced by different types of mechanical treatment under conditions of rapid heating *Combustion, Explosion, and Shock Waves* vol 52 pp 326–28

[2] Abdurakipov S, Gobyzov O, Tokarev M and Dulin V 2018 Combustion Regime Monitoring by Flame Imaging and Machine Learning *Optoelectronics, Instrumentation and Data Processing* vol 54 pp 513–19

[3] Abdurakipov S and Butakov E 2018 Combustion anomalies detection for a thermal furnace based on Recurrent Neural Networks *Journal of Physics: Conference Series* vol. 1105 012043

[4] Gobyzov O, Tokarev M, Abdurakipov S and Lobasov A 2018 Flame state diagnostics using visualization and neural network analysis *AIP Conference Proceedings* vol 2027 040067

[5] Hinton G and Salakhutdinov R 2006 Reducing the dimensionality of data with neural networks *Science* vol 313 pp 504–7

[6] Bengio Y, Courville A and Vincent P, 2013 Representation learning: A review and new perspectives *IEEE Transactions on Pattern Analysis and Machine Intelligence* vol 35 pp 1798–828

[7] Wen L, Gao L and Li X 2017 A new deep transfer learning based on sparse auto-encoder for fault diagnosis *IEEE Transactions on Systems, Man, and Cybernetics: Systems* vol 99 pp 1–9

[8] Xu J, Xiang L, Liu Q, Gilmore H, Wu J, Tang J and Madabhushi A 2016 Stacked sparse autoencoder (SSAE) for nuclei detection on breast cancer histopathology images *IEEE transactions on medical imaging* vol 35 pp 119-30

[9] Goodfellow I, Courville A and Bengio Y 2016 Deep learning *The MIT Press Cambridge*

[10] Smith L 2017 Cyclical learning rates for training neural networks *IEEE Winter Conference on Applications of Computer Vision* pp. 464–72

Journal of Physics: Conference Series **1421** (2019) 012006 doi:10.1088/1742-6596/1421/1/012006

Experimental observation of turbulent exchange between heterogeneous vortex ring and surrounding medium by the shadowgraph method

V V Nikulin and R A Panenko

Lavrentyev Institute of Hydrodynamics SB RAS, pr. ac. Lavrentyeva, 15, Novosibirsk, 630090, Russia

Abstract. A shadow visualization of turbulent exchange between vortex ring containing fluid less dense than that out of it and surrounding medium is performed. A characteristic distance and time of this process are determined and the dependence of time on both vortex velocity and the related Reynolds number is obtained. Using characteristic time, the turbulent diffusion coefficient in the vortex atmosphere is estimated.

1. Introduction

The study of turbulent exchange between a vortex ring and surrounding medium is of fundamental and practical interest. A fundamental importance is that the vortex ring is the example of isolated concentrated vortex. Therefore, the regularities obtained in the study of it are related to a vortex flow pattern and may refer to other concentrated vortices, in particular, to linear ones. A practical importance is due to the fact that vortex rings often appear in different artificial and natural processes taking place, for instance, during explosions, volcanic eruptions, etc. Hence, it is necessary to be able to estimate the characteristics of turbulent exchange as it causes the removal of impurities and heat from the vortex. Moreover, the dynamics of vortex rings, particularly, their expansion coefficient is intimately related to rotation decay owing to dissipation processes caused by turbulence [1].

It is known that a vortex ring is a torus-shaped volume of vortex-type fluid or gas (vortex core) that moves with some surrounding volume of medium called a vortex atmosphere. The study of turbulence in such volumes is essentially complicated due to the presence of non-stationary and vortex pattern of the flow, therefore, the obtaining of turbulent characteristics is an important achievement. By now it has been established experimentally that the exchange takes place mainly between the vortex atmosphere and surrounding medium while the turbulence is suppressed in the core [2]. In [3, 4] a new method of observation of turbulent exchange between vortex ring atmosphere and surrounding medium has been proposed. To this end, vortex rings whose interior fluid density is more than exterior one in the beginning of motion are produced. It is possible to visualize the vortex due to optical inhomogeneities arising during a turbulent exchange of fluids with different optical density inside and outside of it. In [3] the case when the difference of densities has one value has been considered. In [4] the experiments have been performed for three differences of density values. It has turned out that the result was almost independent of the value of this difference at sufficiently large values of velocities of vortex motion.

This work deals with the case when the fluid density in vortex interior is less than that in its exterior in the beginning of motion. A characteristic distance and time of turbulent exchange have been established. Basing on characteristic time, the coefficient of turbulent diffusion in vortex ring atmosphere has been estimated.

2. Experimental results

The experimental setup is shown in figure 1. It consists of uniformly illuminated matte screen (1), vertical chamber with cross-section of 150×150 mm and height 300 mm (2). The chamber is filled with two-layer fluid. The upper layer is water of density ρ_1 = 1g/cm^3 and depth 100 mm. The lower layer is a sugar solution of density ρ_2 = 1.08g/cm^3 and depth 150 mm. A vortex ring is formed by forcing out a pulse jet with the help of a piston from cylindrical tube (3) 21 mm in diameter through a nozzle 12.5 mm in diameter. The piston stroke is 10 mm, so the jet length is 28 mm. The distance from the nozzle to the interface between the layers (4) equals 50 mm. Vortex (5) is formed in the upper layer, moves vertically downward and enters the lower layer containing fluid less dense than the surrounding medium. Vortex ring motion is recorded by the high-speed video camera (6) in the lower layer. The distance between camera and vortex ring is 2.3 m, the pathway passed by the vortex is within the camera view. The record of vortex shadow image is performed perpendicular to vortex motion direction with frequency from 125 to 1000 frames per second. The velocity u and radius of the vortex core R are defined in the lower layer at the distance 20-25 mm from the interface. Kinematic viscosity taken to be equal to 10^{-2} cm^2/s, experimentally measured velocity and radius are used for calculation of the Reynolds number. The temperature of fluids is within the range 20–22 °C. Consequently, the variation of fluid kinematic viscosity is less than 5%. The initial vortex velocity is varied in the experiments. The vortex core of diameter $2R$ is approximately constant and equals (14.3 ± 0.3) mm. The diffusion coefficient of sugar solution in water is $3\cdot10^{-6}$ cm^2/s. The time duration of the process is less than 0.5 s. Then a characteristic length of molecular diffusion is less than $1.2\cdot10^{-3}$ cm and consequently much less than the vortex dimensions. Therefore, molecular diffusion can be neglected.

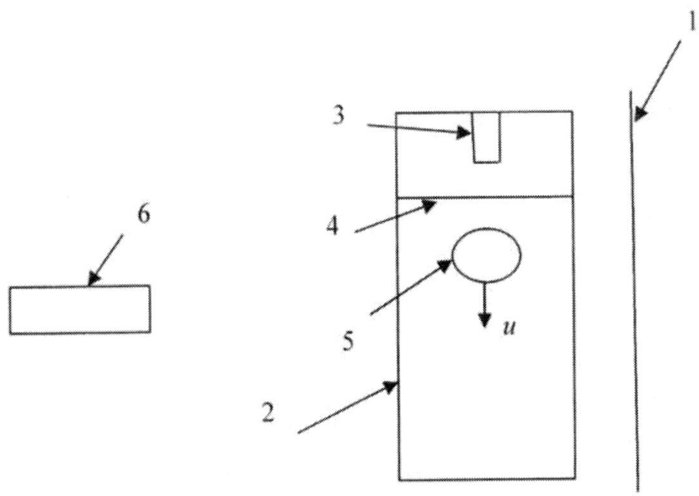

Figure 1. Experimental setup.

The comparison of these experiments with data [3, 4] shows the following. A qualitative difference of shadow image obtained in transition from less dense medium to a denser one from that obtained in transition from a denser medium to less dense one is observed only near the interface. In the second case [3, 4], the image of vortex atmosphere boundary having passed through the interface is observable during the whole transition time. In the process no virtual images exist. During the transition from less dense medium to a denser one before the vortex moves from the interface to a sufficient distance, apart from a real shadow image of atmosphere having passed through the interface, firstly, two virtual images are observed: one of them is reflected from the interface, another one appears due to bended light rays passing through diffusion zone of the interface.

Figure 2 demonstrates shadow images of vortex ring passing through the interface.

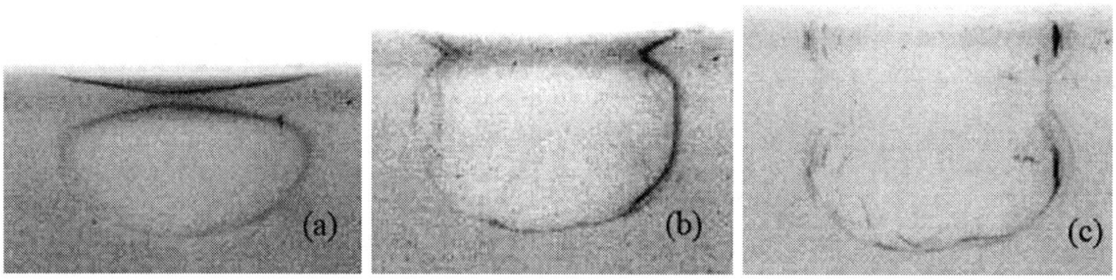

Figure 2. Shadow record of vortex ring transition from upper to lower layer. A real image of front part of the vortex atmosphere and its two virtual images (a), virtual images are partially merged (b), vortex atmosphere and one virtual image (c).

The virtual image disappears when the vortex ring moves from the interface to a distance approximately equal to 20 mm. The further behavior of image is qualitatively identical to that for the motion of vortex ring having a denser fluid than a surrounding one [3, 4]. Figure 3 shows shadow images of vortex ring in the lower layer at various distances from the interface.

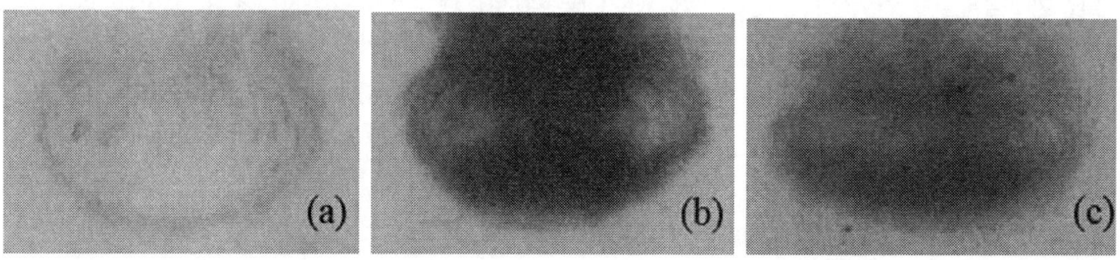

Figure 3. Shadow records of vortex ring at various distances from the interface: 18 (a), 40 (b) and 120 mm (c). Vortex velocity $u = 2.7$ m/s, Re = 19300.

At the distance about 20 mm from the interface, the boundary of the vortex atmosphere is visible, since at such distance turbulent exchange is insignificant; therefore, the fluids inside and outside the vortex are almost uniform, figure 3 (a). As it moves, the uniformity of the media is disturbed and turbulent volumes of one fluid penetrate into another causing the appearance of density gradients and shadow visualization becomes possible. For a vortex ring it looks like an abrupt darkening of the atmosphere. At first, the darkening grows, figure 3 (b), then it starts decreasing at some distance, figure 3 (c). This means that the amount of less dense fluid in the vortex decreases being replaced by the surrounding medium. In the shadow images, the vortex core can be seen, which is brighter than its surroundings due to the effect of turbulence suppression. The centerline diameter of the core torus is determined from the images. The distance from the interface to the centerline was taken as distance z from the interface to the vortex ring, since the core is parallel to the boundary. The average velocity when the vortex passes the distance of 10-20 mm is taken as velocity.

To obtain quantitative characteristic for darkness, the following method is used. According to figure 3 (a), the vortex with its atmosphere is well visible at the distance about 20 mm from the interface. Using such records, the pattern of its atmosphere is constructed. The atmosphere is assumed

Journal of Physics: Conference Series **1421** (2019) 012006 doi:10.1088/1742-6596/1421/1/012006

to be unchangeable during the observation although the outer part of atmosphere may be invisible, figure 3 (c). This assumption is justified by the fact that distance of observation is not more than six vortex diameters. It is known that along such a distance the vortex parameters do not practically vary. The pattern of atmosphere is superposed on the shadow records positioning the pattern to vortex core in the same way. Records are taken at various distances from the interface. Then, average darkness J of the region corresponding to shadow image of vortex ring atmosphere is established.

The typical dependence of average darkness J on distance from interface z is shown in figure 4. Figure 4 illustrates a clearly defined maximum of darkness. The experimental results are shown by the points; solid line is b-spline interpolation of the points. Similar dependence is typical, darkness maximum is observed at all vortex velocities. The value of darkness maximum changes with variation of vortex velocity: the higher velocity the higher darkness maximum.

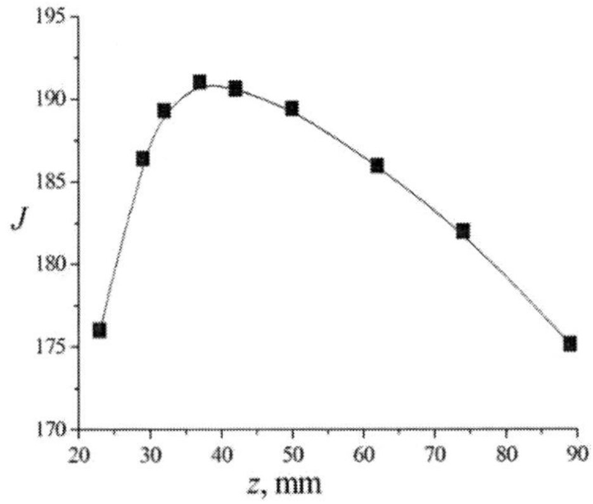

Figure 4. Average darkness of vortex shadow image J versus distance from interface z, u = 130 cm/s, Re = 9300.

As in [3, 4], the distance from the interface to vortex position where darkness has a maximum is characteristic distance of the turbulent exchange process because darkness is related to turbulent exchange. This distance is well defined in every experiment and it is denoted as z_*. So z_* is characteristic distance of the turbulent exchange process between vortex ring and surrounding medium. Using z_* and vortex velocity u, we introduce characteristic time of turbulent exchange as follows: $t_* = z_*/u$. As dimensionless characteristic time we take $\tau = t_* \cdot u/R$, where R is a vortex ring radius. Using expression for t_*, we obtain: $\tau = z_*/R$. In figure 5, the dependence of z_* on Re number is shown.

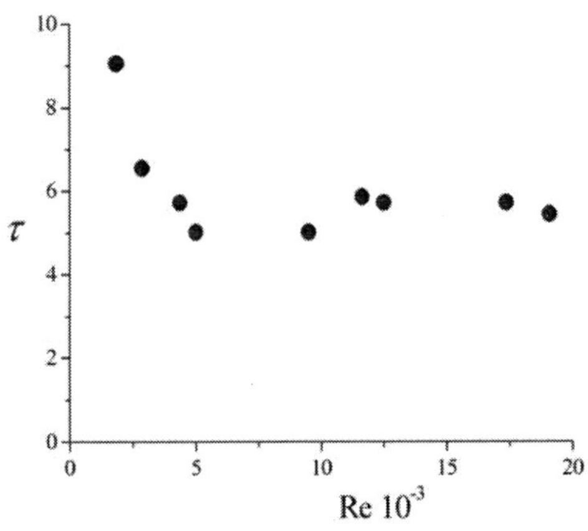

Figure 5. Dimensionless characteristic time of the turbulent exchange process between vortex ring and surrounding medium versus Re number.

According to figure 5, the dimensionless characteristic time initially decreases with an increase in Re number. Then for Re approximately greater than 5000, it is practically independent of Re and equal to 5.5 ± 0.34.

3. Analysis of the results and conclusions

According to [4], vortex radius is 7.5 mm, the average value of z_* and its variance are 32.7 ± 3.8 mm. Then dimensionless characteristic time for data [4] is 4.4 ± 0.5. So turbulent exchange between vortex atmosphere and surrounding medium depends slightly on density differences but the exchange occurs more rapidly when the fluid in the vortex is denser then that out of it.

To estimate the coefficient of turbulent diffusion in vortex atmosphere, we consider diffusion from the volume of spherical form. It has been known that characteristic diffusion time t_d from the sphere of radius R is: $t_d = (R/\pi)^2 \cdot (1/D)$, where D is a diffusion coefficient. In the experiments presented here, characteristic time $t_* = \tau \cdot R/u$. We found diffusion coefficient D such that characteristic diffusion time from the spherical volume will be the same as that from the vortex. Then from equation $t_d = t_*$, we obtain expression for D: $D = (R \cdot u)/(\pi^2 \cdot \tau)$. Since for Re > 5000 τ is approximately constant, then D in this area of Re numbers is proportional to vortex velocity u. The substitution of $\tau = 5.5$, $R = 0.72$ cm, $u = 100$ cm/s into the expression for D yields $D \approx 1.3$ cm^2/s. Thus, the turbulent diffusion coefficient in the atmosphere of experimental vortex is larger than the molecular one by many times.

Acknowledgments

The work was supported by the RFBR grant No. 18-08-00824.

References

[1] Nikulin V V 2014 Analitical model of motion of turbulent vortex rings in an incompressible fluid *J. Appl. Mech. Tech. Phys.* **55** 558–64

[2] Vladimirov V A and Tarasov V F 1973 Structure of turbulence near the core of vortex ring *Dokl. Akad. Nauk SSSR* **245** 1325–8 (in Russian)

[3] Nikulin V V and Panenko R A 2018 Experimental observation of turbulent exchange between vortex ring and surrounding medium of different density *J. Phys.: Conf. Ser.* **1128** 012022

[4] Nikulin V V 2019 Shadow visualization of turbulent exchange between a vortex ring and the environment at different densities of the fluid in the vortex interior and exterior *Tech. Phys.Letters* **45** No.4 8-10

Investigation of temperature field of weakly ionized gas jet by the background oriented schlieren method

I I Korotkikh, Yu I Malakhov and A V Kuchmenko

National Research University "Moscow Power Engineering Institute", Krasnokazarmennaya str., 17, Moscow

E-mail: korotkikhii@mail.ru

Abstract. The paper shows the possibility of using the background oriented schlieren technique to study the temperature field of a weakly ionized plasma jet created as a result of a spark discharge in argon at atmospheric pressure. The design of the spark source, the experimental setup, the experimental method and the processing of experimental data are presented. The temperature field of weakly ionized plasma jet and the temperature distribution around it were obtained.

1. The object of study and the scheme of experimental setup

The object of research in this work is a stream of weakly ionized plasma that is formed during a spark discharge in argon at a pressure close to atmospheric. The study of this object is important in the development and design of medical therapeutic plasma torches, when it is important to know the temperature of the plasma torch. Sources of weakly ionized plasma are increasingly used for the treatment of material surfaces, the application of functional (wear-resistant, electrically conductive, heat-resistant, decorative) coatings (electrical alloying), and the ozone formation.

In modern plasma processes there may be no mechanical contact of the instrument with the workpiece, the effect on the object is carried out only by plasma.

Plasma technologies are increasingly used for surface treatment, plasma cutting, welding; high power plasma torches are used for particle processing, waste disposal, surface activation, repair and hardening of machine parts, tools, dies. To control these technological processes, it is also necessary to know the temperature field inside and around the plasma flows.

The system is a coaxial construction consisting of central and external electrodes. At the end of the outer cylinder it is possible to install nozzles of various configurations for change of the jet geometry and variation of its speed. In addition, in a wide range, it is possible to adjust the input pressure values.

It is possible to install a central electrode of various materials, change the distance between it and the end nozzle. A spark occurs in the gap between the tip of the central electrode and the end of the external electrode (nozzle). The impulse voltage has the form of consecutive trains, with amplitude of 4 kV, followed by a repetition period of 10 ms.

Journal of Physics: Conference Series **1421** (2019) 012007 doi:10.1088/1742-6596/1421/1/012007

The geometry of the discharge region is such that heating, excitation and subsequent ionization of argon atoms occur inside the system, and only neutral and excited atoms that are a source of optical and ultraviolet radiation, enter the open space.

The block diagram of the spark source is presented in figure 1.

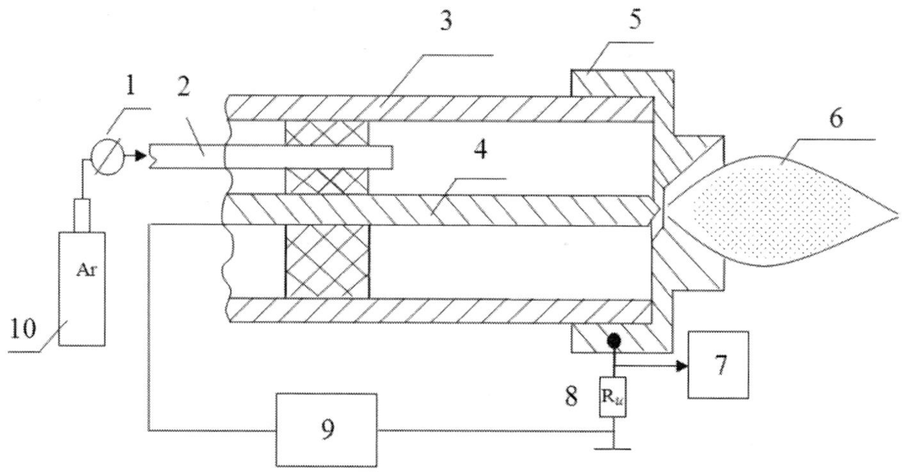

Figure 1. Block scheme of the spark source of low-temperature plasma: 1 – reducer, flow meter and manometer, 2 – gas supply tube, 3 – external protective cylinder, 4 – entral electrode, 5 – external electrode, 6 – outlet gas jet shape, 7 – digital oscilloscope, 8 – resistor for measuring the discharge current, 9 – high-voltage source, 10 – gas cylinder.

The waveform of voltage in discharge gap is presented at figure 2. As can be seen from the waveform, the jet formed by the source represents the periodically repeating pulse trains. The duration of the pulse train is 8 ms, it contains 340 individual pulses and each of them has the duration of 23 μs.

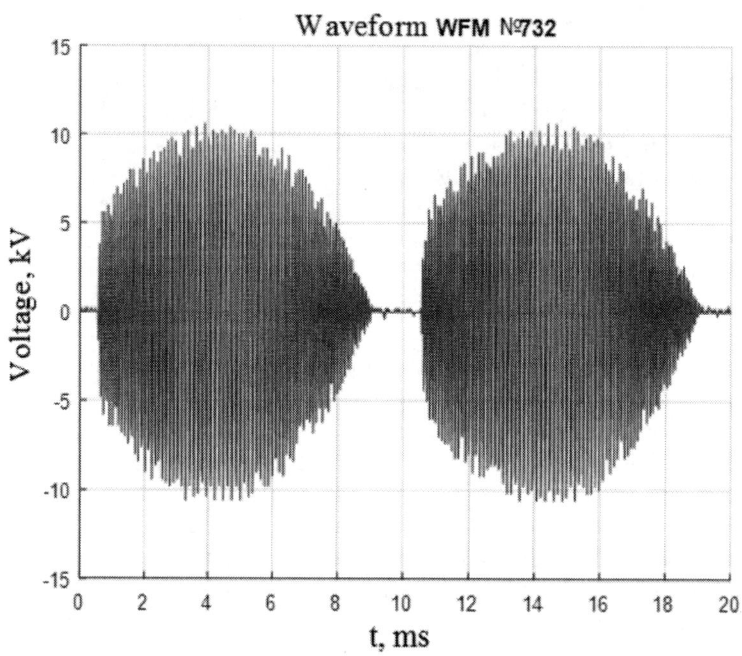

Figure 2. The waveform of voltage between electrodes.

Journal of Physics: Conference Series **1421** (2019) 012007 doi:10.1088/1742-6596/1421/1/012007

The background oriented schlieren method was used for visualization of the temperature field of a jet of weakly ionized gas produced in a spark discharge. The experimental setup for determination of the temperature field is shown in figure 3. The aim of the study was to test the possibility of applying background oriented schlieren method for diagnostics of a plasma stream that changes in time and researching to analyze the dynamics of the flow, so an unstable, and a randomly changing flow was formed. One of the possible flows is presented in figure 4.

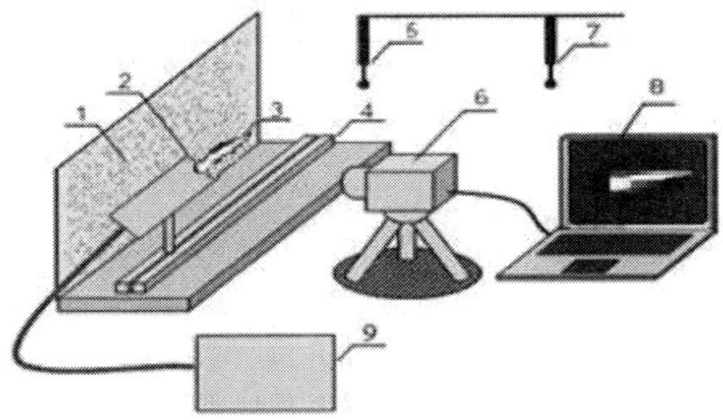

Figure 3. The scheme of the experimental setup for determination of the temperature field of a weakly ionized gas jet by the background oriented schlieren method:

1 – background screen, 2 – spark source of low-temperature plasma, 3 – jet at the exit of the spark source, 4 – optical bench, 5 – measuring thermocouple, 6 – high-speed digital video camera, 7 – control thermocouple, 8 – personal computer, 9 – high-voltage power supply.

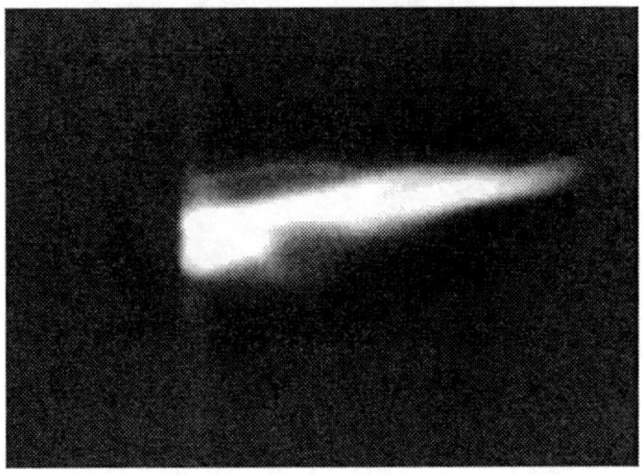

Figure 4. The shape of the plasma torch (exposure time 100 μs).

For the recoording of the jet 3 of a weakly ionized gas formed by the spark source 2 by the background oriented schlieren method, it is nessesary to use the high-speed digital video camera 6, personal computer 8 that is used for storing and processing of images. The video camera is focused on the background screen 1. The spark source is fixed on the optical bench 5; to create a spark discharge, a high-voltage power source 9 of the NP-750-30 type is used. At shooting, a Fastec HiSpec-1 digital video camera was used. The lens Navitar CCTVLENS was used to create a clear image on the plane of matrix. This camera allows shooting of high-speed processes with different frame rates. The kit includes a set of interchangeable lenses, that allows choose the shooting conditions and ensure the best quality of shooting. The aperture of the lens when shooting as closed as possible, it is necessary to ensure the greatest possible depth of field.

The background screen consists of chaotic points of various diameters. Randomness of the points is necessary, since the displacements of points inside and outside the object under study are different. If the location of the points is ordered, then a situation may arise when it is impossible to see the displacement of the points inside the flow under study.

There are the photos of the structured screen used for the background oriented schlieren method: without plasma stream at figure 5 and with plasma stream at figure 6.

Journal of Physics: Conference Series **1421** (2019) 012007 doi:10.1088/1742-6596/1421/1/012007

Figure 5. Photos of the structured screen used for the background oriented schlieren method without plasma flow.

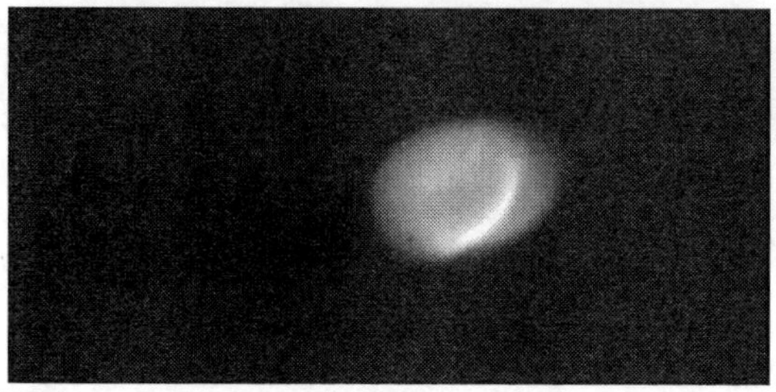

Figure 6. Photos of the structured screen used for the background oriented schlieren method with plasma flow.

The optical axis of the high-speed camera is aligned with the geometrical axis of the plasma torch. Before shooting, the scaling coefficient of the torch image transfer to the plane of the matrix is determined. Next, the images of the object under study (plasma jet) are recorded with frequencies of 1000, 2000 and 3000 frames per second during 2-3 seconds. This time is sufficient to obtain the necessary array of experimental data. Shooting is done with a depth of brightness with a dynamic range of 8 bits. The video camera is focused on the axis of plasma flow under study.

Radiation occurs in a wide range of wavelengths. So various light filters were used at the recording. Each of them corresponds to one of the lines of the argon emission spectrum. The use of light filters makes it possible to isolate narrow regions of the plasma spectrum where there are no argon lines, as well as to avoid the exposure.

2. Determination of the temperature field of plasma jet by background oriented schlieren method and processing of experimental data

The background oriented schlieren method [1, 3] is used for visualization of the temperature field of a weakly ionized gas. The object of study (plasma torch) is located between the high-speed digital video camera and the background screen.

The essence of the background oriented schlieren method is as follows. The screen images of the object under study and without it are recorded using a high-speed digital video camera. The images are randomly structured black dots on a white background. One image is recorded in the absence of a disturbed medium in the image transmission channel. In another picture, the medium is perturbed by the plasma jet that leads to the change in the optical properties of the medium. In the presence of a disturbed medium (thermally excited gas), the position of the background screen points changes. This changes the structure of the image of the background screen.

Displaces of the object points are registered. With the help of cross-correlation processing, changes in the position of points on the structured screen image are analyzed. Further processing is reduced to a change in the coordinates of the background point on screen. It is associated with a change in the refraction of the medium that depends on temperature. Analyzing the obtained images, we can trace the dynamics of the plasma torch development. The field of correlation coefficients is investigated. Correlation equation [1] is

$$B_{fg_{n,m}} \approx \frac{1}{n \cdot m} \sum_{k_1=1}^{n} \sum_{k_2=1}^{m} \left(f_{k_1k_2} - \mu_f\right)\left(g_{k_1k_2} - \mu_g\right), \quad (1)$$

where B – value of correlation coefficients; n – number of rows; m – columns number of the image matrix; $f_{k_1k_2}, g_{k_1k_2}$ – arbitrary random sequences of the values of the background screen point displacements; μ_f, μ_g – average values of these sequences, respectively.

Journal of Physics: Conference Series **1421** (2019) 012007 doi:10.1088/1742-6596/1421/1/012007

Distribution of background screen points in the range of values of correlation coefficients from 0.3 to 1.0 rel. units is presented at figure 7.

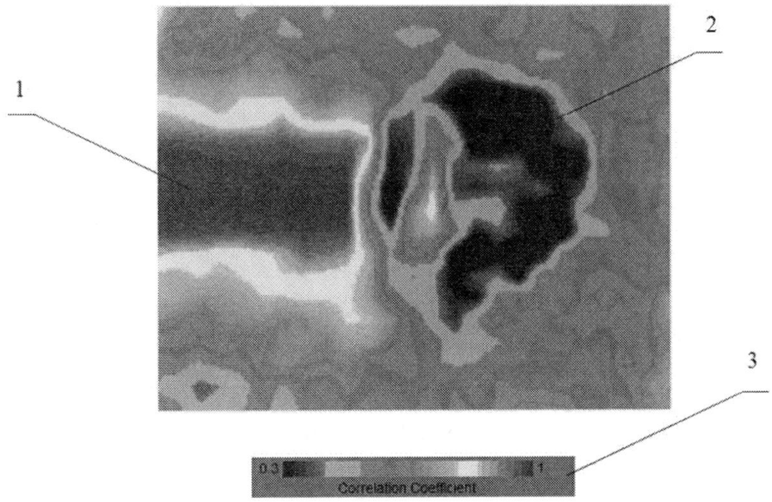

Figure 7. Distribution of background screen points in the range of values of correlation coefficients from 0.3 to 1.0 rel. units: 1 – spark source of low-temperature plasma, 2 – plasma flow under study, 3 – distribution of displacements of the background screen points.

For obtain of the temperature field, it is necessary to determine the normalization factor relating the values of the displacements of the background screen points and the temperature at the reference point. The temperature values are recorded using two thermocouples: a reference one (determining the temperature outside the object under study). The probe one is located in close proximity to the object under study.

Continuous recording of temperature is carried out using the ADC. A sequence of temperature values is formed at the studied point at successive times with steps of 0.1 s.

The algorithm for converting the field of displacements of the points of the background screen in the temperature field is as follows. On the original image there is a vertical or horizontal straight line. The temperature obviously varies evenly along it. Next is the sum of the density gradients along this line.

$$S = \sum_{i=0}^{n-1} \mathrm{Re}(AB_{i,j}), \tag{2}$$

where AB – original complex matrix representing the vector field i, j – row and column numbers, respectively.

In Mathcad the recording of vectors into a matrix is performed in a complex form. The real part of matrix is equal to the horizontal component, and the imaginary part - to vertical component.

To sum up the gradients along a horizontal or vertical straight line, it is necessary to lay down separately the real and separately imaginary components of the matrix elements.

The density gradient is proportional to the gradient of the refractive index and the temperature gradient. Then the temperature difference at the end points of the line (boundary conditions) is divided by the sum of the gradients.

$$x = \frac{T_2 - T_1}{S}, \tag{3}$$

where T_2 and T_1 – temperatures in °C at the opposite ends of the line.

Thus, we get the temperature difference per unit gradient. It is possible to calculate the temperature change from one point of the vector field to another.

After normalization, the temperature field is calculated vertically and horizontally. To do this, take the temperature at any edge of the image (boundary conditions). The temperature of entire field is calculated with the help of sequential addition.

$$TH_{n-1-i,j} = TH_{n-i,j} + x \cdot AB_{n-1-i,j}, \tag{4}$$

where n – the number of rows of the matrix.

A similar procedure is performed in the perpendicular direction.

Then the field values are averaged at this point in the vertical and horizontal direction:

$$T_{i,j} = \frac{TV_{i,j} + TH_{i,j}}{2}, \tag{5}$$

where TV and TH – two matrix of the temperature field calculated in the vertical and horizontal directions, respectively.

The resulting value is equal to the temperature of plasma stream at this point.

The temperature distribution inside and around the plasma flow are shown at figure 8.

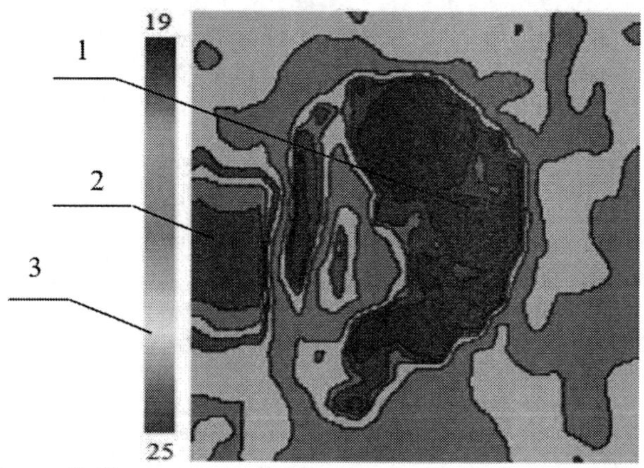

Figure 8. Temperature distribution inside and around the plasma flow: 1 – plasma flow under study, 2 – spark source of low-temperature plasma, 3 – temperature distribution, °C.

The results show that the temperature of plasma jet under study is in the range from 19 to 25 °C. The obtained temperature values indicate that this source forms a stream of cold plasma. The thermodynamic temperature at the outlet of the cylindrical tube is close to room temperature (23 °C) or slightly lower. Since the temperature variation inside the plasma flow is insignificant, it can be concluded that at the exit from the discharge gap there are mainly excited atoms. According to the classical theory, the thermodynamic temperature is related to the density of electrons and atoms ($p = nkT$). To analyze the composition of the particles at the exit from the discharge gap, an estimate of the density of excited atoms in the region under study was made.

It is established that the values of particle density both inside the plasma flow and around it have values of the same order ($n \approx 10^{26}$ 1 / m^3). This fact indicates that at the exit from the discharge gap, excited atoms predominate, and the ionization process takes place inside a cylindrical tube, near the central electrode. Since the density of particles inside the outgoing plasma stream and outside it are practically not differ (the difference does not exceed one order of magnitude), it can be concluded that in this medium the electron density is small, and the field of study consists mainly of excited atoms.

The temperatures of the excited atoms in the outgoing plasma flow decrease slightly compared with the temperatures of the surrounding particles. This effect can be explained by the fact that the ionization of the flow created by this source occurs near the central electrode inside the cylindrical tube. At the exit of the tube, only excited particles can be observed. And their kinetic energy is less than the energy of the surrounding particles.

The obtained experimental data are consistent with the data of other authors [6–15]. The paper shows the possibility of using the background oriented schlieren method for the diagnostics of various plasma medium, including the medium with a high degree of ionization. The considered method allows obtain the temperature field of the plasma flows under study, as well as the concentration field of electrons and excited atoms.

In addition, the use of the background oriented schlieren method allows you to take pictures of various stages of the discharge, analyze the temperature and composition of particles both inside the plasma flow and around it. The method can also help in the analysis of particle drifts and the search for the causes of instabilities of high-temperature plasma, and hence in the search of methods for elimination of these instabilities.

The background oriented schlieren method for plasma streams also allows control the parameters of plasma medium (temperature, concentration, etc.). It can be widely used at various plasma technological processes, such as plasma spraying, plasma particle processing, plasma chemistry and etc.

It should be noted that this method is purely optical, i.e. it is contactless. It does not distort the parameters of the object. It allows exploration of the optical inhomogeneities of small length. The method is simple and universal.

3. Conclusions

In this paper the possibility of applying the background oriented schlieren method to determine the temperature field of a weakly ionized plasma jet is confirmed. A method for recalculating the field of displacements of background screen points in the presence of a plasma stream into the temperature field is presented. Using the background oriented schlieren method it is possible to analyze the temperature and composition of particles inside and around the plasma flow at different points in time. Sources of this type must be stabilized in order to obtain pulses of the same duration and amplitude to ensure accurate determination of the dose of exposure to biological tissue. As an object of study, a construction similar to medical sources of low-temperature plasma was used. At the work with such sources it is necessary to understand what particles are present in the jet of this source (neutral atoms, electrons, ions). For this aim the temperature field of this source jet should be determined. It is necessary to understand the value of UV radiation in this jet, and which of the listed factors is important in assessing the biological effects on human tissue.

In the future, it is intended to move from the temperature field to the density field of particles forming the plasma flow. The results show that in the studied jet charged particles are practically absent. There is intense ultraviolet radiation that is the main factor influencing biological objects. Thermal factor does not play a significant role. The field temperature is slightly different from the environment.

References

[1] Skornyakova N 2011 Background oriented schlieren method and its application *Modern optical methods for the flows studying, Overlay* 93–106

[2] Korotkikh I, Malakhov Yu and Skornyakova N 2017 High-speed recording of discharge formation in inductively coupled plasma *Scientific Visualization* **9** **(3)** 70–6

[3] Glazyrin F, Znamenskaya I, Mursenkova I, Sysoyev N and Czin C 2012 Studies of shock wave flow in the channel by the shadow and background oriented schlieren method *Autometrics* **48** **(3)** 101–10

[4] Mizukaki T 2010 Visualization of compressible vortex rings using the background oriented schlieren method *Shock Waves* **20** **(6)** 531–37

[5] Zare-Behtash H, Gongora-Orozco N and Kontis K 2009 Global visualization and quantification of compressible vortex loops *Journal of Visualization* **12** **(3)** 233–40

[6] Arakeri J, Das D, Krothapalli A and Lourenco L 2004 Vortex ring formation at the open end of a shock tube: a particle image velocimetry study *Physics of Fluids* **16** **(4)** 1008-19

[7] Murugan T, De S, Dora C, Das D and Kumar P 2013 A study of the counter rotating vortex rings interacting with the primary vortex ring in shock tube generated flows *Fluid Dynamics Research* **45** **(2)** 025506

[8] Akiyama T, Yoshimura S, Tomita K, Murakami T and Urabe K 2017 Design of compact dispersion interferometer with a high efficiency nonlinear crystal and a low power CO_2 laser *Journal of Instrumentation* **12** 120–8

[9] Gerasimov S and Trepalov N 2017 Background oriented schlieren method - an optical method for studying of shock waves *Journal of Technical Physics – FSUE Science Publishing House* **87** **(12)** 1802–7

[10] Smolanov N 2017 Electromagnetic and optical properties of microparticles from low-temperature arc-discharge plasma *XLIV International Zvenigorod Conference on Plasma Physics and Controlled Thermonuclear Synthesis – Collection of theses of reports: CJSC STC Plazmaiofan* 342

[11] Malakhov Yu, Nguen K-Shi and Korotkikh I 2015 Features of an automated system for spectral diagnostics of RF plasma torches *Journal of Induction Heating* **1** 21–8

[12] Boselli M, Colombo V, Ghedini E, Gherardi M, Laurita R and Liguori A Characterization of a cold atmospheric pressure plasma jet device driven by nanosecond voltage pulses *IEEE Transactions*on *Plasma Science* **43** **(3)** 713–25

[13] Traldi E, Boselli M, Simoncelli E, Stancampiano A 2018 Shadow visualization methods: a powerful tool for diagnostics of atmospheric plasma *European Physical Journal of Engineering and Tools* **5** **(4)**

[14] Chamorro J, Transvost L, Tsekhas E, Fishfeld G, Kelly X and Mancinelli B 2018 Ambient Species Density and Gas Temperature Radial Profiles Derived from a Schlieren Technique in a Low-Frequency Non-thermal Oxygen Plasma Jet *Plasma Chemistry and Plasma Processing* **38** **(1)** 45–61

[15] Glazyrin F, Znamenskaya I, Koroteeva E, Mursenkova I and Sysoev N 2013 Using the Background oriented schlieren method to study a non-stationary flow with a shock wave *Scientific Visualization – NIIU MEPhI* **5** **(3)** 65–74

The experimental investigation of water spray disperse parameters behind the water injecting system at the aircraft engine inlet by applying PSV

A A Loginova[1,2], O G Chelebyan[1], B I Mineev[1] and A Yu Vasilyev[1]

[1] Central Institute of Aviation Motors, Aviamotornaya 2, Moscow, 111116, Russia
[2] Bauman Moscow State Technical University, 2-ya Baumanskaya 5,105005, Moscow, Russia

E-mail: aaloginova@ciam.ru

Abstract. To provide an aircraft engine certification tests for water ingestion, it is necessary to control accurately the parameters of the water drops generated by the water injecting device simulating rain at the engine inlet. The paper presents the results of autonomous tests of a single-channel centrifugal nozzle for the water injecting device. The tests purpose is to investigate the disperse parameters of the water spray for the typical operation modes of the nozzle as a part of the water injecting system. Tests were carried out at the experimental setup using the PSV (Particle Shadow Velocimetry) technique at three typical nozzle operation modes. As a result of the tests, the distribution graphs of the D_{10}, D_{32} and Dv_{50} diameters, axial velocity (U) and mass flow (M_{abs}) of water droplets over the radius of the spray pattern have been obtained. To assess Dv_{50} mean median diameter of spray droplets the cumulative distribution functions of the droplets mass fraction depending on the size for each of the operation modes of the nozzle have been plotted. On the basis of these functions, the mean median diameters of water spray droplets have been calculated.

1. Introduction

According to the requirements of IAC normative documents [1] it has to be demonstrated that the engine is capable of ingesting simulated rainfall without suffering operating problems. The engine acceptable operation does not allow for flameout, stoppage, prolonged surging or stall. After ingestion the absence of unacceptable mechanical damage, power loss or other adverse effects have to be demonstrated.

To provide an aircraft engine certification tests for water ingestion, it is necessary to control accurately the parameters of the water drops generated by the water injecting device simulating rain at the engine inlet. The autonomous tests of a single-channel centrifugal nozzle for the water injecting device have to be carried out using the PSV technique. The tests purpose is to investigate the disperse parameters of the water spray for the typical operation modes of the nozzle as a part of the water injecting system.

Journal of Physics: Conference Series **1421** (2019) 012008 doi:10.1088/1742-6596/1421/1/012008

2. Operating principle of the PSV technique

The PSV (Particle Shadow Velocimetry) technique is used to visualize particles (e.g. droplets from a spray or bubbles in liquid). The technique is based on high resolution imaging with pulsed backlight illumination (see figure 1). The measurement volume is defined by the focal plane and the depth of field of the imaging system. This technique is independent of the shape and material (either transparent or opaque) of the particles and allows to investigate sizes down to 7 μm using an appropriate imaging system and light source. A double-pulse laser combined with a double-frame camera allows to investigate size dependent velocities. This technique provides information like size distribution, shape and velocity of particles.

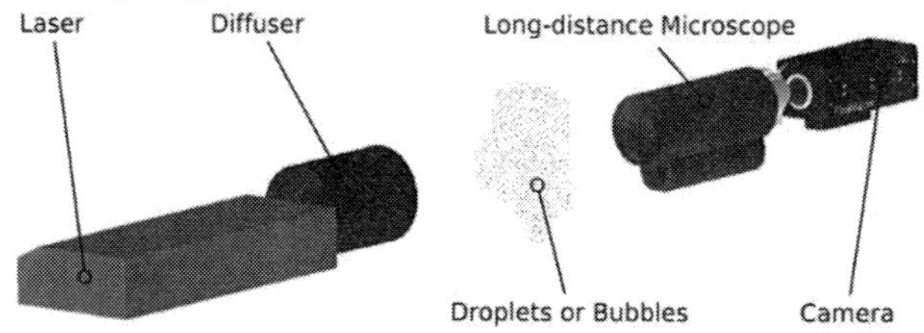

Figure 1. The PSV hardware setup.

The double-pulse laser is located in front of the camera lens. The light from the laser passes through the diffuser and is directed to the area of interest. After the light has passed through the area, its image enters the microscope, and then the magnified image enters the camera matrix and is transmitted to the computer. Using a specialized software, the measurement of particle sizes and their velocities is performed.

For PSV measurements of the characteristics of the spray pattern formed behind the nozzle the optical scheme adjustment and depth of field calibration for the droplet size range from 20 to 1500 μm was performed. The size of the measuring volume after the adjustment is 6.1×4.6×11 mm.

3. Experiment and results

The autonomous tests of a single-channel centrifugal nozzle for the water injecting device have been carried out using the PSV technique. The scheme of the experimental setup is shown in figure 2.

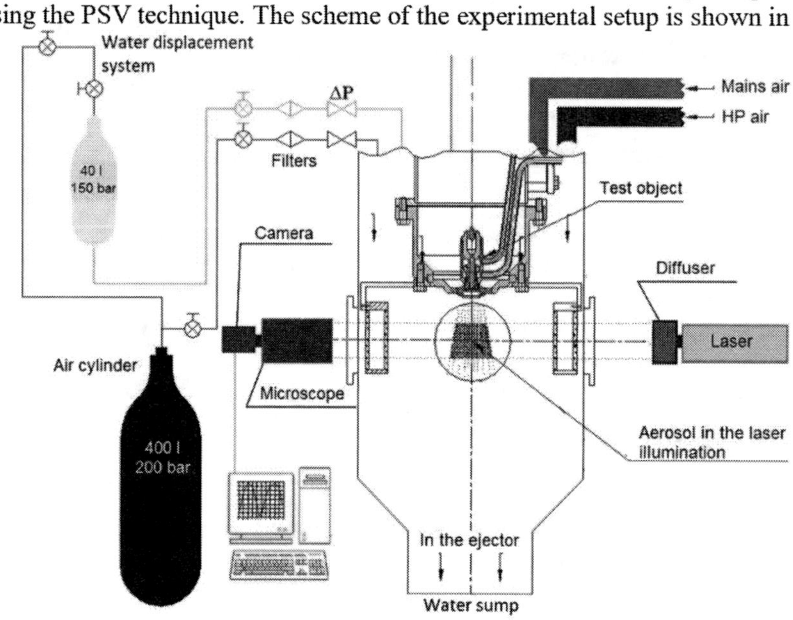

Figure 2. The experimental setup.

Journal of Physics: Conference Series **1421** (2019) 012008 doi:10.1088/1742-6596/1421/1/012008

The measurements have been carried out in cross section along the spray pattern radius, from the axis of the spray jet ($R = 0$) to the distance $R = 170$ mm in increments of 10 mm. In each measurement point 300 images have been shot. Measurements of the spray characteristics have been carried out at a distance of 300 mm from the nozzle at three typical nozzle operation modes:

- $\Delta P_w = 250$ kPa, $G_w = 0{,}162$ l/s;
- $\Delta P_w = 390$ kPa, $G_w = 0{,}203$ l/s;
- $\Delta P_w = 700$ kPa, $G_w = 0{,}263$ l/s.

The nozzle flow characteristics (G_w, l/s) were measured using Krohne mass flow meters. The differential pressure of the water supply (ΔP_w, kPa) was measured by sensors ADZ-SML-25.

The photos of the water torch pattern are shown in figure 3. The shadow images with derived diameters and velocities at the distance $R = 170$ mm from the spray jet axis are shown in figure 4.

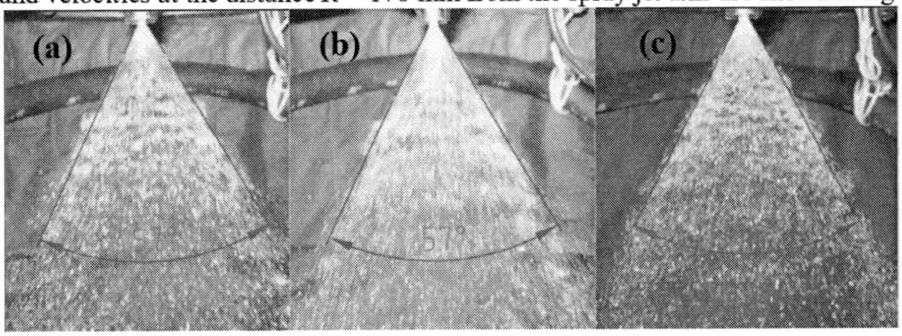

Figure 3. The water torch patterns at operation modes: (a) $\Delta P_w = 250$ kPa, $G_w = 0{,}162$ l/s; (b) $\Delta P_w = 390$ kPa, $G_w = 0{,}203$ l/s; (c) $\Delta P_w = 700$ kPa, $G_w = 0{,}263$ l/s.

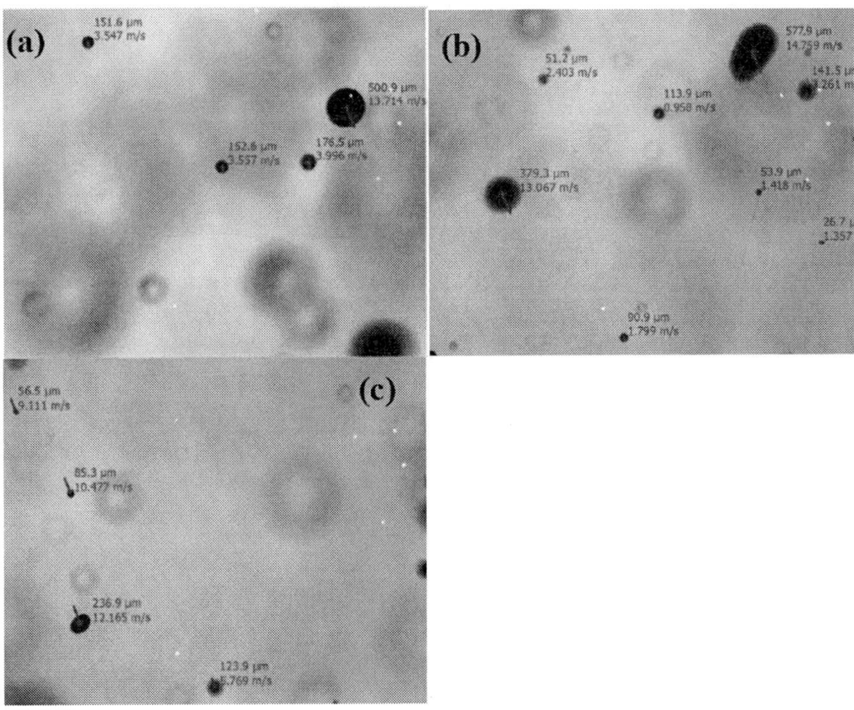

Figure 4. The shadow images with derived diameters and velocities at operation modes: (a) $\Delta P_w = 250$ kPa, $G_w = 0{,}162$ l/s; (b) $\Delta P_w = 390$ kPa, $G_w = 0{,}203$ l/s; (c) $\Delta P_w = 700$ kPa, $G_w = 0{,}263$ l/s.

Journal of Physics: Conference Series **1421** (2019) 012008 doi:10.1088/1742-6596/1421/1/012008

As a result of the tests, the distribution graphs of the D_{10} (arithmetic average of the particle diameter), D_{32} (Sauter mean diameter) and Dv_{50} (mean median diameter), axial velocity (U) and mass flow (M_{abs}) of water droplets over the radius of the spray pattern have been obtained. These graphs are shown in figures 5 and 6 below.

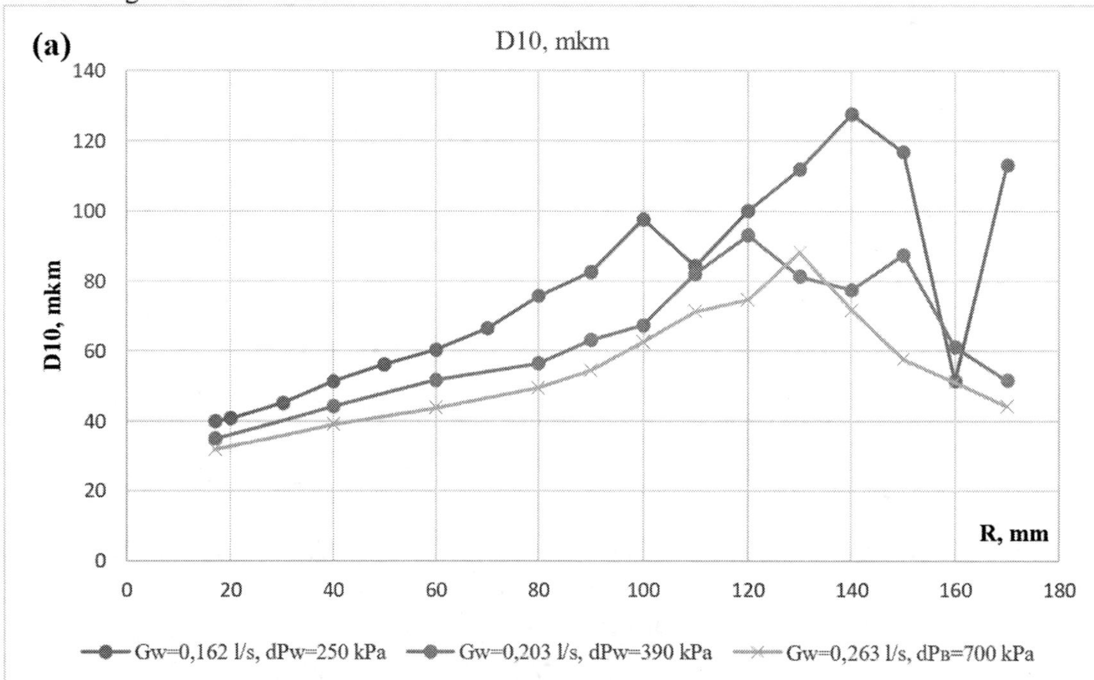

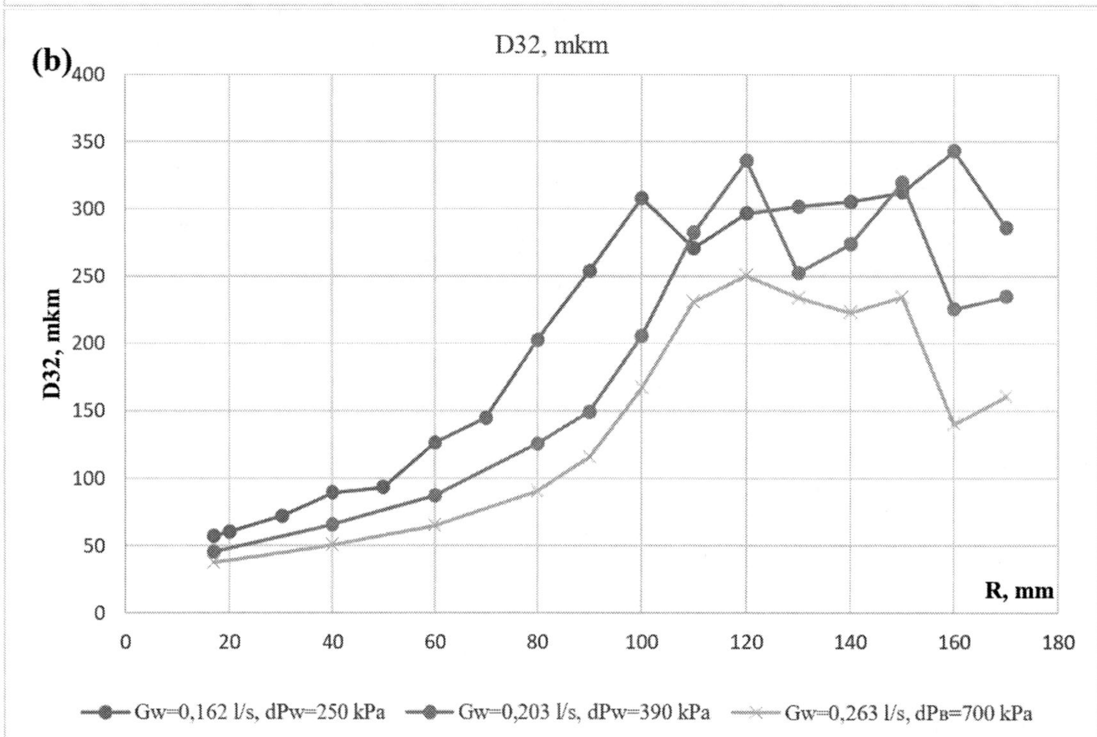

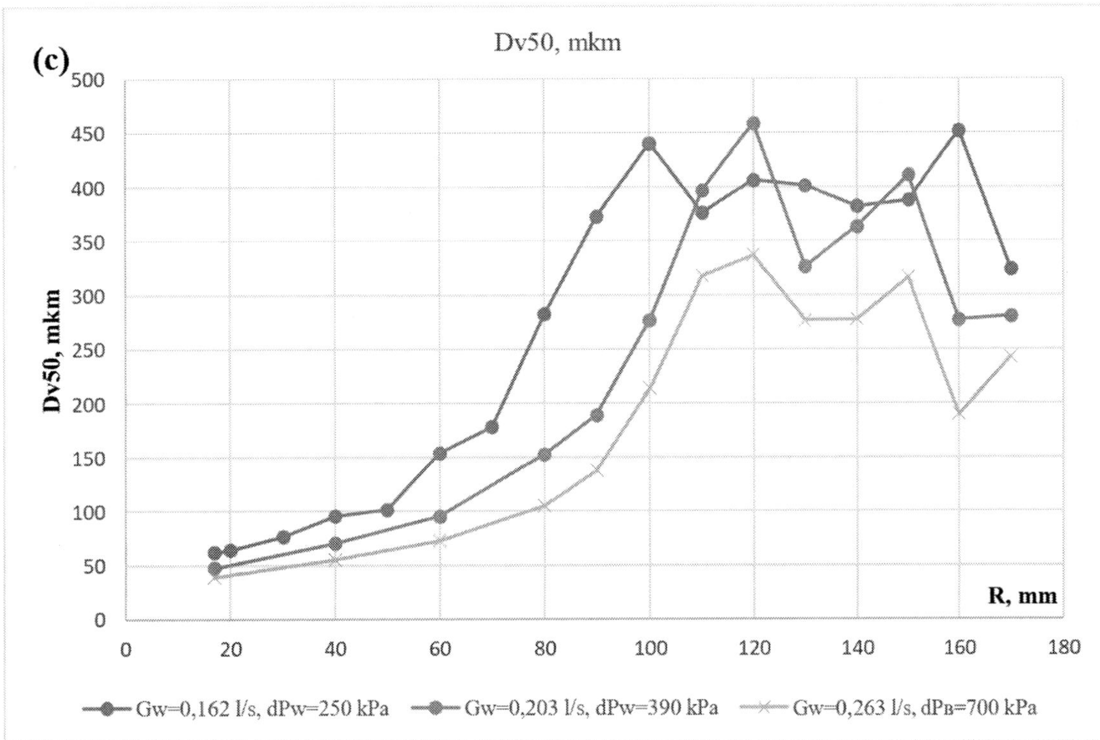

Figure 5. The distribution graphs of the D_{10} (a), D_{32} (b) and Dv_{50} (c) diameters of water droplets over the radius of the spray pattern.

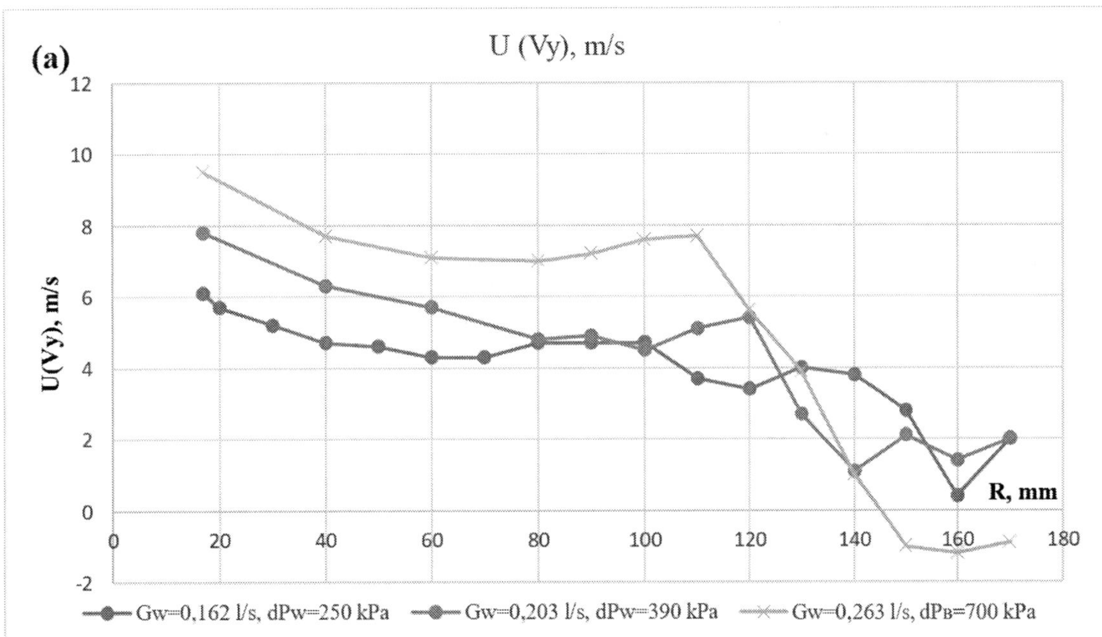

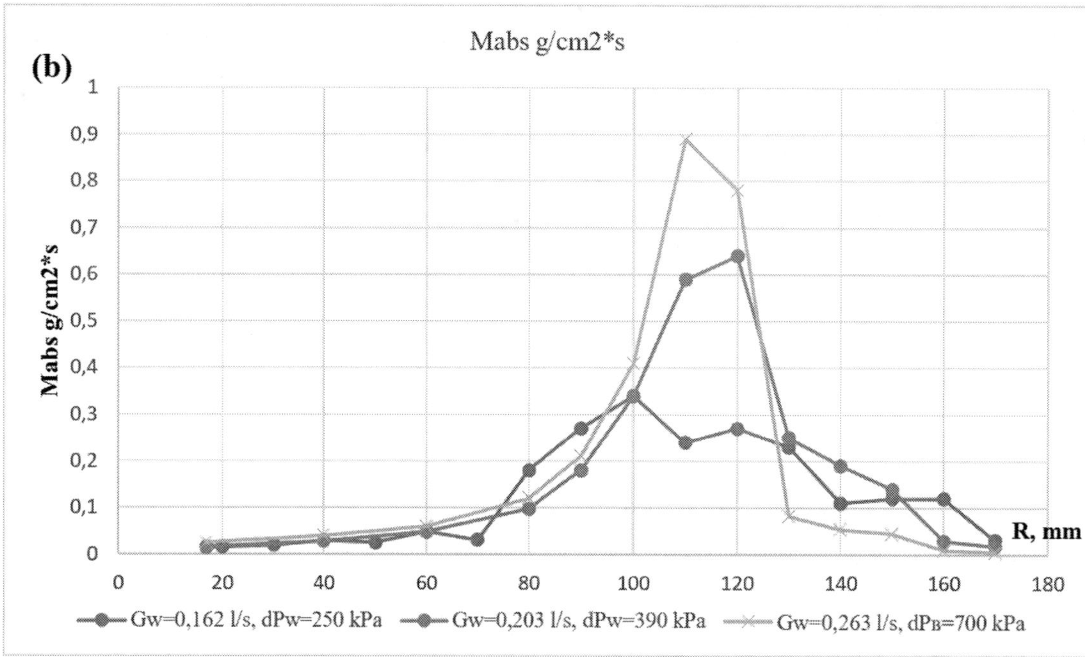

Figure 6. The distribution graphs of the axial velocity U (a) and mass flow M_{abs} (b) of water droplets over the radius of the spray pattern.

The figure 6 (b) shows that major part of the water is distributed in the cross section of the spray torch along the ring at the radii of 100 and 130 mm. The diameter of the spray torch in the measurement plane is 340 mm.

To estimate the mean median Dv_{50} diameter of water droplets in the spray torch cumulative functions of the droplet mass fraction distribution were constructed depending on the size for each of the nozzle operation modes (figure 7). On the basis of these functions, the mean median diameters of water droplets in the spray torch are calculated, the calculation results are presented in table 1.

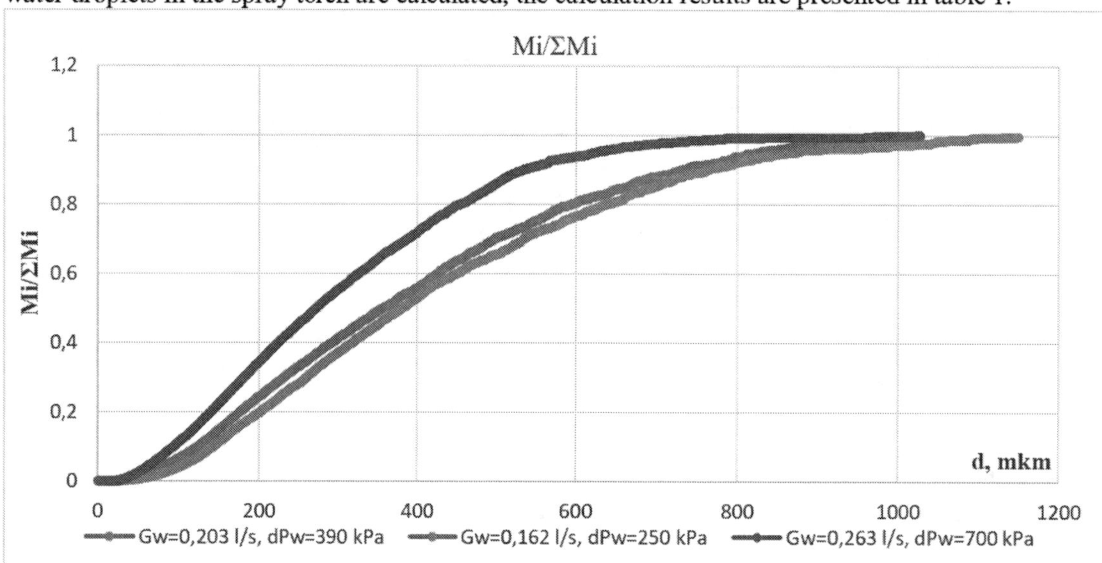

Figure 7. The cumulative functions of the droplet mass fraction distribution.

Table 1. Mean median diameters of water droplets in the spray torch.

Operation mode number	ΔPw, kPa	Gw, l/s	Mean median diameter, mkm
1	250	0.162	380
2	390	0.203	360
3	700	0.263	270

The results obtained in the study allow to choose optimal operating mode and the location of the nozzles in the water injecting device to ensure the required distribution and mean median diameter of the droplets in the stream.

References

[1] Aviation regulations 2018 *Airworthiness standards for aircraft engines* part 33 IAC

Journal of Physics: Conference Series **1421** (2019) 012009 doi:10.1088/1742-6596/1421/1/012009

Researches of two-phase stream by methods of registration of fluorescence of drops of liquid and Shadowgraph

O G Chelebyan, A Y Vasilyev, A A Sviridenkov and A A Loginova

Central Institute of Aviation Motors, Aviamotornaya 2, Moscow 111116, Russia

E-mail: ogchelebyan@ciam.ru

Abstract. In this paper, the fluorescent method is used in conjunction with the shadowgraph method to obtain the concentration distributions and droplet sizes in the spray. A beam of light from a pulsed laser, passing through a longitudinal or cross-section of a jet of atomized fuel, tinted with a fluorescent additive, oxyquinoline, partially scatters on droplets (Mi-scattering), while blue light with a wavelength $\lambda = 488$ nm excites fluorescence in the green range of the wavelength spectrum. A color image of the cross section of the fuel spray by a laser beam is recorded during scanning by a digital camera, the optical axis of which is directed at a right angle to the axis of the laser beam. Thus, at each point of the image of the spray cross section, fluorescence intensities are recorded, proportional to the volume concentration of the droplets and vertically polarized by the component scattered on the light drops with respect to the scattering plane (measurement). The ratio of fluorescence intensity to Mie scattering intensity is proportional to the diameter of the fuel droplets averaged over the Sauter.

1. Introduction

Reducing emissions of pollutants from liquid fuel fired combustion chambers is becoming an increasingly serious problem for industries facing the need to comply with new and upcoming environmental restrictions and regulations. An important condition for meeting these requirements, as well as a critical problem in the development of advanced liquid combustion chambers and high-pressure fuel injectors, is a detailed understanding of the process of mixing fuel with air, droplet size and velocity distribution, and fuel density distribution pattern encountered in high-pressure combustion chambers.

Until recently, the actual flow fields in the combustion chamber were not available for highly sensitive and specific optical sensing methods. The emission characteristics of the combustion chambers of power plants are largely determined by the quality of atomization of fuel injectors. Improving the mixing of fuel with air can reduce NOx emissions, increase the efficiency of the combustion chamber, reduce its size, the instability of combustion, and increase the service life of engines.

The smaller the size of the droplets, the shorter the time of their evaporation, as the total surface of the droplets increases, which accelerates heat transfer. However, on the other hand, very small droplets follow the flow. As a result, the penetration of sprayed droplets into the gas stream may decrease, which will make it difficult to obtain a homogeneous fuel-air mixture. Based on theoretical calculations and experimental work, various researchers recommend optimal values of the average Sauter diameter of the SMD 15-25 μm.

In addition to the average diameter, the maximum droplet size is also important. Large drops reduce spraying and increase fuel penetration, causing irrigation of the combustion chamber walls. To illustrate the droplet size effect, it is easy to calculate that a droplet with a diameter of 40 μm has a fuel mass of eight times the surplus mass of 20 μm droplets and by the time the 20 μm droplets have completely evaporated, the large droplet will still have a diameter of ~ 38 μm.

Therefore, it is important not only to create front-line devices that provide the necessary parameters of the air-fuel mixture, but also to have means of controlling both time-averaged distributions of concentrations and sizes of droplets, and their pulsation values at each point of the combustible mixture. The most modern and informative method for determining these parameters is currently the method based on the fluorescence of a laser-excited dye added to a liquid, or the natural fluorescence of a fuel (kerosene). Together with the measurement of Mie scattering of laser radiation, the method allows to obtain quantitative values of the concentration and size of droplets of sprayed fuel.

Shadowgraph is the only method that can be used to characterize droplets and liquid filaments of arbitrary shape. However, the shadow method, like the PDA method, can only be used in optically dilute aerosols, since the light illuminates both the focused and non-focused drops, and the light rays pass through the entire spray cone before entering the recording chamber. In addition, more complex image processing is required to distinguish between droplets that are in focus and droplets that are out of focus in order to obtain accurate droplet size. While PDA devices use high-power lasers with a continuous wave of radiation, it is advisable to use pulsed lasers that can provide uniform stray light for a shadow image, or use cameras with a very short exposure time.

The fluid mechanics first quantitatively determined the concentration of the LIF method and apparently were applied by Owen in 1976, which measured the concentration and velocity of two coaxial jets flowing into space with a sudden expansion [1]. The LIF / Mie method for determining size droplets has been used in a variety of studies since the study by Yee et al., First published in 1993, in which diesel fuel spraying was studied [2]. The droplets under investigation are spherical. Molecular absorption inside the droplets should be taken into account. The cubic and square dependence of the droplet diameter on the optical signals of LIF and Mie scattering, respectively, should be observed. The dye concentration and laser power remain constant during the experiment. The optical signal LIF is proportional to the power of the laser and at the time of the experiment it does not saturate. The evaporation rates of the fluorescent dye and the liquid medium should be close. Also consider the effects of multiple light scattering and minimize oscillations of the Mie scattering signal.

Research have shown that the use of continuous exciting laser radiation leads to incorrect measurement results near the nozzle, due to secondary scattering on droplets, as well as the presence of non-spherical droplets and liquid ligaments formed at the initial stage of the breakup of fuel film. The use of non-stationary radiation source helps to reduce measurement errors. The paper notes that the contribution of the intensity of scattered light is very difficult to predict, and the observed tendency can change depending on the viewing angle, the transparency of the optical medium, the size of the droplets, their distribution, etc. To calibrate the intensity ratios LIF / Mie SMD data from PDA measurements were used at each measurement point.

In [4] 0.02 g rhodamine 6G dissolved 200 ml of ethanol. This solution was mixed with 4 liters of gasoline and used for experiments. The concentration of rhodamine used (5 mg / l of gasoline) was selected from the literature as the lowest. About 300 images were taken at each time point and averaged over the ensemble after subtracting the background to obtain the final Mie image and fluorescence. The area size of 1 × 1 mm corresponded to 300 × 300 pixels in the images. It is worth emphasizing that, although the uncertainty in the PLIF diameter versus diameter introduces some uncertainty in the calibration factor and, consequently, in SMD, the PLIF method is a quantitative technique, and as such, the PLIF images give the SMD number at any point in the spray. For the standard deviation of the average value of 33%, the error of these measurements, based on a 95%

confidence level, is ± 20%. However, in areas close to the nozzle the assumption of the dependence of PLIF on the diameter is violated due to the presence of ligaments and, therefore, the data in this area will not be accurate. This is, in fact, a limitation of all the methods described in the literature.

Studies of the characteristics of a fuel atomized by a centrifugal nozzle were carried out jointly by the methods of light scattering at small angles and PLIF in [5]. To calibrate the fluorescent method we simultaneously measured the sizes and concentration of droplets in the same cross section of the spray jet of a well-studied (standard) pressure nozzle using the method of light scattering at small angles. The mean values of the Mie intensities of the scattered light and the fluorescent light emission with the surface and volume concentration of the droplets, respectively, were determined from the mean concentrations and droplet sizes in this spray section. These values were used later in the research.

To obtain the concentration distributions and droplet sizes along the radius of the studied cross section of the centrifugal nozzle flame, the Abel transformations of the concentration and droplet size distributions measured along the chords in this cross section by scattering light at small angles were performed. The distributions obtained by the two measurement methods agree satisfactorily with the exception of the region of large concentration gradients in the central part of the spray. This discrepancy may be due to errors that occur during the Abel transform of experimental data representing insufficiently smooth functions.

In the present work, we measured the characteristics of a spray jet of a pneumatic nozzle [6] using the shadow method and the method of recording light on fluorescent kerosene droplets (TS-1) with oxyquinoline dye and irradiated with continuously working laser.

2. Results of investigation

Measurements of the characteristics of the shadow method were carried out at a distance of 30 mm from the nozzle. The droplet speeds, sizes and fuel consumption were measured through the cross section of the spray. Fuel consumption was 1 g / s, air rate 15 g / s.

Figure 1 shows the distribution of the longitudinal component of the velocity of the fuel droplets.

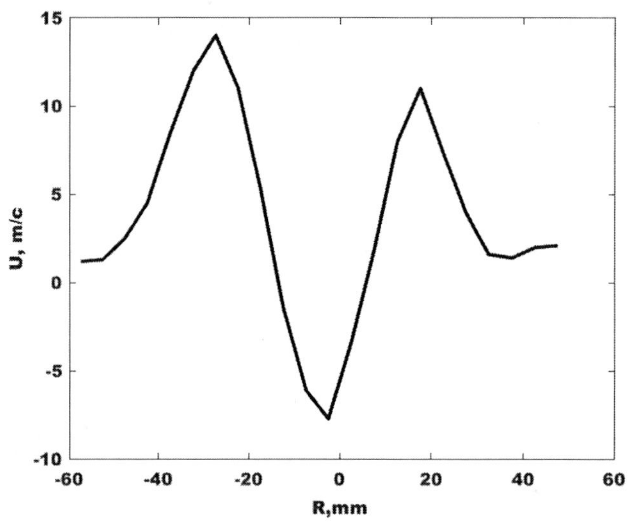

Figure 1. Distribution of the longitudinal velocity component.

As can be seen from the graph of velocity distribution at a distance of 30 mm to the nozzle on the axis of the spray there is an intense recirculation flow, the speed of which reaches 55% of the maximum velocity of the direct flow.

The distribution of the volume flow through the cross section of the spray cone is shown in figure2.

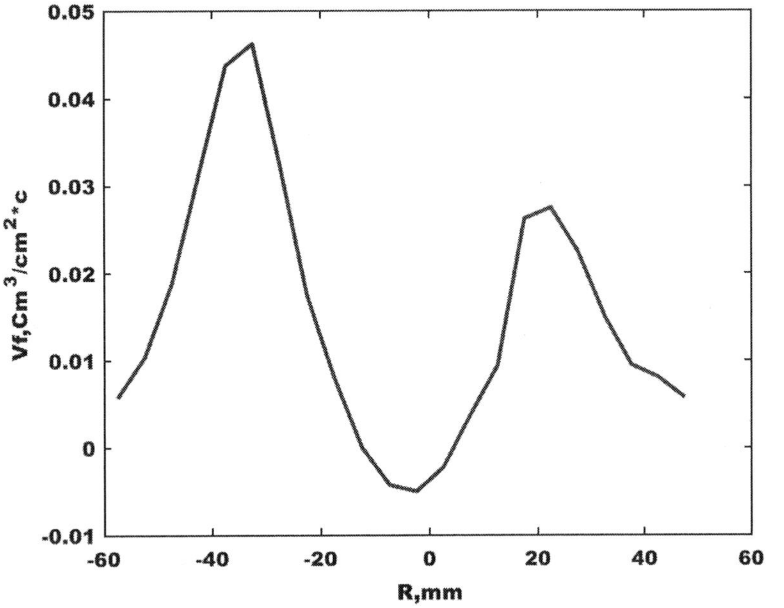

Figure 2. Distribution of the volume flow of fuel droplets in the cross section.

Knowing the velocity distribution of the droplets and their volume flow through the cross sections of the spray, it is possible to determine the volume concentration of the droplets.

With the same operating parameters of the nozzle, measurements of Mie scattering and fluorescence of fuel droplets were carried out. A beam of light from a laser, passing through a longitudinal or transverse (horizontal) cross section of a jet of sprayed fuel tinted with a fluorescent oxyquinoline additive, partially scatters on droplets (Mi-scattering), while blue light with a wavelength $\lambda = 488$ nm excites fluorescence in the green range of the length spectrum waves. A color image of the cross section of the fuel spray by a laser beam is recorded during scanning by a digital photo or video camera, the optical axis of which is directed at a right angle to the axis of the laser beam. Thus, at each point of the image of the spray cross section, the fluorescence intensities I_F and the vertically polarized component I_S Mie of the light scattered on the drops with respect to the scattering plane (measurement) are recorded. Information from a video camera is transfer to a computer. The values $D_{32} = K <D^3> / <D^2>$ *for* $D_{32} >$ 10 μm are calculated from the equations for the intensities of scattered light

$$I_S = \alpha <D^2> N I_0 = 2\alpha\, Cs\, I_0,$$
$$I_F = \beta <D^3> N I_0 = 3\beta\, C_V I_0,$$

Therein N is the numerical concentration of droplets; α and β are constants determined experimentally; I_0 is the intensity of the incident light; Cs and C_V – surface and volume concentration of droplets. Hence, the expressions for the diameter of D_{32} drops, averaged over Sauter, and their C_V concentration are: $D_{32} \sim I_F / I_S$, $C_V \sim I_F$. The constants α and β were determined by comparing the maximum values in the distributions of fuel concentration and droplet sizes at this point, obtained by Shadowgraph and fluorescence. An example of the obtained distributions is shown in figure 3.

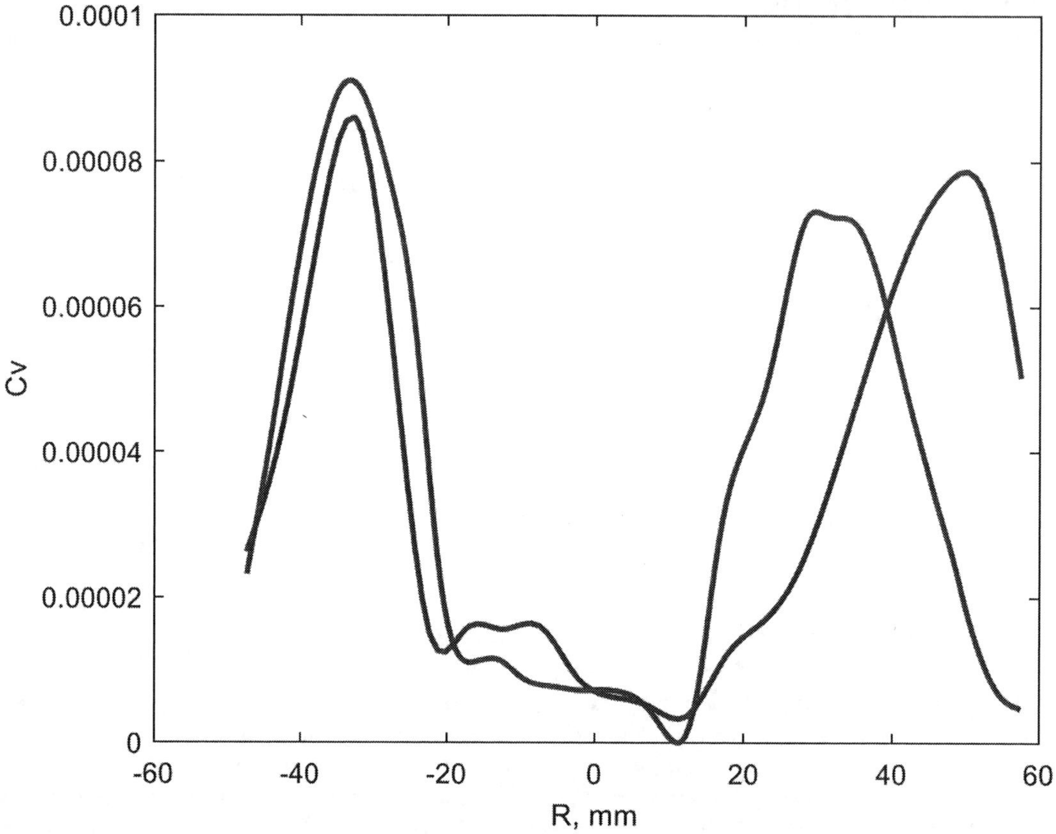

Figure 3. The distribution of the droplets fuel concentration in the cross section of the spray (red curve-Shadowgraph, violet – fluorescence).

Since the proportion of the volume occupied by the sprayed droplets at the exit of the pneumatic nozzle is equal to the ratio of fuel and air rate divided by their density ratio, the average volume concentration of fuel is equal *to* $C_{vmean} = G_f / (G_a \times \rho_f / \rho_a)$. In experiments, this value was measured directly and turned out to be equal to $C_{vmean} = 1.0 \times 10^{-4}$. The characteristics of the spray pattern were measured at a distance of 30 mm from the nozzle. At this distance, the width of the spray is approximately two times larger than directly near the nozzle. Consequently, the average volume concentration of fuel droplets at this distance is approximately 2.5×10^{-5}. Measured as a mean is equal to 3.6×10^{-5}, mean-mean value 2.4×10^{-5}. As can be seen from the figure above, there is a satisfactory agreement of the distributions in the left part of the figure. In the right part of the figure, the width of the spray, measured by the fluorescence method, turned out to be less than that measured by the shadowgraph method. Note also that, when processing the color image of the spray, it was not detected the real fluorescence intensity in the inner area of the spray pattern. This appearance was observed in [3] and is associated with the phenomena of secondary scattering of light by fluorescence. To obtain realistic results, an adjustment was made to the intensity of the fluorescent light. Figure 4 shows the intensities of the green component of the scattered light before and after the adjustment.

Journal of Physics: Conference Series **1421** (2019) 012009 doi:10.1088/1742-6596/1421/1/012009

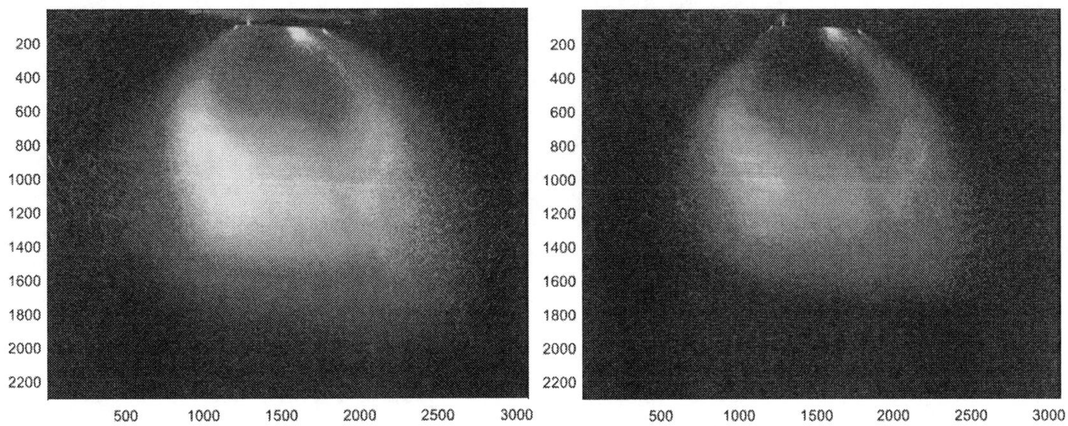

Figure 4. Fluorescence intensity before-and-after updating.

Figure 5 shows the size distribution of droplets in the cross section of the atomization spray. As can be seen from the figure, the difference in droplet sizes in the center of the spray exceeds 100% of the measured local values, although the trend in size distributions is the same.

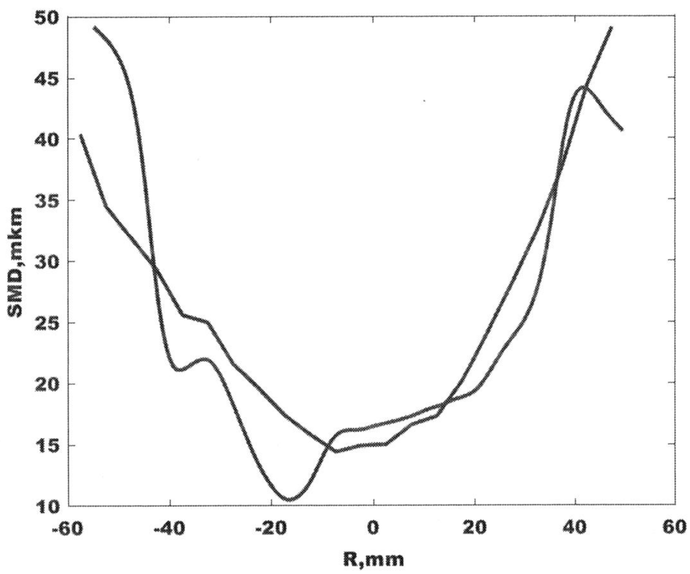

Figure 5. The distribution of the size of fuel droplets Sauter-average in the cross section of the spray (red curve-Shadowgraph, violet – fluorescence).

With a transverse arrangement of a laser sheet, non-uniform distribution of the concentration of fuel droplets in the circumferential direction can be obtained. An example of such a distribution is shown in figure 6.

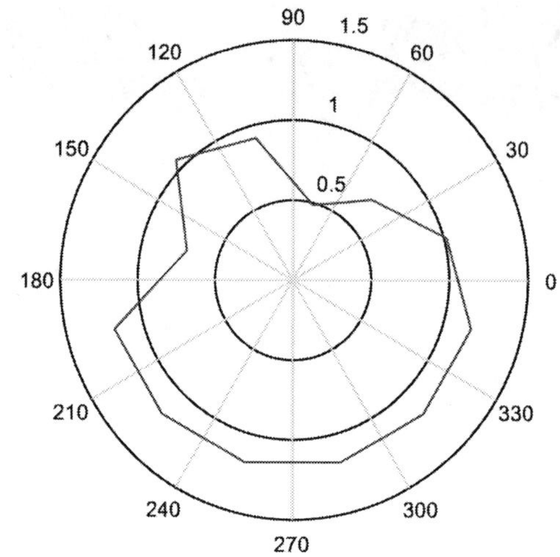

Figure 6. Circumferential irregularity of fuel distribution.

As can be seen from the figure 6 in spray, there is an area with low fuel content. Details of procedure non-uniform calculation are description in [5].

3. Conclusion

A study of the characteristics of spraying fuel in a spray behind a pneumatic nozzle using shadowgraph and fluorescence of droplets of sprayed fuel showed that, despite the informative and high rate production of experimental results, the use of fluorescence requires particularly careful experimentation. This is primarily due to the secondary scattering of light from fluorescent droplets in the spray. It is also necessary to use the minimum concentration of the fluorescent dye to maintain the cubic dependence of the fluorescence intensity on the diameter of the liquid droplets.

References

[1] Owen E 1976 Simultaneous laser measurements of instantaneous velocity and concentration in turbulent mixing flows *J. Agard Appl.* CP **193** (27)

[2] Yeh C, Kosaka H and Kamimoto T 1993 Fluorescence/Scattering Image Technique for Particle Sizing in Unsteady Diesel Spray *Japan. J. Society of Mech. Engineers* B **59** 4008-13

[3] Yogeshwar N 2018 Droplet size, concentration, and temperature mapping in sprays using SLIPI-based techniques *Sweden. University dissertation from Division of Combustion Physics, Lund University*

[4] Anand T, Devendra D, Madan M and Ravikrishna R 2009 Laser-based Spatio-temporal Characterisation of Port Fuel Injection (PFI) Sprays *Int. J. of Spray and Combustion Dynamics vol* 1

[5] Golubev A, Sviridenkov A and Yagodkin V 2012 Researches of two-phase streams dispersion by a method of registration of fluorescence of drops of a liquid and scattering light diffraction *J. Optoelectronics, instrumentation and data processing* **48**(3)

[6] Vasilyev A, Chelebyan O, Zakharov V and Medvedev R 2017 Nozzle module low emission combustonr of a gas turbine engine *Russia. Russian Federation Patent* **170359**

Optical study of the concentration and particle size distribution of octadecylamine in water

N M Skornyakova, A V Anikeev and V A Parshin

V.A. Fabrikant Physics Department, National Research University "Moscow Power Engineering Institute", Krasnokazarmennaya 14, Moscow, 111250 Russia

E-mail: nmskorn@mail.com

Abstract. The objective of this study was to determine the size of the investigated particles in water. The particles studied were octadecylamine inclusions which are invisible to the naked eye. To increase the accuracy of the results obtained, the study was carried out using two optical methods: direct video surveillance in a digital stereo microscope and laser diffraction method.

1. Introduction

Octadecylamine is a chemical compound used for the surface preservation of metals [1]. It is applied in power and other areas. Derivatives and salts of octadecylamine are toxic to bacteria and microorganisms. To implement the process of sorption of octadecylamine molecules on the functional surfaces of structural materials, it is necessary to carry out preparatory operations for the preparation of a highly concentrated aqueous emulsion and for its transportation to the circuit. It is assumed that the method of the emulsion preparation affects the particle size and, accordingly, the properties of the emulsion.

The objective of this study was to determine the size and concentration of the studied particles in water. The particles studied were octadecylamine inclusions. The main difficulty was that visually these inclusions were not observed, since the particles of octadecylamine are transparent [2]. To improve the accuracy of the results obtained, the study was carried out using two optical methods: direct video surveillance in a digital stereo microscope with special LED lighting and a laser diffraction method.

Stereo microscopes belong to the category of optical microscopes. They work on the principle of light reflected from the object being studied. Stereo microscopes are equipped with two eyepieces, the image in which is displayed at slightly different angles. This is what makes it possible to obtain a volumetric enlarged image and makes it possible to evaluate the dimensions of the investigated object.

Stereo microscopes are used in areas where it is necessary to obtain a three-dimensional image, and the perception of contrast and depth is important for the study of the object. In addition, the optical system of the stereo microscope gives a "direct" image of the sample, which allows the use of similar equipment in industry for manipulating microscopic objects. As a rule, stereoscopic microscopes have an upper illumination. However, the presence of a lower backlight is not excluded. Also, manufacturers often make it possible to understaff a microscope with additional sources of illumination: side and ring illumination. For illumination, LED, halogen, fluorescent lamps and even incandescent lamps can be used. Many stereoscopic microscopes show a relatively small magnification of the image: from 10 to 100 times. Although there are instances that can increase the image 200 times.

Laser light (monochromatic, coherent light) interacts with the particles, which have to be characterized in terms of particle size. In dependence of the particles' size, the light waves are scattered by the particles in a characteristic manner: the larger the particles are, the greater is the scattering in forward direction. With particles smaller about 100 nm, the scattering intensity is nearly identical in all directions. Laser diffraction analysis is based on the Fraunhofer diffraction theory, stating that the intensity of light scattered by a particle is directly proportional to the particle size [3]. The angle of the laser beam and particle size have an inversely proportional relationship, where the laser beam angle increases as particle size decreases and vice versa [4].

2. Experimental setup

2.1. Research with stereoscopic microscope

The Altami SM0745 digital stereo microscope was used in this work for direct observation of octadecylamine inclusions in water. In the study, in some cases, an additional ring illuminator was additionally used. Since in this case it was assumed not a single measurement of the particle diameter, but a set of statistical information to obtain the distribution of inclusions by size and determination of concentration, then the obtained images were simultaneously recorded by each method and transferred to a computer.

Together with a stereo microscope, a PL-B621-MF video camera was used, designed to work with microscopes. The PL-B621-MF uses the standard FireWire interface for plug-and-play connection with a host computer. The camera is equipped with a 1.3 MPx CMOS sensor and can provide 30 frames per second at full resolution, or 100 frames per second at VGA resolution. PL-B623-MF is TWAIN, WDM, and IIDC 1.31 (DCAM) compatible.

2.2. Research using a laser diffraction method

Currently, a large number of optical methods for diagnosing flows have been developed: shadow, interference, polarization, holographic. The essence of the laser diffraction method is to determine the size of a particle located in the path of the beam propagation by the period of the emerging diffraction pattern. Laser measurement methods are created to solve problems when the environmental parameters are determined from the measured parameters of the laser radiation at the outlet of the medium under study and the measured (known) parameters of the laser radiation at the input. The laser diffraction method or the small angle method is used. In it, the calculation of the diameter of the present particle is performed according to the formula:

$$d = 1{,}22\lambda \; R/x_{min},$$

where d is the particle diameter; λ is the light wavelength; R is the distance from the investigated plane or lens to the registration plane; x_{min} is the position of the first diffraction minimum.

Experimental setup was developed to solve this problem. It is shown in figure 1. The light from the laser (1) passes through the optical system (2), the elements of which are selected so that the object under study (3) is in the waist of the laser beam. Then the scattered radiation enters the screen (4), the image of it is recorded on the camera (5). The test object is a microliter volume cuvette filled with the test liquid. If the particle under investigation is caught in the path of the laser beam inside the volume under investigation, a diffraction pattern appears on the screen.

As the recording device, a Nikon J1 NIKKOR camera was used. An optical system composed of two collecting lenses was used to obtain the required radius of the waist of the laser beam at the measurement point. The volume of the cuvette used in the experiment was 2 ml. The geometric dimensions of the cell (length, width, height) are equal respectively: 24.1 mm, 7.4 mm and 40.2 mm. Since it was necessary to scan the cross section of the test volume with a laser beam, the cuvette was mounted on a two-coordinate table with micrometric screws vertically and horizontally.

Journal of Physics: Conference Series **1421** (2019) 012010 doi:10.1088/1742-6596/1421/1/012010

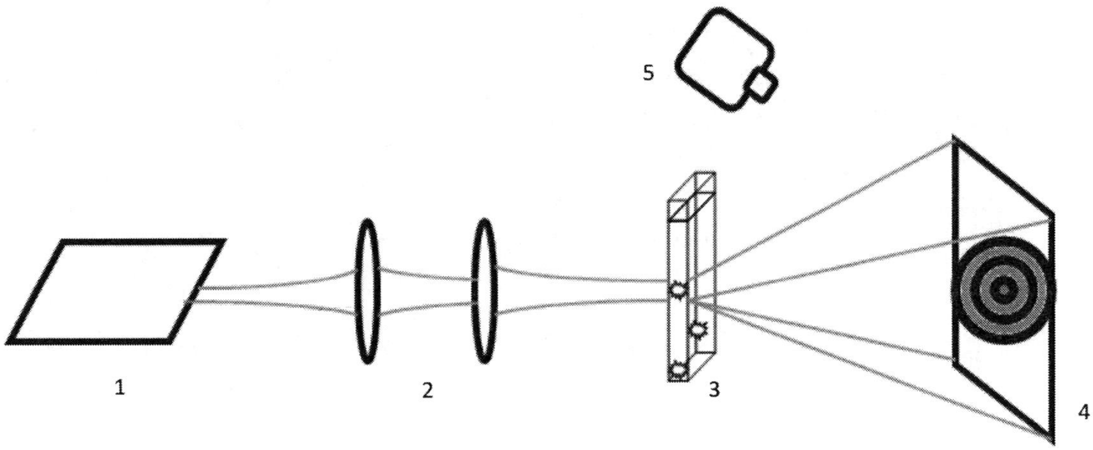

Figure 1. Experimental setup for laser diffraction method.

3. The results of visualization

As a result of visualization of inclusions using a stereo microscope, it was possible to determine that two types of particles are present in the provided emulsion samples. The first type is close in shape to spherical or elliptical particles; hereinafter, it will be called "typical" inclusions. The inclusions of the second type are shapeless, can be threadlike structures, both elongated and looped, in what follows we will call "atypical" inclusions. Examples of observed images are presented in figure 2.

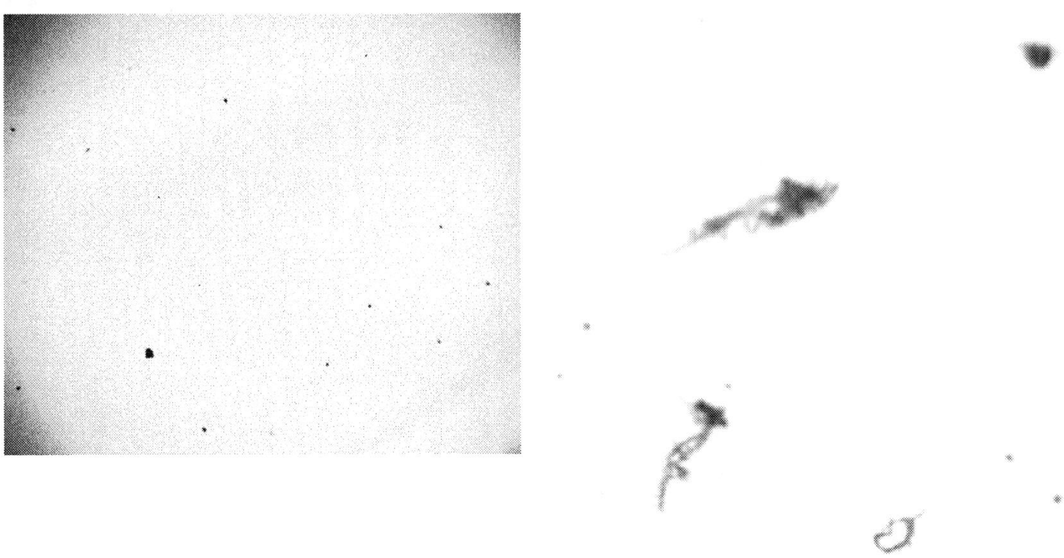

«Typical» inclusions «Atypical» inclusions

Figure 2. The examples of visualization in stereo microscope.

A study of the provided samples by diffraction method was performed at room temperature. Liquid was scanned from the edge of the cell to its bottom for each series of measurements. The movement of the cell in the cross section of the beam was carried out by rotating micrometer screws. The image on the screen was recorded each time a particle fell into the measuring region. Examples of observed diffraction patterns are presented in figure 3.

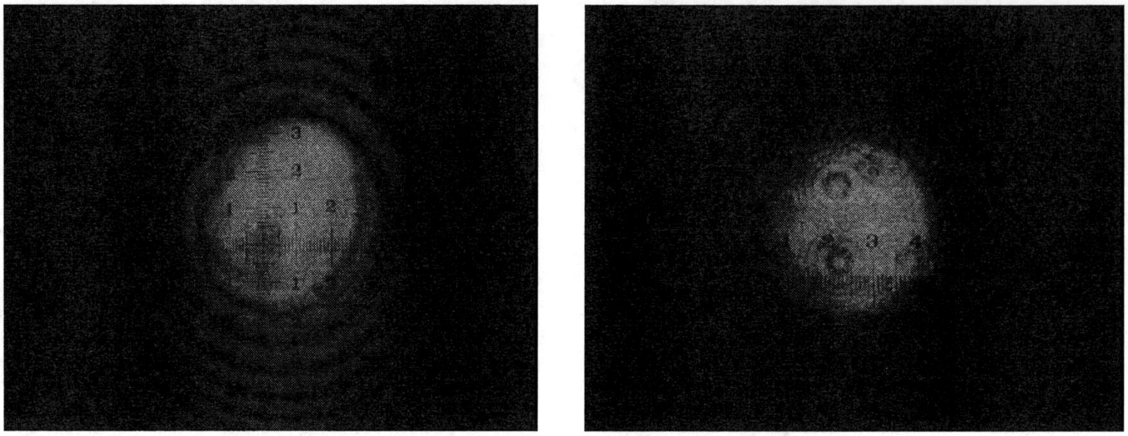

Figure 3. The examples of observed diffraction patterns.

4. The results of measurements

A study was made of the particle size in a suspension of octadecylamine depending on the method of creating the emulsion. The first emulsion was obtained by mechanical stirring of a "powder" of octadecylamine in hot (80°C) water. The second was prepared by dosing the octadecylamine melt through an ejector into a stream of water at 80°C; sampling was carried out immediately after the ejector.

Figure 4 presents the statistical results of the study of the first emulsion. In this emulsion, only "typical" inclusions were observed. On the histogram along the horizontal axis: the diameter of the encountered particles in µm, the left boundary is 3.68 µm, then in increments of 0.7 µm; along the vertical axis: the percentage of particles are given in range of diameters in the liquid (100% – all particles in the test volume). The largest number of particles between diameters of 3.7 µm and 4.4 µm is associated with the resolving power of the microscope, that is, all particles of a smaller diameter also fall into this range.

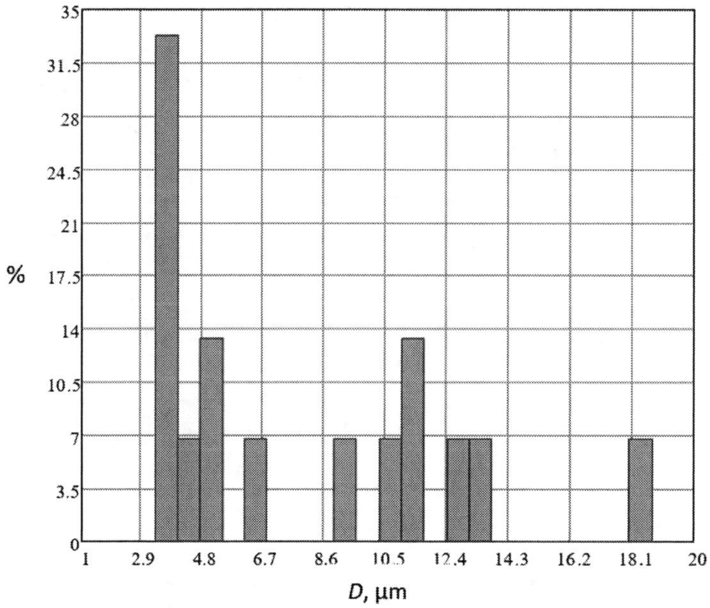

Figure 4. Histogram of particle size distribution in emulsion 1.

By analysing images from a stereo microscope, it is also possible to determine the concentration of inclusions in the emulsion. The total volume of particles per 1 mm^3 of liquid was 4.6 10^{-5} mm^3. The density of octadecylamine is 0.86 g/cm^3 or 0.86 mg/mm^3 [5]. Accordingly, the concentration of 39.56 mg/L.

In the emulsion prepared by the second method, both "typical" inclusions and "atypical" were observed. If for "typical" particles a histogram of the diameter distribution was constructed (assuming that the particles are spherical), then for the "atypical" ones, histograms of the horizontal, vertical, and interspersed areas were constructed.

In figure 5 a bar graph of particle diameter distribution for "typical" particles are presented. On the histogram along the horizontal axis: the diameter of the encountered particles in μm, the left boundary is 1.401 μm, then in increments of 0.42 μm; along the vertical axis: the percentage of particles in a given range of diameters in the liquid (100% – all particles in the test volume).

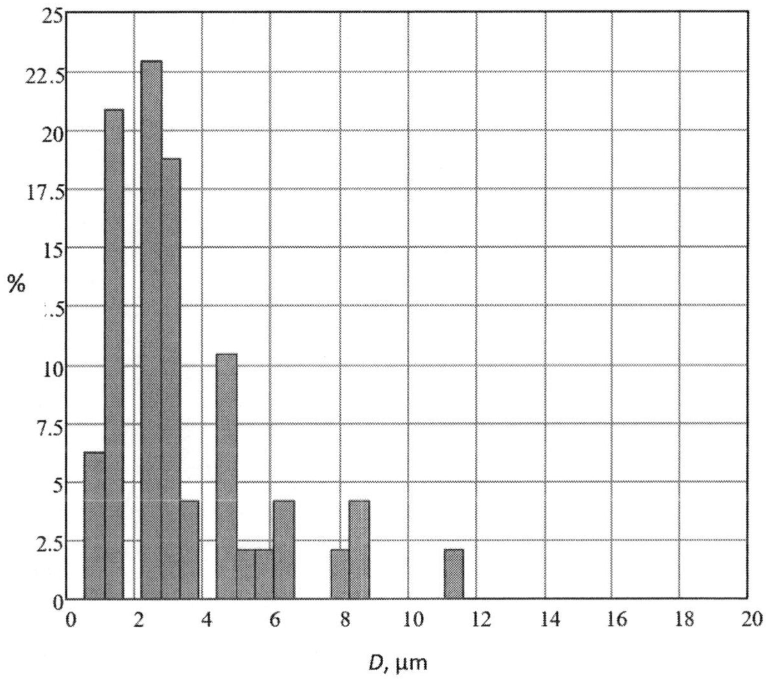

Figure 5. Histogram of "typical" particle size distribution in emulsion 2.

Figure 6 shows the histograms of the distribution of the diameters of the "atypical" inclusions separately horizontally and vertically. Since the particles are far from always oriented horizontally or vertically, the histograms in figure 6 do not give complete information. It will be statistically more correct to compare the inclusions in the liquid over the area. The histogram of the area distribution for "atypical" particles is shown in figure 7. On the histogram along the horizontal axis: the area occupied by particles in square μm, the left border is 27.04 μm^2, then with a step of 14.07 μm^2; along the vertical axis: the percentage of particles in a given range of areas in the liquid (100% – all particles in the test volume).

The total volume of "typical" particles per 1 mm^3 of liquid is 9.6 10^{-6} mm^3. Accordingly, the concentration is 8.26 mg/l (for "typical" particles). The total volume of "atypical" particles per 1 mm^3 of liquid is 8.2 10^{-5} mm^3. Accordingly, the concentration is 70.52 mg/l (for "atypical" particles).

For clarity, the difference between the characteristics of the first and second emulsions, a comparative histogram is shown in figure 8. Blue color (empty rectangles) – liquid 1, red color (filled rectangles) – liquid 2.

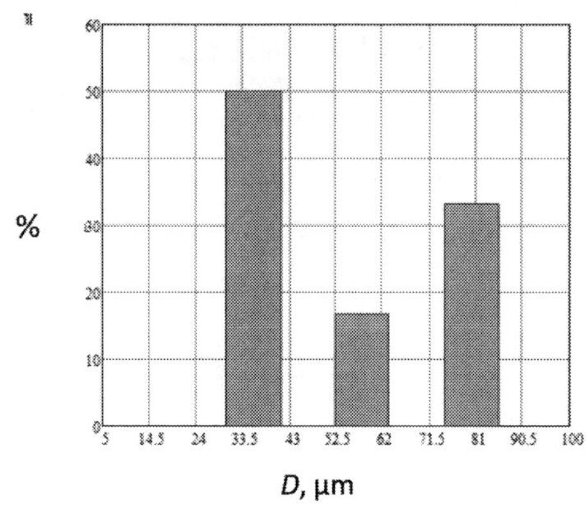

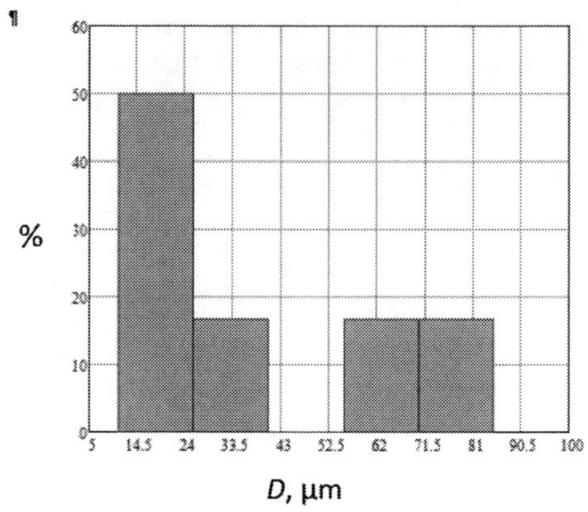

Figure 6. Histogram of "atypical" particle size distribution in emulsion 2.

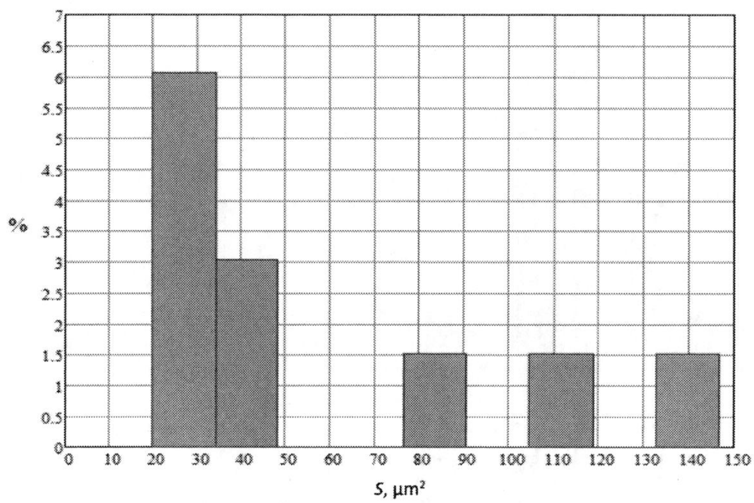

Figure 7. Histogram of the distribution by area of "atypical" particle in emulsion 2.

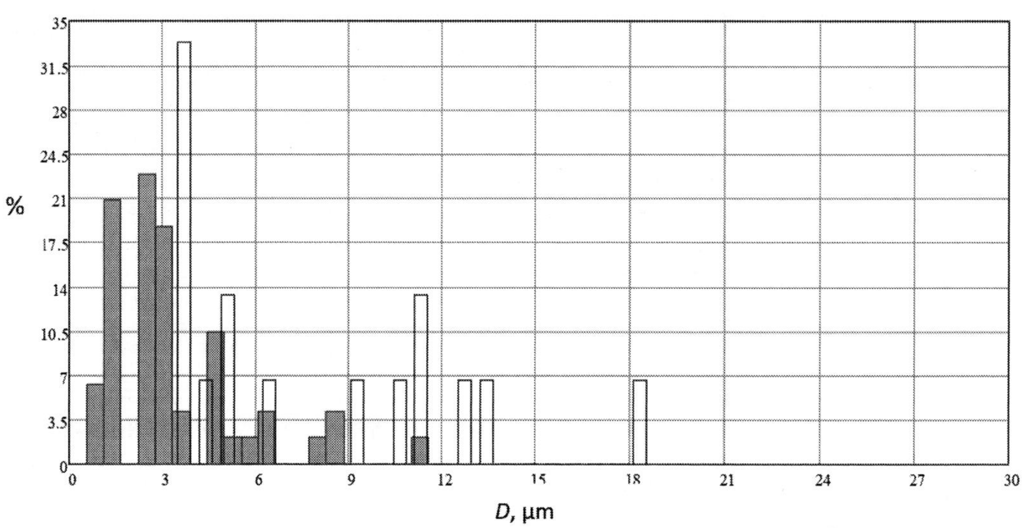

Figure 8. Comparative histogram.

5. Conclusion

The possibility of complex diagnostics by optical methods of the parameters of optically transparent inclusions in water is shown. During the study it was determined that the parameters of the resulting octadecylamine emulsions depend on the method of their preparation. With mechanical stirring of the "powder" of octadecylamine in hot (80°C) water, the particle diameters are obtained in the range from 3.7 to 18.2 μm, while all are of a "typical" type. When preparing the emulsion by dosing the octadecylamine melt through an ejector into a water stream at 80°C, the scatter in the diameters of the "typical" particles is smaller and ranges from 1.4 to 11.8 μm, but the emulsion is an order of magnitude larger (than "typical") the number of "atypical" particles.

6. References

[1] Gosselin R E, Hodge H C, Smith R P and Gleason M N 1976 Clinical Toxicology of Commercial Products (Baltimore: Williams and Wilkins)
[2] Lide D R (ed) 1996 CRC Handbook of Chemistry and Physics (Boca Raton: CRC Press Inc)
[3] Mudroch A 1997 Manual of Physico-Chemical Analysis of Aquatic Sediments (Lewis Publishers)
[4] McCave I N, Bryant R J, CookH F and Coughanowr C A 1986 *Journal of Sedimentary Research* **56 (4)** 561–4
[5] International Chemical Safety Cards (ICSC). http://www.ilo.org/dyn/icsc/showcard.display?p_version=2&p_card_id=1365

Journal of Physics: Conference Series **1421** (2019) 012011 doi:10.1088/1742-6596/1421/1/012011

Diagnosis of low-speed flows by particle shadow velocimetry method

P T Khant and V Yu Barinov

V.A. Fabrikant Physics Department, National Research University "Moscow Power Engineering Institute", Krasnokazarmennaya 14, Moscow, 111250 Russia

E-mail: phyothukhant@gmail.com

Abstract. The method of particle shadow anemometry is one of the PIV modifications. LEDs are used as a radiation source. The application of particle shadow anemometry for diagnostics of low-speed vortex flows is considered in this article. The ability to obtain both visualization of the flow and numerical results was shown.

1. Introduction

Currently, optical methods are widely used in the experiment to measure the parameters of single-phase and two-phase flows. They do not cause disturbances in the flow and have a high temporal and spatial resolution. Low speeds measurement is an integral part of changes, for example, aerodynamic. The aerodynamic force acting on a free-flying model can be determined by measuring the acceleration of the model. If the size of the model does not allow installing the necessary devices on it, the acceleration is found by changing the speed of the model along the path.

The method of particle shadow velocimetry (PSV) is promising in the velocity measurement field, due to its simplicity and low cost relative to laser methods. In case of particle symmetry, the method does not require high-precision adjustment, which makes it faster and easier to obtain the measurement result. This method can be implemented with a single wavelength led and a black and white camera, but the time dynamic range can be increased by using multi-colour LEDs (white) and a colour camera. In this paper, the possibility of using the Scheimpflug adapter for the diagnosis of low-speed flows by particle shadow velocimetry will be considered.

2. Particle Shadow Velocimetry (PSV)

In the method particle shadow velocimetry (PSV) [1-4] LED are used as the light source, with a power is much lower than that of lasers. This technique is presented as a modification of particle image velocimetry (PIV) [5, 6]. PSV neglects the wave nature of light and takes as a basis only the approximation of geometric optics.

Lighting directed in a straight line to the camera matrix is provided by an incoherent source, such as a light-emitting diode (LED). Thus, we are able to distinguish small particles with less energy and time for adjustment.

In an installation based on the PSV technique, it is possible to adjust parameters such as depth of field, field of view and working distance. Such adjustments can be made using the Scheimpflug adapter in addition to the lens.

The attractive side of this new method is the flexibility in the choice of radiation receivers. At the moment, there are no limiting conditions for the selection of the camera, and therefore during the experiment it is necessary to be guided only by the characteristics that are necessary to perform a specific task.

3. Experimental setup

Figure 1 shows an experimental setup for measuring the velocity of particles in water (2). To obtain a shadow of particles in water, dot LEDs (1) with a diverging angle of 120° of different colours (red, green, blue and white) were used as the radiation source. The shooting was carried out using a high-speed camera (4) connected directly to the computer (5). For research, a high-speed camera Fastec Hispec 1 was used with the Scheimpflug adapter [7] and 50 mm objective (3).

The program specially installed on the computer allows not only to select the necessary shooting parameters (frame rate per second, the required survey area, etc.), but also to immediately save the results in the required format and volume.

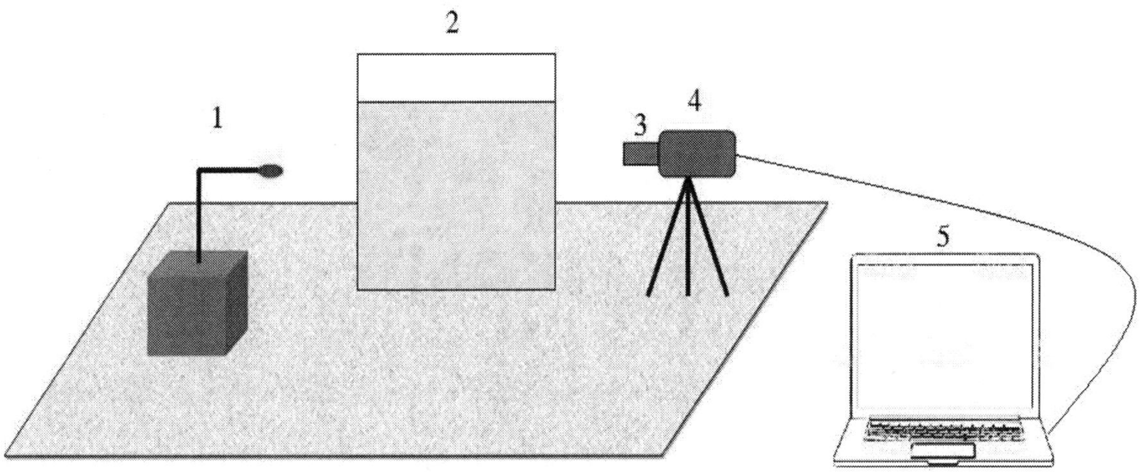

1 – single LED, 2 – cuvette with the object under study, 3 – lens with Scheimpflug adapter, 4 – high-speed camera, 5 – computer with installed software

Figure 1. The scheme of the experimental installation by PSV method.

4. The results of experimental studies

The possibility of flows diagnostics using polystyrene particles with a diameter of 40 μm [8] was investigated. The measurements were carried out in water in a 5×3×1 cm glass cuvette. During the experiment, the survey area was limited to 800×600 pixels with a frame rate per second of 100. As a result of the experiment, sets of images were obtained, which were further processed using the cross-correlation method in the PIVViewDemo program of the PIVTECGmbH company. Processing was carried out with the following parameters: the size of the survey areas was 32×32 pixels with 50 percent overlap. For each case, normalization frames were obtained for the possibility of transition from a pixel to a metric coordinate system. This normalization allows you to get in the program ready-made values of flow rates.

As a result of processing of the experimental images obtained without the use of the Scheimpflug adapter, the vector fields of flow rates shown in figure 2 were obtained.

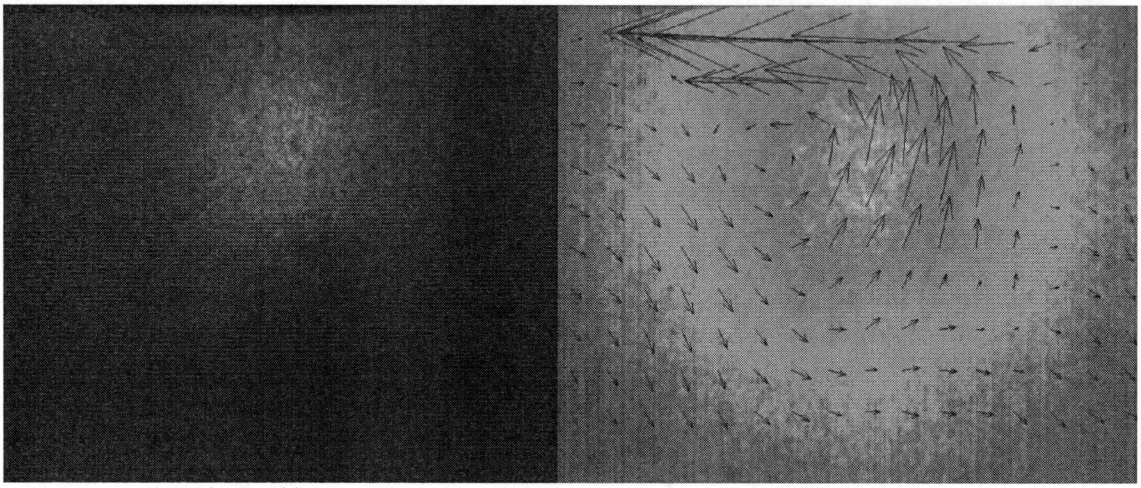

Experimental image Processing result

Figure 2. The results of cross-correlation processing of images obtained without the use of the Scheimpflug adapter.

Also during the experiment the Scheimpflug adapter, which allows to change the depth of field was used. An example of the resulting image and the result of processing in the program PIVView are shown in figure 3.

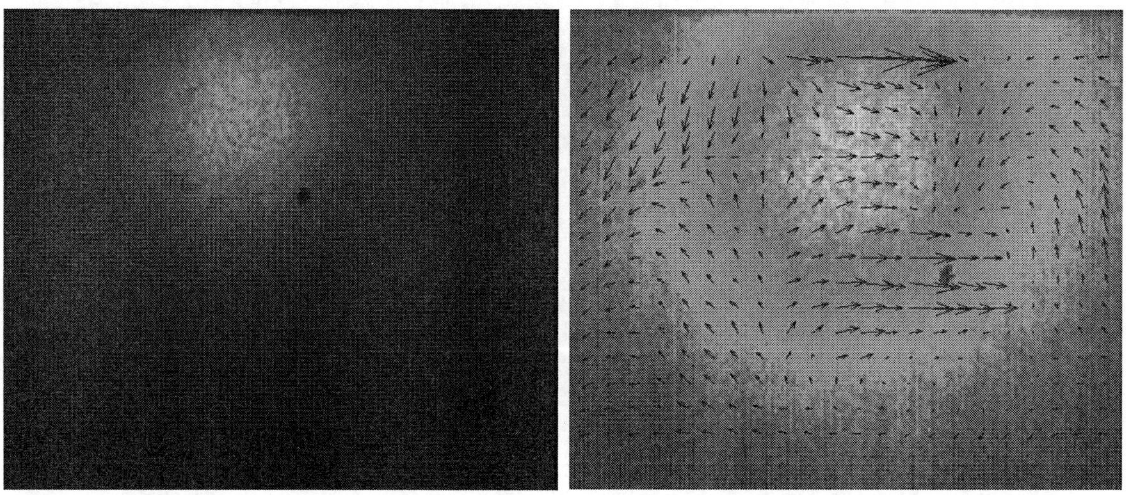

Experimental image Processing result

Figure 3. The results of cross-correlation processing of images obtained using the Scheimpflug adapter.

Image processing in both cases was carried out with the same parameters of the survey areas and overlap. The difference was only in the normalization of pixel / mm, due to the fact that when the lens was tilted due to the adapter, this ratio changed.

Based on the results of processing, the most attractive were vector velocity fields obtained by using a Scheimpflug adapter. As result of the normalization the values of velocity flow rates of order $10^{-3}\div10^{-4}$ m/s were obtained. An example of the obtained value is shown in figure 4.

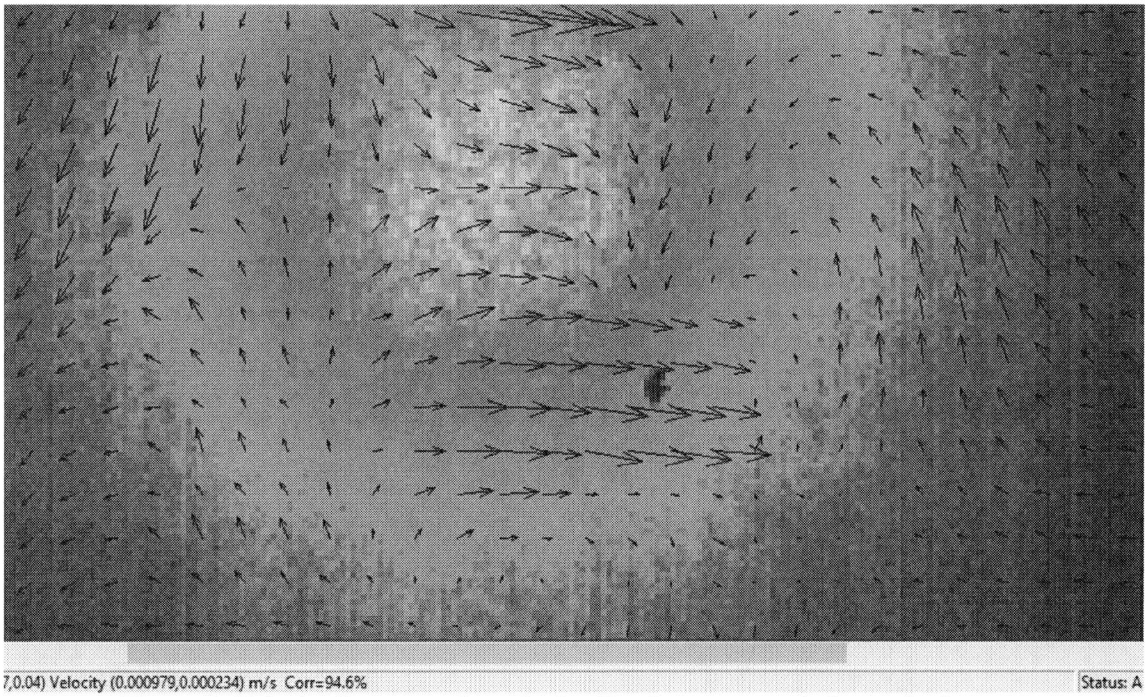

Figure 4. Velocity flow rate values obtained by the cross-correlation method.

5. Conclusion

The vector fields and values of the flow velocity of the order of $10^{-3}\div10^{-4}$ m/s in the process of mixing polystyrene particles at different times using cross-correlation processing are obtained. It was also shown that the use of the Scheimpflug adapter during shooting provides some advantage by changing the depth of field.

References

[1] Estevadeordal J and Goss L 2005 PIV with LED: Particle Shadow Velocimetry (PSV) *Innovative Scientific Solutions* (Inc., Dayton, OH)

[2] McPhail M J, Fontaine A A, Krane M H, Goss L and Crafton J 2015 Correcting for colour crosstalk and chromatic aberration in multicolor particle shadow velocimetry *Measurement Science and Technology* (Measurement Science and Technology, Journal of Physics)

[3] Chetelat O and Kim K C 2002 Miniature Particle Image Velocimetry System with LED In-Line illumination *Measurement Science and Technology, Journal of Physics*

[4] Whitaker S M, Reilly D, Bons J P et al. 2013 A Survey of Airborne Particle Impact Characteristics Using High Speed Particle Shadow Velocimetry DOI: 10.2514/6.2013-2484 Conference: 43rd AIAA Fluid Dynamics.

[5] Modern optical methods of flow investigation 2011 Ed. Rinkevichius B S (Moscow: Overlay)

[6] Raffel M, Willert C E, Wereley S T and Kompenhans J 2007 *Particle Image Velocimetry / A Practical Guide / 2nd ed.* (Springer)

[7] Prasad A K and Jensen K 1995 Scheimpflug stereo camera for particle image velocimetry in liquid flows *Applied Optics*

[8] Kolesnikov S Yu and Skornyakova N M 2018 Comparison of different types of tracers in the particle image velocimetry *Trudy MAI* (Moscow: MAI)

Journal of Physics: Conference Series **1421** (2019) 012012 doi:10.1088/1742-6596/1421/1/012012

The choice of an image processing algorithm for increasing sensitivity of the surface plasmon resonance method

I N Pavlov, A V Vedyashkina and G M Yanina

The V.A. Fabrikant Physics Department, National Research University "MPEI", Krasnokazarmennaya 17, Moscow, Russia

E-mail: inpavlov@bk.ru

Abstract. The paper describes the principles of the surface plasmon resonance imaging method in application to investigation of heat and mass transfer processes. It is shown, that one the main obstacles to increase its sensitivity is low signal-to-noise ratio due to speckles and stray interference structures in the experimental images. An algorithm of image processing is developed to overcome this obstacle. Several ways to do it is discussed in the paper.

1. Introduction

It is obvious that in order to provide the development of almost all scientific areas it is needed to conduct measurements. And the more accurate, sensitive, convenient and demonstrative the used measurement technique is, the more productive results it allows to receive. Thus, the task of development of measurement methods, techniques and devices always remains relevant. The most advanced measurement methods for many characteristics are optical ones. The optical method considered in this paper is the surface plasmon resonance imaging (SPRI) method [1], which allows determining the change in refractive index in a thin boundary layer of investigated medium with thickness of few hundred nanometers. Typically, this method is used to visualize biochemical reactions and to measure the binding constants in them [2]. In this particular work, the method is used for visualization of processes of heat and mass transfer such as evaporation, cooling, mixing liquid droplets [3]. To expand application area and provide more widespread using to this method it was decided to consider some possible ways to increase its sensitivity [4]. And one of them is corresponding algorithm of image processing, the choice of which this paper is devoted.

2. Surface plasmon resonance imaging method

As it is well known, surface plasmons are oscillations of the density of free electrons on a metal surface, excited by an external effect, usually an evanescent wave, produced by the total internal reflection of polarized light inside a glass prism. Under certain conditions (providing the required values of the metal film thickness, wavelength, polarization state and angle of incidence of the exciting light, the ratio of the refractive indices of the glass prism, metal and medium under study), a so-called surface plasmon resonance (SPR) occurs, at which, despite the condition of total internal reflection, the intensity of the reflected light equals to zero. When any of the parameters, for example, the refractive index of the medium, deviates from the resonance value, the efficiency of generation of surface plasmons decreases and reflected light appears, which is a signal of violation of resonant conditions. Thus, if a wide collimated laser beam is used as the exciting light, then on the screen in reflected light one can observe a change in the refractive index of the medium by changing the intensity of the light incident on the screen. A change in the refractive index of a medium may be due to a change in its temperature, phase composition (formation, for example, of ice or gas bubbles in water), concentration of certain substances in a solvent, etc. Knowing the dependence of the refractive index on the changing parameter, it is possible to determine the magnitude of the change in this parameter in the corresponding place of the boundary layer.

In order to demonstrate possibilities of the SPRI method, an experimental setup was developed and created, described in detail in [3]. It consists of a semiconductor laser module with a wavelength of 655 nm and a power of 5 mW, a beam expander, a polarizer, a rectangular glass prism, on the base of which is placed a glass plate with a deposited gold film 50 nm thick, screen or lens and a digital video camera. Investigated drop is placed on the surface of the film illuminated by a wide collimated laser beam and the image of the reflected beam is recorded. As it was shown above, when the refractive index of a thin layer of a drop bordering a film changes, the intensity changes at the corresponding place on the screen. According to the results of processing the images obtained during the process being studied, we can judge about the change in the parameter of interest.

3. Stages of image processing

Each algorithm of image processing consists of several stages. Usually it is preprocessing, filtration, processing itself (calculation of required quantities from image matrix) and sometimes postprocessing (smoothing, plotting, visualization, etc.). Let's consider it sequentially.

3.1. Preprocessing

In our case it consists of selection of desired number of images from received pack of them, then alignment and cropping all selected images in the same way and then renaming in accordance with the resulting sequence. There are no essential features in this stage except for the difficulty of full automation.

3.2. Filtration

The most important part on which we will focus in more detail because that is what determines the quality of information that we get from experimental images. As it can be seen in the examples of experimental images, the main problem is speckle-noises caused by using laser as an excitation light source. This problem can be solved by using moving (e.g., rotating) screen [5], but it entails additional difficulties and is not covered in this paper. Another way of solving this problem is averaging or smoothing. There are few approaches how to do it: use of filters with finite impulse response (FIR-filters), Fourier filtering, wavelet filtering, averaging by few rows or columns of an image and some others. They are needed to be compared in application to our problem.

Speckle-noise is a high-frequency noise that is why we need to use low-frequency filters. Low-frequency spatial filters attenuate high-frequency components (small areas with strong intensity changes) and leave the low-frequency components of the image unchanged. They are used to reduce noise and remove high-frequency components, which improves the accuracy of studies of the content of low-frequency components. As a result of applying these filters, we obtain a smoothed or blurred image. Let us consider the most useful types of low-frequency filters [6].

3.2.1 Arithmetic mean filter. This filter gives the mean value of intensity for a pixel within a local region of an image with help of using mask with equal coefficients. In English literature the name box-filter is also used. The arithmetic mean is defined as

$$G(x\ y)\quad -\sum_{i=1}^{m}\sum\quad F(i,j), \tag{1}$$

where $G(x,y)$ – intensity of a resulted image pixel with coordinates x and y, $F(i,j)$ – intensity of corresponding pixel of an original image within mask, m, n – width and height of a mask correspondingly. An arithmetic mean filter operation on an image removes short tailed noise such as uniform and Gaussian type noise from the image at the cost of blurring the image. The larger the filtering mask becomes the more predominant the blurring becomes and less high spatial frequency detail that remains in the image.

3.2.2. Geometric mean filter. Geometric mean value is calculated by the following formula:

$$G(x,y) = \left[\prod_{i=1}^{m}\prod_{j=1}^{n} F(i,j)\right]^{\frac{1}{m\cdot n}}. \tag{2}$$

The effect of applying this filter is similar to the previous method, however, individual objects of the original image are less distorted.

3.2.3. Harmonic mean filter. This filter is based on the next expression:

$$G(x,y) = \frac{m\cdot n}{\sum_{i=1}^{m}\sum_{j=1}^{n}\frac{1}{F(i,j)}}. \tag{3}$$

The filter suppresses "salt" noise well and does not work with noises like "pepper".

3.2.4. Counterharmonic mean filter. This filter is based on the expression

$$G(x,y) = \frac{\sum_{i=1}^{m}\sum_{j=1}^{n} F(i,j)^{Q+1}}{m\cdot n\cdot\sum_{i=1}^{m}\sum_{j=1}^{n} F(i,j)^{Q}}, \tag{4}$$

where Q is the filter order. The counterharmonic filter is a generalization of mean filters and at $Q > 0$ suppresses noise like "pepper", and when $Q < 0$ – noise like "salt", however, the simultaneous removal of white and black dots is impossible. When $Q = 0$, the filter becomes arithmetic, and when $Q = -1$, it becomes harmonic.

3.2.5. Gaussian filter. At this type of filtration, the two-dimensional Gaussian filter is used:

$$G(x,y) = \frac{1}{2\pi\sigma^2} e^{-\frac{x^2+y^2}{2\sigma^2}}. \tag{5}$$

The larger the parameter σ, the more blurred the image. Commonly, the filter size equals 3σ. In this case the mask size $6\sigma + 1 \times 6\sigma + 1$.

3.2.6. Median filter. This is a non-linear filter which is very useful at suppression of additive noise, especially if the noise is impulse. The method is very simple, does not require customization and therefore has received wide distribution. The procedure of image processing by the median filter is that for each position of the mask, the pixels that fall into it are ordered by increasing intensity. The average count in this ordered sequence (there must be an odd number of them) is called the median and is recorded as the center value of the mask window. As a result of applying this filter, sloping areas and sharp drops are preserved, and impulse noise, whose duration is less than half of the window, will be suppressed. The larger the window, the larger the details will be blurred. There are also other modifications of a median filter, such as adaptive median filter and weighted median filter, but they are usually used in some specific cases and will not be considered in this paper.

Should be added that it has recently become increasingly widespread use of neural networks and deep learning to increase signal-to-noise ratio in corrupted images (e.g., [7]). However, this question is beyond the scope of this work and may be the subject of further research by the authors.

Journal of Physics: Conference Series **1421** (2019) 012012 doi:10.1088/1742-6596/1421/1/012012

Different studies give different ratings and recommendations for choosing filters [8-10], so they are needed to be compared in relation to the SPRI images. The table 1 shows the results of application of these filters to the example of experimental images with three different sizes of mask (3×3, 5×5 and 7×7) and three different amplitude filtering thresholds (50, 100 and 150) for Fourier filtration.

Table 1. The results of application of different filters to an example of experimental images.

Original	Fourier 50	Fourier 100	Fourier 150

Filter	3*3 mask	5*5 mask	7*7 mask
Arithmetic Geometric Harmonic Median Gaussian			

In order to choose the optimal filter, the criterion of the smallest standard deviation in an image area corresponding to homogeneous water droplet was chosen. The table 2 shows the results of its estimation for rectangular area 50×20 pixels in the center of matrix obtained by dividing an image with droplet by the image without it (in other words, matrix of reflectance coefficient). The standard deviation of the same pixel values for non-filtered images is 0.292.

Table 2. Estimation of standard deviations of pixel values in filtered images.

Filter	3×3 mask	5×5 mask	7×7 mask
Arithmetic	0.076	0.069	0.069
Geometric	0.085	0.061	0.060
Harmonic	0.061	0.066	0.066
Median	0.037	0.047	0.047
Gaussian	0.077	0.064	0.068
Threshold	50	100	150
Fourier	0.289	0.206	0.102

Thus, it can be seen that the optimal filter for available images is median filter with mask size 3×3. Apart from minimum standard deviation it allows to get better visual perception and not to lose possible small-scale details in images. For very noisy images an adaptive median filter may be used.

3.3. Processing

After filtration we get images ready for processing. Now it is time to extract quantitative data from these images. For this we take two images obtained with the same experimental parameters (angle of incidence of induced light, intensity distribution across laser beam, recording options, etc.) except absence or presence of investigated liquid on the top of the metal film: in one of them there is an image droplet and the other one is without it (reference image). To calculate reflectance coefficient we divide values of intensity in each pixel by corresponding values in reference image. So we obtain matrix of distribution of reflectance coefficient over the interface between the metal film and investigated medium which corresponds to distribution of refractive index in thin boundary layer in close proximity to this film.

Use of reference image allows us to neglect uneven distribution of intensity by cross-section of a laser beam. But the main difficulty here to provide absolutely the same conditions for registration in order to speckle structure does not change. If we could meet this condition, it would be even unnecessary to do filtration, because it would have been performed automatically. Unfortunately, it is too difficult to fulfill this condition with our technical capabilities, because the SPRI method is very sensitive not only to change of refractive index, but also to the slightest displacements or deformations of parts of the setup. That is why any vibrations or even air flows cause to change in the speckle pattern. This makes it important to choose proper filtering algorithm.

On the other hand, this circumstance can be used to get rid of speckles. Indeed, if we record several images with and without contact of the film and investigated medium with different speckle and stray interference patterns due to vibrations, for example, we can average them and it will have the same effect as physical averaging with help of rotating screen.

In order to get more sensitivity from the same images the controlled contrast enhancement method may be used. This method can allow to artificially increase the reflection coefficient difference between two images of droplets with different refractive indices. The problem is that dependence of reflectance coefficient on refractive index is not linear and for correct interpretation we need to exactly know on which part of this dependence we are working. However, it may be successfully applied for qualitative visualization of change in refractive index of an investigated medium, when it is required.

3.4. Postprocessing

Thus, as a result of processing, we have a matrix of distribution of refractive index in thin boundary layer of an investigated medium in close proximity to the metal film. Sometimes it is needed to reconstruct the distribution of some other value (e.g., temperature, concentration, phase composition, etc.). Then we have to know the dependence of the required value on refractive index of medium and knowing it we can easily recalculate distribution of refractive index into distribution of required value. The next step is plotting smooth graph using experimental points. To do this, we select appropriate values, and sometime spend an average.

4. Conclusion

Based on the above mentioned material, the next algorithm of image processing for experimental images of the surface plasmon resonance imaging method was chosen. At first, selected images should be cropped in exactly the same way so that only the region of interest remains in the image field. Then cropped images should be filtered using median or in some cases adaptive median filter. After filtering the processing begins, which consists of dividing the images by corresponding reference ones in order to get the reflectance coefficient matrices, then averaging by several rows or columns, then calculating a required parameter for each processed image. And finally an approximation curve of dependence of this parameter on time or on coordinates in image should be plotted.

The novelty of the work lies in the fact that all the main modern techniques of image processing were considered and compared in application to the surface plasmon resonance image processing.

It is planned as the nearest future work to implement this algorithm in the Python programming language. The next step may be to use neural networks for image processing in order to reduce noise.

Acknowledgments
This work was financially supported by the Russian Federation President Grant No. MK-6361.2018.8.

References

[1] Wong C L and Olivo M 2014 Surface plasmon resonance imaging sensors: a review *Plasmonics* **9** 809-824
[2] Homola J 2008 Surface plasmon resonance sensors for detection of chemical and biological species *Chem. Rev.* **108** 462-493
[3] Pavlov I N, Rinkevichyus B S and Vedyashkina A V 2016 Application of surface plasmon resonance method for visualization of processes in near-wall liquid layer *J. Phys.: Conf. Ser.* **737(1)** 012044
[4] Pavlov I N, Vedyashkina A V, Raskovskaya I L, Rinkevichyus B S and Tolkachev A V 2018 Laser imaging of physical processes in thin near-wall layer of liquid droplet by surface plasmon resonance *Proc. Int. Conf. Laser Optics (Saint-Petersburg)* 8435196 210
[5] Goodman J W 2007 *Speckle Phenomena in Optics: Theory and Applications* (Roberts and (Company Publishers) p 389
[6] Gonzalez R C and Woods R E 2018 *Digital Image Processing* (Pearson) p 1192
[7] Lehtinen J, Munkberg J, Hasselgren J, Laine S, Karras T, Aittala M and Aila T 2018 Noise2noise: learning image restoration without clean data *Proc. 35th Int. Conf. on Machine Learning* (Stockholm: Sweden) PMLR 80
[8] Hoshyar A N, Al-Jumaily A and Hoshyar A N 2014 Comparing the performance of various filters on skin cancer images *Procedia Computer Science* **42** 32-37
[9] Chandel R and Gupta G 2013 Image filtering algorithms and techniques: a review *Int. J. Advanced Research in Computer Science and Software Engineering* **3(10)** 198-202
[10] Srestha S 2014 Image denoising using new adaptive based median filter *Signal and Image Processing: an International Journal* **5(4)** p 13

Journal of Physics: Conference Series **1421** (2019) 012013 doi:10.1088/1742-6596/1421/1/012013

In-plane interferometry of terahertz surface plasmon polaritons

A K Nikitin[1,2], O V Khitrov[1], V V Gerasimov[2,3], I Sh Khasanov[1] and T A Ryzhova[4]

[1] Scientific and Technological Centre for Unique Instrumentation of RAS, Butlerova str., 15, Moscow, 117342, Russia
[2] Novosibirsk State University, Pirogova str., 1, Novosibirsk, 630090, Russia
[3] Budker Institute of Nuclear Physics of the Siberian Branch of the RAS, Acad. Lavrentieva Pr., 11, Novosibirsk, 630090, Russia
[4] Peoples Friendship University of Russia, Miklukho-Maklaya str.6, Moscow, 117198, Russia

E-mail: alnikitin@mail.ru

Abstract. The paper is devoted to the development of interferometric schemes for determining the complex refractive index $\kappa = \kappa' + i \cdot \kappa''$ κ of surface plasmon-polaritons (SPPs) of the terahertz (THz) spectral range. As the value of κ depends on the dielectric constant ε_m of the metal surface guiding the SPPs it can be used for determining ε_m in the far infrared what other optical methods fail to do due to high reflectivity of metals. We discuss two types of THz SPP interferometers in which the interference pattern is formed as a result of the interaction of SPP beams themselves rather than bulk waves produced by these beams. The first type of interferometers are static devices that enable one to investigate fast processes on the metal surface, while the second type are dynamic ones that make it possible to realize Fourier spectroscopy of the metal surface and its transition layer at THz frequencies. Devices of the both types produce interferograms enabling one to determine the real and the imaginary parts of κ. The results of experiments on the interaction of THz SPPs with flat mirrors and beam–splitting plates, the key elements of the THz SPP interferometers, are presented.

1. Introduction

Surface plasmon-polaritons (SPPs) representing themselves a complex of an evanescent *p*-polarized electromagnetic wave and a wave of the conduction electrons density onto the surface of a material (usually, metal) with negative real part of its dielectric permittivity are widely used for surface control [1]. Being concentrated at the "metal-dielectric" interface SPPs field is very sensitive to the state of the surface and its transition layer. The use of SPPs generated by probing radiation rises up the sensitivity of measurements due to increasing the distance of interaction of radiation with the object and concentration of the field in the near-surface area.

As any EM wave SPPs are characterized by their complex refractive index $\kappa = \kappa' + i \cdot \kappa''$, which determines the measurable characteristics of the SPPs: the propagation length $L = 1/(2k_o \cdot \kappa'')$, SPPs field penetration depth $\delta \sim 1/(k_o \cdot \kappa')$ into air, and their phase velocity $\vartheta_{ph} = c/\kappa'$ (here c is the speed of light; $k_o = 2\pi/\lambda_0$; λ_0 is radiation wavelength in air) [2]. Having measured L (which corresponds to

κ'') and ϑ_{ph} (which corresponds to κ') one can calculate the permittivity of the metal surface or two unknown parameters (thickness and/or the dielectric constant) of the layer covering the surface.

In the visible range SPPs exhibit to a greater extent the properties of a mechanical wave: their ϑ_{ph} is much less than the speed of light and it is usually determined without any problems by the angle of incidence of radiation generating SPPs; while the value of L is only of a few wavelengths and its measurement is problematic.

In the terahertz (THz) region (from 1 to 10 THz), the characteristics of SPPs are very similar to those of a plane wave in free space: their phase velocity ϑ_{ph} is only slightly less than the speed of light (the difference $(\kappa'-1) < 10^{-3}$), while L amounts to tenths of a meter [3]. This complicates the determination of ϑ_{ph}, but makes the assessment of L rather easy.

Interferometry is the most reliable and accurate method for determining the phase velocity of plane EM waves. But this statement is not obvious with respect to surface plasmons.

There are two ways of interferometry of surface plasmons: either to register the result of the interaction of bulk waves (BWs), one (or both) of which is generated by the SPPs, or to record the interferogram formed by the fields of interacting SPP beams in the near-surface region (the in-plane SPP interferometry). To realize the first way one has to change the distance run by the SPPs during the measurements employing a movable coupling element converting the SPPs into BWs [4]; it is a rather time-consuming procedure and the measurements are characterized by the low signal-to-noise ratio as the SPPs-BWs conversion is accompanied by the generation of strong parasitic waves on the coupling element. The second way of SPP interferometry (the in-plane one) is more promising since it does not need the reverse conversion of SPPs into BWs, and the interferogram can be obtained in both convergent and parallel beams; therefore both static and dynamic SPP interferometers can be built (the first of them can be used for studying fast processes, while the second – for creation of surface-plasmon Fourier spectrometers of the THz range).

2. Interaction of THz SPPs with flat mirrors and beam splitters

To create SPP interferometers, it is necessary to master the operations of reflection and splitting of SPP beams. At first glance, such operations can be performed using Bragg gratings formed on the metal surface [5]. The disadvantages of this method of controlling the SPP beams are not only damage to the surface and the inability to control the influence; its main drawback is that interaction of THz SPP beams with diffraction gratings is accompanied by generation of mighty parasitic near-surface bulk waves, illuminating the photodetector and greatly reducing the signal-to-noise ratio. Geodesic prisms also can split and deflect SPP beams [6], but the drawbacks of Bragg gratings are typical for their use as well.

Possible alternative to the planar elements for controlling the THz SPP beams may be their bulk analogous (flat mirrors and beam-splitting plates), since THz SPPs are very close to plane EM waves in their characteristics (see section 1). Below are the results of our experimental studies on the interaction of flat mirrors and partially transparent plates installed perpendicular to the metal surface with THz SPPs generated by radiation of the Novosibirsk free-electron laser (NovoFEL) [7].

2.1. Reflection of THz SPP beams by flat mirrors

The experimental studies on reflection of THz SPP beams by flat mirrors are fully described in [8]. The SPPs were generated by radiation with the wavelength $\lambda_o = 130$ μm having the linewidth of ±1 μm ($\Delta\lambda/\lambda \approx 1\%$). The samples represented themselves flat glass substrates 15×3.5×1.0 cm in size with one optically polished face (15×3.5 cm) coated by a 0.3 μm thick gold layer obtained with the magnetron sputtering technique and overcoated by a zinc sulphide (ZnS) film of various thickness d. The mirror (glass substrate coated by a layer of gold) was installed perpendicular to the sample face in a plane oriented at an angle of 45° to the SPP track. To couple the laser radiation and the SPPs we employed the cylindrical element and the end-fire coupling technique [9, 10]. Scheme of the experimental set up is presented in figure 1; here A is the position of the Goley cell used for detection of the reflected beam intensity. The results of measurements for samples with ZnS films of $d = 0$ and $d = 0.7$ μm are plotted in figure 2.

First it should be noted that THz SPPs retain their nature in interaction with a mirror oriented perpendicular to the sample surface provided the gap between the surface and the mirror edge is less than 2λ; in this case the SPP reflection coefficient was close to unity. But the greater was the angle of deflection α of the mirror from the normal to the surface, the more intensive were the stray bulk waves and for $\alpha \geq 2^\circ$ (this angle limit depends on the surface quality) the reflection coefficient of SPPs tended to zero (see figure 2). Note, that the dependence of reflectivity on the mirror tilt angle α can be used for the real-time control of the SPP beams intensity. Based on this knowledge we developed and tested the device for measuring the propagation length L of THz SPPs with a high signal-to-noise ratio due to immobility of the input and output coupling elements [11].

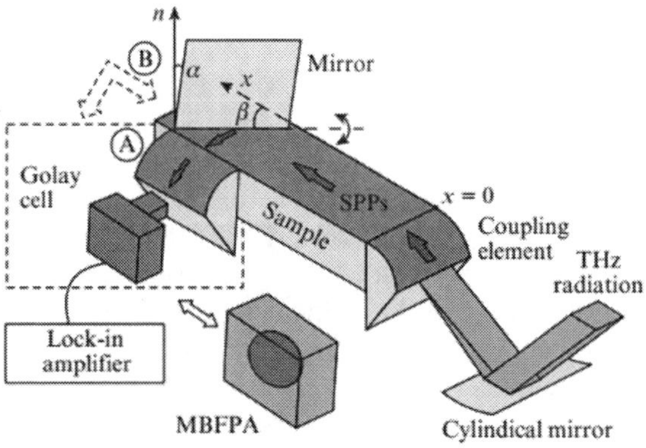

Figure 1. The experimental set up (MBFPA is a microbolometer focal plane array; A and B are the MBFPA positions for the reflection and transmission measurements) [8].

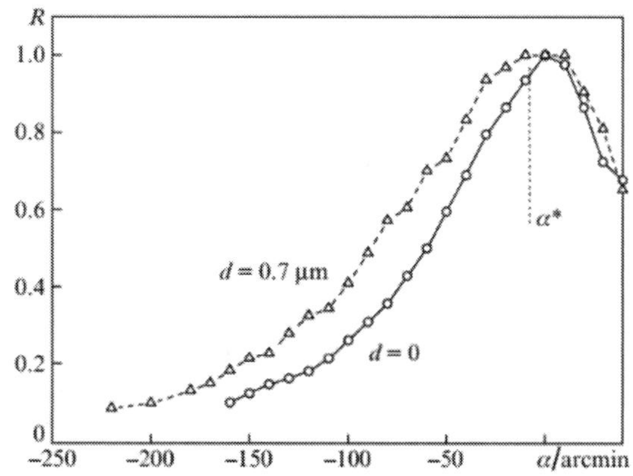

Figure 2. Measured dependencies of the SPP reflection coefficient R on the angle α of the mirror deflection from the normal to the sample surface covered with a *ZnS* layer of thickness d.

2.2. Splitting THz SPP beams by partially transparent flat plates

To test the possibility of splitting THz SPP beams by means of flat plates we performed measurements similar to the ones described in section 2.1 having replaced the mirror with a partially transparent for THz radiation flat plate [12]. As the material of such a plate we chose a kapton tape with thickness of 125 μm; this choice is explained by the value of kapton refractive index ($n \approx 1.88 + i{\cdot}0.02$ [13]) at $\lambda_o = 130$ μm, which provides low absorption of the radiation and approximately the same values of the reflectance R and the transmittance T for a plane THz wave, which is close in its characteristics to the THz SPPs. When measuring T the Golay detector was placed in position B (figure 1).

As a result of the measurements it was found [12] that: 1) THz SPPs are reflected and transmitted by the kapton plate about to the same extent as a plane wave of the same frequency is; 2) the values of R and T for the SPPs, equal to $R = 0.28 \pm 0.03$ and $T = 0.45 \pm 0.05$ at the incidence angle $\beta = 45^\circ$, depend on this angle rather slightly (gradually decreasing and rising with increasing the angle, respectively); 3) the sum of R and T is noticeably less that unity (namely: $R + T \approx 0.8$), that can be explained not only by the absorption of radiation inside the plate, but also by the diffraction losses of the SPPs due to the loose adhesion of the plate to the sample surface and as a result of the SPP passing through of the impedance barrier [14], which is constituted by the plate; 4) the thicker is the dielectric (*ZnS*) film covering the surface of the metale the less is transmittance of the SPPs.

Summing up the results of the experiments, it can be stated that rigid kapton plates of submillimeter thickness can be effectively used for splitting beams of THz SPPs. However, it should be noted, that splitting of THz SPP beams by means of such plates is accompanied by the generation of parasitic bulk waves due to the diffraction of the SPPs on the plate edge. Nevertheless, the intensity of these parasitic waves and the energy losses of the SPPs are significantly less than those occurring on planar diffraction gratings used for the splitting of THz SPP beams.

3. Static two-beam surface plasmon interferometer of the terahertz range

3.1. Chronology of the development of SPP static interferometry

A number of papers on SPP static interferomerty have been published. The earliest of them is [15]. The technique is based on the effect of dramatic change of the phase of radiation generating the SPPs by the method of attenuated total internal reflection (ATR). Later on the idea was picked up by other scientists (e.g. [16, 17]). However, the technique is not applicable in the infrared (IR) and THz ranges due to the macroscopic propagation length of SPPs much exceeding the radiation beam diameter.

A new step in the development of the static SPP interferometry theme is associated with the use of a broadband light source illuminating a slit (or a couple of slits) in an opaque metal layer to generate SPPs propagating normal to the slit(s); on running some distance the SPPs are reflected from a groove parallel to the slit and return to it where they produce bulk radiation interfering with the incident light; the formed interference pattern is analyzed by a spectrometer [18]. Such SPP-interferometers are used mainly for sensor applications and they operate only in the visible or near IR range [19-21].

The first scheme of a static THz SPP interferometer based on the use of flat mirrors was suggested in [22]. On the basis of the scheme, the theory of dispersion Fourier transform infrared SPP-spectroscopy was developed [23]. However experiments to test such interferometers were not carried out due to insufficient knowledge at that time about the interaction of THz SPPs with flat mirrors. Moreover, the operation of such an interferometer could be characterized by a low signal to noise ratio due to diffraction of the SPPs at the edge of the angular mirror splitting the original beam.

3.2. Schematic of the static THz SPP interferometer and its description

Our objective was to simplify both the scheme of the known interferometer for determining the complex refractive index of THz SPPs [22], and the procedure for processing the measurement results.

Simplification of the interferometer scheme is achieved by excluding two angular and three flat mirrors from it. In the new scheme, the splitting of the initial SPP beam into two new ones is realized not by an angular mirror, but by a translucent plane-parallel plate mounted on the waveguiding surface and oriented perpendicular to it and to the SPP track.

Simplification of the procedure for processing the results of measurements is a consequence of the possibility of calculating the imaginary part κ'' of the SPPs' refractive index with respect to the intensity distribution on the photodetector array of each of the interfering beams separately (measured while blocking the other beam with by a flap introduced into the scheme), which makes it possible not only to obtain the calculation formula in an explicit form, but also apply it repeatedly to different points of the interferogram in order to improve the accuracy of κ'' determination.

Figure 3 shows the schematic (top view) of the elaborated device, where the numerals denote: (1) is a source of collimated *p*-polarized (relatively to the SPP guiding surface) monochromatic IR or THz radiation; (2) is an element for coupling the source radiation and the SPPs; (3) is a flat facet of the sample, capable of guiding the SPPs; (4) is a semitransparent flat-parallel plate, oriented at an angle γ with respect to the SPP beam incident on it, and characterized by the coefficient of reflection R and the coefficient of transmittance T as applied to the SPPs; (5) is a plane mirror oriented at an angle $\beta < \gamma$ with respect to the SPP beam incident on it; (6) is a photodetector array adjacent to the facet 3; (7) is an information processing device; (8) is a rotational shutter, allowing to shut alternately the interfering SPP beams. Element (2) represents itself a cylindrical segment with the radius of curvature $r >> \lambda$ and sharp edges of the curved surface. Such form of the coupling element makes it possible not only to use the end-fire coupling method [10], but also to shield effectively the receiver from parasitic exposures [9]. Note that the reflection coefficient of SPPs depends on the angle of deviation of mirror (5) or plate (4) from the normal to the waveguiding surface [8]; this makes it possible to achieve the desired contrast of the interference pattern when tuning the device.

The interferometer works as follows. Radiation (with the wavelength λ_o) from source (1) is directed onto element (2), which converts the bulk radiation into the SPPs. The collimated beam of SPPs propagates along the sample facet (3) and falls on plate (4), which splits the original beam into two secondary ones [12]. The beam, passed through plate 4, falls onto mirror 5 and is reflected by it to

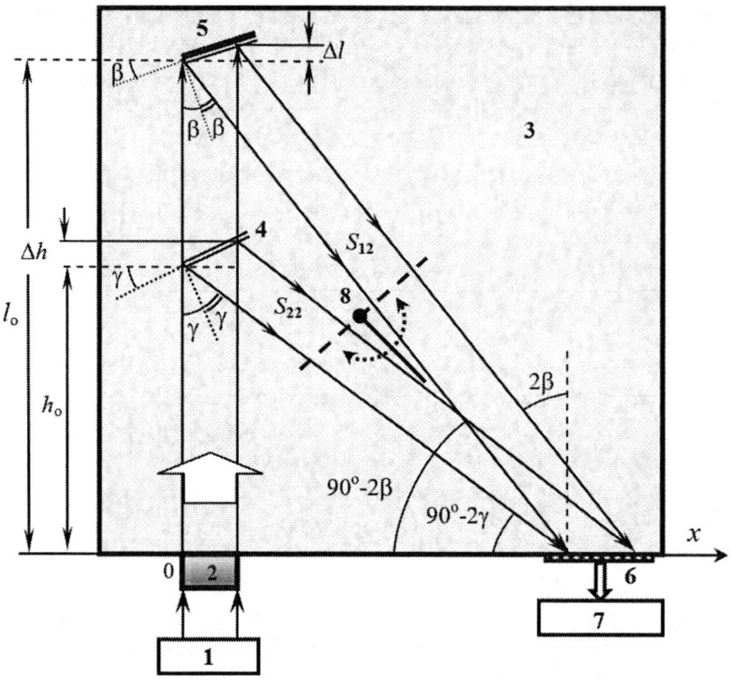

Figure 3. Schematic of the static THz SPP interferometer.

photodetector array (6), where it interacts with the other secondary beam reflected by plate (4). As a result of the interaction of the beams, the interferogram recorded by array (6) is formed. Electric signals from the pixels of array (6) are fed to device (7) which, employing values of the signals in two maxima of the interferogram and the pixel coordinates, calculates the real part κ′ of the SPPs refractive index:

$$\kappa' = \frac{m \cdot \lambda}{\left|\Delta S_j - \Delta S_{j+m}\right|}, \tag{1}$$

here m is the number of periods between the considered maxima of the interferogram; ΔS_j and ΔS_{j+m} are the geometric path differences of the corresponding beams of interfering beams in the j-th and (j+m)-th maxima; $\Delta S_j = (S_1)_j - (S_2)_j$, where the subscripts 1 and 2 correspond to the SPP beams, reflected by mirror (5) and plate (4), respectively; $(S_1)_j$ and $(S_2)_j$ are the distances run by the corresponding rays of the 1 and 2 beams, and calculated by the formulas:

$$(S_1)_j = l_j + (S_{12})_j = l_j \cdot \left(1 + \frac{1}{\cos(2\beta)}\right),$$

$$(S_2)_j = h_j + (S_{22})_j = h_j \cdot \left(1 + \frac{1}{\cos(2\gamma)}\right),$$

here: $l_j = l_o + \Delta l_j = l_o + x_j \cdot \tan(\beta)$ and $h_j = h_o + \Delta h_j = h_o + x_j \cdot \tan(\gamma)$ are the distances travelled by the j-th ray from element (2) to mirror (5) and plate (4), respectively; l_o and h_o are the smallest distances separating mirror (5) and plate (4) from element (2), respectively; x_j is the abscissa of the

j-th ray of the SPP beam on element (2), calculated by the formula: $x_j = x_{j\,int} \cdot \cos(2\beta) - l_o \cdot \sin(2\beta)$, here $x_{j\,int}$ is the coordinate of the same ray on array (6).

Note that formula (1) can be applied many times (for different combinations of j and (j+m) maxima) in processing of this interferogram in order to find the mean value of κ' and to increase the accuracy of its determination.

Value of the imaginary part κ'' of the SPP refractive index can also be determined from the interferogram by solving the nonlinear equation that describes the dependence of the resulting intensity on the coordinate x with respect to κ'' [22]. But we offer an instrument solution to this problem, which makes it easier to process the results of measurements; it implies measuring the intensity distributions $I_1(x)$ and $I_2(x)$ on photodetector array (6) of each of the interfering beams individually, closing the other beam with shutter (8). Then, knowing the pixels abscissas x_{int} of array (6) and intensities of the beams I_1, I_2 on the pixels, the value of κ'' can be calculated by the formula:

$$\kappa'' = \frac{\ln\left[\frac{T}{R} \cdot \frac{I_2(x_{\mathrm{int}})}{I_1(x_{\mathrm{int}})}\right]}{2k_o \cdot \left[(l-h) + \frac{l}{\cos(2\beta)} - \frac{h}{\cos(2\gamma)}\right]}, \qquad (2)$$

here: $l = l_o + x \cdot \tan(\beta)$, $h = h_o + x \cdot \tan(\gamma)$, $x = x_{int} \cdot \cos(2\beta) - l_o \cdot \sin(2\beta)$.

As in the case of determining κ', multiple determination of κ'' contributes to an increase in the accuracy of its sought mean value. We should note that the method described for determining κ'' actually reproduces the known "two-prism" technique for determining the propagation length of SPPs, in which the natural logarithm of the ratio of the intensities of SPPs that have traveled different distances is related to the difference of these distances [2].

3.3. *Numerical modeling*

To illustrate the operation of the interferometer, let us consider the possibility of determining the refractive index of SPPs generated by radiation with $\lambda_o = 130$ µm on a flat surface of a gold sample (100×50 mm^2) disposed in vacuum and containing 0.5 µm thick *ZnS* coating. As beam splitter (4) let's take a kapton plate characterized with $R = 0.28$ and $T = 0.45$. The width of the radiation beam is set equal to 10 mm, the distances $l_o = 80$ mm, $h_o = 50$ mm, $x_o = 30$ mm. In order for interfering beams of SPPs could intersect on array (6), with the chosen values of l_o, h_o and x_o, the angles β and γ should equal 10°17′ and 15°29′, respectively. Recording of the interferogram can be performed by an array of microbolometers (containing 320 pixels 51 µm each) with its total length of 16.32 mm [24].

Figure 4 shows the interferograms recorded in this case on array (6) for different distances h_o spacing the splitting plate (4) from the coupling element (2) under the condition of equal radiation intensity along the cross section of the SPP beam. Note the lager is h_o the greater is the interference pattern period, the higher is the accuracy of κ'determination.

To calculate κ', let us consider the 1-st and 17-th maxima of the pattern depicted in figure 4 (a); on array (6) they have coordinates $x_{1\mathrm{int}} = 30.41$ mm and $x_{17\mathrm{int}} = 40.23$ mm, respectively. On substituting the values l_o, h_o, x_o, λ, β, γ, $x_{1\mathrm{int}}$, $x_{17\mathrm{int}}$ in (1) and setting $m = 16$, we get: $\kappa' = 1.00053$.

To determine κ'' let us select on the interferogram the pixel of array 6 with the abscissa $x_{int} = 36.0$ mm. By shutting alternately the interfering SPP beams with gate (8), we establish that at this point the relative intensities of the beams $I_1/I_o = 0.0322$ and $I_2/I_o = 0.0205$ (I_o is the intensity of the SPP beam at the output of element (2). Then, substituting in (2) the values l_o, h_o, x_o, λ, β, γ, k_{refl}, k_{trans}, x_{int}, I_1/I_o and I_2/I_o, we get: κ''=0.000024.

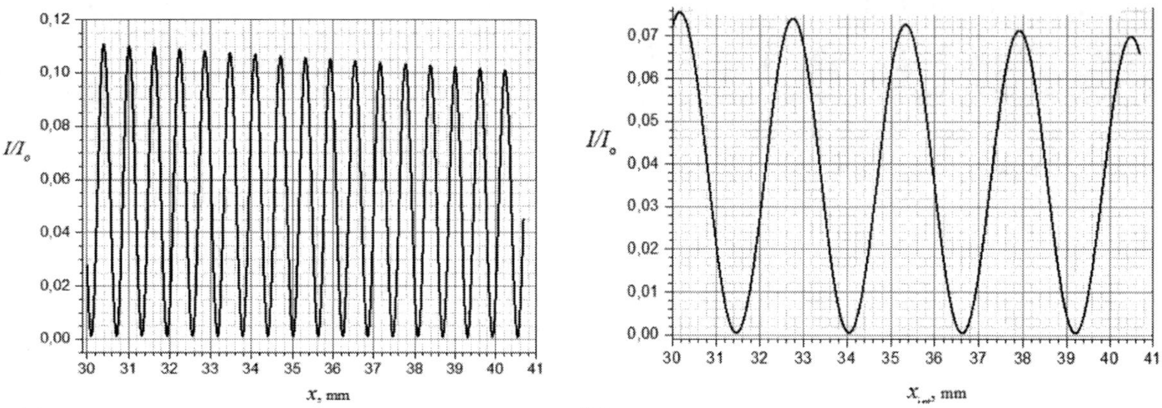

(a) (b)

Figure 4. Interference pattern recorded by the SPP interferometer photodetector array 6 for different distances h_o spacing the splitting plate 4 and the coupling element 2: (a) $h_o = 50$ mm; (b) $h_o = 70$ mm.

Thus, the static interferometer enables one to determine the refractive index of THz SPPs almost instantaneously, which is extremely important for studying fast processes on conducting surfaces.

4. Dynamic surface plasmon interferometer of the terahertz range

4.1. Schematic of the dynamic THz SPP interferometer and its description

The chance of using flat mirrors and beam splitters for reflecting and dividing beams of THz SPPs makes it possible to develop the scheme of a dynamic SPP interferometer, similar to the scheme of the Michelson interferometer for three-dimensional waves.

Figure 5 shows the schematic (top view) of the elaborated THz SPP interferometer, where the numerals denote: (1) is a source of collimated *p*-polarized THz radiation; (2) is an element for coupling the radiation and the SPPs; (3) is a flat facet of the sample capable of guiding the SPPs; (4) is a partially transparent flat-parallel plate oriented at 45° to the incident SPP beam and characterized by the coefficient of reflection R and the coefficient of transmittance T as applied to the SPPs; (5) is an immobile plane mirror oriented normally to the SPP beam reflected by plate (4); (6) is one more mirror, which is oriented normally to the SPP beam transmitted through plate (4) and capable of moving along this beam track; (7) is a single-pixel photodetector; (8) is an information storage and processing device.

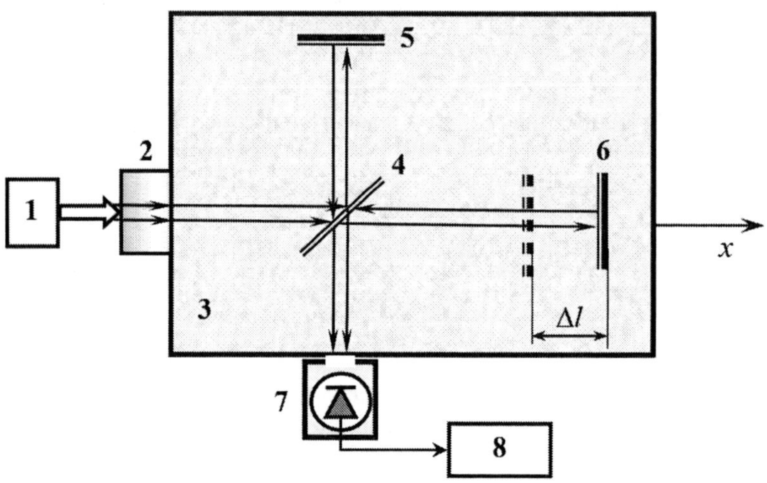

Figure 5. Schematic of the dynamic THz SPP interferometer.

Journal of Physics: Conference Series **1421** (2019) 012013 doi:10.1088/1742-6596/1421/1/012013

The interferometer works as follows. Radiation from source (1) is directed onto element (2), which converts it into the SPPs. The collimated beam of SPPs propagates along the sample facet (3) and falls on plate (4), which splits the original beam into two secondary ones [12]. The beam reflected by plate (4) is reflected from mirror (5) and propagates in the opposite direction [8]. The beam, passed through splitter (4), reaches mirror (6), is reflected from it and returns to splitter (4). The first of the secondary beams partially passes through the plate (4), while the second is partially reflected by it. Having interacted with splitter (4) both secondary beams propagate along the same track (perpendicular to the plane of incidence), reach the edge of the sample facet (3) and, as a result of diffraction on the edge, are converted into bulk waves [25], the intensities of which are proportional to the intensity of the secondary SPP beams. The bulk waves interfere with each other and illuminate photo detector (7). The signal generated by it enters device (8), which stores the value of the signal I and the corresponding coordinate x of mirror (6). Further, mirror (6) is shifted along the x-axis by one "step" Δx and device (8) registers a new couple of values of the signal I and the coordinate $x = x_o \pm \Delta x$ (here x_o is the initial coordinate of the mirror) corresponding to the new position of mirror (6). The procedure of such measurements continues until mirror (6) is shifted to the maximum distance x_{max} from its initial position. The obtained dependence $I_{int}(x)$ is the desired interferogram and it is described by the expression:

$$I_{int}(x) = I_o \cdot R \cdot T \cdot \left[e^{-\alpha\cdot(2b+a)} + e^{-\alpha\cdot(2x+a)} + 2\cdot\sqrt{e^{-2\alpha\cdot(b+x+a)}} \cdot \cos(\Delta\varphi) \right], \tag{3}$$

here I_o is the intensity of the SPP beam incident on splitter (4); $\alpha = 2k_o \cdot \kappa''$ is the attenuation factor of the SPPs; b is the distance from splitter (4) to mirror (5); a is the distance from splitter (4) to the edge of facet (3); x is the distance from splitter (4) to the current position of mirror (6); $\Delta\varphi = k_o \cdot \kappa' \cdot |x_o - x|$ is the phase shift gained by the SPPs on the increment $|x_o - x|$ of the distance spacing splitter (4) and mirror (6).

The interferogram enables one to determine both parts of the refractive index $\kappa = \kappa' + i \cdot \kappa''$ of SPPs. The real part of κ can be estimated by relating the wavelength of radiation in air to the wavelength of the SPPs: $\kappa' = \lambda_0 / \lambda_{SP}$ [1]. To determine the SPP-wavelength (λ_{SP}), one has to divide the double displacement (the change of the distance, run by the SPP-beam) of the movable mirror $\Delta l = 2 \cdot |x - x_o|$ providing monitoring m maxima of the interferogram by the number of these maxima: $\lambda_{SP} = \Delta l / m$. Thus, the final formula for calculating κ' has the form:

$$\kappa' = \frac{m \cdot \lambda_0}{\Delta l} = \frac{m \cdot \lambda_0}{2 \cdot |x - x_o|}. \tag{4}$$

Note that the accuracy of the definition of κ' is proportional to the number of periods m considered.

As in the case of the static SPP interferometer, the value of κ'' can be determined from the interferogram as well. But we again suggest an instrument solution to this problem, which consists in measuring the dependence of the intensity of the secondary beam $I(x)$ passing through splitter (4), sequentially reflected by mirror (6) and by splitter (4) in the absence of mirror (5) onto the facet (3). In this case, for any values I_1 and I_2 measured at the mirror (6) positions with coordinates x_1 and x_2 (where $x_1 > x_2$), respectively, the following ratio is true:

$$\alpha = \frac{\ln(I_2/I_1)}{2 \cdot (x_1 - x_2)}, \tag{5}$$

Equating the right part of expression (5) to the definition formula for $\alpha = 2k_o \cdot \kappa''$, we obtain:

$$\kappa'' = \frac{\ln(I_2/I_1)}{4k_o \cdot (x_1 - x_2)}. \tag{6}$$

Journal of Physics: Conference Series **1421** (2019) 012013 doi:10.1088/1742-6596/1421/1/012013

Multiple calculation of κ''-value for different coordinates x_1 and x_2, in order to find the average value of κ'', improves the accuracy of its determination. Note that the described technique of determining the true value of κ'' actually reproduces the known "two-prism" technique for determining the propagation length of infrared SPPs [26, 2].

The increase in the accuracy of determining both parts the complex refractive by the dynamic SPP interferometer is achieved as a result of: 1) the application of the series method in assessing the value of the interferogram period by recording a greater number of periods compared to the static device due to a greater change in the path difference of the interfering beams in the measurement process; 2) the use of a single-pixel (instead of a multi-pixel) photodetector device characterized by a greater aperture and sensitivity than individual pixels of a photodetector array.

4.2. Computer simulations

As an example, let us consider operation of the dynamic interferometer in which the interferogram is formed by SPPs generated by radiation with $\lambda_o = 130$ µm on the flat face of a gold sample disposed in air and containing 0.5 µm thick ZnS coverage. Let the distances a, b and x_o be the same and equal to 2.0 cm; as the SPP-beam splitting device let us choose the kapton plate (see section 2.2) with $R = 0.28$ and $T = 0.45$. Assuming $I_o = 1$, $\Delta x = 1.0 \mu m$, $x_o = 0$, $x_{max} = 1.0$ cm and substituting the values for κ' and κ'' found in Section 3.3 ($\kappa' = 1.00053$; $\kappa'' = 0.000024$) into the expression (3), the interference pattern for the was calculated. A fragment of it is presented in figure 6. It turned out that the distance x_{max} fit a little more than 153 full periods of the pattern; more precisely, $m = 153$ full periods fit the distance of $(x - x_o) = 9.94$ mm. Substituting values of m, $(x - x_o)$ and λ_o into formula (4) we get the value of $\kappa' = 1.00051 \pm 3 \cdot 10^{-5}$. Note that such a high accuracy of κ' definition is determined not so much by the size of mirror (6) displacement "step" as by the number m of recorded maxima.

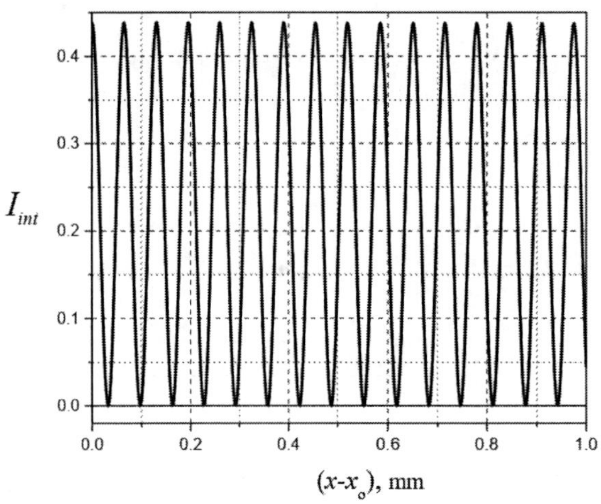

Figure 6. Fragment of the calculated interferogram for the dynamic SPP-interferometer.

For determining the value of κ'' one has to measure the dependence $I(x - x_o)$ obtained for the secondary beam interacting with mirror (6) in the absence of mirror (5). Using the data of measurements and applying formula (6) for many pairs of the coordinate x, one can obtained the average value of κ''. We should note that for a more accurate determination of the attenuation of THz SPPs, it is desirable to choose the maximum displacement of mirror (6) comparable to the propagation length of the SPPs, so that the changes in the intensity of the SPP beam could be measured confidently.

The dynamic SPP-interferometer can be effectively used not only in optical sensing and refractometery of metal surfaces, but also in surface plasmon Fourier spectrometers employing such broadband THz radiation sources as synchrotrons or pulsed lasers.

5. Summary

It has been established that surface plasmon-polaritons (SPPs) of the terahertz (THz) range are reflected by flat mirrors and split by partially transparent flat plates like plane EM waves provided their field penetration depth into air exceeds two wavelengths of the radiation and the mirror or the plate adhere to the surface being oriented along the normal to it. Two types of THz SPP interferometers in which interferograms are formed by the fields of the surface plasmons themselves, rather than by the bulk waves generated by them, are considered: the static and the dynamic ones. The first of these enable one to determine the refractive index of THz SPPs almost instantaneously, which is extremely important for sensing fast processes on conducting surfaces; the second of them are more precise (though require a longer time interval for measurements) and may be the key element for future surface plasmon Fourier spectrometers of the THz range.

Acknowledgements

The experiments were performed using the equipment of the Siberian Synchrotron and Terahertz Radiation Center (the city of Novosibirsk). The reported study was partially funded by the Russian Foundation for Basic Research according to the research project № 19-02-00353\19.

References

[1] Maier S A 2007 *Plasmonics: Fundamentals and Applications* (Springer) 223

[2] Agranovich V M and Mills D L 1982 *Surface polaritons. Surface electromagnetic waves at surfaces and interfaces* (Amsterdam: North Holland Publ. Company) 717

[3] Gomez R J and Berrier A 2013 *Fundumental aspects of surface plasmon polaritons at terahertz frequences* (in "Handbook of terahertz technology for imaging, sensing and communications" Ed. Saeedkia D, Cambridge: Woodhead Publishing Ltd) 688

[4] Nikitin A K et al. 2010 Surface plasmon dispersive spectroscopy of thin films at terahertz frequencies *Proc. SPIE* **7376** 7376 0U

[5] Nazarov M et al. 2008 Surface plasmon terahertz waves on gratings *Comptes Rendus Physique* **9** 232-47

[6] Bogomolov G D, Zhizhin G N, Nikitin A K and Knyazev B A 2009 Geodesic elements to control terahertz surface plasmons *Nuclear Instrum. and Methods in Phys. Research* (*A*) **603** 52-55

[7] Kulipanov G N et al. 2015 Novosibirsk free electron laser - facility description and recent experiments *IEEE Trans. Terahertz Sci. Technol.* **5** 798-809

[8] Gerasimov V V, Knyazev B A and Nikitin A K 2017 Reflection of terahertz monochromatic surface plasmon-polaritons by a flat mirror *Quantum Electronics* **47** 65–70

[9] Gerasimov V V et al. 2016 Growth of terahertz surface plasmon propagation length due to thin-layer dielectric coating *JOSA* (*B*) **33** 2196-2203

[10] Stegeman G I, Wallis R F and Maradudin A A 1983 Excitation of surface polaritons by end-fire coupling *Optics Letters* **8** 386-8

[11] Nikitin A K, Knyazev B A and Gerasimov V V 2018 Terahertz plasmonics: achievements and prospects *EPJ Web of Conferences* **195** 06023

[12] Gerasimov V V, Nikitin A K et al. 2018 Splitting of terahertz surface plasmons by polyimide films *IOP Conf. Series: Journal of Physics: Conf. Series* **1092** 012040

[13] Cunningham P D 2011 Broadband THz characterization of the refractive index and absorption of some important polymeric and organic electro-optic materials *J. Appl. Physics* **109** 043505

[14] Voronko A I, Klimova L G and Shkerdin G N 1987 Reflection of surface polaritons at a dielectric barrier *Solid State Communications* **61** (6) 361-4

[15] Nikitin A K and Tishchenko A A 1991 Phase surface electromagnetic wave microscopy *Techn. Phys. Lett.* **17** 76-79

[16] Nelson S G, Johnston K S, and Yee S S 1996 High sensitivity surface plasmon resonace sensor based on phase detection *Sensors and Actuators* (*B*) **35–36** 187–91

[17] Kabashin A V and Nikitin P I 1997 Interferometer based on a surface-plasmon resonance for sensor applications *Quant. Electronics* **27** 653–4

[18] Temnov V V et al. 2007 Surface plasmon interferometry: measuring group velocity of surface plasmons *Optics Lett.* **32** 1235-7

[19] Gan Q Q, Gao Y and Bartoli F J 2009 Vertical plasmonic Mach–Zehnder interferometer for optical sensing *Optics Express* **17** 20747–55

[20] Ming Y et al. 2012 Surface plasmon interferometer and its sensing applications *IEEE Photonics Journal* **4** 291-9

[21] Melentiev P N et al. 2017 Femtosecond plasmon interferometer *Optics Comm.* **382** 509-13

[22] Bogomolov G D et al. 2009 Determination of the refractive index of IR surface plasmons by static asymmetric interferometry *Bull. Russian Academy of Sciences. Physics* **73** 533-6

[23] Zhizhin G N et al. 2012 Dispersive Fourier-transform spectroscopy of surface plasmons in the infrared frequency range *Optics and Spectroscopy* **112** 545–50

[24] Dem'yanenko M A et al. 2011 Application of uncooled microbolometer detector arrays for recording radiation of the THz spectral range *Optoelectronics, Instrumentation and Data Processing* **47** (5) 109–13

[25] Gerasimov V V et al. 2013 Surface plasmon polaritons launched using a terahertz free electron laser: propagating along a gold-ZnS-air interface and decoupling to free waves at the surface tail end *JOSA (B)* **30** 2182-90

[26] Schoenwald J, Burstein E and Elson J M 1973 Propagation of surface polaritons over macroscopic distances at optical frequencies *Solid State Comm.* **12** (3) 185-9

Long-focus low-cost lens deigned for biomedical hyperspectral camera

V I Batshev[1,2] and A E Gavlina[2]

[1]Laser and opto-electronic Department, Bauman Moscow State Technical University, 2-ya Baumanskaya 5, Moscow, 105005 Russia

[2]Laboratory of Acousto-optic Spectroscopy, Scientific Technological Center of Unique Instrumentation RAS, Butlerova 15, Moscow, 117342 Russia

E-mail: batshev@bmstu.ru, gavlinaae@gmail.com

Abstract. In this work a design of 250 mm F/20 lens for hyperspectral imaging of biomedical micro objects is presented. The designed lens consists of only standard optical elements and does not contain any newly developed component. Its low aperture and small field of view causes that image quality is limited only by diffraction and two optical aberrations – field curvature and chromatic focal shift. That is why an achromatic doublet is utilized as a basic lens component. Its residual field curvature is corrected by a Smith lens – a single plane-concave lens located near the image plane. The components are selected from commercially available ones. A distance between components was the only optimized parameter. Such layout makes the lens cheap and easy to manufacture.

1. Introduction

In recent decades, hyperspectral imaging has been widely used in many different applications, in particular, in biomedicine, for example, in microscopy. There are several ways to obtain a hyperspectral image. For example, in microscopy, tunable filters are utilized for this: liquid-crystal [1] or acousto-optic [2, 3]. An alternative method is to use a special gradient filter [4], which is a thin film attached to the sensor. Thus, an image of an object is formed at the detector, with each part of the object being imaged in a narrow spectral interval passed by the corresponding section of the gradient filter. There are various methods for restoring a hypercube of data; they are discussed in detail, for example, in a presentation [5].

2. Hyperspectral imager layout

Optical scheme of the designed hyperspectral camera is shown on figure 1. Standard microscope objective L1 forms an image at infinity. Lens L2 focuses it to the sensor attached with a gradient filter [4] with 450 nm to 850 nm spectral range. The Axiom Beta camera [6] with 4096×3072 sensor is used in the developed scheme. The sensor size is 22.5 × 16.9 mm, pixel size is 5.5 μm. Microscope objective L1 is removable to obtain different optical magnifications for different objects.

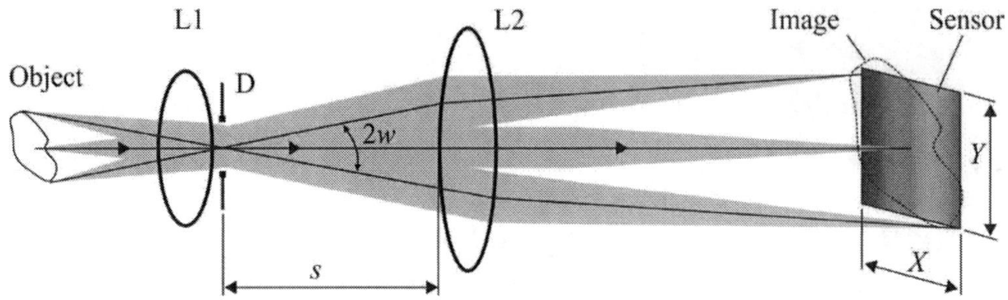

Figure 1. Layout of hyperspectral imaging system: L1 – infinity corrected microscope objective, L2 – designed focusing lens.

The magnification of this system is

$$\beta = f_{L1}/f_{L2} ,$$

there f_{L1} and f_{L2} are the focal lengths of L1 and L2 lenses.

Output parameters necessary to design the L2 focusing lens are slightly the same for most models of infinity corrected microscope objectives: field of view (FoV) $2w \approx 6^{\circ}$ and the exit pupil diameter $D \leq 12$ mm. The aperture stop is located close to the microscope lens L1 that is why the designed L2 lens has an entrance pupil rendered forward for some distance s.

In order for the image to fully fit into the sensor format, the focal length of L2 should be equal to

$$f_{L2} = \sqrt{X^2 + Y^2}/2\tan(w) = 270 \text{ mm}.$$

So, the main optical parameters of the L2 lens are $f_{L2} = 270$ mm, F/# = 20 and FoV = 6°.

In this work it is shown how to design the L2 lens. Off course one can found a standard lens with fixed $f = 270$ mm or zoom-lens and utilize it in this system. But such low aperture gives hope to believe that the lens could be easier, chipper and more compact than any standard lens. So, the mail goal of this work is to show the way to calculate such lens.

3. Lens design

There are two main ways to design a lens. The first is to accurately calculate third order aberrations. The second is the composition of the lens from components with known aberration properties [7].

A low aperture leads to the fact that the aperture aberrations (spherical aberration and coma) do not distort the formed image. Lateral color and distortion are negligible because of small FoV. That is why we can be sure that the optical resolution is restricted by diffraction and chief ray aberrations (field curvature, astigmatism and chromatic focal shift) only.

To make the designed lens not expensive it should be composed of the commercially available components. One of the possible ways to do this and to compensate the residual aberrations is to design the lens consisting of two components. The first one – basic component – is an achromatic cemented doublet. Actually, it is not a problem to found achromatic doublet with the relatively small astigmatism, but the field curvature is significant in all cases even is such small FoV. That is why the second component is a field curvature corrector. It is a so-called Smith lens – a single concave-plane lens located with its flat surface almost close to the sensor. Such location causes that the focal length of the basic component increases only slightly and all values of the aberrations except the image curvature remain the same as those of the base component. So, the optical scheme of the designed lens is shown on figure 2.

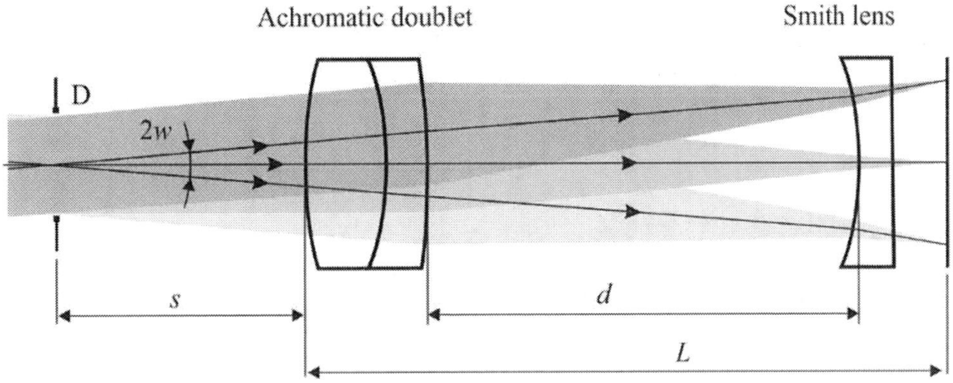

Figure 2. Layout of the L2 lens.

As a first component it is utilized Edmund Optics achromatic cemented doublet coated for visible and near infrared range with 250 mm focal length. The radius of the Smith lens spherical surface was obtained by optimization in ZEMAX software an than selected form available also in Edmund Optics. The glasses utilized in doublet are widely used N-BK7 and N-SF5 from the Schott catalogue, the Smith lens is made of N-BK7.

The distance s should not be too small for the convenience of mounting the lens. On the other hand, this distance should not be large for the light diameter of the components to be acceptable. In our case the dimeters of components are 25 mm that allows us to set $s = 12$ mm.

Finally, optimization was performed, as a result of which the distance d between the components was determined. Table 1 shows us the obtained design parameters. The focal length of the lens obtained is 275 mm, the distance between the first surface and the image sensor (total track) is $L = 255$ mm, so we could consider our lens as a telephoto lens. The entrance pupil position is 15 mm from the first surface. This is the maximum distance where the light diameters of the components do not exceed the physical dimensions of the selected lenses.

Figure 3 shows a tracing model of the L2 lens (a), the obtained spot diagram (b) and the polychromatic average modular optical transfer function (MTF, figure 3 (c)). The results are presented for the different image points: the central point P1 located on the optical axis, the intermediate point P2 and the marginal point P3.

In figure 3 we could see that the image quality is close to the diffraction limit. The image resolution is about 25-30 cycles per mm within the whole FoV. The theoretical (diffraction) limit is also 30 cycles per mm. The largest aberration is chromatic lateral shift. But it is also negligible – it is less than the Airy disc diameter as we can see in figure 3(b).

Table 1. Design parameters of L2 lens.

Surface	Radius, mm	Thickness, mm	Glass	Diameter
Entrance pupil	Infinity	12.0	-	12.0
1	160.73	5.0	N-BK7	13.2
2	-105.64	2.5	N-SF5	13.4
3	-295.75	237.0	-	13.5
4	-51.68	3.5	N-BK7	24.2
5	Infinity	8.9	-	24.5
Image	Infinity	-	-	-

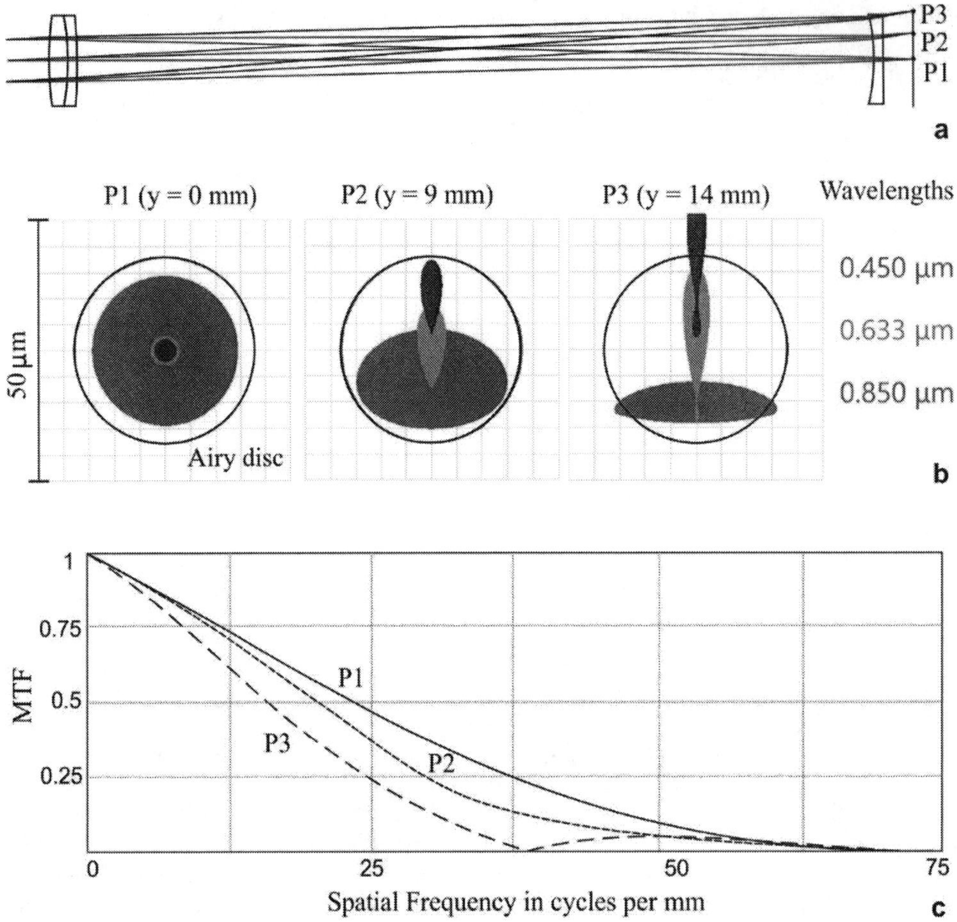

Figure 3. Ray tracing model (a), spot diagrams (b) and modular optical transfer function (c) of the designed L2 lens.

4. Conclusion

In this report it is shown the application of the composition method for lens design. In our case the designed lens consists of the achromatic doublet and the concave-plane singlet as a Smith lens for the image curvature correction. Such an arrangement of the lens made it possible to select the components of their list of standards, commercially available optical elements with known design parameters, which are significantly (about 2 times) cheaper than custom-made elements. In addition, the tolerances of the Smith lens displacements relative to the basic doublet are quite large, which makes the design of the lens mount fairly simple and technological.

References

[1] Beeckman J 2011 *Optical Engineering* **50(8)** 081202.
[2] Machikhin A S, Pozhar V E, Viskovatykh A V and Burmak L I 2015 *Applied Optics* **54(25)** 7508-13.
[3] Polschikova O, Machikhin A, Batshev V, Ramazanova A, Belov A and Pozhar V 2017 Proc. SPIE 105920H
[4] https://www.deltaopticalthinfilm.com/applications/hyperspectral-imaging/
[5] https://vimeo.com/203480245
[6] https://www.apertus.org/axiom-beta
[7] Shannon R R and Wyant J C 1979 *Applied Optics & Optical Engineering*. Academic Press, NY

Applying PIV to study a fluid flow in the vicinity of a circular streamlined cylinder

K G Dobroselsky

Kutateladze Institute of Thermophysics, 1 Ac. Lavrentiev ave., Novosibirsk, Russia

E-mail: dobroselsky@mail.ru

Abstract. The round cylinder streamlining by a turbulent water flow in a hydrodynamic pipe is investigated experimentally. Using the optical method of PIV (Particle Image Velocimetry) the averaged velocity fields near the cylinder are obtained and, on their basis, the geometric characteristics of the vortex zone of the near wake are calculated for the non-cavitation and cavitation flow regimes in the critical region of Reynolds numbers. The use of two mirrors, set at a certain angle to each other, allows obtaining a picture of the velocity fields around the entire cylinder, rather than around half of it, as it is done in most of the works. Using vector patterns of averaged velocity fields, the angles of the boundary layer separation from the cylinder surface along the reverse flow are determined in the considered flow regimes. An asymmetrical separation of the boundary layer from different sides of the streamlined cylinder was obtained. It is shown that the increase in the Reynolds number for the non-cavitation flow regimes leads to a more than two times decrease in the vortex zone behind the cylinder, and, accordingly, to a displacement of the separation angles downstream. Cavitation increases the vortex zone behind the cylinder and displaces the separation angles upstream.

1. Introduction

Cross-streamlined cylindrical surfaces are wide spread in heat exchange, energy and thermal engineering devices. Despite the simple geometry, the flow around the circular cylinder is extremely complex and depends on the flow regime.

Analysis of the literature shows that most of the works on the round cylinder streamlining are associated with the flow of gas and, less often with liquid. There are practically no experimental works on the cavitation flow around the cylinder [1, 2]. Cavitation, in general, is undesirable in many designs, so one needs to know the boundaries of the parameters and the conditions under which it occurs and affects the necessary characteristics.

Conditions of streamlining and the size of the vortex zone (structures) behind the bodies are important for designing the arrangement of both single and groups of streamlined elements in hydraulic and thermal structures. It should be also noted that for building and verifying modern mathematical models, describing the emergence and development of cavitation, it is extremely important to obtain experimental information for cavitation flow regimes of model objects.

This paper, using the PIV method, experimentally studies the velocity field near a round steel cylinder, streamlined by a turbulent water flow, in the critical region with Reynolds numbers over the diameter $Re \approx (2.07\text{-}2.84)\times10^5$ for the non-cavitation and cavitation regimes. Methods for determining the average spatial characteristics of the near vortex wake behind the cylinder are shown.

2. Experimental methodology

The hydrodynamic tube of closed type of Novosibirsk National Research State University was used for experimental studies. Measurements of velocity fields were realized using the PIV system POLIS, developed at the Institute of Thermophysics SB RAS and consisting of: a programmable sync processor; dual solid-state pulse Nd:YAG laser with a wavelength of the laser radiation of 532 nm (pulse energy of 25 mJ, pulse duration of 10 ns, and pulse repetition rate up to 1.3 Hz) with focusing and cylindrical lenses to create a laser sheet; a CCD camera with a resolution of 2048×2048 pixels [3].

Figure 1 shows the main elements of the POLIS and the arrangement of the cylinder in the working area. Perpendicular to the side walls of a rectangular channel with a cross section of 0.08×0.15 m (width × height) there was a round steel cylinder with a diameter d = 0.026 m. The walls of the working area were equipped with viewing windows for optical measurements. The cylinder was streamlined by the water flow with a constant volumetric flow rate. The ratio of the flow rate to the cross-sectional area of the working section served to determine the superficial velocity u of the incoming flow. The flow rate was measured using an ultrasonic flowmeter with a relative volume error of no more than 2%. Pressure and temperature before and after the cylinder were measured by two sensors of pressure and resistance thermometers, respectively, with an error of no more than 0.5% [3]. The working fluid was filtered tap water. All experiments used polyamide tracers with an average size of 50 μm and a density close to that of water.

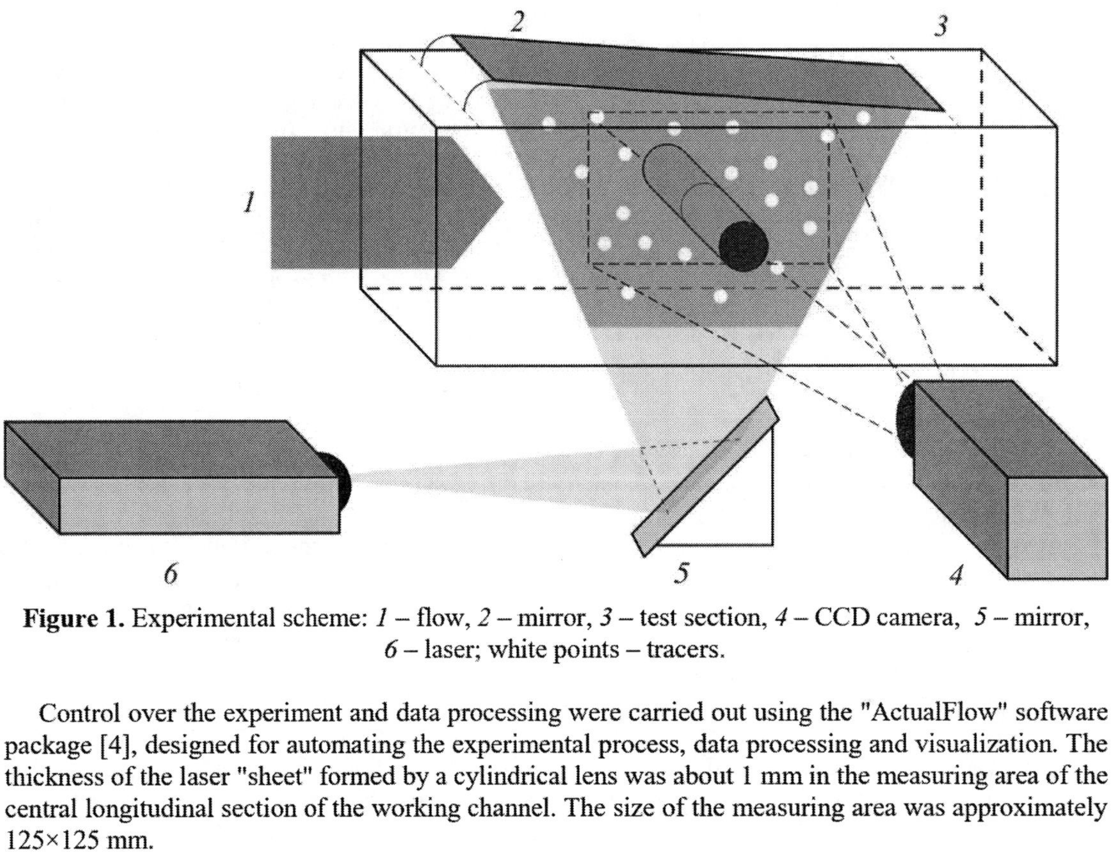

Figure 1. Experimental scheme: *1* – flow, *2* – mirror, *3* – test section, *4* – CCD camera, *5* – mirror, *6* – laser; white points – tracers.

Control over the experiment and data processing were carried out using the "ActualFlow" software package [4], designed for automating the experimental process, data processing and visualization. The thickness of the laser "sheet" formed by a cylindrical lens was about 1 mm in the measuring area of the central longitudinal section of the working channel. The size of the measuring area was approximately 125×125 mm.

Instantaneous velocity fields were calculated using an iterative cross-correlation algorithm with continuous displacement and deformation of elementary computational cells and 75% overlap of computational domains. The sub-pixel interpolation of the cross-correlation peak was realized on three points, using a one-dimensional approximation by the Gauss function. The initial size of the elementary computational domain was 64×64 pix, and the final size was 8×8 pix. The obtained vector fields of instantaneous velocity underwent two validations: by signal/noise ratio with a factor of 2 and by an additive median filter in a region of 7×7 pix. The velocity measurement error depends on the tracer displacement and the number of frames; in our case it was on average 2% and 5% at the displacement by 8 and 2 pixels, respectively [5].

In advance, the velocity field was measured by PIV method in the vertical central section (1000 double images) for various flow rates in a free working area. The velocity profile was found to have a core of constant velocity (the boundary layer on the channel walls was 10-12 mm) with a level of turbulence of the longitudinal velocity component of the order of 0.8-1%.

3. Results and Discussion

Figures 2 and 3 show the fields of averaged velocities v_x and v_y for the non-cavitation regime of the cylinder streamlining in the central part of the channel at a superficial velocity u = 8.8 m/s, the velocity in the free flow core $v_{x\max}$ = 9.8 m/s, the water temperature $t \approx 25$ °C and the corresponding Reynolds number Re $\approx 2.84\times10^5$. Averaging was carried out over 2000 double images. Here x, y are the longitudinal and transverse coordinates related to diameter d, respectively. Figure 3 shows the area with half of a cylinder: x = 0-1.07; y = -0.54-+0.54.

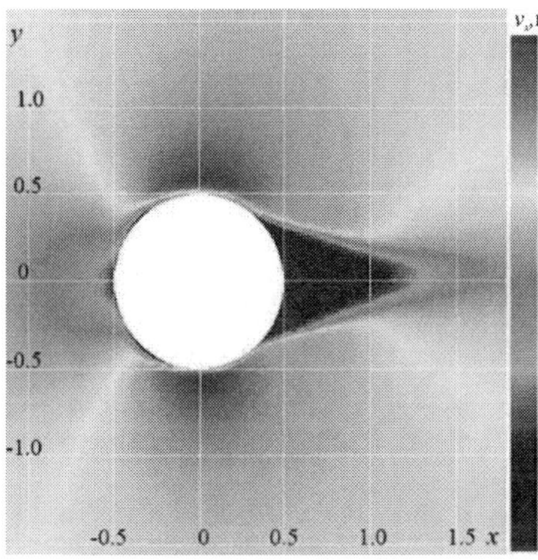

Figure 2. Average velocity field v_x for non-cavitation regime of cylinder streamlining.

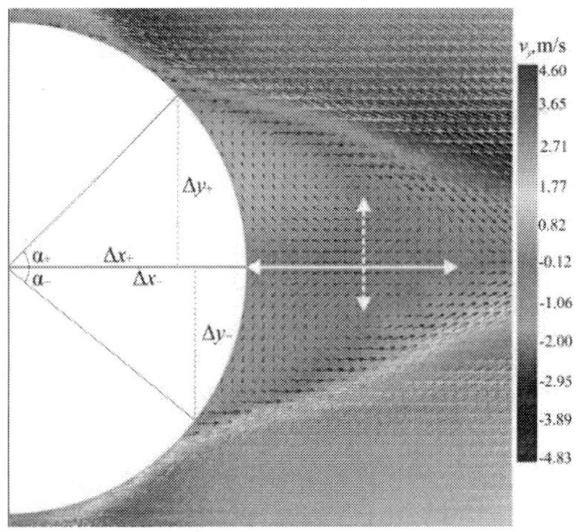

Figure 3. Determination of the main parameters of the near wake behind the cylinder (velocity field v_y).

Just behind the cylinder, a stable turbulent region forms. There opposite rotating vortices appear: one rotates clockwise (from the top), and the second one rotates counterclockwise (from the bottom). Between the vortices, the liquid moves in the direction opposite to the main flow.

In figure 3, white horizontal and vertical dashed lines show the length of the reverse flow l_r and the distance between the vortices Δy_v respectively. The outer boundary of the region of reverse flows is limited by a line, on which the longitudinal velocity component is zero.

On vector pictures of the averaged velocity field, it is possible to determine the final separation angle φ (relative to the frontal point) of the boundary layer along the boundary of the return motion on the cylinder surface. Figure 3 shows the location of the separation angles α_+ and α_- with respect to the upper and lower surfaces of the cylinder and from the rear point, where $\alpha_+ = \tan^{-1}(\Delta y_+/\Delta x_+)$ and $\alpha_- = \tan^{-1}(\Delta y_-/\Delta x_-)$. The final separation angles are determined as $\varphi_+ = 180° - \alpha_+$ and $\varphi_- = 180° - \alpha_-$, respectively. Figure 4 (a-c) shows the average (2000 double images) fields of relative velocities v_x/u at a non-cavitation flow around the cylinder in the central part of the channel for different Reynolds numbers.

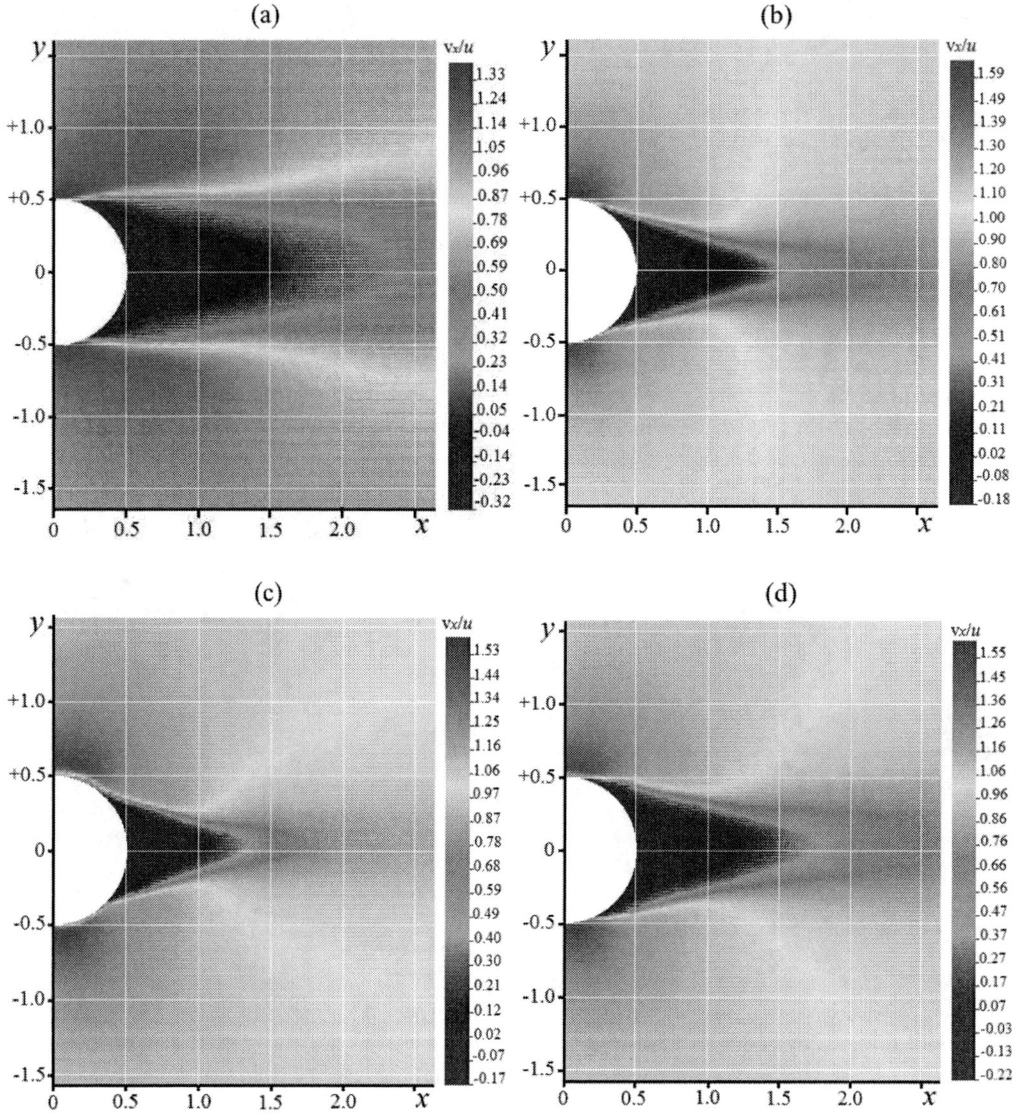

Figure 4. Average velocity field v_x/u: non-cavitation regimes (a) – Re = 2.07×10^5, (b) – Re = 2.54×10^5, (c) – Re = 2.84×10^5; cavitation regime (d) – Re = 2.84×10^5.

The streamlining at the cavitation regime (figure 4 (d)) occurred at the relative cavitation number $\sigma/\sigma_i \approx 0.74$, where σ_i corresponds to the beginning of cavitation. Figure 4 (c) and (d) show comparative pictures of velocity fields at the same Reynolds number, the same temperature, but different pressures before the cylinder, p = 161 kPa and 112 kPa, respectively.

Below, figure 5 presents comparative graphs of $-\tau/q$ from y for different Re, where $\tau = -\rho\overline{u_x u_y}$ is the turbulent stress (u_x, u_y – corresponding pulsations of velocities v_x and v_y; ρ – water density); and $q = \rho u^2/2$ is the average dynamic head of the incoming flow. According to the graphs, it is possible to estimate the transverse size of the vortex zone of the near wake (Δy_w) behind the cylinder, for example, where the solid line comes to zero above and below.

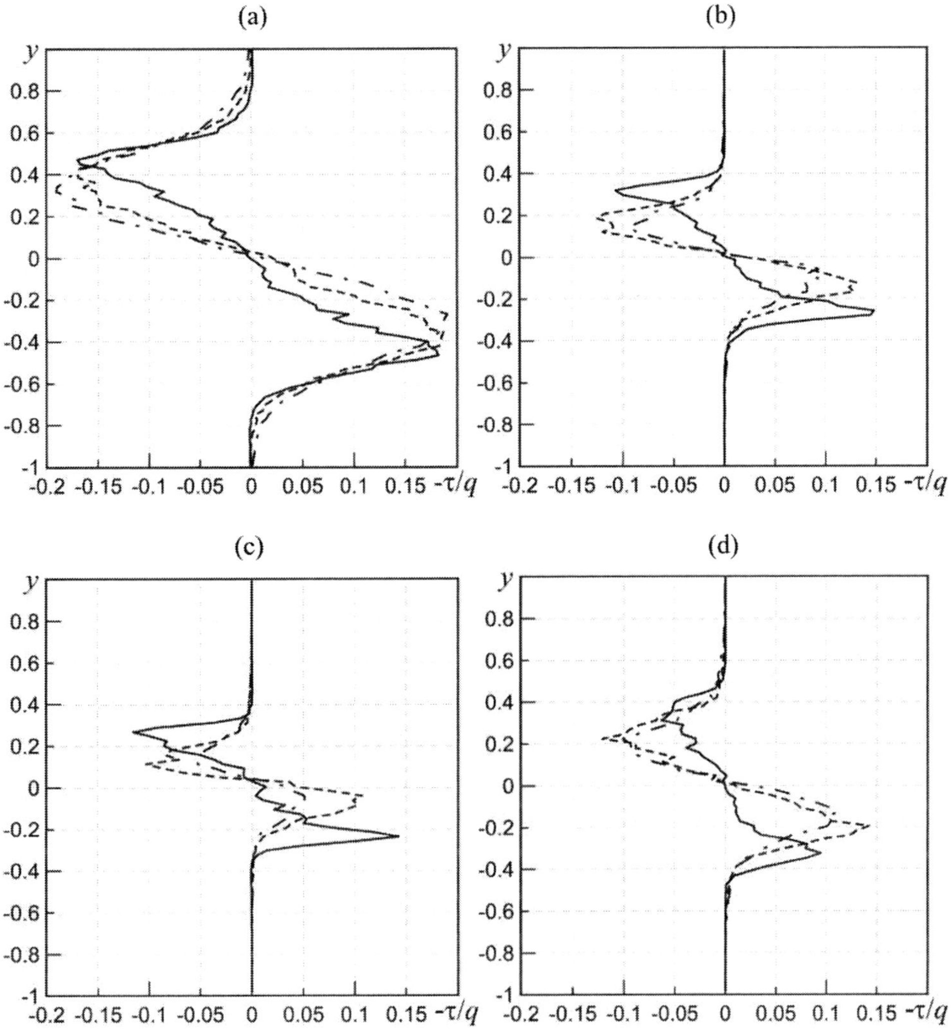

Figure 5. Profiles of relative turbulent stresses: (a) – Re = 2.07×10^5, (b) – Re = 2.54×10^5, (c) – Re = 2.84×10^5, (d) – Re = 2.84×10^5 (cavitation regime). The solid line corresponds to the coordinate x, at which the velocity of the reverse flow behind the cylinder is maximal, the dotted line – x = 1.2, and the dashed line – x = 1.5.

The table 1 below shows the main spatial parameters of the vortex zone behind the cylinder for non-cavitation regimes.

Table 1. Parameters of average vortex structures behind the streamlined cylinder.

Re (10^5)	2.07	2.37	2.54	2.84
Angles of boundary layer separation φ_+, φ_-	105°, 113°	115°, 125°	131°, 140°	134°, 141°
Length of reverse flow l_r (d)	1.00	0.72	0.60	0.45
Width of vortex zone Δy_w (d)	1.6	1.0	0.8	0.7

For cavitation streamlining of the cylinder, the flow parameters were as follows: the separation angles were $\varphi_+ = 129°$, $\varphi_- = 138°$; $l_r \approx 0.65d$; $\Delta y_w \approx 1.0d$. In [6, 7] for Re = $(1–2)\times10^5$ there are given $l_r \approx (0.7–1.05)d$.

The asymmetry of the separation of the boundary layer from the surface of the cylinder is associated with the formation of a laminar separation bubble, which at critical Reynolds numbers (Re $\approx (2\text{-}4)\times10^5$) forms on one side. In this regime, transition to turbulence first occurs in one of the boundary layers and it characterized by the separation with further reattachment of the boundary layer, forming a bubble [8]. The final separation of a turbulent boundary layer occurs at $\varphi \approx 140\text{-}147°$ [8, 9]. The asymmetry and displacement of the near wake upwards can be related to the pressure difference in the flow at the level of the upper and lower parts of the cylinder.

4. Conclusions

The averaged velocity fields obtained by the PIV method allow a reliable investigation of the flow in the vicinity of the streamlined cylinder. It is shown that the kinematic characteristics of the wake behind the cylinder depend on the value of the Reynolds number, as well as on the flow regime.

It has been found that for the non-cavitation flow regimes, with an increase in the Reynolds number, the vortex zone of the near wake behind the cylinder decreases more than two times, and the separation angles of the boundary layer increase by 25%. The vortex zone behind the cylinder is found to increase both in the transverse and longitudinal direction in the cavitation flow regime, relative to the non-cavitation flow.

Acknowledgments

The work was financially supported by the Russian Science Foundation (grant No.19-79-30075).

References

[1] Ihara A and Murai H 1986 Cavitation inception on a circular cylinder at critical and supercritical flow range *Journal of Fluids Engineering* **108** 421

[2] Kumar P, Chatterjee D and Bakshi S 2017 Experimental investigation of cavitating structures in the near wake of a cylinder *International Journal of Multiphase Flow* **89** 207

[3] Dobrosel'skii K G 2016 Flow around a cylinder at the beginning of the critical region *Journal of Applied Mechanics and Technical Physics* **57** No **2** 293

[4] Akhmetbekov E K, Bil'skii A V, Lozhkin Yu A, Markovich D M and Tokarev M P 2006 System of control of experiment and processing of data obtained by the methods of digital tracer visualization (ActualFlow) *Vych. Metod. Programmirov.* **7** 79

[5] Dobroselsky K G 2018 Studying the flow around a cylinder the critical region by the PIV method *J. Phys.: Conf. Series* **1105** 012014

[6] Djeridi H, Braza M, Perrin R, Harran G, Cid E and Cazin S 2003 Near-wake turbulence properties around a circular cylinder at high Reynolds number *Flow, Turbulence and Combustion* **71** 19

[7] Perrin R, Cid E, Cazin S, Sevrain A, Braza M, Moradei F and Harran G 2007 Phase-averaged measurements of the turbulence properties in the near wake of a circular cylinder at high Reynolds number by 2C-PIV and 3C-PIV *Exp. Fluids* **42** 93

[8] Lehmkuhl O, Rodríguez I, Borrell R, Chiva J and Oliva A 2014 Unsteady forces on a critical Reynolds number *Phys. Fluids* **26** 125110

[9] Marr Yu N and Shvegzhda S A 1989 Features of fluid motion near the surface of the cylinder in transverse flow *Izv. Akad.Nauk SSSR, Mekh. Zhidk. Gaza* **4** 65

Journal of Physics: Conference Series **1421** (2019) 012016 doi:10.1088/1742-6596/1421/1/012016

Development of a cylindrical acoustic lens to control the amplitude and phase characteristics of optical radiation

A N Bykhanov[1]

[1]National Research University "MPEI", Krasnokazarmennaya st., 17, Moscow, Russia

Abstract. The promising method of light beam phase control based on beam transmission through the medium layer perturbed with acoustid wave generated by the lateral surface "lens", requives the exact calculation of the acoustic field inside cylindrical cavity of the "lens". We use the comm program Comsol for calculation and demonstrate that for the axial symmetric cylinder, which was used as a test object, the calculated distribution very close coinsides with theoretical form resulted from acoustic equations solutions.

1. Introduction

The trend of modern optoelectronics and photonics is the creation of compact devices for electronic control of the characteristics of optical radiation. Acousto-optic (AO) devices are one of the most common among them. Due to the diffraction of light beams on the gratings created by the acoustic field, it is possible to control their direction of propagation, intensity, polarization and spectral composition [1, 2]. Uniaxial crystals are mainly used as media for the realization of AO interactions. The manufacture of AO cells based on such crystals requires the use of high-tech and time-consuming procedures for growing, orientation, cutting, etc. A limited choice of usable crystals provokes finding for new operating media. One of the directions of further development of AO technologies is the use of liquids as a working medium of AO interaction [3]. Despite the isotropic nature of the medium, AO diffraction in a liquid has significant advantages. One of them is possibility of manufacturing almost any shape of an AO cell, which permit creation of acoustic fields of a complex structure, for example, with the Bessel distribution [4]. The manufacturability and low cost of AO cells production based on liquids give chances to create compact and efficient devices for controlling optical radiation. For example, such AO elements are already used for precision focusing of radiation in various areas of modern microscopy [5-7]. As a physical model rather simple approximation is used so amplitude-phase characteristics of optical radiation has not been completely theoretically and experimentally investigated. Therefore the detailed description of the acoustic field inside AO cell is necessary for the further development of the technology.

This paper is aimed at analytical and computational calculation of the acoustic field inside a pulsating fluid-filled cylinder.

2. Analytical solution

The wave equation in cylindrical coordinates is expressed as:

$$\frac{\partial^2\Phi}{\partial z^2}+\frac{\partial^2\Phi}{\partial r^2}+\frac{1}{r}\frac{\partial\Phi}{\partial r}+\frac{1}{r^2}\frac{\partial^2\Phi}{\partial\varphi^2}=\frac{1}{V^2}\frac{\partial^2\Phi}{\partial t^2}, \tag{1}$$

Journal of Physics: Conference Series **1421** (2019) 012016 doi:10.1088/1742-6596/1421/1/012016

where acoustic field potential $\Phi(r,\varphi,z,t)$ depends on the following variables: r – radial coordinate; φ – angular coordinate (azimuth); z – elevation coordinate (along the axis) and t – time variable; V – sound velocity.

Using the method of separation of variables, we represent the function of the acoustic field in the form: $\Phi(r,z,\varphi,t) = Z(z)\cdot R(r)\cdot G(\varphi)\cdot T(t)$ and rewrite the wave equation with the separation of variables:

$$\frac{1}{R}\frac{\partial^2 R}{\partial r^2}+\frac{1}{r}\frac{1}{R}\frac{\partial R}{\partial r}+\frac{1}{r^2}\frac{1}{G}\frac{\partial^2 G}{\partial \varphi^2}-\frac{1}{V^2}\frac{1}{T}\frac{\partial^2 T}{\partial t^2}=-\frac{1}{Z}\frac{\partial^2 Z}{\partial z^2}=K_z^2 . \quad (2)$$

The solution for the last equation:

$$Z(z) = A'e^{jK_z z} + B'e^{-jK_z z} = (A'+B')\cos K_z z + j(A'-B')\sin K_z z \; .$$

In similar way, we find following dependencies:

$$T(t) = A'\mathrm{e}^{j\omega t} + [B'\mathrm{e}^{-j\omega t}],$$

$$G(\varphi) = A_1\mathrm{e}^{jm\varphi} + B_1\mathrm{e}^{-jm\varphi} = A_2\cos m\varphi + B_2\sin m\varphi = A\cos(m\varphi-\varphi_m)\,, \quad (3)$$

$$R(r,t) = A_m J_m((K-K_z)r) + B_m N_m((K-K_z)r)\,, \quad (4)$$

where ω - the acoustic wave frequency, $\frac{\omega^2}{V^2}=K^2$, $A=\sqrt{A_2^2+B_2^2}$, $\tan\varphi_m=\frac{B_2}{A_2}$, $J_m((K-K_z)r)$ and $N_m((K-K_z)r)$ - Bessel and Neumann functions of the order m. The constant B_m should be set equal to zero, since $N_m(0)=-\infty$, cannot correspond to the actual physical process. But in the case of cylindrical annular channels or sectors in the center of which there is a cylindrical obstacle inaccessible to waves, the second term in (4) should be taken into account. In that case, a particular solution of order m, equation (1) can be written as follows:

$$\Phi_m(r,z,\varphi,t)=[A_m J_m((K-K_z)r)+B_m N_m((K-K_z)r)][A'\mathrm{e}^{jK_z z}+B'\mathrm{e}^{-jK_z z}]A\cos(m\varphi-\varphi_m)\mathrm{e}^{j\omega t}$$

It should be noted that the above approach based on variables separation is definitely applicable if the sound transit time across the cylinder diameter is much less than the oscillation period $2r_0/V \ll 2\pi/\omega$. Let us also assume that the wave process does not depend on z, then $K_z = 0$ and the second bracket in the last equation turns into a constant value. Since the sound pressure and velocity potential are related by the relation $p = j\omega\rho\Phi$ (ρ is the density of the medium inside the cylinder), the general solution of the wave equation for radiation processes (in case of independence from *z*) is written in the form:

$$p(r,z,\varphi,t)=\sum_{m=0}^{\infty}A_m(J_m(Kr)-jN_m(Kr))\cos(m\varphi-\varphi_m)e^{j\omega t}. \tag{5}$$

The radial velocity is obtained from the expression:

$$q_r=-K\frac{\partial\Phi}{\partial Kr}=-K\frac{\frac{\partial p}{\partial Kr}}{j\omega\rho}=j\frac{1}{\rho c}\sum_{m=0}^{\infty}A_m(J'_m(Kr)-jN'_m(Kr))\cos(m\varphi-\varphi_m)e^{j\omega t}$$

Transforming this expression we get:

$$q_r=\frac{A_0C_0(Kr)}{2\rho c}e^{-j\gamma_0(Kr)}+\sum_{m=1}^{\infty}\frac{A_mC_m(Kr)}{\rho c}e^{-j\gamma_m(Kr)}\cos(m\varphi-\varphi_m)e^{j\omega t}. \tag{6}$$

Where $J_0'(Kr)-jN_0'(Kr)=-j\frac{C_0(Kr)}{2}e^{-j\gamma_0(Kr)}$ and $J_m'(kr)-jN_m'(kr)=-jC_m(kr)e^{-j\gamma_m(kr)}$.

The values of C_0, C_m, γ_0 and γ_m are determined by the ratios:

$$\frac{1}{2}C_0\sin\gamma_0=J_1;\qquad \frac{1}{2}C_0\cos\gamma_0=-N_1,$$

Let us determine the sound field created by an infinitely narrow strip of angular width $d\alpha$, located in azimuth $\varphi=\alpha$ and oscillating with the amplitude of velocity $U_0(\alpha)$. The velocity distribution over the surface is written as:

$$u_0(\varphi-\alpha,t)=\begin{cases}U_0(\alpha)e^{j\omega t} \text{ при } -\frac{d\alpha}{2}<\varphi-\alpha<\frac{d\alpha}{2}\\ 0 \quad \text{при } \frac{d\alpha}{2}<\varphi-\alpha<2\pi-\frac{d\alpha}{2}\end{cases}.$$

We decompose the function $u_0(\varphi-\alpha,t)$ in a Fourier series:

$$u_0(\varphi-\alpha,t)=\frac{U_0(\alpha)d\alpha}{\pi}\left[\frac{1}{2}+\sum_{m=1}^{\infty}\cos m(\varphi-\alpha)\right]e^{j\omega t}.$$

The boundary condition $u_0(\varphi-\alpha,t)=q_r(r_0,\varphi,t)$ must be observed on the surface of the cylinder, the speed of sound near the surface of the cylinder coincides with the speed of wall oscillation. In relations (5) and (6) for p and q_r, we introduce the azimuth angle reading from the initial azimuth α; then they will contain φ instead of $\varphi-\alpha$. Equating members with equal indices m in the equality $u_0(\varphi-\alpha,t)=q_r(r_0,\varphi,t)$, we obtain the coefficients of the series:

$$A_0=\frac{U_0(\alpha)d\alpha\cdot\rho c}{j2\pi[J_0'(Kr_0)-jN_0'(Kr_0)]}=\frac{U_0(\alpha)d\alpha\cdot\rho c}{\pi C_0(Kr_0)e^{-j\gamma_0(Kr_0)}},$$
$$A_m=\frac{U_0(\alpha)d\alpha\cdot\rho c}{j\pi[J_m'(Kr_0)-jN_m'(Kr_0)]}=\frac{U_0(\alpha)d\alpha\cdot\rho c}{\pi C_m(Kr_0)e^{-j\gamma_m(Kr_0)}}.$$

The expression A_m through the functions C_m and γ_m is suitable for any values of m, including the value $m = 0$. For the sound field of an oscillating strip located at the azimuth $\varphi=\alpha$, it was found:

$$p=\rho c\frac{U_0(\alpha)d\alpha\cdot e^{j\omega t}}{\pi}\sum_{m=0}^{\infty}\frac{\cos m(\varphi-\alpha)}{C_m(Kr_0)e^{-j\gamma_m(Kr_0)}}[J_m(Kr)-jN_m(Kr)], \quad (7)$$

$$q_r=j\frac{U_0(\alpha)d\alpha\cdot e^{j\omega t}}{\pi}\sum_{m=0}^{\infty}\frac{\cos m(\varphi-\alpha)}{C_m(Kr_0)e^{-j\gamma_m(Kr_0)}}[J'_m(Kr)-jN'_m(Kr)].$$

Due to axial symmetry the sound pressure of the pulsating cylinder will be expressed by the formula (7), in which only the first term of the sum will remain:

$$p(r,t)=\Delta p[J_0(Kr)-jN_0(Kr)]e^{j\omega t}.$$

Since $N_0(0)=-\infty$, this term should be taken into account in the case of cylindrical channels, in the center of which there is a cylindrical obstacle inaccessible to the waves. Therefore $N_0(Kr)=0$.

The general solution of equation (1) will take the following form:

$$p(r,t)=\Delta p J_0(Kr)e^{j\omega t}.$$

The result is a complex wave. Since only the real part has physical meaning, we will rewrite the solution in the form:

$$p(r,t)=\Delta p J_0(Kr)\cos(\omega t), \quad (8)$$

$\Delta p = \dfrac{2\rho V U_0}{C_0(Kr_0)} e^{j\gamma_0(Kr_0)}$ is the pressure amplitude, $C_0(Kr) = \sqrt{\dfrac{2}{\pi Kr}}$, $\gamma_0(Kr) = Kr - \dfrac{\pi}{4}$, U_0 is the amplitude of the sound wave.

The harmonic of the medium pressure according to (8) leads to the formation of the distribution of the refractive index in it

$$n(r,t) = n_0 + \Delta n J_0(Kr)\cos(\omega t), \tag{9}$$

where n_0 is the refractive index of the undisturbed medium. As shown in [4], when a constant frequency of sound f is applied, such a component behaves with respect to optical radiation with a wave vector k as a lens with optical power $\left(K^2 v/2k\right)\cos(\omega t)$, where v is the Raman-Nath parameter [8].

3. Numerical calculation

The expression (9) inspire us to calculate and simulate more complex acoustic fields. This technique for controlling the characteristics of optical radiation can be significantly expanded by applying a modulated acoustic signal, specifying not only a cylindrical, but also a more complex surface of revolution and analyzing other factors.

Figure 1 shows the dependences of the acoustic pressure inside a cylindrical radiating surface at a frequency of 300 kHz, obtained by means of modeling (a) and direct numerical calculation (b).

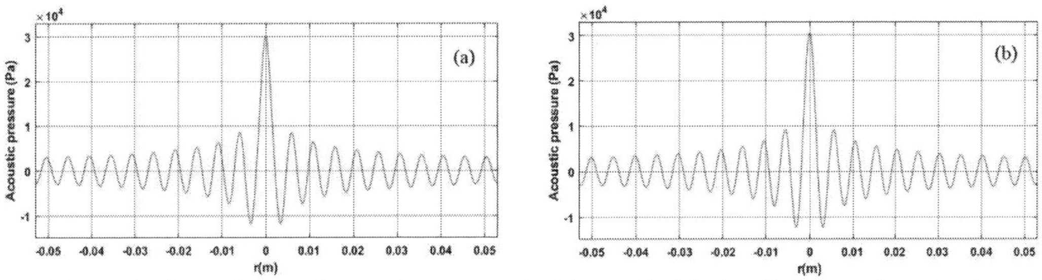

Figure 1. Dependence of acoustic pressure inside a cylindrical radiating surface at a frequency of 300 kHz, obtained by modeling (a) and calculating (b).

These dependences were obtained for a cylinder with r_0 = 0.053 m and the oscillation speed of the inner wall of the cylinder U_0 = 0.001 m/s. It can be seen that a standing wave is formed inside the cylindrical surface, which is formed by the interference of waves emitted by opposite walls. This fact allows us to consider a cylindrical surface filled with liquid as an acousto-optic cell. Comparing dependencies on figure 1 (a) and (b), we can conclude that the field modeled in the Comsol Multiphysics software fully corresponds to the field calculated using (9) under the same conditions. This gives us an opportunity for modeling complex fields with use of this program.

4. Conclusion

The main result obtained is that models derived with use of the built-in Comsol Multiphysics modules correlates very well with analytical calculations. This fact open the way for systematic study of the acoustic fields inside the radiating cylindrical surface, which, in turn, provides more possibilioties for light beams control in an isotropic medium with different cell geometries.

References

[1] Goutzulis A and Rape D 1994 *NY: Dekker*
[2] Molchanov V, Kitaev Yu, Kolesnikov A, Never V N, Rosenstein A Z, Solodovnikov N P and Shapovalenko K G 2015 *Theory and practice of modern acoustooptics* Moscow: MISIS
[3] Gamalath K and Jayawardena G 2012 *Int. Let. Chem., Phys. and Astr.* **4** 39
[4] Grulkowski I and Kwiek P 2006 *Opt. Commun.* **267** 14
[5] Akemann W 2015 *Optics Express* **23** 28191
[6] Friedman N, Kaplan A and Davidson N 2000 *Optics Letters* **25** 1762
[7] Duocastella M, Vicidomini G and Diaspro A 2014 *Optics Express* **22** 19293
[8] Grulkowski I, Szulzycki K and Wojtkowski M 2014 *Optics Express* **22** 31746

Journal of Physics: Conference Series **1421** (2019) 012017 doi:10.1088/1742-6596/1421/1/012017

Prospects of applying the elastic light scattering effect for the development of an optical method to the colloidal solutions diagnositic

M V Sapronov, N M Skornyakova

National Research University "Moscow Power Engineering Institute", Krasnokazarmennaya str. 14, Moscow, 111250, Russia

E-mail: maks-sapronov@yandex.ru, nmskorn@mail.ru

Abstract. In the paper the possibility of developing an optical method for colloidal solutions diagnostic based on elastic light scattering was investigated. The scheme of the experimental setup for light scattering indicatrix registration was presented, and the results of its work were demonstrated. The algorithm and the program of experimental data processing have been developed. Experimental light scattering indicatrices with 434, 546 and 656 nm wavelengths in a colloidal sulfur solution have been obtained. The scattering indicatrix models dynamics with changes in the scattering center size were presented. Qualitative comparison of the experimental and model results was carried out.

1. Introduction

The colloidal solution consists of dispersion medium and dispersed phase, which is suspended particles with the size in range from 1 to 1000 nm. The scope of application and use of colloidal solutions are numerous and diverse due to the unique nanoscale structures properties of the dispersed phase.

Traditionally, various colloidal solutions are widely used in medicine, pharmacology [1] and cosmetology. At the present stage of scientific and technical development substance nanoscale structures have been used in various narrow special areas. For example, on the basis of colloidal quantum dots receivers of thermal infrared radiation were developed [2], organic semiconductors for photovoltaic cells were created [3], colloidal nanoscale crystals of aluminum hydroxide AlOH are used for spectral radiation conversion [4], sulfur and carbon nanostructures are used for the production of lithium battery electrodes [5, 6, 7].

The challenges variety in which colloidal solutions are used requires qualitative and quantitative diagnostics methods of the colloidal solutions properties and parameters, such as the size, shape and concentration of the dispersed phase particles, the preparation time and the solution stability time . Among the existing diagnostic methods, the following are the most common. Nephelometry method [8, 9, 10], which allows to measure the dispersed phase concentration in a colloidal system based on the scattered light intensity measurement in it, optical [11] and electron [12] microscopy methods used to determine the nanometer-sized particles size and shape, a method based on dynamic light scattering, which is used to search for the size of nanoparticles [13, 14]. It should be noticed that these methods don't allow simultaneous concentration and size of the nanostructures measurement, in the present time, this is the most frequently used methods combination based on scattering G. the Mi and analytical ultracentrifugation [15, 16].

This paper discusses the prospects of using the elastic light effect to develop a method that will determine the particle size of the colloidal solution dispersed phase. The relative angular distribution of the scattered light intensity the scattering indicatrix is supposed to be used as the measured informative parameter.

2. Methodology

2.1. The experimental setup

For the light scattering indicatrix experimental determination by the colloidal solution dispersed phase, a model of the setup was developed; the scheme is shown in figure 1.

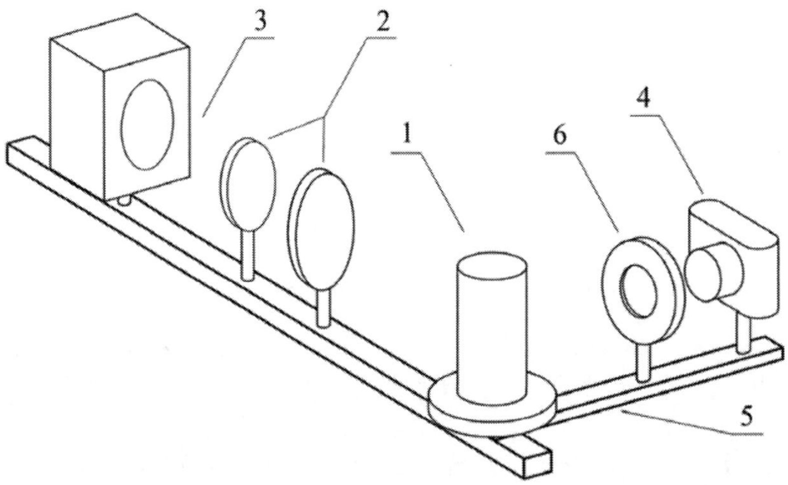

Figure 1. The experimental setup scheme: 1 – cylindrical vessel with the investigated medium, 2 – the lens system, 3 – light source, 4 – matrix photodetector, 5 – movable rod, 6 – light filter.

The incoherent light source (3) is a light-emitting diode lamp, which is placed in a light-tight housing with a round shape single outlet and adjustable size. The source spectral power density (3) has three maxima located in the blue, green and red regions of the visible spectrum. A narrow probing light beam is formed by the lens system (2) and directed to a cylindrical vessel (1) with colloidal solution of sulfur in water.

Recording of the scattered radiation is carried out on a camera (4) Nikon 1J5 [20]. The camera is mounted on a rod (5), which can rotate around the cylindrical vessel axis, with the optical axis of the lens crosses the axis of the vessel at any position of the rod. The lens focal length is chosen so that the image plane contains the axis of the cylindrical vessel. The experimental setup also provides a holder for band-pass filters (6), due to which the spectral selectivity of the detected light is provided. Three light filters with the maximum transmittance $\lambda_1 = 656$ nm, $\lambda_2 = 546$ nm, $\lambda_3 = 434$ nm wavelengths are provided for.

2.2. The experiment methodology

Previously, it was necessary to find the ratio between the values of the probing radiation spectral power density at wavelengths corresponding to the used light filters ($\lambda_1 = 656$ nm, $\lambda_2 = 546$ nm, $\lambda_3 = 434$ nm). At first the light source images in the scattering medium absence through each of the filters were recorded, and then for each image the average pixel brightness was calculated. The obtained brightness values were used as normalization coefficients, taking into account the difference in the probing beam spectral power density, the photo detector sensitivity and the transmittance coefficients of light filters at different wavelengths.

At the experiment main stage, a colloidal solution was produced. For this, in 200 ml of distilled water H_2O was added 0,2 g sodium thiosulfate $Na_2S_2O_3$, there was a usual salt solution. Then 0,5 ml of 12% hydrochloric acid solution HCl was added to the obtained solution. After some time, sulfur began to be released in the suspended molecular clusters form – the dispersed phase, and the solution began to become cloudy.

The scattering indicatrix recording technique was to obtain a series of the digital probing beam images in scattered light at different camera positions. The light filter was placed in the holder, the camera was installed in such a way that the axis of the lens coincided with the axis of the probing radiation formation system. This position of the camera was taken as the reference angle beginningof the rod rotation. Then a series of images was recorded, in which each next frame differed from the previous one by that the rod with the camera rotates around the axis of the cylindrical vessel by five degrees clockwise from the start of the reference. A similar series of images was obtained with each of the three filters. All measurements were carried out in the external lighting absence. The dynamics of experimental probing beam images in scattered light with wavelengths $\lambda_1 = 656$ nm, $\lambda_2 = 546$ nm, $\lambda_3 = 434$ nm with increasing rod rotation angle is shown in figures 2-4, respectively.

2.3. The experimental results processing

First of all, the radiation source images were processed and the normalization coefficients of the scattered light relative intensity for different wavelengths were determined. The cylindrical vessel with the colloidal solution has an optical force and the scattered light refraction occurs on the vessel surfaces. The scattered radiation doesn't change its propagation direction after leaving the vessel, if it falls on its side wall normally. Therefore, only the part of the experimental image that contains the vessel axis was processed. In such area of each experimental image, the average pixel brightness was calculated, which is proportional to the scattered radiation intensity. The average value of the pixel brightness, taking into account the corresponding normalization factor, was compared with the rod rotation angle with the camera, at which the processed frame was obtained. After processing all the series images, light scattering indicatrix of certain wavelength is obtained.

3. Results

In the course of the experiment, after 30 and 60 minutes from the producing of the solution has been obtained three series of the probing beam images in scattered light, all three series were done with different filters. As the processing result of experimental images the scattering indicatrix presented in figures 5 and 6 were constructed.

The rod rotation angle with the camera when receiving each next shot in the series was chosen equal to 5 degrees. The constructive features of the setup allowed measurements in the angle range from 10 to 160 degrees, so the scattering indicatrices in figures 5 and 6 aren't closed and have two discontinuities.

4. Direct problem solution of elastic scattering in the framework of the G. Mi theory

The modeling 2D and 3D light scattering indicatrix program in the framework of G. Mi theory was developed by the authors [17, 18, 19]. It allows obtaining the scattering indicatrix of monochromatic unpolarized or linearly polarized radiation in dielectric or slightly absorbing spherical particles.

The particle shape limitation can be attributed to the calculating disadvantages of the scattering by the G. Mi formulas, but this calculation method has a significant advantage. The set of values that can take the initial data (the relative complex refractive index m and the radius r of the particle, the wavelength of the probing radiation λ) is practically limited only by the computer computational resources. The use of other methods for solving the direct problem of elastic light scattering is limited by the initial data permissible values. At the expense of the G. Mi theory noted advantage received program were universal, allowing to trace the light scattering dynamics when changing the relative size of $x = 2\pi \cdot r/\lambda$ or the particles refractive index in a wide range.

The scattering indicatrix 2D-models dynamics of nonpolarized monochromatic light for three wavelengths $\lambda_1 = 434$ nm, $\lambda_2 = 546$ nm, $\lambda_3 = 656$ nm on a particle with the relative refractive index $m = 1{,}532$ when changing its radius r in the range from 1 to 11 nm is shown in figure 7, in the range from 11 to 22 nm – in figure 8, in the range from 22 to 33 nm – in figure 9, in the range from 50 to 90 nm –

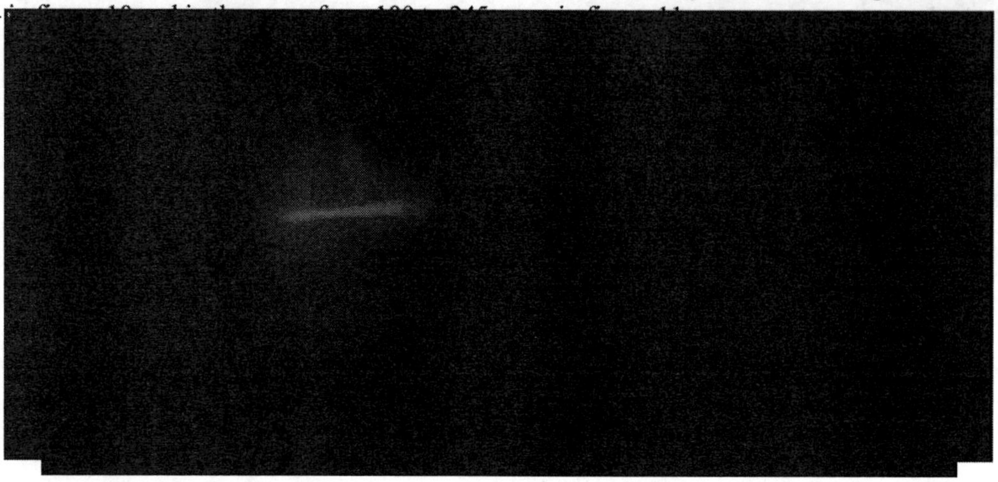

Figure 2. Dynamics of the experimental probing beam images in scattered light with 434 nm wavelength (avi file).

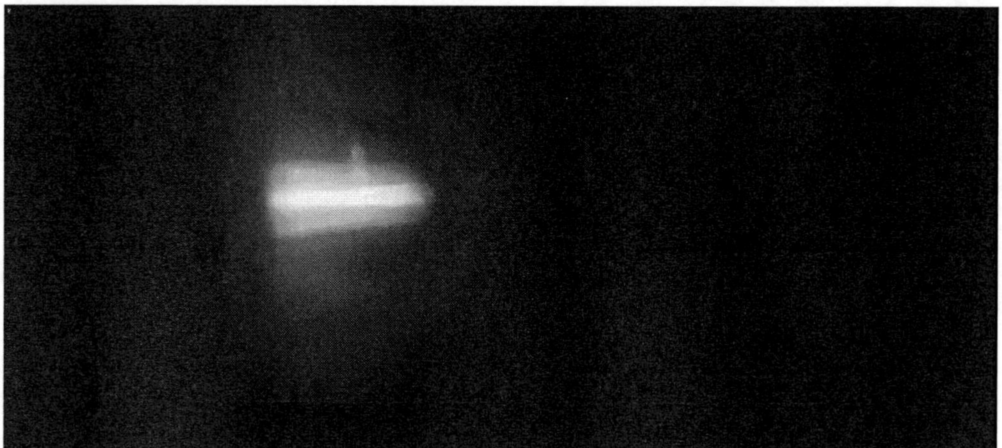

Figure 3. Dynamics of the experimental probing beam images in scattered light with 546 nm wavelength (avi file).

wavelength (avi file).

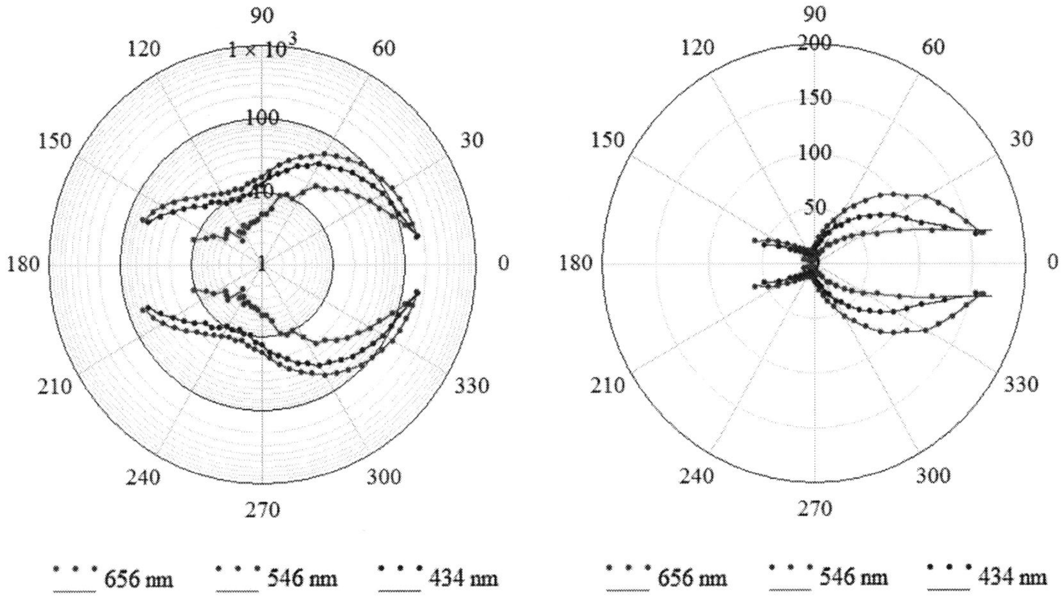

a) b)

Figure 5. Experimental light scattering indicatrices with wavelengths $\lambda_1 = 434$ nm, $\lambda_2 = 546$ nm, $\lambda_3 = 656$ nm by colloidal sulfur particles 30 minutes after the solution was produced: a) in a grid with conventional scale; b) in a grid with logarithmic scale.

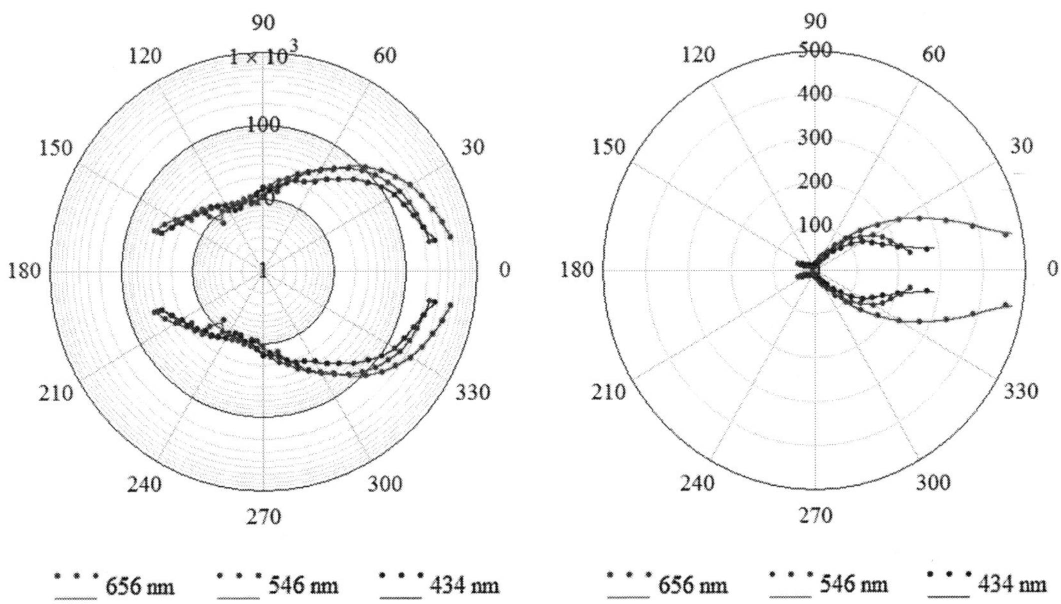

a) b)

Figure 6. Experimental light scattering indicatrices with wavelengths $\lambda_1 = 434$ nm, $\lambda_2 = 546$ nm, $\lambda_3 = 656$ nm by colloidal sulfur particles 60 minutes after the solution was produced: a) in a grid with conventional scale; b) in a grid with logarithmic scale.

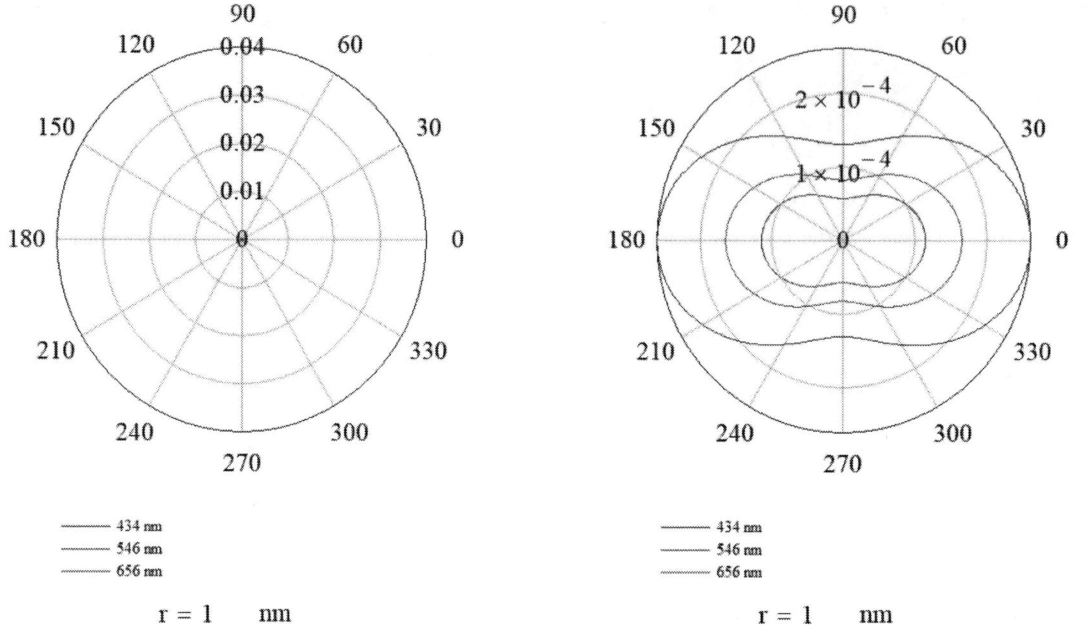

a) b)

Figure 7. Two-dimensional light scattering indicatrix models with wavelengths $\lambda_1 = 434$ nm, $\lambda_2 = 546$ nm, $\lambda_3 = 656$ nm on a particle with the relative refractive index $m = 1{,}532$ when changing its radius r from 1 to 11 nm: a) in a grid with constant scale; b) in a grid with variable scale (avi file).

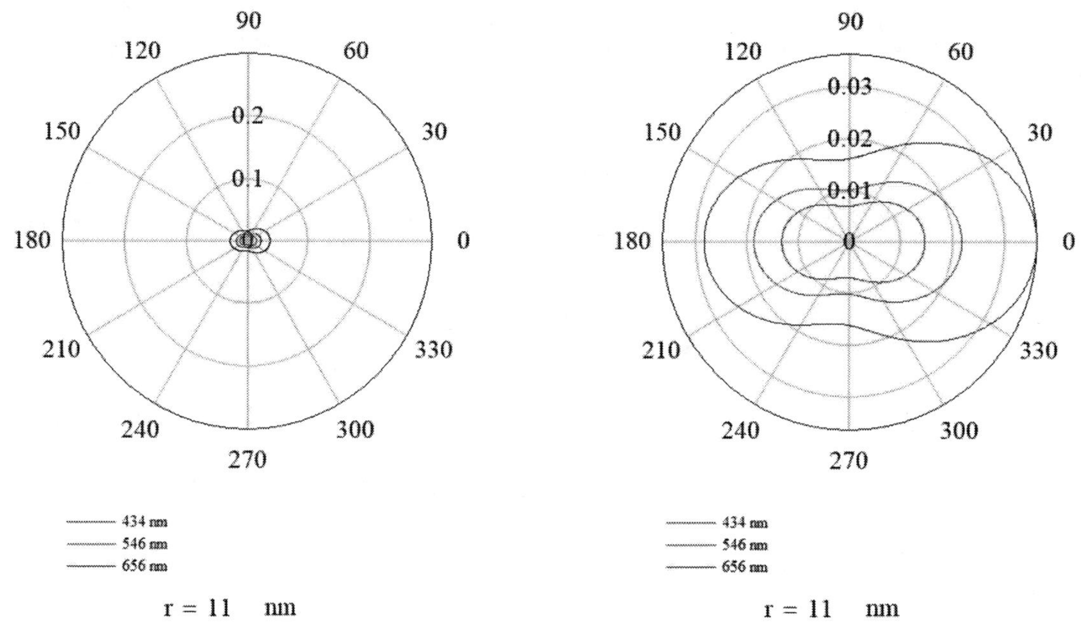

Figure 8. Two-dimensional light scattering indicatrix models with wavelengths $\lambda_1 = 434$ nm, $\lambda_2 = 546$ nm, $\lambda_3 = 656$ nm on a particle with the relative refractive index $m = 1{,}532$ when changing its radius r from 11 to 22 nm: a) in a grid with constant scale; b) in a grid with variable scale (avi file).

Journal of Physics: Conference Series **1421** (2019) 012017 doi:10.1088/1742-6596/1421/1/012017

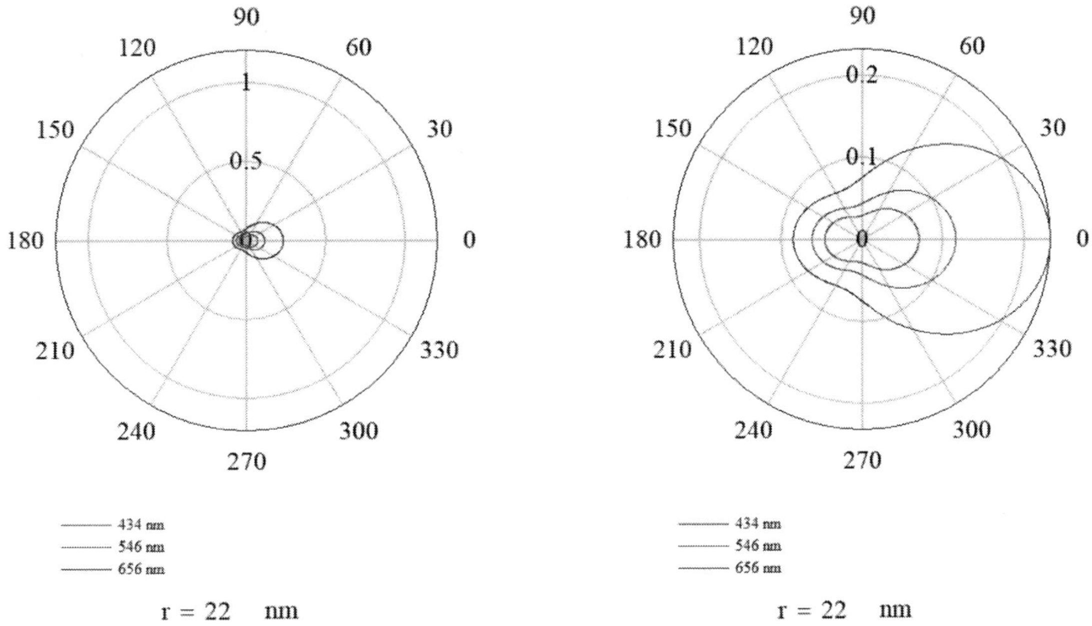

a) b)

Figure 9. Two-dimensional light scattering indicatrix models with wavelengths $\lambda_1 = 434$ nm, $\lambda_2 = 546$ nm, $\lambda_3 = 656$ nm on a particle with the relative refractive index $m = 1{,}532$ when changing its radius r from 22 to 33 nm: a) in a grid with constant scale; b) in a grid with variable scale (avi file).

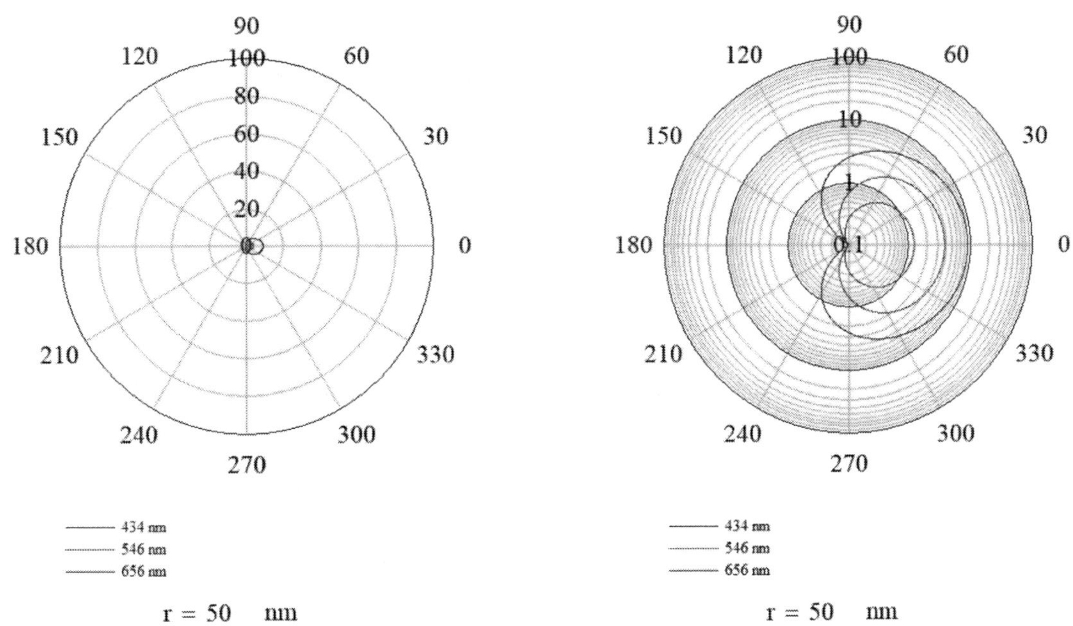

Figure 10. Two-dimensional light scattering indicatrix models with wavelengths $\lambda_1 = 434$ nm, $\lambda_2 = 546$ nm, $\lambda_3 = 656$ nm on a particle with the relative refractive index $m = 1{,}532$ when changing its radius r from 50 to 90 nm: a) in a grid with constant scale; b) in a grid with logarithmic scale (avi file).

Conventionally, three particle sizes ranges can be distinguished, for each of which the scattering features have significant differences. When the particle is small ($x << 1$), we talk about the light elastic scattering in the Rayleigh approximation. This case corresponds to the scattering indicatrix in figures 7 and 8. Presented in figures 9 and 10 of the scattering phase function correspond to the Rayleigh–Gans approximation ($x \sim 1$), the features of which can be attributed to the pronounced scattering forward and the emergence of a small number the interference of the "petals" in the scattering phase function. Figure 11 shows the scattering indicatrices corresponding to the approximation of large particles ($x >> 1$), when the interference of scattered light is manifested in the form of a significant maxima and minima number in the angular intensity distribution. The number of interference "petals" increases with increasing particle radius.

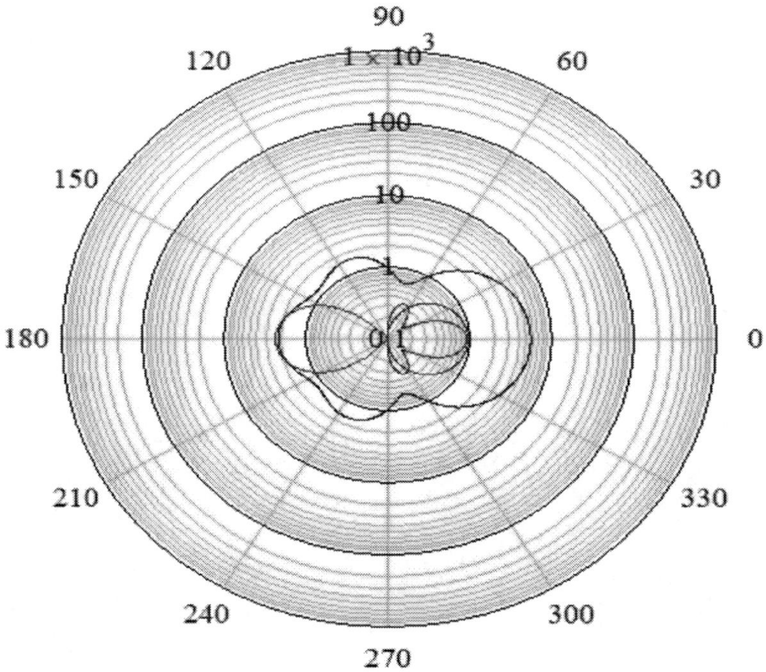

—— 434 nm
—— 546 nm
—— 656 nm

r = 100 nm

Figure 11. Two-dimensional light scattering indicatrix models with wavelengths $\lambda_1 = 434$ nm, $\lambda_2 = 546$ nm, $\lambda_3 = 656$ nm on a particle with the relative refractive index $m = 1{,}532$ when changing its radius r from 100 to 245 nm in a grid with logarithmic scale (avi file).

5. Discussions

The size increase of the colloidal sulfur particles over time after the solution was produced should have caused a change in the scattering indicatrix shape . Indeed, the light scattering nature of different wavelengths after 30 (figure 5) and 60 (figure 6) minutes after the solution was produced is different. Figure 5 shows that in the entire angles range at which measurements were made, the light with 546 nm wavelength experiences the greatest scattering, and the light with a wavelength of 656 nm undergoes the least scattering. From the figure 11 it can be seen that at angles from 10 to 50 degrees, light with a wavelength of 656 nm is more scattered, and light with wavelengths of 434 and 546 nm is scattered weaker, at angles from 85 to 160 degrees, the intensity of the scattered light at each wavelength differs slightly. Shown in figure 10 the scattering phase function doesn't have intersection points, while the scattering phase function in figure 11 cross each other.

The light scattering indicatrix shape of different wavelengths in each of the figures 10 and 11 is also different. But all the curves have common features: a pronounced maximum in the "forward" direction; a dip near the angle of 100 degrees. The "petals" at the experimental distributions aren't available, but the interference effects are manifested in the fact that most of the scattering light isn't experiencing the shortest wavelength, which is scattering characteristic in the absence of interference phenomena. All these features indicate the scattering in the Rayleigh-Gans approximation, qualitatively the experimentally obtained scattering phase function correspond to the models presented in figure (10).

6. Conclusion

On the comparison basis of experimental and calculated light scattering indicatrices it was concluded that their qualitative coincidence, which means that theelastic light scattering phenomenon can be used to develop an optical method for the diagnosis of colloidal solutions.

The paper shows the difference between experimental light scattering indicatrix with different wavelengths on the same particle system, as well as their change over time. Provided that the wavelengths of the detected scattered light are known, the scattering indicatrix can be used as measured informative parameters to determine the scattering particles size. If on the basis of the considered experimental scheme to produce an automated system for measuring the scattering indicatrix, it will be possible to trace the dynamics of the scattering indicatrix in time and, based on it, to determine the preparation time and the stability time of the colloidal solution.

Acknowledgments

The work was supported by the Ministry of higher education and science of the Russian Federation (project 3.8009.2017/BCh).

References

[1] Wongmekiat A, Tozuka Y, Moribe K, Oguchi T and Yamamoto K 2007 *Chem. and Pharmac. Bull.* **5** 359-363
[2] Hafiz S, Scimeca M, Sahu A and Ko D-K 2019 *Nano Converg.* **6** 1
[3] Luo K, Wu W, Xie S, Jiang Y, Liao S and Qin D 2019 *Appl. Sci. (Switzerland)* **9** 1885
[4] Kim M, Lee D, Woo H-Y, Kim T-H, Paik T and Kim K-S 2019 *Materials and Design* **175** 107800
[5] Zheng G, Yang Y, Cha J J, Hong S S and Cui Y 2011 *Nano Let.* **11** 4462-67
[6] Guo J, Xu Y and Wang C 2011 *Nano Let.* **11** 4288-94
[7] Ji L, Rao M, Aloni S, Wang L, Cairns E J and Zhang Y 2011 *Energy and Envir. Sci.* **4** 5053-59
[8] Jakmunee J, Udnan Y, Morrison R, Beckett R, McKinnon I and Grudpan K 2003 *Analytical Sci.* **19** 1495-98
[9] Matsuzaki K, Murase O, Sugishita K, Yoneyama S, Akada K, Ueha M, Nakamura A and Kobayashi S 2000 *Biochimica et Biophysica Acta – Biomembranes* **1467** 219-226
[10] Gimeno-Furio A, Hernandez L, Navarrete N and Mondragon R 2019 *Renewable Energy* **140** 493-500
[11] Attota R, Kavuri P P, Kang H, Kasica R and Chen L 2014 *Appl. Phys. Let.* **105** 163105
[12] Boyd R D, Gunnarsson R, Pilch I and Helmersson U 2014 *J. of Phys.: Conf. Series* **522** 012065
[13] Beliciu C M and Moraru C I 2009 *J. of Dairy Sci.* **92** 1829-39
[14] Nobbmann U and Morfesis A 2009 *Materials Today* **12** 52-54
[15] Mächtle W 1999 *Biophysical J.* **76** 1080-91
[16] Lechner M D 2005 *J. of the Serbian Chem. Society* **70** 361-369
[17] Sapronov M V and Skornyakova N M 2017 *Scientific Visualization* **9** 42-53
[18] Sapronov M V and Skornyakova N M 2017 *Optical Methods of Flow Investigation: Proc. XIV Sci.-Tech. Int. Conf. (Electronic Materials)* (Moscow: «Pero») pp 243-250

Application of computer vision and neural network analysis to study the structure and dynamics of turbulent jets

S Abdurakipov[1,2,*], M Tokarev[1,2], E Butakov[1,2] and V Dulin[1,2]

[1] Kutateladze Institute of Thermophysics SB RAS, Novosibirsk, Russia
[2] Novosibirsk State University, Novosibirsk, Russia

***E-mail:** s.s.abdurakipov@gmail.com

Abstract. In this paper we solved the problem of detecting and detailed analysis of the characteristics of coherent structures in non-swirling and swirling turbulent jets from the 3D PIV experiment and DNS modelling database. To solve the problem intellectual approaches of computer vision based on deep neural networks were applied. It is shown that with the use of a generative competitive neural network, it is possible to reconstruct the structure and dynamics of a turbulent flow with an increased spatial resolution, which is important for analysis and interpretation. The approach is based on a modification of the loss function, which minimizes the residual part of the hydrodynamic equations (the continuity equations and the Poisson equation) for the reconstructed data. It is shown that generative models are more effective in the reconstruction of high moments of turbulent flow than conventional methods POD and DMD. Using a fully convolutional neural network, an automatic segmentation of the turbulent jet flow region and the external environment based on instantaneous velocity and pressure fields was carried out. For a swirling jet, the area of reversed flow is identified. It is shown that the complex of the developed machine learning algorithms successfully copes with the localization of coherent structures, the detection of their trajectory, characteristic size and phase propagation velocity. The obtained new fundamental information is important for a deeper understanding of the role of vortex structures in mixing processes in a jet flow. The developed complex of algorithms for the identification and analysis of characteristics of coherent structures can be applied to a database of measurements and numerical simulation of a wide class of hydrodynamic flows.

1. Introduction

It is well known that large-scale vortex structures play an important role in the processes of heat and mass transfer in jet flows. In this regard, the study of their characteristics is an important task both from a fundamental and from a practical point of view. The use of the non-contact optical method PIV (Particle Image Velocimetry) and the method of direct numerical simulation DNS (Direct Numerical Simulation) allows to obtain detailed information on the distribution of the instantaneous velocity field in turbulent flows. However, often, a large amount of data makes it difficult to analyze and interpret the results. Thus, the obtained primary information requires a detailed analysis using automatic procedures for the identification of the characteristic flow features. The development of intelligent data analysis methods and the growth of computing power made it possible to use neural network algorithms to study the structure and dynamics of turbulent flows. There is a large number of examples of machine learning methods being used, for example, in many aviation problems of simulation of empirical phenomena, malfunction diagnosis, and control [1-4]. In this paper, we solved the problem of detecting and detailed analysis of the characteristics of coherent structures in non-swirling and swirling turbulent jets from PIV experiment and DNS modeling database. To solve the problem, intellectual approaches of computer vision based on deep neural networks were applied. Now, this actual topic is practically not developed in the literature.

In many literary sources computer vision problem statements of classifying and detecting objects are known. In the task of classifying objects, we want the model to determine which individual object or scene is present in the image. Therefore, the classification operates on a high level of abstraction. The task of detecting objects is also well known, where we are trying to find and classify several objects in the image, drawing bounding boxes around them, and then classifying what is in this field. Therefore, detection takes place at the middle level, where we have quite useful information, but it is still rather rough, because we define only the bounding box and do not get an accurate idea of the object shape. Semantic segmentation is the most informative of these three tasks, because we try to classify each pixel in the image. In the literature there are many works on semantic segmentation, for example [5-11]. One of the efficient algorithms for semantic segmentation are fully-convolutional neural networks (FCN) [5]. A fully convolutional network (FCN) learns the mapping from pixels to pixels without extracting proposals of regions unlike region-based models such as Mask R-CNN [7]. The FCN pipeline is a continuation of the classic convolutional neural networks (CNN) (see figure 1). The basic idea is to force classic CNN to take an image of arbitrary size as input. The limitation of CNN is to take images of a fixed size due to the use of fully connected layers that are fixed.

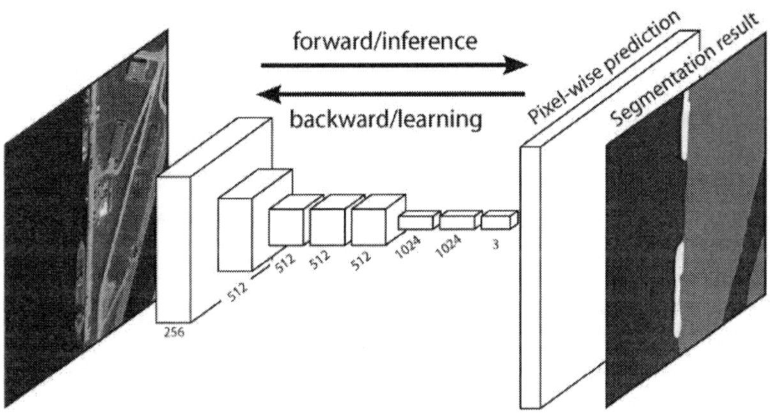

Figure 1. The architecture of fully-convolutional network (FCN) for pixel-wise prediction [5].

In contrast, FCN have only convolutional and pooling layers, which enable them to make predictions at inputs of arbitrary size. To solve this problem, many more advanced FCN-based approaches have been proposed, including SegNet, DeepLab-CRF and Dilated Convolutions [8, 9]. In this paper, the architecture of the deconvolution network [5] was used to segment the velocity and pressure fields.

Generative adversarial neural networks (GANs) are widely used to create photorealistic images [10-13]. However, in the case of applying such generative models to real physical data, the basic physical equations for the new examples generated by the model may not be satisfied. In this paper, we applied an approach to training the generative model for the reconstruction of turbulent flow fields [13]. The approach is based on the modification of the loss function, which is aimed at minimizing the residual part of the hydromechanics equations (continuity equation and the Poisson equation) for the generated data. This mathematical operation can significantly improve the properties of the model to reconstruct the physical dynamics of the turbulent flow. We applied and refined the algorithm for the turbulent flow of a submerged round jet. Accurate modeling of turbulent flow is extremely difficult due to the large number of degrees of freedom of a turbulent flow (proportional to $Re^{9/4}$). In [14], it was shown that GAN can demonstrate better results in the reconstruction of higher order moments than other popular machine learning approaches, such as the principal component method - in the literature on sections of hydromechanics better known as Proper Orthogonal Decomposition (POD). In a recent paper [15], neural networks that take into account the physical equations were developed to solve ordinary differential equations (ODE). In their solution, the authors used the root-mean-squared error function with the contribution not only of the solution error, but also with residual part of the differential equation. The approach was implemented and tested for a one-dimensional problem and cannot be extended for three-dimensional unsteady turbulent flows. In [12] generative models were used to create high-resolution photorealistic images. Despite the fact that the results are extremely encouraging, the approach cannot be applied to solving the current problem due to the absence of guarantees of physical realism of synthetic fields. In [14] generative models were used as an alternative to the well-known principal component analysis approach (PCA) to create complex geological models. It is shown that GAN reproduces the distribution of several physical quantities simultaneously.

This paper is devoted to the analysis of the applicability and creation of a generative adversarial neural network model, which allows learning from low spatial resolution data and reconstructing the relevant physical dynamics of the turbulent flow. This model can be used for many engineering applications. For example, for the reconstruction of low-dimensional modeling data, including for an approximate engineering calculation of the structural loads on a wind turbine, etc., the reconstruction of experimental data limited by the spatial resolution of the sensor. In this paper, we applied a generative adversarial neural network by analogy with [13] to reconstruct the structure and dynamics of a turbulent flow with an increased spatial resolution, which is important for analysis and interpretation.

The purposes of this work were to explore the possibilities of modern deep learning methods to study the structure and dynamics of complex turbulent flow.

2. Experimental and numerical database

The database of jet flow velocity fields was obtained as a result of 3D PIV experiment and DNS simulation. The non-swirling and swirling jet flows were organized in a closed hydrodynamic circuit, which included a water tank, pump, flowmeter, and test section. The rectangular test section ($200\times600\times200$ mm^3) was made of Plexiglas (see figure 2). A contraction nozzle in the bottom of the test section produced the jet flow. The nozzle exit diameter d was 15 mm. A vane swirler was mounted inside the nozzle to organize jets with swirl. Details on the nozzle and mean velocity data can be found in [16]. The measurement volume of $40\times40\times40$ mm^3 (approximately $2.5d\times2.5d\times2.5d$) was recorded by four high-speed CMOS cameras (Photron FASTCAM SA5). Polyamide seeding particles (50 μm in size) in the measurement volume were illuminated by a high-repetition pulsed Nd:YAG laser (Photonix DM100-532) with the average power of 100 W. The illumination area was organized as a vertical cylindrical beam of 40 mm in diameter. The in-house "ActualFlow" software, developed in the Institute of Thermophysics, was used to measure and process the acquired data. The acquisition frequency was 2 kHz. During each run 2000 images were recorded. Three independent runs were carried out for each swirl case to verify the consistency of the results for a single run. The separation between neighboring vectors was 0.84 mm. The spatial and temporal resolution was limited by approximately 3 mm and the 5 ms, respectively. Pressure fields were derived from velocity fields by solving the Poisson equation [16].

Direct numerical simulation data of a turbulent jet [17] performed with an in-house high-order finite-difference/pseudospectral code that solves the compressible Navier-Stokes equations using 400 million grid points. The computational domain was divided into subdomains to use a set of processor cores using MPI protocol and OpenMP. The computational grid used about 400 million nodes, which was sufficient to resolve even the smallest turbulent scales up to the size of Kolmogorov. We consider a jet at Reynolds number Re = 5940 based on the bulk velocity U_b in the pipe and its diameter D generated by a fully turbulent pipe flow entering a uniform stream with a coflow velocity $u_{co} = 0.27U_b$. The physical region had a longitudinal coordinate along the flow up to x/D = 40. Together with a sufficient spatial resolution, the computation features a very long-time realization with around $400D/U_b$ time units, allowing one to statistically assess even the far field of the jet.

(a) (b)

Figure 2. Photographs of (a) experimental setup and (b) illuminated volume.

3. Flow fields reconstruction. Generative Adversarial Network

Three-dimensional PIV data on the velocity and pressure fields of dimension 12×12×12 are fed to the input of the generative model. The model allows you to increase the frequency of spatial discretization up to 44×44×44, while maintaining the physical properties of the generated distributions. The PIV database contained 2,500 velocity fields for 4 separate runs (10,000 total). Additional fields for training generative model were generated by using Reduced Order Model (ROM) based on POD technique [17]. Such an approach makes sense of augmentation, which is often used when training a neural network in order to increase the generalizing ability of a model. The full dataset contained about 15,000 velocity and pressure fields. Pressure fields were derived from velocity fields by solving the Poisson equation [16].

The generator is a ResNet residual network (residual neural network) architecture, each block of which includes convolutional layers with batch normalization. The discriminator is a deep convolutional neural network with fully-connected output layers for binary classification. The architecture of the generator and discriminator is shown in figure 3.

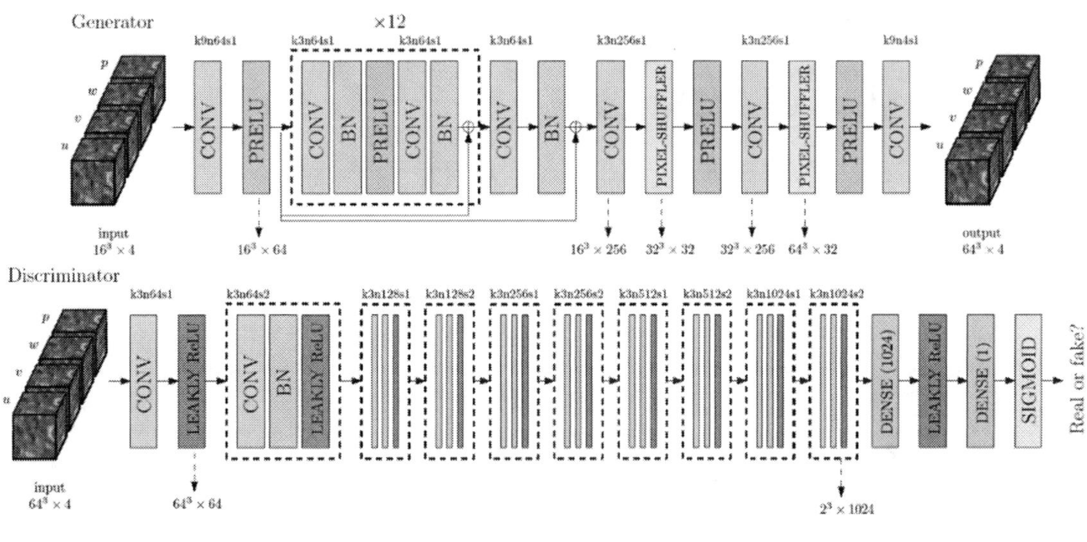

Figure 3. The architecture of generator and discriminator of generative adversarial neural network. In figure k is the convolutional kernel size, n is the number of channels, s is the stride for each convolutional layer [13].

The velocity and pressure fields must satisfy the continuity equation and the Poisson equation: $\nabla \mathbf{u} = 0$ and $\Delta p = -(\mathbf{u}\nabla)\,\mathbf{u}$, where $\mathbf{u} = (u, v, w)$ is a three-component velocity vector, p is pressure. However, these equations may not be exactly satisfied by the fields of the generative model. To overcome this drawback, the residual part of these equations can be used as a regularizer for the model. The total loss function of Q_{model} includes 5 component terms:

$$Q_{\mathrm{model}} = \alpha Q_{\mathrm{resnet}} + (1-\alpha) Q_{\mathrm{adver}},$$
$$Q_{resnet} = \beta Q_{\mathrm{field}} + (1-\beta) Q_{\mathrm{gove}},$$
$$Q_{\mathrm{field}} = \gamma Q_{\mathrm{MSE}} + (1-\gamma) Q_{\mathrm{enst}},$$
$$Q_{\mathrm{gove}} = \delta Q_{\mathrm{pois}} + (1-\delta) Q_{\mathrm{cont}}.$$

The loss function, which is minimized for the generator, is a linear combination of two loss functions Q_{field} and Q_{gove}, which makes it possible to more accurately describe the physically relevant dynamics of the turbulent flow. Q_{field}: Q_{MSE} is the standard deviation between the actual high-resolution velocity fields and the generated fields from GAN model. Q_{enst} is the mean squared error in the resulting enstrophy $\Omega = \boldsymbol{\omega}\boldsymbol{\omega}$, where $\boldsymbol{\omega} = \nabla \times \mathbf{u}$ is the vorticity. Q_{gove} is the residual part of the continuity and Poisson equations, considered by analogy with [13]. Q_{adver} is the logistic loss function, defined by analogy with [13]. To train the discriminator, we used the logistic loss function for the classifier problem, which distinguishes between real and generated data examples. We used the generator model based on the residual neural network ResNet without the content part to check the convergence and tuning of the hyperparameters (the number of ResNet blocks and the parameters of the physical loss function Q_{gove}).

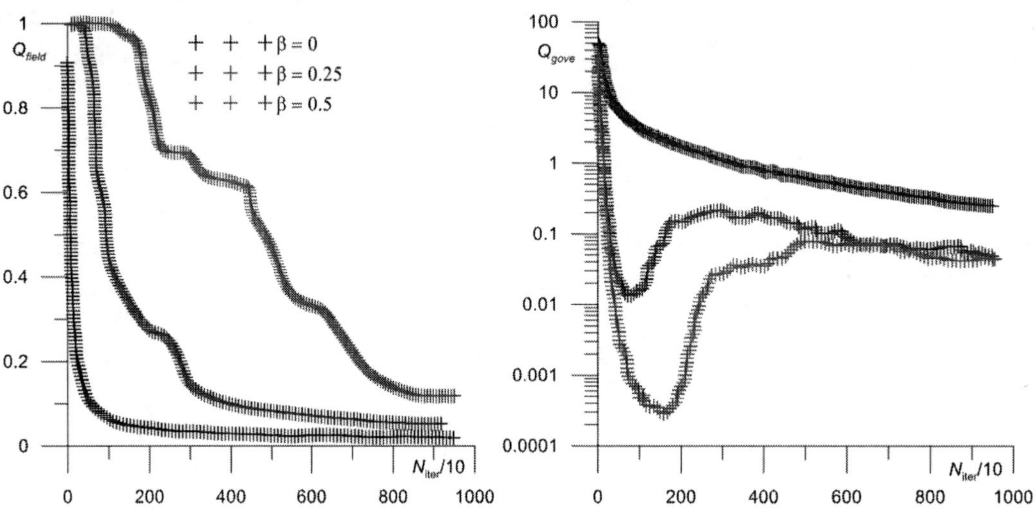

Figure 4. Graphs of content (Q_{field}) and physical loss (Q_{gove}) functions depending on the number of iterations (N_{iter}) for different values of the coefficient β.

The used model of generative adversarial neural network includes a discriminator and a generator, shown in figure 3. The initial generator weights were initialized from the pre-trained ResNet model, the He-Xavier initialization was used for the discriminator weights. At the first ~ 100-300 iterations, only the discriminator is trained to compensate the advantage of the generator due to pre-selected weights. Then the discriminator and the generator were trained together in a competitive manner, until the loss functions come to a plateau, and the discriminator output reaches ~0.5 (see details in [13]). Adam (adaptive inertia optimizer) was used to update the weights and train both neural networks. The size of the batch was 5 and was chosen to train both the ResNet model and the generative adversarial neural network due to the limitations of the available GPU memory. We chose the learning rate α = 0.0001 in the process of optimization of the hyperparameters (iteration over the grid). In the process of learning, the coefficient for the physical and content parts of the loss function was analyzed. The coefficient β = 0.9 was selected based on a compromise between a decrease in the physical part and an increase in the content part of the loss function. To increase the sustainability of the generative model in the learning process, we reduced the speed of learning. The figure 4 shows the convergence curves of the loss function with increasing number of learning iterations for various values of β coefficient.

Figure 5 shows a comparison of the velocity fields in the jet cross-section according to PIV data. For example, one cross-section y/d=2.0 of the jet is shown. Fields with a reduced resolution of 12×12×12 were fed to the input of the generative model, while the original was about 4 times higher (44×44×44). In figure 5 it can be seen that the reconstructed image with good accuracy repeats the original velocity field in high resolution. This result was achieved by changing the loss function and adding the term responsible for the physical content.

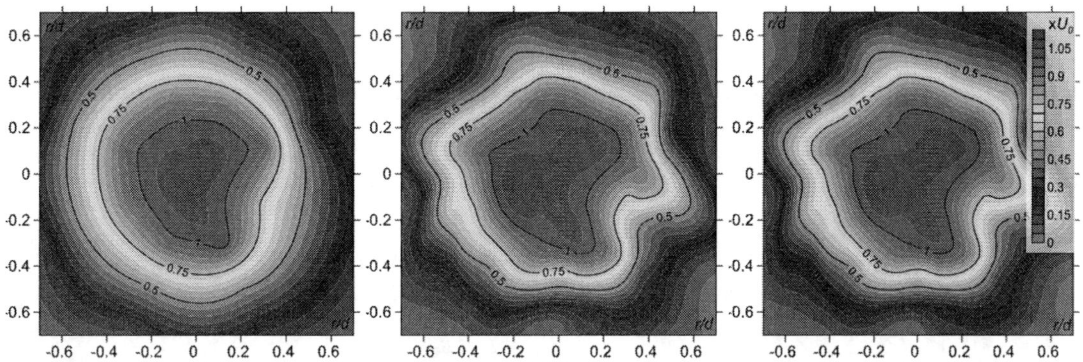

(a) (b) (c)

Figure 5. Comparison of velocity fields (from left to right): low resolution (a), from GAN model (b) and high resolution (c). The color represents the axial velocity component of the round turbulent jet in cross-section (y/d = 2.0).

4. Flow fields segmentation. Fully-Convolutional Neural Network

In this work, a fully-convolutional neural network (FCN) was used to solve the problem of segmentation of velocity vector fields of a turbulent jet. Similarly, each vector of three velocity components was considered by analogy with a separate RGB pixel in image segmentation. The deep learning model is based on the pre-trained neural network of the VGG-16, as was done in [5]. In our FCN implementation, we reuse the pre-trained VGG-16 layers numbered 3, 4, and 7, and then apply convolutions with the 1x1 filter to these layers. This stage resembles an encoder that extracts features from images. The encoding is followed by a decoding step using multiple layers of upsampling. Increasing the sampling rate occurs through the use of transposed convolution (inverse convolution operation). Figure 6 shows the structure of the encoder and decoder.

The network has been trained on using the following hyperparameters: learning rate = 0.0001, keep probability = 0.25, epochs = 25, batch size = 16, initial std = 0.01, number of classes = 2, image shape= (64, 64). We used GAN model described below to increase the spatial resolution to 64x64x64. Neural network weights optimization was performed by using Adam optimizer, and minization of cross-entropy loss function. The cross-entropy loss is computed against the correct ground truth image. Image labeling was performed on the basis of criteria for vorticity to determine the region of turbulent flow and axial velocity to determine recirculation zone.

For example, figure 7 shows the results of segmentation of instantaneous velocity fields of a swirling jet using a fully-convolutional neural network. For example, one cross-section y/d = 1.0 of swirling jet is shown. The green color shows the identified region of the turbulent flow of a swirling jet (with sufficient vorticity). The blue area in the center visualizes a segmented area of reversed flow. The gray area visualizes negative axial velocity values. You may notice that the segmentation area of the

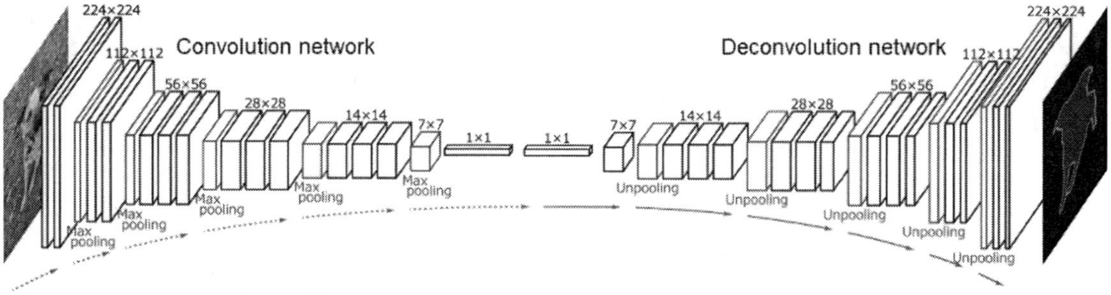

Figure 6. The architecture of fully-convolutional neural network (FCN) [6].

recirculation zone is slightly larger than the contour of the negative values of the axial velocity component. The figure shows a few typical situations. This is due to the fact that 3D velocity fields separated by slices are used when training fully-convolutional neural network. Extracting 3D flow structure information directly using 3D convolutions and increasing number of epochs during training process can improve segmentation accuracy. However, it is a fairly time-consuming in the computational point of view.

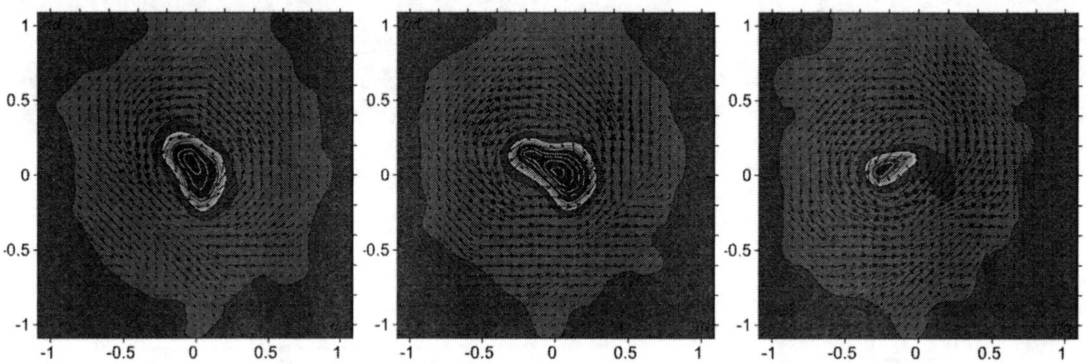

Figure 7. Examples of segmentation of instantaneous velocity fields of swirling jet in cross-section (y/d = 1.0) by using fully-convolutional neural network. The green contour shows region of turbulent flow, blue contour in the central part shows the segmented area of reverse flow, gray contour indicates negative values of the axial velocity in cross-section.

5. Conclusions

In this work, computer vision technology was tested on the basis of deep neural networks to solve problems of reconstruction of velocity and pressure fields of a turbulent flow with high resolution and to identify the characteristic flow features. In the process of research, two important tasks were considered. The first task concerned the construction of a model for the reconstruction of flow velocity fields with an increased spatial resolution, which is often important for the accompanying data analysis and modeling tasks. The second task was aimed at building a deep learning model for automatic detection of the characteristic turbulent flow features in instantaneous velocity fields. To solve these problems, an extensive databases of 3D PIV experiments and DNS modeling were used. To increase the dataset, an augmentation procedure based reduced order modeling was applied. It is shown that the generative adversarial neural network can be effectively used to increase the spatial resolution up to four times. The result was achieved by applying an approach based on modification of the loss function, which minimizes the residual part of the hydrodynamic equations (the continuity equations and the Poisson equation) for the reconstructed data. It is shown that generative models are more effective in the reconstruction of high moments of turbulent flow than conventional methods POD and DMD. The results show that the fully-convolutional neural network can be used for automatic segmentation of flow features, such as turbulent flow region, recirculation zone, coherent vortex structures, etc. This approach can be effectively applied to localize, determine the characteristic size and trajectory of vortex structures. The developed deep learning models can be effectively used to develop systems for automating the process of studying hydrodynamic flows.

Acknowledgements
This work was financially supported by the Russian Science Foundation, grant number 18-79-00138.

References
[1] Lee C, Kim J, Babcock D and Goodman R 1997 Application of neural networks to turbulence control for drag reduction *Physics of Fluids* **9** 1740-1747
[2] Abdurakipov S, Gobyzov O, Tokarev M and Dulin V 2018 Combustion Regime Monitoring by Flame Imaging and Machine Learning *Optoelectronics Instrumentation and Data Processing* **54** 513-519
[3] Abdurakipov S, Tokarev M, Pervunin K and Dulin V 2019 Modeling of the tonal noise characteristics in a foil flow by using machine learning *Optoelectronics Instrumentation and Data Processing* **55** 123-131
[4] Abdurakipov S and Butakov E 2018 Combustion anomalies detection for a thermal furnace based on recurrent neural networks *Journal of Physics: Conference Series* vol 1105 pp 012043
[5] Long J, Shelhamer E and Darrell T 2015 Fully convolutional networks for semantic segmentation. *Proceedings of the IEEE conference on computer vision and pattern recognition* pp 3431-3440
[6] Noh H, Hong S and Han B 2015 Learning deconvolution network for semantic segmentation *Proceedings of the IEEE international conference on computer vision* pp 1520-1528
[7] He K, Gkioxari G, Dollár P and Girshick R 2017 Mask r-cnn *Proceedings of the IEEE international conference on computer vision* pp 2961-2969
[8] Badrinarayanan V, Kendall A and Cipolla R 2017 Segnet: A deep convolutional encoder-decoder architecture for image segmentation *IEEE transactions on pattern analysis and machine intelligence* vol 39 pp 2481-2495
[9] Chen L, Papandreou G, Kokkinos I, Murphy K and Yuille A 2018 Deeplab: Semantic image segmentation with deep convolutional nets, atrous convolution, and fully connected crfs *IEEE transactions on pattern analysis and machine intelligence* vol 40 pp 834-848
[10] Goodfellow I, Pouget-Abadie J, Mirza M, Xu B, Warde-Farley D, Ozair S, Courville A and Bengio Y 2014 Generative adversarial nets *Advances in neural information processing systems* pp 2672-2680
[11] Radford A, Metz L and Chintala S 2015 Unsupervised representation learning with deep convolutional generative adversarial networks *arXiv preprint* 151106434
[12] Ledig C, Theis L, Huszár F, Caballero J, Cunningham A, Acosta A and Shi W 2017 Photo-Realistic Single Image Super-Resolution Using a Generative Adversarial Network *CVPR* **2** 4
[13] Nimmagadda S, Subramaniam A, Wong M, Borker R and Lele S 2018 Turbulence Enrichment using Generative Adversarial Networks *Bulletin of the American Physical Society* **63**
[14] Chan S and Elsheikh A 2017 Parametrization and Generation of Geological Models with Generative Adversarial Networks *arXiv preprint* 170801810
[15] Raissi M, Perdikaris P and Karniadakis G 2017 Physics Informed Deep Learning (Part I): Data-driven solutions of nonlinear partial differential equations *arXiv preprint* 171110561

[16] Alekseenko S, Abdurakipov S, Hrebtov M, Tokarev M, Dulin V and Markovich D 2018 Coherent structures in the near-field of swirling turbulent jets: A tomographic PIV study *International Journal of Heat and Fluid Flow* **70** 363-379
[17] Mullyadzhanov R, Sandberg R, Abdurakipov S, George W and Hanjalić K 2018 Propagating helical waves as a building block of round turbulent jets *Physical Review Fluids* **3** 062601

Peculiarities of the acousto-optical interactions of Bessel light beams in crystals

V N Belyi[1], G V Kulak[2], G V Krokh[2], P I Ropot[1] and O V Shakin[3]

[1]LOD, Institute of Physics of NASB, 68 Nezalezhnasti ave., Minsk, Belarus 220072
[2]Physical engineering faculty, State Pedagogical University,28 Studencheskaia str. Mozur, Belarus 247760
[3]State University of Aerospace Instrumentation, 67 Bolshaya Morskaya str., Saint-Petersburg, Russia 190000

E-mail: [1]v.belyi@dragon.bas-net.by, [2]g.kulak57@mail.ru, [3]oshakin@mail.ru

Abstract. The noncollinear acousto-optical filtration of diffracted Bessel light beams in uniaxial paratellurite crystals has been investigated. With the use of the method of overlap integrals it is shown that independently on the order of Bessel light beams in the conditions of transverse phase-matching of diffracted waves in the range of optical spectrum of 0.4-1.1 nm in paratellurite crystals the bandwidth of transmission of ~0.52 nm is reached due to detuning from Bragg's synchronism only. It is shown that in uniaxial gyrotropic crystals polarization-independent diffraction of Bessel light beams is possible, i.e. Bragg diffraction efficiency is independent of the polarization state of the incident beam. It has been stated that the physical reason of such diffraction is simultaneous realization of two processes of anisotropic scattering under which the Bragg synchronism conditions are realized for the orthogonally polarized elliptical Bessel beams.

1. Introduction

Acousto-optic (AO) diffraction of light waves in plane-wave approach or Gaussian light beams is quite well studied [1]. Features of the collinear acousto-optic transformation by ultrasound of Bessel light beams (BLB) are investigated [2, 3]. Owing to narrow diffraction structure of BLB of the highest orders and narrow dark central spot, Bessel beams of high order can be used for steering of atoms at long distances and focusing of cold atoms. AO tunable filters (AOTF) have emerged in present time as convenient, all-electronic, light weight and higher speed means of spectral filtering [4]. AOTF are based on the interaction of acoustic and optical waves in certain crystals, the most popular of them is paratellurite (TeO_2) [5].

In this paper with the use of the method of overlap integrals the noncollinear AO filtration of polychromatic Bessel light beams of higher orders at propagating in the uniaxial crystals is considered. At the same time as an example, the AO interaction of azimuthally uniform BLB of zero order ($m = 0$) and the azimuthaly non-uniform different orders ($m \neq 0$) in a paratellurite crystal is considered. In this work the features of the AO Bragg polarization diffraction of BLB propagating nearly to the optical axis of a gyrotropic uniaxial crystal on ultrasonic traveling wave is investigated theoretically and experimentally.

2. Theoretical investigations and discussion

We will consider the geometry of the AO interaction for which slow shear ultrasonic (US) wave propagates in TeO_2 a crystal in the direction which has angle α to axis $X \| [110]$ and also occupies space between the plane $z = 0$ and $z = l$. The axis of incident BLB is located in the XZ plane at an angle θ_o to X. At the same time in the direction of a diffraction order the longitudinal phase-matching conditions of diffracted waves as $\vec{k}_o + \vec{K} = \vec{k}_e + \Delta\vec{k}$ (anisotropic diffraction of o-type wave into e-type wave) takes place, where $\vec{k}_o$, $\vec{k}_e$ are the wave vectors of the incident and diffracted waves; $\vec{K}$ are the wave vectors of US wave; $\Delta\vec{k}$ represents the phase velocity mismatch between the incident and diffracted optical wave (longitudinal phase mismatch). An AOTF in imaging systems usually requires a large angular aperture that is achieved under the noncritical phase matching conditions. These conditions in the wave vector diagram are realized when the tangents of the wave vector surfaces, corresponding to incident and diffracted light are parallel. The gyrotropy property of crystal is not taken into account for such interaction geometry. The axis of incident BLB is located in the *XZ* plane at an angle φ_1 to the front of the ultrasonic wave. The cross section of the surface of wave vectors of a gyrotropic uniaxial crystal by the *XZ* diffraction plane and the location of the incident $\vec{k}_o$ and $\vec{k}_e$ and diffracted ($\vec{k}_o^{'}$ and $\vec{k}_e^{'}$) wave vectors Bessel beams are used. At the same time two types of vector interactions, namely, $\vec{k}_o+\vec{K}_1=\vec{k}_e^{'}, \vec{k}_e+\vec{K}_2=\vec{k}_o^{'}$, where $\vec{K}_1$, $\vec{K}_2$ components of the US wave, are seen to contribute to the diffracted field. In further calculations a simplified model of the acoustic beam was adopted, which is a superposition of two plane waves with wave vectors $\vec{K}_1$ and $\vec{K}_2$, propagating at a small angle to the *X* axis α.

The diffraction efficiency is calculated using the expression

$$\eta=\frac{\chi^2\sin^2\left[l_d\sqrt{\chi^2+(\Delta k_z/2)^2}\right]}{\chi^2+(\Delta k_z/2)^2}, \tag{1}$$

where

$$\chi=\frac{\pi n_o^4 g_m p_{ef}}{2n\lambda_0\sin(\theta_o+\alpha)\sin(\theta_e+\alpha)}\sqrt{\frac{2P_a}{l_1 l_2\sigma\upsilon^3}},$$

$$\Delta k_z=b\left\{\sqrt{\left[(\xi(\theta_o)-\eta)^2+2\eta\xi(\theta_o)(1-\cos(\theta_o+\alpha))\right]}-1\right\}$$

$$\xi(\theta_o)=\frac{n_e}{\sqrt{n_o^2\cos^2\theta_o+n_e^2\sin^2\theta_o}},\eta=\frac{\lambda f}{n_o\upsilon},\ b=\left(-\frac{2\pi n_o\Delta\lambda}{\lambda_0^2}\right),$$

and $p_{ef}=[(p_{12}-p_{11})-p_{44}\sin(2\alpha)/8]$ (p_{11}, p_{12}, p_{44} is the photoelastic constants, σ is the density of a crystal, υ is the phase speed of US wave, λ_0 is the central length of a light wave in vacuum; $\Delta\lambda$ is the detuning of the wavelength from central Bragg λ_0; f is the frequency of ultrasound; g_m is the overlap integral; l_d is the length of the AO interaction taking into account the direction of US beam group velocity; $n=(n_o+n_e)/2$, where n_o, (n_e) is the ordinary (extraordinary) index of refraction of a crystal; P_a is the power of a US wave, l_1 is the length of a piezoelectric transducer along a direction $X||[110]$, l_2 is the width of a piezoelectric transducer). We must consider overlap integral corresponding to *o-e* type Bessel light beam transformation from the relation

$$g_m=\frac{\int_0^{2\pi}\int_0^{R_B}|(e_o^*e_e)|\,\rho d\varphi d\rho}{\sqrt{\int_0^{2\pi}\int_0^{R_B}|e_e|^2\,\rho d\varphi d\rho}\sqrt{\int_0^{2\pi}\int_0^{R_B}|e_o|^2\,\rho d\varphi d\rho}}, \tag{2}$$

where vector-function polarizations have the form

Journal of Physics: Conference Series **1421** (2019) 012019 doi:10.1088/1742-6596/1421/1/012019

$$\vec{e}_o = e_{o1}\,\vec{e}_1 + e_{o2}\,\vec{e}_2 + e_{o3}\,\vec{e}_3, \qquad (3)$$
$$\vec{e}_e = e_{e1}\,\vec{e}_1 + e_{e2}\,\vec{e}_2 + e_{e3}\,\vec{e}_3,$$

and

$$e_{o1} = \mathrm{i}q_{o\perp}\cos\theta_o'(J_{m-1}(q_{o\perp}\rho)e^{-\mathrm{i}\varphi} + J_{m+1}(q_{o\perp}\rho)e^{\mathrm{i}\varphi})/2 + \mathrm{i}q_{o||}\sin\theta_o' J_m(q_{o\perp}\rho),$$
$$e_{o2} = q_{o\perp}\cos\theta_o'(J_{m-1}(q_{o\perp}\rho)e^{-\mathrm{i}\varphi} - J_{m+1}(q_{o\perp}\rho)e^{\mathrm{i}\varphi})/2\sqrt{1+a^4\mathrm{tg}^4\theta_o'} +$$
$$+ q_{o\perp}\sin^2\theta_o'(J_{m+1}(q_{o\perp}\rho)e^{\mathrm{i}\varphi} - J_{m-1}(q_{o\perp}\rho)e^{-\mathrm{i}\varphi})/2,$$
$$e_{o3} = q_{o\perp}\cos\theta_o'(J_{m-1}(q_{o\perp}\rho)e^{-\mathrm{i}\varphi} - J_{m+1}(q_{o\perp}\rho)e^{\mathrm{i}\varphi})a^2\mathrm{tg}\theta_o'/2\sqrt{1+a^4\mathrm{tg}^4\theta_o'} +$$
$$+ q_{o\perp}\sin 2\theta_o'(J_{m+1}(q_{o\perp}\rho)e^{\mathrm{i}\varphi} - J_{m-1}(q_{o\perp}\rho)e^{-\mathrm{i}\varphi})/4,$$
$$e_{e1} = k^{-1}n_o^{-2}n_e^{-2}\{[aq_{e\perp}^2\sin^2\theta_e'(J_{m-2}(q_{e\perp}\rho)e^{-\mathrm{i}2\varphi'} - J_{m+2}(q_{e\perp}\rho)e^{\mathrm{i}2\varphi'})/4]$$
$$+ [aq_{e\perp}q_{e||}\cos\theta_e'((J_{m+1}(q_{e\perp}\rho)e^{\mathrm{i}\varphi'} - J_{m-1}(q_{e\perp}\rho)e^{-\mathrm{i}\varphi'})/2]\},$$
$$e_{e2} = \mathrm{i}k^{-1}n_o^{-2}n_e^{-2}\{[aq_{e\perp}^2\sin 2\theta_e'(J_{m+2}(q_{e\perp}\rho)e^{\mathrm{i}2\varphi'} + J_{m-2}(q_{e\perp}\rho)e^{-\mathrm{i}2\varphi'})/4] +$$
$$- [aq_{e\perp}q_{e||}\cos 2\theta_e'((J_{m-1}(q_{e\perp}\rho)e^{-\mathrm{i}\varphi'} + J_{m+1}(q_{e\perp}\rho)e^{-\mathrm{i}\varphi'})/2] -$$
$$- a(k^2 - 3q_{e\perp}^2/2)\sin 2\theta_e' J_m(q_{e\perp}\rho)\},$$
$$e_{e3} = \mathrm{i}k^{-1}n_o^{-4}\{[-q_{e\perp}^2\sin 2\theta_e'(J_{m+2}(q_{e\perp}\rho)e^{\mathrm{i}2\varphi'} + J_{m-2}(q_{e\perp}\rho)e^{-\mathrm{i}2\varphi'})/4] +$$
$$+ [q_{e\perp}q_{e||}\sin 2\theta_e'((J_{m-1}(q_{e\perp}\rho)e^{-\mathrm{i}\varphi'} + J_{m+1}(q_{e\perp}\rho)e^{-\mathrm{i}\varphi'})/2] +$$
$$+ q_{e\perp}^2(1+\cos^2\theta_e' + q_{e||}^2\sin^2\theta_e')J_m(q_{e\perp}\rho)\}.$$

Here $\vec{e}_1, \vec{e}_2, \vec{e}_3$ is the single vector of crystallographic coordinate X_1, X_2, X_3; ρ, φ are the cylindrical coordinates; $J_p(\xi)$ ($p = m$, $m\pm1$, $m\pm2$) is the Bessel functions.

Figure 1 presents the dependences of the diffraction efficiency η on the transverse phase mismatch parameter $q_n=\Delta q/q_{o\perp}$ ($\Delta q=|q_{e\perp}-q_{o\perp}|$) for different BLB mode value.

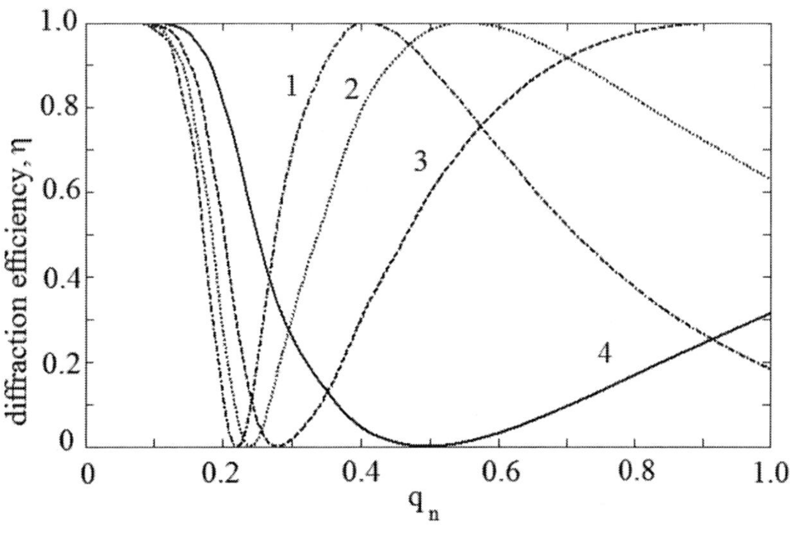

Figure 1. Dependence of η on q_n for mode $m = 0(1)$, $1(2)$, $2(3)$, $3(4)$ (TeO_2 crystal is considered; $\alpha = 10^0$, $\gamma_o = \gamma_e = 0.5^0$, $R_m = 6$ mm, $P_a = 24$ mW, $f = 159$ MHz, $\lambda_0 = 0.75$ μm, $l_1 = 1$ cm, $l_2 = 0.5$ cm).

Under longitudinal and transversal conditions the diffraction efficiency reaches maximal value of $\eta = 1$. For US intensities reaching the same values for longitudinal phase synchronism, the diffraction efficiency became lower till zero value and then we see enhancement of it till the maximal value. With the use of the relations (1)-(3) the physical characteristic of the acousto-optical filtration in the optical spectral region 0.4-1.1 μm has been studied. The central optical wavelength would be $\lambda_0 = 0.75$ μm. The index of refraction is $n_o = 2.24$, $n_e = 2.237$. For the length of a piezoelectric transducer $l_1 = 1$ cm along a direction $X||[110]$ and the width of a piezoelectric transducer $l_2 = 0.5$ cm the maximal diffraction efficiency is $\eta = 1$ for $P_a = 24$ mW.

Figure 2 presents the dependence of the diffraction efficiency η on the transmission bandwidth $\Delta\lambda$ of the AO device, in the region of 0.4-1.1 nm.

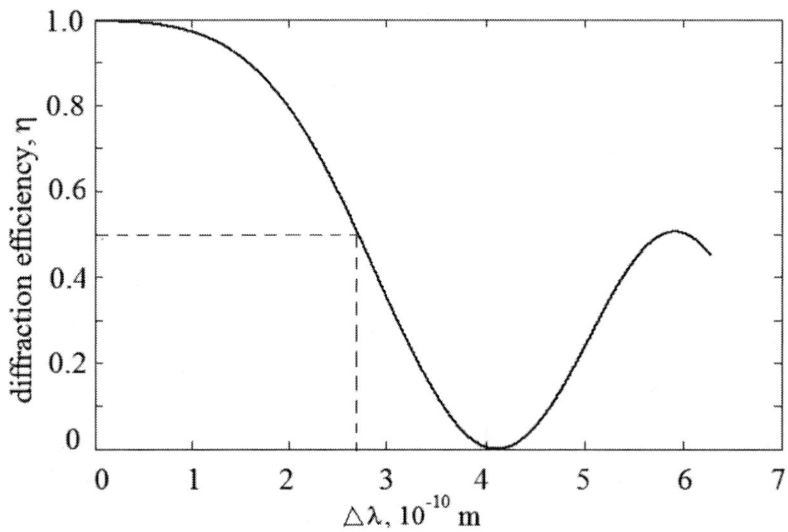

Figure 2. Dependence of η on Δ under condition of cross phase synchronism ($q_n = 0$) for different BLB modes $m = 0 \div 30$ (TeO_2 crystal is considered; $\alpha = 10^0$, $\gamma_o = \gamma_e = 0.5^0$, $R_m = 6$ mm, $P_a = 24$ mW, $f = 159$ MHz, $\lambda_0 = 0.75$ μm, $l_1 = 1$ cm, $l_2 = 0,5$ cm).

The full width of the filter passband has been calculated at the level of 50% from maximal value of diffraction efficiency. In the conditions of transverse phase synchronism filter passband is $\Delta\lambda_{1/2} = 0.52$ nm. Under maximal diffraction efficiency and absence of transversal phase-matching the acoustooptical filtration was not realized during the higher level of sideband maxima in diffracted orders.

Let BLB with τ ellipticity and azimuth polarization ψ (orientation of a big polarization axis of the ellipse relative to the diffraction plane) incident on the boundary of the crystal. The crystal active BLB are represented as own elliptically polarization waves with ellipticity parameter τ_o and τ_ε. Then, taking into account both *e-o* and *o-e* anisotropic diffraction, we obtain the following expression for diffraction efficiency:

$$\eta = \frac{1}{1+\tau^2}\left[(\cos^2\psi + \tau^2\sin^2\psi)\sin^2(g_m\chi_{oe}) + (\sin^2\psi + \tau^2\cos^2\psi)\sin^2(g_m\chi_{eo})\right], \quad (4)$$

where

$$\chi_{oe} = \Delta\left[(\tau_o\tau_e\cos\varphi_2 + \cos\varphi_1)\Delta_1 + (\sin\varphi_1 + \tau_o\tau_e\sin\varphi_2)\Delta_2\right],\ \Delta_1 = [n_e^4(p_{11} - p_{12})]/\tau_{oe},$$
$$\Delta_2 = (n_o^2 n_e^2 p_{44})/2\tau_{oe},\ \Delta = \pi U_{12}/\left[2\lambda_0(n_o n_e\cos\varphi_1\cos\varphi_2)^{1/2}\right]$$

Here strain tensor component $U_{12} = |\nabla_2 U| = \left|2I_a / \sigma \upsilon^3\right|^{1/2}$, $\tau_{oe} = \left[(1+\tau_o^2)(1+\tau_e^2)\right]^{1/2}$; p_{11}, p_{12}, p_{44} photoelastic constants of the crystal.

3. Experimental results and discussions

For an AO cell on TeO_2 crystal, BLB diffraction with a wavelength of 632.8 nm was considered during the propagation of beams near the optical axis of the crystal. Here linearly polarization of Gaussian beam from *He-Ne* laser with the power of 25 by means of the plate turned into circularly polarized first, and then using a conical lens (an axicon with an angle at the base of 2 degrees) a zero-order Bessel light beam was formed. Film polarizer at the entrance to the crystal sets up necessary states of linear polarization of incident (ψ) Bessel beam. The intensity distribution of the diffracted BLB in the far-field zone is shown in figure 3.

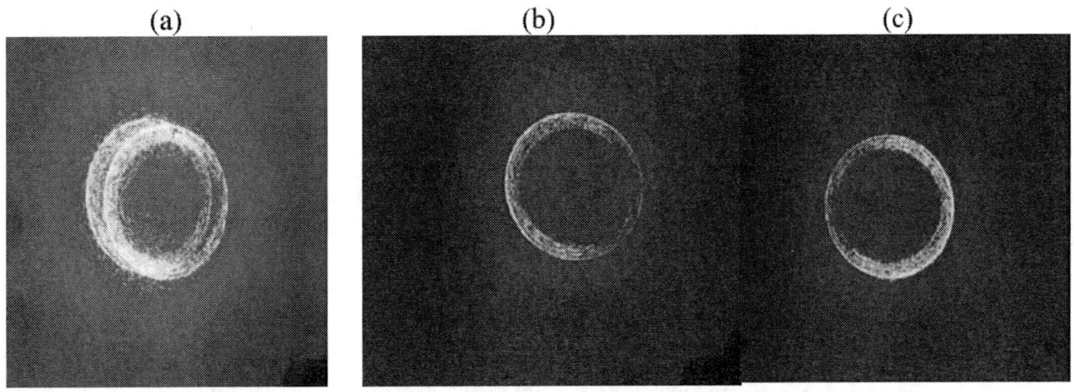

Figure 3. Ring intensity distribution of two diffracted BLBs in the far field when a light beam is incident at a small angle φ_I to the optical axis Z of a TeO_2 gyrotropic uniaxial crystal: (a) the total field of the diffracted light beam at the crystal output; (b), (c) the diffracted field after the polarizer is installed in front of the camera in two mutually orthogonal positions, respectively.

It can be seen that in the far-field zone the diffracted beam is spatially divided into two annular fields, corresponding to *o-e* and *e-o* AO interactions. The ring character of the fields indicates that the diffracted field is a superposition of two Bessel beams with the angles of the cones coinciding with the angle of the cone of the incident BLB. The location of the film polarizer directly behind the exit face of the AO cell showed that the ring fields observed in the far-field zone have elliptical polarization orthogonal to each other (figure 3) in this case, the ring intensity distribution corresponds to *e-o* diffraction in figure 3 (b), and diffraction in figure 3 (c). This fact confirms the simultaneous implementation of two processes *(e-o* and *o-e)* of the interaction of Bessel beams with approximately equal diffraction efficiency. The diffraction efficiency η from the polarization azimuth ψ of a linearly polarized incident BLB was investigated. Figure 4shows the experimental dependence of the diffraction efficiency η of the polarization azimuth ψ of incident linearly polarized Bessel beam on ultrasound at different intensities I_a.

Acoustic power was estimated considering loss power in coordination with the piezoelectric transducer and generator losses on gluing with sound-line. The maximum width h of the piezoelectric transducer used to excite an ultrasonic beam with divergence 2α satisfies the equation $h \approx \Lambda_0 / 2\alpha$, where Λ_0 is the length of the ultrasonic wave. For the considered geometry of the AO interaction and the central frequency f_0 = 30 MHz of the ultrasonic wave in a TeO_2 crystal, the value $h \approx 11$ mm. In the experiment a piezoelectric transducer with a width of h_0 = 8.5 mm was used. Therefore, an acoustic beam with the divergence of $\Delta\Theta \approx 0.26^0$that exceeds the angle value of $2\alpha \approx 0.22^0$ Consequently, adopted in the calculation model of the acoustic beam with low divergence is adequate to experiment and allows theoretical description of the complex phenomenon of diffraction in gyrotropic uniaxial crystal as a superposition of two processes of anisotropic scattering of elliptically polarizing BLB in terms of Bragg synchronism.

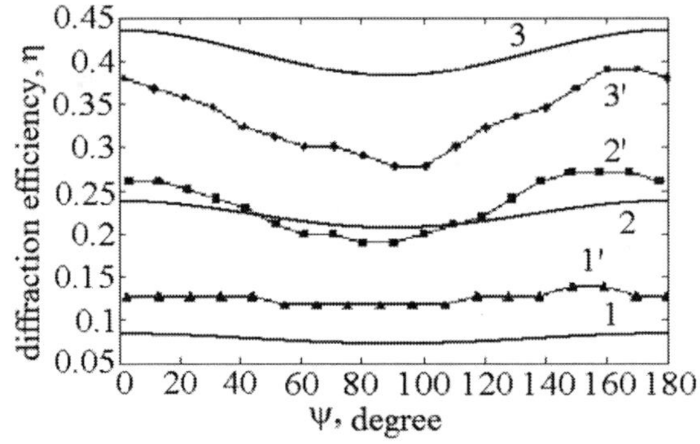

Figure 4. The dependence of the diffraction efficiency η on the azimuth of the incident light polarization ψ at different intensities of ultrasound I_a: 1-10, 2-30 , 3-60 mW/cm^2 ($l = 8.5$ mm , $n_o = 2.26$; $n_e = 2.1$; $\varphi_1 = 0.7^0$, $\varphi_2 = 0.4^0$, $\gamma = 0.5^0$, $\tau = 0$; $\alpha = 0{,}11^0$, $f_0 = 30$ MHz, $p_{11} = 0.0074$, $p_{12} = 0.87$, $p_{44} = -0.17$; $\tau_o = 0.92$, $\tau_\varepsilon = 0.9$; $\sigma = 5.72$ kg/cm^3, $R_B = 1$ mm, $g_0 = 1$, TeO_2; 1-3 is the theory, 1'- 3' is the experiment).

From figure 4 it follows that with these parameters of the AO interaction there is a dependence of the diffraction efficiency of the BLB on the polarization state of the incident light. The change in diffraction efficiency due to the polarization of the incident light is ~25%. By reducing the triple level supplied from the signal generator the polarization independence of diffraction (curve 1) has been obtained. In this case, the dependence of diffraction efficiency on the polarization state of the incident light is in the level of error and is less than 4%. From the above figure 4 dependence implies that optimum conditions for polarization-independent diffraction can be realized due to the choice of a suitable length of AO interaction and intensity of ultrasonic waves. It should be noted that bandwidth investigated for anisotropic AO diffraction was $\Delta f \sim f_0$, where f_0 is the center frequency of the ultrasound source. On the basis of formula (4) dependence of diffraction efficiencies η on the azimuth of the polarization of the incident BLB was theoretically investigated. Figure 4 shows the calculated dependence of the diffraction efficiency η on polarization azimuth ψ of linearly polarized incident Bessel beam of ultrasound at different intensities I_a. The results obtained at low ultrasound intensities are in good agreement with the experimental data presented in figure 4 (curves 1, 2 - theory and curves 1', 2' - experiment).

It should be noted that polarization-independent AO BLB diffraction can be implemented in other gyrotropic uniaxial crystals (for example, in quartz and tellurium [1], [4], [5]) based on perspective geometries of AO interaction. Polarization-independent acousto-optical modulation of the nearly-non-diffracted BLB of infrared spectral range, traveling at the vicinity of tellurium optical axis on the slowly shear ultrasonic waves is investigated. It is shown that under any ultrasonic intensity for *XZ*-cut of tellurium crystals less dependence of diffraction efficiency of BLB from the light polarization is taken place. Polarization-independent modulation of BLB under the Bragg diffraction by ultrasound in *YZ* crystal plane on the shear ultrasonic wave polarized under -63^0 to *Z* axis in a *YX* plane is stated.

4. Conclusion

With the use of the method of overlap integrals it is shown that independently on the order of vortex Bessel light beams in the conditions of transverse phase-matching of diffracted waves in the range of optical spectrum of 0.4-1.1 nm in TeO_2 crystals the bandwidth of transmission of ~0.52 nm is reached due to detuning from Bragg synchronism. The features of AO diffraction of Bessel light beams propagating at small angles to the optical axis of a gyrotropic uniaxial crystal were theoretically and experimentally investigated. The distinctive feature of the polarization-independent modulation is the independence of the diffraction efficiency on the polarization state of the incident BLB. Such polarization-independent AO devices for Bessel beams are promising for applications in laser technology (in particular for laser cutting and welding), for probing, absorbing, and scattering media to great depth, laser location, high-resolution microscopy and other areas, as well as for solving problems in laser control and diagnostics, micro- and nanotechnology.

References

[1] Balakshy V I, Parygin V N and Chirkov L E 1985 *Physical bases of an acoustooptics* (Moskov: Radio and communication)
[2] Belyi V N, Kulak G V, Krokh G V and Shakin O V 2014 *Opt. and Spectroscopy* **121** 458-65
[3] Belyi V N, Kazak N S, Khilo P A, Petrova E and Khilo N A 2015 *Universal Journal of Physics and Application* **9** 220-24
[4] Denes J L, Gottlib M S and Kaminsky B 1989 *Opt. Eng.* **37** 1262-67
[5] Gupta N, Voloshinov V B, Knyazev G A, and Kulakova L A 2012 *Journal of Optics* **14** 1-9

Modeling pictures of the flow past rough and smooth surfaces in the program comsol multiphysics

Win Thu and D V Ilin

National research University Moscow power engineering Institute „MPEI", Krasnokazarmennaya street., 14, Moscow, 111250 Russia,

E-mail: winthu.thu51@gmail.com

Abstract. The aim of this work was to obtain simulated flow patterns of rough surfaces by aerodynamic flow at different speeds for comparison with the results obtained by anemometry from images of particles in a laboratory wind tunnel. The results obtained by the program Comsol Multiphysics, which is used to simulate complex high-turbulent flows characteristic of gas flows in aircraft engines and gas turbine plants.

1. Introduction

When describing the gas flow near a solid surface, the models of interaction of gas particles with the surface play a decisive role. Any real surface for a falling gas particle is rough. Therefore, one of the most important factors that need to be taken into account in the process of interaction is roughness [1, 2]. Roughness as a property of a real non-smooth surface is manifested through a set of individual irregularities that form the microstructure of the surface. It is associated with such surface properties as microhardness, friction, Aero - and hydrodynamic resistance, etc. the Problem of boundary conditions for the gas-dynamic task often depends on the ignorance of the real surface structure, which interact with the gas molecules. At the same time, experiments show that the surface roughness at the atomic level and above has a significant effect on the energy and momentum transfer during the flow of solids by gas flows.

To date, both analytical methods – series expansions, approximation theory, asymptotic expansions – and numerical methods (calculation of multiple integrals) are used to study the interaction of a gas with a rough surface.

In this work, a numerical simulation of the flow pattern of rough surfaces by an aerodynamic flow at different speeds is carried out for comparison with the results obtained by anemometry from images of particles (AIP). Anemometry by images of particles (AIP) enables contactless measuring of the velocity of the motion of gaseous, liquid and solid media containing light-scattering heterogeneities. This method is now widely used in scientific research and technical applications. Anemometry by images of particles (AIP) can solve a wide range of problems: from the study of slow directional movements in capillaries and living cells to remote measurements of the turbulent velocity of gas flows in supersonic tubes and wind speed in the atmosphere. The measured velocities range from several μm/s to several km/s [3].

In the application of optical methods of flow diagnostics to specific tasks often raises the question of confirmation of the correctness of the results. The solution of the problem is obtained using the comsol Multiphysics software package, which uses the finite element method for the numerical solution of systems of partial differential equations [4].

The program contains various modeling and simulation tools, which include tools for geometric constructions, a finite element mesh generator, various solvers for linear and nonlinear problems, and post-processing tools.

2. The calculation of the modeling pictures of the flow around rough surfaces using the program comsol multiphysics

To solve the problem of air flow along the rough surface (Rz = 160 μm) in Comsol Multiphysics, a physical interface (module) of the turbulent flow was used, taking into account the properties of the viscous boundary layer near the solid wall. Of the many models of turbulent flow offered by Comsol Multiphysics, the v2-f model was chosen. This model is low-Reynolds and can be used to calculate the flow over the entire thickness of the boundary layer, including the viscous sublayer and buffer layer.

The equations solved by the Turbulent Flow interface, v2-f, are the Reynolds-averaged Navier-Stokes (RANS) equations for momentum conservation and the continuity equation for mass conservation [5-7].

The Navier-Stokes Equations – it is a system of differential equations that describes the motion of a viscous Newtonian fluid or gas (1 and 2):

$$\begin{cases} \rho \dfrac{d\mathbf{u}}{dt} + \rho(\mathbf{u}\cdot\nabla)\mathbf{u} = \nabla\cdot\left[-p\mathbf{I} + \mathbf{K}\right] + \mathbf{F}, \\ \rho\nabla\cdot(\mathbf{u}) = 0, \end{cases} \qquad (1)$$

where: ρ – fluid density, t – time, p – pressure, **u** – fluid velocity, ∇ – Nabla operator, **F** – vector field of mass forces, **I** – intensity of turbulent flow and vector **K** is calculated as the product of the sum of viscosities (dynamic and turbulent) and velocity operators:

$$\mathbf{K} = (\mu + \mu_T)(\nabla\mathbf{u} + (\nabla\mathbf{u})^T), \qquad (2)$$

where μ – dynamic viscosity, μ_T – turbulent dynamic viscosity.

The first equation (1) in the system is the equation of motion itself. In the left part are the product of the density of the corresponding acceleration. In the right part – the product of density on the forces of pressure and internal friction. The second equation (2) is the continuity equation. Its physical meaning is the preservation of mass for the flow of liquid.

The expression for **K** is nothing but a substantial derivative (also called a complete derivative). It shows how the acceleration of a material point that moves in a stationary gas medium changes. It shows the change in the properties of a point over time, as if it were stationary. A term $\nabla\mathbf{u}$ is called a convective derivative describing the evolution of properties at a fixed point due to the fact that a gas medium flows through it at a speed.

It should be noted that the system of Navier-Stokes equations gives very accurate solutions if the laminar gas flow is considered, but in turbulent flow the equations are very sensitive to the values of the coefficients: a change in the Reynolds number by 0.05% can lead to a radically different result.

For simplicity, consider an example in which the gas flow rate is considered constant, etc. $d\mathbf{u}/dt=0$. In this case, the system of Navier Stokes equations takes the form (3):

$$\begin{cases} \rho(\mathbf{u}\cdot\nabla)\mathbf{u} = \nabla\cdot\left[-p\mathbf{I}+\mathbf{K}\right]+\mathbf{F}, \\ \rho\nabla\cdot(\mathbf{u}) = 0. \end{cases} \tag{3}$$

For the v2-f model, the turbulent viscosity is based on velocity fluctuations normal to the current lines.This makes it possible to model the anisotropy of turbulence and to separate the effects with a low Reynolds number and take into account the effect of the walls. The nonlocal influence of pressure fluctuations on turbulent fields is taken into account by applying elliptic relaxation to the pressure deformation term [8]. In COMSOL Multiphysics, this effect is formulated for normalized velocity oscillations $\xi = \frac{u^2}{k}$ and the elliptic mixing function α, which is used to combine the effects associated with the walls in the far field. Thus, the complete model for compressible gas flow has for our case the form (4):

$$\begin{cases} \rho(\mathbf{u}\cdot\nabla)k = \nabla\cdot\left[\left(\mu+\frac{\mu_T}{\sigma_K}\right)\nabla k\right]+P_K-\rho\varepsilon, \\ \rho(\mathbf{u}\cdot\nabla)\varepsilon = \nabla\cdot\left[\left(\mu+\frac{\mu_T}{\sigma_\varepsilon}\right)\nabla\varepsilon\right]+\frac{1}{\tau}\left(C'_{\varepsilon1}(\zeta,\alpha)\right)P_K-C'_{\varepsilon2}(k,\varepsilon,\alpha)\rho\varepsilon), \\ \rho(\mathbf{u}\cdot\nabla)\zeta = \nabla\cdot\left[\left(\mu+\frac{\mu_T}{\sigma_K}\right)\nabla\zeta\right]+\frac{2}{k}\left(\alpha^3\mu+\frac{\mu_T}{\sigma_K}\right)\nabla k\cdot\nabla\zeta+(1+\alpha^3)f_w+\alpha^3 f_h-\frac{\zeta}{k}P_K, \end{cases} \tag{4}$$

where k – turbulent kinetic energy, σ – stresses in fluid flow, ε – turbulent dispersion rate. Since the equation of turbulent kinetic energy k and turbulent scattering velocity ε is derived directly analytically, it uses a number of auxiliary coefficients that provide the required dimension of the input quantities, including these coefficients $C_{\varepsilon1}$ = 1,5, $C_{\varepsilon2}$ = 1,9, C_μ = 0,09, σ_k = 1,4. This model introduces two important concepts – generation $P_k = \mu_T\left[\nabla u : \left(\nabla u + (\nabla u)^T\right)\right]$ and dissipation ε. The physical meaning of generation of turbulence P_k is in the generation of new vortices and ripples that make up the turbulence. Dissipation ε, on the contrary, is the dispersion of large vortices into smaller ones, which leads to averaging of the flow and reducing turbulence. The elliptic mixing function α is related to the turbulence length $L = C_L \max\left[\frac{k^{3/2}}{\varepsilon}, C_\eta\left(\frac{\nu^3}{\varepsilon}\right)^{1/4}\right]$, according to the expression 5:

$$\alpha - L^2\nabla^2\alpha = 1, \tag{5}$$

where $\nu = \mu/\rho$ - kinematic viscosity of the gas.

Turbulent dynamic viscosity μ_T in equation (4) and parameters and parameters f_w and f_h in equation (4) are defined according to expressions (6-8):

$$\mu_T = \rho C_\mu k\zeta\tau, \tag{6}$$

$$f_w = -\zeta\frac{\varepsilon}{k}, \tag{7}$$

Journal of Physics: Conference Series **1421** (2019) 012020 doi:10.1088/1742-6596/1421/1/012020

$$f_h = -\frac{1}{\tau}\left(C_1 - 1 + C_2\frac{P_K}{\rho\varepsilon}\right)\left(\zeta - \frac{2}{3}\right), \tag{8}$$

where $\tau = \max\left[\frac{k^{3/2}}{\varepsilon}, C_\tau\sqrt{\frac{\nu}{\varepsilon}}\right]$.

The stresses in the gas flow in the wall area σ_w and distance l_w are found from the solution of boundary conditions for the flow (9-10):

$$\nabla G \cdot \nabla G + \sigma_w G(\nabla \cdot \nabla G) = (1 + 2\sigma_w)G^4, \tag{9}$$

$$l_w = \frac{1}{G} - \frac{l_{ref}}{2}, \tag{10}$$

where G – inverse distance to the wall, l_{ref} – the distance from the wall at which there is a linear change in the air velocity (viscous sublayer).

3. Computational models

In figure 1 shows a General view of the geometry of the computational domain. The air flow is moving from left to right. The flow rate varies in steps from 0.1 and 0.5 m/s to 5.5 m/s in increments of 0.5 m/s. the lower boundary of the computational domain is represented as a curved periodic structure, repeating the irregularities of the test sample of the grinding skin.

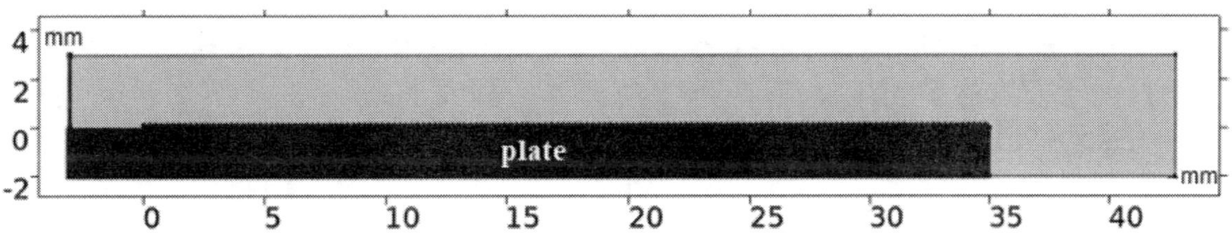

Figure 1. General view of the computational domain of the model.

Now let's move on to the construction of the grid. The most widely used in computational practice are piecewise linear finite elements of triangular shape (figure 2.), which correspond to the approximation of the approximate solution on each triangle by a linear function. In the convection and diffusion problems we obtain analogs of schemes with central-difference approximation of convective terms. In addition, triangular-shaped meshes allow to obtain a more accurate and fast-converging solution on complex areas, such as on the roughness considered in this problem. Special attention was paid to the construction of a dense grid near the solid wall, which used a special section of the grid generator settings Boundary Layers. The result of meshing in the boundary layer is shown in figure 3.

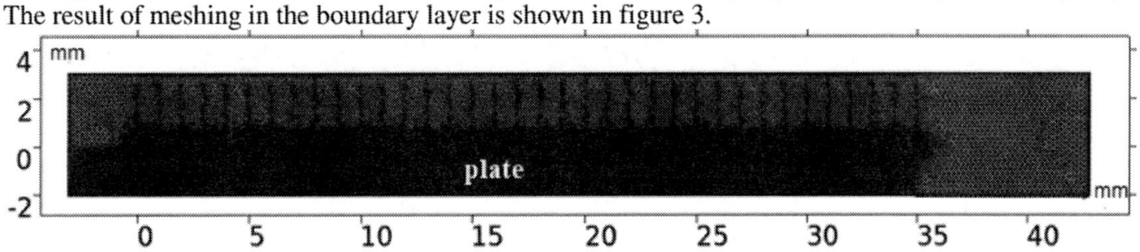

Figure 2. Finite element mesh.

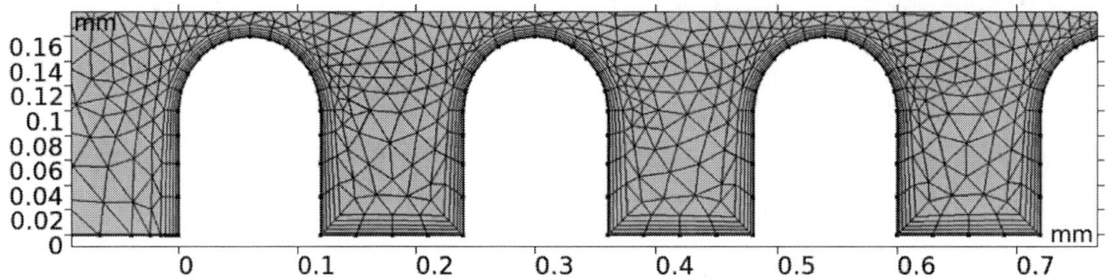

Figure 3. Finite element mesh in boundary layer.

4. The results are simulated pictures of the flow around rough surfaces and smooth surfaces by an aerodynamic flow with different speeds

As a result of the calculation, the distribution pattern of the air flow velocity along the curved surface is obtained. In figures 4-7 shows the final characteristic velocity field for different air flow rates at the input boundary of the model.

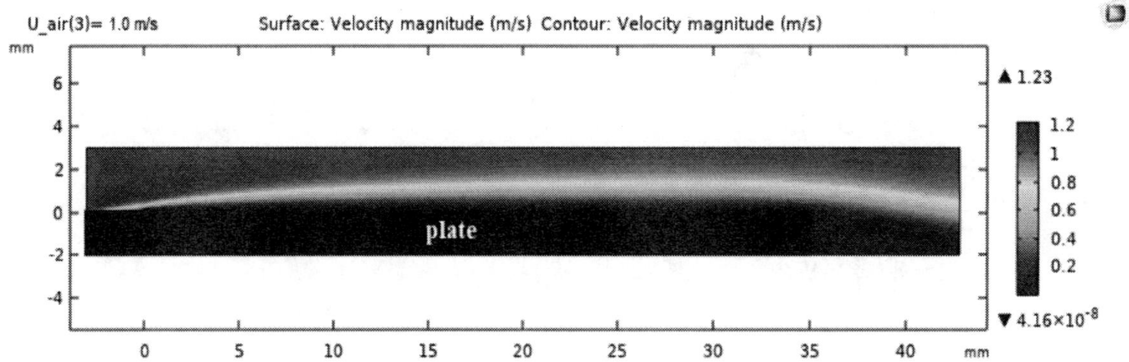

(a)

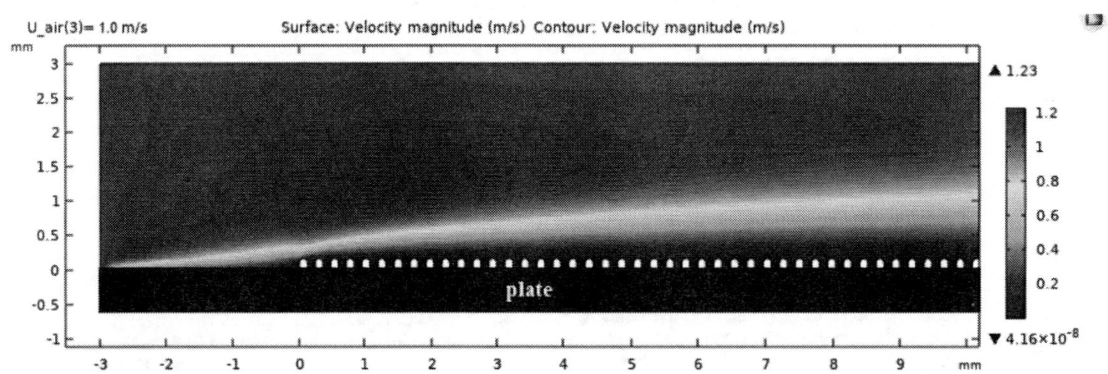

(b)

Figure 4. (a) Flow velocity field at the input flow velocity of 1.0 m/s in rough surface. (b) The input flow velocity field with a flow velocity of 1.0 m/s brings the viewer closer to the rough surface for comparison with 1.0 m/s smooth surface.

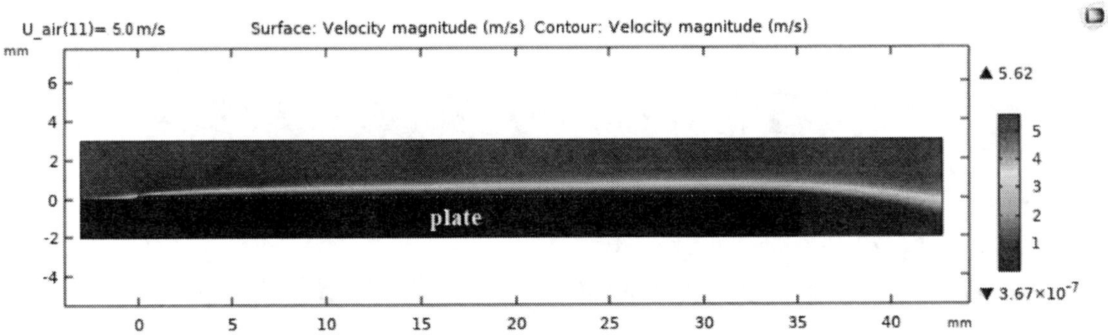

(c)

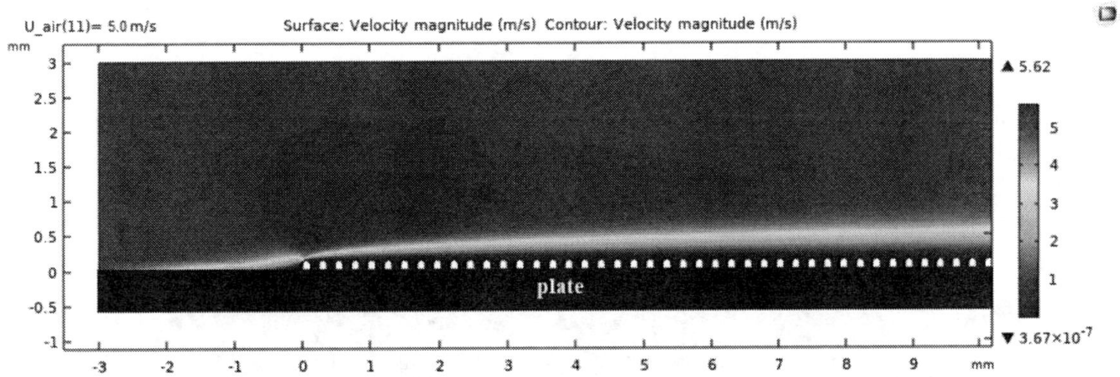

(d)

Figure 5. (c) Flow velocity field at the input flow velocity of 5.0 m/s in rough surface. (d) The input flow velocity field with a flow velocity of 1.0 m/s brings the viewer closer to the rough surface for comparison with 5.0 m/s smooth surface.

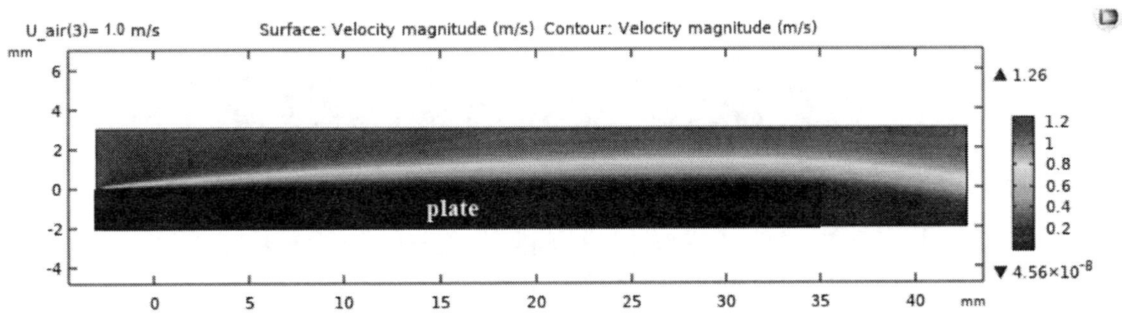

(e)

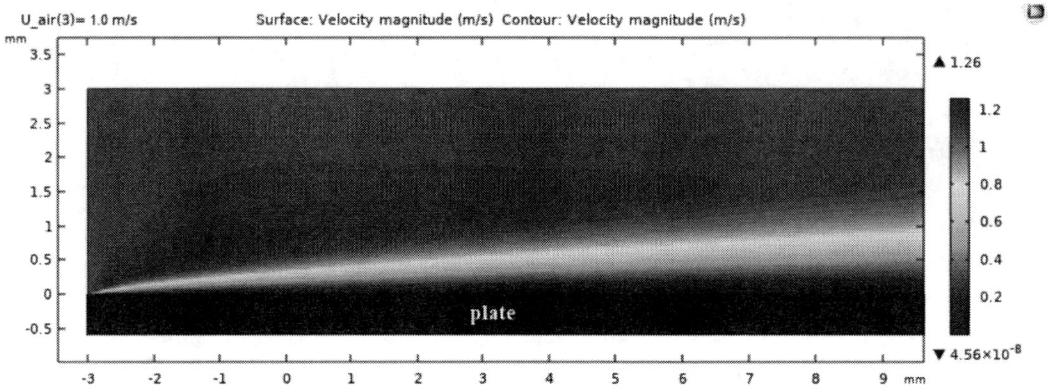

(f)

Figure 6. (e) Flow velocity field at the input flow velocity of 1.0 m/s in smooth surface. (f) The input flow velocity field with a flow velocity of 1.0 m/s brings the viewer closer to the smooth surface for comparison with 1.0 m/s rough surface.

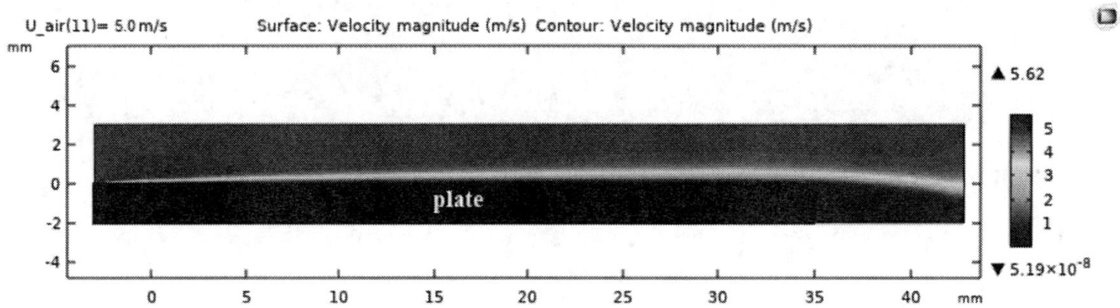

(g)

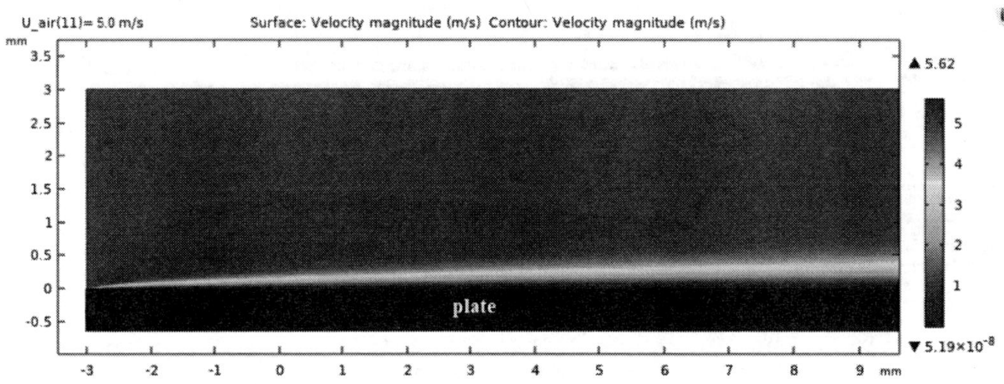

(h)

Figure 7. (g) Flow velocity field at the input flow velocity of 5.0 m/s in smooth surface. (h) The input flow velocity field with a flow velocity of 5.0 m/s brings the viewer closer to the smooth surface for comparison with 1.0 m/s rough surface.

With an increase in the input flow rate, a decrease in the transition zone width between the lower region with a minimum flow rate near the curved surface and the upper region with a maximum flow rate and the developed turbulent flow region is observed. It is obvious that this effect is due to the increase in pressure in the upper region of the flow.

For the convenience of assessing the impact of the input flow rate on the picture of the air flow near the solid curved wall, a graph was constructed showing the value of the flow rate at a height of 0.25 mm (figure 8 (i)) above the lowest point of the surface under study and a smooth surface (figure 8 (j)).

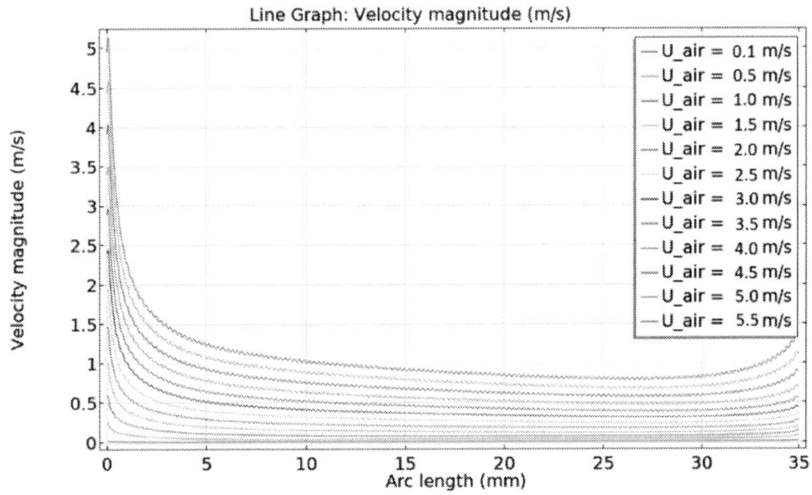

(i)

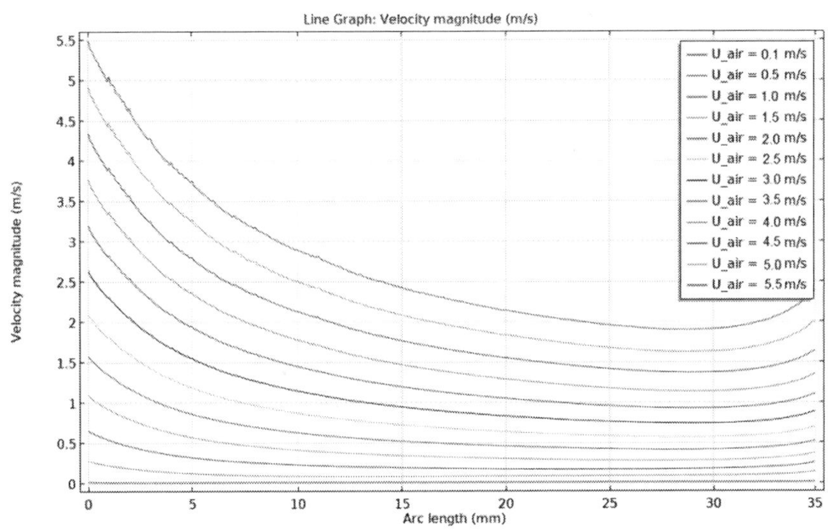

(j)

Figure 8. (i) Influence of the input flow rate on the nature of the envelope flow in rough surface. (j) Influence of the input flow rate on the nature of the envelope flow in smooth surface.

The result is shown in the figure 8. As noted earlier, with the increase in the flow rate, the pressure in the near-wall lower region increases and the flow of irregularities becomes noticeable with the repetition of the nature of the relief. When comparing the results for smooth and rough surfaces it can be seen that the flow of rough surfaces velocity is very much reduced.

5. Conclusion

The paper presents the calculation and modeling of flow patterns of rough surfaces in the program Comsol multiphysics at different speeds from 0.1 and 0.5 m/s to 5.5 m/s in increments of 0.5 m/s. As a result of the calculation, the distribution pattern of the air flow velocity along the curved surface under study is obtained.

With an increase in the input flow rate, a decrease in the transition zone width between the lower region with a minimum flow rate near the surface and the upper region with a maximum flow rate and the developed turbulent flow region is observed.

References

[1] Miroshin R N 1981 *The intersection curves of the Gauss processes* (publishing House Leningrad state University named after Zhdanov A A) 212
[2] Khalidov I A 2014 The use of poly-Gaussian stochastic processes to the modeling of flow over rough surfaces flow of rarefied gas *Vestnik St. Petersburg University* **1(59) 3** 428-37
[3] Rinkevichius B S 1978 *Laser anemometry* (Moscow: Energy) 160
[4] Krasnikov G E, Nagornov O V and Starostin N V 2012 *Modeling of physical processes using COMSOL Multiphysics* (Moscow: national research nuclear University MEPhI) p 184
[5] Kharitonov V P 2012 *Fundamental equations of fluid mechanics* (MGTU пamed. N. Eh. Bauman) 65
[6] Durbin P A 1993 Application of a near-wall turbulence model to boundary layers and heat transfer *Int. J. Heat and Fluid Flow* **14(4)** 316-23
[7] Frick P G 1998 *Turbulence: models and approaches* (Course of lectures Part 1 Perm State technical University Perm) 108
[8] Smirnov E M and Habakuk A V 2010 *The flow of viscous fluid and turbulence model: the methods of calculation of turbulent flows* (lectures Saint-Petersburg state Polytechnical University) 127

Journal of Physics: Conference Series **1421** (2019) 012021 doi:10.1088/1742-6596/1421/1/012021

Compact interferometer for precise shape testing of large-size convexaspherical mirrors

A E Gavlina[1], D A Novikov[2] and M V Sergeeva[1]

[1]Laboratory of Acousto-optic Spectroscopy, Scientific Technological Center of Unique Instrumentation RAS (STC UI RAS), Butlerova 15, Moscow, 117342 Russia;
[2]All-Russian research institute of metrological service, Ozernaya 46, Moscow, 119361 Russia

E-mail: gavlinaae@gmail.com, dnovikov@vniims.ru, applesin17@mail.ru

Abstract. A model of an interferometer for testing of convex aspherical mirrors up to 250 mm in diameter based on the scheme of orthogonal rays is presented. An algorithm for interferogram processing has been developed. The results of measurements of the profile of a convex spherical mirror on the presented interferometer are presented.

1. Introduction

The development of testing methods for convex aspherical and spherical mirrors of large diameter is an urgent task today because when testing convex mirrors by traditional methods, it is necessary to use precision optical elements with a diameter larger than the diameter of the test surface [1-3].

Professor D.T. Puryaev proposed the basic scheme of orthogonal rays for testing convex aspherical and spherical optical surfaces in 1982 (figure 1). The scheme of orthogonal rays is a scheme of illumination by a collimating beam of a testing surface, which is directed along the normal to its axis of symmetry. An interference method for testing convex optical surfaces based on an orthogonal ray scheme was also proposed by D.T. Puryaev [4]. Therefore, the proposed test method is applicable for small mirrors only.

We offer an improved interference method for testing convex spherical and aspherical optical parts, which is suitable for testing parts, both small and large diameters [5-7]. We analyze the entire tangential section of the interference pattern, which makes it possible to test the shape of the tangential profile of the surface, in contrast to the method proposed by D.T. Puryaev. The dimensions of the analyzed tangential cross section of the interference pattern are smaller than the diameter of the test mirrors; therefore, the method is applicable for testing mirrors of large diameter.

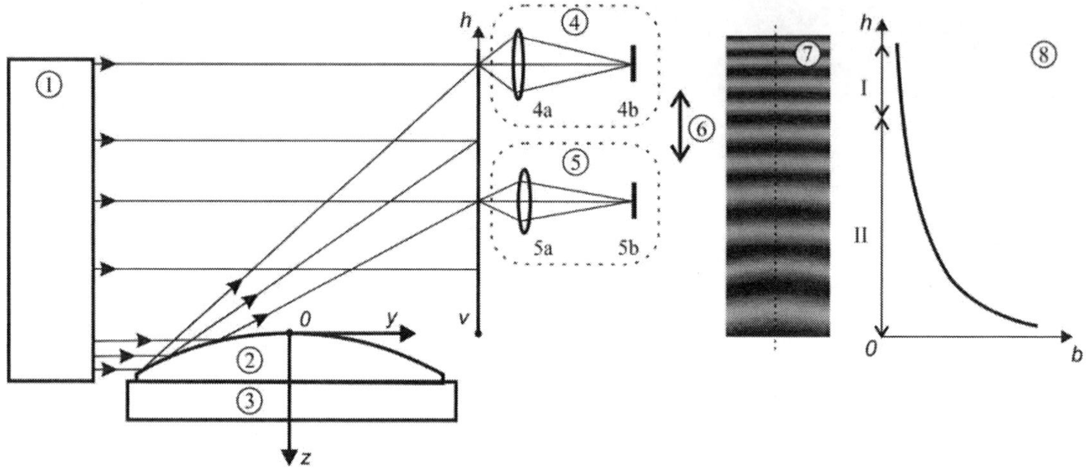

Figure 1. The layout of improved interference method for testing convex mirrors.

2. The experimental setup

The scheme of orthogonal rays has been implemented for the first time in practice in the STC UI RAS. The optical scheme of the interferometer is shown in the figure 1. The collimated beam formed by a collimator (1) directed onto tested mirror (2). The reference and object beams interfere in the plane of the object of the micro-objectives (4a), (5a). The images of the interferogram are projected onto video cameras (4b), (5b). To register the entire interference pattern, we move the microscopes (4) and (5) consisting of a micro-objective and a video camera along the registration plane using a linear translator (6). The information contained in the registered interference pattern (7) contains information about the profile of the test surface in the tangential section. To check the entire surface, we rotate it around the axis of symmetry 0z of the tasting mirror 0z using the turntable (3).

The width of the interference fringes decreases with increasing coordinate in the registration plane. The most frequent interference bands are formed on the edge of the interference pattern. For this reason, the layout for testing convex mirrors contains two micro-objectives with different magnifications.

Figure 1 shows the plot 8 of the width of the interference band (b) versus the coordinates of the centers of maximum of the interference band (h). The applications of micro-objects (4a) and (5a) are also shown in figure 8. Micro-objective (4a) with high magnification $\beta = 40$ and numerical aperture A = 0.6 is used at the field I for registration the most frequent interference bands. Micro-objective (5a) with magnification $\beta = 10$ is used at the field II.

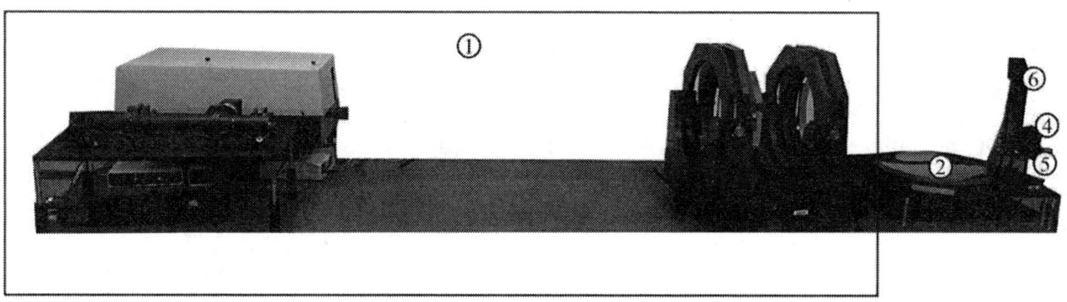

Figure 2. The experimental setup.

The created interferometer for testing convex aspherical mirrors is included in the setup of the Fizeau interferometer [8-9] located at the All-Russian research institute of metrological service. The interferometer forms a reference collimated beam of 300 mm diameter. Figures 2 shows: Fizeau interferometer (1), testing spherical mirror (2) and the system of interferogram registration (4)-(6).

3. Data processing

The plot of the width of the interference band versus the coordinates of the centers of maximum of the interference band $b(h)$ contains information about test mirror profile $z(y)$ in the tangential section. The coordinate system of the test surface is shown in figure 1.

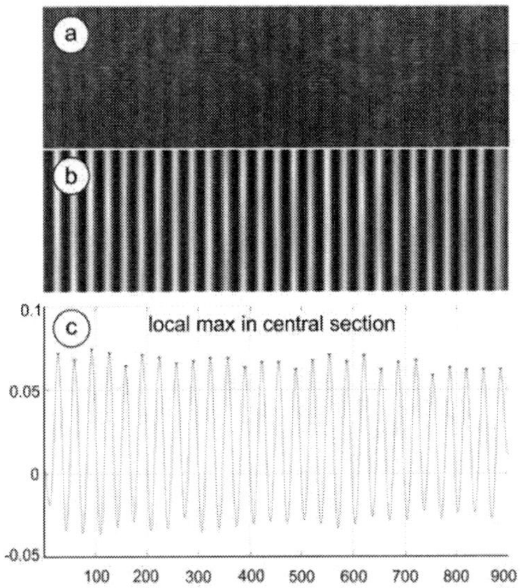

Figure 3. The data prossecing of the obtained interference pattern.

Thus, the test procedure is reduced to measuring the coordinates of the interference bands maxima, building the dependence $b(h)$, based on experimental data. The experimental plot is compared with the theoretical one.

Figure 3 shows an example of a registered fragment of the interference pattern (a), an image of the same fragment (b) filtered using a spatial Fourier transform and an intensity distribution in the filtered image with the coordinates of the centers of interference maxima superimposed on it depending on the coordinates of the pixel (c).

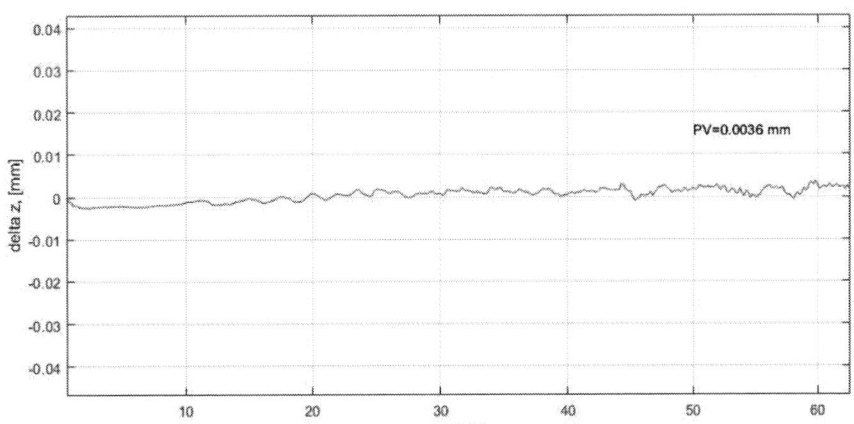

Figure 4. The mirror profile test result in the tangential section.

The figure 4 shows the profile test result in the tangential section obtained with the created interferometer. The maximum shape deviation of the test mirror (Δz) is 0.0036 mm.

4. Conclusion

An interferometer has been designed and manufactured for testing convex aspherical mirrors, which does not require precision large optical components in their work. The experimental data obtained using an interferometer are presented.

Acknowledgments

The results of the work were obtained using the equipment of the Center for Collective Use of the Scientific and Technological Center for Unique Instrumentation Making of the Russian Academy of Sciences (STC of the Russian Academy of Sciences) [10].
The authors are grateful to Professor Daniil Trofimovich Puryaev for his help with the work.

References

[1] Burge J H and Zhao C 2012 *Proc. SPIE* 8450
[2] Yan F, Fan B, Hou X, Wu F and Lei B 2014 *Proc. SPIE* 9282
[3] Burge J H, Su P and Zhao C 2008 *Proc. SPIE* 7018
[4] Puryayev D T Invention patent N3467407.25-28, "Testing Method of the optical surface profiles", (in Russian), author certificate 1044969 USSR, MKI4G 01 B П/24 (1982)
[5] Goncharov A V, Druzhin V V and Batshev V I 2009 *Proc. SPIE* 7389
[6] Druzhin V V and Batshev V I 2010 *Proc. IONS* 8
[7] DOI: 10.1117/12.2504364
[8] Vishnyakov G N, Zolotarevskii S Y and Novikov D A 2012 *Metrologia*
[9] Novikov D A, Kononogov S A, Zolotarevskii S Yu, Lyaskovskii V L, Vishnyakov G N, Gusev A S and Levin G G 2013 *Measurement Techniques* № 9
[10] http://ckp.ntcup.ru

Application of acoustic emission and laser optoacoustics at various stages of defect formation during friction stir welding

K Stepanova[1], I Kinzhagulov[1] and Y Yakovlev[2]

[1] ITMO University, Kronverksky Pr. 49, St. Petersburg, 197101, Russia
[2] Baltic State Technical University, 1st Krasnoarmeyskaya, 1, St. Petersburg, 190005, Russia

Abstract. This paper covers basic aspects of laser ultrasound and acoustic emission non-destructive testing methods in application of monitoring friction stir welds during its deformation and fracture. Stress-strain curves for such welds were obtained and the mechanical characteristics of welded joints of aluminum-magnesium alloy were determined. Using acoustic elastic effect levels of mechanical stresses were measured. Signals of acoustic emission, stimulated by one-axis loading, were registered and analyzed. It was established that acoustic emission during deformation and fracture of samples of welded joints is associated with the course of the process of plastic flow of the material. The nature of the distribution of acoustic emission parameters in the destruction of welded joints with defects has distinctive features. In conditions of one axis quasi static loading rate event count acoustic emission is the most informative parameter.

1. Introduction

Friction stir welding (FSW) is widely introduced among different industries because of high technological and constructional requirements for products. This technology is effectively applied in aircraft, rocket production and shipbuilding [1].

One of the main reasons for defect formation during FSW is the difference between real and required values of technological welding parameters. This difference may cause significant decrease in quality of final welds and, therefore, whole products [1]. Thereby, active multiparameter control of welding process is needed to obtain desired quality of welded joints.

The purpose of the presented research is an assessment of acoustic emission (AE) method for defect formation control during the process of aluminum alloys FSW.

The behavior of butt welded aluminum-magnesium alloy joints during the elastic and plastic deformation was investigated for development of acoustic emission control principles for FSW control in situ. The investigation revealed specific features of acoustic emission parameters which appeared during the tensile test of joints' specimens.

The method of laser optoacoustics was used for actual stress measurement in the area of weld root in specimens with different structure of joints.

2. Description of work

2.1. Materials

Specimens were made of butt welded joints produced by FSW. Material of specimens was high strength aluminium magnesium alloy. Specimens No. 1, No. 2 had the same geometry but was made of base material (without weld). Specimens Nos. 3 A, 4A, 6A, 9Ad, 10 Ad were joints with weld in the middle part (figure 1). Joints in specimens No. 9Ad and No. 10Ad were made from joints which were produced with welding process parameters deviated from optimal ones. Geometric characteristics of these specimens are presented in table 1.

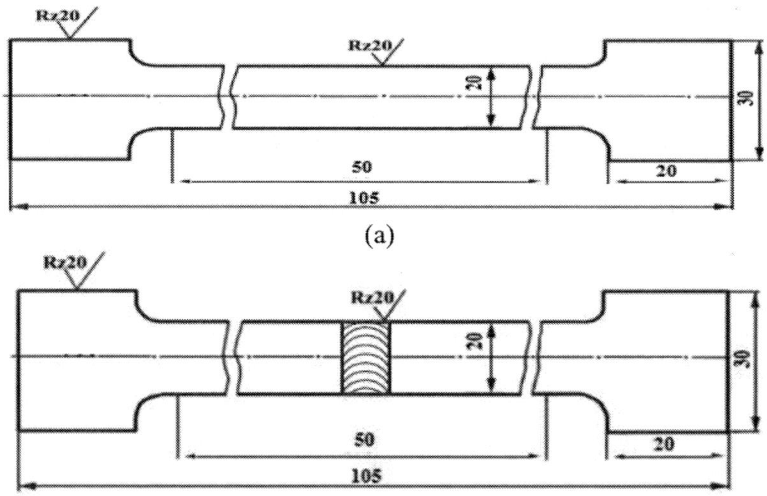

(b)

Figure 1. Drafts of specimen: (a) - Nos. 1,2 (base metal); (b) - Nos. 3A, 4A, 6A, 9Ad, 10Ad (weld joint).

Table 1. Geometric characteristics of specimens.

No.	Initial thickness, a_0 (mm)	Initial width, b_0 (mm)	Gauge length, l (mm)	Initial cross-sectional area, F_0 (mm^2)	Alloy
1	7.4±0.2	20.0±0.2	50.0±0.2	148.0 ±0.4	
2	7.4±0.2	20.0±0.2	50.0±0.2	148.0 ±0.4	
3A	7.4±0.2	19.5±0.2	50.0±0.2	144.3 ±0.4	
4A	7.4±0.2	20.0±0.2	50,0±0.2	148.0 ±0.4	AMg6
6A	7.4±0.2	19.8±0.2	50.0±0.2	146.5 ±0.4	
9Ad	7.4±0.2	19.9±0.2	50.0±0.2	147.3 ±0.4	
10Ad	7.4±0.2	20.0±0.2	50.0±0.2	148.0 ±0.4	

2.2. Stress measurement using laser ultrasound method

Measurements were carried out with use of acoustoelastic effect. Ultrasound waves were generated thermooptically by laser ultrasound method. This method provides an opportunity to generate nanosecond length waves [2]. The informative parameter for stress measurement in metal constructions is the speed of longitudinal subsurface wave (LSW) [3].

Generation and receiving of LSW are provided by dual element transducer. Acoustic contact is obtained with additional pressing through the thin layer of contact fluid between a transducer and specimens.

2.3. Acoustic emission during uniaxial tension specimen testing

Acoustic emission data were recorded under static loading of specimen. Static uniaxial tensile tests were performed with a constant loading rate of 0.2 kN·s^{-1} at a temperature of +23 °C. Analyzed Informative AE parameters: count event N_Σ, rate event count N'_Σ, AE signal rise time Δt. These parameters were compared with the level of the applied load P and the relative deformation ε. AE was recorded by piezoelectric acoustic emission transducers (AET) with a passband of 100-700 kHz. The

AE signals received by AET were amplified with a 40 dB preamplifier. Elements of the test bench are shown in figure 2.

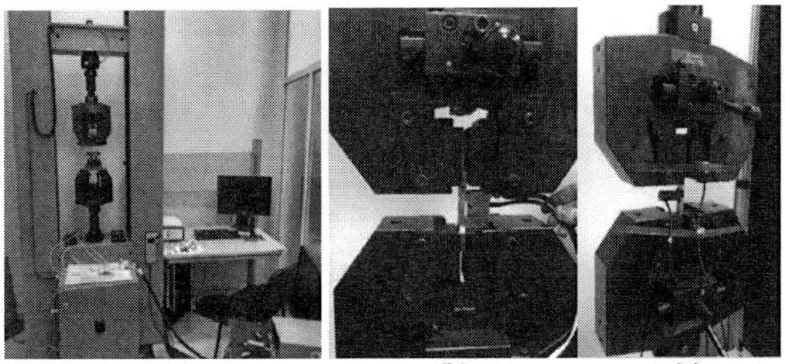

(a) (b) (c)

Figure 2. Elements of the test bench: (a) – apparatus for uniaxial tension; (b) – placement of the laser-ultrasonic transducer on the specimen; (c) – placement of the AET.

3. Results & discussion

3.1. Results of stress measurements

Before the actual measurement it was necessary to perform graduation of measuring tools to get the value of acoustoelastic coefficient K_υ which is specific for concrete material, alloy or group of alloys. This coefficient allows to estimate mechanical stresses using the measured LSW speed. Calculation of K_υ was made according to equation [3]:

$$K_\upsilon = \sigma \frac{\upsilon}{\upsilon - \upsilon_0}, \quad (1)$$

where σ – mechanical stress in a specimen (according to actual loading), MPa; K_υ – coefficient of acoustoelasticity of the specimen material, kg·m^{-1}·s^{-2}; υ – speed of head subsurface ultrasound wave in specimen material during the loading, m·s^{-1}; υ_0 – speed of head subsurface ultrasound wave in specimen material without the loading (zero actual stress), m·s^{-1}.

To derive graduation characteristics υ_0 and υ_0 were measured. After that, the value of K_υ was calculated using specimens No. 1 and No. 6A. The speed was measured according to the scheme (figure 3).

The graduation characteristics (figure 4) here is the linear approximation of measured speed-stress pairs of values on chosen levels of stresses. The average value of K_υ is based on results of two loading cases – unloading in the elastic area of specimen No 1 and loading in the elastic area of specimen No 6A.

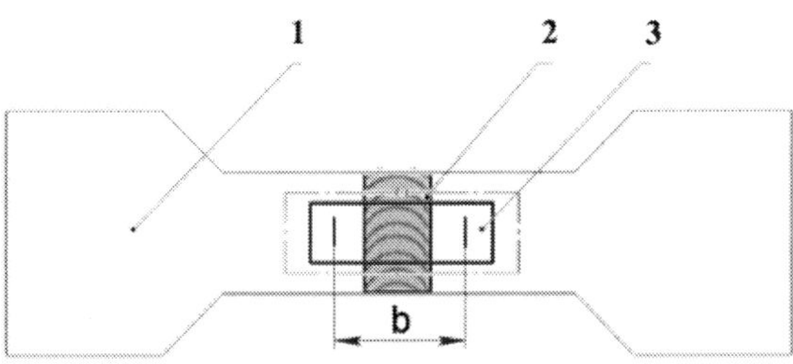

Figure 3. The scheme of measurements of the velocity of propagation of the head ultrasound (1 – specimen; 2 – measurement area at the root of the weld, 3 – transducer; *b* – base length of the transducer).

According to the measurement results, K_υ was calculated and the graduation characteristic of the dependence of the stresses σ on the velocity υ was built. The average value of the coefficient K_υ was calculated from the results of two loading cycles – unloading in the region of elastic deformations of specimen No. 1 and under load in the region of elastic deformations of specimen No. 6A. Figure 4 shows the results of the calibration performed on samples No. 1 and No. 6A.

Using the obtained coefficient K_υ, stresses σ were measured in specimens with joints during the loading (table 2, figure 5).

The relative error $\Delta\bar{\sigma}$ of the measurement result σ was determined by the equation (2):

$$\bar{\Delta}\sigma = \frac{\Delta\sigma}{\bar{\sigma}} \cdot 100\%, \tag{2}$$

where $\Delta\sigma$ – absolute error of the result of measurement σ, MPa.

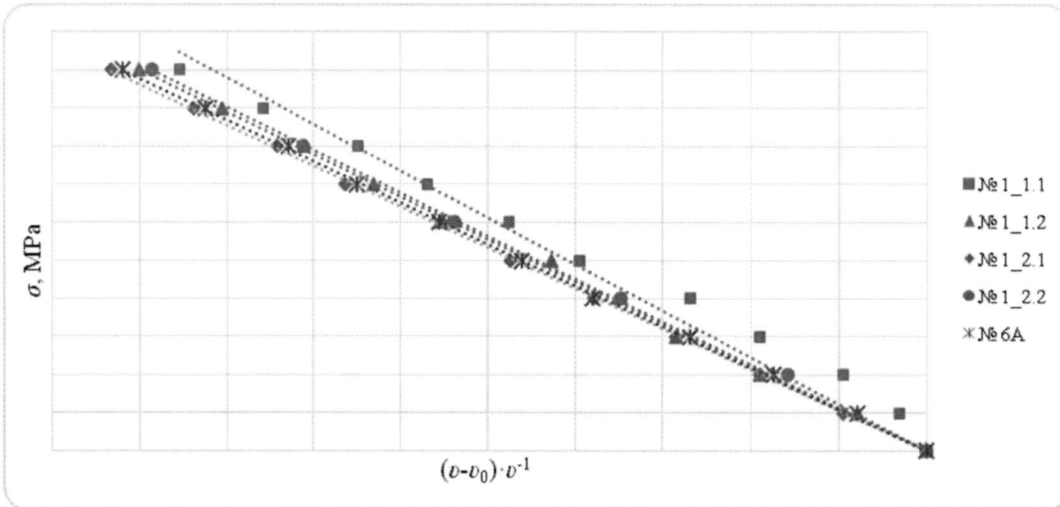

Figure 4. Graduation characteristic for aluminium-magnesium alloy.

The true value of σ_p was taken as the calculated value of the stresses in the cross section of the specimen, defined as the ratio of the applied load P to the cross-sectional area F_0.

Table 2. Results of stress measurements in specimens.

Load		Results of stress measurements					
		Specimen No. 6A		Specimen No. 9Ad			
				Load 1		Load 2	
P (kN)	σ_p (MPa)	σ (MPa)	$\bar{\Delta}\sigma$ (%)	σ (MPa)	$\bar{\Delta}\sigma$ (%)	σ (MPa)	$\bar{\Delta}\sigma$ (%)
0.0	0	0	0	0	0	0	0
2.8	20	18	10	6	72	26	30
5.6	40	40	1	24	41	39	2
8.4	60	61	2	49	19	61	1
11.2	80	86	8	67	16	78	3
14.0	100	104	4	93	7	97	3
16.8	120	126	5	107	11	121	1
19.6	140	148	5	139	1	141	1
22.4	160	165	3	142	11	158	1
25.2	180	187	4	173	4	170	5
28.0	200	209	4	233	17	191	5

As a result of comparison of applied and measured stresses, the relative measurement error on specimen No. 6A did not exceed 10%. The convergence of the results suggests that the stress distribution in the root region and the heat-affected zone of the welded joint of sample No. 6A under the action of tensile forces occurs similarly to the development of the base metal stresses, and the effective stresses in the heat-affected zone are zero.

This fact allows us to conclude that the stress-strain state of the joints made by FSW changes under the action of operating loads in the same way as in the base metal. This conclusion is valid for products from aluminum alloys, physically justified by the fact that the maximum temperature in the welding area is lower than the melting point of material. It reduces welding stresses and makes it possible to create joints of equal strength.

An applied load corresponding to a stress of σ = 200 MPa led to the occurrence of microplastic deformations in the root region of the weld of sample No. 9Ad. This can be seen in the non-linear part of the diagram (figure 5) with load 1 of specimen No. 9Ad.

The magnitude of the errors at the stages of loading-unloading of sample No. 9Ad allow to conclude about the change in the nature of the stress-strain state in the measurement zone of the welded joint, made with the deviation of the welding parameters.

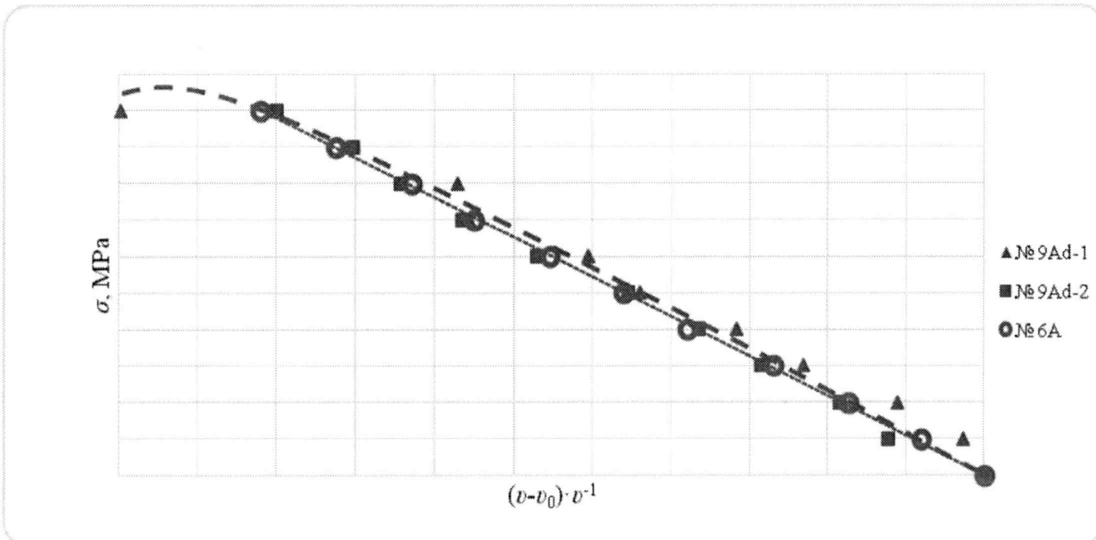

Figure 5. Stress-speed curves for specimen alloy (9Ad-1 – loading, 9Ad-2 – unloading).

3.2. The results of uniaxial static tensile tests

The structure of FSW joints has a number of characteristic features. In the center of the weld (core), ultrafine grains are formed, the formation of which is the result of dynamic or collective dynamic polygonization. The core structure is called the onion ring because of its ring-shaped structure (figure 6 (a) (4)). The zone of thermomechanical influence (TMAZ) (figure 6 (a) (1)) is intermediate between the zone of thermal influence (TAZ) (figure 6 (a) (2)) and the core and represents the transition from one type of structure to another – from a grain textured structure base metal (figure 6 (a) (3)) to the layered structure of the central part of the weld [4, 5].

In FSW joints specimens, the plane of destruction was perpendicular to the layers of friction formed by the sliding shoulder of the welding tool on the surface.

In the tensile test of specimen of welded joints, it was noted that the destruction of samples Nos. 3A, 4A, 6A occurred in two stages. The primary crack was formed from the root of the seam, then the destruction occurred along the border between TMAZ and HAZ (figure 6) and ended with a break in the upper part of the welded joint. In the lower part, the seam is less durable due to insufficient mixing of the metal, due to the increased heat sink to the welding substrate, which reduces ductility.

The destruction of specimens No. 9Ad and No. 10 Ad took place differently. The failure of specimen No. 9Ad and No. 10 Ad occurred according to the weld material. Macrocrack originated at the root of the welded joint in the core zone. The fracture surface is visually divided into two parts - zone I with a rough relief characteristic of static destruction of ultrafine materials, and upper zone II with a dimpled relief similar to secondary fracture (figure 6) [5].

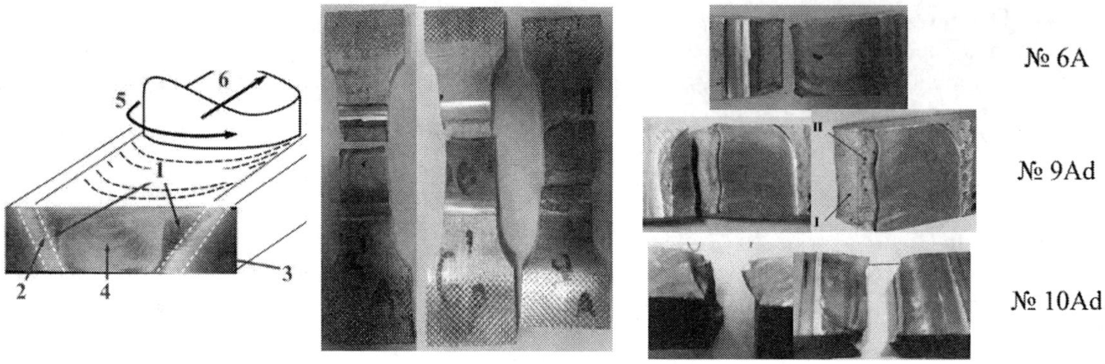

(a) (b) (c)

Figure 6. Specimens after tests: (a) – structure zones (1 - TMAZ; 2 – HAZ; 3 – base metal; 4 – weld core; 5 – direction of a tool spin; 6 – direction of weld); (b), (c) – specimens after tests.

The destruction of specimens No. 9Ad and No. 10 Ad took place differently. The failure of specimen No. 9Ad and No. 10 Ad occurred according to the weld material. Macrocrack originated at the root of the welded joint in the core zone. The fracture surface is visually divided into two parts - zone I with a rough relief characteristic of static destruction of ultrafine materials, and upper zone II with a dimpled relief similar to secondary fracture (figure 6) [5].

The destruction of the core of the weld is not typical for the conditions of tensile tests. This circumstance is due to the presence of defects in the structure of the welded joint. The nature of the fracture qualitatively reflected the original structural state of the welded joint. The results of static tests on uniaxial loading are shown in table 3.

Table 3. The results of uniaxial tensile tests.

No. specimen	Strain, ε_{max} (%)	Average strain, ε_{max} (%)	Ultimate strength, σ_{uts} (MPa)	Reducing of strength, $\Delta\sigma_{uts}$ (%)
1	16.2	16.4	398	Average value $\sigma_в$ =389
2	16.6		380	
3A	17.8	17.4	363	6.8
4A	17.5		354	9.0
6A	17.0		356	8.6
9Ad	14.5	12.3	337	13.5
10Ad	10.1		341	12.3

The results of uniaxial tensile tests are shown in figure 7.

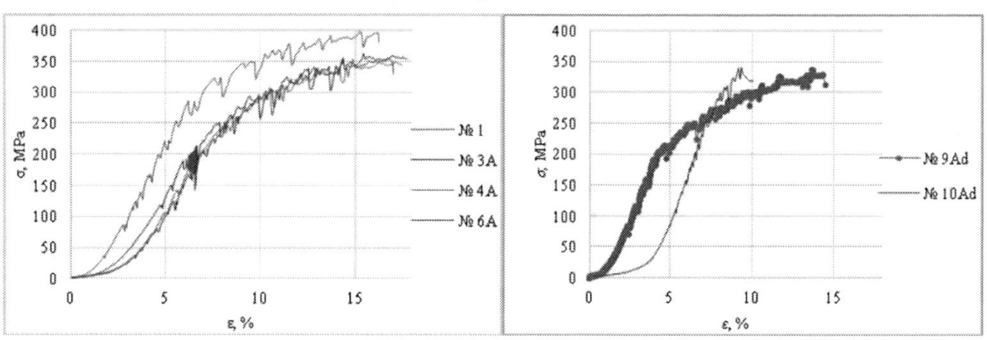

(a) (b)

Figure 7. Stress-strain curves: (a) – base metal and welded joint without defects; (b) – welded joints with defects.

The modulus of elasticity of tested welded joints differs by no more than 2-3%, while the modulus of elasticity of the base metal is greater than the elastic modulus of welded joints by 5-7%, which is explained by the greater ductility of the weld material due to the grinding of polycrystals of the material during welding.

The reduced ductility and strength of specimen No. 10Ad is associated with the presence of an angular displacement type defect [6]. During loading, both tensile stresses and bending stresses occurred in the specimen, caused by the pressure of the grippers of the testing machine on the fixed parts of the sample.

The reduced ductility and strength of sample No. 9Ad is explained by the fact that the weld was formed in two passes of the welding tool, which led to the formation of a double-recrystallized structure of the weld.

3.3. The results of the acoustic emission parameters analysis

During AE control a linear type of location was implemented using three AET (figure 2). The parameters of the signals received in the zone between the AET, corresponding to the zone of the subsequent destruction of the specimen, were analyzed. Thus, signals caused by friction in the grips of the testing machine were excluded from the processing of AE.

The amplitude filtering criteria is determined based on the ratio of the useful signal amplitude, which is more than 10 times greater the noise level. According to the rise time of the forefront Δt, the signals characterizing discrete AE were chosen, since many authors attribute this type of AE to the evolution of macroscopic defects [7]. The results of the AE count event N_Σ distribution and the AE rate event count N'_Σ distribution during testing of the base metal samples (No.1, No.2) are shown in figure 8.

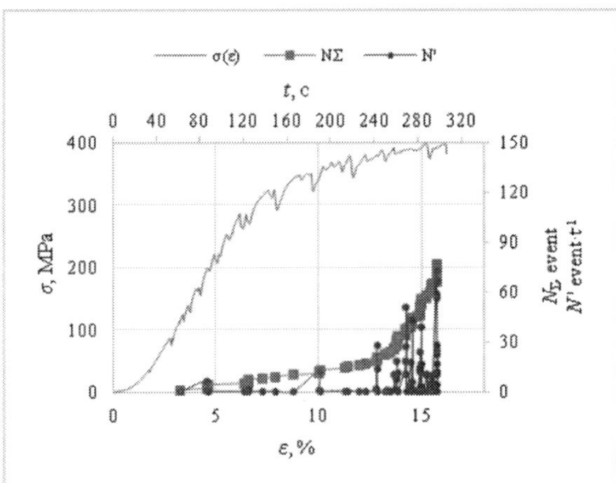

(a)

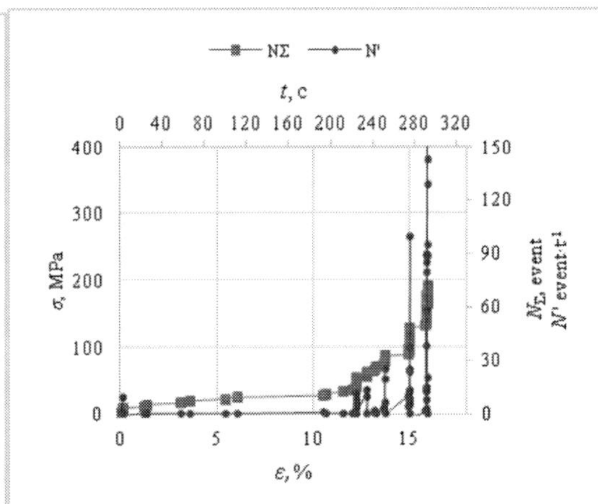

(b)

Figure 8. Diagrams of AE parameters distribution of the base metal samples: (a) - No. 1; (b) - No. 2.

From the standpoint of the kinetic approach, a decrease in the AE rate event count at the stage of plastic flow of No. 1, No. 2 samples can be caused by a slowing down of the movement of dislocation clusters, due to their high density.

The active growth of the AE count event with increasing load is due to the processes of combining dislocation clusters, the destruction of dislocation loops with the formation of destruction foci (microcracks), leading to the formation of macrocracks [8, 9].

The results of the AE count event N_Σ distribution and the AE rate event count N'_Σ distribution during testing of the welded joints samples (Nos. 3A, 4A, 6A) are shown in figure 9.

The microstructure of welded joints, in terms of acoustic emission, has its own characteristics, due to the specificity of a material plastic flow process [10].

For samples Nos. 3A, 4A, 6A, local maxima of AE rate event count fall within the yield strength of the material, which, due to the dislocation theory of plastic deformation of crystalline materials, is explained as the formation of dislocation clusters, their migration, annihilation and the release of dislocations to the surface [11].

The count event recorded at the plastic flow stage of welded joints material (Nos. 3A, 4A, 6A) exceeds this parameter for base metal specimens (No. 1, No.2), which indicate the presence of additional AE sources related to the structure of the weld: the generation of dislocations at the boundaries of the core zone and thermomechanical zone (TMZ) in particular.

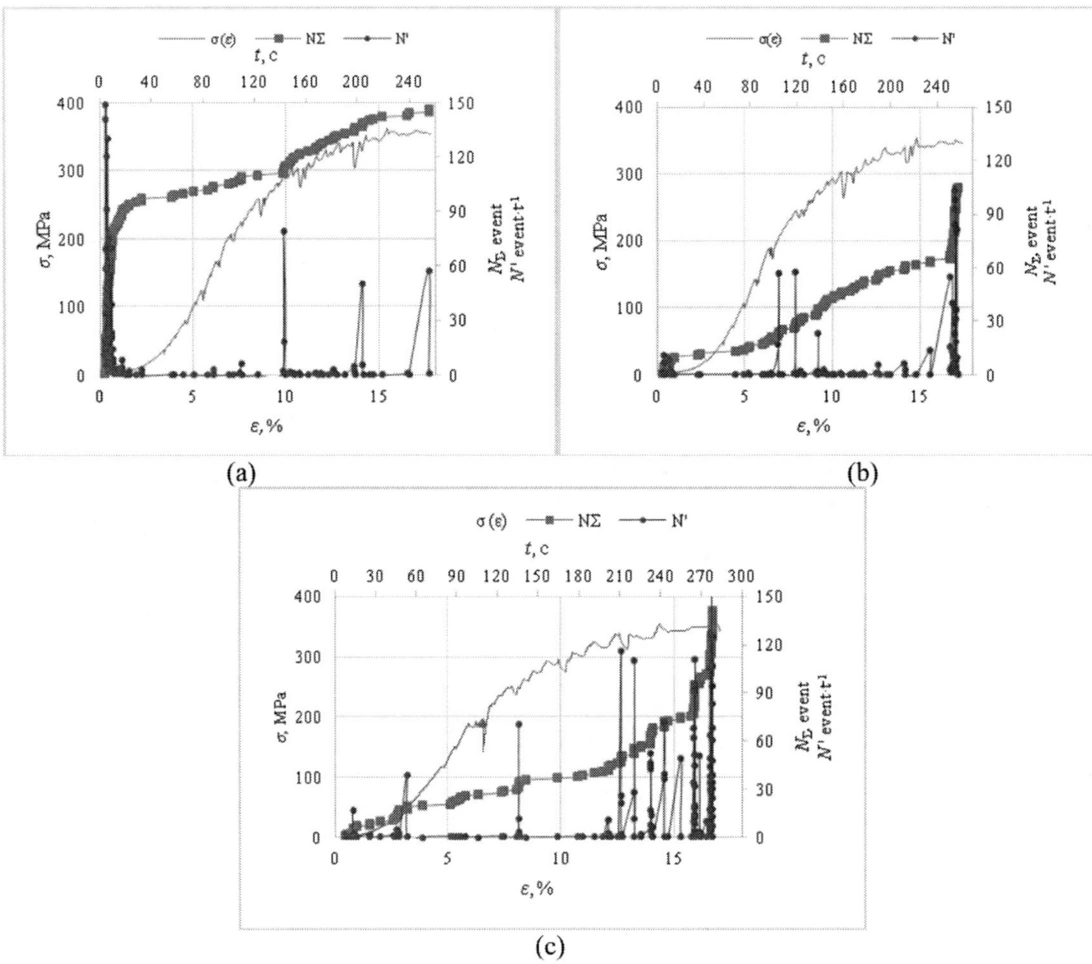

Figure 9. Diagrams of AE parameters distribution of welded joints samples: (a) - No. 3A; (b) - No. 4A; (c) - No. 6A.

For samples No. 4, No. 6A, the abrupt accumulation of AE signals (the maximum of AE rate event count) preceding the destruction of the samples is characteristic. Sources of AE in this case were the processes of combining dislocation clusters with the formation of localized microcracks and their subsequent merging into a macrocrack, the growth of which led to destruction [11].

Sample No. 6A is characterized by the presence of local maxima of AE activity at a load level (0.85÷0.9) σ_{uts}, probably related to the accumulation of dislocations, overcoming obstacles with the release of stored energy in the form of AE radiation, at the boundaries of the core zone and TMZ.

The results of the AE count event N_Σ distribution and the AE rate event count N'_Σ distribution during testing of the welded joints samples (Nos. 9Ad, 10Ad) are shown in figure 10.

For samples No. 1, Nos. 4A, 6A, 9Ad in the area of elastic deformation (up to 200 MPa), low AE activity was observed due to the Kaiser effect caused by the preloading of these samples to determine the acoustoelastic coefficient.

The distribution of AE parameters for samples No. 9Ad, No. 10Ad, made with a deviation of the values of technological parameters of welding, had a number of features.

For sample No. 9Ad, the first local maximum of AE rate event count corresponds to a stress level of $\approx 0.7 \cdot \sigma_{uts}$. For a given sample with a double-recrystallized structure, probably a discrete AE is associated with twinning processes, the emergence and growth of slip bands, with cracking mechanisms that cause spatial and temporal heterogeneity of deformation [9].

For sample No. 10Ad, the first local maximum of AE rate event count was registered at a stress level of $\approx 0.4 \cdot \sigma_{uts}$. This level of stress is in the area of elastic deformations of samples without structural defects. This feature is caused by the initial deformation of specimen No. 10Ad (a defect of the type angular displacement of edges). The uniaxial load applied during the tests led to the formation of additional bending stresses in the weld zone in the sample.

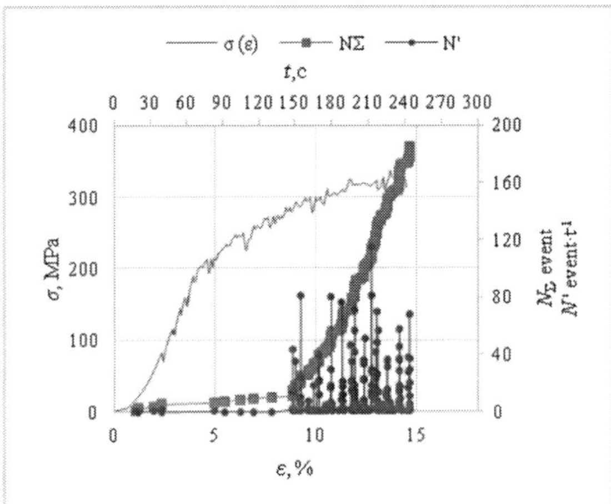

(a)

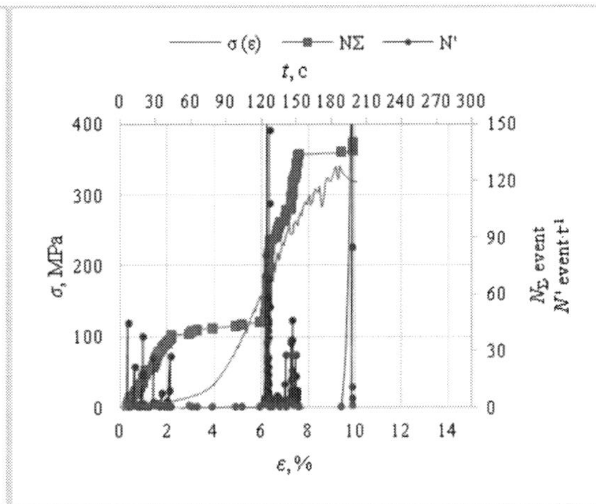

(b)

Figure 10. Diagrams of AE parameters distribution of welded joints samples with defects: (a) - No. 9Ad; (b) - No. 10Ad.

The correlation of the AE signals parameters with the parameters of fracture and plastic deformation of welded joints samples, made with deviation of the technological parameters of welding and without them, will allow to proceed to the study of the processes of defect formation during the welding.

4. Conclusion

The results of experimental studies have shown the potential applicability of the acoustic emission method for the defect formation control during FSW of products from aluminum alloys.

During the process of a welded joint forming, AE sources are associated both with the formation of structural inhomogeneities zones (defects) and with a technological change in the stress-strain state of the material.

For reliable interpretation of the registered AE signals, it is advisable to monitor the parameters of the stress-strain state of the base metal and the weld after welding by the laser-ultrasonic method. The difference in the values of the parameters characterizing the stress-strain state will indicate the presence of structural inhomogeneities in the welded joint. Control supplemented in this way will allow obtaining the most informative analysis of the AE information.

References

[1] Lohwasser D and Chen Z 2009 *Friction Stir Welding. From Basics to Appl.* (New Delhi:Woodhead Publishing Ltd) p 15
[2] Karabutov Jr A A, Sapozhnikov O A and Karabutov A A 2010 *J. Phys. of Wave Phenomena* **18** 297-302
[3] Marusina M Y, Fedorov A V, Berkutov I V and Bychenok V A 2017 *Russian J. Measurement Techniques* **59** 1165–69
[4] Schneider J A and Nunes A C 2004 *J. Met. and Mat. Trans.* B **35** 777–783
[5] Sizova O V, Kolubaev A V, Kolubaev E A, Zaikina A A and Rubtsov V E 2014 *J. Met. Work. and Mat. Science* **64** 14–20
[6] American Welding Society 2016 *Specification for friction stir welding of aluminum alloys for aerospace applications* (AWS - D17.3/D17.3M:2016) Retrieved from https://standards.globalspec.com/std/13041310/aws-d17-3-d17-3m
[7] Tetelman A S 1971 *J. Mater. Res. And Stand.* **11** 13–16
[8] Petrov M G and Ravikovich A I 2006 *J. of Applied Mechanics and Technical Phys.* **47** 143-151
[9] Lebyodkin M A 2014 *J. Materials Science Forum* **783-786** 204–209
[10] Polyakov V V, Egorov A V, Salita D S and Kolubaev E A 2015 *J. Izvestiya of Altai State University* **85** 50-54
[11] Kok S, Bharathi M S, Beaudoin A J, Fressengeas C, Ananthakrishna G, Kubin L P and Lebyodkin M A 2003 *J. Acta Materialia* **51** 3651–62

Optical-acoustic diagnostics of cracks on the surface of solid bodies

N S Kazak[1], G V Kulak[2], T V Nikilaenko[2], A G Matveeva[2] and O V Shakin[3]

[1]LOD, Institute of Physics of NASB, 68 Nezalezhnasti ave., Minsk, 220072, Belarus
[2]Physical engineering faculty, State Pedagogical University,28 Studencheskaia str. Mo zur, 247760, Belarus
[3]State University of Aerospace Instrumentation, 67 Bolshaya Morskaya str., Saint-Petersburg, 190000, Russia

E-mail: [1]lod@dragon.bas-net.by, [2]g.kulak57@mail.ru ,[3]oshakin@mail.ru

Abstract. The features of optical-acoustic diagnostics of defects in the form of cracks on the surface of solid bodies in a pulsed mode are investigated. It has been established that pulsed surface Rayleigh waves should be used to detect cracks. The largest amplitude of the scattered surface acoustic wave is achieved at the optimum width of the laser spot, directed to the center of the crack.

1. Introduction

Optical-acoustic (OA) sources of ultrasonic (US) waves have several advantages over traditional (piezoelectric and electromagnetic-acoustic), including the absence of contact with the medium, the ability to easily change the geometric parameters of an acoustic antenna, diagnose objects moving at any speed [1]. For the excitation of short acoustic pulses, the application of OA methods under pulsed laser exposure is promising [2, 3]. The excitation of a Rayleigh SAW upon absorption of nanosecond laser pulses in a fused silica material was studied in [4]. The features of diagnostics of inhomogeneities in the form of strips of gold on the surface of fused quartz are considered.

2. Theoretical results and discussion

The surface defect in the form of a crack has a width l and is located at a distance of ρ from the center of the laser beam and at a distance of ρ' from the receiver of the scattered SAW. The excitation region of the US wave has the shape of a rectangle with the area $S_f = ab$, where a is the width of the rectangle, b is its length. In our case $a << b$ [5]. A laser pulse propagates along the Z axis and excites high-frequency ultrasonic waves of different polarization and spatial-angular distribution. Suppose that OA excitation is carried out in a linear mode [1], and the laser pulse has a Gaussian amplitude distribution in time $f(t) = \exp(-t^2/\tau^2)$. Then the Gaussian acoustic pulse has a duration of τ_a exceeding the duration of the laser pulse. The pulse of ultrasonic SAW displacements has a width of the spectrum

$\Delta\Omega \sim 1/\tau_a$ and a central frequency $\Omega \sim \Delta\Omega$. The frequency spectrum of a Gaussian acoustic pulse excited on the surface of a solid body has the form $F(\Omega) = \tau_a\sqrt{\pi}\exp(-\tau_a^2\Omega^2/4)$, where Ω is the circular frequency of the ultrasonic wave. The frequency spectrum of the components of the scattered Rayleigh SAW polarized along the *Z* and *X* directions has the form [5]

$$\tilde{U}_{Rz}(\Omega) = \frac{P_f S_{noл} F(\Omega)\eta^2}{4\mu}\sqrt{\frac{2k_R}{\pi\rho}}\frac{\sqrt{1-\gamma^2\eta^2}}{C_0}\frac{\sin(A_R)}{A_R}\frac{\sin(B_R)}{B_R}e^{i\phi_R}, \tag{1}$$

$$\tilde{U}_{Rx}(\Omega) = \frac{\tilde{U}_{Rz}(\Omega)(1-\eta^2/2)}{\sqrt{1-\gamma^2\eta^2}}, \quad A_R = ak_R/2,\ B_R = b(k_R\sin\alpha)/2, \tag{2}$$

$S_f = ab$, where *a* (*b*) is the size of the excitation strip along the *X* (*Y*) axis; *α* is the azimuth angle measured from the *X* axis (in the *XY* plane), P_f is the pressure on the surface of a solid in the region of laser irradiation; *ρ* is the distance from the excitation region to the crack; $\phi_R = (k_R\rho' - \Omega t + \pi/4)$; $k_R = \Omega/\upsilon_R$, υ_R is the phase velocity of the SAW, $\eta = (0{,}87 + 1{,}12\nu)/(1+\nu)$ and ν is Poisson's ratio; $\gamma = \mu/(\lambda + 2\mu)$; λ, μ are the constant Lame; $C_0 = 2(2-\eta^2) - C_1/C_2 - C_2/C_1 - 2C_1C_2$ where $C_1 = \sqrt{1-\eta^2}$, $C_2 = \sqrt{1-\eta^2\gamma^2}$. Note that the frequency-angular distribution of the surface ultrasonic waves excited by laser pulses is determined by the diffraction function on the rectangular excitation region and the frequency spectrum of the incident acoustic wave, i.e. $[F(\Omega)\sin(A_R)/A_R][\sin(B_R)/B_R]$.

The incident Rayleigh ultrasonic wave with frequency *Ω* is directed to the defect and scatters on it in the forward direction. Using the two-dimensional Green function in the X_1X_2 plane (in the $X_1X_2X_3$ coordinate system), the amplitude of the scattered SAW can be found from the expression [6]

$$U_{Ri}^{sc}(\mathrm{r}') = -\int_s n_p(\mathrm{r}')C_{jpmn}\frac{\partial}{\partial x_n}G_{im}(\mathrm{r},\mathrm{r}')[U_{Rj}(\mathrm{r})]dS_r, \tag{4}$$

where n_p are the components of the unit vector normal to the crack surface (CR), C_{jpmn} are the elastic moduli of the material; $[U_{Rj}(r)]$ – components of displacement in the region of an open crack, induced by Rayleigh SAW; the Green function for displacement in the excited wave has the form [6]

$$G_{im}(\mathrm{r},\mathrm{r}') = \sum_{\alpha=L,T}\left(\frac{i}{4\pi\mu}\right)\left(\frac{k_L}{k_T}\right)^2\int_{-\infty}^{+\infty}(1-\xi^2)^{-1/2}d_i^\alpha(\xi)d_m^\alpha(\xi)e^{i(\alpha p(\mathrm{r}-\mathrm{r}'))}d\xi, \tag{5}$$

where k_L, k_T are the wave numbers of the longitudinal and shear bulk ultrasonic waves in the material; $\vec{d}^L = \vec{p}$, $\vec{d}^T = [\vec{e}_3\,\vec{p}]$, where $\vec{\mathrm{p}} = (\xi, \sqrt{1-\xi^2})$, $\vec{\mathrm{e}}_3 \| X_3$.

After completing the asymptotic estimate of the integral (5) and substituting the resulting expression in (4), we can integrate over the inner surface of the defect by obtaining the expression for the components of the displacement vector of the scattered SAW of the form

$$U_{Rx,z}^{sc} = \sqrt{\frac{2}{\pi\rho' k_{L,T}}}\mathrm{e}^{-i\pi/4}\,|U_{Rz,x}|\,Q_{L,T}I_{L,T}, \tag{6}$$

where

$$Q_L = -k_L^3\lambda/4k_T^2\mu,\ \Omega_T = k_T/4,\ I_{L,T} = \sin[l(k_{L,T}+k_R)/2]/(k_{L,T}+k_R),$$

moreover, ρ' is the distance from the scattering center to the receiver of ultrasonic waves.

The temporal form of the scattered acoustic pulse is determined by the integral [1], [3]

$$U^{sc}_{Rx,z}\left(\tilde{t}_R\right)=\int_{-\infty}^{+\infty}U^{sc}_{Rx,z}(\Omega)e^{-i\Omega\tilde{t}_R}d\Omega, \tag{7}$$

where $\tilde{t}_R = t-\rho'/\eta\upsilon_s$. Substituting expressions (6) into (7), and, completing the integration by numerical methods, we obtain the temporal form of the longitudinal and transverse components of the scattered acoustic pulse of Rayleigh ultrasonic waves in the receiver region. The fast Fourier transform method was used in the numerical calculations. The calculations were performed for the Rayleigh SAW propagating along the free surface of a material made of steel (*Fe*). It was assumed that $P_f = 100$ MPa, $\upsilon_l = 5900$ m/c, $\upsilon_s = 3200$ m/c, $\alpha = 0^0$, $\lambda = 4.9\cdot10^{10}$ Pa, $\mu = 7.84\cdot10^{10}$ Pa. It was assumed that a model defect in the form of a crack of width l has a flat boundary and is oriented along the axis *OX*. It is shown that for small widths of the laser spot, the shape of the SAW pulse in the form of a "kink" takes place for the incident pulse of the Gaussian shape. For large spot widths $a \sim 6$ mm, the pulse shape has the form close to *S*-shaped with the presence of a "precursor" with small amplitude of oscillations in time. The oscillograms of the scattered SAW in the forward direction for the normalized amplitude when scattered on a defect in the form of a crack is investigated. It follows that the form of the oscillograms changes with a change in the spot width of the laser spot a. The amplitude value of the scattered SAW pulse varies significantly with increasing laser spot width a. The greatest amplitude in time is achieved by a pulse with a laser spot width $a = 2.5$ mm. The width of the laser spot is a multiple of the integer number of SAW wavelengths that fit on it.

Figure 1 shows the dependences of the amplitude of the Z component of the scattered SAW on the position of the laser excitation source x_1. It was assumed that there is a small deviation Δx_1 due to the experimental features imposed on the accuracy of the orientation of the center of the laser spot.

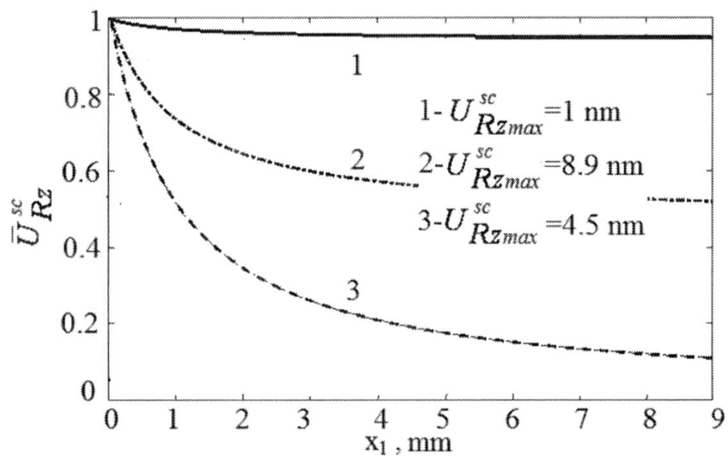

Figure 1. Dependence of the amplitude of the transverse component of the pulse of Rayleigh SAW when the position of the excitation point x_1 is changed relative to the crack at different widths of the laser excitation a: 1-0.25, 2-2.5, 3-5.5 mm (*Fe*, τ_a=12 ns, $\alpha = 0$, ρ'=10 mm, l=0,2 mm, b=30 mm, $\Delta x_1 = 0.01$ mm).

As can be seen from figure 1, when the laser beam moves relative to the crack, there is a pronounced maximum reached when the center of the laser beam is located in the vicinity of the crack ($x_1 \approx 0$). It is determined that the amplitude of the scattered SAW in the receiver region increases with increasing width of the excitation strip. An increase in the pulse amplitude with increasing US frequency f is associated with the dispersive properties of a scattering crack for the transverse component of the incident Rayleigh SAW. With increasing frequency $f$the frequency of oscillations increases significantly with a simultaneous increase in the amplitude of the SAW reaching the receiver. When $a = 5.5$ mm amplitude $U^{sc}_{Rz} \approx 4.5$ nm.

3. Conclusion

Registration surface defects in the form of a crack can be detected using scattering effects of Rayleigh surface acoustic wave in the pulsed mode and the second measurement time pulse shape scattered by a surfactant defect. A significant decrease in the amplitude of the scattered SAW takes place when the receiver location area deviates from the crack location area. By choosing the spot width of the laser spot is possible to achieve the maximum value of the signal of the scattered SAW in the receiver region.

References

[1] Gusev V E and Karabutov A A 1991 *Laser optoacoustics* (Moscow: Science)
[2] Tam A C 1984 *Phys. Lett.* **45** 510-12
[3] Zang S Y, Paul E M, Fassbender S, Schleichert U and Arnold W 1991 *Res. Nondestr. Eval.* **2** 143-55
[4] Lin H N, Maris H J and Freund L B 1993 *Appl. Phys.* **73** 37-45
[5] Budenkov G A and Nedzvtskaya O V 2004 *Dynamic problems of the theory of elasticity* (Moscov: Fizmatlit)
[6] Hirose S 1997 *J. Faculty of Environmental Science and Technology* **2** 89-98

Separate BOS-visualization of temperature and pressure distributions in a solid medium

D G Sychev

Physics Department, Moscow Power Engineering Institute, 17 Krasnokazarmennaya street, Moscow, Russia

E-mail: sychdmitrij@yandex.ru

Abstract. Background Oriented Schlieren (BOS) is a technique widely used for visualizing disturbances in optically transparent objects. This method makes possible to visualize the variation of the refractive index, which depends on such physical parameters as temperature, pressure and concentration. Therefore, obtaining information about these parameters is also possible. In the course of the work, an experiment was conducted to visualize the optical inhomogeneity formed by changes in pressure and temperature in plane-parallel glass plate. Processing of experimental data has shown that under certain conditions it is possible to obtain information about the individual parameters of the environment which forms a complex inhomogeneity.

1. Introduction

Flows of gases and liquids are very common phenomena, both in natural and industrial conditions. Necessity to visualize gas flows in industrial aviation has provided optical methods with wide recognition in the field of experimental gas dynamics, since they allow the flow to be diagnosed without introducing additional disturbances into it [1].

Various refractometric methods (such as schlieren, shift-interferometry) are most often used in the study of optically transparent inhomogeneities. These methods are based on the use of the same phenomenon-the reaction of the light beam to refractive index alterations [2]. Refractometric methods are usually used to investigate the distribution fields of the refractive index or its gradients, which are able to be converted into quantities related to the refractive index itself.

Following the evolution of digital technology, techniques which based on the main principles of the above-mentioned methods began to develop and at the present times have reached the mark of an independent measuring technique; for example - Background Oriented Schlieren which application will be discribed in this work. Currently, it is widely used for qualitative and quantitative visualization of wind tunnels flows [3], in full-scale flight tests [4], and in explosive engineering [5]. The significant advantage of this method is the simplicity of its practical implementation and the modest requirements for technical base which are lower in comparison with analogues.

The main objective of this work is to study the possibility of obtaining information about each of the environment characteristics involved in the formation of complex optical inhomogeneity. In this paper, we will consider the BOS-visualization of optical inhomogeneities caused by the heating and deformation of the glass plate.

2. Background Oriented Schlieren in quantitative diagnostics

As already noted, the basis of BOS is the phenomenon of refraction: due to the violation of the optical homogeneity of the medium in the path of propagation of light rays, the trajectory of their movement is distorted. Violation of the optical homogeneity of the medium is expressed in the inconstancy of the refractive index *n* over its volume. Pressure, temperature and concentration are the main parameters of the medium, affecting its refractive index. For clarity, consider the dependence of the refractive index of air on its pressure and temperature [6]:

$$n_{t,p} - 1 = (n_0 - 1) \cdot \left[\frac{p \cdot (1 + (1.049 - 0.0157 \cdot t) \cdot 10^{-6} \cdot p)}{720.883 \cdot (1 + 0.003661 \cdot t)} \right]. \quad (1)$$

Here n_0 – is the refractive index of air at initial conditions (t_0= 15 °C, p_0= 760 mm. Hg.), p – pressure in mm. Hg., t - temperature in Celsius degree. Figure 1 shows the corresponding graphs.

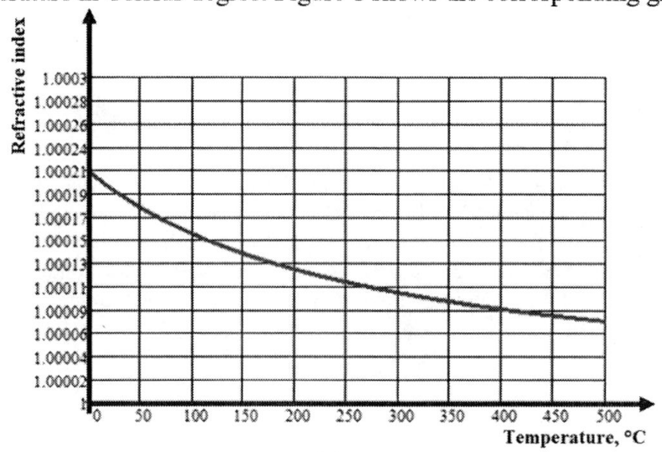

(a)

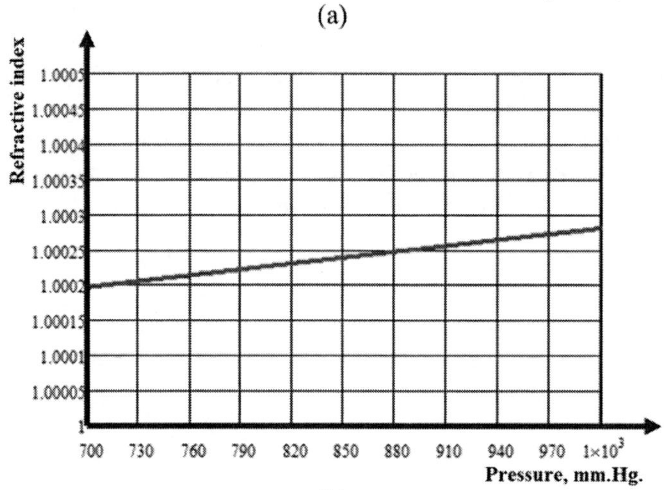

(b)

Figure 1. The dependence of the refractive index of air on its pressure and temperature. (a) – dependence of the refractive index of air on its temperature; (b) – dependence of the refractive index of air on its pressure.

Journal of Physics: Conference Series **1421** (2019) 012024 doi:10.1088/1742-6596/1421/1/012024

From figure 1 it is clear that temperature has a significantly greater effect on the refractive index in comparison with pressure. Therefore, if the problem in question does not require exceptional accuracy in its solution, then small changes in pressure can be neglected. Information about the studied medium is obtained by analyzing a series of consecutive images of the background screen, distorted by small gradients of the refractive index. In addition to distorted images of the background screen, there must be at least one image in which these distortions are absent. This is necessary to perform cross-correlation processing, the result of which will be the vector field of the image displacements, which is, in fact, the visualization of optical inhomogeneity. The vector field of displacements can be recalculated into the surface of the distribution of the index of refraction. If the refractive index gradient was caused by a change in any one of the characteristics of the medium, then the distribution of the refractive index can be recalculated into the distribution of this characteristic itself. As an example, figure 2 shows the temperature distribution in the air flow from an industrial dryer.

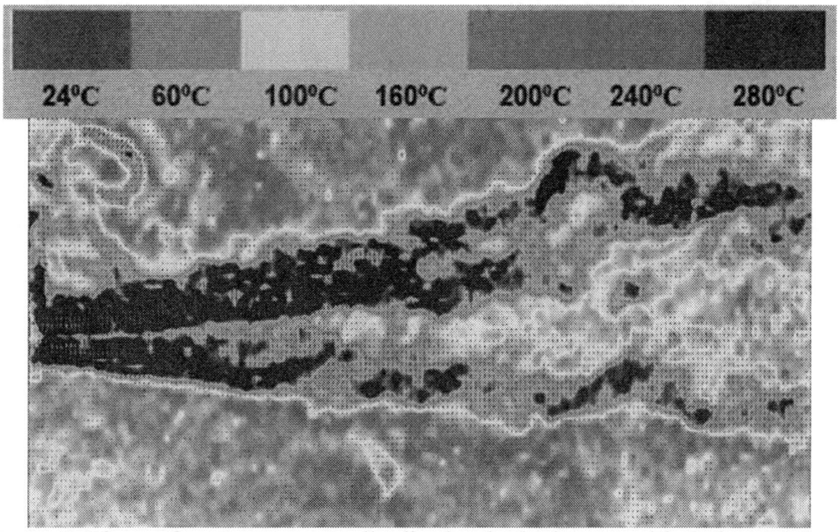

Figure 2. The surface temperature distribution in the jet assembly industrial dryer.

It is worth to be noted that this result was obtained without taking into account changes in pressure (the gradient of the refractive index arising from a change in pressure was considered negligible). However, such neglect is not always permissible. Therefore, the issue of the possibility of obtaining separate information on several environmental parameters that have formed complex optical inhomogeneity is highly relevant.

3. Experiment

The main goal of this work was to obtain and visualize optical inhomogeneity formed by changes in pressure and temperature. Due to the fact that in laboratory conditions it is rather difficult to create a pressure gradient in the air that is perceptible for BOS, it was decided to perform an experiment using a solid. The experimental setup scheme was a standard BOS-scheme, in which a plane-parallel glass plate with internal defects caused by residual mechanical stresses acted as a test object. A schematic representation of the experimental setup is presented in figure 3.

The experiment was concluded to the registration of several series of images of the background screen. Each subsequent series differed from the previous one by introducing additional optical inhomogeneity. The plane-parallel plate was installed normally to the optical axis of the camera lens on the metal surface, with which it was fixed using a metal clamp. With the help of these clutches inside the plate, the occurrence of mechanical stresses was provoked, due to which an optical inhomogeneity was formed due to the pressure gradient. After that, without unclamping the plate under study, its lower interface was heated with a candle flame. Thus, an optical inhomogeneity

Journal of Physics: Conference Series **1421** (2019) 012024 doi:10.1088/1742-6596/1421/1/012024

caused by a temperature gradient arose inside the plate. The heating temperature of the metal surface on which the plate was located was measured using a chromel-copel type thermocouple connected to a digital multimeter. Images were recorded at temperatures of 75 °C, 100 °C, and 125 °C on a JAI RMC-2030GE camera with a high-aperture lens. The frame-rate was 16 fps at a resolution of 1600 × 1200 pixels. The distance from the edge of the lens to the background screen was 0.515 m, the plate itself, 5 mm thick, was located at a distance of 0.2 m from the background screen.

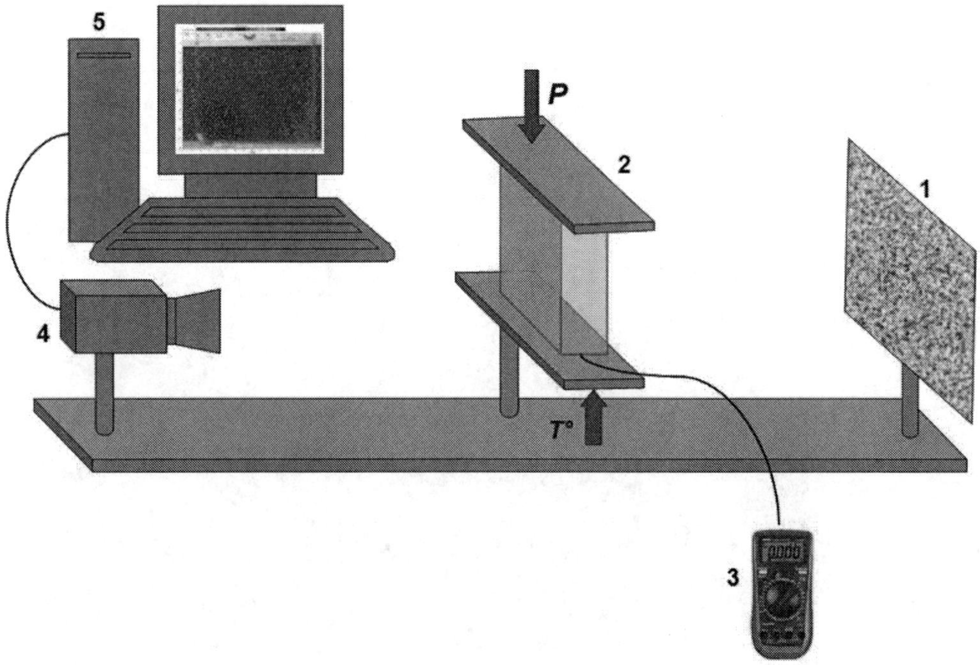

Figure 3. Schematic of the experimental setup: 1 – background, 2 – test-object (plane-parallel glass plate), 3 – digital multimeter, 4 – camera, 5 – computer.

4. Experimental results

The processing of the obtained images was performed in the PivViev 2C v.3.8.0 (demo version) with the same settings for all image sequences.

Table 1. PIV Evaluation setup.

Window size	64×64 px
Step size	16×16 px
Correlation mode	Standart (FFT) correlation
Interrogation mode	Multiple-pass interrogation; 2 passes
Sub-pixel peak fit	3-point Gauss fit

At the first stage of processing, the background image without plane-parallel plate between the camera and the screen on which camera was focused was used as a reference frame. An example of the processing results is shown in figure 4.

Journal of Physics: Conference Series **1421** (2019) 012024 doi:10.1088/1742-6596/1421/1/012024

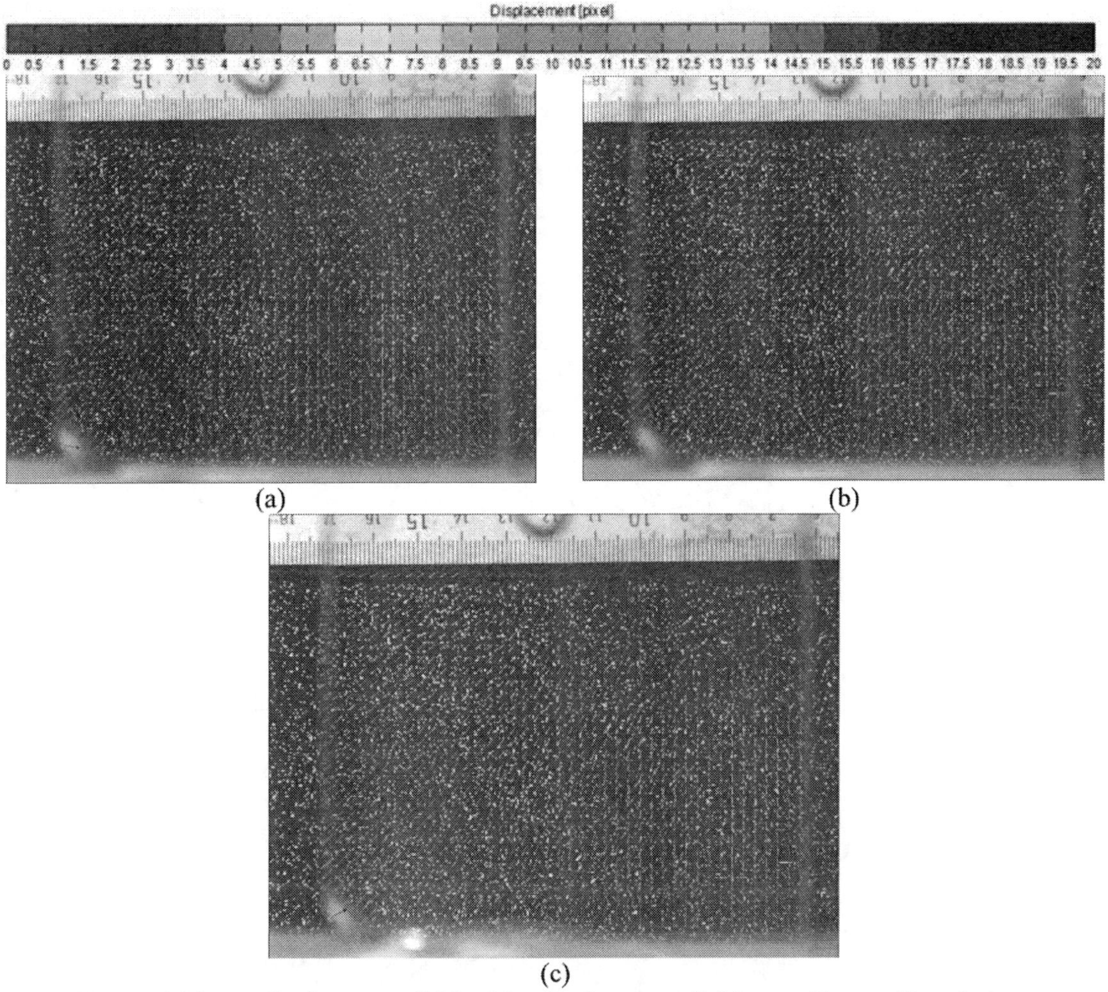

Figure 4. Vector displacement fields: (a) – displacement field caused by unaffected plate, (b) – displacement field caused by deformed plate, (c) – displacement field caused by deformed and heated plate.

From these results, it can be concluded that the mechanical stresses inside the plate cause significantly greater distortion of the image in comparison with the distortions caused by changes in pressure and temperature. However, analyzing the processed images for each series, it was noted that during the passage from frame to frame, the displacement fields differ slightly from each other. These minor differences are precisely due to the inhomogeneities formed due to compression and heating of the plate. For this reason, processing was performed for each sequence using background image with a plane-parallel plate between the camera and the screen as a reference. The corresponding results are shown in figure 5.

It can be seen from this figure that the image displacement values for this experimental scheme are rather small and lie in the range from 0.5 to 1.5 pixels. Nevertheless, visualizing the inhomogeneities caused by changes in pressure and temperature is possible.

Since, during the experiment the plate was first deformed and then heated up, it becomes possible to separate the effects of pressure and temperature during visualization. To do this, for processing a series of frames with heating of the plate, the image on which the plate is deformed but not yet heated should be used as a reference frame. As a result, the image displacement field will be obtained, due only to a change in temperature. An example of such processing is shown in figure 6.

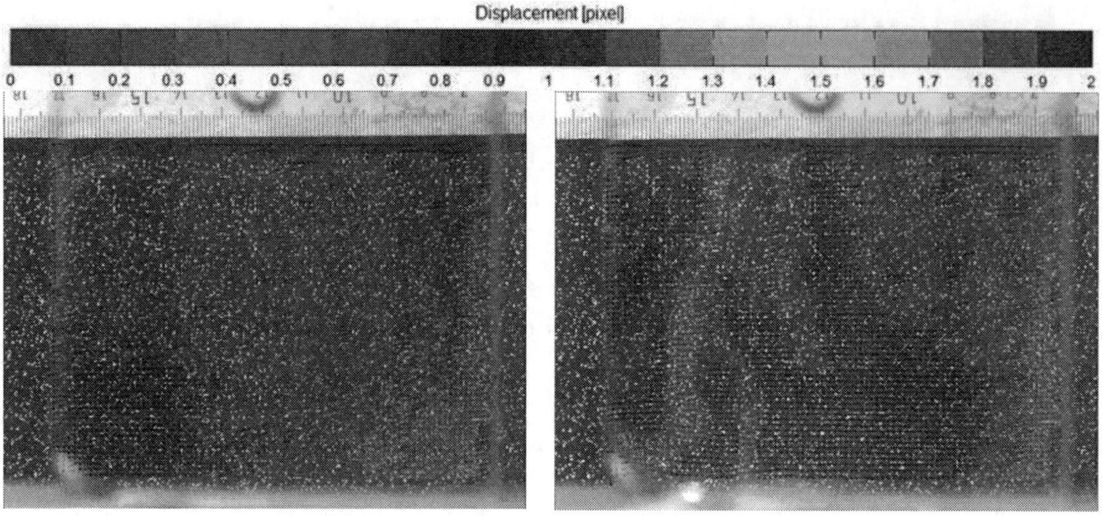

Figure 5. Processing results: (a) – displacement field due to deformation, (b) – displacement field due to temperature changes and deformation.

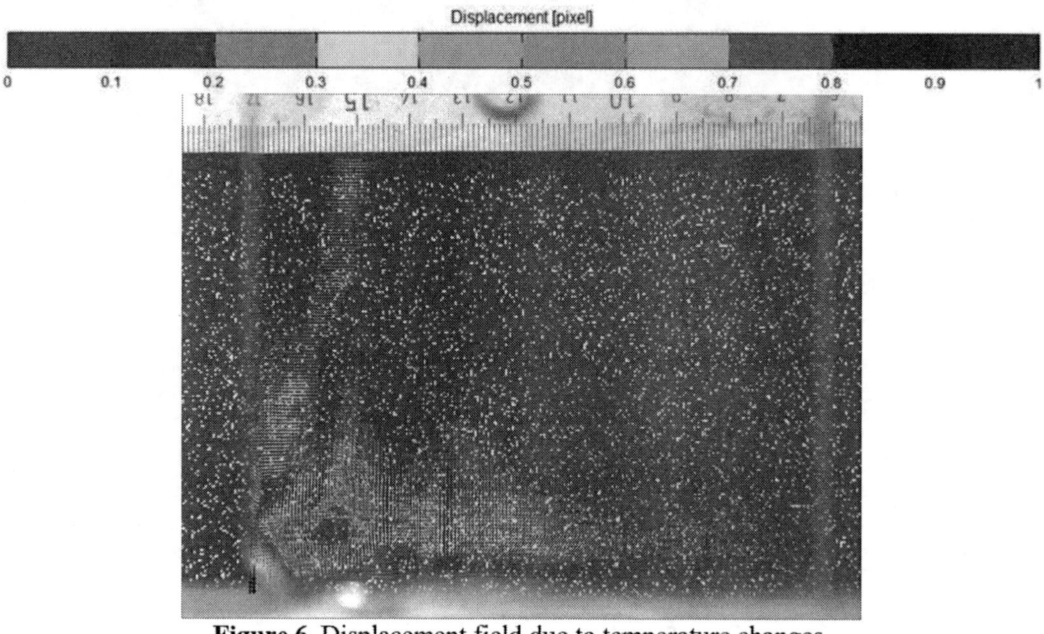

Figure 6. Displacement field due to temperature changes.

5. Conclusion

One of the most commonly used refractometric methods for visualizing optical inhomogeneities is BOS. This method has a certain potential as a measuring technique for obtaining the distributions of various environment characteristics related to its refractive index. However, to obtain sufficiently accurate data, it is necessary to obtain information on each of the factors that formed inhomogeneity separately.

In the course of the work, an experiment was carried out to investigate the optical inhomogeneity formed in a glass plate by the influence of temperature and mechanical stresses. Processing of experimental images showed that it is possible to visualize the distribution field of each parameter in the presence of another if their changes do not occur at the same time or the distribution of one of them is known.

References

[1] Belozerov A F 2014 Modern technologies for creating systems of optical and physical measurements of gas-dynamic fields with sizes up to 1000 mm *Optoelectronics, Instrumentation and Data Processing* **5** 50 429–41

[2] Inshakov S I 2007 Visualization of gas flows structures by mean of schlieren and interferometry methods *Vestnik Samarskogo Gosudarstvennogo Aerocosmicheskogo Universiteta im. S.P. Koroleva* **2** 13 108–13

[3] Ota M, Leopold F, Noda R and Maeno K 2015 Improvement in spatial resolution of background-oriented schlieren technique by introducing a telecentric optical system and its application to supersonic flow *Experiments in fluids* **56** 48 1–10

[4] Bauknecht A, Ewers B, Wolf C, Leopold F, Yin J and Raffel M 2015 Three-dimensional reconstruction of helicopter blade-tip vortices using a multi-camera BOS system *Experiments in fluids* **56** 1866 1–13

[5] Venkatakrishnan L, Suriyanarayanan P and Jagadeesh G 2013 Density field visualization of a micro-explosion using background oriented schlieren *Journal of Visualization* **16** 177–80

[6] Cox A N 2000 *Allen's Astrophysical Quantities* (New York: Springer)

Development of the system for quantitative gas-liquid flow visualization

A V Kuchmenko, D A Koshelev, N M Skornyakova and P P Shcherbakov

National Research University «Moscow Power Engineering Institute», Krasnokazarmennaya 17, Moscow, 111250, Russia

E-mail: ann.131@yandex.ru

Abstract The paper is concerned with the developing of an optical-electronic complex for visualization and quantification of the gas-liquid flow from the nozzle using the particle image velocimetry method depending on various flow parameters. Selection of optical elements for experimental research was based on the possibility of visualizing the flow, determining its type, opening angle, and also obtaining vector velocity fields in the plane under study, on the basis of which the gas-liquid flow drops velocities dependencies on the coordinates in given sections of the vector field were constructed.

1. Introduction

For a long time, numerous applications of nozzles in industry have been known. To date, there are methods for calculating nozzles and the results of theoretical and experimental studies used in various fields of science and industry, but it is often quite difficult in experimental studies to determine the flow parameters by contact methods [1-8]. The solution to these difficulties can be the use of contactless optical methods which are widely used due to modern high-speed video cameras, lenses with various parameters, software and the rapid optical methods development for researching liquid and gas flows.

The particle image velocimetry technique (PIV) is one of the optical contactless methods of fluxes research, allowing to recorder instantaneous velocity fields in the measurement plane. The experimental data obtained using the PIV method will also make it possible in the future to simply determine the flow structure from the nozzle (full or hollow cone), as well as the jet opening angle [9-12]. The use of the purposed technique can significantly reduce the time of the experiment, which often plays an important role in the complex technological processes.

The main aim of the work is to develop a contactless optical-electronic complex for the quantitative visualization of a dispersed gas-liquid flow, the production and processing of experimental images. Based on the images obtained, the nozzle type (full or hollow cone), the opening angle of the jet were determined, and the vectorial velocity field was obtained, the dependences of the velocities of the flow drops in various sections were plotted.

2. Experimental setup

For the implementation of the experimental setup to the flow study using the method of particle image velocimetry, the presence of a laser sheet and tracers is necessary [13-15]. The light from the laser module passes through a cylindrical lens, which forms a diverging laser sheet, exposing the flow in a certain section. A blue diode laser with 450 nm wavelength and 1000 mW was used as a laser module.

To determine the flow rates, tracers are introduced using the spraying system for oils and other liquids, the speed of which determines the flow rate, but in this experiment drops from the gas-liquid flow itself appeared as tracers. The main criteria related to tracer are small size, spherical shape, and smooth surface. The recording of the stream propagation was carried out using a high-speed camera which registers the illuminated stream against a darkened background with a certain frame rate. We used a monochrome high-speed camera Fastec HiSpec 1 with a CMOS matrix, which allows obtaining images at speeds up to 506 frames per second at a resolution of 1280 × 1024 pixels. This model of a video camera has high sensitivity - 3200 ISO and the ability to shoot up to 112000 frames per second. For the correct construction of the object under study vector velocity field, the number of recorded frames per second is at least 500. On the basis of the first series of experiments, the necessary parameters of the lens to obtain high-quality images were determined. The parameters studied at this stage include a sufficient lens aperture ratio (f1.2-close) and a focusing range (0.3 - infinity). A lens that satisfies all of the above parameters is the Navitar DO1212 with a focal length of 12 mm. The main task of the measurement system, the collection and processing of information during the experiment is to receive data from the camera, as well as registering the data (recording in computer memory). In addition, in real-time mode, the stream is displayed on the screen to select the required shooting parameters (recording speed, light quality, etc.). Further processing of the obtained data and analysis of the results of the experiment. Image processing allows calculating the displacement of particles between two successively recorded frames using cross-correlation analysis, which uniquely determines the flow rate in a given area.

3. Experimental data processing

Using experimental images (visually), it is possible to determine an equally important parameter - the type of jet from the nozzle. Figure 1 (a), (b) shows photographs of the flows from two different nozzles.

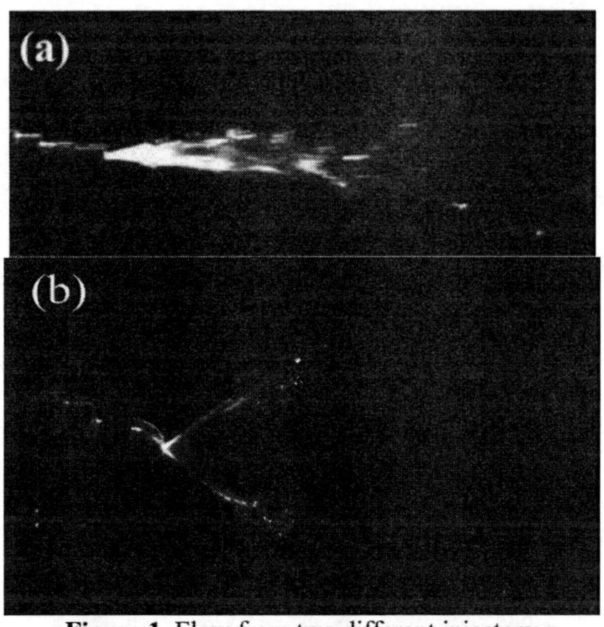

Figure 1. Flow from two different injectors:
a – nozzle type full cone; b – hollow cone type nozzle.

Journal of Physics: Conference Series **1421** (2019) 012025 doi:10.1088/1742-6596/1421/1/012025

It was obtained that the injector being investigated is of a full cone type. Such nozzles are used to a greater extent for cooling in view of the fact that the force of the blow flow from it is significantly less than that of other types, and at the same time it allows the liquid to be evenly distributed on the surface of the cooled object.

The angle of jet opening was determined from the geometry of the jet due to the program JMicroVision. The method of determining the angle of jet aperture from the nozzle is shown in figure figure 2.

Figure 2. Determination of the jet aperture angle from the nozzle.

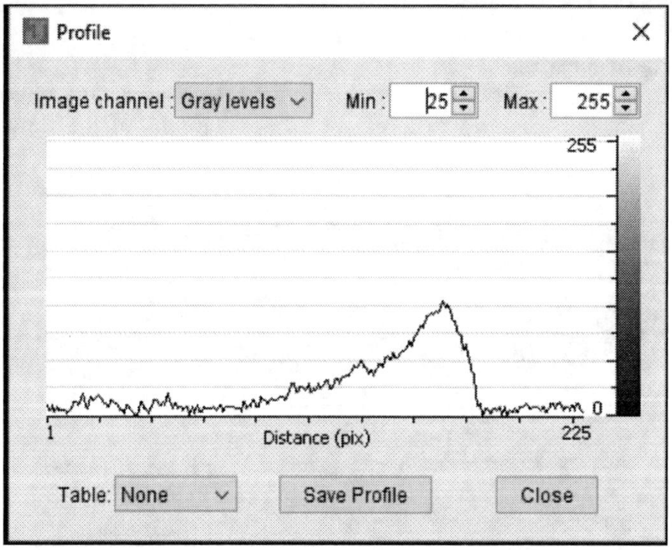

Figure 3. The flux brightness distribution (the yellow line indicates the brightness threshold of the flux structure).

To determine the reliable width of the flow, the JMicroVision program allows building a brightness distribution, which allows you to consider the brightness threshold of the flow structure to correctly determine the aperture angle of aperture In this case, the threshold corresponded to 25 (≈ 10 % of maximum brightness 255). And according to the graph of the brightness distribution in the section, the jet geometry was determined. Table 1 shows the obtained values of the jet aperture angle at pressures of flow in 2 and 4 bars.

The jet opening angle from the nozzle for 2 and 4 bar is not significantly different and has values of 17 and 18 degrees, respectively. When conducting experimental studies, processing was performed to determine the quantitative flow characteristics from the nozzle. Using cross-correlation processing, the velocity field was constructed in the cross section of the gas-liquid flow from the nozzle. Processing was carried out using the PIVview program with the following parameters: polling area - 64 pixels; step - 32 pixels, 50% overlap, the algorithm for determining the correlation peak - 3-point Gaussian approximation. The function of eliminating erroneous vectors was also applied. Baseline experimental data are shown in figure 4.

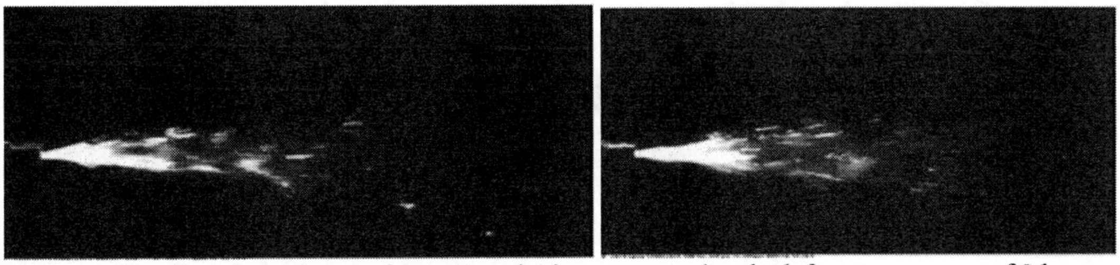

Figure 4. The flow from the nozzle at a 2 and 4 bar pressure (on the left - at a pressure of 2 bar, on the right - at a pressure of 4 bar).

The result of constructing the vector velocity field is shown in figure 5

Figure 5. The result of constructing the flow velocity field from the nozzle at a pressure of 2 and 4 bar (on the left - at a pressure of 2 bar, on the right - at a pressure of 4 bar)

Figure 6 shows the lines for constructing the values of velocity in the cross section X and Y of the gas-liquid flow.

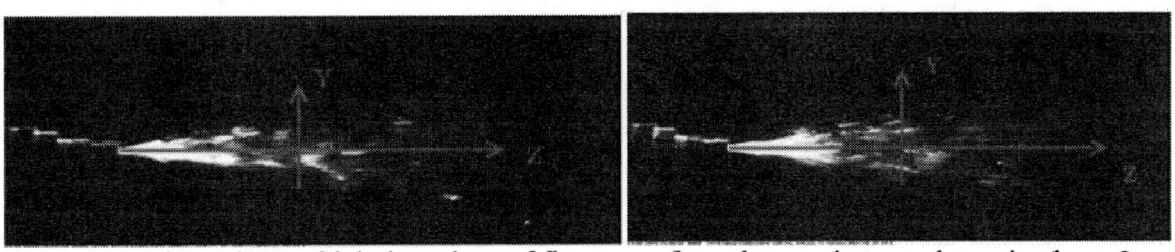

Figure 6. Sections along which the values of flow rates from the nozzle were determined at a 2 and 4 bar pressure (to the left - at a pressure of 2 bar, to the right - at a pressure of 4 bar).

According to the obtained values of the velocities of the flow drops, the dependences of the average velocity of the drops on the Z and Y coordinates were plotted.

Journal of Physics: Conference Series **1421** (2019) 012025 doi:10.1088/1742-6596/1421/1/012025

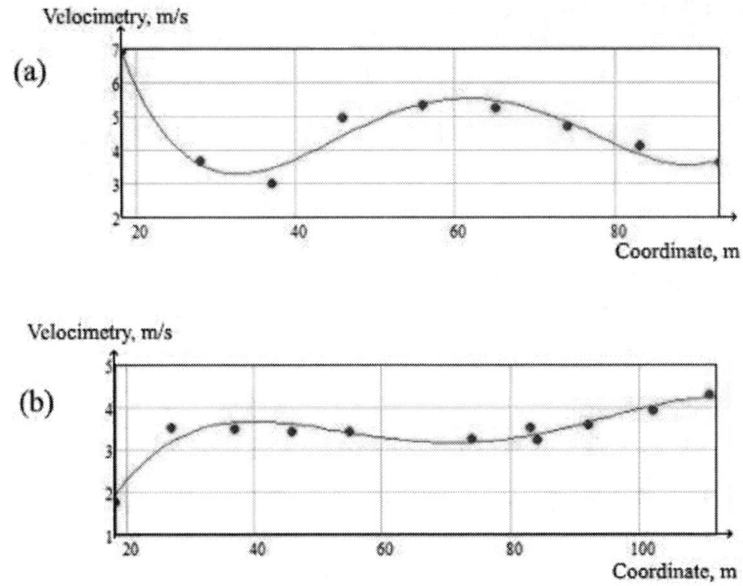

Figure 7. The dependence of the velocity of the particle flow from the nozzle from the coordinates Z, Y (pressure 2 bar):
a – dependence of particle flow velocities on Z coordinate; (b) – dependence of particle flow velocities on Y coordinate.

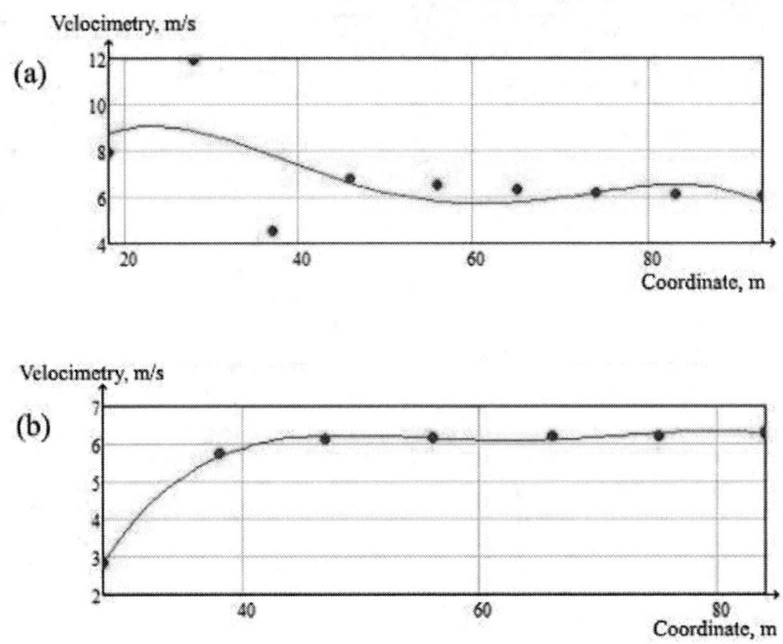

Figure 8. The dependence of the velocity of the particle flow from the nozzle from the coordinates Z, Y (pressure 4 bar):
a – dependence of particle flow velocities on Z coordinate; (b) – dependence of particle flow velocities on Y coordinate.

The dependences obtained are in good agreement with the data given in [16]. Dependencies for 2 and 4 bars pressure turned out to be similar. Considering the dependence of speed on the Z coordinate, it is possible to conclude that at the initial stage the flow is a zone of continuous liquid flow, then small particles slow down and begin to fall. Medium particles combine with each other and move, gradually acquiring greater speed, and then slow down and also fall down. The dependence of the velocity on the Y coordinate shows the asymmetry of the jet, since the velocity is different at different points of this section.

4. Conclusion

An optical-electronic device was developed to enable the dispersed flow quantitative visualization. The efficiency of the installation of high-speed visualization and quantification was tested at two flow pressures. Test and experimental images were obtained that are necessary for carrying out the algorithm of the PIV method. After analyzing the experimental data, it was found that the best for the experiment is a high-aperture lens. For this case, experimental images were processed, which were used to determine the jet opening angle, nozzle type, flow velocities distribution at two pressures, and dependences of the velocities of the flow drops in various sections were plotted. It should be stressed that this installation will need to be upgraded to study various nozzles types, which is related to the size of the droplets themselves, as well as the type of jet from the nozzles, but these improvements will not be global.

Acknowledgments

The work was carried out with the financial support of the RNF (project 16-19-10457-P).

References

[1] Hasselmann K, Aus Der Wiesche S.and Kenig E Y 2019 Optimization of Piecewise Conical Nozzles: Theory and Application *Journal of Fluids Engineering, Transactions of the ASME* **141(12)** 121202

[2] Bang B-H, Kim Y-I, Jeong S, Yarin A L and Yoon S S 2019 Theoretical model for swirling thin film flows inside nozzles with converging-diverging shapes *Applied Mathematical Modelling* 76 607-16

[3] Gregorio E, Torrent X, Planas S and Rosell-Polo J R 2019 Assessment of spray drift potential reduction for hollow-cone nozzles: Part 2. LiDAR technique *Science of the Total Environment* 687 967-77

[4] Maly M, Sapik M, Cejpek O, Katolicky J and Jedelsky J 2019 Effect of spill orifice geometry on spray and control characteristics of spill-return pressure-swirl atomizers *Experimental Thermal and Fluid Science* **106**, 159-70

[5] Persico G, Rodriguez-Fernandez P and Romei A 2019 High-Fidelity Shape Optimization of Non-Conventional Turbomachinery by Surrogate Evolutionary Strategies *Journal of Turbomachinery* **141(8)**, 081010

[6] Oh S Y, Shin J S, Kim T S, Chung C-M and Lee J 2019 Effect of nozzle types on the laser cutting performance for 60-mm-thick stainless steel *Optics and Laser Technology* 119,105607

[7] Feng S, Xiao L, Ge Z, Du X and Wu H 2019 Parameter analysis of atomized droplets sprayed evaporation in flue gas flow *International Journal of Heat and Mass Transfer* **129** 936-52

[8] Mirshahi M, Yan Y and Nouri J M 2015 Influence of cavitation on near nozzle exit spray *Journal of Physics: Conference Series* **656(1)**, 012093

[9] Maly M, Sapik M, Cejpek O, Katolicky J and Jedelsky J 2019 Effect of spill orifice geometry on spray and control characteristics of spill-return pressure-swirl atomizers *Experimental Thermal and Fluid Science* **106**, 159-70

[10] Hillamo H, Anttinen T, Aronsson U, Andersson O and Johansson B 2011 Flow field measurements inside a piston bowl of a heavy-duty diesel engine SAE Technical Papers

[11] Sergeev D A 2011 etry using particle images (PIV-methods) in laboratory modeling of geophysical flows *Bulletin of Nizhny Novgorod University. N.I. Lobachevsky* **4 (2)** 522 - 24.
[12] Knysh Yu A, Redkin E S and Dmitriev D N 2011 Experimental study of a swirling gas jet by the method of digital tracer visualization *Bulletin of the Samara State Aerospace University. Academician S.P. Korolova* **5(29)** 11-17
[13] Varaksin A Yu, Protasov M V., Marinichev D V and Vasiliev N V 2015 Analysis of tracer particle parameters for optical diagnostics of vortex flows *Measuring equipment* **6** 46-9
[14] Bernhard Wieneke 2015 PIV uncertainty quantification from correlation statistics *Measurement Science and Technology* **26(7)**
[15] Tokarev M P, Markovich D M and Bilsky A V 2007 Adaptive particle image processing algorithms for calculating instantaneous velocity fields *Computational Technologies* **12 (3)**
[16] Eskov A V, Svistula A E, Cherepov O D and et all 2006 The method for determining the dispersed composition of droplets of a jet of sprayed liquid: Patent RU 2277442 № 16

Micro-PTV technique application to velocity field measurements in immiscible liquid-liquid plug flow in microchannels

A V Kovalev[1,2], A A Yagodnitsyna[1,2] and A V Bilsky[1,2]

[1]Kutateladze Institute of Thermophysics, 1, Lavrentev Ave., Novosibirsk, 630090, Russia
[2]Novosibirsk State University, Novosibirsk, 2, Pirogova str., Novosibirsk, 630090, Russia

E-mail: therfmig@gmail.com

Abstract. The present paper presents the results of micro-PTV technique application to velocity fields measurements in plug flow of immiscible liquids in T-shaped straight and curved microchannels. The data processing algorithms are described; the procedure for velocity circulation assessment was proposed and applied to measure velocity circulation in aqueous plugs at different flow rates. The comparison with micro-PIV technique application results is performed.

1. Introduction

Microfluidic devices are of great interest in the last years due to multiple advantages such as high-intensity heat and mass transfer, small volumes of reagents, the possibility to control local parameters and safety of the processes carried out [1]. Immiscible liquid-liquid flows in microchannels are widely used in chemical and biomedical applications and allow to enhance the outcome of extraction processes and other chemical reactions [2,3]. Plug flow when a series of plugs of one liquid surrounded by liquid film and separated by slugs of another liquid is essential for microfluidic devices because of its unique properties. The circulations inside plugs enhance mixing and increase mass transfer through the dimeric interface. There is a need to measure velocity fields inside plugs and assess velocity circulation in order to develop and optimize microfluidic devices. Although Microresolution Particle Image Velocimetry (micro-PIV) has become an established technique for velocity field measurements in microflows since 1998 [4], it has some drawbacks in application to immiscible liquid-liquid plug flow. Due to confinement, the velocity distribution inside plugs has the areas with high velocity gradients which impose constraints on the maximum interrogation area. On the other hand, typically flow seeding density is quite low in microflows which leads to large size interrogation areas or velocity fields averaging. Latter has the meaning only for quasi-stationary flows which is not the case for plug flow. The alternative technique to micro-PIV is Microresolution Particle Tracking Velocimetry (micro-PTV). The technique is based on tracking of individual particles on the image frame rather than the correlation of interrogation areas containing a group of particles; thus it can be applied to the high gradient velocity fields.

In the present work, the micro-PTV technique was applied to velocity field measurements in plug flow of immiscible liquids in T-shaped microchannels. The procedure was developed to assess velocity circulation inside plugs.

2. Experimental setup

Velocity field measurements were carried out in plug flow regime of immiscible liquids in T-shaped straight and curved microchannels. The top view of the microchannels is presented in Figure 1. The width of the inlet channels was 200 μm; the width of the outlet channel was 400 μm while the depth of the channels was equal to 200 μm. The curvature radius of the curved channel was 600 μm. Castor oil was a carrier liquid; distilled water played a role of the dispersed phase. Flow rates of dispersed and carried liquids were varied in the range 0.695μl/min $<Q_d<$ 4.17 μl/min and 0.695 μl/min $<Q_c<$ 8.34 μl/min by means of KD Scientific double syringe pump. The dispersed phase was seeded by 2 μm fluorescent particles with 0.07% concentration by volume. The scheme of the experimental setup is shown in figure 1. The microchannel was mounted on the Zeiss microscope stage. The velocity measurements were performed at a distance of 8.5 mm from a T-junction and the curved part. The focal plane was set to the central plane of the channel. The flow was illuminated through the microscope lens by double Nd:YAG laser with 532 nm wavelength. The light emitted by particles was directed to a dichroic mirror and captured by a double exposure CCD camera with 4 Mpix resolution. The spatial resolution achieved was 0.8 μm/pixel.

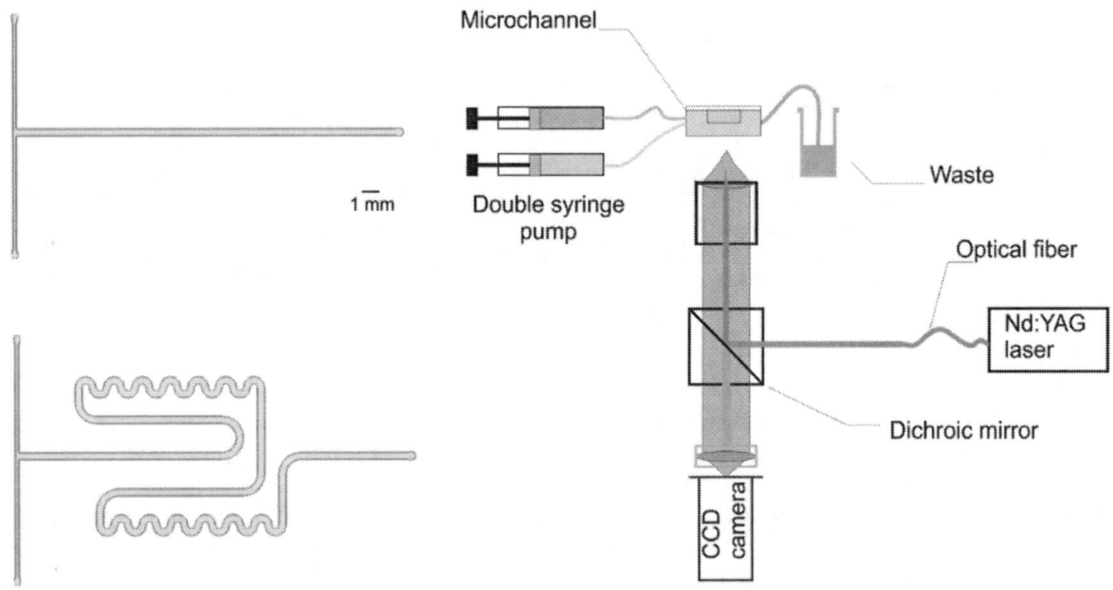

Figure 1. Left: the top view of straight and curved T-shaped microchannels. Right: the scheme of the experimental setup for micro-PTV measurements.

3. Results and discussions

3.1. Velocity field calculation

All calculations were performed using our in-house ActualFlow software, where PTV algorithm was realized. For PTV calculations a pair of tracer images with a short time delay between them is required similarly to PIV. The algorithm detects individual particles at both the first and the second frame and then matches them to obtain the most probable displacement of each particle. The algorithm includes three main stages: *particle identification, relaxation algorithm, and result correction.* At the first stage – *particle identification* – a raw image of tracer particles (Figure 2 (a)) is filtered using the Ricker wavelet convolution with a Fourier transform of an initial image. This operation allows to single out characteristic structures and to decrease noise. Further, local maxima at the image are found. A particular pixel is defined as local maxima if it has maximum intensity in the area of 3×3 pixels around it. Finally, at this stage, we validate if local maxima correspond to tracer particles. To perform this validation, in the vicinity of each local maximum we calculate average intensity I_{avg} and compare the normalized difference between local maximum intensity I_{max} and I_{avg} with some threshold value P:

$$\frac{I_{max}-I_{avg}}{I_{max}} \geq P \quad (1)$$

The threshold parameter P should be chosen for particular experimental data because images of tracer particles can differ significantly. Too high values of P can lead to a small amount of identified particles, whereas too small values lead to misidentification and spurious particles. The example of particle identification is presented in Figure 2 (b). In this work, P was varied from 0.9995 to 0.9998.

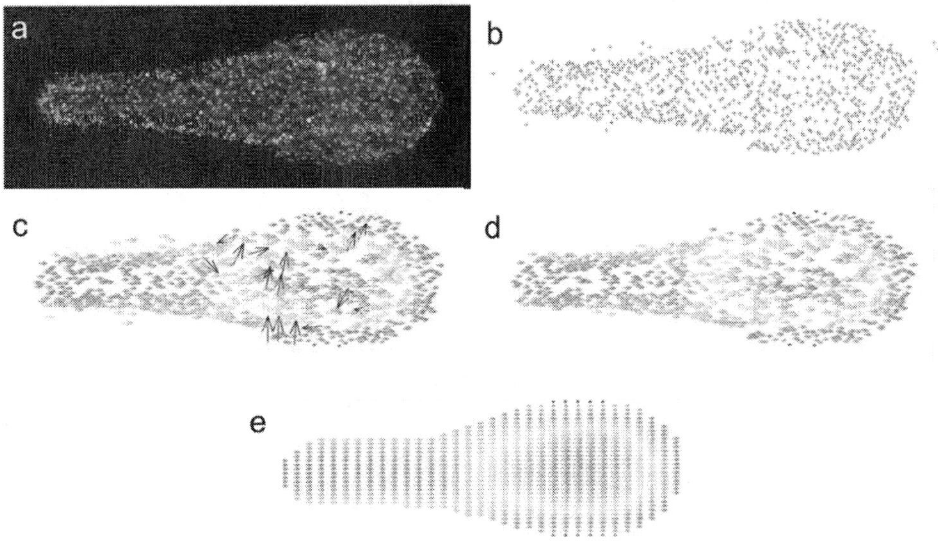

Figure 2. Stages of velocity field calculation using the PTV algorithm: a) a raw image of tracer particles; b) particles found at the image by the PTV algorithm; c) a raw PTV vector field; d) an irregular PTV field after range validation and mask imposition operations; e) a regular velocity field interpolated from PTV data.

The second stage is the relaxation algorithm which estimates first frame particle probabilities of displacing in corresponding particles in the second frame. This algorithm is based on two main principles: maximum particle velocity (or displacement) is defined, and continuity of velocity field is satisfied (local groups of particles moves in the approximately same direction). When the most probable displacement is determined, the corresponding velocity vector is calculated for each pair of particles. A more detailed description of the relaxation algorithm is presented in [5].

At the final stage, the correction of results is performed using the correlation function maximum [6]. Each particle from the second frame is shifted according to the previously calculated displacement and correlation between particle from the first frame, and the shifted particle are calculated. Since a maximum of the correlation function is known, corrections to the initial displacement are made. An example of the resulting irregular velocity field is presented in Figure 2 (c). If spurious or irrelevant vectors appear and it is impossible to eliminate them just by varying threshold value P, additional procedures such as image masking or velocity vector range validation should be applied. Irregular field after mask imposition and range validation is shown in Figure 2 (d). Eventually, irregular fields were interpolated into a regular grid using second order local polynomial algorithm for further calculation of derivatives (Figure 2 (e)). Amount of nodes in the grid was approximately equal to the number of vectors in the irregular field. For experiments in this paper typical grid resolution was 8×8 μm^2..

3.2. Velocity circulation assessment

To assess circulations inside plugs for liquid-liquid plug flow it is convenient to use vorticity fields calculated from velocity derivatives. Often, the PIV algorithm provides unsatisfactory results and resolution of velocity field insufficient to calculate the vorticity field. For instance, this case takes place for serpentine microchannels, where velocities inside plugs have different directions. In Figure 3 the comparison between vorticity fields obtained from PIV and PTV data is presented. PIV velocity field was calculated using 64×64 pixel grid without overlapping, further refinement of the grid was impossible due to initial image quality and insufficient flow seeding. As can be seen in Figure 3, the vorticity field obtained from PIV data, even the main vorticies in the front part of the plug are resolved poorly. Besides, small values of velocity vectors near the plug boundary introduce significant error into the resulting vorticity field. By comparison, using interpolated PTV results, adequate vorticity field can be calculated even for a curved channel.

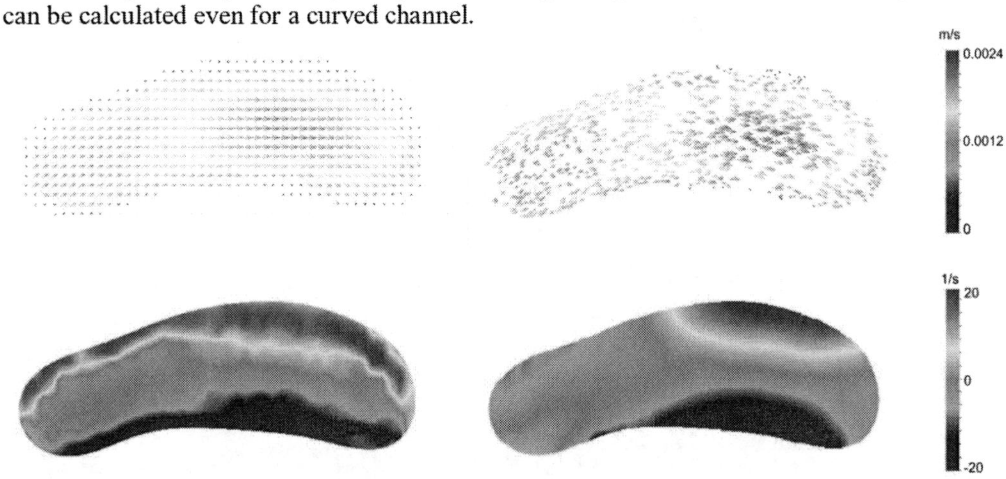

Figure 3. Comparison between PIV (left) and PTV (right) measurements of velocity and vorticity fields inside the aqueous plug in the curved microchannel.

For the assessment of velocity circulation inside plugs, the following procedure has been developed. From calculated velocity derivatives, the vorticity component, directed orthogonally to the measurement plane, were obtained. The total circulation in the plugs was calculated from vorticity distributions using the following equation:

$$\Gamma = |\Gamma_+| + |\Gamma_-| = \iint_S \left|\frac{\partial v_y}{\partial x} - \frac{\partial v_x}{\partial y}\right| dxdy \approx \sum_{i=0}^{N} |\omega_z|_i S_i \quad (2)$$

We approximated the area integral with the sum of the z-th component of vorticity ω_z multiplied by the cell area S for each cell in the circulation areas. Circulation areas were defined as areas with vorticity values greater than 25% of the maximum for a given plug. It allowed us to eliminate areas with low vorticity from consideration. To take into account the circulation regions with both positive and negative vorticity values, the values of ω_z were taken in absolute value.

Using the described approach to the total circulation assessment experiments with castor oil - water plug flow in the straight microchannel were conducted. The plot of the total circulation versus $Ca_{\text{bulk}}Q_d/Q_c$ parameter, where Ca is the capillary number, is shown in figure 4. Total circulation depends linearly on $Ca_{\text{bulk}}Q_d/Q_c$. This result is in agreement with the work [7], where $Ca_{\text{bulk}}Q_d/Q_c$ was shown to describe plug shape and velocity fields inside aqueous plugs in castor oil.

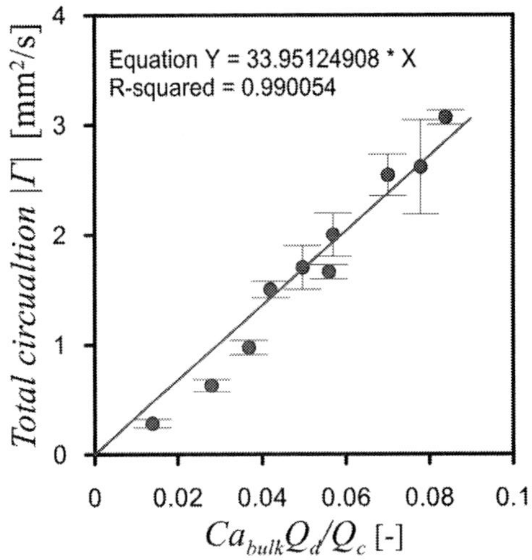

Figure 4. Velocity circulation in aqueous plugs depending on $Q_d/Q_c \cdot Ca_{bulk}$.

4. Conclusion

In the present study, micro-PTV technique was applied to velocity field measurements in plug flow of immiscible liquids in microchannels. Two types of microchannels were used: straight T-shaped microchannel and curved T-shaped microchannel. Velocity field and velocity circulation calculation procedures were described. It was demonstrated that the micro-PIV technique does not take satisfactory results for velocity circulation assessment in case of curved plug due to the restriction of seeding density on the minimum size of the interrogation area. Velocity circulation at different flow rates was calculated in the T-shaped microchannel, its linear dependence on $Q_d/Q_c \cdot Ca_{bulk}$ was shown.

Acknowledgments

This work was carried out under a state contract with IT SB RAS.

5. References

[1] Squires T M and Quake S R 2005 Microfluidics: Fluid physics at the nanoliter scale *Rev. Mod. Phys.* **77** 977–1026

[2] Šalić A, Tušek A and Zelić B 2012 Application of microreactors in medicine and biomedicine *J. Appl. Biomed.* **10** 137–53

[3] Xu C and Xie T 2017 Review of Microfluidic Liquid–Liquid Extractors *Ind. Eng. Chem. Res.* **56** 7593–622

[4] Santiago J G, Wereley S T, Meinhart C D, Beebe D J and Adrian R J 1998 A particle image velocimetry system for microfluidics *Exp. Fluids* **25** 316–9

[5] Theunissen R, Stitou A and Riethmuller M L 2004 A novel approach to improve the accuracy of PTV methods *12th Int. Symp. Appl. Laser Tech. to Fluid Mech. Lisbon, Port.* 12–5

[6] Akhmetbekov Y K, Markovich D M and Tokarev M P 2010 Study of the PTV method with individual particle correlation correction *Comput. Technol.* **15** 57–72

[7] Kovalev A V, Yagodnitsyna A A and Bilsky A V 2018 Flow hydrodynamics of immiscible liquids with low viscosity ratio in a rectangular microchannel with T-junction *Chem. Eng. J.* **352**

Journal of Physics: Conference Series **1421** (2019) 012027 doi:10.1088/1742-6596/1421/1/012027

Elimination of distortions on images in the existence of turbulence

S Yu Belov

Physics Department, Moscow Power Engineering Institute, 17 Krasnokazarmennaya street, Moscow, Russia

E-mail: BelovSY@mpei.ru

Abstract. The aim of the study is to eliminate the influence of the turbulence model on the photographed images. This problem is solved by computer processing of images containing such distortions. In the course of the work, images with simulated turbulent distortions were obtained, and they were also post-processed using cascade filters.

1. Introduction

The problem of turbulence is quite common in everyday scientific life. Often, when conducting field experiments, a photo or video is taken. The environment surrounding the object under study has heterogeneities in the refractive index, which leads to a distortion of the results obtained. In normal photography, this is noticeable in the quality of the recorded picture. Suppose you make an experiment on photofixation and the subsequent digitization of a car license plate located at a considerable distance from the camera. In the presence of moderate or severe turbulence, distortion will occur, resembling a blurring. Due to this distortion, it may happen that a car license plate cannot be recognized, which is applicable to the digitization task inevitably leads to an error. When visualizing through the atmosphere, in addition to defocusing, a distortion of the geometry of the observed object may also occur. This well shows that the elimination of the influence of turbulence is a fairly promising direction for research [1]. The methods of turbulence research are the Background Oriented Schlieren method, Particle Image Velocimetry (PIV) method, Particle Trace Velocimetry (PTV) method, and others. Each method has its limitations. So, the Background Oriented Schlieren method is essentially integral, and with the help of it information about the volume containing heterogeneity is obtained. PIV and PTV methods allow to obtain information in the cross section of the volume, but the experimental setups are more expensive and more complex in their implementation [2-5].

Elimination of image distortions is carried out by computer filtering. In the case of elimination of a certain type of distortion, there is often a model of this distortion, which allows us to solve the inverse problem and determine the parameters of the necessary filtering. Currently, a lot of work is being done on modeling turbulent flows using numerical simulation methods, but such calculations are usually approximate or require large computational power. Such simulations make it possible to understand what the distribution of the air flow velocity or pressure will be in a particular object under study, for example, in a turbine [6-9]. In the general case, when the distortion model is not known or the reverse solution is not found, sometimes you can use the basic filtering operations in order to achieve image enhancement.

Journal of Physics: Conference Series **1421** (2019) 012027 doi:10.1088/1742-6596/1421/1/012027

2. Modeling of distortions and filtering of distorted images

One of the variants of modeling the distortions introduced by turbulence on images obtained from a camera while photographing is presented below. The distortion model in [10] is based on the consideration of the turbulent properties of the atmosphere. This model is presented below:

$$H(u,v) = \exp(-k \cdot (u^2 + v^2)^{5/6}), \qquad (1)$$

where the coefficient k describes the turbulent properties of the atmosphere. So, at $k = 0.001$ there will be moderate turbulence, and at $k = 0.00025$ it will be weak. Up to a factor of 5/6 in the exponent, this expression coincides in its form with a Gaussian low-frequency filter [11].

To simulate the distortion, an algorithm was implemented by software that performs convolution of the original image with a filter. Figure 1 shows the distortion with different used coefficient of the turbulent atmosphere.

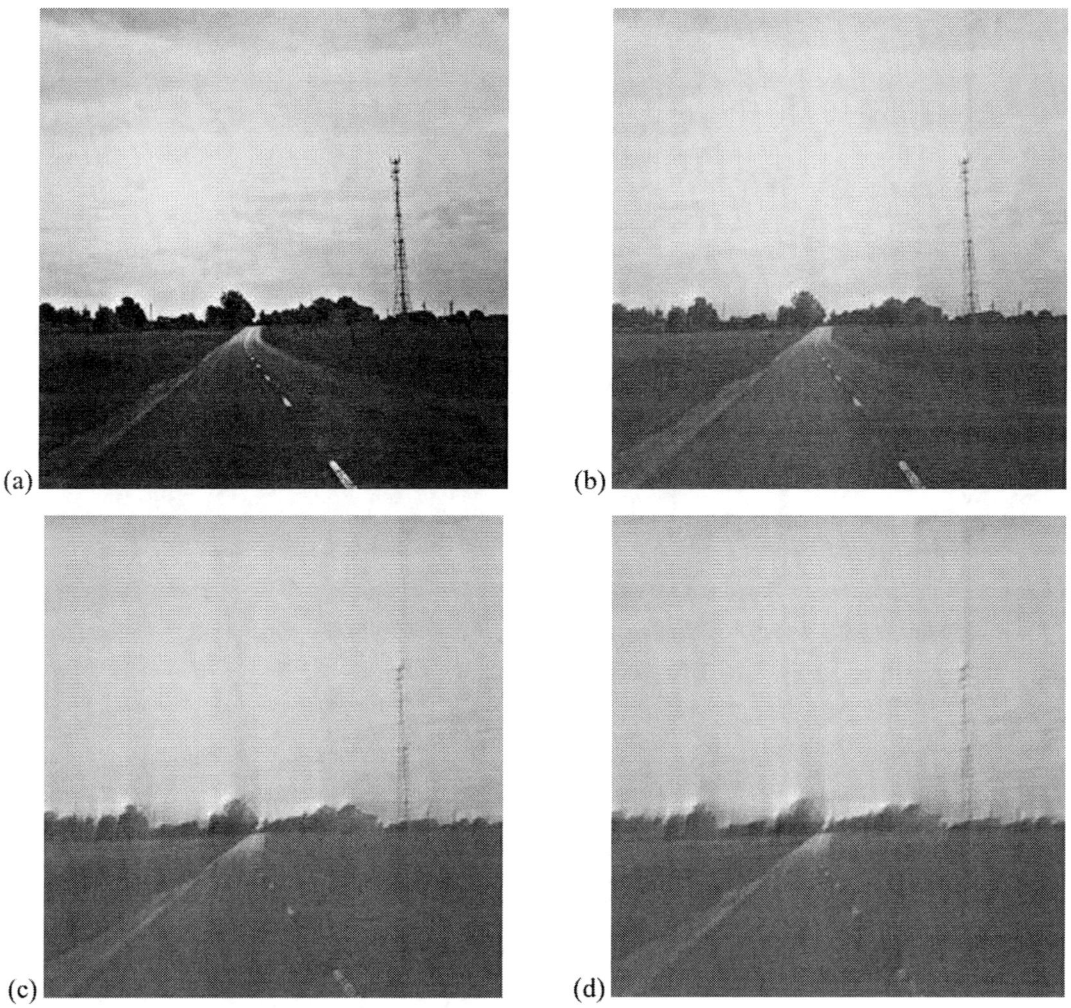

Figure 1. Simulation of turbulent distortion in images: (a) – $k = 0$; (b) – $k = 0.0001$; (c) – $k = 0.0002$; (d) – $k = 0.0003$.

The distorted images were shifted 5 pixels to the right in order to determine the relative error at different turbulence coefficients. Then the relative error δ can be defined as the ratio of the difference between the theoretical and experimental shift to the theoretical shift. The experimental shift was determined using the PivView 1.7 Demo program. This program performs correlation processing of images. For a pair of images in the program, you can see not only the distribution of the correlation coefficient, but also the average displacement along the axes of the image. Two images are loaded into the program – one reference and the other with distortion. An example of downloaded images is presented in figure 2 [12-13].

(a)

(b)

Figure 2. Downloadable in the program PivView 1.7 Demo frames: (a) – reference image; (b) – Image with the parameter of introduced inhomogeneity $k = 0.0001$ and offset by 5 pixels.

Then, using cross-correlation processing, we construct the vector field of displacements on the images. To do this, each image is divided into survey windows with a given size, and then a cross-correlation maximum is searched for each of the survey windows. The vector field for the data in figure 2 can be seen in figure 3.

Figure 3. Constructed vector field.

The average image offset can be viewed in the "PIV" → "PIV-Statistics" menu. Here the program shows the average horizontal and vertical displacement, which is required for estimating the error. If we process a set of images with different effects of atmospheric distortion, we can construct a graph estimating the error δ depending on the coefficient k (figure 8).

When an object is shot on a video camera, the influence of atmospheric distortion has its own fluctuation. A simple averaging of, say, three frames here does not help to increase the quality of the image, which will be shown below. For example, three images with simulated distortion were taken. Between themselves, the images differed in the degree of turbulence influence (coefficients $k_1 = 0.0001$; $k_2 = 0.0003$; $k_3 = 0.0005$). When calculating the average of the average image will likely get a blur, but does not improve in visual perception (figure 4). Therefore, this technique is not worth using.

Figure 4. Averaged frame.

When shooting through the atmosphere, the image of the object under study becomes blurred. This may be due both to the shaking of the camera itself or the object, or because of, for example, a strong wind in the space between the object and the camera. Given that turbulent distortions in their effects on the recorded image resemble a Gaussian low-pass filter, it is logical to assume that when filtering, use a filter that has the opposite effect [14]. For the best visual perception, an image containing the result of superimposing turbulence on a useful signal can be processed by increasing the sharpness. One way to increase the sharpness is to use a convolution of the original image with a Laplace mask. Below is a formula for image processing [15].

$$f'(x,y) = f(x,y) - k_L \cdot z(x,y), \tag{2}$$

where k_L is the sharpness gain (selected experimentally), $z(x, y)$ is the convolution of the original image with the Laplace mask. This formula will be valid with a mask, the central element of which has a negative value. An example of such a mask can be a 3 × 3 matrix with a central element equal to –8, and the remaining elements equal 1. Possible variants of the Laplace mask are presented below:

$$L_1 = \begin{pmatrix} 0 & 1 & 0 \\ 1 & -4 & 1 \\ 0 & 1 & 0 \end{pmatrix}, \; L_2 = \begin{pmatrix} 1 & 1 & 1 \\ 1 & -8 & 1 \\ 1 & 1 & 1 \end{pmatrix}, \; L_3 = \begin{pmatrix} 2 & -1 & 2 \\ -1 & -4 & -1 \\ 2 & -1 & 2 \end{pmatrix}.$$

For example, an image with a simulated distortion of $k = 0.0002$ was taken. In figure 5 you can see the result of the increased sharpness. During processing, we must not forget that as a result of filtering, the brightness values of pixels may change in such a way that they go beyond the original brightness range of pixels (for example, in the MathCAD program, the brightness varies from 0 to 255). In this case, it is worthwhile to further optimize the image – that is, to maintain its dynamic range.

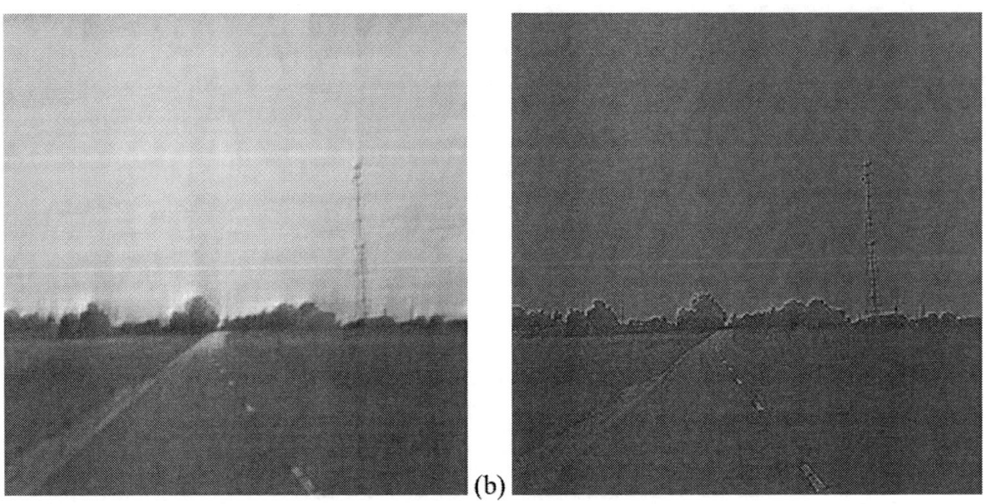

(a) (b)

Figure 5. Result of sharpening: (a) – the original image; (b) – processed image.

The result will be better if in addition to the increase in sharpness add an increase in contrast (figure 6). The most favorable result is achieved precisely by the sequential application of these filtrations.

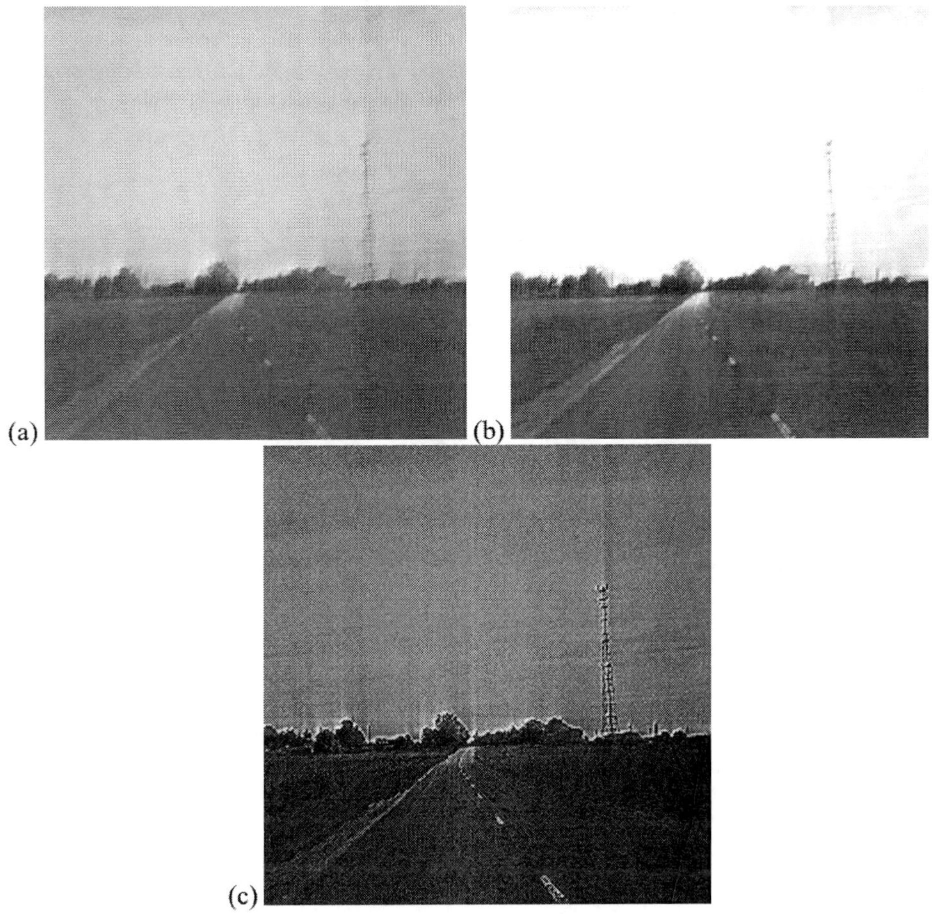

Figure 6. Image Processing Results: (a) – source image with $k = 0.0002$; (b) – contrast enhancement; (c) – sharpening and contrast enhancement.

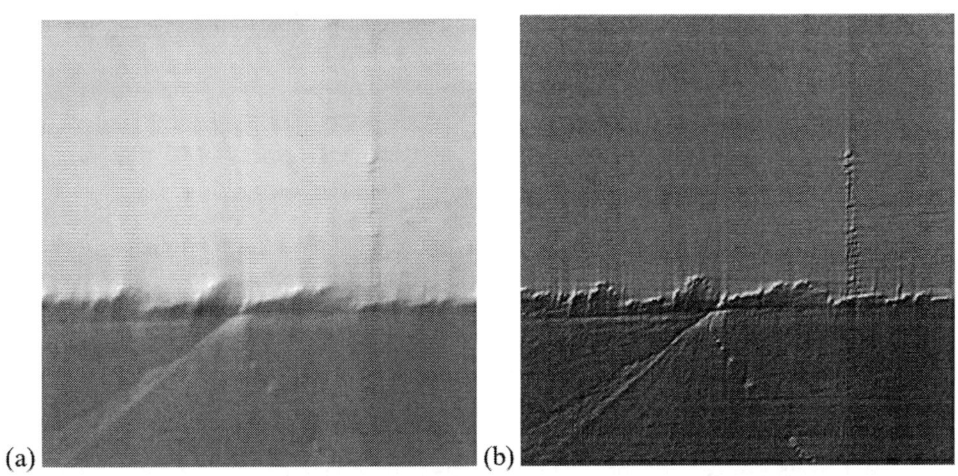

Figure 7. Image Processing Result: (a) – source image with $k = 0.0006$; (b) – processed image.

However, this algorithm is far from universal – it is suitable only for images with a low influence of turbulence. Figure 7 shows the result of processing in the case of a stronger distortion ($k = 0.0006$). It is clearly seen that when processing minor distortions intensified, which was to be expected.

For the processed images, an error estimate was also made (figure 8), which showed that the cascade of filters used in the case of small turbulence actually improves the image.

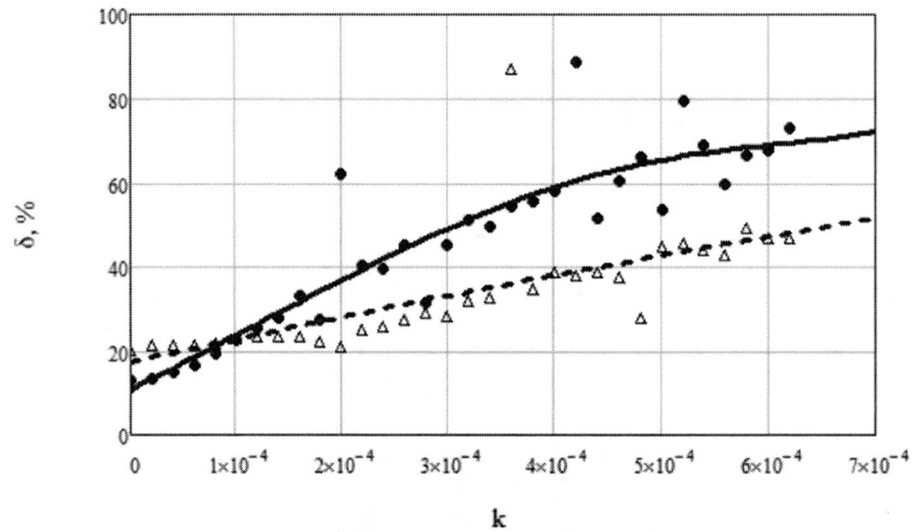

– – – The dependence of the error on the coefficient *k* for the processed images
– The dependence of the error on the coefficient *k* for the original images

Figure 8. Error estimate.

3. Conclusion

According to the results of processing, it can be said that the selected algorithm is suitable only for distortions corresponding to weak turbulence. When processing it is worth paying attention to the noise of the original image, because if you use sharpening, then in the presence of noise they will also stand out. To eliminate the effects of stronger turbulence in images, it is still worth solving the inverse task when using the distortion model.

Acknowledgments

This work was funded by Russian Foundation for Basic Research (project 19-07-00921A).

References

[1] Tahtali M, Lambert A J and Fraser D 2010 Graphics processing unit restoration of non-uniformly warped images using a typical frame as prototype *Proceedings of SPIE – The International Society for Optical Engineering* **7800** №78000H

[2] Cando E H, Luo X W, Hidalgo V H, Zhu L and Aguinaga A G 2016 Experimental study of liquid-solid two phase flow over a step using PIV *IOP Conference Series: Materials Science and Engineering* **129 (1)** №012054

[3] Cameron S M, Nikora V I and Marusic I 2019 Drag forces on a bed particle in open-channel flow: Effects of pressure spatial fluctuations and very-large-scale motions *Journal of Fluid Mechanics* **863** 494-512

[4] Mikheev N I, Dushin N S and Saushin I I 2017 Technique for velocity vector field dynamics measurement on the basis of smoke visualization of flow *Journal of Physics: Conference Series* **899 (6)** №062003

[5] Liang Z, Liu X, Jiang F and Wang C 2010 Coordinate positioning of turbulence tracing particle based on cross correlation matching and image projection transformation *3rd International Symposium on Test Automation and Instrumentation, ISTAI 2010* 586-91
[6] Sandoval J, Escauriaza C, Mignot, E. and Mao L 2018 Turbulent flow dynamics and mass transport processes in a natural surface storage zone using field data and numerical simulations *E3S Web of Conferences* **40** №05064
[7] Liang Z-W, Liu X-C, Wang Y-J and Wang C 2010 Calculation of turbulence kinetic energy based on image light stream vectors and k-ε differential equations *ICCASM 2010 - 2010 International Conference on Computer Application and System Modeling, Proceedings* **1** №5619081 V1721-5
[8] Thomas Scarlett G and Viola I M 2019 Unsteady hydrodynamics of tidal turbine blades *Renewable Energy* **146** 843-55
[9] Li Z, Wen B, Wei K, Yang W, Peng Z and Zhang W 2019 Flexible dynamic modeling and analysis of drive train for Offshore Floating Wind Turbine *Renewable Energy* **145** 1292-305
[10] Hufnagel R E and Stanley N R 1964 Modulation Transfer Function Associated with Image Transmission Through Turbulent Media *J. Opt. Soc. Amer.* **54** 52-61
[11] Gonzalez R C and Woods R E 2008 Digital Image Processing 3rd ed. Upper Saddle River NJ: Prentice Hall 978-0-13-168728-8
[12] Van der Kindere J W, Laskari A, Ganapathisubramani B and de Kat R 2019 Pressure from 2D snapshot PIV *Experiments in Fluids* **60 (2)** 32
[13] Mukaro R and Govender K 2013 Experimental study of turbulence and water levels in shoaling and breaking waves using digital image processing techniques *Journal of the South African Institution of Civil Engineering* **55 (2)** 59-74
[14] Sieberth T, Wackrow R and Chandler J H 2014 Influence of blur on feature matching and a geometric approach for photogrammetric deblurring *International Archives of the Photogrammetry, Remote Sensing and Spatial Information Sciences – ISPRS Archives* **40 (3)** 321-6
[15] Soloviev N V and A M Sergeev 2010 Improving the quality of bitmaps: Tutorial. - SPb.: SPbSU ITMO p 158

Laser diagnostics of the instability onset in the confined vortex flows

I V Naumov and M A Tsoy

Kutateladze Institute of Thermophysics, SB RAS, Academician Lavrentyev pr., 1, Novosibirsk 630090, Russia

E-mail: naumov@itp.nsc.ru

Abstract. This paper focus is the application of optical methods of pulsating vortex flow diagnostics on the example of swirling flow instability in a cylinder with rotating lid and aspect ratio $h = 3.3, 3.4$. To obtain a complete picture of auto-generated pulsations development, long-term LDV measurements of the tangential velocity component were carried out with a continuous method of the Reynolds number variation from 2000 to 3500 over 8 hours (30,000 seconds). In the view of the smallness of the Reynolds number increase rate $\Delta Re / \Delta t = 0.05$ s^{-1}, this process can be considered as quasi-equilibrium. A comparison with the widely used discrete method of the Reynolds number variation is made, given range of the Reynolds numbers was studied with a step $\Delta Re = 50$ for 20 minutes per point. It was found that at the same h, Re different vortex structures are formed depending on the Reynolds number variation method. In the case of discrete the Reynolds number variation, the development of pulsations may require up to 2000 rotating lid revolutions or 500 seconds, it should be considered conducting the experiments.

1. Introduction

The effective use of closed vortex centrifugal mass-and heat exchangers enormous potential in the first place is affected by our understanding of vortex motion depending on the intensity and swirl of the flow and ability to predict it [1]. It is of immense importance to control vortex flow instability for improving mass transfer processes in closed bio-and chemical reactors. In such reactors a complex fluid flow is formed, characterized simultaneously by a vortex and circulation movement [2]. Modeling those flows, cylindrical and polygonal cross-section containers with rotating disk inscribed in the end wall are used [3-6]. The development of instability usually is investigated by optical methods (LDV and PIV) after the establishment of a given flow regime [4-8]. Such optical methods allow to avoid external flow disturbance, since instabilities development is very sensitive [9]. Both the experiment and the numerical simulation show with good agreement that significant cylinder elongations lead to the flow stability loss due to multiplets formation, inducing velocity pulsations grow [10, 11].

However, the study of particular steady-state regimes does not allow to obtain a complete picture of pulsations development with an increase in the Reynolds number. For example, in [7, 8], it was found that with swirl increase in the flow, with certain geometric parameters, the flow becomes expectedly pulsating. With a further vortex motion intensity increase, period of velocity pulsations also increases up to hundreds or even thousands of disk rotation periods, and, with an even further increase in the angular velocity of rotation of the disk, the flow stabilizes. Another interesting fact for the first time was recorded in [12], where two possible instability development scenarios were experimentally found in an intense swirling flow: the increase or decrease of the amplitude of velocity pulsations with increasing flow swirl. Moreover, it was found that in the meridional cross-section the maximum amplitude of radial and axial velocity component pulsations expectedly grows with the Reynolds number increase, if the frequency of secondary disturbances is lower than of the primary ones, or, conversely, unexpectedly decreases, if greater. Usually, experimenters fix the aspect ratio h and discretely change Re during the experiment, accounting for the time necessary for the flow relaxation in the entire volume (usually 100÷200 disc revolutions). When the regime is changed, transition can influence the development of flow instability. It is possible to overcome this factor if disk rotation speed and corresponding Reynolds number are changed linearly at a rate small enough to neglect influence of transient processes.

2. Experimental setup and experimental methods

The development of instability in a well-studied swirling flow generated in a closed stationary cylinder with rotating lid is investigated in current work. The main motion of the fluid is toroidal rotation around cylinder axis with axial motion towards the lid in near-axis zone and in opposite direction near the walls. Flow structure in this simplest configuration depends on two parameters: the ratio of cylinder height to the radius $h = H / R$ and the Reynolds number $Re = \Omega R^2/\nu$, where Ω is the angular velocity of rotation of the lid and ν is the kinematic viscosity of the fluid.

Experimental investigation of pulsations development was carried out in a vertical transparent glass cylinder of optical quality with an inner diameter of 150 mm and length 1000 mm with bottom rotating lid (figure 1 (a)). Upper end wall position can be adjusted in order to change aspect ratio h. The rotating disk radius is R= 74.5 mm. The gap between disk and the cylinder wall is negligibly small - 0.5 mm, less than 0.5% of the disk diameter. The reference point of the coordinate system is set in the center of rotating lid, r-axis passes through bottom plane and z-axis coincides with the vertical container axis. Water-glycerin solution (75% glycerin) was used as a working fluid. Polyamide spheres of neutral buoyancy with an average diameter of 10 microns were used as light-scattering particles for LDV measurements. To minimize optical aberrations and temperature fluctuations test section was placed inside a rectangular transparent glass container with dimensions 500x500x1000 mm filled with water.

(a)

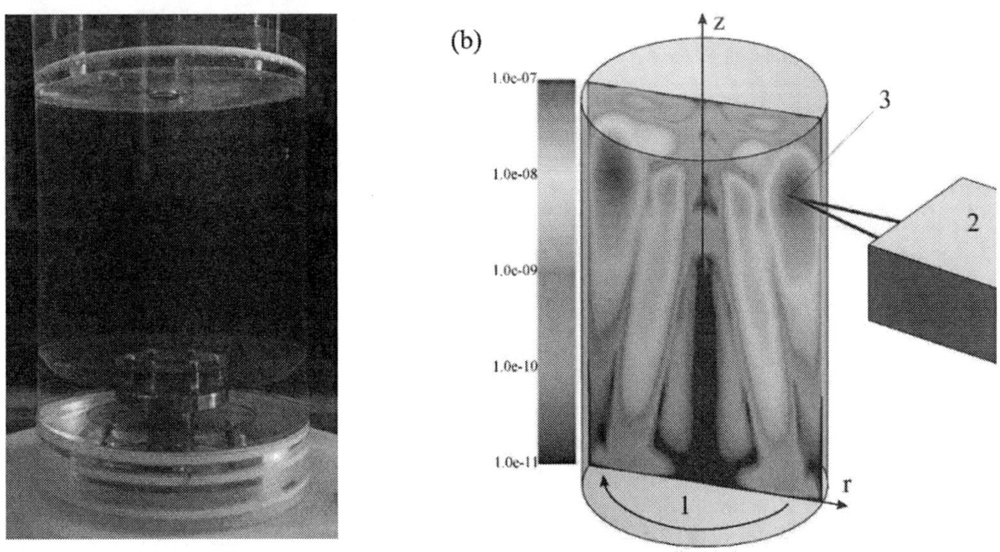

Figure 1. (a) – Photo of cylindrical cavity, (b) – Example of numerical computation and LDV measuring point, h =3.4, Re = 3200. Variance of velocity magnitude, m^2/s^2. 1 – rotating lead, 2- LDV, 3 – LDV measuring point at $z = 2.6R$, $r = 0.75R$.

Journal of Physics: Conference Series **1421** (2019) 012028 doi:10.1088/1742-6596/1421/1/012028

The temperature dependence of working fluid viscosity was approximated by the second order polynomial. As the viscosity of the working fluid is sensitive to temperature, the fluid temperature was measured with an accuracy of 0.1 °C, thus reducing the uncertainty of viscosity determination to 0.2%. The total error of *Re* did not exceed ±10 in the range of investigated Reynolds numbers.

A two-component LDV system (LAD-06i) developed at the Kutateladze Institute of Thermophysics SB RAS, based on a Mitsubishi ML1013R semiconductor laser (with 70 mW power and 684 nm wavelength) and operating in a backscattering configuration, is used to measure the tangential velocity component. A two-frequency differential optical configuration is applied with a frequency shift of 80 MHz based on a Bragg cell to eliminate the directional ambiguity. Split by the Bragg cell, two laser beams are then focused by the optical head with a focal distance of 0.5 m and measurement volume of 0.05×0.05×1 mm (in air).

Rotating disk was driven by a stepper motor. Digital sinusoidal control signal was generated on PC, converted to analog form by DAC and supplied to the driver. Speed of the motor (and, accordingly, the Reynolds number of the flow) is proportional to the frequency of the signal, transmitted to the driver. Thus allowing us to programmatically set the Reynolds number schedule for the whole experiment.

Loss of stability and the development of flow pulsations lead to emergence of clear peaks on the spectrogram of velocity pulsations (figure 2), which corresponds to a change in vorticity – the formation of azimuthal wave modes or rotating vortex multiplets (figure 3) [4, 7, 11]. Since velocity pulsations are not equal in the flow volume, it is necessary to determine the area of flow where pulsations will be most pronounced. During the preliminary CFD calculations, the velocity dispersion distribution was obtained and a measuring point of $z = 2.6{\cdot}R$, $r = 0.75{\cdot}R$ (figure 1 (b)) was assigned, which correlates with the area of maximum of velocity pulsations [7, 8].

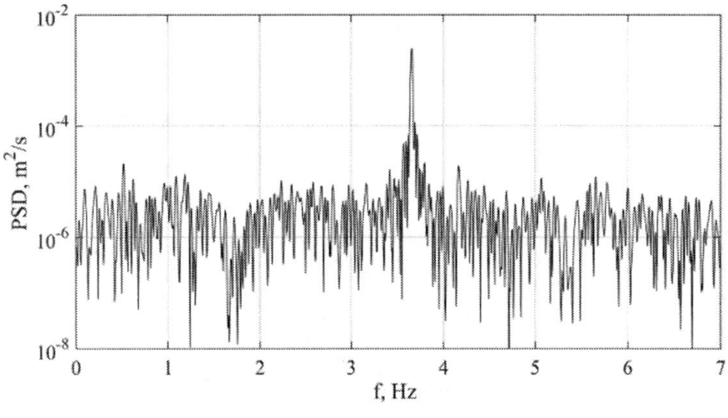

Figure 2. Velocity pulsations spectrum.

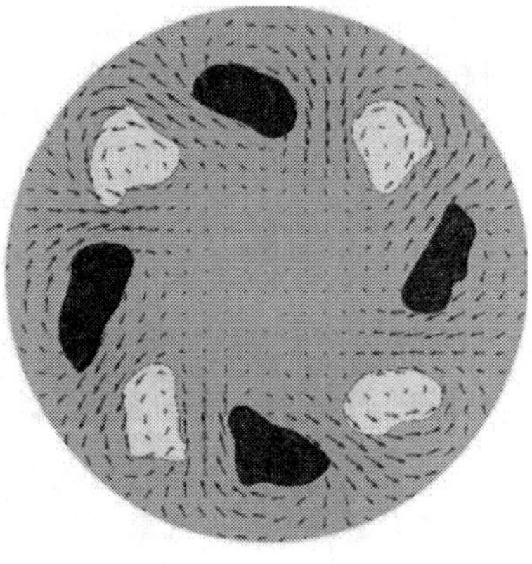

Figure 3. Iso-contours of phase averaged axial vorticity, rotating multiplet [7].

To experimentally study auto-generated flow pulsations two different methods of the Reynolds number variation were used and compared.

For the continuous method of the Reynolds number variation, the Reynolds number was linearly increased from 2000 to 35000 during 30000 seconds with growth rate $\Delta Re / \Delta t = 0.05$ s^{-1}. During the experiment, tangential velocity component was continuously measured by LDV and recorded.

For the discrete method of the Reynolds number variation, the range of the Reynolds numbers 2000÷3500 was divided into a sequence of discrete steps with the step size $\Delta Re = 50$. At each step, measurements were performed at a constant disk rotation speed during the time $T_S = 1200$ s (figure 4).

After that at even intervals of 10 seconds the disc was decelerated, held fixed and then accelerated to a new speed value corresponding to the next step during $T_P = 30$ sec.

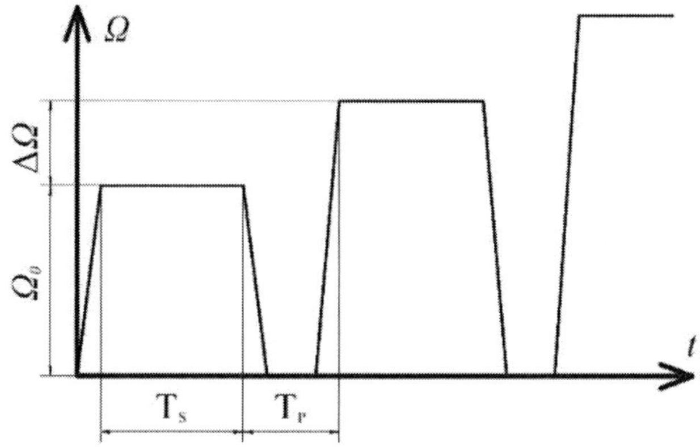

Figure 4. Discrete method of the Reynolds number variation chart.

For this study aspect ratios $h = 3.3$, 3.4 and the Reynolds number range 2000÷3500 were chosen due to the fact that at $h = 3.3$, according to previous works [4, 7, 8], with increasing angular velocity of disk rotation an anomalous stabilization of the vortex motion appears. Of additional interest is the fact that the dimensionless pulsation frequency $\omega_m \approx 0.3$ found in the experiments correspond to the rotation of the $m = 3$ vortex structure, while numerical calculations show that the transition to the non-stationary flow manifests itself as a vortex structure formation with wave modes $m = 3$ and 2 with frequencies $\omega_m \approx 0.3$ and 0.2 respectively.

3. Results

Time-dependent power spectral density of the tangential velocity component measured by LDV is presented on Figure 5 for aspect ratios $h = 3.3$ and 3.4 for both continuous and discrete method. Threshold was set to hide irrelevant noise. Using our Reynolds number variation schedule we can map the Reynolds number values on the time axis and normalize frequency of pulsations f by the frequency of disk rotation f_D.

$$f(t) \xrightarrow{Re(t)} \omega_m(Re); \qquad \omega_m = \frac{f}{f_D}; \qquad f_D = \frac{Re \cdot \nu}{2 \cdot \pi \cdot R^2}.$$

It can be noted that for both $h = 3.3$ and 3.4, the ranges of the Reynolds numbers where velocity pulsations are registered do not depend on the method of the Reynolds number variation. At $h = 3.3$, the pulsations are found at $2125 < Re < 2300$, and at $Re > 2700$. At $h = 3.4$, the pulsations arise at $Re = 2065$ and exist till the limit of investigated range $Re = 3500$. Additional wave modes are presented at $Re > 3100$. These results completely agree with previous experimental studies [7, 8].

However, there are significant differences in the values of dimensionless frequencies ω_m near the transition to instability border. In the case of continuous method, the first-appeared mode $m = 3$, has a frequency $\omega_m \approx 0.3$ and permanently exists with a further increase in the Reynolds number. The arise of secondary modes $\omega_m \approx 0.22$, 0.45, 0.5 at $Re > 3200$ does not destroy this main mode containing most of the energy.

Using a discrete method, there are certain regimes where pulsation frequencies differ from continuous method. These differences are presented in table 1. In these regimes at the same h and Re, depending on the Reynolds number variation method, different wave modes may arise.

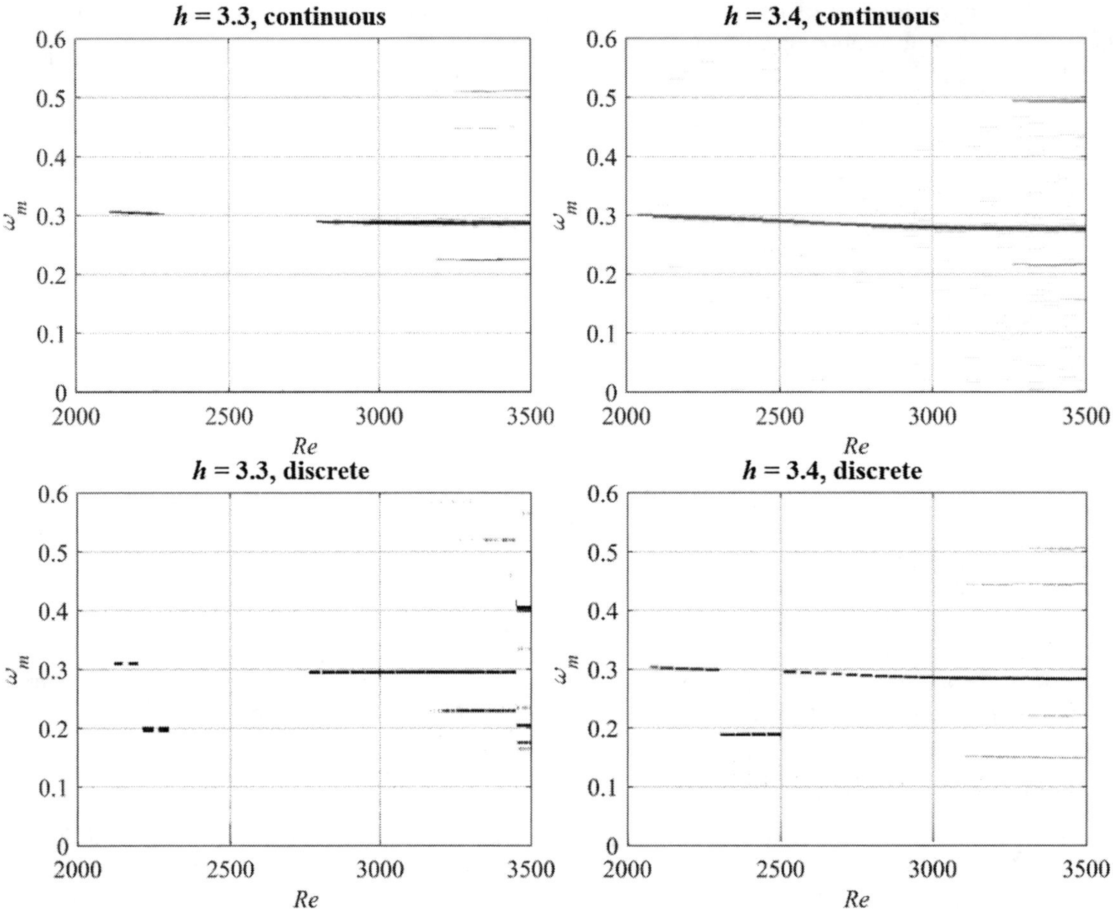

Figure 5. Spectrograms of LDV velocity measurements.

With continuous method the first first-appeared mode $m = 3$, $\omega_m \approx 0.3$, becomes dominant and, holding the energy in itself, suppresses the appearance of other modes with a further increase in *Re*. In the case of a discrete method the rotating disk is stopped after each step, viscous friction leads to energy dissipation with flow relaxation and subsequent vortex structure collapse. Thus, setting new regime at next step, there is not a trace of vortex structure in the flow at the previous step. The “new” flow develops from the rest, and at the same time, another wave mode with $\omega_m \approx 0.2$ may become dominant. It is possible that acceleration of the disk affects energy transferring to one or another azimuthal wave mode during transition process when the disk is accelerated to a given speed. This fact requires additional experiments with variation of the time taken by the disk to accelerate to a given speed.

Table 1. Velocity pulsation frequencies.

h	*Re*	2100	2150	2200	2250	2300	2400	2500	2600	2700	2800	3000
3.3	ω_m Cont.	0.31	0.31	0.3	0.3	-	-	-	-	0.29	0.29	0.29
	ω_m Discr.	0.31	0.31	**0.19**	**0.19**	-	-	-	-	0.29	0.29	0.29
3.4	ω_m Cont.	0.3	0.3	0.3	0.3	0.3	0.3	0.3	0.29	0.29	0.28	0.28
	ω_m Discr.	0.31	0.3	0.3	0.3	**0.19**	**0.19**	**0.19**	0.29	0.29	0.28	0.28

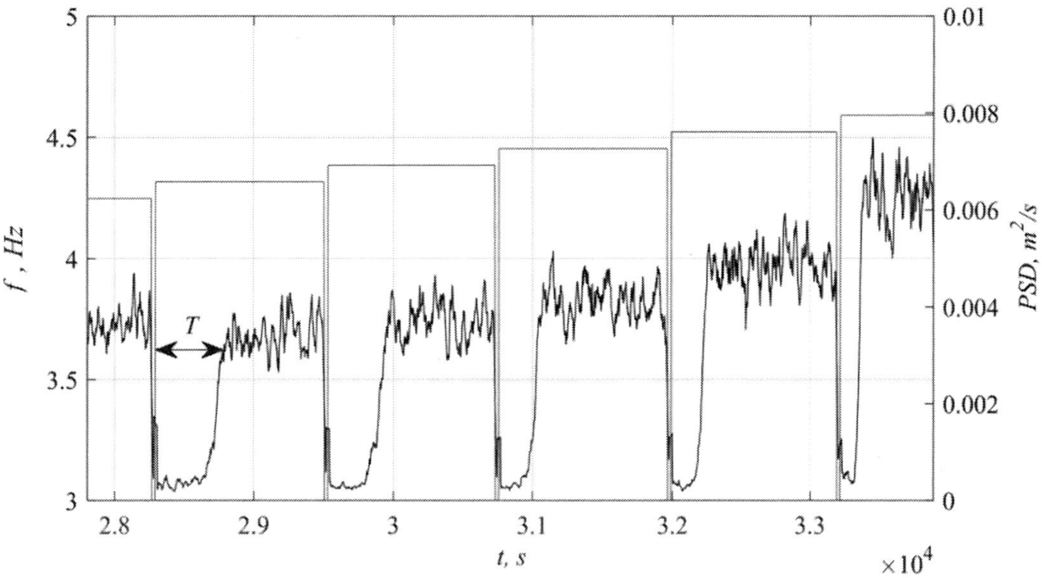

Figure 6. Disk rotation frquency and velocity pulsations PSD during experiment at $h = 3.4$, $2600 < Re < 2850$.

One can notice small gaps in the spectrograms for the discrete method in the range of *Re* 2100÷2300 for $h = 3.3$ and 2100÷3000 for $h = 3.4$, looking like a dashed line with varying steps. To explain their origin power spectral density of velocity pulsations with and schedule of disk rotation frequency f_D are presented in figure 6 for $h = 3.4$ and $2600 < Re < 2850$. It can be seen that after setting disk rotation to a constant rate there is some delay *T* in pulsation development with order of 10^2 seconds. The dependence of delay *T* normalized on the disk rotation period $T_D = 1/f_D$ over the Reynolds number for $h = 3.4$ is presented on figure 7. It can be noted that near the region of bifurcations associated with instability onset, T_D reaches a value up to 2000. Corresponding time interval required for complete flow development in the entire volume reaches up to 500 seconds. Thereby there exist regimes requiring significantly more time for the flow to develop than widely used interval of 60-100 seconds.

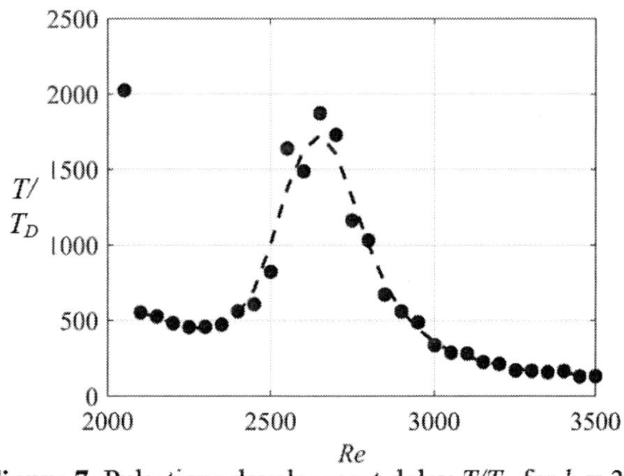

Figure 7. Pulsations development delay T/T_D for $h = 3.4$.

It can be assumed that slow vortex structure development speed is determined by regimes proximity to the border between steady and unsteady flow on the *h*, *Re* plane. This is evidenced by the large value of the delay at $Re = 2050$, and also in the range 2500÷2800, which is close to the range where at $h = 3.3$ there are no pulsations at all. With a further increase of the Reynolds number there is a slow decrease of delay, approaching to the usually used value of 100-200 revolutions of the disk. Additionally, it can be noted that the change of dominant modes at $Re = 2300$, 2500 does not affect this delay. Extended study of flow transition to unsteadiness over wider range of aspect ratios h is required.

4. Conclusion

To study the development of vortex flow instability during a long-term experiment we applied two different methods for the Reynolds number variation over 2000÷3500 range: continuous with linear change of the Reynolds number and discrete. Regimes to study were chosen due to the specific features of instability development in a cylinder of high elongation. It was found that using continuous method, to obtain a complete picture of auto-generated velocity pulsations development, LDV measurements during 30000 seconds, requiring the minimum involvement of an experimenter are valid. Due to the small rate of change of the Reynolds number $\Delta Re / \Delta t = 0.05\ s^{-1}$, such a process can be considered as quasi-equilibrium. The obtained results are compared with the common used discrete method of the Reynolds number variation, where a given range of the Reynolds numbers was studied with a step $\Delta Re = 50$, for 20 minutes per step. Variation of rotating disk velocity by discrete steps was automated by use of programmatically set schedule.

Regimes were found where various vortex structures are formed and dominate, depending on the used method. It was found that in the case of discrete method, the development of vortex flow close to instability border may require up to 2000 disk revolutions or 500 seconds. This fact should be accounted studying instability onset region borders.

Acknowledgments

Research was supported by the Russian Science Foundation (Project № 19-19-00083).

References

[1] Gupta A K, Lilley D G and Syred N 1984 Swirl Flows *Tunbridge Wells, Kent: Abacus Press* 475
[2] Shtern V N 2012 Counterflows *New York: Cambridge University Press* 488
[3] Escudier M P 1984 *Exp. in Fluid* **2** 189-96
[4] Gelfgat A Y, Bar-Yoseph P Z and Solan A 2001 *J. Fluid Mech.* **438** 363-77
[5] Naumov I V, Dvoynishnikov S V, Kabardin I K and Tsoy M A 2015 *Phys. Fluids* **27** 124103
[6] Naumov I V and Podolskaya I Yu 2017 *J. Fluid Mech.* **280** 263-83
[7] Sørensen J N, Naumov I V and Mikkelsen R F 1984 *Exp. in Fluids* **41** 425-40
[8] Sørensen J N, Gelfgat A Y, Naumov I V and Mikkelsen R F 2009 *Phys. Fluids* **21** 054102
[9] Thompson M C and Hourigan K 2003 *J. Fluid Mech.* **496** 129–38
[10] Sørensen J N, Naumov I V and Okulov V L 2011 *J. Fluid Mech.* **683** 430-41
[11] Herrada M A, Shtern V N and Torregrosa M M 2015 *J. Fluid Mech.* **766** 590-610
[12] Naumov I V, Sørensen J N and Okulov V L 2007 *Tech. Phys. Letter* **33** 775-78

High speed visualization of self-oscillation of axisymmetric cavitating cavity in a square vortex chamber

S G Skripkin[1], M A Tsoy[1], P A Kuibin[1], Zhigang Zuo[2] and Shuhong Liu[2]

[1]Kutateladze Institute of Thermophysics SB RAS, Lavrentyev ave., Novosibirsk,630090, Russia
[2]Department of Energy and Power Engineering, State Key Laboratory of Hydro Science and Engineering, Tsinghua University, Beijing, 100084, China

Abstract. An experimental investigation of the multiphase turbulent flow in a square vortex chamber was carried out to examine the self-oscillation of a cavitating cavity in a wide range of flow rates and pressure conditions. Such phenomenon is observed behind turbine runner when the hydraulic units operate at full load regime at high flowrate, and leads to significant flowrate and pressure pulsations in the entire hydraulic system. The dynamics of the cavity over time, and the freaquency of the volumetric oscillations frequency was studied via high speed visualization. The grey-scale images of the vortex cavity were binarized. By assuming an axially symmetric flow, the volume of cavity for each frame was integrated, and further investigated using FFT. Quantitative information on the structure of a single-phase flow was also obtained using the PIV technique.

1. Introduction

The swirling flows can be found in many technical devices and applications: cyclone separators to separate solid particles from the gaseous phase, a vortex generators to improve stability of the flame in combustion chambers, and hydraulic turbines operating at part load regimes with extensive recirculation zone. The variety of effects associated with the flow swirl (formation of recirculation zone, stationary and non-stationary vortex structures) attracts close attention of researchers for many years. Dealing with multiphase swirling flow the study becomes much more complicated. The two-phase flow resulting from cavitation processes, greatly complicates the task of research and the analytical description of the swirling flows and the coherent vortex structures arising in them. The occurrence of a vapor (gas) volume corresponds to the introduction of a compliant element into the system, which is an additional source of unsteady pulsations. The first paragraph after a heading is not indented.

One of the problems caused by cavitation appearance in fluid machinery is the interaction with the flow path. It is generally recognized that cavitation has a significant impact on the power and operational characteristics of hydraulic units [1-3]. Compared with partial load conditions, when pressure pulsations are caused by the precession of a spiral vortex rope, in hydraulic turbines operating at full load at high discharge conditions cavitation oscillation of axisymmetric vortex rope volume is observed in the draft tube cone. This in turn leads to significant flowrate and pressure pulsations in the entire hydraulic system. Rheingans [4] was one of the first to draw attention to the problem of non-stationary processes in flow systems in the presence of cavitation. He associated the power fluctuations of a hydroelectric power plant with non-stationary phenomena occurring in the flow path, and also made attempts to classify and describe them. A more detailed systematic description was presented a few decades later [5, 6]. An analytical study of the characteristics of non-stationary cavitation in a wide frequency range was performed by Tsujimoto et al. [7]. Otsuka et al. [8] investigated the cavitation characteristics on the basis of a closed cavity model, in which variations in the cavity length are allowed. It was found that this model smoothly transforms into quasistationary calculations at the lower frequency limit, in contrast to the model with a fixed cavity size. It is also shown that the influence of the cavitation number on the cavitation compliance and the flow mass gain factor depends on the frequency of perturbations. Comparing with experimental data, it was found that the model describes non-stationary cavitation processes only at a qualitative level. In the swirl free conditions, Streeter [9] developed a method for calculating the cavitating flow in pipes. The model describes the passage of a rarefaction wave through a liquid and the formation of a cavitation cavity. After that, the speeds of each phase are calculated separately. The cavitating flow is enhanced by "shock" fronts of the fluid. Equations are derived for determining the velocity of shock fronts and increasing the pressure in them.

To account for the interaction of the liquid column with the cavitation (gas) cavity in the work of Chen et al. [10] developed a one-dimensional model. It was found that the effect of the diffuser of the conical section destabilizes the regime at all flowrates. While the swirl effect stabilizes/destabilizes the system at high/low flow rates than at the flow rate at which the integral swirl parameter is close to zero. In both cases, the oscillation frequency is determined from the cavitation compliance and the size of the conical section. It has been shown that in the general case the amplitude of the flow rate fluctuation is much larger in the downstream of the runner as compared to that in the upstream. Later, the Chen model was refined by Kuibin et al. [11]. In their work, it was noted that the Bernoulli equation gives an averaged pressure in the cross section of the channel but not pressure on the wall. Moreover, the pressure correction due to the swirl of flow should depend on the radius of the cavitation cavity, which was not considered in the Chen model. Also, an analytical model of a cavitating vortex rope was developed, and various vorticity distributions for axisymmetric vortices were considered. It is shown that the swirl effect on the condition of flow stability in a hydraulic system is rather weak, and its effect decreases with increasing volume of the cavitation cavity. The developed semi-empirical model of a two-phase helical vortex rope predicts frequencies that are in good agreement with the experimentally measured data at different gas contents. Among the latest experimental works, it should be pay attention to the study by Muller et al. [12, 13]. They showed that the characteristic breathing motion of the cavitation vortex rope is governed by an important variation of the flow swirl in the draft tube. It was found that the torque on the runner shaft is synchronized with the cavity oscillation. Based on the analysis of the wall pressure synchronized flow visualizations they proposed potential governing mechanism, identifying the swirl variation due to the development of cavitation on the runner blades as the key factor.

Due to the complexity of conducting experiments on full-scale hydro turbines or model stands, as well as a large number of factors affecting the vortex flow patterns, it has been attempted to isolate the experimental modeling of the formation of an axisymmetric vortex cavitation cavity in a square vortex chamber. To create a flow swirl, a combination of stationary and rotating swirlers is used, which allows us to obtain velocity distributions, similar to the distributions behind a real turbine runner. Such an approach made it possible to very carefully control the main parameters: flow rate, swirl intensity and cavitation number for an isolated study of their influence on the dynamics of the cavitating vortex cavity.

2. Experimental setup and techniques

Measurements are carried out at the test facilities of the Kutateladze Institute of Thermophysics, Laboratory of Ecological Problems of Heat Power Industry, on a vertical closed-loop experimental test rig with the square vortex chamber made of Plexiglas (figure 1), providing an optical access for PIV measurements and high-speed visualization technique. The swirl generator (combination of stationary and rotating vane swirlers) designed to simulate flow in a Francis turbine operated at different discharge conditions partially reproduces the swirl generator designed at Politehnica University of

Timisoara, Romania [14]. The swirl intensity is varied by changing two main parameters: flow rate and runner rotational speed. The detailed description of experimental setup can be found in [15]. A sealed expansion tank with a free surface and a vacuum pump connected to it is installed in the upper part of the hydraulic test rig, this allows varying pressure level in vortex chamber controlling cavitation inception number.

To perform the visualization of the vortex flow pattern, a high-speed CMOS camera PCO 1200.hs is used at 820 frames per second. The high temporal resolution allowed to resolve most of the frequencies when using the fast Fourier transform to the imaging data.

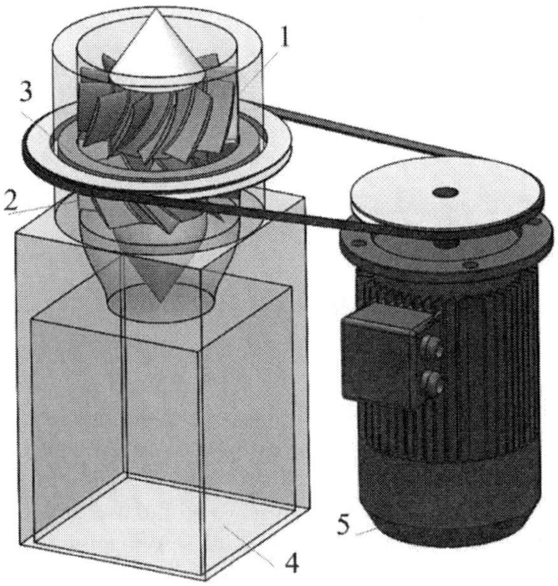

Figure 1. View of experimental setup, the main flow direction is downwards, the runner is rotated clockwise; 1 – stationary swirler, 2 – runner, 3 – magnetic couple, 4 – square vortex chamber, 5 – electric motor supply runner speed up to 3000 rpm; max Re = 10^6.

During visualization, the searchlight was installed opposite the camera behind the working area, visualizing the cavitation cavity to the light. With the selected lighting scheme, due to the scattering of light at the interface, the obtained images contained a uniform light background with darker areas corresponding to the caitating cavities. For each received half-tone eight-bit frame, a procedure to increase the contrast was carried out - one percent of the lower and upper brightness values of the pixels were discarded, and the remaining values were scaled to a range of 0-255. To determine the boundaries of the vapor-gas cavity, images were binarized – each pixel was assigned to one of two classes – the cavitation cavity or the background. The Otsu method was used to calculate the binarization threshold for the halftone image, this algorithm allows to divide the pixels into two classes, calculating such a threshold that the intraclass dispersion is minimal. Assuming that the cavitation cavity is quasi-symmetric with respect to the axis of the vortex chamber, the image area was integrated over the angle and the cavity volume was calculated, which after calibration was converted to cubic centimeters. Based on the temporal realization of the cavitation cavity volume for each mode, spectrograms were calculated using fast Fourier transform. Also, PIV system is used to obtain velocity distribution in vertical cross section in cavitation free conditions. For each investigated regime the 5000 images have been collected.

3. High-speed visualisation and PIV results

Flow similar to the flow behind the runner of the Francis turbine operated at a full load has been experimentally reproduce in a square vortex chamber. Figure 2 shows a series of images obtained from high-speed flow visualization with a shooting frequency of 850 frames per second. The selected images cover a half cycle and shows the different stages of the oscillation of the cavitation cavity at a low-frequency. It can be seen that fragment “D” has a significantly larger cavity volume compared to “A”. As the volume of the vortex cavity increases in the flow path, the hydrodynamic resistance increases, which is accompanied by significant pulsations of the liquid column and its effect on the runner of the model turbine. Also, as the diameter of the cavitation cavity increases, the local section decreases and the local parameter of the flow swirling increases. The study of the mechanism of such self-excited oscillations is an important task in the safe operation of natural hydraulic units.

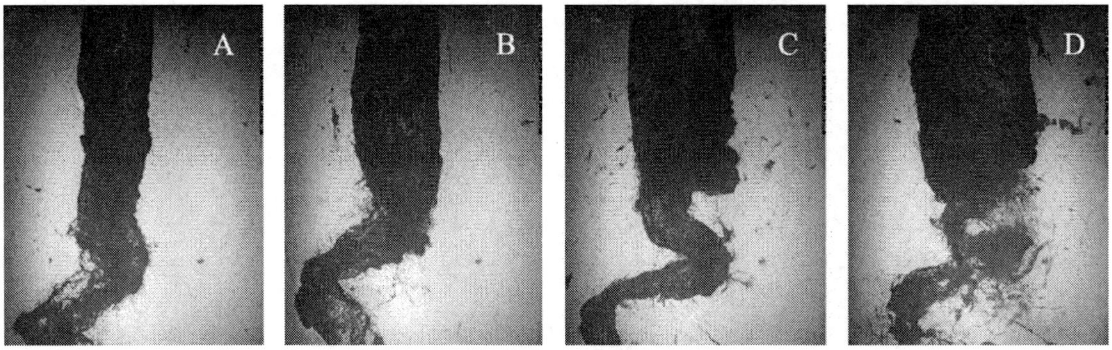

Figure 2. Evolution of vortex volume in time, $N = 860$ rpm, $Q = 112$ m^3/h.

Figure 3 shows the spectrum obtained using the fast Fourier transform; it can be seen that several frequencies dominate the spectrum: 3.9Hz, 9.9Hz, 30.5Hz and 121.8Hz. By decomposing the original signal at these frequencies, it is possible to obtain images of the change in volume at each frequency. The frequency of 3.9 Hz is dominant and corresponds to low-frequency “breathing” motion with the highest amplitude. Other frequencies 9.9 and 30.46 are presumably large-scale modulation of low-frequency oscillations of the volume, and the frequency 121.8 corresponds to waves propagating along the vortex, resulting in the formation of small local constriction reducing the cavity volume.

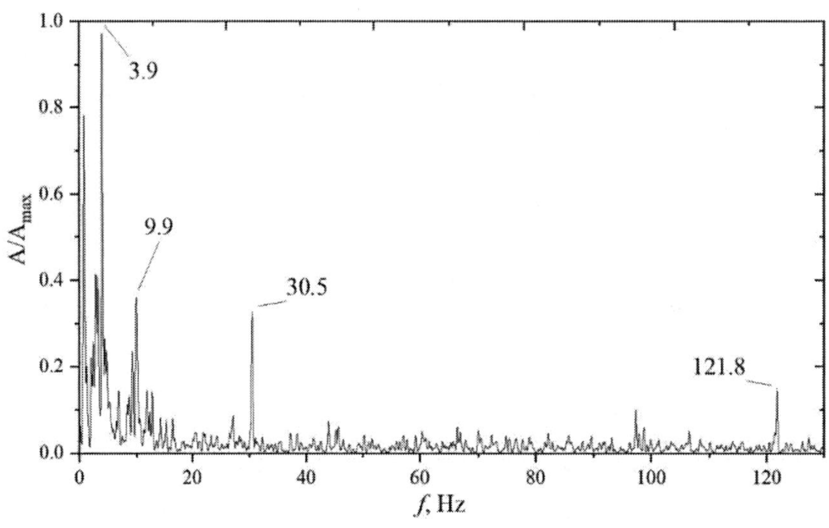

Figure 3. Power spectral density of volume pulsation, $N = 860$ rpm, $Q = 112$ m^3/h.

It is important to study the effect of the pressure variation (cavitation number) on the characteristics of the phenomenon of self-excited oscillations. A series of experiments was carried out at a fixed flow rate and runner rotational speed and a variation of the additional vacuum in the tank connected to the hydraulic test rig. Figure 4 shows two spectrograms corresponding to different pressure conditions. The d*P* value in the figures corresponds to the additional vacuum in the working section relative to the atmospheric pressure without considering the liquid level. It can be seen that, with a smaller pressure drop, d*P* = -33 kPa, the vortex cavity has much smaller volume. Analyzing the spectral characteristics, it can be found that large-scale volume pulsations at low frequency 3.9 Hz disappear, but pulsations at a frequency of 30 Hz remain and dominate.

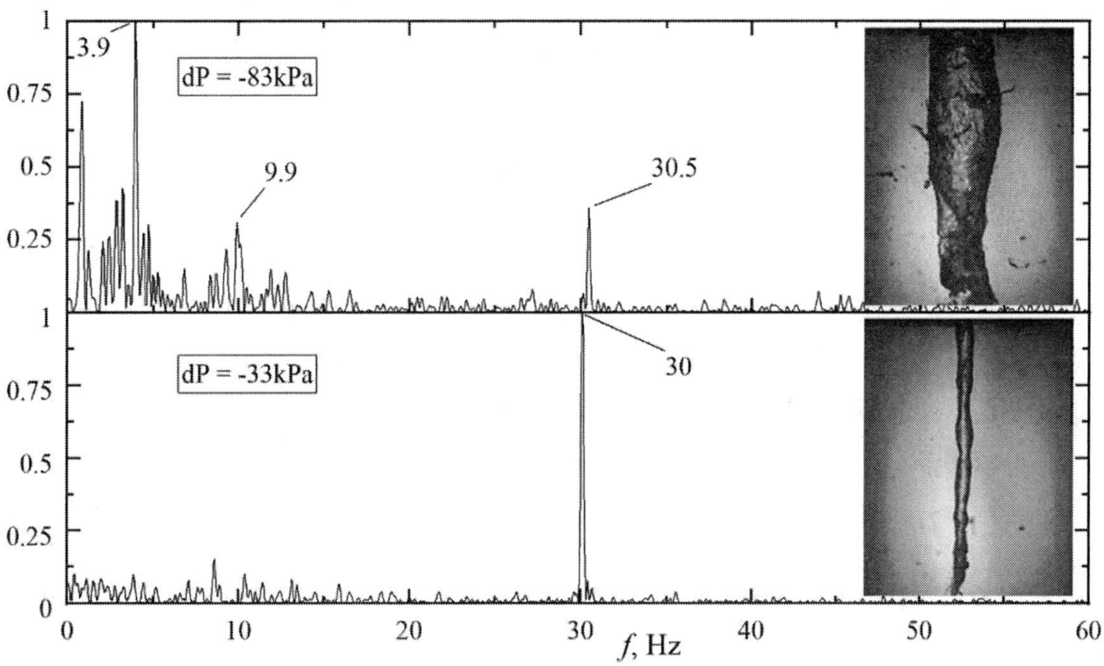

Figure 4. Comparison at volume oscillation spectrum normalized on maxima of spectrum amplitude at different pressure level., d*P* – pressure drop relative to atmosphere measured in vacuum chamber connected with upper part of test rig, $N = 860$ rpm, $Q = 112$ m^3/h.

To obtain quantitative information on the flow structure, a series of PIV measurements were carried out in the central vertical cross section of the vortex chamber. Figure 5 shows the averaged distribution of the axial velocity component, streamlines are also plotted. It can be seen that the main flow is concentrated in the central region, and in the near-axial zone a deceleration area forms and there is a significant deficit in axial velocity. A torus-shaped vortex is presumably formed in the near-wall region, which is associated with a sudden expansion of the swirling flow at the entrance to the vortex chamber. The accumulated PIV data are planned to be used in the future for POD and DMD analysis, which will allow a comparison between the obtained frequencies and the vortex structures.

4. Conclusion

Using the methods of experimental modeling in a square vortex chamber it has been reproduced qualitatively similar flow pattern as observed in a draft tube of Francis turbine operated under full load conditions. This study attempts to correlate these observations with a full load surge phenomenon in Francis turbine. Large-scale vapor-gas cavities is observed, the volume of which can oscillate significantly in time, which is an additional source of unsteady flowrate and pressure pulsations in the hydrodynamic path. Using high-speed visualization and digital image analysis allow us to obtain four different oscillation frequencies: 3.9 Hz, 9.9 Hz, 30.5 Hz, 121.8 Hz at $Q = m^3/h$. Reducing the diameter of the cavitation cavity by varying the pressure in the experimental test rig suppresses low-frequency volume oscillations. Gathered PIV data can be used for further POD analysis or as case for CFD study.

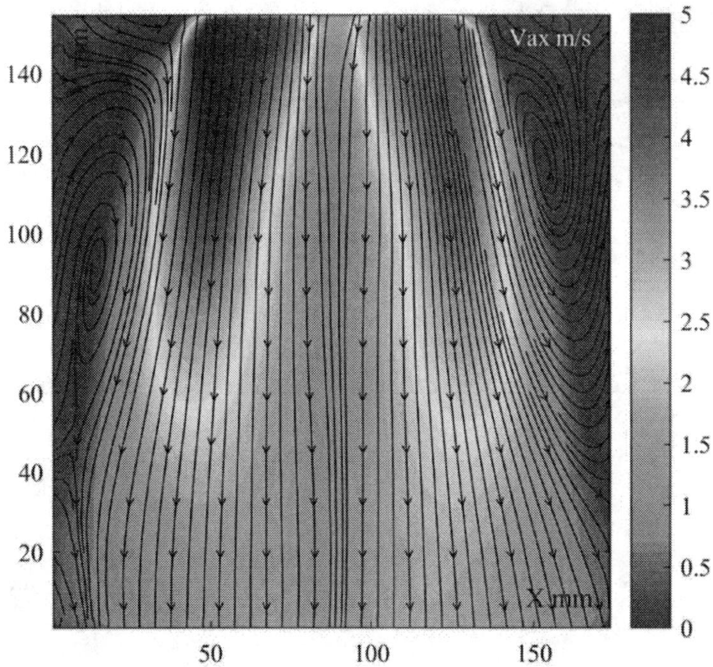

Figure 5. Streamlines and axial velocity distribution in central vertical cross section.

Acknowledgements
The study of cavitation influence on characteristic of vortex cavity was financially supported by RFBR (project 18-58-53052), the study of flow structure by PIV measurements was carried out under state contract with IT SB RAS.

References

[1] Favrel A, Müller A, Landry C, Yamamoto K and Avellan F 2016 LDV survey of cavitation and resonance effect on the precessing vortex rope dynamics in the draft tube of Francis turbines *Exp. Fluids* **57**

[2] Lu G, Zuo Z, Sun Y, Liu D, Tsujimoto Y and Liu S 2017 Experimental evidence of cavitation influences on the positive slope on the pump performance curve of a low specific speed model pump-turbine *Renew. Energy* **113** 1539–50

[3] Alligne S, Nicolet C, Tsujimoto Y and Avellan F 2014 Cavitation surge modelling in Francis turbine draft tube *J. Hydraul. Res.* **52** 399–411

[4] Rheingans W J 1940 Power swings in hydroelectric power plants *Trans. ASME* **62** 171–84

[5] Braisted D M and Brennen C E 1980 Auto-Oscillation of Cavitating Inducers *Polyphase Flow and Transport Technology* 157–66

[6] Nishi M, Kubota T, Matsunaga S and Senoo Y 1980 Study on swirl flow and surge in an elbow type draft tube *10th IAHR Symposium on Hydraulic Machinery and Cavitation* **1** 557

[7] Tsujimoto Y, Kamijo K and Yoshida Y 1992 Theoretical analysis of rotating cavitation in rocket pump inducers *28th Joint Propulsion Conference and Exhibit* **115** (Reston, Virigina: American Institute of Aeronautics and Astronautics) 135–41

[8] Otsuka S, Tsujimoto Y, Kamijo K and Furuya O 1996 Frequency Dependence of Mass Flow Gain Factor and Cavitation Compliance of Cavitating Inducers *J. Fluids Eng.* **118** 400
[9] Streeter V L 1983 Transient Cavitating Pipe Flow *J. Hydraul. Eng.* **109** 1407–23
[10] Chen C, Nicolet C, Yonezawa K, Farhat M, Avellan F and Tsujimoto Y 2008 One-Dimensional Analysis of Full Load Draft Tube Surge *J. Fluids Eng.* **130** 041106
[11] Kuibin P A, Pylev I M and Zakharov A V. 2012 Two-phase models development for description of vortex-induced pulsation in Francis turbine *IOP Conf. Ser. Earth Environ. Sci.* **15** 022001
[12] Müller A, Favrel A, Landry C and Avellan F 2017 Fluid–structure interaction mechanisms leading to dangerous power swings in Francis turbines at full load *J. Fluids Struct.* **69** 56–71
[13] Müller A, Favrel A, Landry C, Yamamoto K and Avellan F 2014 On the physical mechanisms governing self-excited pressure surge in Francis turbines *IOP Conf. Ser. Earth Environ. Sci.* **22**
[14] Bosioc A I, Susan-Resiga R, Muntean S and Tanasa C 2012 Unsteady Pressure Analysis of a Swirling Flow With Vortex Rope and Axial Water Injection in a Discharge Cone *J. Fluids Eng.* **134** 081104
[15] Skripkin S, Tsoy M, Kuibin P and Shtork S 2019 Swirling flow in a hydraulic turbine discharge cone at different speeds and discharge conditions *Exp. Therm. Fluid Sci.* **100** 349–59

Visualization of plasma fluxes by a measurement of own rotation of a dust particle

E S Dzlieva, V Yu Karasev, I R Krylov, I Ch Mashek, L A Novikov and S I Pavlov

Saint-Petersburg State University, Universitetskaya nab. 7/9, Saint-Petersburg, 199034, Russia

E-mail: v.karasev@spbu.ru

Abstract. A research of own rotation of single dust particle which is in a dust trap in the stratified glow discharge under imposed magnetic field is carried out. It is experimentally shown that in the scale of Debye length an action of azimuthal torque from the fluxes of ions and electrons is mutually compensated; an angular velocity of rotation of dusty particle remains invariable up to magnetic fields of 200 G. The found effect was given theoretical interpretation on a basis of consideration of ambipolar diffusion of positive ions and electrons to the particle surface.

1. Introduction

The visualization of plasma fluxes in low-temperature plasma is a relevant, but difficult problem, in particular, because of low ionization degree of plasma (up to 10^{-7}). So the dusty plasma [1-2] which is intensively developing in the last 20 years offers a new technical capability of the diagnostics. At the same time the dust particle works as "the optical probe", which mechanically responds to action of plasma flows and thereby visualizes them. Having information about the plasma fluxes, it is possible to control actively them in discharge chambers and technological systems. Similar researches in the area of heterogeneous electric field, near walls of discharge chambers as well as in magnetic field are especially relevant.

An own rotation of dust particles around the center of inertia was revealed experimentally by N. Sato [3-4]. So far its detailed research is absent for different discharge conditions and particles of any shape. A lot of information on own rotation of dust particles was given by work [5] in which the method of the coordinate tracing allowing to take effective measurements with transparent spherical particles was offered. The dust particles with a diameter up to 40 microns, which have a charge up to 10^6 electrons on a surface and rotate with angular velocity up to 10^4 rad/s, generate magnet moment up to 10^5 Bohr magnetons and define the magnetic properties of complex plasma [6-7]. Dynamic, structural, electrical and thermodynamic properties of complex plasma are rather well studied now, but magnetic properties after emergence of works [4-7] only begin to be studied.

An observation of own rotation of dust particles is a difficult technical problem which demands a significant optical multiplication and application of video recording with a high speed when the observed dust particle is located in a trap in plasma. Observations show that all dust particles possess own rotation (the magnetic moment). However, this rotation appears as threshold effect depending on magnetic induction and the power input [8-9].

Journal of Physics: Conference Series **1421** (2019) 012030 doi:10.1088/1742-6596/1421/1/012030

In the present work the dependence of angular velocity of rotation of the single dust particle on magnetic field is discussed. The experiment shows extremely weak change of angular velocity in magnetic field. The physical interpretation of rotation taking into account the fluxes of the electrons and ions going on the dust particle in the course of maintenance of its stationary charge in magnetic field is offered.

2. The scheme of the experiment and the optical registration of rotation

The registration of rotation of dusty particles was made by a coordinate tracing method [5-6] which consists in the following. Spherical transparent hollow glass particles (microspheres with the density of 0.4-0.9 g/cm^3 with a diameter of 10-120 μm, wall thickness 1–2 μm) were used. On their surface there are spot defects which scatter the light passing through the particle. The setup described in [5] was applied to creation of complex plasma. Particles were injected in the dust trap one by one. Their illumination was carried out sideways by laser radiation; the video recording was implemented from above by video camera with the CCD matrix with parameters: resolution of 1 megapixel and the frequency up to 240 frames per second. Tracing of the image was carried out by reciprocating motion of the system of optical registration (it consisted of a non-magnetic optical head part of microscope MBS-10 providing the a hundredfold increase rigidly connected with the video camera). Illumination and tracing were made in the horizontal plane in mutually perpendicular directions. The scheme is shown in figure 1.

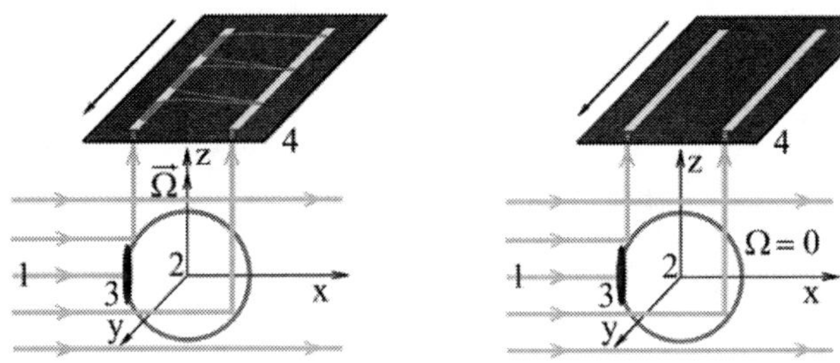

Figure 1. The scheme explaining the method of coordinate tracing. 1 – laser illumination, 2 – transparent particle, 3 – the schematic image of defect on the surface, 4 – the CCD matrix. When the matrix is moved (it is shown by an arrow) and the particle is rotated (it is shown by Ω) the registered signal is a cycloid which parameters depend on characteristics of the movement.

From the obtained results we will give the main one: the extremely weak dependence of angular velocity of own rotation of the particle on vertical magnetic field in the range of 0-200 of G is observed. Examples of dependences for two particles are given in figure 2. The result is of interest because it was expected the fast rotation of dust particles [10-11]. For example in [10], when imposing magnetic field, own rotation of dust particles with an angular velocity up to $\Omega \approx 6{\cdot}10^6$rad·s^{-1} was expected.

Without the magnetic field the direction of rotation of the particle and the value of angular velocity are random values. In magnetic field the majority of particles behave similar to atoms of paramagnetic: angular velocity almost does not change, its direction is opposite to vector of magnetic field, and the magnetic moment of a particle is directed along the magnetic field. In some cases the own rotation as function of magnetic field, has threshold character, more detail it was described in [9-10].

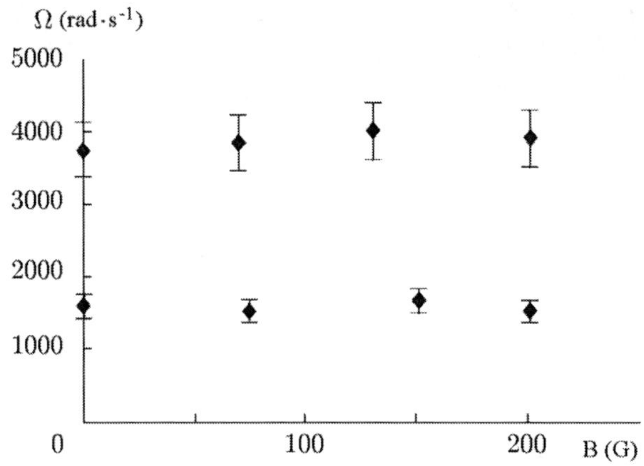

Figure 2. Example of dependence of angular velocity of own rotation on magnetic induction for two particles of spherical shape of 20 microns in size. Conditions: gas neon, pressure is 0.15 Torr, current 1.3 мA.

3. Interpretation of results

In the considered plasma the vertical fluxes of ions and electrons, blowing in the levitated not quite symmetric dust particle, spin it without any magnetic field as windmill blades twist. In this work the dust particles of spherical shape were used. Nevertheless, without magnetic field the considerable angular velocity of rotation up to thousands rad/s were observed. It occurs because the surface charge generally gathers on cambers of the particle and enhances its accidental asymmetry. Probably, also the threshold nature of rotation observed for some particles is explained by asymmetry of electric field of a particle and the field of its environment. In the presented work we are interested only the change of rotation of a dust particle under the influence of external magnetic field. Let's consider fluxes of positive ions and electrons on a particle surface, figure 3.

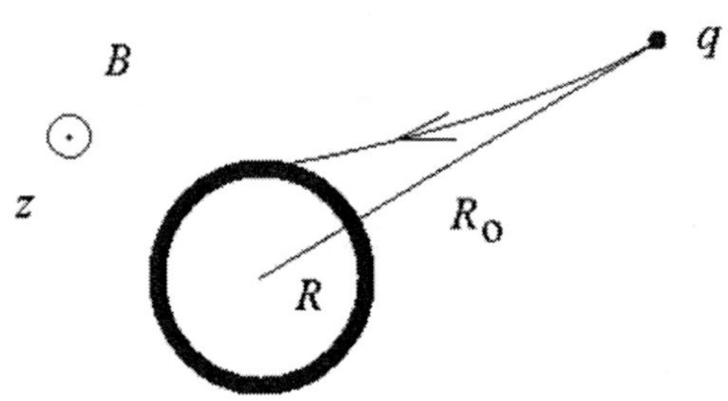

Figure 3. Illustration of the movement of the positive ion to the dust particle in magnetic field.

To simplify the analysis, we will consider that nothing depends on vertical coordinate; therefore we will consider a task only in coordinates of the horizontal plane, and we will look for only a vertical component of angular velocity of particle rotation. Let's consider the movement of the positive ion to the surface of the dust particle with zero starting velocity. In vertical magnetic field the positively charged ion gets the angular momentum in the system connected with the center of the particle. At a collision with the particle this angular momentum is transferred to the particle and twists it. Let r is the distance from the ion to the center of the particle; φ is a corner between the initial direction on the ion from the center of the particle and the current direction on the ion. The projection of ion velocity to the

Journal of Physics: Conference Series **1421** (2019) 012030 doi:10.1088/1742-6596/1421/1/012030

direction to the center of the particle is $V_{-r} = -\frac{\mathrm{d}r}{\mathrm{d}t} > 0$. Let's present the ion velocity as the sum of two components $\vec{V} = \vec{V}_{-r} + \vec{V}_{\varphi}$; two components of Lorentz force correspond to them: $F_{\varphi} = qV_{-r}B$ and $F_{-r} = -qV_{\varphi}B$. The second component of the force slows down the movement of the ion to the particle, but it will not interest us as the moment of this force is equal to zero. Moment of force $F_{\varphi} = qV_{-r}B$ is equal

$$M = M_z = F_{\varphi}r = qV_{-r}Br = -qB\frac{\mathrm{d}r}{\mathrm{d}t}r.$$

The angular momentum is connected with it:

$$L = L_z = \int M_z \mathrm{d}t = -qB\int \frac{\mathrm{d}r}{\mathrm{d}t} r\,\mathrm{d}t = -qB\int r\,\mathrm{d}r.$$

Limits of integration are defined by initial removal of the ion from the center of the particle R_0 and final removal equal to particle radius R:

$$L = L_z = -qB\int_{R_0}^{R} r\,\mathrm{d}r = \frac{qB}{2}\left(R_0^2 - R^2\right).$$

This angular momentum twists the dust particle.

We found the angular momentum transferred to the particle by one positive ion. Now we will consider the continuous flux of positive ions on the particle surface, this flux is connected with ambipolar diffusion. The angular momentum transferred in unit of time is equal to the moment of forces:

$$M = M_z = \frac{\mathrm{d}L_z}{\mathrm{d}t} = \frac{\mathrm{d}}{\mathrm{d}t}\left\{\frac{qB}{2}\left(R_0^2 - R^2\right)\right\} = \frac{B}{2}\left(R_0^2 - R^2\right)\frac{\mathrm{d}q}{\mathrm{d}t} = \frac{B}{2}\left(R_0^2 - R^2\right)I,$$

here I is the current of positive ions on the particle surface. As a result, the moment of forces acting on the particle from positive ions does not depend on weight, it is defined by the current I of these ions

$$M = \frac{B}{2}\left(R_0^2 - R^2\right)I.$$

In the given consideration we considered that the initial velocity of the charge moving to the dust particle is equal to zero. It is quite fair for positive ions, but it is not fair for electrons. The matter is that electrons are slowing down at the movement to the negatively charged particle therefore their initial velocity is obviously other than zero. The movement of electrons with different initial values of impact parameter will be, as shown in figure 4.

Different electrons untwist the dust particle in different directions, and on average the electrons untwist the particle a little. For an electron at its movement to the particle the same expressions for the Lorentz force, moment of force and change of the angular momentum of the particle, are the same as for the positive ion, only the charge and sizes proportional to it will change the sign. Respectively, change of the angular momentum of an electron during its movement will be equal to the same value on the module

$$\Delta L = \Delta L_z = -\frac{q}{c}B\int_{R_0}^{R} r\,\mathrm{d}r = \frac{qB}{2c}\left(R_0^2 - R^2\right) = -\frac{eB}{2c}\left(R_0^2 - R^2\right).$$

If to consider the electron which hits in the particle perpendicular to its surface (that is the final angular momentum of the electron relatively to vertical axis passing through the center of the particle is equal to zero), then the initial angular momentum of an electron:

$$L_0 = \frac{eB}{2c}\left(R_0^2 - R^2\right).$$

Thus, the average initial angular momentum of the electrons reaching the particle surface is approximately equal $L_0 = \frac{eB}{2c}\left(R_0^2 - R^2\right)$.

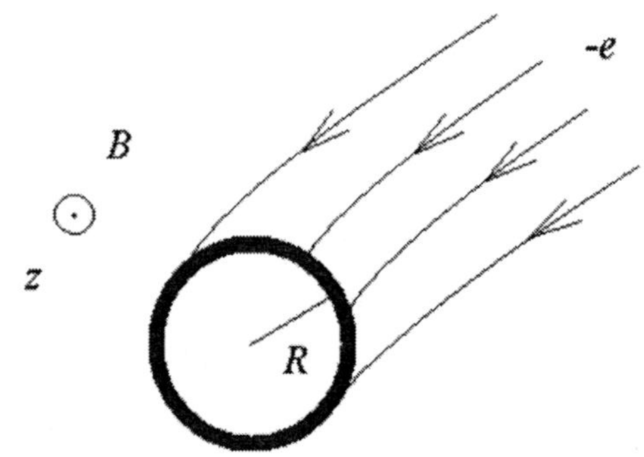

Figure 4. Illustration of the movement of electrons to the dust particle in magnetic field.

Electrons on sufficient removal of R_0 from the particle have isotropic distribution of velocities in the horizontal plane and the zero average angular momentum relatively the particle. A part of electrons with the angular momentum $L_0 = \frac{eB}{2c}\left(R_0^2 - R^2\right)$ then sticks to the dust particle and drops out of consideration of the remained electrons. As a result, in a layer of electrons at a distance of R_0 from the particle from each vanishing electron not compensated angular momentum $-L_0 = \frac{eB}{2c}\left(R_0^2 - R^2\right)$ is formed. Not compensated angular momentum which is formed in unit of time is equal to the moment of forces

$$M = \frac{B}{2c}\left(R_0^2 - R^2\right)I,$$

which in accuracy coincides with the moment of forces acting on the particle from current of positive ions, but differs in the sign because of other sign of current I.

It is possible to assume that the flux of positive ions twists the dust particle; and the flux of electrons twists in the counter direction not the particle, but the gas surrounding it at some distance R_0 from the center of the particle. As a result, the particle has to rotate under the influence of the flux of positive ions, but in an experiment it is not observed. The matter is that positive ions begin the movement not with zero initial velocity as it was considered above, and with some accidental dispersion of initial velocities. As a result of this dispersion the flux of positive ions on the dust particle is similar to figure 4 for electrons, but only with a deviation to the opposite side. Further reasoning for the flux of positive ions is similar to reasoning for the flux of negative electrons. As a result, both fluxes twist not the particle, but the plasma surrounding it in the counter directions, and the twisting of the particle is weak.

Thus, Ampere force acts on the current, flowing to the surface of the dust particle in vertical magnetic field and twist a mote and the plasma surrounding it. At ambipolar diffusion current of positive ions and current of negative electrons on the surface of the mote are equal on the module. Respectively, two Ampere forces untwist not the mote, but surrounding it plasma, with an identical force in the counter directions. Total current of ambipolar diffusion is equal to zero therefore the total moment of forces acting on this current is equal to zero. As a result, the rotation of the mote observed in an experiment is not enough.

4. Conclusion

It is experimentally shown that in the scale of Debye length the action of the azimuthal moments of forces from fluxes of ions and electrons is mutually compensated, the angular velocity of rotation of the dust particle remains almost invariable up to magnetic fields in 200 G. Experiments were carried out in the glow discharge in the neon in the stratified mode. Hollow transparent particles about 20 μm in the diameter were used. The original method of coordinate tracing was applied to diagnostics of their own rotation.

The found effect of extremely weak dependence of angular velocity of rotation of the dust particle on magnetic field was given the theoretical interpretation on the basis of consideration of ambipolar diffusion of positive ions and electrons on the particle surface. In magnetic field the charges at the movement to the surface of the particle get the angular momentum. It is given to the particle and the plasma surrounding it and does not depend on the mass of charged particles of the flux. Respectively, the angular momentum of the flux of ions is generally compensated by the angular momentum of the flux of electrons.

Acknowledgments

The work was supported by RFBR, grant № 18-02-00113. The equipment and materials used were partially acquired through the program of the Development of the Material and Technical Base of Saint-Petersburg State University.

References

[1] Fortov V E and Mofill G E 2010 *Complex and dusty plasmas: from laboratory to space* (NewYork: Taylor & Francis Group)
[2] Vladimirov S V, Ostrikov K and Samarian A A 2005 *Physics and Applications of Complex Plasmas* (London: Imperial College Press)
[3] Fukagawa K, Uchida G, Iizuka S and Sato N 2001 *Proc. XXV Int. Conf. on Phenomena in Ionized Gases* vol 3 (Nagoya, Japan) p 37
[4] Sato N 2005 *AIP Conf. Proc.* **799** 97
[5] Karasev V Yu, Dzlieva E S, Eikhval'd A I, Ermolenko M A, Golubev M S and Ivanov A Yu 2009 *Phys. Rev.* E **79** 026406
[6] Karasev VYu, Dzlieva E S, Ermolenko M A, Golubev M S and Ivanov A Yu 2011 *Contr. Plasma Phys.* **51** 509
[7] Tsytovich N V, Sato N and Morfill G E 2003 *New Jour. Phys.* **5** 43
[8] Dzlieva E S, Karasev V Yu and Petrov O F 2012 *JETP* **114** 167
[9] Karasev V Yu, Ermolenko M A, Dzlieva E S, Pavlov S I, Novikov L A and Mashek I Ch 2016 *Tech. Phys.* **61** 618
[10] Krasheninnikov S I 2006 *Physics of Plasmas* **13** 114502
[11] Ishihara O and Sato N 2001 *IEEE Trans. Plasm. Scien.* **29** 179

Compensation of spectral image shift in AOTF-based system

A A Bykov[1,2], P V Zinin[2], K M Bulatov[2], D D Khokhlov[1,2] and I B Kutuza[2]

[1]National Research University Moscow Power Engineering Institute, Krasnokazarmennaya 14, Moscow 111250, Russia
[2]Scientific-Technological Center of Unique Instrumentation of the Russian Academy of Sciences, Butlerova 15, Moscow 117342, Russia

E-mail: bykval@bk.ru

Abstract. Recently, it was shown that the use of a tandem acousto-optic tuneable filter (TAOTF) in combination with a camera allows one to measure the temperature distribution at the focus of a laser beam on an object's surface. The TAOTF advantage is a reduction of aberrations and spatial distortions. Unfortunately, a TAOTF significantly reduces the intensity of the transmitted light beam, requires longer exposure time and often hampers temperature measurements below 1200 K. A single-crystal AOTF is free from this drawback and allows increasing the intensity of the incident radiation to the camera sensor. The paper presents experimental data for comparing measurements of the temperature distribution at the laser focus, obtained using single-crystal AOTF and TAOTF. A strong chromatic drift caused by AOTF-based system was compensated using the digital processing of the spectral images.

1. Introduction

Measurement of the temperature distribution on the surface of laser heated solids is an important task of geophysics and materials science [1]. In earlier studies [2], it was shown that the use of a tandem acousto-optic filter (TAOTF) in combination with a camera allows measuring the temperature distribution in focused high power laser beam. Temperature measurement is carried out by recording the spectrum of thermal radiation of object and further comparing the theoretical Planck curve with experimental data [3]. The TAOTF was used to avoid strong chromatic drift of the AO spectroscopic image in AOTF imaging system. Unfortunately, a TAOTF significantly reduces the intensity of the transmitted light beam, which makes it impossible to measure temperatures below 1200 K and also with a low intensity of the object's radiation, a long exposure time of the camera is required.

The goal of this work is to demonstrate that the use of AOTF results in a twofold increase in the intensity of multispectral images as compared to a TAOTF. A strong chromatic drift of the AO spectroscopic imaging system was compensated by the use of developed software allowing the alignment of all the images recorded at different wavelength. As a result, the threshold for measuring the minimum temperature is reduced, and the exposure time is also reduced, which allows several times to increase the speed of shooting hyperspectral images. The paper presents experimental data for comparing measurements of temperature distribution in ribbon filament lamp heated by current, obtained using AOTF and TAOTF.

2. Experimental Method

The use of acousto-optic filtering makes it possible to investigate the spectral images of the object under study in a fairly wide range of wavelengths in a certain part of the optical spectrum. The main advantage of acousto-optic tunable filters in the field of optical studies is their high speed and arbitrary positioning of the filter transmission line, compactness, simplicity of design, the absence of moving parts, and characteristic small control voltages. The optical scheme of an AOTF is shown in figure 1. The principle of operation of an AOTF is based on the effect of light diffraction in an optically anisotropic crystal on traveling acoustic waves excited by a piezoelectric transducer [4]. A crystal of paratellurite is usually used as a transparent optical medium. A feature of AOTFs is a specific transmission function, which, in addition to the main line, has a set of low intensity side lobes described as sinc^2-function.

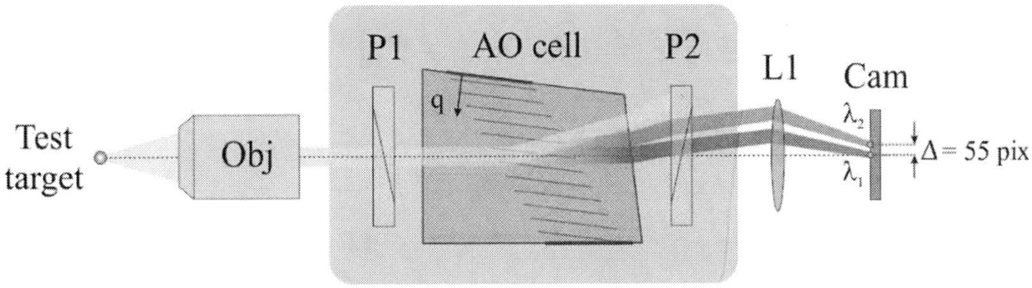

Figure 1. Image shift (Δ) on interval from λ_1 = 640 nm to λ_2 = 740 nm in AOTF. P1, P2 – crossed polarizes; L1 – lens; AO cell – tuneable AO crystal; Obj – objective; Cam – video camera with CCD;.

The main problem of using AOTF for imaging is the transverse chromatic aberration (figure 1) [5-7]. These aberration lead to a transverse displacement of the image occurring in the diffraction plane. To reduce the effect of transverse chromatic aberrations in the manufacture of AOTFs, the output face of the filter is tilted at a certain angle, and the input face is orthogonal to the transmitted beam to reduce reflection losses.

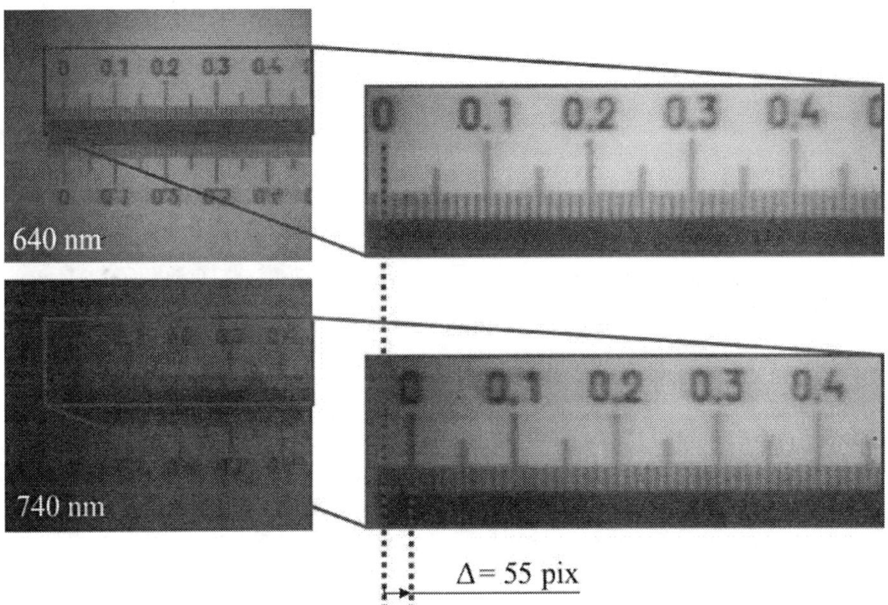

Figure 2. The image shows the displacement on 55 pixels (189 μm) of transmitted radiation in wavelength diapason from 640 nm to 740 nm.

Series of images was taken (figure 2) to evaluate the displacement of the image on AOTF filter. Images taken at 640 nm and 740 nm are shown in figure 2. The relative shift from the image taken at 640 nm to that taken at 740 nm found to be 189 μm or 55 pixels on the camera sensor. There was used an Imaging Source camera with pixel size of 3,75 μm. Resolution test target was used to measure the shift (figure 2). Figure 3 shows the results of the measurements of the transverse shift of TAOTF and AOTF. They show a linear dependence on the wavelength. The displacement of the image is small for TAOTF and is only 7,5 μm for the wavelength range from 640-740 nm. The dependence of the transverse

Journal of Physics: Conference Series **1421** (2019) 012031 doi:10.1088/1742-6596/1421/1/012031

$$z(\lambda) = 2.53(0.89) + 0.572(0.014) \cdot \lambda$$

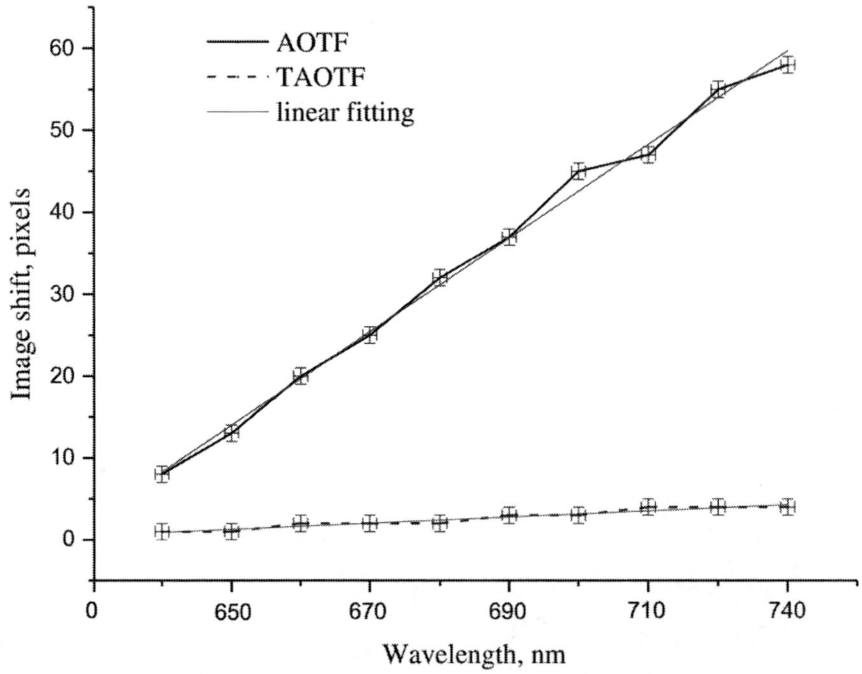

Figure 3. Dependence of image displacement on wavelength for TAOTF and AOTF. Linear least square fits are shown as red lines.

3. Results

To compare results obtained by the TAOTF, and AOTF and resistivity methods were used to conduct an experiment similar to that described in [3, 8] in a current configuration (figure 4). A tungsten lamp was heated by an electric current (10A) and the temperature of the heated tungsten plate was measured using AOTF and TAONF. A lamp with a flat uniformly heated tungsten tape was used as a radiating heated object. The temperature of lamp was determined by fitting [3] the actual signal to Planck's equation at each point of the specimen's surface (figure 4):

$$I(\lambda,\varepsilon,T) = \varepsilon g(\lambda,T), \quad g(\lambda,T) = \frac{c_1}{\left\{\lambda^5\left[\exp(\frac{c_2}{\lambda T}) - 1\right]\right\}}$$

where I is the spectral intensity, ε is the sample emissivity, λ is wavelength, T is the temperature, and c_1 and c_2 are physical constants ($c_1 = 2\pi c^2$, $c_2 = h{\cdot}c/k = 0.01432$ m·K). Because the temperature of the heated body should be uniform over the surface the variation of the temperature (~40 K) over the surface is related to the error of the AOTF method.

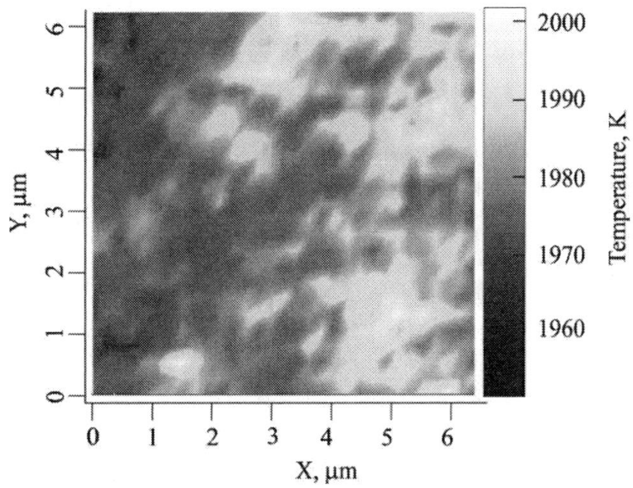

Figure 4. Temperature distribution on heated ribbon filament lamp.

When the current was 10 A, the temperature measured by AOTF was appeared to be 1969(60) K, and that obtained by TAOTF was 2012(81) K. The number in the brackets are 95% confidence interval described in [3]. The temperature of the lamp measured by TAOTF and AOTF are in an agreement within experimental errors. We also mention that the confidence interval determined AOTF is 30% less than that determined by TAOTF. It might be related to the fact that time of the exposure for the AOTF system is nearly three times as low as that for TAOTF system. It was established experimentally that the transmittance of AOTF is 2.8 times higher than that of TAOTF. This allows reducing the exposure time on the camera while shooting a series of images and as a result data collection speeds up.

4. Conclusions

The paper demonstrates the use of two types of acousto-optical spectral filters with respect to the problem of measuring the temperature of a heated object. A lamp with a flat uniformly heated tungsten tape was used as a radiating heated object. It was established experimentally that the transmittance of AOTF is 2.8 times higher than that of TAOTF. This allows reducing the exposure time on the camera while shooting a series of images and as a result data collection speeds up. We demonstrate that using AOTF also as a TAOTF allows measuring temperature of heated objects. The temperature of lamp was determined by fitting [6] the actual signal to Planck's equation at each point of the specimen's surface.

Acknowledgements

This work was supported by the program "Fundamental scientific research of the National academy of sciences for 2013-2020" (№0069-2016-0004). This work was performed using the Unique Scientific Instrument "Laser heating at diamond anvil cell" [http://ckprf.ru/usu/507563/].

References

[1] Bassett W 2001 *Rev. Sci. Instrum.* vol 72, p. 1270
[2] Machikhin A, Zinin P, Shurygin A and Khokhlov D 2016 *Opt. Lett.* **41** 901
[3] Zinin P, Bykov A, Machikhin A et al 2019 *High Pressure Res.* **39** 131
[4] Xu J and Stroud R 1992 *Acousto-optic devices: principles, design, and applications* (New York: Wiley)
[5] Voloshinov V, Yushkov K and Linde B 2007 *J. Opt. A* **9** 341
[6] Machikhin A and Pozhar V 2009 *Instrum. Exp. Tech.* **52** 847
[7] Ryu S, You J, Kwak Y and Kim S 2008 *Opt. Express* **16** 17138
[8] Bulatov K, Mantrova Y, Bykov A et al 2017 *Computer Optics* **41** 864

A review of non-polar liquids as materials for bulk acousto-optic devices operating with terahertz radiation

P Nikitin[1,2]

[1]Scientific and Technological Center of Unique Instrumentation of the RAS, Butlerova str. 15, Moscow 117342, Russia
[2]Lomonosov Moscow State University, Leninskie Gory 1, Moscow 119991, Russia

E-mail: nikitin.pavel.a@gmail.com

Abstract. In this paper, the acoustic and optical properties of a number of liquids transparent in the terahertz range are systematized. The acousto-optic quality is calculated for the quasi-orthogonal and collinear geometry of the acousto-optic interaction. The parameters of acousto-optic deflectors and filters of terahertz radiation are determined. The diagrams linking the resolution for the filter and the maximum number of resolvable elements for the deflector with the diffraction efficiency are given.

1. Introduction

Acousto-optic (AO) devices are widely used to control electromagnetic radiation in the visible, infrared and ultraviolet ranges [1-3]. The most effective in terms of electrical power consumption are devices operating with shortwave radiation. This is due to the fact that the ratio of the intensity I_1 of diffracted radiation to the intensity I_0 of radiation incident on an AO cell is inversely proportional to the square of the radiation wavelength λ:

$$\frac{I_1}{I_0} \propto \frac{M_2 P_a}{\lambda^2}, \tag{1}$$

where P_a is the sound wave power, M_2 is the AO figure of merit, depending on the acoustic, optical and photo-elastic properties of the medium.

To control laser THz radiation, metallic and silicon diffraction elements are used. Thus, the AO effect can be used for real-time processing of information in the THz range in areas of science and technology that do not require large values of the intensity of diffracted radiation. The well-known AO crystals, such as paratellurite (TeO_2) and lithium niobate ($LiNbO_3$), are opaque in the THz range [4, 5] and the transparent crystals, such as germanium (Ge) and silicon (Si) have relatively small values of M_2. Therefore liquids become the most suitable media for AO interaction. To date, there are several review papers about the choice of the AO interaction medium in different spectral ranges of electromagnetic waves. Liquids were used as the AO interaction medium only in the visible range and an overview of their AO properties can be found in the works of Uchida, Watanabe and Hoff [6-8]. The purpose of this work is to select liquids that are most suitable for creating AO devices operating with THz radiation.

2. Acoustical, optical and photo-elastic properties of liquids

To obtain acceptable values of the diffraction efficiency, the length of the AO interaction region have to be of the order of 1 cm. Therefore, only relatively transparent liquids with a radiation absorption coefficient of no more than $\alpha < 5\ \text{cm}^{-1}$ were selected. It was assumed that the radiation intensity I decreases exponentially with the distance l

$$I \propto \exp(-\alpha l)\,. \tag{2}$$

As noted above, the suitability of the material from the point of view of acousto-optics is determined by its AO figure of merit M_2, which for liquids is determined by the following relation [6]

$$M_2^{(\text{liquid})} = \left(\rho \frac{\partial n}{\partial \rho}\right)^2 \frac{4}{\rho V^3} = \left[\frac{(n^2-1)(n^2+2)}{6n}\right]^2 \frac{4}{\rho V^3}\,, \tag{3}$$

where n – refractive index, V – sound velocity, ρ – density.

Table 1 shows the values of n and α at $\lambda = 140\,\mu\text{m}$ for a number of non-polar liquids, as well as for such polar liquids as alcohols. Since the values of the elasto-optical constant $\rho(\partial n/\partial \rho)$ in the THz range are unknown, table 1 was supplemented with the values of M_2 calculated by using the equation (2). Acoustic properties of the liquids are also shown in the table 1. Attenuation the sound was assumed exponential and its power P_a is related to distance l in the following way

$$P_a \propto \exp(-\alpha_s l)\,. \tag{4}$$

Table 1. Acoustical, optical and acousto-optical properties of selected liquids.

liquid	ρ (g/cm^3)	V (km/s)	$\alpha_s/(2F^2)$ (10^{-17} s^2/cm)	n	α (cm^{-1})	M_2 (10^{-15} s^3/kg)
C_6H_{12}	0.774 [9]	1.255 [9]	190 [10]	1.421 [11]	0.37 [11]	604
C_6H_{14}	0.655 [12]	1.077 [12]	60 [6,10]	1.372 [13]	0.69 [13]	847
$C_{16}H_{34}$	0.770 [12]	1.338 [12]	100 [12]	1.428 [13]	0.69 [13]	520
C_6H_6	0.881 [14]	1.298 [14]	850 [10]	1.496 [11]	5.5 [11]	709
C_8H_{10}	0.880 [10]	1.360 [10]	63 [10]	1.492 [15]	5.1 [15]	605
C_2H_5OH	0.789 [10]	1.155 [10]	54 [10]	1.498 [16]	130 [16]	1135
$C_5H_{11}OH$	0.814 [10]	1.294 [10]	97 [10]	1.477 [16]	45 [16]	705
CCl_4	1.584 [6]	0.922 [6]	550 [6]	1.487 [11]	1.4 [11]	1053
$CHBr_3$	2.877 [6]	0.920 [6]	290 [6]	1.595 [17]	6.0 [17]	960
CS_2	1.256 [6]	1.141 [6]	6100 [6]	1.61 [18]	3.5 [18]	1228

The liquids given in table 1 have approximately the same refractive index $n \approx 1.4$ in the THz range. Among non-polar liquids, cyclohexane (C_6H_{12}) is the most transparent. Non-polar liquids, such as alcohols, absorb THz radiation much more strongly. This is due to the resonant stretching of hydrogen bonds. The acoustic properties of the listed liquids are very similar: the density is about $\rho \approx 0.8$ g/cm^3, and the speed of sound is $V \approx 1.2$ km/s. It should be noted that at relatively high sound frequencies F value of the attenuation coefficient α_s of sound becomes comparable or even higher than radiation absorption coefficient α. Therefore value of some combination of α_s and α will be a criterion for using a non-polar liquid as an AO interaction medium.

3. Quasi-orthogonal geometry of acousto-optic interaction

With the quasi-orthogonal geometry of the AO interaction, the diffracted radiation propagates approximately in the same direction as the transmitted. This configuration is shown in figure 1 and can be used to create THz radiation modulators and deflectors. The angle of deviation of the diffracted radiation is equals to twice the Bragg angle $\theta_B \ll 1$ and depends linearly on the frequency of the sound F [19]

$$\sin\theta_{\mathrm{B}} = \frac{\lambda F}{2V}. \tag{5}$$

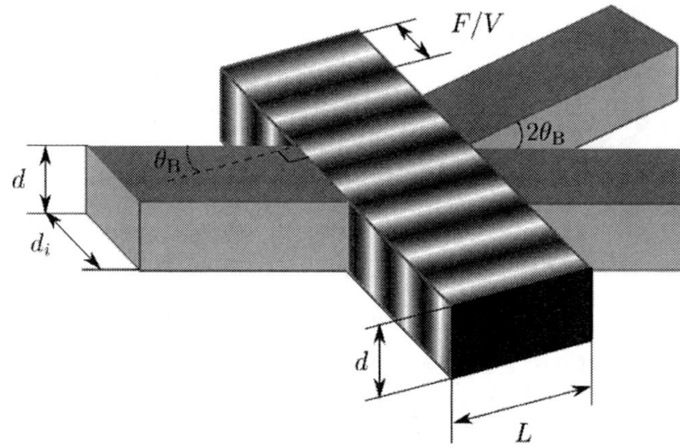

Figure 1. Scheme of the quasi-orthogonal diffraction of radiation on the phase grating, induced by sound wave.

Since the angle of deflection of the diffracted radiation is much less than unity, then during calculations we set Bragg angle equal to $\theta_{\mathrm{B}} = 0.1$ (or approximately 5.7°). The AO interaction will be observed at the ultrasound frequency F

$$\sin\theta_{\mathrm{B}} \approx \theta_{\mathrm{B}} = 0.1; \quad F = 2\sin\theta_{\mathrm{B}}\frac{V}{\lambda} = 0.2\frac{V}{\lambda}. \tag{6}$$

As established in the paper [20], in an absorbing medium at low diffraction efficiency the ratio of the intensity $I_1(L)$ of the diffracted radiation at the output of an AO cell to the intensity $I_0(0)$ of the radiation incident on an AO cell is determined by the following expression

$$\frac{I_1(L)}{I_0(0)} = \frac{\pi^2}{2\lambda^2}\frac{M_2 P_{\mathrm{a}}}{d} L\exp(-\alpha L), \tag{7}$$

where L is the length of the AO interaction region, approximately equaled to the width of the sound beam, and d is the width of the sound beam in the plane orthogonal to the AO interaction plane (see Figure 1).

From the expression (7) it follows that with increasing L, the diffraction efficiency linearly increases due to the interaction of radiation with a large number of periods of the sound wave on the one hand and decreases exponentially due to the absorption of radiation in the medium on the other hand. Thus, there is an optimal length L_{opt} of AO interaction, at which the maximum diffraction efficiency is achieved [20]

$$L_{\mathrm{opt}} = \frac{1}{\alpha}; \quad \frac{I_1(L_{\mathrm{opt}})}{I_0(0)} = \frac{\pi^2}{2\lambda^2}\frac{M_2 P_{\mathrm{a}}}{d\alpha e}. \tag{8}$$

It was assumed that the diffraction efficiency at the wavelength of $\lambda \pm \Delta\lambda/2$ is 2 times less than the diffraction efficiency at the wavelength λ. Under the quasi-orthogonal geometry of the AO interaction, the following proportion between the bandwidth of the radiation wavelength and the bandwidth of the sound frequency is valid: $\lambda / \Delta\lambda = F / \Delta F$ [21].

As is known from the literature, the ultrasound frequency bandwidth ΔF, in which an effective AO interaction is observed (at the edges of this region, the diffraction efficiency decreases by 2 times) is calculated as [22]

$$\Delta F = \frac{1.8nV^2}{\lambda FL}. \tag{9}$$

The number $N = \Delta\theta / \delta\phi$ of resolvable elements is determined by the ratio of the angle bandwidth $\Delta\theta$, in which the effective scanning of the diffracted beam is realized, to the diffraction divergence $\delta\phi = \lambda / (nd_d)$ of the radiation in the AO interaction plane. Here d_d is the width of the diffracted radiation beam in the AO interaction plane, which is assumed to be equal to the width d_i of the beam incident on the AO cell (see figure 1).

The Bragg interaction is resonant in nature and requires the fulfilment of the condition of Bragg synchronism. When light is diffracted into a + 1 diffraction order, this condition has the form $\mathbf{k}_1 = \mathbf{k}_0 + \mathbf{K} + \Delta\mathbf{K}/2 + \boldsymbol{\eta}$, where $\Delta K \equiv |\Delta\mathbf{K}| \ll |\mathbf{K}|$ and $|\boldsymbol{\eta}| \ll |\mathbf{K}|$. Here $\mathbf{K}$ is the wave vector of sound corresponding to the Bragg matching condition; the value of ΔK corresponds to the bandwidth of sound frequency from the relation (9); $\mathbf{k}_0$ and $\mathbf{k}_1$ are the wave vectors of the incident and diffracted radiation and their moduli are the same $|\mathbf{k}_0| = |\mathbf{k}_1| = k$. In addition, the notation $\boldsymbol{\eta}$ is used for the mismatch vector, which is directed orthogonally to the sound beam [23].

The relation between the wave vectors of the interacting waves can be shown on the wave vector diagram. Since the wave surface has the shape of a circle and the Bragg angle is small $\theta_B \ll 1$, then $|\boldsymbol{\eta}| \ll \Delta K$. Therefore, the bandwidth of angles $\Delta\theta$ in the medium can be calculated as the ratio of the bandwidth of sound vectors ΔK, in which the AO interaction is observed, to the module of the wave vector of light k. Thus, the number N of resolvable elements is proportional to the width d_i of the light beam and inversely proportional to the length L

$$N = \frac{\Delta\theta}{\delta\phi} = \frac{\Delta K / k}{\lambda / (nd_i)} = \frac{(2\pi\Delta F / V)/(2\pi n / \lambda)}{\lambda / (nd_i)} = \frac{1.8nVd_i}{\lambda FL} = \frac{0.9}{\sin\theta_B}\frac{d_i}{L}n. \tag{10}$$

4. Collinear geometry of acousto-optic interaction

A feature of the collinear AO interaction in liquids is that the diffracted radiation propagates towards the incident one. In the literature, the term “backward” collinear diffraction is used to describe this type of AO interaction. From the vector diagram of phase matching, it follows that collinear diffraction is observed when the wave vector of the sound is 2 times larger than the wave vector of light. The required ultrasound frequency is determined by the following relation [19]

$$F = \frac{2nV}{\lambda}. \tag{11}$$

Since the refractive index of liquids $n \approx 1.5$ is small, the maximal of angle of incidence of sound on the input optical window is limited by approximately 40°. Therefore, it becomes possible to construct an AO cell so that the path travelled by the sound to the input optical window is much less than the length of the AO interaction, and the attenuation of sound in this region can be neglected. In this case, the ratio of the intensity of diffracted radiation to the intensity of the incident radiation, as well as the optimal length of the AO interaction are calculated from the following equations [24]

$$L_{\text{opt}} = \infty, \qquad \frac{I_{-1}(0)}{I_0(0)} = \frac{\pi^2}{2\lambda^2}\frac{M_2 P_a}{S}\left(\alpha + \frac{\alpha_s}{2}\right)^{-2}, \tag{12}$$

where S is the cross-sectional area of the sound beam, equal to the area of the piezoelectric transducer.

The value of the two-sided band Δk by the wave vectors of light is determined from the expression for ΔK taking into account the fact that $\Delta K = 2\Delta k$

$$\Delta K = 2\Delta k = 2\left(\alpha + \frac{\alpha_s}{2}\right). \quad (13)$$

Thus, the spectral resolution R of the AO filter depends mainly on the absorption coefficient of the radiation α and the attenuation coefficient of the ultrasound α_s

$$R = \frac{k}{\Delta k} = \frac{2\pi n}{\lambda} \frac{1}{\alpha + \alpha_s / 2}. \quad (14)$$

5. Results and discussion

It was assumed that $d_i = 10$ mm and $d = 5$ mm. When the width of the sound beam in the AO interaction plane is equal to the optimal length $L_{opt} = 1/\alpha$, and the ultrasonic power is $P_a = 1$ W, the values of the diffraction efficiency I_1/I_0 and the number N of solvable elements of AO deflectors are calculated and given in table 2.

Table 2. Properties of acousto-optic deflectors based on liquids.

liquid	L_{opt}, cm	F, MHz	I_1/I_0, 10^{-4}	N
C_6H_{12}	2.7	2.5	3	5
C_6H_{14}	1.4	2.1	2.3	9
$C_{16}H_{34}$	1.4	2.9	2.7	9
C_6H_6	0.18	3.0	0.5	70
C_8H_{10}	0.20	3.1	0.4	70
C_2H_5OH	0.008	2.7	0.03	1800
$C_5H_{11}OH$	0.022	2.9	0.06	600
CCl_4	0.7	2.1	2.7	19
$CHBr_3$	0.17	2.3	0.6	90
CS_2	0.29	2.8	1.3	50

When calculating the parameters of AO filters, it was assumed that the piezoelectric transducer area is $S = 5\times 5$ mm^2, and the sound power is $P_a = 1$ W. The results of the calculations are given in table 3.

Table 3. Properties of acousto-optic filters based on liquids.

liquid	F (MHz)	α_s (cm^{-1})	I_{-1}/I_0, (10^{-4})	R (10^3)
C_6H_{12}	25.4	2.5	2.4	0.4
C_6H_{14}	21.1	0.5	9.3	0.6
$C_{16}H_{34}$	27.2	1.5	1.5	0.4
C_6H_6	27.7	13	0.05	0.06
C_8H_{10}	29.0	1.1	0.19	0.12
C_2H_5OH	24.7	0.7	0.001	0.005
$C_5H_{11}OH$	27.3	1.4	0.003	0.014
CCl_4	19.6	4.2	0.9	0.19
$CHBr_3$	20.9	2.2	0.19	0.10
CS_2	26.2	47	0.017	0.027

Based on the obtained results, a diagram was constructed for the AO deflectors (see figure 2), which interlinks the number N of resolvable elements and the diffraction efficiency I_1 / I_0 for the liquids considered at the radiation wavelength $\lambda = 140\,\mu\text{m}$.

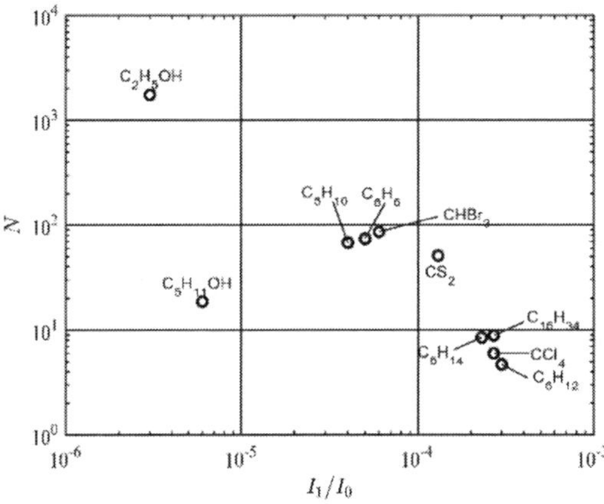

Figure 2. Dependence of the number of resolvable spots on the diffraction efficiency.

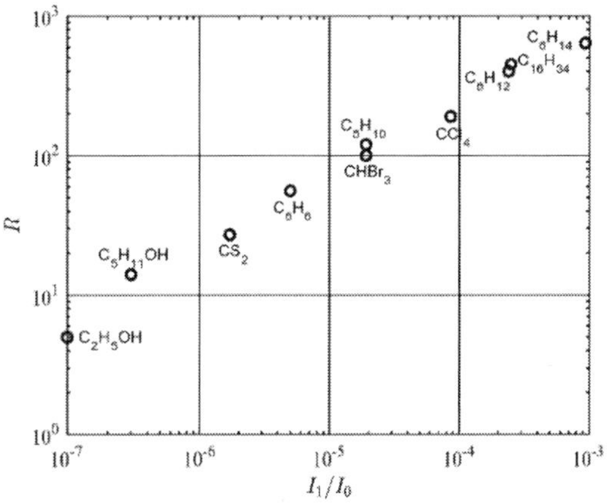

Figure 3. Dependence of the AO filter spectral resolution on the diffraction efficiency.

From figure 2, it follows that AO deflectors with a higher diffraction efficiency will be characterized by a smaller number of resolvable elements. That could be explained by the fact that the diffraction efficiency is proportional to the length of the AO interaction L, while the number of solvable elements is inversely proportional to L. Therefore, the isolines (corresponding to a certain liquid) in this diagram will be parallel to the diagonals of the squares, since the value of the product $N \cdot (I_1 / I_0)$ remains constant as L changes.

A similar diagram (see figure 3) was also constructed for AO filters. It interlinks the resolution R of AO filters and the diffraction efficiency I_1 / I_0 for the liquids considered. From the diagram a trend is visible: the higher resolution R of the devices in which liquid is used as the interaction AO medium, the larger maximum achievable diffraction efficiency I_1 / I_0. Moreover, as I_1 / I_0 increases by 2 orders of magnitude, R increases by an order of magnitude. It can be explained in the following way. As follows from the relations (12) and (14), the diffraction efficiency is inversely proportional to the square of the combination $(\alpha + \alpha_s / 2)$ of the sound attenuation coefficient and the radiation absorption coefficient, while the resolution is inversely proportional to the first power of the specified combination. This can be qualitatively explained as follows. The larger number of sound wave periods radiation interacts with, the higher diffraction efficiency and resolution. In this case, the length of the region of effective AO interaction decreases with increasing attenuation of the sound and absorption of the radiation.

6. Summary

The first time systematization of the acoustic and optical properties of liquids made it possible to estimate the value of their AO figure of merit in the THz range. It is shown that due to the low diffraction efficiency, the length of the piezoelectric transducer should be equal to the reciprocal of the absorption coefficient of the radiation. Correlation between resolution and diffraction efficiency for the AO filter as well as correlation between the number of resolvable elements and diffraction efficiency for the AO deflector based on the liquids were estimated. It is established that for the manufacture of these devices it is advisable to use saturated hydrocarbons as the medium of AO interaction.

Acknowledgments
Results have been obtained under support of the Russian Science Foundation (RSF) grant No.18-12-00430.

References
[1] Zhang C, Wang H, Sheng Z, Zhang X, Yuan J, Shi L and Wang P 2018 *Applied Optics* **57** 8595–8599
[2] Gupta N and Voloshinov V 2014 *Journal of Optics* **16** 035301
[3] Ward J and Valle S 2018 *Fiber Lasers and Glass Photonics: Materials through Applications* vol 10683 ed Taccheo S, Ferrari M and Mackenzie J (Proc. SPIE) p 1068327
[4] Unferdorben M, Szaller Z, Hajdara I, Hebling J and Palfalvi L 2015 *Journal of Infrared, Millimeter, and Terahertz Waves* **36** 1203–1209
[5] Unferdorben M, Buzady A, Hebling J, Kiss K, Hajdara I, Kovacs L, Peter A and Palfalvi L 2016 *Journal of Infrared, Millimeter, and Terahertz Waves* **37** 703–709
[6] Uchida N 1968 *Japanese Journal of Applied Physics* **7** 1259
[7] Fukumoto A and Watanabe A 1970 *Japanese Journal of Applied Physics* **9** 662–665
[8] Hoff F 1971 *Japanese Journal of Applied Physics* **10** 269–270
[9] Gonzalez B, Gonzalez E, Calvar N, Dominguez I and Canosa J 2010 *J. Chem. Eng. Data* **55** 2294–2305
[10] Kikoin I (ed) 1976 *Tables of physical values* (Moscow: Atomizdat)
[11] Pedersen J and Keiding S 1992 *IEEE J. Quant. Electron.* **28** 2518–2522
[12] Djerdjev A and Beattie J 2008 *Phys. Chem. Chem. Phys. 10* 4843–4852
[13] Laib J and Mittleman D 2010 *J. Infrared, Millimeter, and Terahertz Waves* **31** 1015–1021
[14] Abulencia J and Theodore L 2009 *Fluid flow for the practicing chemical engineer* (Hoboken, NJ: John Wiley & Sons)
[15] Jin Y, Kim G, Shon C, Jeon S and Kim J 2008 *Journal of the Korean Physical Society* **53** 1879–1885
[16] Chamberlain J, Afsar M, Hasted J, Zafar M and Davis G 1975 *Nature* **255** 319–321
[17] Zafar M, Zafar F, Shamim A and Arif Z 1984 *Infrared Physics* **24** 505–509
[18] Yu B, Zeng F, Xing Q and Alfano R 2003 *Appl. Phys. Lett.* **82** 4633–4635
[19] Royer D and Dieulesaint E 1999 *Elastic Waves in Solids II* (Springer-Verlag Berlin Heidelberg)
[20] Nikitin P, Voloshinov V, Gerasimov V and Knyazev B 2017 *Technical Physics Letters* **43** 635–637
[21] Voloshinov V, Molchanov V and Mosquera J 1996 *Optics & Laser Technology* **28** 119–127
[22] Gordon E 1966 *Proceedings of the IEEE* **54** 1391–1401
[23] Voloshin A and Balakshy V 2015 *Acta Physica Polonica A* **127** 60–62
[24] Nikitin P and Voloshinov V 2015 *Physics Procedia* **70** 712–715

Journal of Physics: Conference Series **1421** (2019) 012033 doi:10.1088/1742-6596/1421/1/012033

Special features of laser diagnostics of gas flows in a thick-walled glass pipe

K M Lapitskiy, D A Poyatsyka, I A Lapitskaya, A Yu Poroykov and N S Ilina

National Research University "Moscow Power Engineering Institute", Krasnokazarmennaya 14, Moscow, Russia

Abstract. The article presents application of optical gradient methods for the research of inhomogeneity of gas flows through a thick-walled glass cylinder. The influence of not only inhomogeneous gas medium, but also cylinder walls on the process of laser radiation propagation was researched. Methods for describing the refraction of the laser sheet for various cases of its orientation in space in geometrical optics approximation were worked out. The analysis of displacement of the laser sheet's sections on the screen after penetration through inhomogeneity was conducted. Calculations of the laser sheet's shape distortion during penetration through a transparent hollow cylinder were performed. Theoretical distributions of the laser sheet's sections on the screen were constructed. Various cases of the laser sheet orientation in space were considered. The influence of an optically inhomogeneous medium on the laser sheet's shape were considered.

1. Introduction

The research of liquid and gas flows has a great practical value in designing of long-distance transportation systems. Particularly, it is necessary to investigate the influence of temperature gradient, density, pressure and other parameters on the strength characteristics of a pipeline.

Notwithstanding the fact that in most cases pipelines are made of opaque materials, there is a task to conduct a test research of the flow during its propagating through the pipeline made of a transparent material, such as glass. In laboratory conditions it is possible to produce the model of the pipeline in question entirely from glass, and for practical purposes to make a small part of the pipeline transparent for testing and control.

Optical methods of inhomogeneous media research, in comparison with other methods of measurement, have significant advantages [1]. Such methods make it possible not to introduce mechanical disturbances into the medium under research, provide remotability and the ability to determine the local characteristics of the inhomogeneous flow. The mentioned optical methods include, for example, refractive gradient methods based on the use of optical radiation of incoherent and coherent light sources.

In the present paper, the special features of the propagation of the laser sheet (LS) through a hollow glass cylinder are investigated, methods for calculation of the LS refraction for different cases of LS penetration through the cylinder are considered. By the means of computer modelling, the pictures characterizing the LS profile after penetration through the cylinder in the presence of a homogeneous and inhomogeneous medium in its cavity were obtained. The influence of not only inhomogeneous gas medium, but also cylinder walls on the process of laser radiation propagation was researched.

2. Simulation diagram

The present research investigates the possibility of application of laser refractography method for diagnostics of flows in pipelines. The figure 1 demonstrates the general diagram of laser refractography method for the research of refraction of structured laser radiation (SLR) in inhomogeneous medium [2]. It consists of the laser ()1 and SLR generation block (2). In this particular case, SLR is represented by laser sheet (3), which is penetrated through the medium under investigation (4). The laser sheet profile registered by the video camera (6) is observed on the screen (5).

In the present diagram the laser and the laser sheet generation system can be moved in space for the purpose of scanning of the medium under investigation.

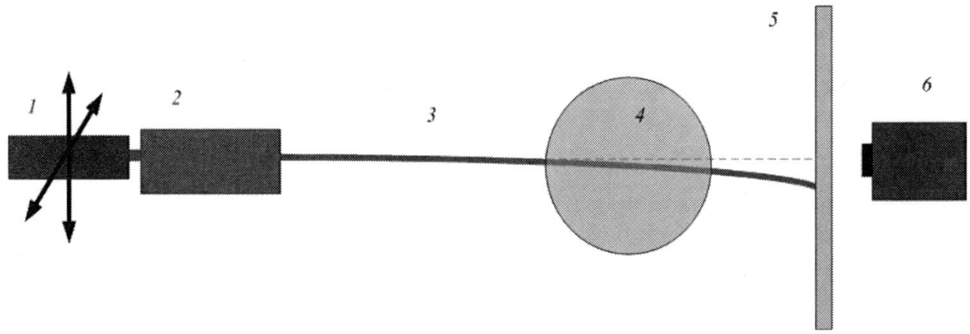

Figure 1. Diagram of laser refractography installation: 1 – laser, 2 – SLR generation system, 3 – laser sheet, 4 – medium under investigation, 5 – screen, 6 – receiving system (video camera + PC).

The system pictured in the figure 2 was used to calculate the effect of a glass hollow cylinder on the laser sheet. In the present system structured laser radiation penetrates through the glass cylinder and the laser sheet profile is observed on the screen. Parameters of the cylinder and dimensions of the system were selected based on the availability of test objects of the research: the radiuses of the cylinder $R_1 = 50$ mm, $R_2 = 54$ mm, the height of the cylinder $x = 100$ m, refraction index of glass $n = 1.5$, the distance to the screen $L_3 = 150$ mm. Insofar as the cylinder under investigation is a cylindrical lens, the refraction of the beams will occur only in a plane perpendicular to the axis of the cylinder.

An assemblage of geometrical-optical beams propagating parallel to each other within the width of the laser sheet is used in the capacity of the laser sheet's model. To calculate the penetration of the laser sheet through the cylinder, formulas for real beams penetrating through the lens system were used, which take into consideration the rules of signs adopted in geometric optics.

The cylinder under examination contains four refractive surfaces, in which the radii of curvature of the first two surfaces are respectively equal to the radii of curvature of the second two modulo. The task is to determine the profile of the laser sheet observed on the screen. Let's consider the laser sheet's model as an assemblage of geometrical-optical beams. It's necessary to calculate the penetration of each beam through the system. However, since the calculation must be performed for the laser sheet, it is necessary to calculate the course of assemblage of beams propagating parallel to each other through small distances to create a model of the laser sheet. The width of the laser sheet is numerically equal to the diameter of the smallest cylinder.

Therefore, it is possible to derive a function to calculate the displacement on the screen depending on the distance between the point of incidence of the beam on the system and the axis *OZ*. For this purpose, it is necessary to calculate the path of the beam through each refractive surface.

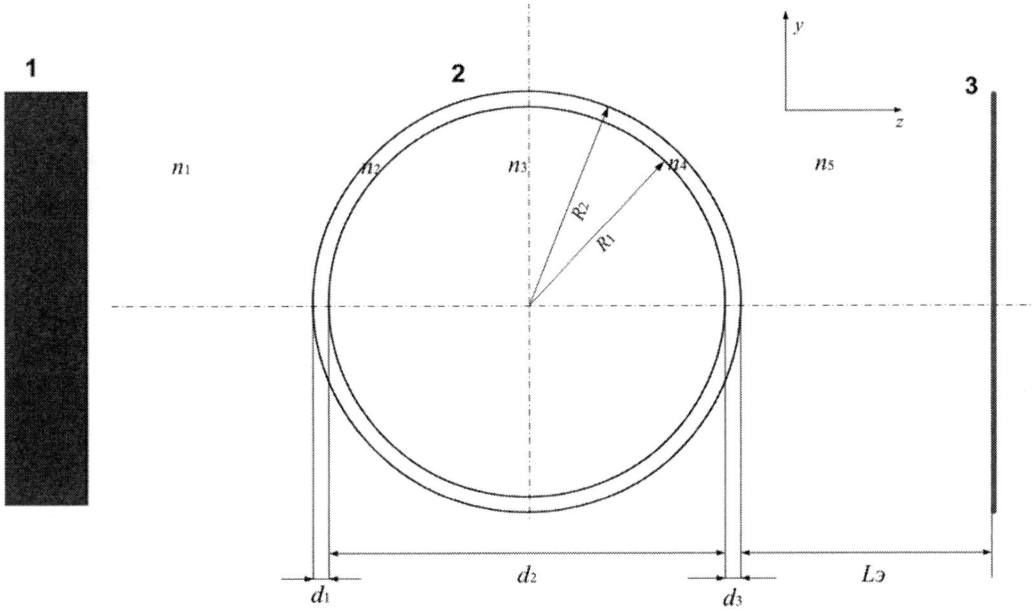

Figure 2. System for calculation of the laser sheet profile: 1 – laser sheet generation system, 2 – thick-walled cylinder, 3 – screen for the laser sheet observation.

3. Calculation of the LS trajectory subject to the existence of inhomogeneous medium in the cylinder's cavity

An important task is to research into the flow parameters during the transportation of natural gas. The base of natural gas is methane (CH_4), usually it also includes heavier hydrocarbons, methane homologues: ethane (C_2H_6), propane (C_3H_8), butane (C_4H_{10}) and various non-hydrocarbon impurities.

Let the cylinder's cavity contains methane with a refractive index n_2 = 1.000441 (for λ_D = 589.3 nm) [7]. In the first approximation, let us assume that in the cylinder's cavity the properties of the medium change rather slowly along the radial coordinate, thus the law of change of the refractive index from the radial coordinate is described by a linear function. It is valid to say that this refers to radially stratified inhomogeneous medium. Let the refractive index changes according to linear law:

$$n(r) = 2 \cdot 10^{-6} \cdot r + 1.000441$$

Formulas for spherically inhomogeneous medium were used to calculate the beam trajectory in the geometric optics approximation [1]. The formulas describe the trajectory before and after the deflection point, and the angle value for the deflection point (figure 3).

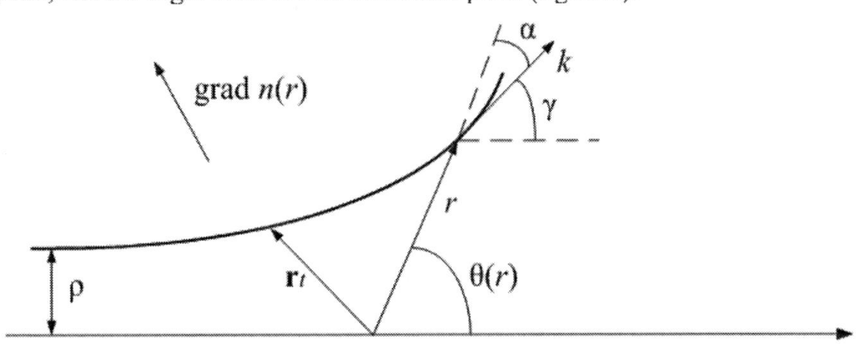

Figure 3. Beam trajectory in spherically stratified medium.

Journal of Physics: Conference Series **1421** (2019) 012033 doi:10.1088/1742-6596/1421/1/012033

Beam trajectory has two symmetrical branches: before and after r_t. For calculation we will use formulas for beam trajectory in spherically stratified medium 6 - 9.

$$\theta_1(r) = \theta_0 + \int_{r_0}^{r} \frac{n_0 \cdot \rho}{r \cdot \sqrt{r^2 \cdot n^2(r) - \rho^2 \cdot n_0^2}} dr, \tag{1}$$

where the angle θ_0 characterizes the direction of the beam at the entry to the medium and is determined as:

$$\theta_0 = \frac{\pi}{2} + arctg \frac{z_0}{\rho}, \tag{2}$$

The impact parameter for a beam penetrating parallel to *OZ* is the distance between the beam's entry point into the inhomogeneity and the *OZ* axis, that is, the distance x_0 in the accepted coordinate system.

The beam equation after the deflection point:

$$\theta_2(r) = \theta_t + \int_{r}^{r_t} \frac{n_0 \cdot \rho}{r \cdot \sqrt{r^2 \cdot n^2(r) - \rho^2 \cdot n_0^2}} dr, \tag{3}$$

where θ_t - the angle corresponding to the deflection point of the beam and is determined as:

$$\theta_t = \theta_0 + \int_{r_0}^{r_t} \frac{n_0 \cdot \rho}{r \cdot \sqrt{r^2 \cdot n^2(r) - \rho^2 \cdot n_0^2}} dr, \tag{4}$$

By virtue of the fact that the above formulas must satisfy the condition that the beam enters the inhomogeneous medium in a parallel way to the propagation axis, and in this case, the original beam propagates in a parallel way, it is necessary to determine a new coordinate system. The angle of entry into the inhomogeneity is determined by the angle of exit of the beam after passing the second refractive surface.

Recalculation to the new coordinate system (figure 4) was executed using geometric relations. For the selected model of the refractive index dependence on the radial coordinate, the distortion will occur only in the area of the deflection point and this will cause the beam's displacement when it impinges on the third refractive surface.

Through the instrumentality of the above formulas and taking into account the recalculation to the new coordinate system, it is possible to construct the beams' trajectories in the inhomogeneous medium. The figure 5 illustrates the trajectories of the assemblage of the beams describing the laser sheet for several heights $|h| = 10, 20, 30, 40, 45$ mm.

The trajectories shown on the figure 5 demonstrate that a thick-walled glass cylinder, like a dispersing lens, changes the shape of the laser sheet relative to the initial one, and also changes its divergence angle. The beams penetrating above a certain distance from the axis, experience total internal reflection and are not included in the limits of the cylinder's cavity.

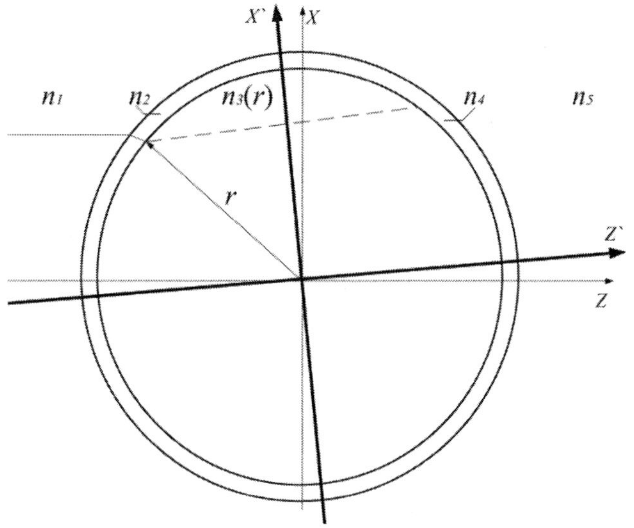

Figure 4. New coordinate system for calculation.

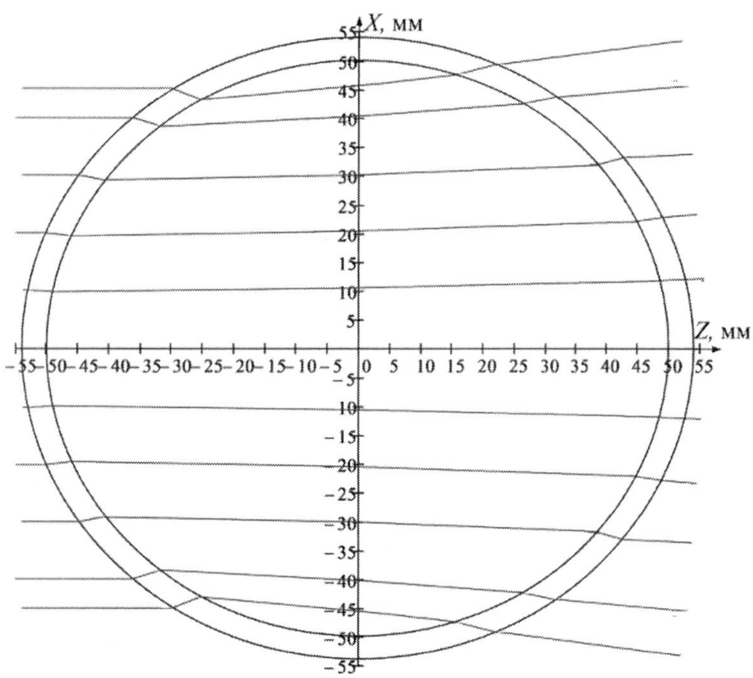

Figure 5. Trajectories of the beams propagating in inhomogeneous medium.

By virtue of the fact that the nature of inhomogeneity can change in the cylinder's cavity both perpendicularly and along its axis, then propagating the laser sheet at an angle to the cylinder axis it is possible to obtain its profile in the presence or absence of inhomogeneity. If the LS will penetrate at an angle to the cylinder's axis, then the screen will demonstrate a displacement, the position of elements of which depends not only on the initial distance from the axis, but also on the position along the cylinder's axis. The figure 6 illustrates the profile for the angle 45° and its behaviour in this situation.

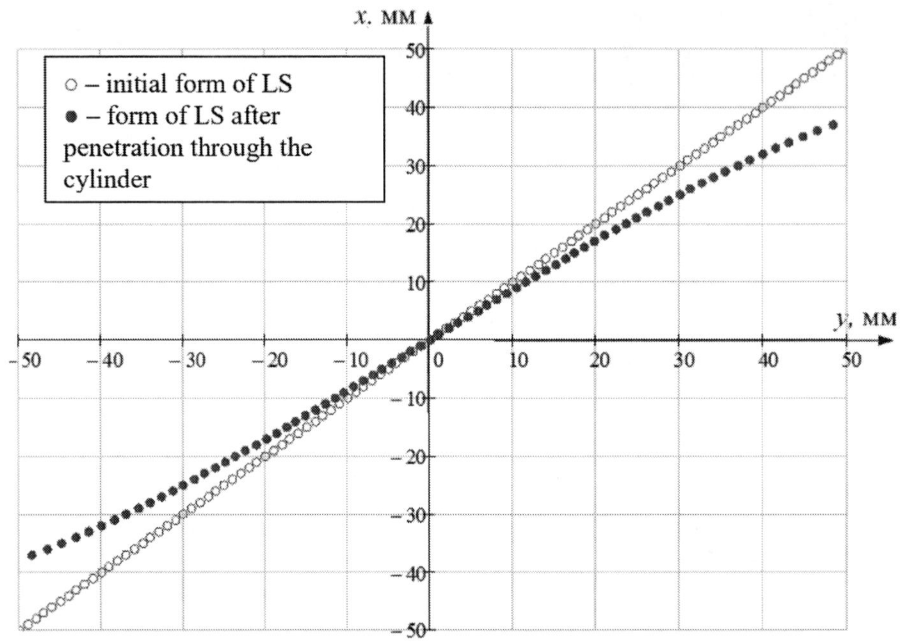

Figure 6. Laser sheet at an angle α = 45°.

4. Conclusions

In the present paper, the influence of the hollow glass cylinder on the laser sheet for different cases of its propagation in space, as well as in the presence of various inhomogeneities inside the cylinder, were investigated. The distortion of the laser sheet shape during its propagation through the cylinder in the case when there is air in the cylinder's cavity was estimated. The width of the laser sheet observed on the screen increased after the penetration through the cylinder, which made it possible to make an estimate of the presence and nature of inhomogeneity in the gas flow.

References

[1] Evtikhieva O A, Raskovskaya I L and Rinkevichius B S 2008 *Laser refractography* (Moscow: FIZMATLIT) 176
[2] Soyfer V A 2003 *Methods of computer-based optics* (Moscow: FIZMATLIT) 687
[3] Evtikhieva O A, Imshenetskiy A I, Rinkevichius B S and Tolkachev A V 2004 Computer-laserrefractive method of research of optically inhomogeneous flows *Measurement equipment* **6** 15-8
[4] Rinkevichius B S 1990 *Laser diagnostics of flows* (Publishing house MPEI) 288
[5] Raskovskaya I L, Rinkevichius B S and Tolkachev A V 2007 Laser refractography of opticallyinhomogeneous mediums *Quantum electronics* **12** 1176-80
[6] Rabinovich V A and Khavin Z Ya 1977 *Quick-reference chemical book* (Chemistry) 160
[7] https://www.calc.ru/594.html

Development of a series of interchangeable optical tips for a video endoscope

V I Batshev[1,2,3], A S Machikhin[2,3], A V Gorevoy[2,3] and D D Khokhlov[2,3]

[1] Laser and Opto-electronic Department, Bauman Moscow State Technical University, 2-ya Baumanskaya 5, Moscow, Russia;

[2] Laboratory of Acousto-optic Spectroscopy, Scientific Technological Center of Unique Instrumentation RAS, Butlerova 15, Moscow, Russia;

[3] Department of Electrical Engineering and Introscopy, Institute of Automatics and Computer Engineering, National Research University "Moscow Power Engineering Institute", Krasnokazarmennaya 14, Moscow, 111250 Russia

E-mail: batshev@bmstu.ru, machikhin@ntcup.ru, gorevoy.av@ntcup.ru, khokhlov.dd@ntcup.ru

Abstract. In this work a design of four different optical tips for video endoscope is presented. In particular, the report presents the results of the calculation of optical systems with normal (50°) and wide (80°) field of view (FoV) for direct (0°) and side (90°) view. All tips have simple to manufacture design and produce good image quality. The results of image simulation are presented.

1. Introduction

Nowadays stereoscopic machine vision systems are widely used for various industrial and medical applications [1-4]. Devices of this type provide object visualization as well as measurements of its geometrical parameters. Industrial applications of stereoscopic vision include remote visual inspection [5-7] of different hard-to-reach objects, for example tubes, engines, heat exchangers, turbines.

In papers [8-10] we presented the developed miniature stereoscopic machine vision system for non-destructive testing that provides high-quality images and the novel calibration algorithm that allows accurate three-dimensional (3D) shape calculation. The optical layout of the designed system and the photo of its prototype are shown on figure1.

The system consists of the basic objective (1) and the removable stereoscopic tip (2). The important feature of the designed system is that the aperture stop A is placed between the objective 1 and the tip (2). Constructively it is part of the objective. This allows us to minimize the diameters of the optical components. The prism is used to form two images of the same observing object from different points of view on the single sensor.

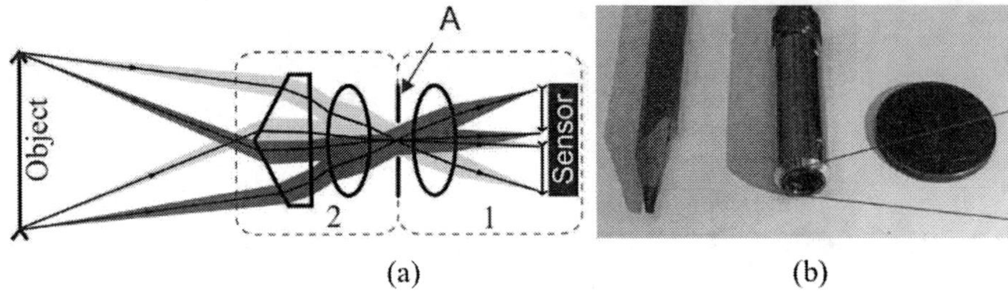

Figure 1. Prototype of the stereoscopic endoscopic optical system: optical scheme (a) and photo (b).

The main optical characteristics of this system are presented in table 1. The obtained image quality is enough to reconstruct 3D structure of the tested object and to get the accuracy of geometrical measurements about 50 μm.

Table 1. Optical parameters.

Stereoscopic system parameters	
Working distance (WD)	5 – 40 mm
FoV (in a single channel)	45°
Magnification	0.25 – 0.06
F/#	11
Resolution (@ WD = 15 mm)	40..100 lines per mm
Basic objective parameters	
Working distance	40 – 100 mm
FoV	15°
Focal length	4 mm
Magnification	0.1 – 0.04
F/#	8.5
Resolution (@ WD = 45 mm)	50..90 lines per mm
Sensor parameters	
Resolution	1280 x 720
Pixel size	1.4 μm x 1.4 μm
Image area	1.819 x 1.033 mm (1/9")

The formation of a stereoscopic image of inspected object and the measurement of its geometrical parameters is a function, in many cases necessary, but it is not so often used in endoscopic testing as a simple visual inspection. In this regard, the task of expanding the functionality of the developed video endoscope seems to be relevant. For this reason, several interchangeable optical tips are designed, which, when used in conjunction with the basic endoscope objective, provide various of its optical characteristics.

2. Interchangeable optical tips design

We have developed four types of tips for direct and side-view with different FoV: 50° and 80°. They are presented in figure 2. T0-15 (figure 2 (a)) is direct-view tip with angular magnification of 0.3 that provides the endoscopic FoV of 50°. This tip consists of a single quasi-afocal thick lens. In combination with the right-angle prism, it becomes a side-view tip T90-50 (figure 2 (c)).

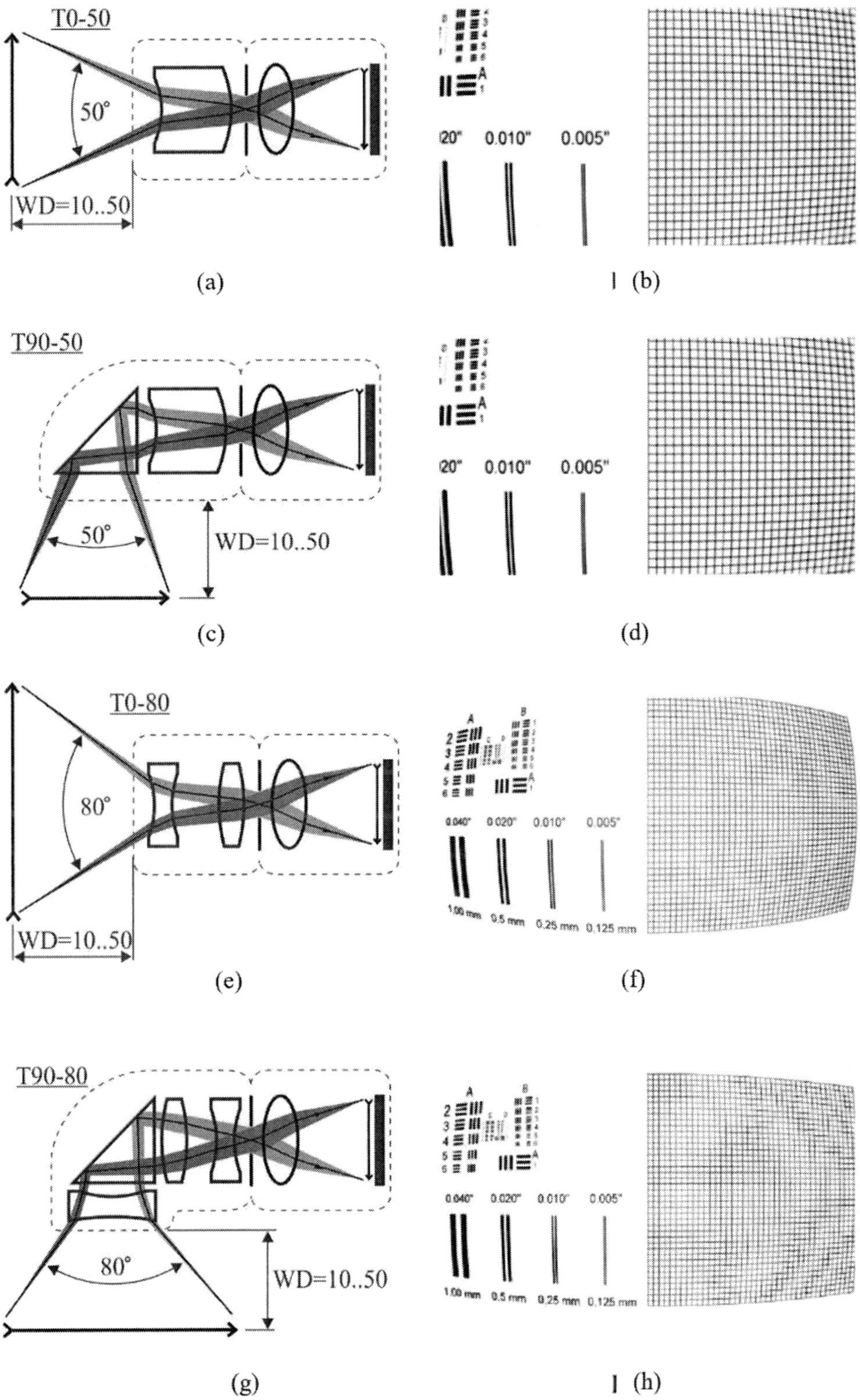

Figure 2. Designed endoscopic tips (a, c, e, g) and corresponding image simulations (b, d, f, h).

Actually, it is possible to obtain more than 50° FoV with single lens, but it causes a significant complication in its manufacture because of its small radii. That is why the next direct-view wide-angle tip T0-80 (figure 2 (e)) has a two-lens design. It is also quasi-afocal.

T90-80 (figure 2 (g)) is a side-view wide-angle 80° tip with angular magnification of 0.188. It is the most complicated of the designed tips. To avoid the large size of the prism, the diverging lens is located in front of it.

All tips have a working distance WD = 10-50 mm which is convenient for inspecting most types of technical objects. Actually, all designed tips could be re-adjusted ore recalculated without significant changes for another working distance if necessary.

Figure 2 (b),(d), (f), (h) shows the results of image simulation for the endoscopic system using the corresponding designed tips. Simulation was performed in ZEMAX software. The object was placed at the distance WD = 20 mm. We could see that all tips produce good image quality. Images obtained with tips T01-50 and T90-50 look the same because of there is no difference between these tips except the prism. T90-80 consists of 3 lenses and has better aberration correction than T0-80, therefore images in figures 2 (f) and 2 (h) show different distortion.

Optical resolution of the endoscope using the designed tips is almost the same for all tips. For image center, the resolution varies from 70 to 100 lines per mm within the working distance range. For the image margin, the resolution is from 45 to 70 lines per mm. It is quite enough for the tasks of visual inspection.

3. Conclusion

Presented interchangeable optical tips produce high image resolution (up to 100 lines per mm) and significantly expand the functionality of the developed endoscopic device for non-destructive testing of technical objects and, as a result, make it commercially more attractive.

Acknowledgments

The study was supported by Russian Science Foundation (project #17-19-01355).

References

[1] Batchelor B G 2012 *Machine Vision Handbook* (London: Springer)
[2] Davies E R 2012 *Computer and Machine Vision: Theory, Algorithms, Practicalities* (4th Edition) (London: Academic Press)
[3] Zhu Y K, Tian G Y, Lu R S and Zhang H 2011 *Sensors* **11** 7773-98
[4] East J E et al. 2016 *Endoscopy* **48(11)** 1029-45
[5] Ng A 2015 *Proceedings of NDT Malaysia* http://www.ndt.net/events/MINDTCE-15/app/content/Paper/33_Ng.pdf
[6] Geng J and Xie J 2014 *IEEE Sensors Journal* **14(4)** 945-60
[7] https://www.aero-mag.com/exploring-latest-advances-remote-visual-inspection-3d-stereo-measurement
[8] Machikhin A S, Batshev V I, Gorevoy A V, Khokhlov D D and Bykov A A 2019 *Proc. SPIE* 110410M
[9] Batshev V I, Machikhin A S and Kachurin Y Y 2017 *Proc. SPIE* 104664D
[10] Machikhin A S, Batshev V I, Gorevoy A V, Khokhlov D D, Naumov A A and Kuznetsov A O 2019 *Optik* **185** 1172-81

Journal of Physics: Conference Series **1421** (2019) 012035 doi:10.1088/1742-6596/1421/1/012035

Measurement of high-speed gas-droplet flow characteristics by pulse photography method

N I Bokuchava[1] and F M Inochkin[2]

[1] I.I. Polzunov Scientific and Development Association on Research and Design of Power Equipment (NPO CKTI), Atamanskaja st., 3/6, 191167, St. Petersburg, Russia
[2] Saint Petersburg National Research University of Information Technologies, Mechanics and Optics, Kronversky pr., 49, 197101, St. Petersburg, Russia

Abstract. The subject of this work is the optical probe measurement system for the measure of droplets parameters in high-speed two-phase flow. A system with dark-field illumination including two pulse lasers as illumination sources is used for droplets registration. Rays from lasers intersect at the angle of 120° and form a measurement volume. When a droplet crosses the volume, it refracts a part of the light. Refracted rays pass through the optical system and create two highlights on the sensor. Distance between two highlights is equal to the droplet diameter. System for measurement of coarse droplets parameters in wet steam flow, based on the described method, is developed. Main components of the system are the optical probe and PC with software. The system allows determining the following parameters of droplets: size distribution (20–600 μм), velocity distribution (0.1–300 m/s), movement direction. The results of measurements of the system in two-phase flow are presented in this work. The system is successfully operated as part of the complex for investigation of two-phase flow parameters on the model turbine.

1. Introduction

Measurements of wet phase characteristics is an essential component of wet-steam flows investigation. Optical methods can be used for carrying out such type of measurements.

The subject of this work is the system for measure structural-kinematic parameters of high-speed two-phase flow in steam turbine.

Operation of this system is based on a pulsed photography method. The method allows measuring the follow parameters: droplets size distribution, droplets velocity distribution and droplets movement direction.

2. Particularities of investigated environment

Wet phase in wet-steam flow is amount to mix of two fractions: fine fraction (80–90% of mass) with droplet diameter 0.05-2 μm and a coarse fraction (10-20% of mass) with droplet diameter 10-1000 μm.

Velocity and movement direction of fine droplets correspond to these of main flow. However, velocity and movement direction of coarse droplets may differ significantly from those of the main flow.

Various methods (mainly optical methods) are used for measure parameters of fine fraction. There are many measurements results on the model and the full-scale turbines.

At contrast, the question about measurements of coarse fraction parameters is still open. There are not generally accepted results from measurements of size and velocity of erosion dangerous droplets.

Photography registration of moving droplets is one of the first method used for investigation of two-phase flow disperse structure. The photo demonstrates the shape and size of droplets with high confidence. This property of photo is important when preliminary information about the droplets is absent.

Nevertheless, using this method for investigation of wet steam flow in a turbine can cause some difficulties:

- droplets velocity may be up to 300 m/s. Illumination source with very short pulses should be used for registration such values;
- stringent requirements are met for dimensions of a measurement device;
- there is not preliminary information about the droplets, which requires an expansion of measurement range;
- there is statistically less than one droplet in a measurement volume (figure 1), which requires large quantities of data in order to obtain a sufficient representative measurement.

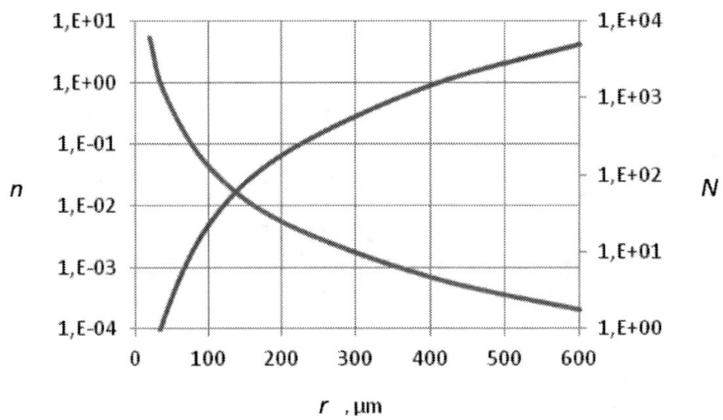

Figure 1. The expected concentration of droplets in the flow part of the turbines n – number of droplets in measurement volume, N – number of batches needed for registration of 1 droplet.

3. Illumination system

Choose a type of illumination system is an essential aspect of the realization of a pulse photography method.

Usually, the following illumination system is used for a photoregistration: measurement volume is located between illumination source and registration device. In that case, an image of droplet looks like a black-out area on a light field.

This illumination system's design is simple, but it has several significant disadvantages. The primary disadvantage is the uncertainty of the registered volume depth estimate, which influences the calculation of droplets concentration. Besides, the regime of multiple expose cannot be applied because of image contrast loosing.

Illumination system described in the inventor's certificate [1] is different from mentioned above by illumination sources location off-axis of the registered system. Light from sources goes to the area of measurement (measurement volume) under angle β to the optical axis of registered system. This system (so-called dark-field illumination system) is realized in the optical probe (videoprobe) described in this paper.

Journal of Physics: Conference Series **1421** (2019) 012035 doi:10.1088/1742-6596/1421/1/012035

In that case, translucent spherical droplet plays the role of a small wide-angle lens when it crosses the measurement volume. Images of illumination sources as two bright highlights against the dark field are formed on registration plane.

Distance between the highlights is equal of droplet diameter if the value of angle β is chosen correctly. Dependence of highlights location relative of droplet size on value angle β and rays' way in a droplet (n = 1,33) are demonstrated on figure 2. This dependence is presented both for reflected and refracted ray.

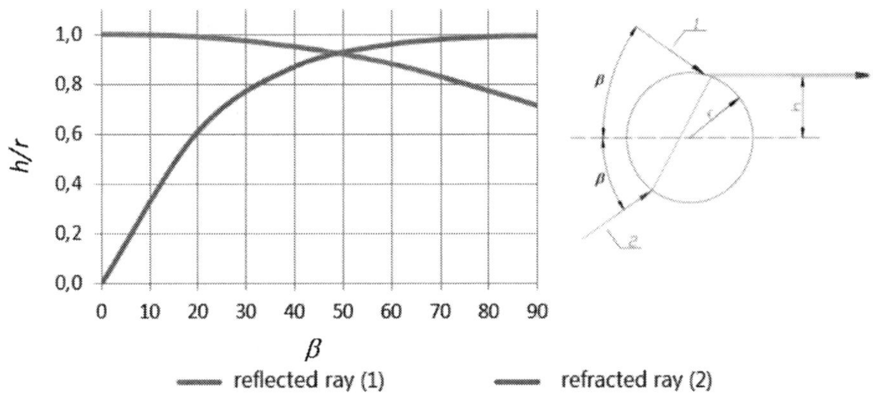

Figure 2. Way of rays in a droplet: h – the distance between droplet image center and highlight, r – droplet radius.

If $\beta = 50^0$, values of h/r for reflected and refracted rays are equal. It means, two highlights pair (from opposite illumination sources) superimpose on one another. Highlight formed by the refracted ray is significantly brighter then highlight formed by the reflected ray.

Value of angle β was chosen 60^0 , when the videoprobe was designed. In that case, the distance between highlights is equal to 0.98 droplet diameter.

Image of droplet with a diameter 1 mm is shown on figure 3. This image was made by the videoprobe. Highlights formed by refracted and reflected rays are clearly visible.

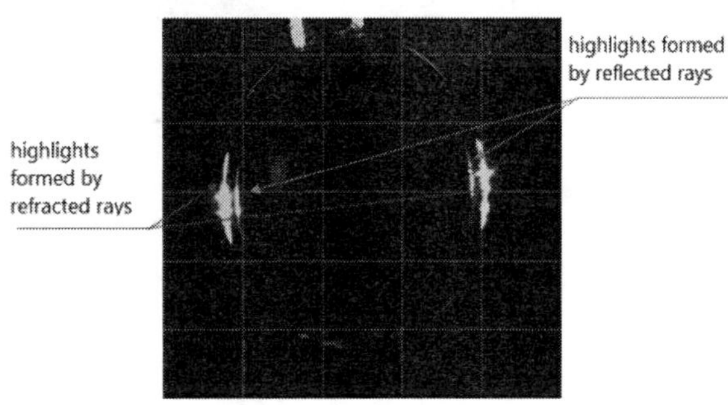

Figure 3. Droplet image.

Dark-field illumination system has two advantages:

- using multiple exposure regime recording without picture contrast loss (regime of information storage); multiple exposure is amount to increase of measurement volume in corresponding numbers of times. This advantage is critical when droplets concentration is very low;
- scale constancy on measurement volume depth; distance between highlights centers is kept and does not depend on droplet location relative of the focal plane.

Optical probe for fractional analysis of coarsely dispersed gas-droplet flow based on this illumination system was designed in [2].

4. Videoprobe. Optical head

The main component of the videoprobe is an optical head containing all optical and electronic components (figure 4). This design is capable of short illumination pulse generation and high-resolution image acquisition.

Illumination sources are based on pulsed laser diodes with a central wavelength of 905 nm. In total there are 4 diodes in the videoprobe, two diode pairs are located on both sides of the videoprobe. These diodes provide illumination pulses with minimal duration of 30 ns and optical output power up to 90 W. In case of elevated image brightness, the pulse width can be increased up to 100 ns. High speed of applied diodes provides opportunity to acquire several batches of pulses for a single video frame. This results in ability to determine size of droplets being measured and at the same time their velocity and movement direction.

An image sensor is a 5-megapixel monochrome CMOS with a pixel pitch of 2.2 μm. In order to improve image capture performance for pulsed illumination mode, the sensor operates in global reset release (GRR) mode [3].

Video signal and setup commands for illumination pulse generator and image sensor are transmitted over USB interface via a high-speed controller integrated into the electronic board. Precision generation of control pulses for the illumination source is implemented by the programmable logic (FPGA).

Configuration of the optical head is shown on figure 4. Rays go through guiding prisms and form measurement volume. Droplet refracts part of light when it crosses the volume. Refracted rays pass through optical system and form an image of the droplet as two bright highlights on matrix.

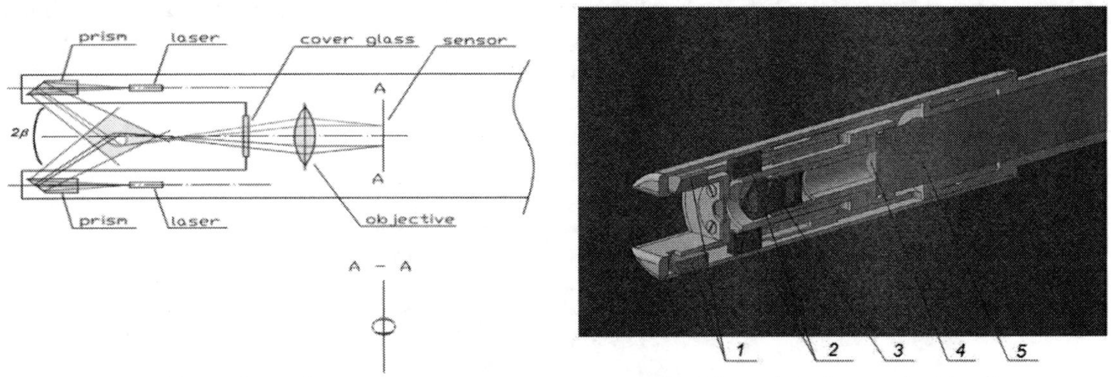

(a) Optical scheme (b) 3D model of an optical head

Figure 4. Optical head. 1 – guiding prisms, 2 – laser diodes, 3 – objective, 4 – matrix, 5 – the electronic board.

Measurement volume can be separate into two areas (figure 5):

- two-side lighting area, all droplet parameters (size, velocity, movement direction) can be detected if the droplet is in this area;
- one-side lighting area, only velocity and movement direction can be detected if droplet is in this area.

Journal of Physics: Conference Series **1421** (2019) 012035 doi:10.1088/1742-6596/1421/1/012035

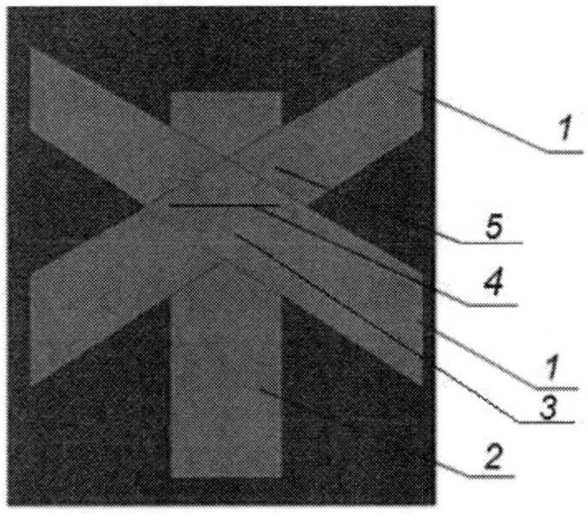

Figure 5. Scheme of measurement volume forming. 1 – rays from illumination sources, 2 – view field of objective, 3 – measurement volume, 4 – focal plane, 5 – one-side lighting areas.

Value of measurement volume should be known for calculation of droplets concentration in a flow. Since the measurements of droplets size are carried out in a wide range, dependence of value of measurement volume on the droplet size (figure 6) should be considered for calculation of the concentration.

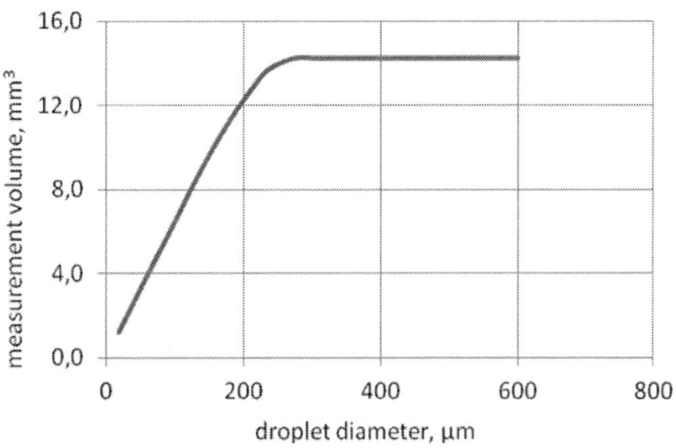

Figure 6. Dependence of value of measurement volume on the droplet diameter.

5. System for measuring structural-kinematic parameters of high-speed two-phase flow

System for measuring of droplets parameters in wet-steam flow in turbine (figure 7) consists of the follow parts:

- videoprobe;
- control PC with installed software for image acquisition and data processing;
- system for air blowing and cooling;
- traverse system for moving the probe in the turbine;
- power supply.

Journal of Physics: Conference Series **1421** (2019) 012035 doi:10.1088/1742-6596/1421/1/012035

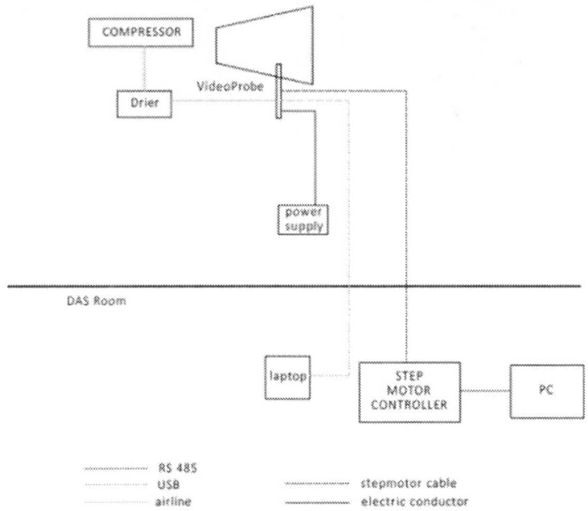

Figure 7. System for measurement of droplet parameters in a steam turbine.

The system measures the follow parameters:

- droplets diameter: 20 – 600 μm;
- velocity: 0.1 – 300 m/s;
- movement direction: ±45° relative to the angle of videoprobe installation.

Assembled videoprobe is made as optical head installed on a bar with diameter 20 mm (figure 8). Electronic and power cables, and channels for air delivery lay in the bar. Elements of traverse system (step motors) are installed on the bar (figure 8).

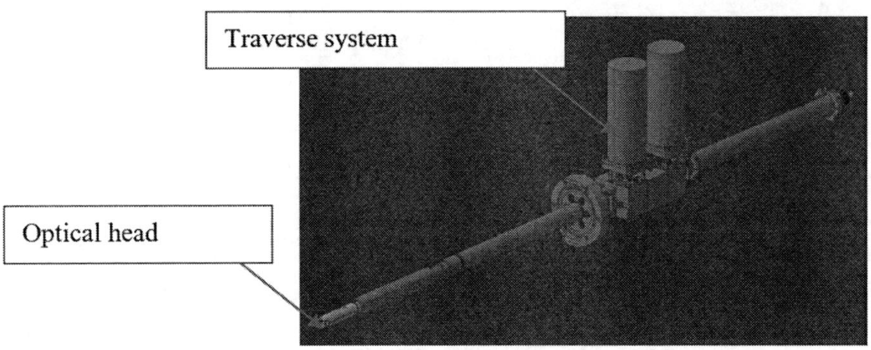

Figure 8. Assembled videoprobe.

Constant air blowing of power-on videoprobe is performed for two reasons:

- optical elements protection from contamination and water film. Guiding prisms and protection glass have slits for air outlet for this purpose;
- electronic elements cooling. The temperature inside of the videoprobe should be less than 60°C according to operation conditions. Resistance-temperature is used for temperature control. Value of temperature is displayed in a window of image acquisition software and recorded in the output data file.

Specific software was developed for videoprobe operation. It consists of two components:

- image acquisition software;

- data processing software.

Image acquisition software controls illumination sources and image sensor, and also enables video stream and single-frame capture. Video stream is stored in a specially developed format, without image compression involved in capture process. Each output frame contains image sensor and illumination sources parameters.

The software is capable of adjusting image sensor readout parameters and of increasing frame rate by decreasing the number of sensor pixels being read. This can be achieved by limiting image area or image resolution.

While illumination pulses are generated, multiple images of objects are exposed on the photosensitive area of image sensor.

In order to increase droplets detection probability, several batches of image pulses are generated for each frame. At the end of exposure image readout is started. An example of timing diagram is shown a figure 9. There are 2 frames with 3 batches of illumination pulses and 4 pulses in each batch for the demonstrated example.

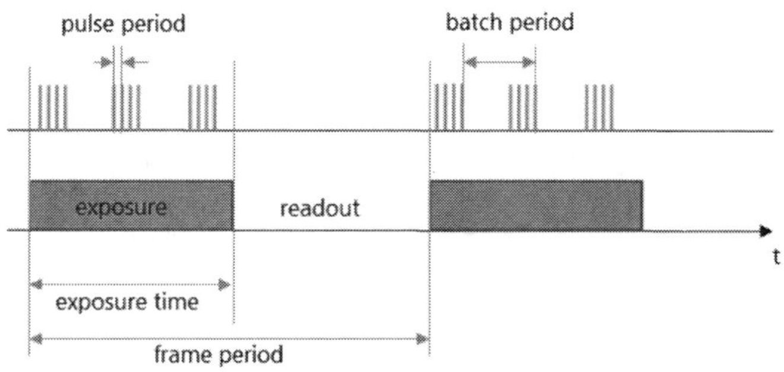

Figure 9. Videoprobe timing diagram example.

Described illumination design renders image of each droplet as a sequence of highlights pairs (figure 10), which can be analyzed to determine droplet parameters: size, velocity and movement direction.

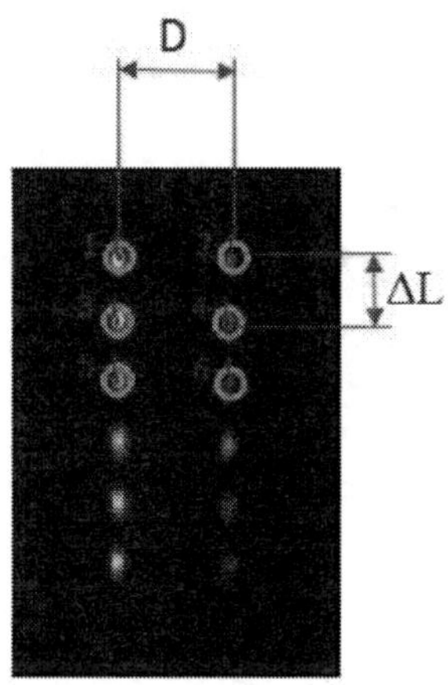

Figure 10. Moving droplet image example.

Data processing software was developed for droplet parameter estimation. Droplets diameter D is estimated by distance between highlights.

Droplets velocity: $V=\Delta L/\tau$, τ - pulse period, ΔL – the distance between pairs of highlights. As a result of processing the acquired data, the following information is retrieved:

- droplets size distribution;
- droplets velocity distribution;
- droplets movement direction.

6. Examples of measurements results

Described measurement system is applied in NPO CKTI for various problem solving. For example, results of droplets in spray nozzle parameters measurements are presented on figure 11.

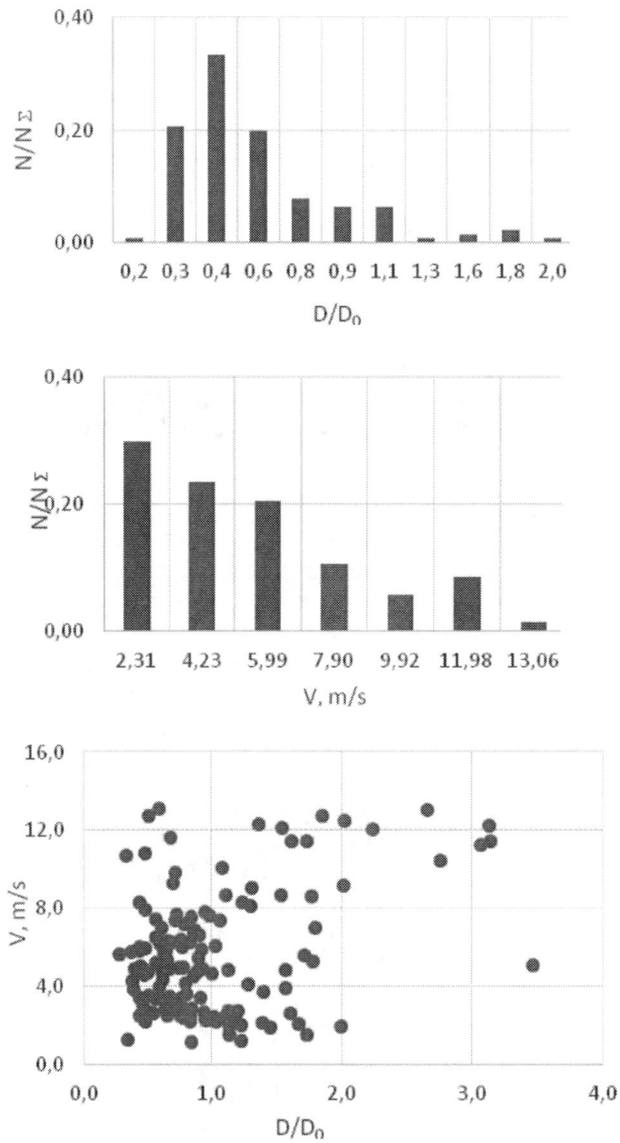

Figure 11. Results of droplets in spray nozzle parameters measurements. D – droplet diameter, µm, D_0 – average droplets diameter, µm, N – number of droplets, N_Σ - total number of droplets, v – droplet velocity, m/s.

Moreover, the measured parameters of droplets in the steam turbine are shown on figure 12.

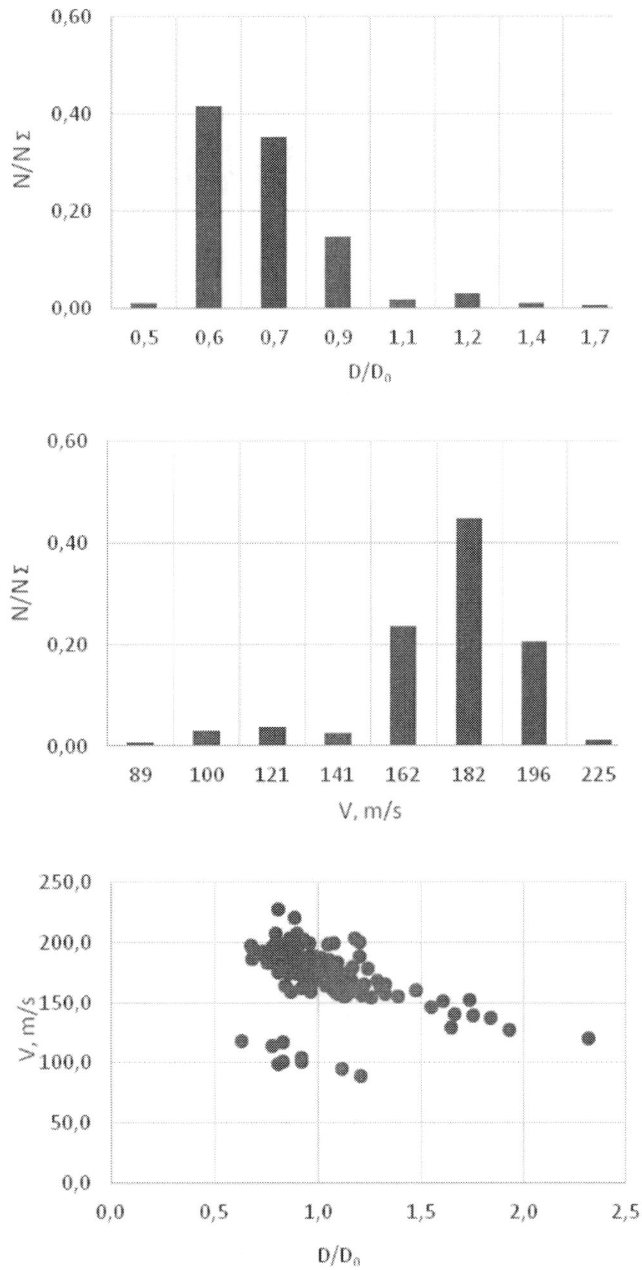

Figure 12. Results of measurements of parameters of droplets in the model steam turbine. D – droplet diameter, μm, D_0 – average droplets diameter, μm, N – number of droplets, N_{Σ} - total number of droplets, v – droplet velocity, m/s.

Measurements by the described system allow getting correlation between all measured parameters (figure 13).

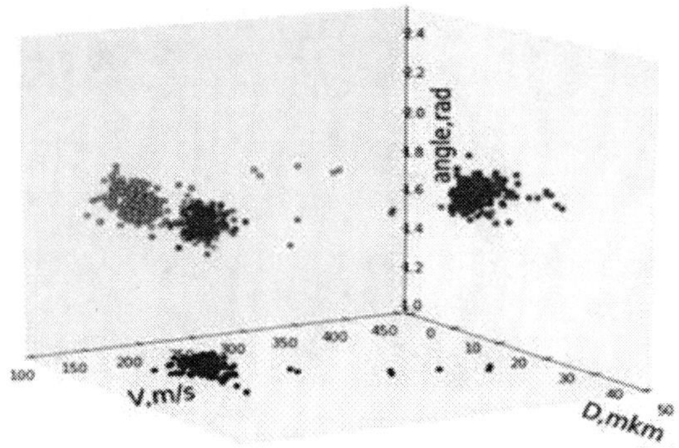

Figure 13. Example of measurements result on a model turbine (relation between diameter, velocity and movement direction).

7. Conclusions

The system for measurement of high-speed erosion dangerous droplets parameters in wet-steam flow is designed. Operation principle of the system is based on the pulse photography method.

The system allows getting droplets size distribution, droplets velocity distribution and movement direction of droplets in the flow.

Specific software for recording photo and video files and processing them was developed. This software based on modern image recognition methods.

References

[1] Dobkes A L and Feldberg L A 1989 Fotogrammetricheskij zond dlya registracii kapel' vlagi v potoke para ili gaza. Opisanie izobreteniya k AS №SU1485079A1 07.08.89 №21

[2] Feldberg L A and Dobkes A L 1990 *Study of dispersion structure of gas-drop flows in heat-and-power-engineering* vol 9 (Moscow: Tyazhmash Energeticheskoye mashinostroyeniye) chapter 3

[3] Kruglov S K, Bronstein I G, Vasiliev V B and Inochkin F M 2015 A compact videoprobe for ultrafast phenomena imaging in high-speed gas-liquid flows *Voprosy radioelektroniki. Seriya (Tekhnika televideniya* vol 6) 141-5

LED illumination module for fluorescence and spectroscopic studies in microscopy

V I Batshev[1,2,3], A A Blokhin[4] and A S Belov[4]

[1] Laser and Opto-electronic Department, Bauman Moscow State Technical University, 2-ya Baumanskaya 5, Moscow, 105005 Russia;
[2] Laboratory of Acousto-optic Spectroscopy, Scientific Technological Center of Unique Instrumentation RAS, Butlerova 15, Moscow, 117342 Russia;
[3] Department of Electrical Engineering and Introscopy, Institute of Automatics and Computer Engineering, National Research University "Moscow Power Engineering Institute", Krasnokazarmennaya 14, Moscow, 111250 Russia;
[4] Digital Medicine LLC, 42 Bolshoy boulevard, Skolkovo Innovation center, Moscow, 143026 Russia

E-mail: batshev@bmstu.ru, aab@chance2live.com, asb@chance2live.com

Abstract. In this report we produce a microscopic illuminator consisting of four light emitting diodes (LEDs) with different wavelength. The radiation power of each LED could be controlled independently. The dichroic mirrors are used as a beam combiner. So, it is possible to create a highlight of the required spectral composition on the object under study. This illuminator is useful both for fluorescent analysis or spectral imaging.

1. Introduction

Nowadays, to study some types of micro-objects using a microscope, it is often not enough to obtain their image. There are various methods to increase the information content of research. In particular, for the study of biological objects using the technique of phase contrasting, fluorescent analysis, spectral imaging and other methods [1]. Spectral imaging requires either a wavelength-tunable light source [2, 3] or a wavelength-tunable optical filter, for example an acousto-optical [4], which allows us to take a series of images at required different wavelengths and thus analyze the spectrum of the light coming from each point of the object.

To excite the fluorescence in the object under study, narrow-band short-wave optical radiation should be directed at it. Typically, such light is generated using a filter that cuts out a portion of the spectrum of the radiation from a broadband source. Sometimes laser, LED of other sources are used for this technique [1].

There are a large number of different markers introduced into the object of study. They fluoresce when illuminated at different wavelengths. Recently, the technique of using multiple markers simultaneously, which are introduced into different parts of the object or mixed. This technique is applied first of all for living cells study. It allows different areas fluoresce at different spectral ranges.

Journal of Physics: Conference Series **1421** (2019) 012036 doi:10.1088/1742-6596/1421/1/012036

For this technique it is more useful to implement a specialized light source that generates a radiation in several narrow spectral ranges. Such a source may also contain a wideband source and several light filters however it is more efficient if the lighter consists of several LEDs with different spectral ranges emitting simultaneously [5].

In our laboratory, tasks often arise when it is important not only to obtain luminescence, but also to measure the spectrum of luminescent radiation at each point of the object. This requires both a fluorescent illumination module and a spectral imager. Such systems for «mixed» fluorescent analysis and spectral imaging were previously created [6, 7], but their applicability is very limited in the number of spectral channels either in the lighting or in the registration branch.

To study of a large number of objects with minimal human involvement in this process it is required to automate the measurement process and synchronize the devices used: spectral imager, video camera, fluorescent module, microscope stage.

For spectral imaging we utilize a home-made acousto-optical spectral imaging module developed and fabricated specially for the microscope. It is described in detail in [8]. But there is no commercially available fluorescent illumination module which can be synchronized with our acousto-optical module. That is why we developed a LED illuminator for fluorescent microscopy of modular construction and with independent control of the radiation power of each LED source. In the near future we plan to synchronize it with other nodes of the microscope.

2. Designed LED illuminator

We used four sources LED1-LED4 of 3 W power with different spectra, as shown in figure 1 (a). Their radiation indicatrix are shown on figure 1 (b).

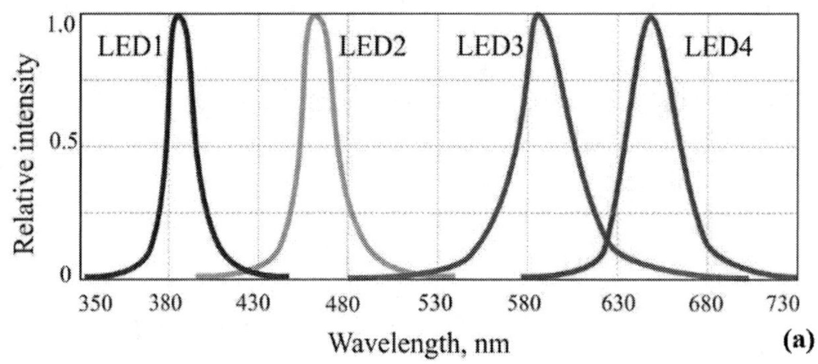

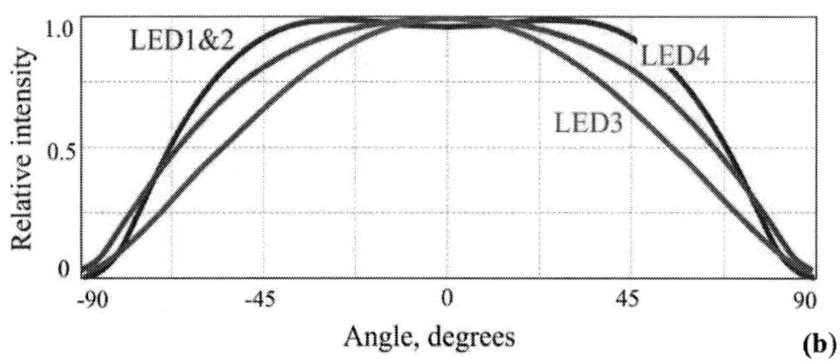

Figure 1. LEDs spectra (a) and radiation indicatrix (b).

The optical layout of the developed fluorescent illuminator is shown in figure 2. Lenses L1-L4 are used for beam collimation. We utilize equal collimating lenses, but their positions relative to the corresponding LEDs are adjustable to compensate the difference of LEDs indicatrix (figure 1 (b)) and to ensure maximum uniformity of illumination of the object under study. Actually, lens positioning should be done once before the measurements. The combination of collimated beams is carried out by dichroic mirrors DM1-DM3.

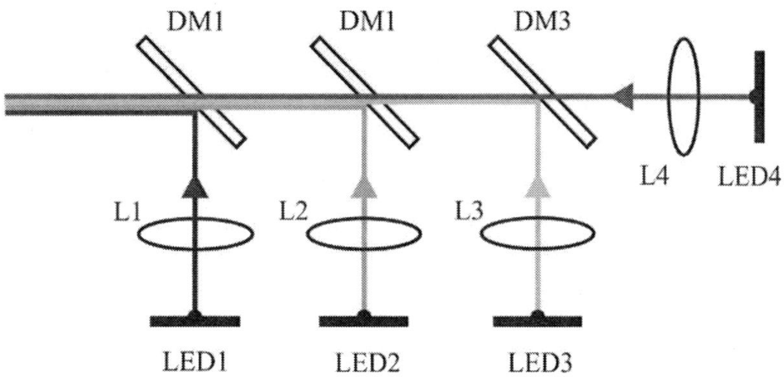

Figure 2. Optical scheme of the designed illuminator.

The microscope attached with the designed fluorescent LED illumination module and acousto-optical spectral imager AOS is presented in the figure 3.

The designed fluorescent illumination module represents a beam combiner unit (1), to the housing of which two LED sources (2), (3) can be attached, or one LED source and a similarly designed beam combiner unit with a longer wavelength border. This allows us to apply simultaneously as many LED sources with different wavelengths as we need. In the case when only one LED source is utilized, it can be attached directly to the microscope. Thus, the modular principle of the developed fluorescent LED illuminator is implemented.

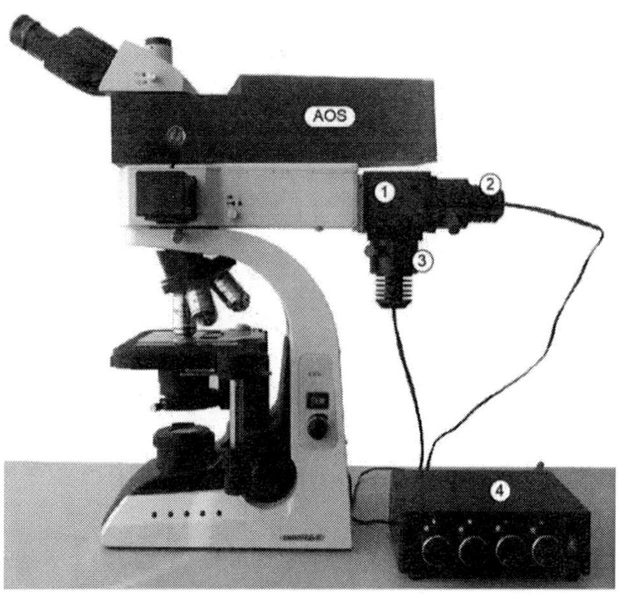

Figure 3. Microscope with acousto-optic spectral imager (AOS) and designed illuminator: 1 – beam combiner unit; 2, 3 – LED sources; 4 – control unit.

Each LED source is connected to a single control unit (4) that allows us to control the intensity of each LED source separately. By this time, we have manufactured a control unit in which manual control of the power of each LED is implemented. In the future, software control and synchronization will be realized.

3. Conclusion

We developed a fluorescent LED illumination module which is convenient to apply with acousto-optical spectral imaging module to obtain a device for mixed fluorescent analysis and spectral imaging in microscopy of biological objects.

The modular design of the developed fluorescent illuminator allows the use any number of LEDs with spectral characteristics required for specific object studies. The power of each LED is graduated smoothly and independently of the others. Synchronization of all microscope devices will allow us to create an automated tool for the study of a large number of samples with minimal human involvement in the measurement process.

The designed illuminator could find application not only in microscopy. It can also be used in colorimetry tasks, for spectral visualization of objects with their non-destructive testing and in other applications.

Acknowledgment

This development was performed using the equipment of the Center for Collective Use of the Scientific and Technological Center of Unique Instrumentation of the Russian Academy of Sciences (STC UI RAS) [9].

References

[1] Murphy D B 2001 *Fundamentals of light microscopy and electronic imaging* (NY: Wiley-Liss) p 357
[2] Hall H, Bridges M, Leavesley S J and Robinson J P 2007 *Proc. SPIE* 66680W
[3] Batshev V, Machikhin A, Pozhar V and Khokhlov D 2017 *Proc. SPIE* 1046639
[4] Levenson R and Mansfield J R 2006 *Cytometry Part A* **69(8)** 748-58
[5] Aswani K K, Jinadasa T and Brown C M 2012 *Microscopy today* **20(04)** 22-28
[6] Lou X, Yoon E 2011 *15th International Conference on Miniaturized Systems for Chemistry and Life Sciences* (USA: Seattle Washington) 598-600
[7] Haraguchi T, Shimi T, Koujin T, Hashiguchi N and Hiraoka Y 2002 *Genes to Cells* **7(9)** 881-87
[8] Polschikova O, Machikhin A, Batshev V, Ramazanova A, Belov A and Pozhar V 2017 *Proc. SPIE* 105920H
[9] http://ckp.ntcup.ru

Application of the particle trajectory imaging for modelling dusty plasma levitation on the Moon

A Yu Poroykov[1], S A Bednyakov[2], A V Zaharov[2], G G Dolnikov[2], A N Lyash[2], I A Shashkova[2] and I A Kuznetsov[2]

[1] National Research University "Moscow Power Engineering Institute", Krasnokazarmennaya 14, Moscow, Russia
[2] Space Research Institute of the Russian Academy of Sciences, Profsoyuznaya 84/32, Moscow, Russia

Abstract. The future robotic and human lunar landing missions to the Moon has many factors which turn them into difficult technology and science tasks. One of these factors is the influence of the dust. Upper layers of the regolith are an insulator. The regolith exposed to permanent bombardment by micrometeorites and acquire a charge due to solar wind fluxes and solar UV radiation. These factors create a charge distribution on the surface of the Moon: positive on the illuminated side and negative on the night side. On the day side of the Moon near the surface layer exists possibility of formation an electric field. Charged dust particles of micron and submicron sizes can take off and levitate over the surface. The aim of the work is to get visualization of the dynamic of dust particles under a charged surface with simultaneously estimating its parameters as accurately as possible. The experimental setup based on vacuum chamber for physical modelling of dusty plasma levitation is described. For visualization of the dust particles trajectory a stereo system of two cameras with a laser as source of illumination is used. Image processing techniques for estimating the particle trajectory in three dimensional coordinates and examples of processing results are presented.

1. Introduction

Recently, an interest in exploring the Moon has been reborn. Missions for this aim are declared by the USA, China, India, and the European Union. Russia also prepares two landing missions: Luna-Glob and Luna-Resource. But the future robotic and human lunar landing missions to the Moon have many factors which turn them into difficult technology and science tasks. One of these factors is the influence of lunar dust particles. Meteorites bombardment of the lunar surface, accompanied by shock-explosive phenomena causes disintegration and mix of the lunar soil in depth and on area simultaneously [1]. Under the influence of these factors, dust particles become one of the most active and aggressive components of lunar regolite, as well as regolith of any airless body of the solar system. Such particles, acquiring an electric charge under the influence of ultraviolet radiation from the Sun and solar wind, can rise above the surface of the Moon and levitate, creating a near-surface plasma-dust exosphere [2].

Properties of the regolith and the dust particles (density, temperature, composition, charge etc.) as well as near-surface lunar exosphere depend on solar activity, lunar local time and position of the Moon relative to the Earth's magnetotail. Upper layers of the regolith are an insulator and it is charging by the solar UV radiation and the solar wind. These factors create a charge distribution on the surface of the Moon: positive on the illuminated side and negative on the night side [3]. Charge distribution depends on the local lunar time, latitude and the electrical properties of the regolith. On the day side of the Moon near the surface layer exists possibility of formation an electric field. Dust particles of micron and submicron sizes are able to overcome the gravity force and the adhesive van der Waals forces in the electric field, take off and levitate above the surface [4]. An altitude of levitation depends on the near-surface electric field, charge and size of dust particles.

Important part of investigation of the lunar surface is study of dynamics dust particles under the regolith surface. Studying the properties of the lunar dust is important both for scientific purposes of lunar exosphere investigation [5] and for technical safety of robotic and manned missions [6]. PmL [7] a dust analysis device for future Russian flights is intended to study the dynamics of the dusty plasma near the lunar surface.

There are some recent articles [8-10] which present several approaches for modeling of the dusty plasma. Also, in some articles [8, 11] physical experimental modeling of this levitation is presented. The aim of our work is to get visualization of the dusty plasma with simultaneously estimating its parameters as accurately as possible.

2. Experimental setup

For conducting physical experiments on modelling the dusty plasma levitation an experimental setup was created (figure 1). It is based on a vacuum chamber. It consists of two parts. The first part is for creating conditions for the dusty particles levitation and it is including a system of supply and control of vacuum, a nonconductive substrate with dust particles on it, a steel grid above particles and a source of high voltage. High voltage from the source (2-12 kV) was applied to conductor on the bottom of the substrate. The particles of the dust with different materials and sizes were loaded on the substrate. The steel grid is located above the particles and connected to the ground of the high voltage source. Distance from the grid to substrate were changing in different experiments and was in range 3-8 mm.

The second part is optical visualization of the particles levitation. This part includes a vacuum viewport, a laser as a source of light, an optical system for illumination of an investigation the chamber volume, and a stereo system of CMOS cameras.

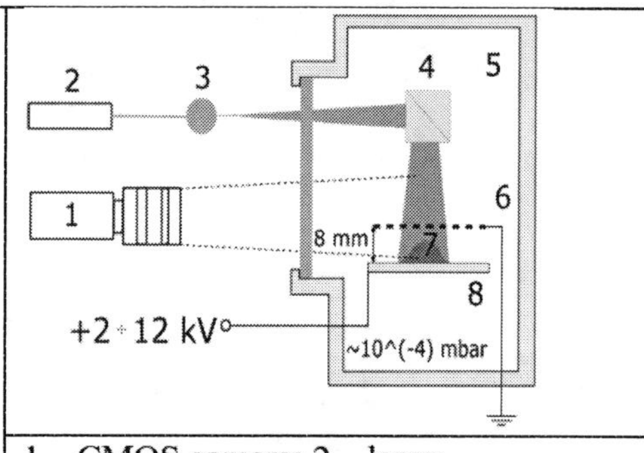

1 – CMOS camera; 2 – laser;
3, 4 – optics for laser illumination of dust particles; 5 – vacuum chamber;
6 – steel mesh; 7 – dust particles (1 μm, Fe; 40-100 μm, SiO_2);
8 – conductive substrate

Figure 1. Picture and scheme of the experimental setup for investigation the dusty plasma levitation (one of the chamber flanges is open).

In the experimental setup a laser diode with wavelength 532 nm and power of 300 mW was used. The light from laser goes through the optical system of several lenses. The optical system expands the beam and makes its cross section close to rectangular. The expanded beam goes through the vacuum viewport, reflects on mirror inside vacuum chamber on angle near 90° and illuminate the investigation area above the dusty particles. A volume of the investigation area was 5×5×3 mm^3.

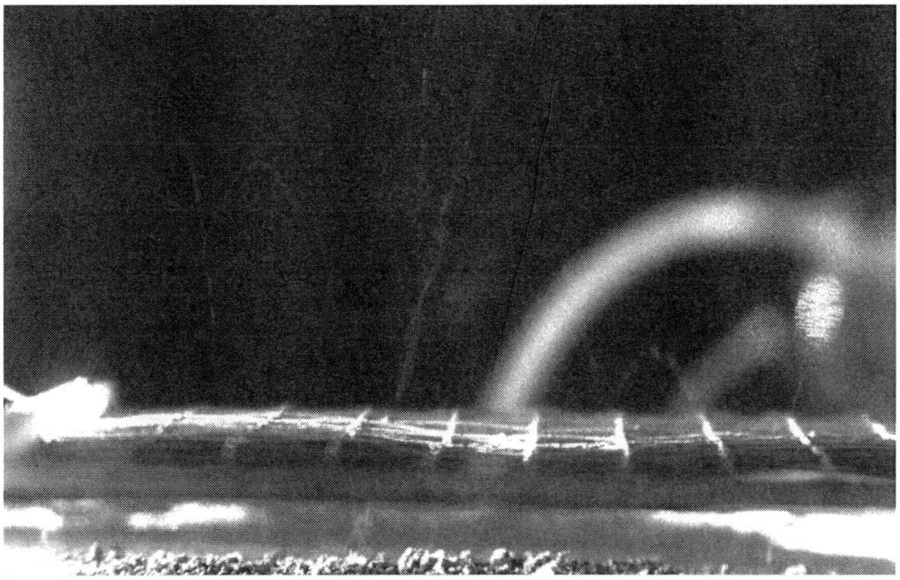

Figure 2. Direct visualization of the flight of SiO_2 particles with size of 40-100 μm (the thin lines on the picture).

Registration of images of the illuminated dusty particles was performed using the stereo system of two cameras. Baumer VLG-24M cameras with Sony IMX249 CMOS sensor, resolution 1920x1200 pixels and size of pixel 5.86×5.86 μm^2 were used.

(a) (b) (c) (d)

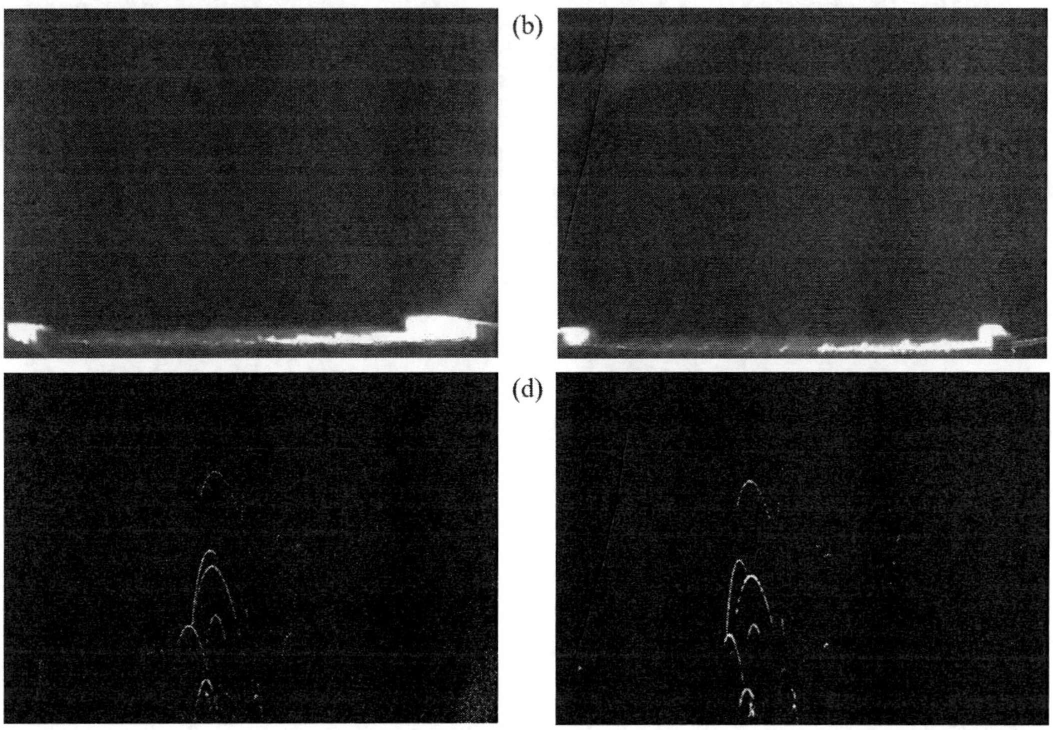

Figure 3. The first stage of the stereo system images processing: (a), (b) – raw registered images; (c), (d) – images with noise reduction and thresholding.

Journal of Physics: Conference Series **1421** (2019) 012037 doi:10.1088/1742-6596/1421/1/012037

3. Image processing technique

Image processing after the experiment is used to get complete information about particle trajectory. Trajectory should be reconstructed in three dimensional coordinates for estimating dust particles parameters as accurately as possible. A main problem of experimental images is lack of light in the investigation volume. Maximum power of the light source is limited by the possibility of an additional influence on the dust particles.

An image processing technique consists of several sequential operations. For each camera in the stereo system we:

- remove noise from images by subtracting each experimental image with a reference image without particles;
- binarize images with fixed threshold (figure 3 (c) and (d));
- detect connected components and filter them by their area and form.

Adaptive thresholding and Otsu method didn't give appropriate results after processing due to high noise rate. So binarization with fixed and relative low threshold value (below 10-15) was used. Some images was filtered by performing morphology opening.

After these operations we process two corresponding images from the stereo system cameras:

- for each filtered connected component from the first camera we calculate the best match component on image from second camera by criteria of maximum normalized correlation coefficient (figure 4 (a) and (b));
- calculate two dimensional coordinates of corresponding points for two matched components (figure 4 (c) and (d));
- calculate three dimensional coordinates of the corresponding points (figure 5).

Last operation gives us 3d trajectory of particle. It can be fit to parabola for estimation of particle parameters.

(a) (b) (c) (d)

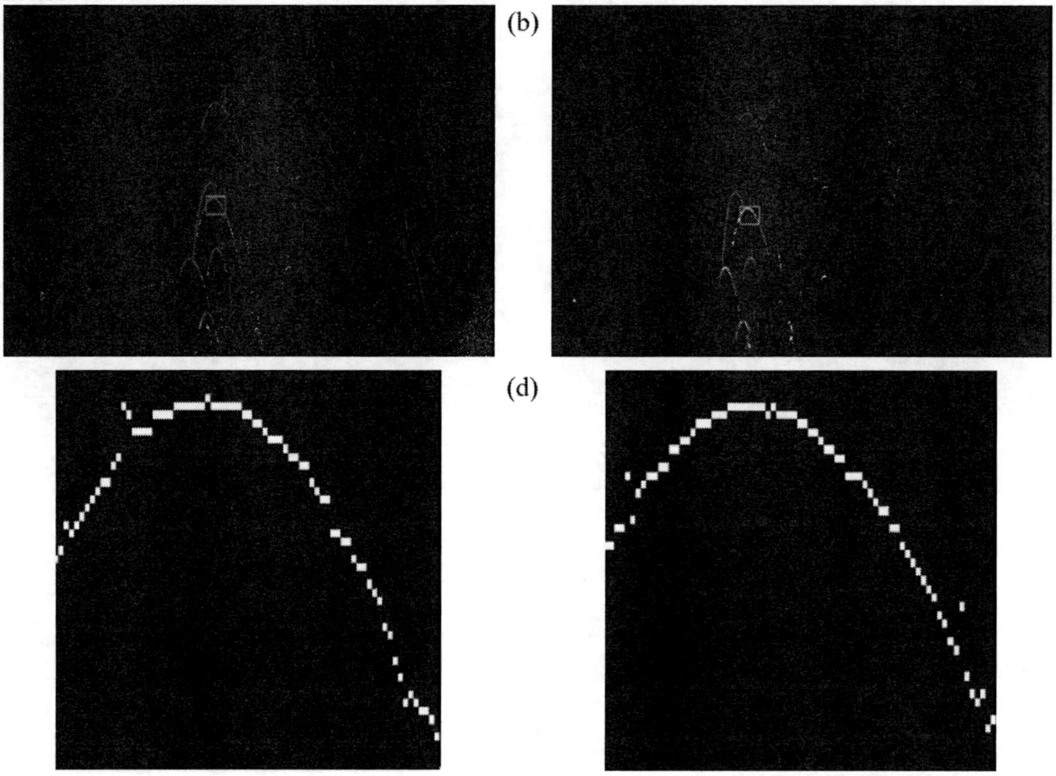

Figure 4. The second stage of the stereo system images processing: (a), (b) – two matched contours on different images by criteria of maximum normalized correlation coefficient; (c), (d) – corresponding points for two matched contours.

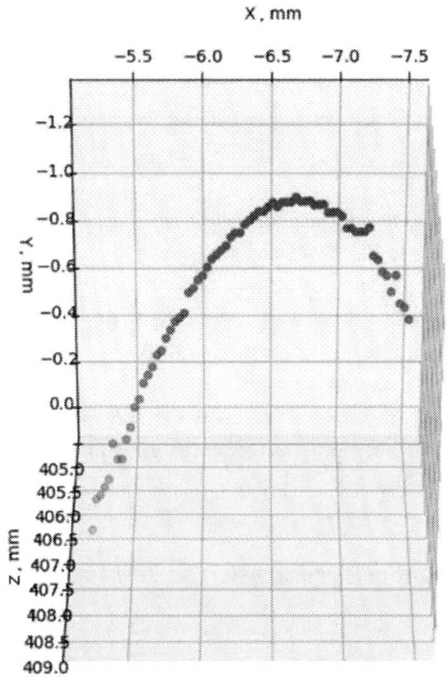

Figure 5. Particle trajectory obtained after the image processing.

For reconstruction three dimensional coordinates classic pinhole camera model was used. Before registration of experimental images cameras calibration with chessboard was done. Chessboard contains 8×7 squares with size of 5 mm. For each calibration 50 pairs of chessboard images in different positions were used. RMS error of calibration for single camera was below 0.5 pixel. RMS for stereo system calibration was in range 0.5-0.7 pixel.

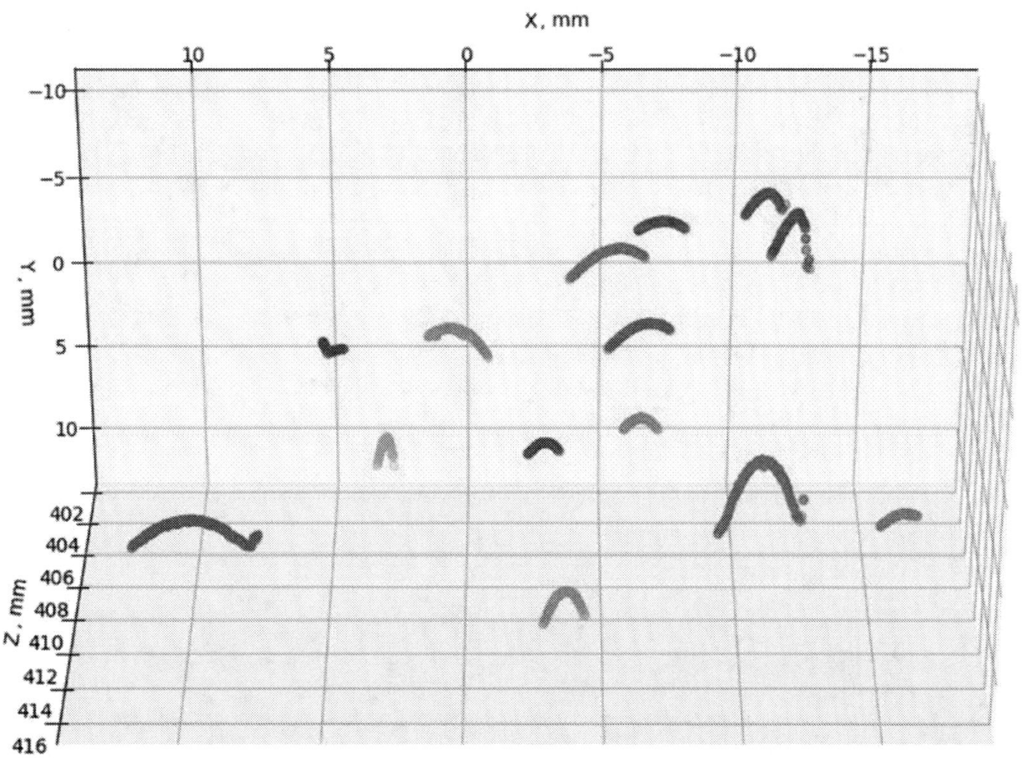

Figure 6. Particles' trajectories obtained after processing.

4. Conclusions

Visualization of the dusty plasma with simultaneously estimating its parameters is presented in this work. The experimental setup based on vacuum chamber for physical modelling of dusty plasma levitation is described. Visualization of the dust particles levitation is performed by the stereo system of two cameras. The particles were illuminated by a laser beam which is formed by special optic system.

Main aim of the work was determination of a trajectory of particle levitation in 3d coordinates. Trajectories was obtained by classical approach – calculating 3d point coordinates from 2d coordinates of points from different cameras of the stereo system. Maximum power of the light source was limited due to the possibility of an additional influence on the investigating particles. Therefore a lack of light in the investigation volume led to the need to use special image processing technique.

The proposed approach and image processing technique allow not only to visualize, but to obtain quantitative values of levitation parameters. These parameters make a valuable contribution to physical experiments on modelling the dusty plasma levitation and increase the likelihood of success of future robotic and human lunar landing missions to the Moon.

Acknowledgments

This work was supported by the Russian Scientific Foundation (the grant № 17-12-01458).

References

[1] Hörz F, Grieve R, Heiken G, Spudis P and Binder A 1991 *Lunar Sourcebook* (Cambridge University Press) p 61
[2] Manka R H 1973 *Photon and Particle Interactions with Surfaces in Space* ed Grard R J L (Holland: Reidel Publishing Co) pp 347–361
[3] Freeman, J W, Fenner M A and Hills H K 1973 *J. Geophys. Res.* **78** 4560–7
[4] Stubbs T J, Vondrak R R and Farrell W M 2006 *Adv. Space Res. 37 59–66*
[5] Colwell J E, Batiste S, Horanyi M, Robertson S and Sture S 2007 *Rev. Geophys.* **45** RG2006
[6] Afshar-Mohajer N, Wu Ch-Y, Curtis J S and Gaier J R 2015 *Adv. Space Res.***56** 1222–41
[7] Kuznetsov I A, Zakharov A V, Dolnikov G G, Lyash A N, Afonin V V, Popel S I, Shashkova I A and Borisov N D 2017 *Sol. Syst. Res.* **51 (7)** 611–22
[8] Wang X, Schwan J, Hsu H W, Grün E and Horányi M 2016 *Geophys. Res. Lett.* **43 (12)** 6103–10
[9] Orger N C, Alarcon J R C, Toyoda K and Cho M 2018 *Adv. Space Res.* **62 (4)** 896-911
[10] Popel S I, Zelenyi L M and Horányi M 2017 *JETP Lett.* **105 (10)** 635–40
[11] Orger N C, Toyoda K, Masui H and Cho M 2019 *Adv. Space Res.* **63 (10)** 3270-88

Journal of Physics: Conference Series **1421** (2019) 012038 doi:10.1088/1742-6596/1421/1/012038

Calculation of the optical variosystems with tunable optical power lenses

P A Nosov, D E Piskunov, M A Vinogradov, V O Tigaev and A A Yablokova

Bauman Moscow State Technical University, ul. Baumanskaya 2-ya, 5/1, Moscow, 105005, Russia

Abstract. Modern optical variosystems include lenses with a fixed and tunable focal length, which can also move relative to each other according to a certain law. The method of dimensional synthesis of the variosystems converting incoherent radiation and laser Gaussian beams is considered. Examples of the calculation of optical variosystems are presented.

1. Introduction

Optical vario- or zoom-systems have become widespread in various fields of science and technology. In their development, different approaches are used to allow a required change in the optical characteristics of the system. The most common approach is to move the components of the optical system according to a certain law [1-3]. Recently lenses with tunable optical power have begun to be used [4-6]. To change the optical power of the lens, it is necessary to change either its shape or the index of refraction. Liquid optical media are used for these purposes, therefore in most cases lenses with variable optical power are meant to be liquid lenses. The shape of the lens surface can be changed both mechanically and electrically. Today, there are several technologies for creating liquid lenses, two of them are based on the effect of electrowetting, as a result of which the curvature of the contact surface of two immiscible liquids changes, and on the basis of a change in the curvature of the elastic membrane limiting volume. These technologies have been commercialized by Varioptic and Optotune. Liquid crystal lenses are also perspective. The tunable refractive index of the liquid crystal lens allows to create the necessary phase profile and adjust the optical power of the lens.

The calculation of variosystems with liquid lenses includes the same steps as for optical systems with moving lenses [1, 3]. The main feature of the calculation of an optical variosystem based on liquid lenses appears at the stage of structural and dimensional synthesis. Moreover, the development of optical systems for the transformation of laser Gaussian beams should be carried out on the basis of the laser optics theory [7-9]. Let's consider the features of the development of optical variosystems of different types and methods for their automated calculation.

2. Optical variosystems calculation

In the general case, the problem of calculating of a variosystem is reduced to solving a system of nonlinear equations [3, 10]:

$$P(d,\varphi)=P_k, \tag{1}$$

where $\boldsymbol{P}$ is the vector of paraxial values, which should be equal to the required values (e.g. the focal length and the length of the system), $\boldsymbol{d}$ is the vector of the distances between the components, $\boldsymbol{\varphi}$ is the vector of optical powers of the components, $\boldsymbol{P}_k$ is the vector of required values of the paraxial values for the k-th position.

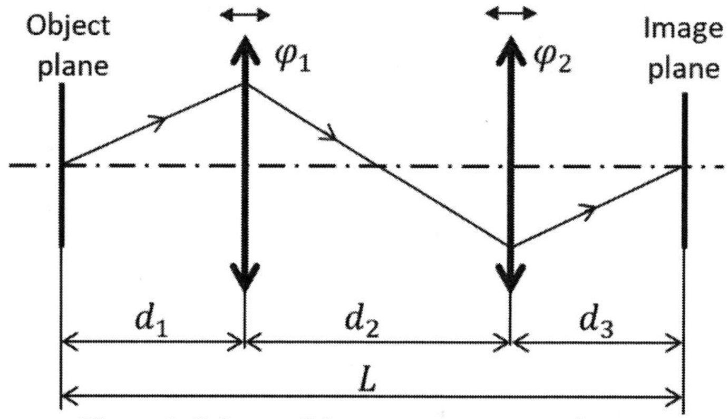

Figure 1. Scheme of the two-component variosytem.

For a two-component optical variosystem that converts incoherent homocentric radiation, the system (1) has the following form (figure 1) [10]:

$$\begin{cases} d_1 + d_2 + d_3 = L, \\ 1 - d_2\varphi_1 - d_3\varphi_1 - d_3\varphi_2 + d_2 d_3 \varphi_1 \varphi_2 = \beta, \\ 1 - d_1\varphi_1 - d_1\varphi_2 - d_2\varphi_2 + d_1 d_2 \varphi_1 \varphi_2 = 1/\beta, \end{cases} \tag{2}$$

where φ_i is the optical power of the component ($i = 1, 2$), d_1 and d_3 are the distances from the object plane to the first component and from the second component to the image plane, respectively, d_2 is the distance between the components, L and β are the length and transverse magnification of the system, respectively (they form the vector of the paraxial values $\boldsymbol{P}_k$).

For incoherent homocentric radiation in the analytical solution of the system of equations (2) in the case of a variosystem with moving components, it is assumed that the vector $\boldsymbol{\varphi}$ is known, i.e. the problem is reduced to determining the laws of components moving. One of approaches to solving the system (1) with the known vector $\boldsymbol{\varphi}$ is its transformation to a power equation. From (2) it is possible to obtain the law of displacement of the components of a two-component variosystem [6, 11]:

$$d_2 = \frac{L}{2}\left[1 \pm \sqrt{1 - 4\left(\frac{\varphi_1 + \varphi_2}{L\varphi_1\varphi_2} + \frac{(\beta - 1)^2}{L^2 \beta \varphi_1 \varphi_2}\right)}\right]. \tag{3}$$

Substituting the solution of (3) to the system of equations (2), one can determine the distances d_1 and d_3.

If the change in the magnification in the variosystem is carried out using liquid lenses, then it is assumed that the separations are known and it's necessary to determine the laws of change in the optical powers. In this case, for a two-component system, the solution of the system (2) is [6]:

$$\varphi_1 = \frac{\beta(d_1 - d_2) - d_3}{\beta d_1 d_2}, \quad \varphi_2 = \frac{d_3 + d_2 - d_1\beta}{d_3 d_2}.$$

In [3], it was shown that for convenience of numerical calculations, the system of equations (1) could be converted to a system of the form:

$$\boldsymbol{P}(\boldsymbol{a},\boldsymbol{b}) = \boldsymbol{P}(m),$$

where m is the argument of the basis function that determines the position of the components or optical power; $\boldsymbol{a},\boldsymbol{b}$ are vectors of coefficients of expansion into basis functions.

The laws of displacement $d(m)$ and changes in the optical power $\varphi(m)$ of individual components, which are presented in the form of expansion into basis functions, are written as follows:

$$d(m) = \sum_{i=0}^{N} a_i F_{di}(m),\ \varphi(m) = \sum_{i=0}^{N} b_i F_{\varphi i}(m), \tag{4}$$

where a_i, b_i are expansion coefficients from $\boldsymbol{a},\boldsymbol{b}$ vectors; N is the number of expansion members.

As a result, the calculation of a variosystem is reduced to the determination of the expansion coefficients a_i, b_i. Such an approach makes it possible to calculate both traditional systems with moving components and with liquid lenses, as well as combined ones, i.e. including either moving components or optical components with tunable optical power.

As a rule, the system of equations (1) has several solutions. For example, in the case of a two-component system, we have two solutions (3). This means that two paths of movement of the components are possible. In numerical calculations, the solution is defined as individual points on the curves (3), which determine the corresponding position of the components. A problem may arise when these points belong to different curves. This means that the magnification of the system will change with a jump while the movement of the components is smooth. If we define a solution in the form of the expansion (4), then the problem of matching the roots for a smooth change in magnification does not arise, since the solution is defined as a curve (of the law of displacement), but not a single point on it.

Now let's consider the calculation of optical variosystems for the transformation of Gaussian laser beams. Let the variosystem be designed to form a waist of a Gaussian beam of a constant diameter at different distances from the input waist (of the laser). This problem can be solved by an optical system in which components move along the optical axis [12], or can change their optical power. Figure 2 shows an optical system with a lens of variable optical power and its longitudinal movement to form a waist of a Gaussian beam of a constant diameter at different distances from the input waist.

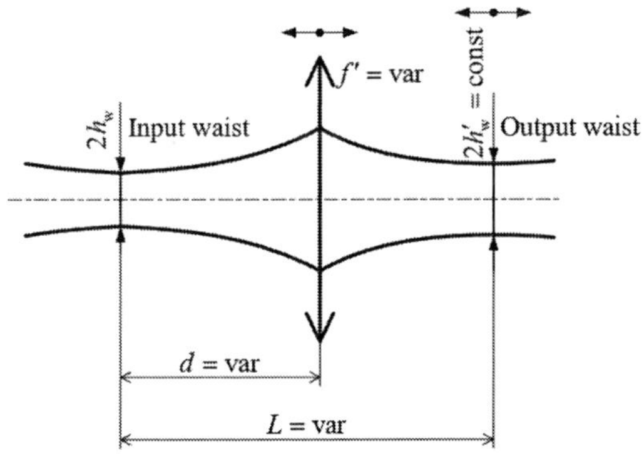

Figure 2. Single-component laser variosystem with tunable optical power lens and its longitudinal movement: d is the distance from the waist of the input beam to the lens, L is the distance from the waist of the beam at the lens input to the output waist, f' is the lens focal length.

Journal of Physics: Conference Series **1421** (2019) 012038 doi:10.1088/1742-6596/1421/1/012038

When the Gaussian beam is transformed by the optical system, its invariant is performed for its spatial parameters [7-9, 12]: $2h_w 2\theta = 2h'_w 2\theta' = \text{const} = 4M^2 \lambda/\pi$ and $(2h_w)^2/z_R = (2h'_w)^2/z'_R = \text{const} = 4M^2 \lambda/\pi$, where $2h_w$, 2θ and z_R are the diameter, angular divergence and Rayleigh range of the beam, λ is a beam wavelength, M^2 is the quality factor of the beam. The parameters of the transformed beam are denoted with a prime symbol. From this expression it follows that when the Gaussian beam is formed with a constant waist diameter at different distances from the input waist, the angular divergence of the output beam also remains constant.

Analysis of the transformation of the Gaussian beam of the laser optical system gives the following result [7]. The angular divergence of the beam on the optical system output is equal to $2\theta'_k = 2h_{F_k}/|f'_k|$, where $2h_{F_k}$ the diameter of the beam at the input of the *k*-th optical system component in its front part, f'_k is the back focal length of the *k*-th component of the optical system.

Then the parameters of the beam and variosystem for forming the waist of a constant diameter at different distances from the input waist are connected by the expression:

$$\frac{2h_F}{|f'|} = \text{const} = \frac{2M^2\lambda}{\pi h'_w} = 2\sqrt{\frac{M^2\lambda}{\pi z'_R}}.$$

From this condition, one can obtain the law of change in the focal distance of the lens from its position.

The position of the waist of the Gaussian beam transformed by the lens and the distance from the input to the output waist are determined by longitudinal magnification in the near zone α_G [7-9, 12]:

$$\alpha_G = \frac{f'^2}{z_w^2 + z_c^2},\ z'_w = -z_w \alpha_G,\ L = -z_w(1+\alpha_G) + 2f',$$

where z_w and z'_w are the distances that determine the position of the waist of the input and transformed beam relative to the front and rear foci of the optical system, respectively. For the distances z_w and z'_w, the following rule of signs holds true: to the right of the foci – positive, to the left – negative.

3. Examples of variosystems calculation

Example 1. Let's consider as an example the results of calculating of a 20^x variolens with a focal lengths range from 15 to 300 mm, consisting of four components. The change in focal length is provided by moving of the 2nd, 3rd and 4th components of the optical system along the optical axis. As the basis functions in the formula (4) we choose power functions. Then the distances between the components of the variolens can be represented as an expansion in the power series of the following form:

$$d(f') = a_0 + a_1 f' + a_2 f'^2 + a_3 f'^3, \tag{5}$$

where as a parameter *m* from (4) the focal length f' of the variolens can be chosen.

Table 1 presents the paraxial parameters of the components of the calculated variolens, and table 2 presents the expansion coefficients for the corresponding air distances. The laws of the moving components are shown in figure 3

Table 1. Paraxial parameters of the variolens.

Component #	Focal length (mm)	Distance	Relative aperture
1	78.700	d_1	0.600
2	-18.850	d_2	0.900
3	32.170	d_3	0.700
4	-44.000	d_4	0.200

Table 2. Expansion coefficients of the air distances of the variolens in (5).

	a_0 (mm)	a_1 (mm)	a_2 (mm^{-1})	a_3 (mm^{-2})
d_1	18.893	0.18064	-0.00030755	1.95E-07
d_2	85.994	-0.22193	8.37E-05	-5.51E-07
d_3	41.88	-0.053718	5.21E-05	2.13E-07
d_4	3.2334	0.095008	0.00017175	1.44E-07

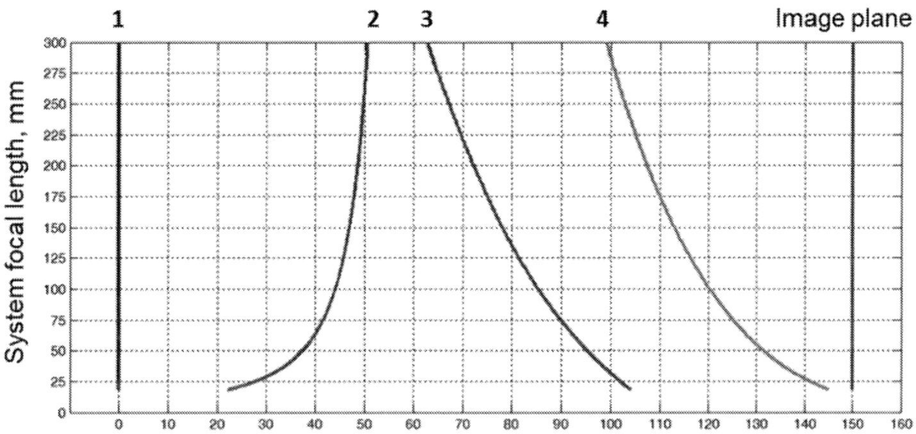

Figure 3. The laws of displacement of the components of the variolens.

The apertures of the components (table 1) were obtained with the following system parameters: a linear field in the image space is 15 mm, the f-number at the minimum focal length is 2.8, and 6 at the maximum focal length. The resulting system can be used for subsequent aberration synthesis.

Example 2. Let us now consider the results of calculating a 2^x variolens with liquid lenses with a focal length range of 17..34 mm (figure 4). The variolens consists of 6 components, two of which, the second and fifth are liquid lenses. The laws of change of the focal length of the *i*-th component f_i' are determined as an expansion in powers of the focal length f' of the entire lens:

$$f_i'(f') = a_0 + a_1 f' + a_2 f'^2 + a_3 f'^3 + a_4 f'^4. \tag{6}$$

Table 3 presents the expansion coefficients, and Figure 5 shows the laws of change in the focal length of liquid lenses (figure 4).

Table 3. The expansion coefficients of focal lengths of the components of the variolens in (6).

	a_4 (mm^{-3})	a_3 (mm^{-2})	a_2 (mm^{-1})	a_1	a_0 (mm)
f_2'	-0.0011	0.1051	-3.4208	35.6882	100.0000
f_5'	0.0003	-0.0083	-0.0845	5.1280	-11.8967

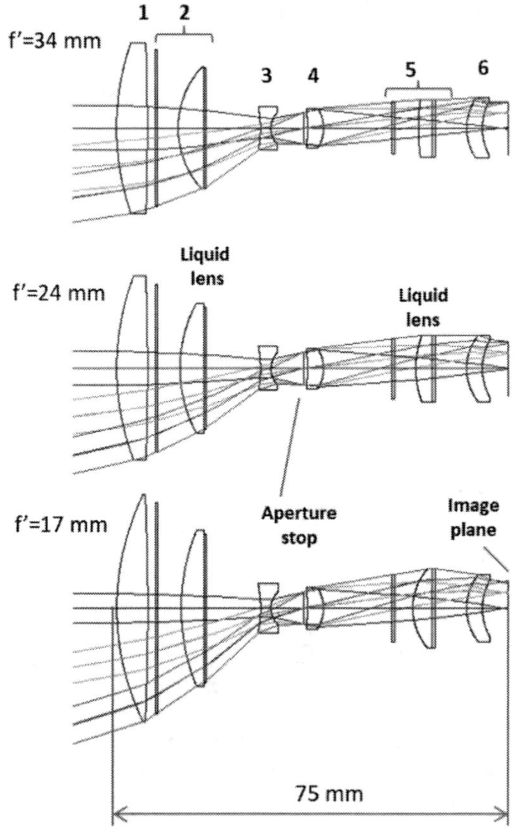

Figure 4. The scheme of the variolens with liquid lenses for 3 positions.

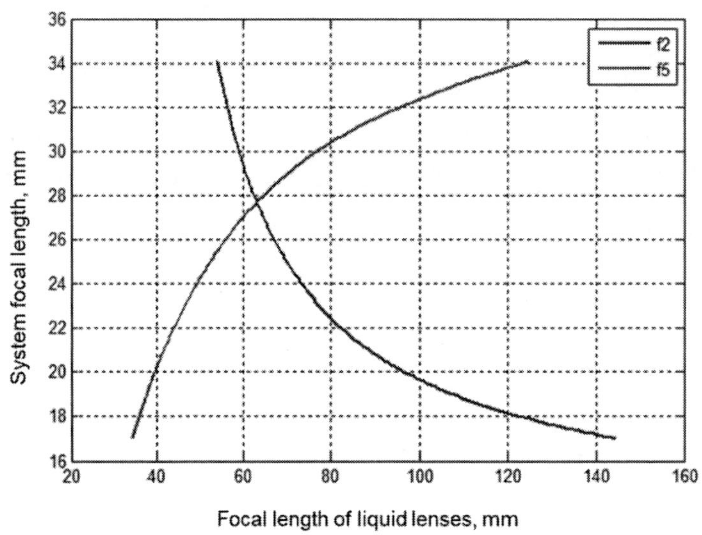

Figure 5. The laws of the focal length change of the tunable components.

The parameters of a variolens after the transformation to a system of lenses of finite thickness are presented in tables 4 and 5.

Table 4. The optical characteristics of the components.

Component #	f_i' (mm)	$s'_{F'}$ (mm)	s_F (mm)	$s'_{H'}$ (mm)	s_H (mm)	d (mm)
1	82.5	78.88	-82.17	-3.63	0.34	6.36
2*	80.6	76.79	-80.64	-3.85	0	10.42
3	-6.5	-6.73	7.6	-0.21	1.08	6.93
4	13.6	13.98	-11.34	0.42	2.22	17.63
5*	56.5	53.67	-56.47	-2.8	0	5.88
6	83.5	73.31	-93.04	-10.17	-9.56	5

Note: * tunable components, s_F and $s'_{F'}$ are the front and rear focal distances of the component, s_H and $s'_{H'}$ are positions of the front and rear principal points of the component, d is the air distance.

Table 5. Parameters of the tunable components.

Focal length of the variolens (mm)	Focal length of the component (mm)	
	# 2	# 5
17.0	120.67	35.76
24.0	80.64	56.47
34.0	52.84	122.28

The field of view of the calculated system of the visible spectrum range varies in the range of 17°...33°. The f-number at the minimum focal length is 2.8, at the maximum is 4. The illumination drop across the field does not exceed 25%. The resulting system is a good initial approximation for subsequent optimization.

Example 3. Let laser radiation have the following parameters: wavelength 1.03 μm, waist diameter $2h_w = 150$ μm, beam quality factor $M^2 = 1.05$. It is required to calculate the parameters of the laser variosystem, which forms a beam with a waist diameter of $2h'_w = 75$ μm and provides a change of $L > 50$ mm. The results of calculating of the variosystem for the initial, intermediate, and extreme positions of the lens with tunable optical power are presented below (for the notation, see figure 2):

f' (mm)	d (mm)	L (mm)
17.06	47.0	71.54
25.08	72.5	109.43
33.34	98.0	147.51

4. Conclusion

The considered methods and the above expressions make it possible to carry out an automated synthesis of optical variosystems of different types. Examples of calculating the variosystems show the effectiveness of the considered methods, and the resulting systems are a good initial approximation for subsequent optimization. The use of lenses with tunable optical power made it possible to abandon the mechanical movements in classical systems and reduce the number of moving components of the laser variosystem. Variosystems can be used in video camera lenses, computer vision systems with a rapid change in magnification, in laser systems for manipulating micro-objects (optical tweezers), and other areas [13-20].

Acknowledgments

The reported study was partially supported by RFBR, research project No. 18-38-20155.

References

[1] Pakhomov I I 1976 *Zoom Systems* (Moscow: Mashinostroenie)
[2] Tanaka K 1982 Paraxial analysis of mechanically compensated zoom lenses. 1: Four-component type *Appl. Opt.* **21** 2174–83

[3] Rozhkov O V, Piskunov D E, Nosov P A, Pavlov V Y, Khorokhorov A M and Shirankov A F 2018 Bauman MSTU scientific school "zoom lens design": Features of theory and practice *Computer Optics* **42(1)** 72–83
[4] https://www.optotune.com/
[5] Blum M, Büeler M, Grätzel C and Aschwanden M 2011 Compact optical design solutions using focus tunable lenses. *Proc. of SPIE* **8167**
[6] Miks A and Novak J 2010 Analysis of two-element zoom systems based on variable power lenses *Opt. Express* **18** 6797–810
[7] Pakhomov I I, Rozhkov O V and Rozhdestvin V N 1982 *Optoelectronic quantum devices* (Moscow: Radio i Svyaz')
[8] Pakhomov I I and Tsibulya A B 1988 Computational Methods for Laser Optical Systems Design *J. Soviet Laser Res.* **9(3)** 321–429
[9] Anikanov A G, Pakhomov I I and Shirankov A F 2010 Structural synthesis of laser optical systems when their parameters are restricted *J. Opt. Technol* **77(2)** 101–6
[10] Pakhomov I I, Piskunov D E and Khorokhorov A M 2012 Numerical technique of variable magnification systems with an arbitrary number of variable components *Engineering Journal: Science and Innovation* **9(9)**
[11] Khorokhorov A M, Piskunov D E and Shirankov A F 2016 First-order method of zoom lens design by means of generalized parameters *J. of the Opt. Soc. of America* A **33(8)** 1537–45
[12] Nosov P A, Piskunov D E and Shirankov A F 2019 Laser varifocal system synthesis for longitudinal Gaussian beam shifting *Appl. Opt.* **58** 3347–53
[13] Gulina Y S, Koliuchkin V Y and Trofimov N E 2018 Mathematical model of the human visual system *Optical Memory and Neural Networks (Information Optics)* **27(4)** 219–34
[14] Khatsevich T N and Volkova K D 2017 Stabilization of the exit pupil in an optical zoom sight with variable magnification *J. of Opt. Technol.* **84(9)** 605–12
[15] Druzhkin E V and Khatsevich T N 2013 Compact thermal viewers *J. of Opt. Technol* **80(6)** 343–8
[16] Kovalev M S, Krasin G K, Odinokov S B, Solomashenko A B and Zlokazov E Yu 2019 Measurement of wavefront curvature using computer-generated holograms *Optics Express* **27(2)** 1563–8
[17] Bobrinev V I, Kovalev M S, Odinokov S B and Sagatelyan G R 2016 Investigation of the properties of a beam reconstructed from volume holographic optical elements used in optical observation devices *Russian Physics Journal* **58(10)** 1457–66
[18] Machikhin A S, Batshev V I, Gorevoy A V, Khokhlov D D, Naumov A A and Kuznetsov A O 2019 Compact stereoscopic prism-based optical system with an improved accuracy of 3-D geometrical measurements *Optik* **185** 1172–81
[19] Machikhin A S, Batshev V I, Pozhar V E, Naumov A A and Gorevoy A V 2018 Acousto-optic tunable spectral filtration of stereoscopic images *Optics Letters* **43(5)** 1087–90
[20] Greisukh G I, Ezhov E G, Kalashnikov A V and Stepanov S A 2012 Diffractive-refractive correction units for plastic compact zoom lenses *Appl. Opt.* **51** 4597–604

Journal of Physics: Conference Series **1421** (2019) 012039 doi:10.1088/1742-6596/1421/1/012039

Photothermocapillary detection of conductive track ruptures on a printed circuit board coated with a protective film

A Yu Zykov[1], **N A Ivanova**[2]

[1] Department of Physics, Methods of Control and Diagnostics, Tyumen Industrial University, Volodarskogo 38, Tyumen,625000, Russia
[2] Photonics and Microfluidics Laboratory, Tyumen State University, Volodarskogo 6, Tyumen, 625003, Russia

Email: aleksandr.zykov@inbox.ru; n.ivanova@utmn.ru

Abstract. The photothermocapillary (PTC) method has been proposed to detect ruptures of the copper tracks of a printed circuit board coated with a black protective film absorbing the pump laser radiation. The ruptures were simulated by obtaining parallel tracks of different lengths by etching in ferric chloride solution. The method consisted of the scanning with the pump laser of the tracks along a line perpendicular to them and passing through their midpoints, and plotting a scan-profile of the diameter of PTC signal as a function of the scan coordinate. The scan-profile was a sequence of maxima and minima the diameters of the PTC signal corresponded to the beam position in the middle between adjacent tracks and in the midpoint of the track, respectively. It was found that with an increase in the track length above a critical value, dependent on the laser beam power, the diameter of PTC signal reaches saturation. With increasing power by 1.85 times, the sensitivity of the PTC method increases by 2.57 times. The spatial resolution of the PTC method reaches tens of micrometers. Experimental results have shown that the proposed PTC method is simple and efficient for detecting of hidden defects of the copper tracks on the PCB.

1. Introduction. Photothermocapillary effect

Increasing the compactness of printed wiring and reducing the width of conductors (conductive tracks) and the gaps between them make it necessary to control of the continuity of the electrical circuits. To detect such defects as ruptures of conductors, cavities in soldered contacts, insulation strength of disconnected circuits, insulation thickness heterogeneity and detachments from the surface of the printed circuit boards (PCB) are often not available for standard visual inspection. The control of these defects is complicated when the tracks are covered with a nontransparent protective film. More often, to check the continuity of the printed circuits, contact methods based on measuring the electrical conductivity by using ohmmeters or electric calls are used. However, in recent years, thermal methods allowing the detection of the entire complex of the PCB defects have become increasingly applicable. Among these are optical techniques such as thermoreflectance (infrared microscopy) [1-4], liquid crystal thermography [5] and scanning thermal microscopy [6]. However, the implementation of these methods typically requires both complex experimental setup, highly precise and very expensive instruments and complex algorithms of data analysis that does not allow for realization of cheap, compact and easy-to-use devices. In present work, we propose a novel and simple express-method for the control of defects on PCBs, which is based on using of the photothermocapillary (PTC) convection in a thin liquid layer on a tested solid substrate. The mechanism of the PTC convection is as follows [10]. When a laser beam hits an absorbing substrate covered with a thin transparent liquid layer a heat source at a solid-liquid interface arises. The heat distributes into the substrate and into the bulk of liquid. In the latter case, the temperature of the free liquid surface increases, that leads to a local decrease in its surface tension, $\gamma = \gamma_0 + \gamma'_T(T - T_0)$, where γ_0 is the surface tension of the liquid at temperature T_0; γ'_T is the thermocapillary coefficient, which is negative for most pure liquids. As a result, a surface tension gradient along the free liquid surface arises and induces the radially outward thermocapillary liquid flows, which create the concave deformation in the liquid layer in the laser spot (see, figure 1 (c)). The capillary pressure below the deformed surface causes the bottom flow towards the heated spot, wherein it conjoins with the surface flows resulting in the stationary thermocapillary deformation. The deformed liquid interface represents a concave mirror. A probe laser beam reflected from this mirror forms a circular interference pattern on a remote screen, called as the photothermocapillary (PTC) signal, which has a bright and wide outer ring (see, figure 1 (c)). Recently we have shown [7-10] that the PTC diameter is very sensitive to variations of thermal conductivity of substrates, hence the PTC effect can be considered as a simple and easily implemented method for nondestructive control of hidden defects of the PCB. To validate the proposed method a test sample of PCB with parallel conductive tracks of different length, which serve as a model of the ruptured tracks, was used.

2. Experimental part

2.1. The sample preparation

The tracks of different length (l = 2, 4, 8, 16, 32 and 64 mm) and 1.4 mm of width, located with a gap (insulation) of 1.3 mm (the gap between the last two tracks was 1.9 mm) were etched out on a one-sided PCB in ferric chloride solution using a mask. After obtaining the desired copper tracks, the surface of the PCB was cleaned from dust and treated with acetone. Then the PCB was glued to a duralumin disc, which works as a heat sink and the PCB stiffness amplifier. After the adhesive had solidified, a ring of 70 mm in diameter was glued to the PCB surface, and then this cell was filled with a liquid black nitrocellulose varnish (Zapon lacquer). After the solvent evaporated a coating film about 30 μm thick was formed on the copper tracks. A cross-section of the sample and the map of the tracks location under coating film are shown in figure 1 (a), (b). To perform the scanning experiment a layer of silicon oil (μ = 4.56 mPa·s) of 300 μm thick was deposited on the sample.

2.2. Experimental setup and method

The experimental setup is shown in figure 1 (c). The pump laser beam (1) (P_e = 21.3 mW, λ = 632.8 nm, d_e = 2.5 mm) passing through the optical system (2)-(4) (figure 1(c)) and focused with a lens (20) (d_e^* = 0.5 mm) falls almost vertically on the silicone oil layer (5), with the thickness controlled with the calibrated wire (6). The liquid layer is protected from dust by a transparent lid (9). The probe laser beam (15) (P_b = 0.3 mW, λ = 632.8 nm, d_p = 5 mm) is expanded with a spherical mirror (16) and directed to the TC depression in the liquid layer with the mirror (17). The reflected from the liquid surface the probe beam forms on a remote screen (19) the PTC signal. The horizontal position of the sample is provided by the system consisting of a massive base (11), on which a Teflon cell (12) with a solidified layer of Wood's alloy (13) is placed. A plane-concave lens (14) was laid down with a concave surface onto the Wood's layer. To scan the PCB sample, the cell is shifted horizontally using the micrometric screw (21). The diameter of the PTC signal, D, is measured with a ruler (22). In experiments, the power of the pump beam was chosen according to the following requirements: (i) the absence of the liquid layer rupture, and (ii) the clearly visible external wide ring of the PTC signal, which can be measured with sufficient accuracy. The power of the pump laser beam was changed using a set of light filters.

The scanning of the PCB sample was performed with a step of 100 μm starting at the track of 2 mm length in the direction perpendicular to the tracks and passing through their midpoints (figure 1 (b)). The scan-profiles as the dependency of D on x were plotted, where x is the horizontal scan coordinate.

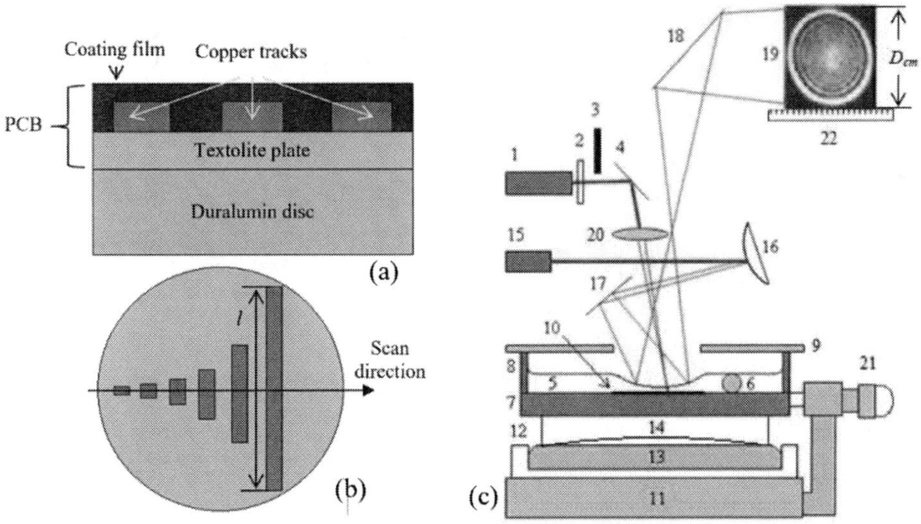

Figure 1. (a) Schematic cross-section view of the printed circuit board test sample. (b) A map of the location of the copper tracks of different lengths (the horizontal arrow indicates the scan direction). (c) The experimental setup: 1, 15 – the pump and probe lasers, 2 – attenuator, 3 – shutter, 4, 17, 18 – mirrors, 5 – liquid layer, 6 – calibrated wire, 7 – test substrate, 8 – ring, 9 – cover, 10 – absorbing film, 11 – massive base, 12 – Teflon cell, 13 – solidified Wood's alloy, 14 – plane-concave lens, 16 – spherical mirror, 19 – PTC signal on a screen, 21 – micrometric screw, 22 – ruler.

3. Results and discussion

Figure 2 (a) shows the scan-profiles of the PCB sample obtained at the powers of the pump laser beam 2.8 and 5.2 mW. The maximum, D_{max}, and minimum, D_{min}, values of the PTC diameter correspond to the position of the beam in the midpoints between the copper tracks (the isolation gaps) and in the midpoints of the conductive tracks, respectively. The decreasing D on the copper tracks is related with a sharp increase of the thermal losses due to dissipation into the copper track due to its high thermal conductivity in comparison with the textolite ($k_c/k_t \cong 1176$), where k_c and k_t are the thermal conductivity of copper and textolite, respectively. It turned out that the increasing l of the copper tracks leads to a slight decrease in D_{max} that is caused by the influence of the size of the heat-conducting tracks on the amount of dissipated heat into the substrate, i.e., by the increasing of the so-called "effective" thermal conductivity of the substrate. However, for the track of 64 mm length, an opposite behavior is noticed, that is an increase in D_{min} and D_{max}, which can be related to the larger isolation gap with the adjacent track (l = 32 mm) on the one side and the absence of heat-conducting tracks on the other side.

To analyze the obtained results the dependency of D_{min} on the length of the copper tracks was plotted in the semi-logarithm scale (figure 2 (b)). As is seen, with increasing l of the tracks, D_{min} of the PTC signal decreases and at a critical value of l_c reaches a constant value. Thus, the limit of applicability of the PTC method of the scanning of the tracks is determined by the reaching of $D_{min}(l)$ saturation. The critical l_c is determined as a cross point of the straight lines fitting the decreasing and stationary parts of $D_{min}(l)$ dependency (figure 2 (b)). For the pump laser power of 2.8 and 5.2 mW, the values of l_c are 4.6 and 5.0 mm, respectively. The constant value of D_{min} implies that the tracks in the region $l > l_c$ can be considered as the thermally "long". With increasing the power of the pump laser beam, there is a tendency to increase the value of l_c and this fact is the starting point for further research and increasing the range of applicability of the photothermocapillary method. The sensitivity of the method is defined as the slope $\Delta D_{min}/\Delta l$ of the $D_{min}(l)$ dependence. As can be seen in figure 2 (b) the sensitivity $\Delta D_{min}/\Delta l$ depends on P_e and takes on values of 16 and 43 at 2.8 and 5.2 mW respectively, i.e., with increasing P_e by 1.9 times $\Delta D_{min}/\Delta l$ grows by 2.7 times. Taking into account that the error of D_{min} measurement is around 1mm, one may conclude that the PTC method allows detecting the minimal changing of the copper track length down to 20 and 60 μm, respectively.

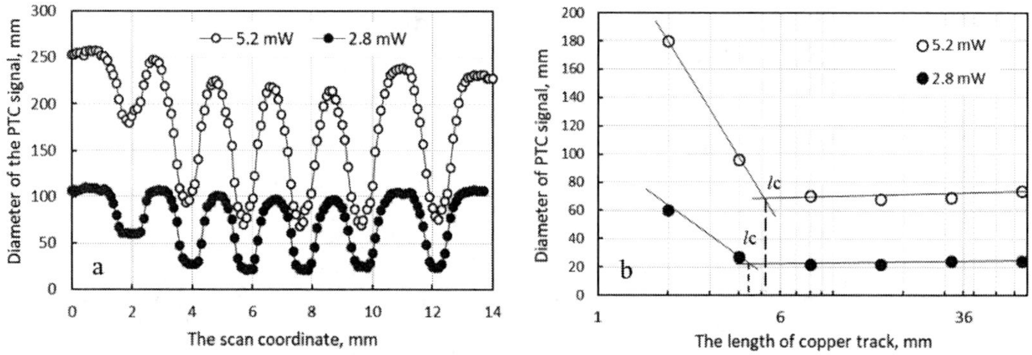

Figure 2. (a) Scan-profiles of the copper tracks at powers of the laser beam: ○ - 5.2 mW, ● - 2.8 mW. (b) The PTC signal of the copper track as a function of the track length.

4. Conclusions

The photothermocapillary method for detecting ruptures of the copper tracks of the PCB coated with a black protective film has been proposed. The method involves the covering of the PCB sample with a liquid layer, the scan of the copper tracks along a line perpendicular to tracks and the plotting of a scan-profile as the diameter of PTC signal versus the scan coordinate. The minima diameter of the PTC signal on the scan-profile, corresponding to the position of the pump beam in the midpoints of tracks, decreases with an increase in the track length until reaching a constant value. The latter determines the limit of applicability of this method in terms of the track length. The resolution of the PTC method reaches tens of micrometers. The method is simple in implementation and does not require special equipment and software for processing the optical signal.

Acknowledgements

The work was supported by the Russian Foundation for Basic Research (grant No. 17-08-00291).

References

[1] Voigt P, Hartmann J and Reichling M 1996 *J Appl Phys* **80**(4) 2013 – 18
[2] Christofferson J and Shakouri A 2004 *Microelectronics Journal* **35** 791 – 6
[3] Langer G, Hartmann J and Reichling M 1997 *Rev. Sci. Instrum* **68** (3) 1510–13
[4] Lo H and Ram R J 2012 *J Electronic Materials* **41**(6) 1332 – 36
[5] Cahill D G, Goodson K and Majumdar A 2002 *J Heat Transfer* **124**(2) 223
[6] Goodby J W, Collings P J, Kato T, Gleeson H, Tschierske C, Vill V and Raynes P 2014 *Handbook of Liquid Crystals* (John Willey & Sons)
[7] Bezuglyi B A, Zykov A Yu and Semenov S V 2008 *Russian J Nondestruct Testing* **44** 391–4
[8] Bezuglyi B A, Zykov A Yu and Semenov S V 2008 *Tech. Phys. Lett* **34** 743–6
[9] Bezuglyi B A and Zykov A Yu 2009 *Tech Phys Lett* **35** 650–2
[10] Zykov A Yu and Ivanova N A 2017 *Applied Physics B* **123**(9) 2–7

Journal of Physics: Conference Series **1421** (2019) 012040 doi:10.1088/1742-6596/1421/1/012040

Effect of the nanoparticle composition on the fluorescence of carbon nitride coatings

A B Bogomolov[1,2], P V Zinin[1], S A Kulakov[2], V A Kutvitsky[2], M F Bulatov[2] and I B Kutuza[1]

[1]Scientific and Technological Center for Unique Instrumentation RAS, Moscow, Butlerova str., 15, 117342, Russia
[2]Russian Technological University "MIREA", Moscow, Vernadskiy ave. 78, 119454, Russia

E-mail: bogomolov1395@gmail.com

Abstract. In this report we present results on synthesis of graphitic carbon nitride (g-C_3N_4) coating on various nanoparticles. Carbon nitride coatings were grown on of SiO_2, TiO_2, and Fe_2O_3 nanoparticles by the decomposition of the melamine at 300 °C. Relative fluorescence quantum yield was determined by comparing the fluorescence intensity of coated nanoparticles with that of the fluorescein in NaOH solution excited a green laser (532 nm). It was demonstrated that the g-C_3N_4 coated TiO_2 nanoparticles have the highest quantum yield of the fluorescence among investigated nanoparticles.

1. Introduction

Recently the detection of intense fluorescence in some graphitic carbon nitride phases and especially its strong dependence on the nanostructure of these phases [1] has added a new facet to the already known extraordinary properties of graphitic C-N phases, emerging from their unique electronic structure. With medium band gap as well as thermal and chemical stability in ambient environment, near-stoichiometric graphitic C_3N_4 (g-C_3N_4) became one of the most promising photocatalytic materials [2]. In addition, considerable attention has been paid on its photoelectronic application, such as light emitting devices, photocathodes, optical sensors, etc. [3]. However, systematic studies of the effect of the atomic structure, nanostructure and chemical bonding on optical and elastic properties of the C-N compounds have not been carried out yet.

Since a pioneering work on the synthesis of graphitic (g-C_3N_4) and hollow spherical (s-C_3N_4) modification of carbon nitrides by solid-state reaction between cyanuric chloride lithium nitride [4, 5], an abundance of methods for the synthesis of g-C3N4 phases has been developed and reviewed elsewhere [6-8]. In [6] an overview of graphitic carbon nitride materials for the use as catalysts is given. Several different preparation routes of the synthesis of graphitic carbon nitride materials from molecular precursors were reviewed. This review also reveals that the term “graphitic carbon nitride” covers a wide range of different compounds, with microstructures ranging from amorphous to nanocrystalline and from compact to nanoporous, depending on the synthesis path. A lot of the reported compounds have compositions close to C:N=3:4 and are named g-C_3N_4. Structural model molecular clusters for the g-C_3N_4 showed on figure 1. Model is based on carbon vacancy order leading to an orthorhombic cell.

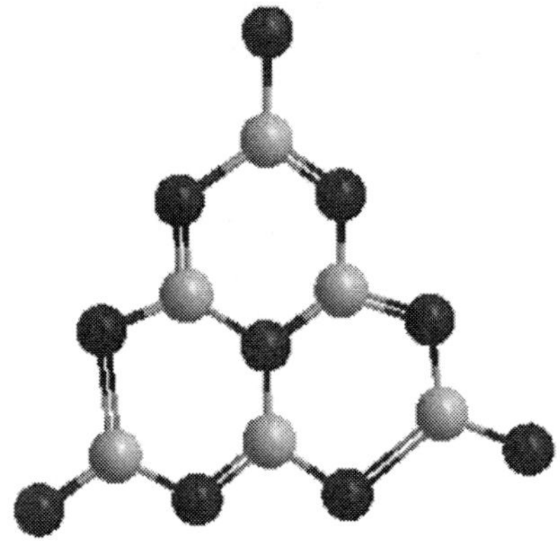

Figure 1. Atomic structure of the g-C_3N_4 layer. Blue atoms are nitrogen, gray atoms are carbon.

There are many ways to obtain g-C_3N_4 phases. In this report we developed a modification of the method of the synthesis of luminescent particles described in [9]. It is based on the thermal decomposition of melamine. The process of thermal decomposition of melamine can be carried out in any laboratory without use of any tools for work with toxic substances.

In this report we present results on synthesis of carbon nitride coating on various nanoparticles. The coating was obtained by the decomposition of the melamine at 300 °C. Carbon nitride coatings were grown on of SiO_2, TiO_2, and Fe_2O_3 nanoparticles.

2. Material and Methods

For synthesis of the g-C_3N_4 coated nanoparticles we used the melamine with a purification 99.7%, SiO_2 nanoparticles with a size from 5 to 15 nm produced by Sigma Aldrich (Purification 99%), TiO_2 nanoparticles with a size from 50 to 150 nm (Purification 99%); Fe_2O_3 particles with a size of 2 μm (purification 98.7%), Fluorescein (purification 98.7%), sodium hydroxide (purification 99%), and deionized water. The g-C_3N_4 coating was prepared by the thermal decomposition of melamine. The melamine powder was grinded in an agate mortar for 15 minute. Then powder was heated to 300°C with a ramp rate of 10 °C/min inside a sealed glass reactor. After 6 hours melamine completely transformed into homogeneous g-C_3N_4 white solid powder

$$C_3N_6H_6 \xrightarrow{300^\circ C} g - C_3N_4 + 2NH_3 \tag{1}$$

g-C_3N_4 coating on SiO_2 and TiO_2 nanoparticles were obtained by a similar technique. The mixture of the melamine powder and nanoparticles with a ratio of 1:3 after 30 minute grinding were heated at 310 °C for 6 hours in a sealed glass reactor.

The choice of material of nanoparticles for substrates was made taking into account their chemical and thermal stability, and toxicity [10]. TiO_2 is an effective photocatalyst in the ultraviolet (UV) region. We believe together with g-C_3N_4 it will increase photocatalic activity [11]. The microparticles Fe_2O_3 (E172 are of interest because their coloring properties, as this will enable the production of a biocompatible dye with small fluorescent properties. The SiO_2 nanoparticles with size 5-15 nm were used because fullerene-like microstructure can be grown on their surfaces [4]. As a reference for the quantum yield was used a fluorescein as substance with the known fluorescence output. The absolute quantum yield for fluorescence of 0.1N NaOH is 92.5% [12]. The fluorescein solution was prepared by dissolving 0.01 g powder fluorescein in 100 ml 0.1N sodium hydroxide. Sodium hydroxide was prepared by dissolving 0.4 g of NaOH in 100 ml of deionized water. We used a low concentrations of the fluorescein to avoid quenching of luminescence.

Journal of Physics: Conference Series **1421** (2019) 012040 doi:10.1088/1742-6596/1421/1/012040

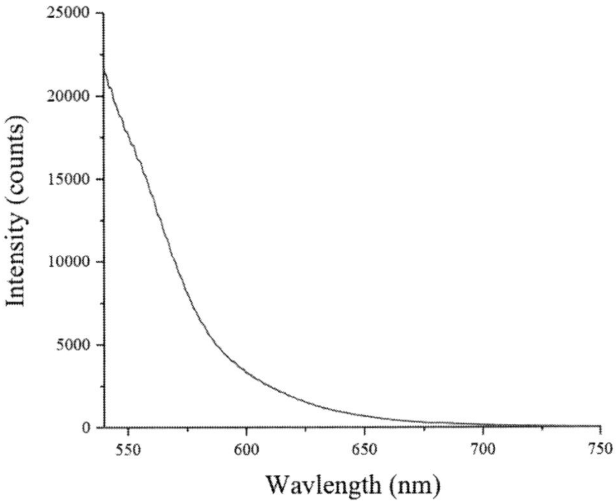

Figure 2. Fluorescence spectrum of the fluorescein solution excited by 532 nm laser. The power of the laser was 20 mW. Exposure time was 0.02 s.

Luminescence spectra were excited by a green lase (532 nm, Oxxius, LCX-532L-150-CSB-PPF) at room temperature. The spectra were collected by Solar Laser Systems M266, with detector U2C16H10426; 532 nm. Calibration was done with a neon lamp. A relative quantum yield, Q_y, was measured by comparing with etalon values using the following expression.

$$Q_y = Q_{et} \frac{S_{dye}}{S_{et}} \quad (2)$$

where Q_{bt} is the quantum yield of the fluorescence of the fluorescein in sodium hydroxide, S_{dye} and S_{et} are the areas under fluorescence peaks of the fluorescein and C_3N_4 coated nanoparticles. The value of quantum yield of the fluorescence is 92% when the blue laser is used for the fluorescence excitation. In our experiments we use the green laser (figure 2). Under such an excitation (532 nm) only half of the intensity of the fluorescence intensity of the fluorescein is excited [13]. Therefore we set the value Q_{bt} =0.46 in (2).

3. Resultant and discussion

Figure 1 shows the spectrum of fluorescence of the fluorescein`s solution. The fluorescence spectra of coated nanoparticles are showed on figure 3. The g-C_3N_4/ TiO_2 nanoparticles have the highest intensity of the fluorescence, and the g-C_3N_4/Fe_2O_3 ones have the lowest. The fluorescence intensity of the g-C_3N_4/ SiO_2 nanoparticles is lower than that of g-C_3N_4/Fe_2O_3 nanoparticles, but is lower than that of the g-C_3N_4/ TiO_2 ones. The results of the relative of quantum yield measurements are shown in table 1.

We do not know why the intensity of the fluorescence of the g-C_3N_4/ TiO_2 nanoparticles is the highest among investigated set. In [1], the high value of the fluorescence quantum yield for spherical *s*-C_3N_4 nanoparticles was attributed to the spherical shape of the phase. However, titanium dioxide can also generate localized plasmon resonance [14]. Nanoscale metallic systems are modifying collective electronic excitations that increase the quantum yield. It also is possible that Purcell effect increases the likelihood of spontaneous transitions and decreases lifetime of molecules in the excited state [15].

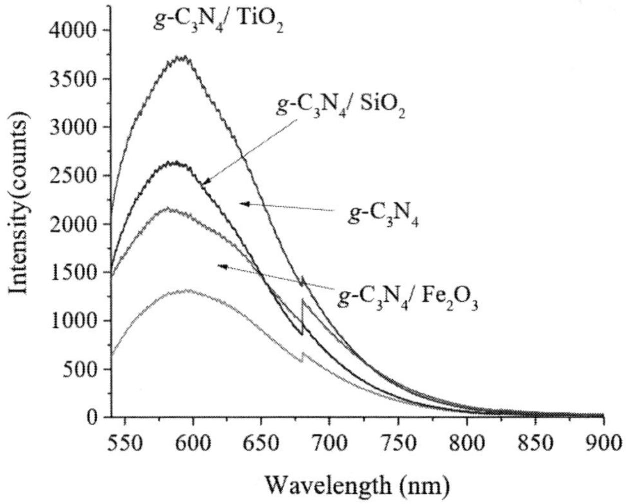

Figure 3. Fluorescence spectra of the solutions of different nanoparticles: $g\text{-}C_3N_4/TiO_2$; $g\text{-}C_3N_4/SiO_2$; $g\text{-}C_3N_4$; $g\text{-}C_3N_4/Fe_2O_3$. The power of the laser was 20 mW. Exposure time was 0.02 s.

The high level of the fluorescence of $g\text{-}C_3N_4$ coated nanoparticles can be used as fluorescent labels serving to identify the presence or the spatial position of the molecule under study. The fluorescent label must be chemically stable and exhibit stable fluorescence, which is independent of external factors and minimally varies with time. Thus, g-C3N4 acts as a passive beacon, signaling the location of the molecule to which it can be attached.

Table 1. Relative fluorescence quantum yield.

Material	S_{dye}	Q_y
$g\text{-}C_3N_4/Fe_2O_3$	90 500	0.03
$g\text{-}C_3N_4$;	180 200	0.06
$g\text{-}C_3N_4/ SiO_2$	324 000	0.09
$g\text{-}C_3N_4/ TiO_2$	469 900	0.14

4. Conclusion

We successfully synthesized $g\text{-}C_3N_4$ coatings on SiO_2, TiO_2 and Fe_2O_3 nanoparticles and microparticles by thermal melamine decomposition. The relative quantum yield of the $g\text{-}C_3N_4/ TiO_2$ nanoparticle found to be the highest among synthesized $g\text{-}C_3N_4$ coated particles. The relative quantum yield of the $g\text{-}C_3N_4/ TiO_2$ nanoparticle was measured to be 14%. The low level of the fluorescence of $g\text{-}C_3N_4/Fe_2O_3$ particles may be related to the micron size of the Fe_2O_3 particles. The results show the possibility of using $g\text{-}C_3N_4$ coated nanoparticles in photoelectronic devices, such as light emitters, photocathodes and biomarkers.

Acknowledgment

This work was performed using the Unique Scientific Instrument "Laser heating at diamond anvil cell" [http://ckprf.ru/usu/507563/].

Reference

[1] Zinin P V, Ryabova A V, Davydov V A, Khabashesku V, Boritko S, Sharma S K, Pominova D V and Loshenov V 2015 *Chem. Phys. Lett.* **633** 95
[2] Wang X C, Maeda K, Thomas A, Takanabe K, Xin G, Carlsson J M, Domen K and Antonietti M 2009 *Nat. Mater.* **8** 76
[3] Dong G P, Zhang Y H, Pan Q W and Qiu J R 2014 *J. Photochem. Photobiol. C-Photochem. Rev.* **20** 33
[4] Khabashesku V N, Zimmerman J L and Margrave J L 2000 *Chem. Mat.* **12** 3264
[5] Zimmerman J L, Williams R, Khabashesku V N and Margrave J L 2001 *Nano Lett.* **1** 731
[6] Thomas A, Fischer A, Goettmann F, Antonietti M, Muller J O, Schlogl R and Carlsson J M 2008 *J. Mater. Chem.* **18** 4893
[7] Xie H Q, Fu Y, Zhang Q Z, Yan K, Yang R, Mao K H, Chu P K, Liu L Z and Wu X L 2019 *Talanta* **196** 530
[8] Yang L Y, Wu X X, Luo L, Liu Y and Wang F 2019 *New J. Chem.* **43** 3174
[9] Hatamie A, Marahel F, and Sharifat A 2018 *Talanta* **176** 518
[10] Mary Ealias A and M P S 2017 A review on the classification, characterisation, synthesis of nanoparticles and their application vol. 263
[11] Yan Y Y, Kuang W C, Shi L J, Ye X L, Yang Y H, Xie X B, Shi Q S and Tan S Z 2019 *J. Alloy. Compd.* **777** 234
[12] Magde D, Wong R and Seybold P G 2002 *Photochem. Photobiol.* **75** 327
[13] Panchompoo J, Aldous L, Baker M, Wallace M I, and Compton R G 2012 *Analyst* **137** 2054
[14] Dahlman C J, Agrawal A, Staller C M, Adair J and Milliron D J 2019 *Chem. Mat.* **31** 502
[15] Choy W C H, Chen X W, He S L, Chui P C and Aoe 2008 The Purcell Effect of Silver Nanoshell on the Fluorescence of Nanoparticles (Shenzhen: Aoe)

Computer modeling of acoustic standing waves in the coolant of nuclear power plants

K N Proskuryakov[1], A V Anikeev[2], and E Afshar[3]

[1,2,3]Department of Nuclear Power Plants, National Research University 'MPEI', Moscow, Russia

E-mail: [1]ProskuriakovKN@mpei.ru, [2]AnikeevAV@mpei.ru, [3]AfsharI@mpei.ru

Abstract. In nuclear power plants (NPPs), the dynamic interactions of the coolant with the structures are among the main factors determining the dynamic loads on the equipments, its service life and reliability. The most hazardous is the resonant interaction of vibrations of the equipment with acoustic standing waves (ASW) that arise in accidental conditions.It is pointed out that the study of accident conditions on a full-scale object model is not possible and, accordingly, the development of adequate computer models to simulate the conditions is of prime importance. The analysis of acoustic systems with single-phase and two-phase fluid media is based on the theory of elastic wave propagation in liquids and gases.The analysis is based on the equations of fluid state, equations of motion, continuity equation and the equation expressing the law of conservation of energy. It is shown that in view of the unity of the differential equations of acoustic and electrical systems, the study of the propagation of the volume flow rate of one-dimensional pulsating flow of a compressible single-phase or two-phase fluid in the acoustic system can be replaced by the study of the propagation of electric current in the lines of electrical circuits.The paper presents an acoustic scheme of the NPP coolant circuit formed by acoustic elements considering their inherent thermal-hydraulic and geometric parameters. It is shown that the results of calculations of the frequencies of the ASW are confirmed by the measurements carried out at the NPP.

1. Introduction

The operational experience of nuclear power plants (NPP) shows that the fluid flowcauses vibrations of equipment and their elements. These processes are among the main factors determining the dynamic loads on the equipments, their service life and reliability. The most hazardous case occurs due to resonant interaction of vibrations of equipment with acoustic standing waves (ASW) arising in the coolant [1].

One of the most important mechanical systems that determine the reliability and safety of nuclear reactors are fuel assemblies (FA). To improve the economic performance of the nuclear fuel cycle and achieve the required level of safety, it is necessary to increase the service life and reliability of fuel assemblies and fuel rods (FR). The solution of these problems is largely related to the vibration processes in the structural elements upon the impact of the coolant flow.

To predict the reliability of existing FA and substantiate the reliability of new FA designs, research have been carried out on vibration and hydrodynamic processes, including test loops, computational modeling and vibration control during commissioning tests and in operating reactors. The FA under operating conditions, is a hydro elastic oscillatory system consisting of two interacting subsystems – mechanical and hydrodynamic. Currently, the mechanical subsystem is the most studied part and for which there already exists a sufficiently reliable computational analysis.The hydrodynamic subsystem, due to the complexity of the description of the processes of formation of random hydrodynamic loads on streamlined surfaces, and due to the influence of thermal-hydraulic and acoustic characteristics of the flow on the occurrence of self-excited oscillations, has been studied to a large extent, insufficiently.

It is very important to be able to predict adequately FA reliability, since its experimental study in an operating reactor is impossible, and it is also not possible to reproduce the interrelations of thermal-hydraulic, neutron-physical and vibroacoustic processes that determine the complicated dynamic loads.

The principles of mathematical modeling of mechanical systems are known, which allows to analyze both the oscillations of FA (usually, low-frequency) and the oscillations of individual fuel rods. As it is known, the main sources of excitation of oscillations, are non-stationary hydrodynamic processes in the coolant (acoustic waves, turbulence, vortex formation, cavitation, etc.) and the imbalance of rotating mechanisms, primarily circulation pumps (mechanical forces). In this case, the vibrational characteristics of the acoustic system of the coolant are not reduced to the sum of the properties of its individual elements, since the interaction of vibrational processes in these elements gives the system new properties that are not present in individual elements.

2. The method of electro-acoustic analogies

In modern acoustics, methods of solving problems that are borrowed from electrical engineering are widely used. This became possible due to the fact that in many fields of physics, including acoustics and in electrical engineering, many problems are described by the same differential equations. The rapid development of electrical and radio engineering has led to a more complete study of similar problems in electrical systems

Combination of individual acoustic elements can lead to formation of complex devices, which are similar in their action to various electrical (or mechanical) systems. If the geometrical dimensions of these devices are small compared to the length of acoustic waves, then such devices could be considered systems with concentrated rather than distributed parameters. The analysis of acoustic systems with single-phase fluid media is based on the theory of elastic wave propagation in liquids and gases [2, 3]. The analysis is based on the equations of fluid state, equations of motion, continuity equation and the equation expressing the law of conservation of energy.

An important stage in the development of methods for analyzing acoustic systems of coolant in nuclear power plants was the justification of the validity of electroacoustic analogies for one-dimensional pulsating flow of a two-phase medium, both with single valued and multi-valued hydrodynamic characteristics [4]. Taking into account the assumption that the sound pressure Δp at all points along the channel length is only a function of time, the equations of continuity and conservation of momentum of a one-dimensional pulsating flow of a two-phase medium are obtained in the form of a system of linear differential equations

$$\begin{cases} \dfrac{\partial W}{\partial x} + C_a \dfrac{\partial \Delta p}{\partial t} + G_a \Delta p = 0 \\ \dfrac{\partial \Delta p}{\partial x} + m \dfrac{\partial W}{\partial t} + R_a W = 0 \end{cases}. \quad (1)$$

Here Δp is the pressure change, which occurs during the passage of a sound wave in a fluid, in steam or in a mixture environment (sound pressure). The sound wave when propagating in a fluid, forms compression and rarefaction, which create pressure changes in relation to the average static pressure p.

W is volumetric flow rate of the coolant in the pipe; L is the length of the pipe; S is the cross section of the pipe; $C_a = \frac{L \cdot S}{k \cdot p} \cong \frac{L \cdot S}{\rho_{lv} a^2}$ is the acoustic compliance of the medium; a is the speed of sound; k is the adiabatic factor; R_a is the active acoustical resistance of fluid flow; $G_a = \frac{1}{k} \frac{W \cdot L}{p}$ is the wave conductivity of the medium; $m = \frac{\rho_{lv} \cdot L}{S}$ is the acoustic mass; ρ_{lv} is the density of the two-phase (liquid and vapor)medium.

These equations are known in the literature under the name of Telegraph equations [5]. The solution of the system of linear differential equations (1) gives functional dependences of sound pressure and volume flow of the compressed medium in the pipeline with distributed constant acoustic compliance of the medium, acoustic mass, wave conductivity of the medium, and active resistance of the medium flow on variables *x* (distance along the pipeline axis) and *t* (time).

The solution of system of linear differential equations (2) gives functional dependences of electric current (*i*) propagation along the cable and voltage (*u*) with distributed constant capacitance (*C*), inductance (*L*), conductivity (*G*) and ohmic resistance (*R*)

$$\begin{cases} \frac{\partial i}{\partial x} + C\frac{\partial u}{\partial t} + Gu = 0 \\ \frac{\partial u}{\partial x} + L\frac{\partial i}{\partial t} + Ri = 0 \end{cases}. \tag{2}$$

Due to the unity of the differential equations of acoustic (1) and electric (2) systems, the study of the propagation of the volume flow of a compressible single-phase or two-phase fluid in the acoustic system can be replaced by the study of the propagation of an electric current along the cable [4].

From the joint consideration of systems (1) and (2), a number of analogies between acoustic and electrical quantities follow. These analogies are shown in table1.

Table1. Analogies between acoustic and electrical parameters.

Acoustic system			Electrical system		
Parameter	symbol	measurement unit	parameter	symbol	measurement unit
Sound pressure	Δp	Pa	Voltage	u	V
Volume flow rate	W	m^3/s	Current	i	A
Acoustic compliance	C_a	m^4s^2/kg	Capacitance	C	F
Acoustic mass	m	kg/m^4	Inductance	L	H
Active acoustical resistance	R_a	kg/m^5s	Active resistance	R	Ω

The acoustic systems are introduced in the form of a set of electrical circuits and the resulting electrical circuit (equivalent) is analyzed by methods developed in electrical engineering. As shown in [6], in the system of electroacoustic analogy, a longitudinal element of a pipe is equivalent to a passive quadric pole made in the form of a power transmission line segment. Thus, the self frequency of liquid pressure fluctuations in the pipe is determined by the formula

$$f_0 = \frac{1}{2\pi\sqrt{mC_a}}. \tag{3}$$

For any segment *i* satisfying the conditions in [4, 6]

$$L_i \ll \frac{a_i}{f_{0i}}, \tag{4}$$

$$C_{ai} = \frac{L_i \cdot S_i}{\rho_i \cdot a_i^2}, m_i = \frac{\rho_i \cdot L_i}{S_i}, f_{0i} = \frac{a_i}{2\pi \cdot L_i}, \tag{5}$$

where a_i, m/s – the speed of sound in the liquid in the pipe; ρ_i, m/s – the density of the liquid; L_i, S_i – the length, and the cross-sectional area of the i-th section, respectively.

For n segments connected in series:

$$C_a = \sum_{i=1}^{n} C_{ai} \qquad m = \sum_{i=1}^{n} m_i \qquad f_0 = \frac{1}{2\pi\sqrt{m \cdot C_a}} \tag{6}$$

The problem of developing acoustic models of coolant motion in the main NPP equipments, its individual components and in a set of structural elements was first solved in [4]. Developed methods of modeling and calculation of acoustic waves in the coolant of nuclear power plants allowed to their author K. N. Proskuryakov create an acoustic model of a nuclear reactor and organize its testing and verification at a nuclear power plant with a light water reactor.

The radiation of elastic waves by the coolant flow was investigated in paper [7]. A method of acoustic measurements, has been developed to identify hazardous modes associated with the processes of spontaneous vortex formation and self excited swirling of the streams of coolant.The experimental method is based on the measurement of the amplitude-frequency characteristics of sound vibrations in a swirling jet. In contrast to traditional methods of non-destructive testing of thermal-hydraulic parameters, such as ultrasonic, radiation, and magnetic, a new method of acoustic resonances is used in this research work. This method allows to register the radiation of elastic waves by fluid caused by local dynamic reformation of its structure and to predict the appearance of dangerous periodic fluctuations, the frequencies of which are close to the natural frequencies of the NPP components.

It is experimentally established that when the maximum flow rate of the swirling flow is reached, a spontaneous restructuring of the amplitude-frequency characteristics of the acoustic oscillations of the hydromechanical systems occurs.When the frequency of vortex motion coincides with the natural frequency of oscillations of the elastic end surface of the vortex chamber, resonance (or critical) mode of flow mode occurs. The critical mode is characterized by a sharp increase in the amplitude of sound vibrations at the resonant frequency, and the suppression of oscillations at all other frequencies.

It was found that in the resonant flow mode, the flow decomposes into separate spiral vortices, the rotational frequency of which coincides with the natural frequency of bending vibrations of the end surface of the vortex chamber.

Some special series of experiments have shown that in the critical transition region corresponding to the frequency range of acoustic oscillations generated during the formation of a system of stable coherent spiral-vortex structures, the sound pressure recorded by the amplitude-frequency characteristics, increases by 2 to 3 orders of magnitude. Therefore, when the vortex frequencies approach to the natural frequencies of the hydro-mechanical system, the probability of detecting a dangerous resonance mode increases due to a sharp increase in the amplitude of the signal. The research results are aimed at the development of theoretical foundations and instrument complex to provide the technology of vortex diagnostics of the coolant flow states of nuclear power plants.

3. Development of methods for predicting and preventing acoustic oscillations of thecoolant,that lead to amplification of fuel rod and fuel assembly vibrations

As a result of the conducted research studies, some methods and algorithms of calculations have been already developed to predict and to prevent the conditions for the increase of vibrations of fuel rods and fuel assemblies due to acoustic vibrations of the coolant. The dependence of the bandwidth on the operation mode of the reactor plant is investigated. For this purpose, an acoustic model of the coolant in the core is used, in which the presence of the gas phase formed during radiolysis, is taken into account.

The algorithms for calculation of the quality factor for the coolant oscillating circuit and the bandwidth are given in paper [8], that makes it possible to specify the location of the fuel assemblies with an increased level of vibration in the investigated mode of operation of VVER-1000 core. It should be emphasized that the appearance of acoustic vibration frequencies and vibrations in the bandwidth is possible only at a certain value of the average temperature of the coolant in the fuel assembly.

The developed methods and algorithms can be used for quantitative estimates of the natural frequency of coolant pressure fluctuations, quality factor and bandwidth of the acoustic elements of the primary circuit and their connections, for existing and for projected power plants such as VVER, VVER-SCP (reactor cooled by water in super critical pressure) [8]. It is shown that in order to prevent resonance of coolant and fuel assembly vibrations, it is necessary and sufficient to bring the natural frequency of fuel assembly oscillations outside the coolant bandwidth.

3.1. Prediction of operating conditions that lead to vibro-acoustic resonances of the coolant with vibrations of FA and fuel rod

The data related to parameters of hydrodynamic loads, vibro dynamic behavior and stress-strain state of the reactor internals (RI) of VVER-1000 equipments in stationary and transient conditions are provided in paper [9]. These data were obtained during the implementation of a set of laboratory, test loop and in-core studies of theRI and FA of VVER-1000 (including physical models of various scales, as well as the full scale assembly and in-core components). The factors determining the hydrodynamic loads and vibro dynamic behavior of the RI and FA of VVER-1000, were identified as a result of many years of experimental and computational work. The studies have confirmed that the main factors determining the parameters of hydrodynamic loads and vibro-dynamic behavior of the equipments are the thermal-hydraulic characteristics of the primary circuit, the combination of operating main circulation pumps, as well as those factors not considered at design stage.

It is noted in [10], that the problem of vibrations at the NPP arose due to the fact that at the design stage of the first reactors, the dynamic impact of the coolant flow on mechanical vibrations of structural elements had been underestimated. However, ensuring the vibration strength of the equipments and pipelines of the reactor plant continues to be an actual problem. There are several reasons that make it difficult to solve this problem, the main of which, is the impossibility of forecasting of exciting vibration impacts with acceptable accuracy at the design stage, as well as due to following reasons: the lack of reliable data on the characteristics of the cyclic strength of structural materials for the hypercycle zone of the fatigue curve, high sensitivity of vibrations to small changes in design, external influences, the variety of structural forms of equipment elements, the reasons behind vibration excitations and their mechanisms.

The formation of acoustic field of the reactor plant, is mainly caused by self-oscillations of the coolant flow that are generated in the acoustic elements of the primary circuit. The methods for the determination of ASW, are provided in [11].

The acoustic elements of the primary circuit of VVER 440 NPP [12] are shown in the figure 1. The results of the calculation of the frequencies of ASW in the first circuit of a VVER – 440 NPP, are given in table 2 [12].

The software package of the vibration control system of the main equipment named SUS, is currently implemented and used in Novovoronezh NPP successfully; it consists of a program for automatic process of spectra and a program for automatic selection of peaks in the vibration spectra [13].

The results of the calculation of the frequency of ASW for all stages are compared with the measured data of vibration of the main equipment of the primary circuit that are provided in [14]. The results of the comparison indicate that in the studied stages of start-up modes, resonances of the ASW with vibrations of the equipments can occur in the following elements of NPP equipments:

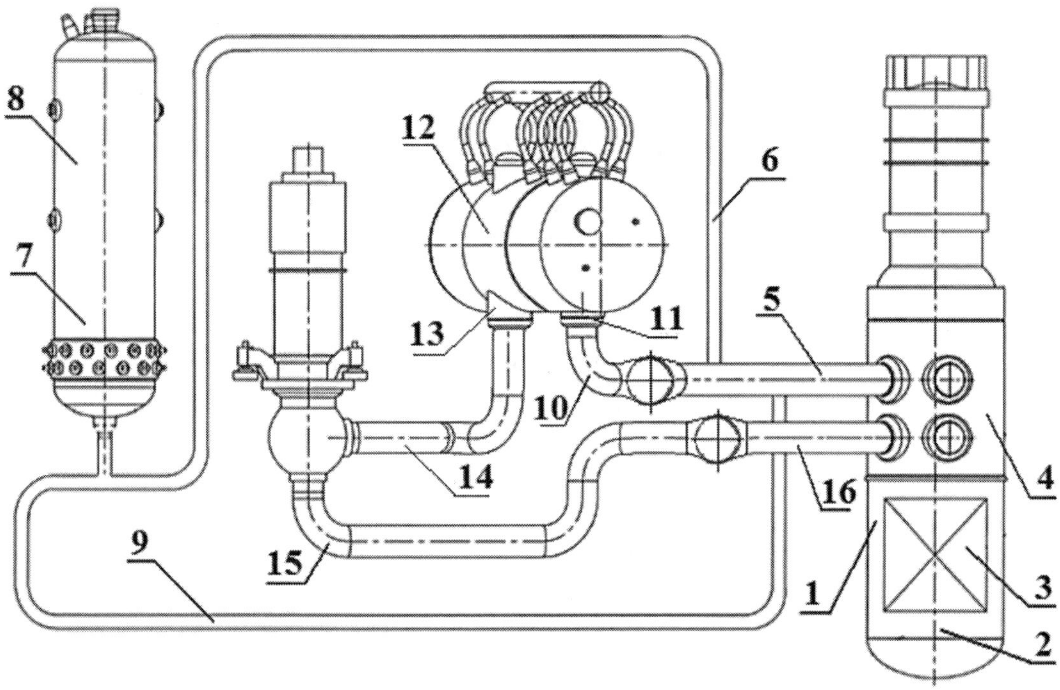

1 – downcomer region of the reactor; 2 – lower grid;3 – reactor core; 4 – upper grid ; 5 – section of the main circulation circuit (MCC) from the reactor to the main gate valve (MGV); 6 – surge line from the hot leg to the pressurizer (P); 7 – pressurizer (water side); 8 – pressurizer (steam side); 9 – surge line from the pressurizer to the hot leg; 10 – section of the hot leg ofMCC from the MGV to the hot collector; 11 – hot steam generator(SG) collector ; 12 – heat exchange surface of the SG; 13 – cold SG collector; 14 – section from cold SG collector to main circulation pump (MCP); 15 – cold leg section of MCC from MCP to MGV; 16 – cold legsection of MCC from MGV to the entrance of downcomer region of the reactor

Figure 1. Acoustic scheme of VVER-440 NPP.

Table 2. The results of the calculations of ASW frequencies in sections of acoustic scheme of first circuit.

Parameters of mode 1	$T = 256\ ^0C$, $P = 12{,}7\ kg/cm^2$, $H=4.34\ m$							
Section No.	1	2	3	4	5	6	7	8
Frequency, Hz	18.3	44.1	71.7	28.6	20.4	7.1	23.7	11.5
Section No.	9	10	11	12	13	14	15	16
Frequency, Hz	6.7	43.6	42.5	19.3	42.6	20.7	56.9	36.3
Parameters of mode 2	$T = 268\ ^0C$, $P = 125\ kg/cm^2$, $H=3.37m$							
Section No.	1	2	3	4	5	6	7	8
Frequency, Hz	17.7	43.1	69.2	27.6	19.5	6.8	29.5	9.9
Section No.	9	10	11	12	13	14	15	16
Frequency, Hz	6.4	41.8	40.6	18.5	41.3	20.1	55.3	35.3

1.5 - 3.0 Hz – frequency of vibrations of FA with one fixed end are observed:
• in combination circuit consisted of several numbers of sections, (16+1+2+3+4), from cold leg of MGV to core exit in the range of temperature values, at the reactor outlet, of 281 - 296 °C and pressure values of 12.1 - 12.3 MPa;
• in combination circuit consisted of several numbers of sections, (16+1+2+3+4+5), from cold leg to hot leg of MGV in the range of temperature and pressurevalues , at the reactor outlet ,of 60 - 296 °C and 1.8 - 12.3 MPa, respectively.
4.0 - 6.0 Hz – frequency of vibrations of FA with two fixed ends are observed:
• in the combinational circuit consisted of several numbers of sections, (16+1+2), from MGV cold leg to the bottom of the core in the range of temperature and pressure values, at the reactor outlet, of 60 - 296 °C and 1.8 - 12.3 MPa, respectively;
• in the combination circuit consisted of several numbers of sections, (1+2+3+4), from the beginning of the downcomer region of the reactor to the reactoroutlet in the range of temperature and pressure values, at the reactor outlet, of 60 °C - 136 °C and 1.8 - 4.8 MPa,respectively;
8.0 - 12.0 Hz – 2nd mode oscillations of FA with two fixed ends are observed:
• in the combination circuit consisted of several numbers of sections, (1+2+3), from the beginning of the downcomer region of the reactor to the top of the core in the range of temperature and pressure values,at the reactor outlet , of 136 - 296 °C and 4.8 - 12.3 MPa, respectively;
• in the combination circuit consisted of several numbers of sections, (1+2+3+4), from the beginning of the downcomer region of the reactor to the exit from the reactor in the range of temperature and pressure values,at the reactor outlet, of 60 - 136 °C and 1.8 - 4.8 MPa, respectively.

4. Prediction of vibroacoustic resonances in the reactor cores of VVER NPP in case of loss of coolant accidents

To calculate the self-frequency of coolant pressure oscillations in the steam-generating channel, the following relation is obtained in [4]

$$f = \frac{a_{lv}}{2\pi}\sqrt{\frac{\rho_{lv}}{\rho_l}}(L_l \cdot L_{lv})^{-0.5}, \quad (7)$$

where ρ_{lv} is the density of the coolant in two-phase region; ρ_l is the density of the coolant in single-phase region; L_l is the length of section in single-phase coolant; L_{lv} is the length of section in two-phase coolant; a_{lv} is the speed of sound in two-phase medium; f is self frequency of coolant pressure oscillations.

In paper [15] it is shown that the calculated values of the frequencies of the ASW obtained by using the relation (7), that was developed for analysis of acoustic systems with two-phase fluidmedium [4], coincide with the results of measurements of the pressure fluctuations of the coolant in the reactor core.The experimental evidence that was given in [15], and the justification of the use of the method of electroacoustic analogies for the calculation of the frequencies of ASW in boiling reactors, that was developed in [4], are now of particular importance for improving the competitiveness and safety of nuclear power plants. Prediction and control of frequencies of ASW at different stages of accidental conditions in boiling condition of core coolant, will optimize the emergency cooling system of the reactor and manage severe accidents and reduce capital costs for the construction of NPPs. The

developed methods and algorithms of calculations are applied in [8] to quantify the range of change of self frequency of coolant pressure oscillationsin VVER-1700 (reactor with fast-resonance neutron spectrum) and VVER-1200.
In paper [9] it is shown that the interdisciplinary approach is necessary for the solution of actual current problems as for calculation of the ASW frequencies of the coolant in a reactor core in different conditions of maneuverable, transitional and accidental modes.

One of the ways to suppress the self-oscillations of the coolant is the use of dampers, which are devices that filter the acoustic vibrations of the working fluid and dissipate their energy. The Helmholtz resonator is one of the most effective dampers. When designing a Helmholtz resonator, it is important to take into account the effect of the vortex flow in its path. For this purpose, it is advisable to use the techniques of finite element modeling, with which the calculated justification of the damper scheme and the optimal values of its acoustic characteristics are determined. This method is used by Areva company [16], which in 2013 developed a complex program for the comprehensive study and analysis of the vibro-stressed state of NPP equipments, its internals and pipelines. The program includes measurements of the vibro-stressed state of equipments and pipelines, and determination of the safety margin for material strength in the stationary state and transient conditions.

In this program, it is noted that in assessing and predicting the service life, it is necessary to take into account the hydrodynamic and acoustic parameters of the coolant flow, as well as the current state of the equipment contacts with the RI and with the construction structures. The disadvantage of this approach is the dependence of the results on the current state, which makes it difficult to predict the service life, and the complexity and high cost of evaluation calculations.

In contrast to this approach, in paper [17] it is proposed an alternative option to ensure carrying ability of the nuclear installation equipment design, which provides for preventing the occurrence of conditions which lead to the maximum possible dynamic loads caused by vibration-acoustical resonance.

As the most effective means of preventing vibration-acoustical resonance, it is proposed to use an acoustic filter of the Helmholtz resonator type. This installation will provide suppression of acoustic fluctuations of the coolant coinciding with natural frequencies of fluctuations of fuel assemblies and fuel rods, and the excited fluctuations of pressure caused by the main circulation pump rotation or external and seismic impacts.

5. Conclusions

It is proved that the pressure fluctuations of a one-dimensional pulsating flow of a compressible single-phase or two-phase fluid used as a coolant of a nuclear reactor can be replaced, through investigation, by the similar voltage fluctuations in electric circuit lines.

The developed mathematical models are versatile, quite simple and, as evidenced by the results of measurements at NPPs, are effective for determining the acoustic properties of complex systems withanaccuracy sufficient to solve a number of important practical problems.

Developed methods of modeling and calculation of acoustic waves in the coolant, allowed to their author K. N. Proskuryakov to create an acoustic model of a nuclear reactor and organize its testing and verification at a nuclear power plant with a light water reactor.

The obtained results can be used to predict the occurrence of vibration-acoustic resonances in NPP equipments under operating and accidental conditions, as well as under seismic and shock impacts.

References

[1] Proskuryakov K N 2005 Vibroacoustic certification of nuclear power plants-an urgent problem of nuclear power plants *Thermal engineering*(Moscow, Russia) **12** pp 30-34

[2] Charny I A, 1951 *Unsteady Movement of a Real Fluid in Pipes* (Moscow: State publication of Technical –Theory Literature)

[3] Lependin L F 1978 *Acoustics The Manual for Higher Educational Institutions* (Moscow: High School Publishimg House) p 448

Journal of Physics: Conference Series **1421** (2019) 012041 doi:10.1088/1742-6596/1421/1/012041

[4] Proskuryakov K N 1984 *Thermal-hydraulic Excitation of Vibrations of the Coolant in ReactorInternals of the Nuclear Power Plants* (Moscow: MPEI) p 67
[5] Atabekov G I 1970 *Theoretical Foundations of Electrical Eengineering Part 1* (Moscow: Energia)
[6] Olson G 1947 *Dynamic Analogies* (Moscow: State publishing house of foreign literature)
[7] Pozdeeva I G and Mitrofanova O V 2018 Diagnostics of the vortex structure of coolant flows based on the method of acoustic resonances *Thermophysics and Aeromechanics* (Russia , Obninsk) p 182
[8] ProskuryakovKN and Novikov K S 2010 Determination of the region of vibro-acoustic resonances of the coolant and fuel assemblies in future high-power reactor *Journal of Atomic Energy* (Russia, Moscow) **3** pp 151 – 155
[9] Dragunov Y G, Dranchenko B N, Abramov V V and Khairetdinov V U 2007 *Proc. of Con. On Safety Assurance of VVER NPP* (Podolsk, Russia) Vibro dynamic research in support of design decisions for VVER, Materials of the conference " Safety provision in VVER NPPs" JSC EDB "GIDROPRESS", section No. 2
[10] Shary N V 2008 Calculation methods of substantiation of the strength and dynamics of reactor plant structures for VVER NPP author's abstract of thesis submitted for scientific degree of Ph. D (Podolsk, Russia)
[11] Proskuryakov K N 1999 *The Use of Vibro Acoustic Noise for the Diagnosis of Technological Processes in Nuclear Power Plants* (Moscow: MPEI publishment) p 68
[12] Proskuryakov K N, Fedorov A I, Zaporozhets M V and Volkov G Y 2016 Study of acoustic standing waves in the primary circuit of NPP VVER-440 NPP in the start-up modes *Global Nuclear Safety* (Russia, Volgodonsk) **2** (19) pp 59 – 69
[13] Slepov M T 1999 Development of methods and interpretation of data with regard to the noise diagnostics of the reactor units of Novovoronezh NPP,author's abstract of thesis submitted for scientific degree of Ph. D(Obninsk, Russia)
[14] Arkadov G V, Pavelko V I and Usanov A I 2004 *Vibration Diagnostics of VVER* (Moscow: Energoatomizdat) p 344
[15] Fomichev M S 1989 *Experimental Hydrodynamics of Nuclear Power Units* (Moscow: Energoatomizdat) p 247
[16] ANP-10306NP 2013 Comprehensive vibration assessment program for internal parts of U.S. EPR reactor ,Technical report
[17] Proskuryakov K N and Zaporozhets M V 2016 Study of acoustic vibrations in reactors and prospects of their use for the justification of residual life *Vestnik MEI* **5** pp 19-24

Journal of Physics: Conference Series **1421** (2019) 012042 doi:10.1088/1742-6596/1421/1/012042

Determination of parameters of the compelled capillary disintegration of liquid jets

A V Boukharov and S I Kukanov

Department of low temperatures, National Research University «MPEI», Russia, 111250 Moscow, Krasnokazarmennaya, 14

E-mail: boukharov@mail.ru

Abstract. The technique of the automated precision parameters diagnostics of the compelled capillary disintegration of liquid jet is developed for the solution of heat-physical problems of receiving the stable monodisperse drop streams. The analysis of the images received from two digital cameras, located at the angle ninety degrees to each other is the basic of this procedure. One of cameras fixes the image of a drop flow in plane *X*, another - in plane *Y*. The minimum time of storing images is ~0.1 microsecond. By means of specially developed software in the "on line" mode it is possible to process images and to determine the following parameters: jets speed, drops speed, jets length, jets diameter and drops diameter. Testing on objects with the known geometrical sizes confirmed working capacity and efficiency of the technique and the software. The relative error of parameters determination does not exceed 0.5% for jets with length of several millimeters. The developed technique and the software allow to increase the accuracy and reliability of parameters determination of compelled capillary disintegration of liquid jets and drop streams and expand possibilities of parameters control of setups for receiving monodisperse streams.

1. Introduction

The basic trend of modern development of disperse technologies is passes from disperse systems with wide spacing in parameters to the monodisperse systems consisting of spherical particles flows of a different phase status with small dispersion by the sizes and the speed. The small dispersion by the sizes and speed allows not only to improve considerably existing technological processes, but also to create new technologies. Now the monodisperse streams and technologies developed on their base are used in the following fields of science and technology: scientific instrument making, space technologies, power and electronics, biology and medicine, ecology. The detailed description of the technologies using the monodisperse streams is submitted in [1-7].

Structurally and technologically the installations for receiving the monodisperse streams can differ from each other, however, they should have the following necessary elements: the monodisperse drops generator with a nozzle, the working chambers and sluices connecting working chambers. The description of some types of the installations is submitted in [8, 9]. A theoretical basis of receiving stable monodisperse drops stream is the theory of compelled capillary disintegration of Rayleigh-Weber (the theory of CCDJ) [10].

There are many methods of definition of separate characteristics of CCDJ: jets length; jets speed and drops speed; jets diameter and drops diameter; deflection angle of jets from the vertical. However for receiving the stable monodisperse streams it is necessary to know not separate characteristics, but all characteristics of CCDJ and drop streams at the same time. The technique of the automated definition of CCDJ and drop streams characteristics was developed.

2. Technique of determination of the CCDJ and drops stream parameters

Underlying the technique is processing the image of the streams and drop flows. The full technique consists of the following parts: image acquisition, image processing and definitions of CCDJ and drop streams characteristics.

2.1. Image acquisition technique

Two digital video cameras PCO-PixelFly of Pro.imaging located at an angle ninety degrees to each other and the illumination system consisting of two pulsed light sources are used for the definition of CCDJ characteristics (figure 1). One of cameras fixes characteristics of CCDJ and monodisperse drops in plane X (camera X), another — in plane Y (camera Y).

There are following elements: CCD matrix, analog-digital converter and fast internal memory in each video camera. Cameras connect to the computer by means of RS-232 interface.

The illumination system consists of two pulsed light sources. Synchronous operation of the cameras and light sources is carried out by means of the external or internal generator which start the video cameras and the illumination system at the same time. The full cycle of cameras operation consists of two parts: mode of picture storage and reading mode.

In the storing mode the image of a jet, drop stream and additional objects is focused on matrixes of video cameras, and are remembered in fast memory in the form of separate frames. Usually additional objects which image is remembered together with the image of jets and drops are: an output nozzle of the drops generator and the sluice through which drops pass between working chambers (figure 1). Storing time of a frame is ~0.1 microsecond. Thanks to short storing time the jet and drops are displaced slightly in space, and the image turns out bright and contrast.

In the reading mode the frames are consistently read out from an internal memory of cameras and sent to memory of the computer. The minimum reading time of a frame is 40 msec.

2.2. Image processing technique

Read frames are located to the respective area of memory X (folder X) or Y (folder Y) in the computer. For selection of information on the CCDJ parameters, each frame is exposed to special processing which consists of the following stages: contrasting, selection of the closed dark bodies, identification of the selected bodies and scaling.

Contrasting of the image is made at an initial stage. Contrasting consists in the analysis of brightness value of the image and the background in each point of a frame. If point brightness is less than background brightness, then this point is defined as black. Otherwise, the point is defined as white. This operation is necessary for receiving a clear boundary between a background and the studied objects: jets, drops and other support entities.

The search of the closed dark bodies which are in limits of one frame begins after contrasting. The identification of separate bodies: the nozzle, a jet, drops, the sluice is made after their selection.

The nozzle is understood as the body touching the upper bound of the frame and having the relation of depth dimension to lateral dimension smaller 1.

The sluice is understood as the body touching the lower bound of the frame and having the relation of depth dimension to lateral dimension more than 1.

If the lower bound of a nozzle has a sharp reduction of width, then all lower part is recognized as a jet.

Objects in the drop form have no contacts with frame borders, for their selection the additional selection condition – rotundity coefficient is imposed. The rotundity coefficient of a drop indicates proximity of the found body circuit to an ellipse with the same vertical and horizontal size.

The final stage of processing is scaling. As a result of the made experiments the determination of scaling ratio by the known sizes of two bodies (the nozzle and the sluice) and the fixed distance between them was the most optimum. In this case borders of all bodies are defined steadily and at single coefficient of increase the error of delimitation makes ±3 microns.

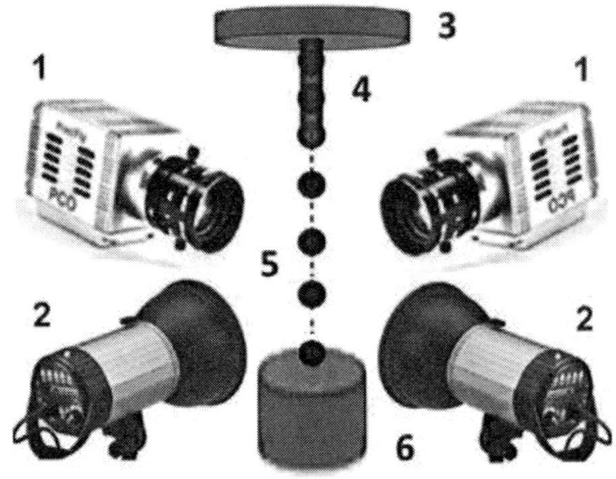

Figure 1. Arrangement of the optical system elements by definition of CCDJ and monodisperse drops characteristics: 1 – digital video cameras, 2 – pulsed light sources, 3 – the nozzle, 4 – a jet, 5 – drops, 6 – the sluice.

2.3. Technique of definition of CCDJ and drop streams characteristics

Definition of CCDJ and drop streams characteristics: jet lengths, deflection angle of jets from the vertical, diameter of a jet and drops, speeds of a jet and drops occurs as follows.

The beginning of a jet is accepted the line with sharp change of image width. At first the right and left borders of a jet is defined for this line, and then – position of the jet center. Jet diameter is equal to the difference of coordinates between the right border and left. The end of a jet is accepted the line in which the right and left borders of the image coincide. Jet length is defined as the difference between the coordinate of the initial center of a jet and the coordinate of a point of borders coincidence.

Definition of deflection angle of jets axis from the vertical for each of the planes comes on the following algorithm. At first the right and left borders of a jet are defined for each line of the image. Then the position of the jet center is defined for every line. The deflection angle from the vertical is defined as a result of approximation of all jet centers coordinates.

The bodies which do not have contacts with image borders are accepted to drops. At first the right and left borders are defined for each such object line by line in each plane. Then boundary points for each plane are approximated by means of the least-squares method by circles. As a result of approximation we receive the following characteristics: diameter of a drop and coordinate of each drop center. Coordinates of next drop centers are used for determination of drop speed in planes X and Y and full speed of a drop.

3. Description of the program of definition of CCDJ and drop streams parameters

The technique described above was a basis for creation of the program for image processing and definition of CCDJ and drop streams parameters in "on line" mode [11, 12]. At the same time the program processes the images from two cameras which are conditionally called X and Y. The block – scheme and the working window of the program are submitted in figure 2 and 3.

The block– scheme of the program includes the following stages: task of initial parameters, obtaining the image from the camera, definition of transition point from light to dark for every line, the analysis of transition points and definition of full border of all dark bodies, definition of bodies type, determination of jet and drops parameters, an output of results.

At each camera it is possible to configure the following settings: permission in horizontal direction, permission in vertical direction, sensitivity, type of start (external or internal), digit capacity, exposure, scale, interval between frames, operation mode (framing or video mode), image recording, image processing parameters. The choice of necessary parameters is made in the corresponding window of the program (figure 3).

In the upper window of the program it is possible to configure the following settings: permission, sensitivity, a type of start, digit capacity in bits and exposure.

The buttons with plus signs (increase) and minus (reduction) are used for change of the image scale in program window.

In the lower right window of the program it is possible to select an operation mode of cameras. “Frame” button allows to receive the single image, “Video” button - the image from cameras through a certain interval. The time interval between frames is specified in the lower part of the window.

The “Processing Parameters” mode is selected in the upper working program window for setup of parameters of image processing. In this mode it is possible to configure the following settings: preprocessing, parameters, work with objects.

Preprocessing allows to cut out the most informative fragment of the image from a full frame and to use only this fragment in the course of further image processing. For convenience of postprocessing this fragment can be transformed in addition. The following opportunities are provided in the program: turn of a fragment on a corner, to multiple 90 degrees, and reflection of this fragment in horizontal or vertical direction.

“Parameters” button allows to select the level of a binding to object border in the contrasting mode.

After contrasting of the image the program carries out search of the closed dark bodies and their identification. For correct algorithms execution it is necessary to specify the following parameters in the section "Work with Objects": minimum area, squareness coefficient, body width relation to its height, coefficient of change of body width, rotundity coefficient.

The “Calibration” mode is intended for determination of scaling ratios on each plane. In addition it is necessary to specify the liquid density and exciting frequency for determination of weight, volume and speed of drops. The scaling ratio decides by comparison of the known sizes of the nozzle, the sluice and the fixed distance between them in microns on the corresponding sizes in pixels. The sizes of the nozzle, the sluice and the fixed distance between them are entered in the editor located below on a tab. If necessary it is possible to set scale manually, for this purpose it is necessary to put “Specify Manually” checkbox. The received scaling ratios are used for determination of jet length, the diameter of jet and drops, the speed of a jet and drops, deflection of nozzle axis and sluice axis, deflection of jet axis and sluice axis.

A set of tabs in the left lower part of a primary window of the program allows to select the current output mode of results of a program runtime. The type of tabs for some modes is presented in figure 3. “Viewing” tab allows to browse the images received from cameras without their processing. By means of “Disk writing” button it is possible to write the specified number of image files on a disk in the specified folder (the folder is selected by a folder image button). Names of files form from date and time of obtaining the image. The plane in which the camera is located (X or Y) is specified at the end of a name. It is possible to interrupt the sequence of record with “Interrupt” button.

The “Adjustment of Cameras” mode is intended for alignment of images and scales of both cameras. In this mode the following data are programmatically defined and output: data on nozzle height in pixels, nozzle thickness in pixels, stream width in pixels, deflection angle of a jet or a nozzle from the vertical. At coincidence of the sizes on images of both cameras and corners equal 0, cameras are considered as aligned.

In “Nozzle Adjustment” mode the values of space shift of nozzle axis from sluice axis are programmatically defined and displayed.

In "Jet Adjustment" mode the values of space shift of jet axis from sluice axis are programmatically defined and displayed. Besides, the average length of not broken up part of a jet is defined and displayed.

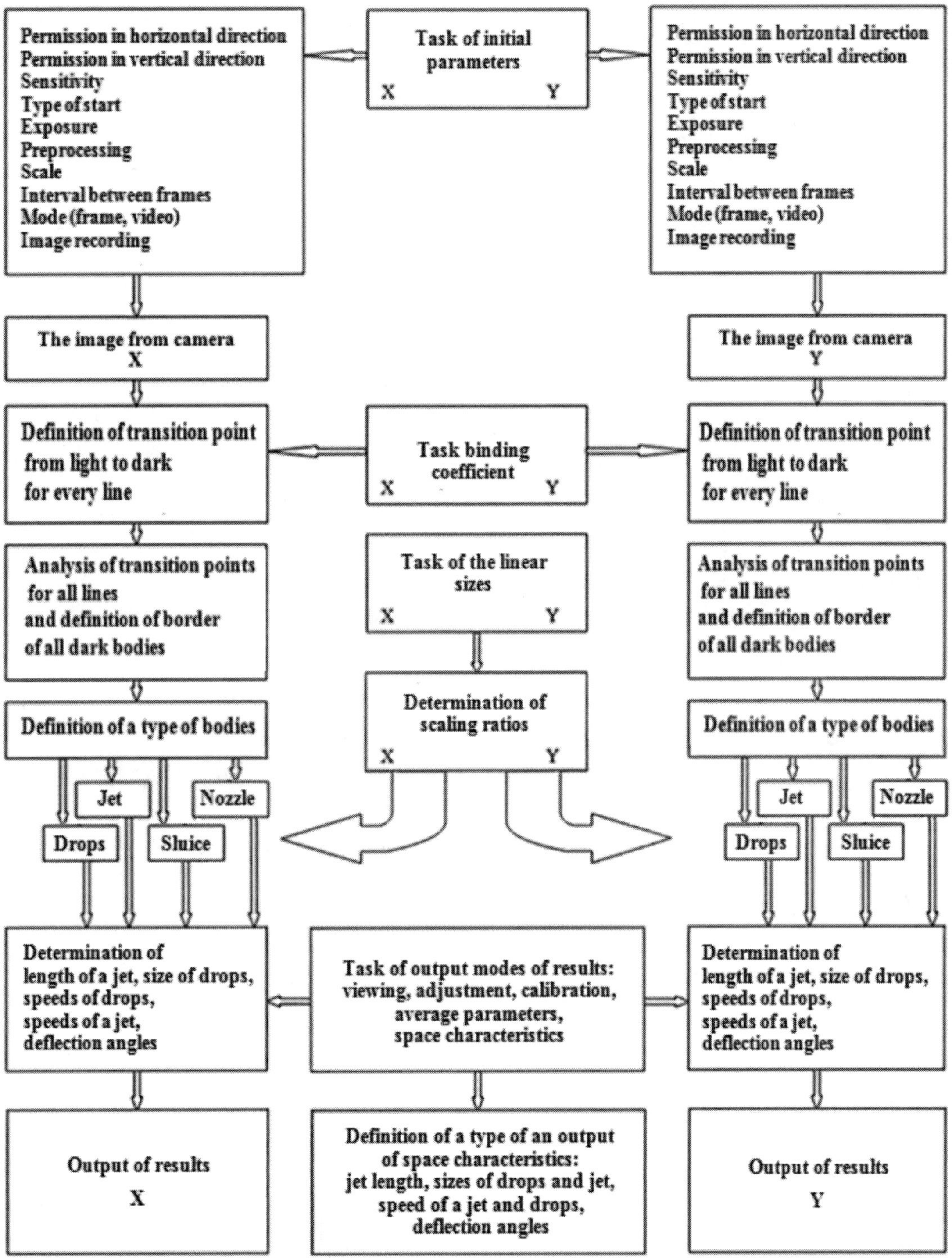

Figure 2. Block scheme of the program for determination of CCDJ and drop streams parameters.

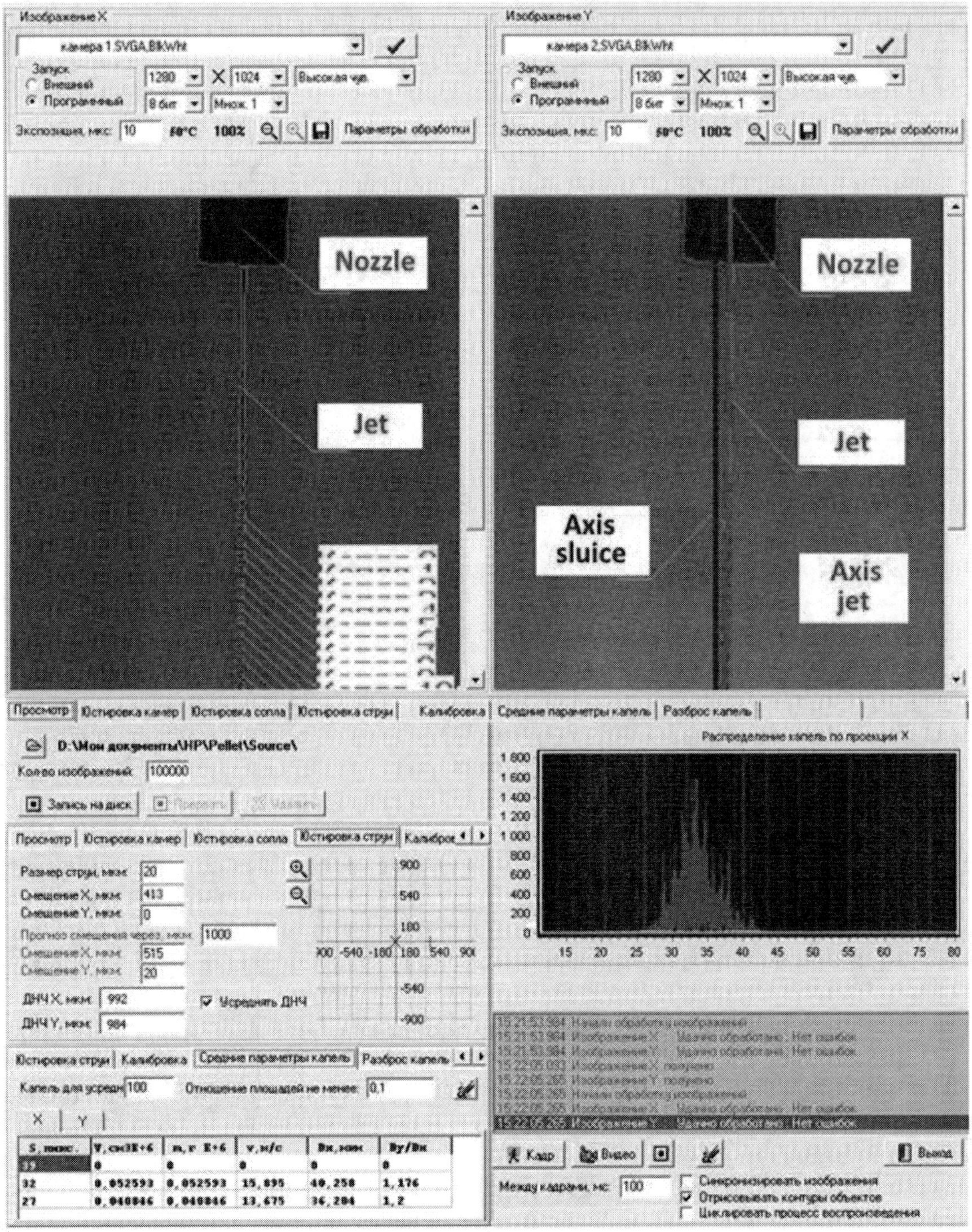

Figure 3. Working window of the program for determination of CCDJ and drop streams parameters.

"Average Parameters of Drops" mode is intended for definition of mean values of drops: volume, weight, speed and diameter. For determination of drops mass it is necessary to specify liquid density in addition. The quantity of drops for averaging is set in the corresponding editor.

"Dispersion of Drops" mode is intended for accumulation of statistics and determination of dispersion on drops diameters. The quantity of drops for averaging is specified in the corresponding editors. Preliminary results are displayed in the right lower part of a primary window of the program (figure 3).

"Draw Circuits of Objects" checkbox in the lower part of a primary window of the program allows to apply circuits and axes of the received bodies on the image.

"Exit" button is intended for completion of work with the program.

4. Testing of the technique and software

For testing of the technique and the software the object with the known sizes was located in focus of a lens of video cameras, and its geometrical sizes experimentally were defined. Knowing the difference between experimentally measured geometrical sizes of the object and its exact sizes, it is possible to define an error of the technique and the software. The glass fibers of different diameter (from 100 μm to 400 μm) and different length (from 0.5 mm to 1 mm) were use as experimental objects. The error of determination of length and diameter made ±0.1 μm. The upper end of the fiber fastened to an output nozzle of the drops generator, and lower end connected to a small load.

By means of the special justified screws installed on the drops generator in "Adjustment of Cameras" mode the maximum coincidence of cameras axes to drops generator axis and to fiber axis was carried out. Then in "Jet Adjustment" mode were defined: average diameter of fiber, length of fiber and measurement error.

As a result of processing of experimental data it was established that the relative error of determination of length for jets over one millimeter long does not exceed ±0.5%. For shorter jets the error increases and for jets 100 μm long reaches ±5%.

Thus, testing confirmed working capacity and efficiency of the developed technique of definition of CCDJ and drop jets characteristics.

5. Conclusion

The technique of the automated precision diagnostics of CCDJ and drop streams parameters is developed for the solution of heat-physical problems of receiving the stable monodisperse drop streams.

The analysis of the images received from digital cameras, located at the angle ninety degrees to each other is the basic of this technique. In the "on line" mode it is possible to define the following characteristics by specially developed software: jet speed, drops speed, jet length, jet diameter, drops diameter, an angular deviation of jet axis from the vertical.

Testing of the technique and the software on objects with the known geometrical sizes confirmed their working capacity and efficiency. For jets of several millimeters the relative error of determination of jet length, drops speed, drops size and jet speed does not exceed 0.5%. The error of definition of an angular deviation of jet axis from the vertical does not exceed 0.003 rad.

The developed technique and the software allow to increase the accuracy and reliability of determination of CCDJ and drop streams parameters and significantly expand possibilities of control of installations parameters for receiving monodisperse streams.

References

[1] Ametistov E V and Dmitriev A S 2002 *Monodisperse Systems and Technologies* (Moscow: Moscow Power and Engineering Institute Press) p 390

[2] Gorodov A K, Bukharov A V, Ginevsky A F and Dmitriev A S 1996 Cryodisperse technology of receiving new materials *J. Conversion* **8** 28

[3] Bukharov A V and Dmitriev A S 2013 *Cryogenic Corpuscular Targets in Power* (Moscow: Moscow Power and Engineering Institute Press) p 144

[4] Bukharov A V, Ametistov E V, Ginevsky A F and Bukharova M A 2013 Heat-physical prob lems of cryogenic corpuscular targets *Questions Atomic Science and Technology (Ser.Nucl. and Phys. Res.)* **6** 205

[5] Bukharov A V, Buscher M and Ginevsky A F 2010 Use of a cryogenic corpuscular target for experiments on laser acceleration of the charged particles *Questions Atomic Science and Technology (Ser.Nucl. and Phys. Res.)* **2** 60
[6] Mattick A and Hertzberg A 1981 Liquid droplet radiators for heat rejection in space *J. of Energy* **6** 387–93
[7] Koroteev A A 2008 *Drop Refrigerators - Radiators of Space Power Installations of New Generation* (Moscow: Mechanical Engineering Publishing House) p 159
[8] Bukharov A V and Bludov A 2006 *J. Vestnik MPEI* **4** 11
[9] Bukharov A V and Buscher M 2006 *J. Vestnik MPEI* **2** 16
[10] Rayleigh Lord 1892 On the Instability of Cylindrical Fluid Surface *Phil. Mag.* **34** 177–87
[11] Bukharov A V and Kukanov S I 2013 *Proc. 19th Int. Conf. students (Moscow)* vol 4 (Moscow: Moscow Power and Engineering Institute Press) p 55
[12] Bukharov A V and Kukanov S I 2015 Program for definition characteristics of liquid streams and drops "JET-1" (Moscow: Moscow Power and Engineering Institute) 2015616003/69

Visualization and PIV-study of the formation of a cooling film when flowing around the leading edge of the gas turbine nozzle blade

S V Veretennikov and O A Evdokimov

Department of general and technical physics, Soloviev Rybinsk State Aviation Technical University, Pushkin st.173a, Rybinsk, 152934, Russian Federation

E-mail: serveret@yandex.ru

Abstract. The results of particle image velocimetry near the perforated leading edge of the nozzle vane of gas turbine are presented. Visualization of the formation of cooling film on the surface of the blade with determination of its thickness is carried out. On the basis of the obtained results, the influence of the blown cooling air mass flow rate on the formation of the film is analyzed.

1. Introduction

One of the main problems of organization of effective cooling leading edges of nozzle blades of the first stages of gas turbines is providing a stable cooling film, which pushes the flow of hot gases from the surface of the blade. It is known that during film cooling, the behavior of the coolant jets significantly depends on such parameters as the blowing angle of the jet, its velocity and density. The last two parameters are taken into account by the blowing parameter m.

The real conditions of discrete blowing coolant through the rows of holes require preventing contact of the flow of high-temperature combustion products with the surface of the blade and escaping its burnout, therefore it is necessary to provide mixing of jets within the boundary layer and to avoid deep penetration of the coolant jets into the core of the blowing flow.

In case of cooling of a leading edge having the advanced perforation in the form of several rows of holes, the organization of effective blowing of the coolant is complicated by significant gradients of static pressure and gas velocity in regions of blowing the coolant. This is due to the peculiarities of the main flow on the blade. In this case, the holes in the adjacent rows are in significantly different conditions for the available pressure gradient and therefore the speeds of the coolant and the blowing parameters m will be different. This problem can be solved by redistributing the flow of the coolant through the holes on the basis of varying the hydraulic resistance of the supply paths. This will provide the required values of the blowing parameter corresponding to the optimal range of effective film cooling.

Until recently, the main method of experimental study of the film cooling of the blades was thermal bench tests. However, the temperature conditions of the walls, as well as the presence of burnout, soot, etc., can be considered as practically the only criteria for the cooling efficiency of the blades according to the results of such tests. At the same time, the reasons of the defects are not always obvious, since the thermal state of the blades is subject to changes due to non-stationary modes of operation of other engine elements, in particular the combustion chamber.

Possible solution of the problem of the formation of high-quality thermal protection of the blade surface is detailed study of the flow structure near the leading edge of the blade with determination of velocity vectors and visualization of the convective film in different modes by the blowing parameter m. Such studies cannot be carried out by traditional gas-dynamic measurement methods based on direct contact of sensors with the diagnosed flow since the introduction of even small sensors into the flow can completely change its structure. For this reason, these measurements should be carried out by non-contact optical methods, which have become widespread abroad, and in the last decade in Russia. The most famous among them are the methods of particle image velocimetry (PIV), particle tracking velocimetry (PTV), as well as high-speed video and optical visualization by inert impurities.

PIV method belongs to the class of non-contact optical methods for measuring velocity in flows [1, 2]. Among other tools for the study of the flow structure, it occupies a special place due to the ability to record instantaneous spatial velocity distributions. This advantage is particularly important in the study of flows containing large-scale vortex structures, information about which is partially lost when using single-point diagnostic methods.

The study of gas-dynamic parameters of flow in turbines using PIV has been actively developing since the early 2000s. At the initial stage, non-contact measurements of the velocity vectors were carried out during the flow around the blade models. However, the full advantages of non-contact measurements were revealed in the study of eddy formation in edge traces, the structure of secondary flows and features of interaction with the coolant jets.

A sufficiently large number of PIV studies have been performed to study stator-rotor interaction in turbine stages [3-5]. These papers contain the results of measurements of instantaneous and averaged velocities, as well as the parameters of the vortex created by the separation of the flow from the trailing edge of a blade. Some papers [6, 7] show the possibility of applying PIV to identify the secondary flow in the interblade channels. The measurements of the velocity allow to estimate intensity of secondary flows and their influence on aerodynamic losses. The paper [8] presents fairly detailed results of PIV studies of the formation of coherent vortex structures in the boundary layer on the surface of a low-pressure turbine blade.

Detailed PIV measurements of the velocity near the model of the trailing edge of the blade with film cooling for the purpose of studying the mechanism of generating aerodynamic losses are given in studies [9-11]. It should be noted that almost all such studies are limited to the use of 2D PIV, and measurements are carried out on scaled models of blades to facilitate the formulation.

In the experimental study [12], which is closest to the subject of this paper, the local velocity field for a package of nozzle blades of a high-pressure turbine with film cooling was analyzed using the methods of flow visualization and PIV measurements. The visualization was performed near the flow deceleration point of the leading edge of the blade for a more detailed understanding of the flow structure, while the PIV method was used to measure the velocity field of the coolant jets at the outlet of the perforation holes in the 2D vertical plane.

This paper deals with an experimental study of flowing around the leading edge of the nozzle blade of a high-pressure turbine with an advanced perforation of film holes at different flow rates of the coolant. The main purpose of the work is to identify the structure of the main flow in terms of its interaction with the coolant jets blown through the film holes in the leading edge of the blade, and determine the thickness of the cooling film at the surface of the blade.

2. Experimental Stand and Object of Research

In order to carry out PIV research, experimental was developed (figure 1). This stand consists of a package of three blades (1), side channel blower SKS-1000 (2), controlled by a variable frequency drive (VFD). To illuminate the tracers a double pulse Nd:YAG laser Quantel Ever Green 70 was used. Registration of tracers was carried out by a cross-correlation 4MP digital camera Bobcat ImperX B2020. Synchronization of equipment during the experiment was provided by the processor "Polis" (5) using the software ActualFlow. The seeding particles were titanium dioxide powder and glycerin vapor.

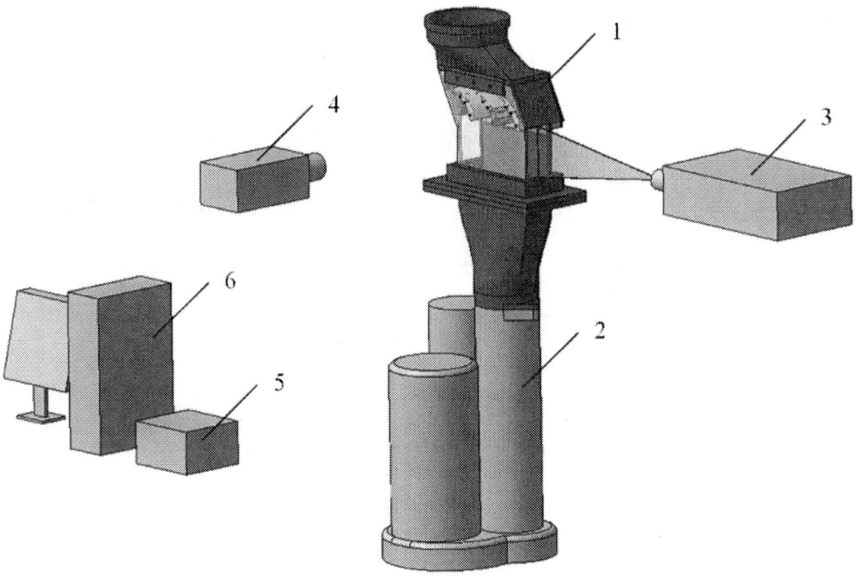

Figure 1. Scheme of the experimental stand: 1 – package of three blades; 2 – side channel blower; 3 – double pulse Nd:YAG laser; 4 – cross-correlation 4MP digital camera; 5 – synchronizing processor "Polis"; 6 – PC.

The package of blades (figure 2 and 3) consists of an input diffuser, a flow stabilization section, a test object, and an output section. The test object is a curvilinear channel with optical access on three sides, in which a flat package of three blades is placed. The left and the right blades are made without internal cavities and designed to create conditions close to original flow around the central blade. The central blade is removable and has internal cavity for inserting the deflector. During the tests, metal deflector from full-scale blades was used. This deflector was slightly shortened for installation inside the internal cavity of the central blade. All the tested blades were manufactured using 3D printing from polyamide. After manufacturing, a consistent grinding of the surfaces of the blades and its platforms, as well as drifting of the film holes to the nominal diameter, were performed. The condition of the blade surface and the holes is shown in figure 4. A photo of the test stand is shown in figure 5.

The remaining elements of the stand are made by 3D printing of ABS plastic. Additionally, two deturbulating and balancing nylon nets are installed in front of the inlet diffuser and in front of the flow stabilization section.

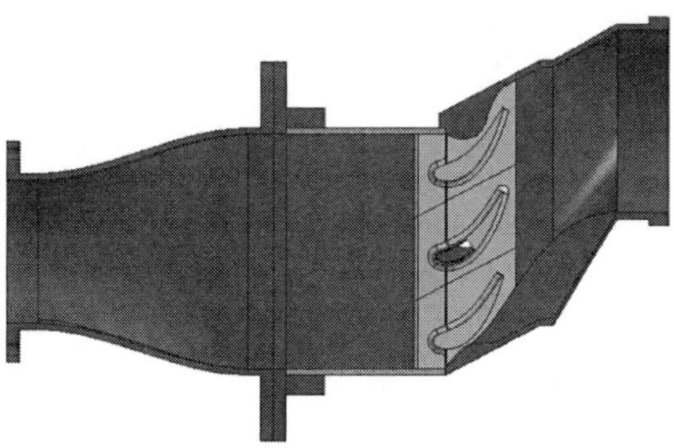

Figure 2. Longitudinal section of the package of the blades.

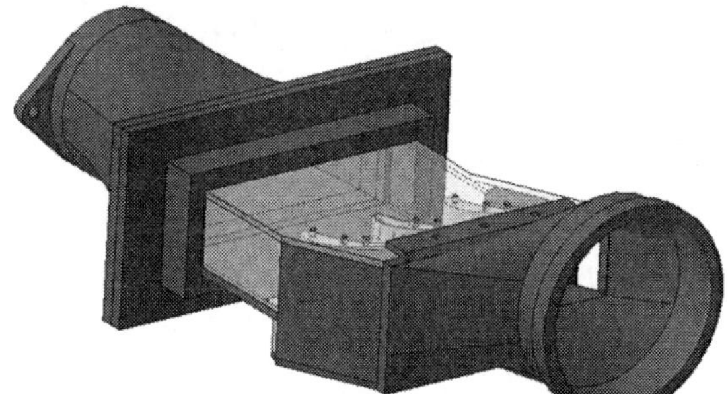

Figure 3. General view of the package of the blades.

During the experiments, there were made measurements of temperature and pressure of surrounding air as well as the temperature of the main flow created by the blower, temperature, pressure and mass flow rate of the air supplied into the internal cavity of the blade by a piston compressor. PIV measurements of the velocity field were carried out in sections shown in figure 6.

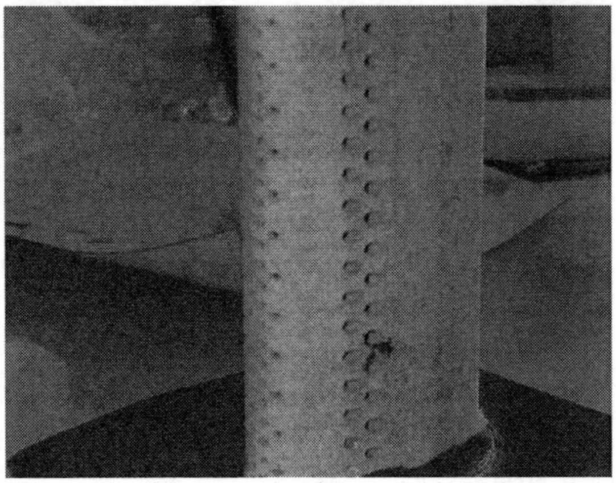

Figure 4. Photograph of the test blade from the leading edge after the grinding process.

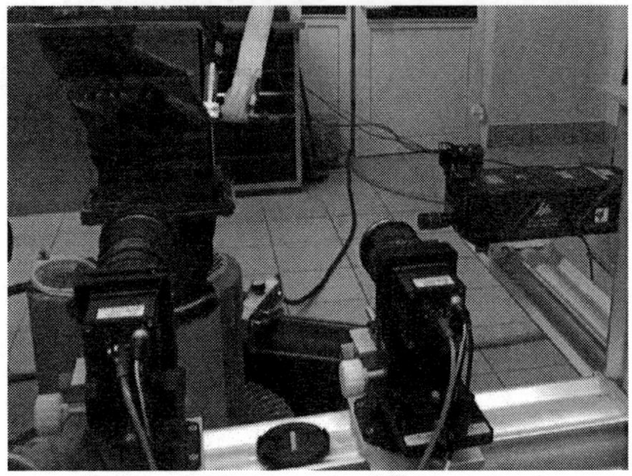

Figure 5. Photograph of the test stand.

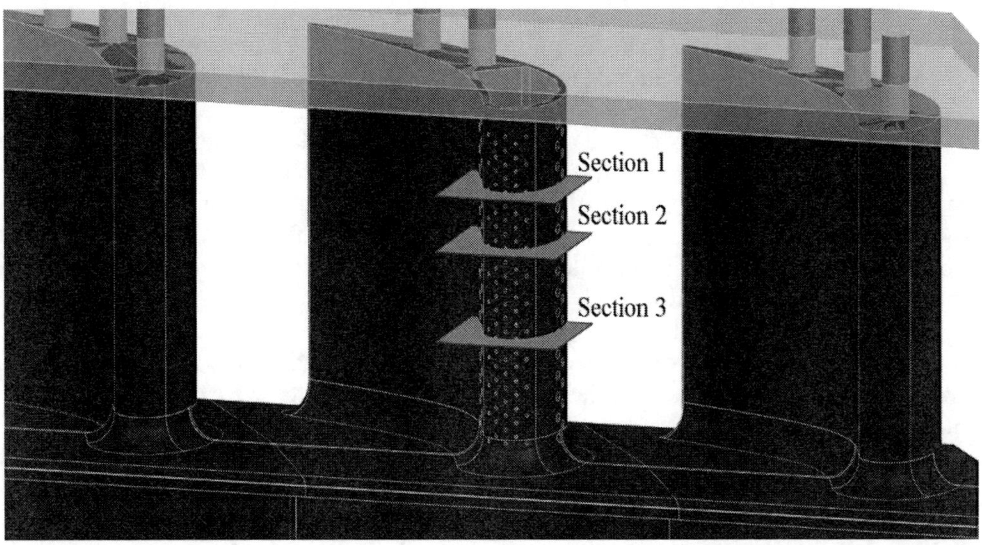

Figure 6. The layout of the control sections for PIV measurements and visualization.

3. Results and Discussion

The results of experimental PIV studies as well as visualization of the formation of the cooling film on the surface of the blade are shown in figures 7-14. These results correspond to the main modes with value of the relative coolant mass flow rate G_c equal to 1.6%, 3.2%, 4.8% of the main air flow rate. It should be noted that due to the necessity to organize the laser plane as well as because of the geometric features of the experimental model, measurements were carried out only at the leading edge and part of the suction side of the central blade. The remaining areas in the figures are blacked out (in figure 13 – in blue), which corresponds to the absence of measurements.

For all three cross sections studied, increase in the blowing parameter m in the range from 0.5 to 2.5 leads to an intense increase in the thickness of the cooling film. However, the rate and nature of the increase for different cases are different. In particular, it was found that in the studied range of m, the forming film is quite non-uniform. Operation modes with relative coolant mass flow rate $G_c > 4\%$ near the leading edge define deep penetration of cooling jets into the main stream. Drift of such jets by flowing air is observed only at a length of about 10..15 diameters of the perforation hole. This is accompanied by the formation of a recirculation zone behind the jet, in which, obviously, an intensive mixing of the coolant with the main air occurs. Obviously, these modes should be avoided by setting the range for the blowing parameter from 1.0 to 1.5, which corresponds to the relative coolant mass flow rate 2.5..4.0%. Studies have shown that in this case a uniform cooling film can be formed on the surface of the blade.

Cooling film has the smallest thickness in the zone corresponding to the impingement point of the main air flow. At point 2 (figure 13), there is a deep penetration of cooling jets into the main flow observed practically for all the operation modes. This indicates a large value of the blowing parameter m for this row of holes. In this zone, the film can be detached from the surface of the blade with the formation of a suction zone behind the hole. It forms favorable conditions for potentially hot gases to flow from the main stream into this zone.

It is important to understand the formation of the cooling film on the surface of the blade is a significantly unsteady process. Analysis of flow structure in the central row of perforation holes (figure 14) indicates its high instability. This is realized in the form of alternate deflection of the jet to the pressure and suction sides.

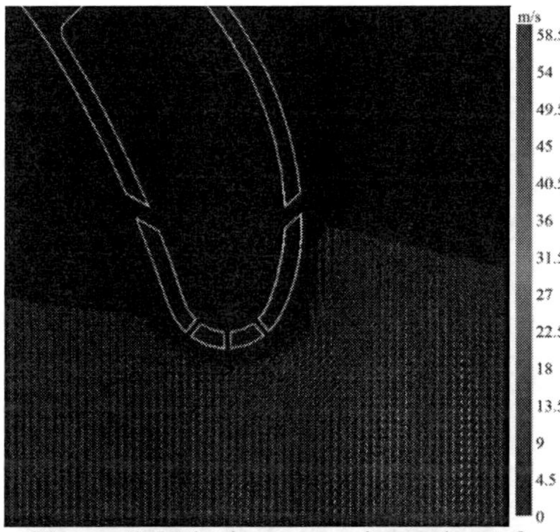

Figure 7. Velocity vectors in section 1 for operation mode $G_c = 1.6\%$

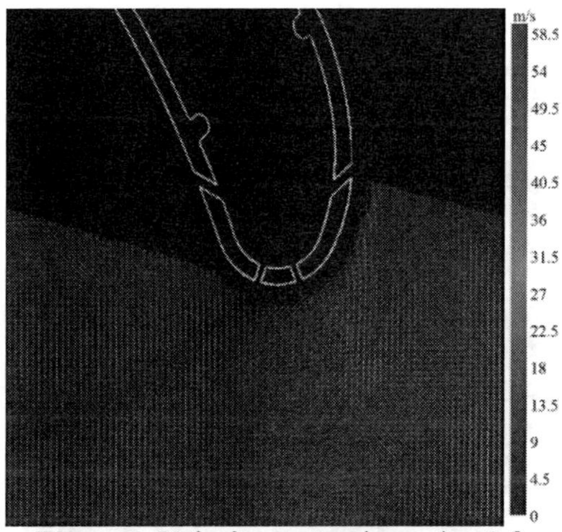

Figure 8. Velocity vectors in section 2 for operation mode $G_c = 1.6\%$

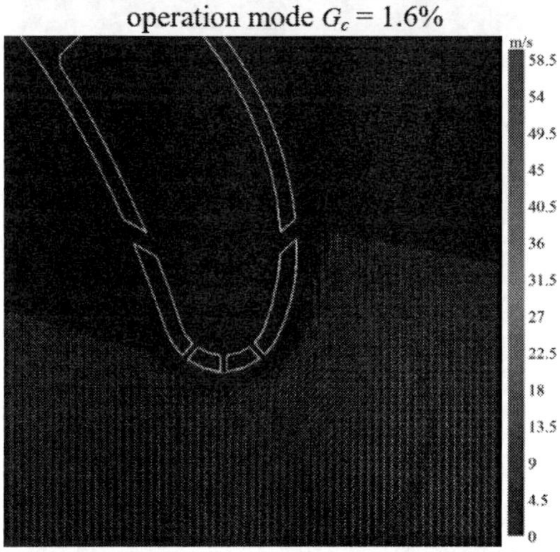

Figure 9. Velocity vectors in section 1 for operation mode $G_c = 3.2\%$.

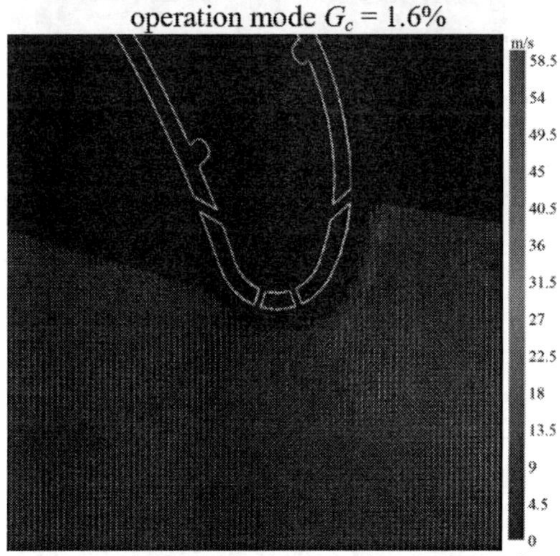

Figure 10. Velocity vectors in section 2 for operation mode $G_c = 3.2\%$.

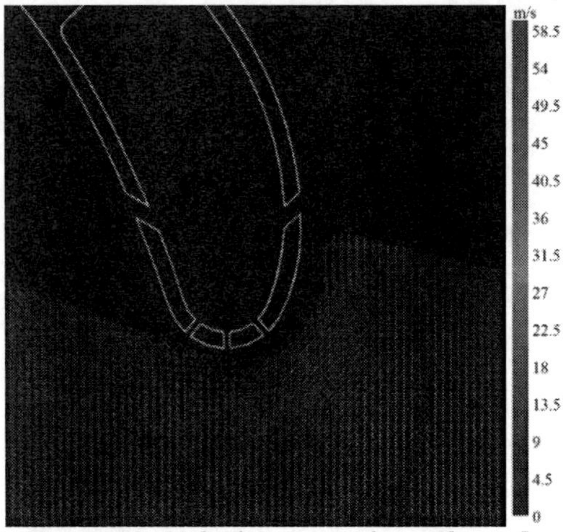

Figure 11. Velocity vectors in section 1 for operation mode $G_c = 4.8\%$.

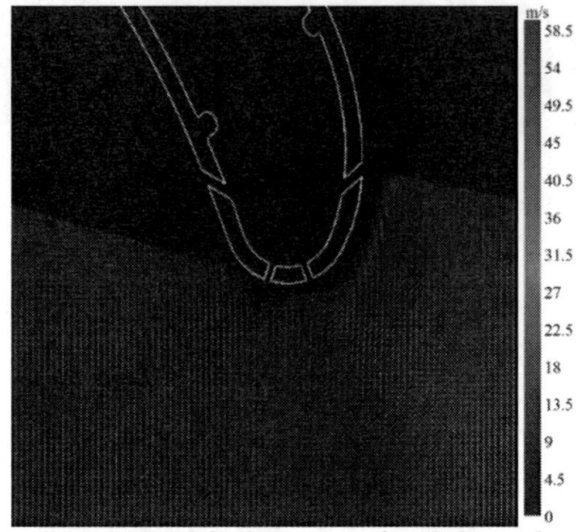

Figure 12. Velocity vectors in section 2 for operation mode $G_c = 4.8\%$.

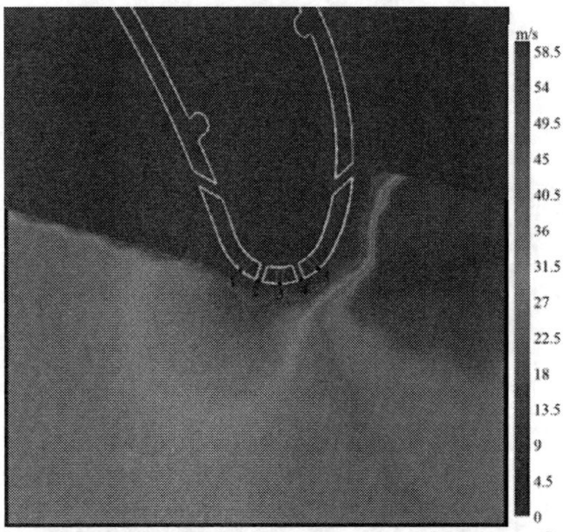

Figure 13. Velocity contours and layout of control points for operation mode $G_c = 4.8\%$.

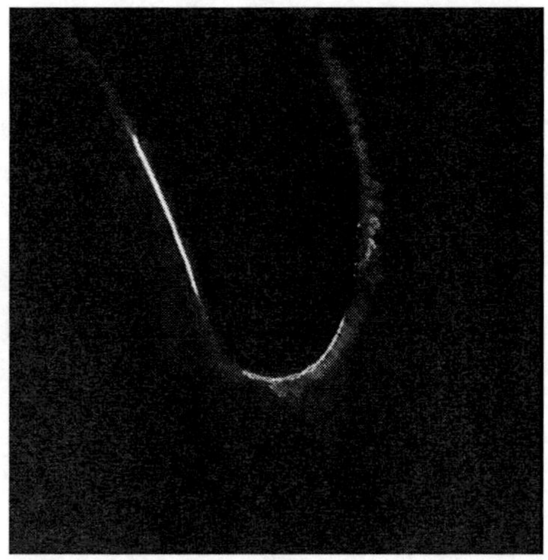

Figure 14. Visualization of the formation of cooling film on the surface of the blade in section 3 for operation mode $G_c = 4.8\%$.

4. Conclusion

The experimental stand for velocity measurement and flow visualization near leading edge of the blade under conditions of formation of cooling film is developed and made.

PIV study of the flow around the perforated surface of leading edge are carried out for 3 different sections and variable relative coolant mass flow rate blown through the film holes in the range from 1,6% to 6,2%. Based on the obtained velocity fields and the flow visualization, it is shown that at point 2 (figure 13) almost all operation leads to deep penetration of the cooling jets (from 3 mm to 4.2 mm) into the main air stream. This indicates a large value of the blowing parameter m.

The analysis of the obtained results allowed to determine the most favorable operating range for the film cooling of the leading edge by the blowing parameter m from 1.0 to 1.5. These values correspond to relative coolant mass flow rate from 2.5% to 4.0%. Studies shown that in this case sufficiently uniform cooling film is formed on the surface of the blade.

Acknowledgments

This work was financially supported by RSF (Russian Science Foundation) grant No. 18-79-00180 "Thermophysics of combined reacting jets and impact flows."

Journal of Physics: Conference Series **1421** (2019) 012043 doi:10.1088/1742-6596/1421/1/012043

References

[1] Akhmetbekov E, Bilsky A, Lozhkin Yu, Markovich D and Tokarev M 2006 Control system of experiment and processing data obtained by methods of particle image velocimetry (Actual Flow) *Computational methods and programming* **7** 79-85

[2] Westerweel J 1993 *Digital Particle Image Velocimetry - Theory and Application* (Delft: Delft University Press)

[3] Lang H, Mørck T and Woisetschläger J 2002 Stereoscopic particle image velocimetry in a transonic turbine stage *Experiments in Fluids* **32** 700-9

[4] Woisetschläger J, Pecnik R and Göttlich E, 2008 Experimental and numerical flow visualization in a transonic turbine *Journal of Visualization* **11** 95-102

[5] Woisetschläger J, Mayrhofer N and Hampel B 2003 Laser-optical investigation of turbine wake flow *Experiments in Fluids* **34** 371-8

[6] Senoo A, Mizuki S and Tsujita H 2000 Investigation of internal flow in ultra - highly loaded turbine cascade by PIV method *Journal of Thermal Science* **9** 193-8

[7] Tian Y, Ma H and Ma R 2017 Stereoscopic PIV measurements of the flow field in a turbine cascade *Journal of Thermal Science* **26** 89-95

[8] Davide L, Simoni D and Ubaldi M 2017 Coherent Structures Formation During Wake-Boundary Layer Interaction on a LP Turbine Blade *Flow Turbulence Combustion* **98** 57-81

[9] Chen Y, Matalanis C G and Eaton J K 2008 High resolution PIV measurements around a model turbine blade trailing edge film-cooling breakout *Experiments in Fluids* **44** pp 199-209

[10] Uzol O and Camci C 2001 Aerodynamic loss characteristics of a turbine blade with trailing edge coolant ejection: part 2 – external aerodynamics, total pressure losses, and predictions *Journal of Turbomachinery* **123** 249–57

[11] Raffel M and Kost F 1998 Investigation of aerodynamic effects of coolant ejection at the trailing edge of a turbine blade model by PIV and pressure measurements *Experiments in Fluids* **24** 447-61

[12] Kukutla P and Prasad B 2017 Secondary flow visualization on stagnation row of a combined impingement and film cooled high-pressure gas turbine nozzle guide vane using PIV technique *Journal of Visualization* **20** 817-32

Estimating the quality of stereoscopic endoscopic systems

A A Naumov[1,2], A V Gorevoy[1,2,3] A S Machikhin[1,3], V I Batshev[1,2], and V E Pozhar[1,2]

[1]Scientific Technological Center of Unique Instrumentation, Russian Academy of Science, Moscow , 15 Butlerova, Russia
[2]Department of Laser and Opto-Electronic Systems, Bauman Moscow State Technical University, 5/1 Baumanskaya 2-ya, Moscow, Russia
[3]Department of Electrical Engineering and Introscopy, National Research University "Moscow Power Engineering Institute", 14 Krasnokazarmennaya, Moscow, Russia

Abstract. Stereoscopic video endoscopic systems are widely used for the control of hard-to-reach objects at the stages of production, assembly and testing of various devices. Using a stereoscopic endoscope as a measuring tool requires its geometrical calibration. The effectiveness of such systems is largely determined by the quality of an image, based on modulation transfer function (MTF) of the optical system. In this paper, we describe and test an algorithm for automatic calculating the MTF for images of a plane test object with chessboard pattern. Experimental data was obtained using two self-developed prism-based endoscopic stereoscopic systems. Coincidence of computer simulation results and results of applying the proposed algorithm shows its effectiveness for estimating the image quality provided by stereoscopic endoscopic systems.

1. Introduction

Currently, stereoscopic video endoscopic systems are a key tool for non-destructive testing of hard-to-reach elements in complex technical systems, such as aircraft engines, steam generators, etc. Due to small diameter of the endoscopic probe, stereo images in such devices are usually formed by the prism, which creates two images of the inspected surface on the camera sensor [1].

To use a stereoscopic endoscope as a measuring tool, it is necessary to calibrate it as a stereo system [2]. Known methods require a plane test object with chessboard pattern, located in several positions inside the working volume of the system [3, 4]. Additionally, it is necessary to estimate the quality of the image obtained by the system, because it affects the accuracy of 3D surface reconstruction. The image quality depends on the noise generated by the sensor and the electronic circuits as well as on the distortions introduced by an optical system. According to the standard [5], the slanted-edge method of MTF measurement is used to estimate the performance of the optical system.

We propose to use the same plane test object with chessboard pattern to perform calibration and assess image quality for the stereo endoscope. Such an object contains both areas of inclined boundaries required for measuring the MTF and areas of constant intensity suitable for estimating noise parameters. It will significantly reduce the required time and cost of auxiliary equipment, because estimations of MTF, distortion and noise, as well as calibration can be performed simultaneously.

In this paper, we experimentally demonstrate MTF measurements for two self-developed endoscopic stereoscopic probes using the images of chessboard patterns, like for calibration procedure. We compare the results obtained using these systems at three working distances and theoretical dependencies calculated for these optical systems in optical design software Zemax.

2. Instruments and methods

We carried out MTFs estimation for two self-developed prism-based stereo endoscopic systems, which technical characteristics are presented in table 1. We further denote these systems as OS1 and OS2.

Table 1. Technical issue for used endoscopic stereoscopic systems.

Feature	OS1	OS2
Matrix diagonal	1/9"	1/6"
Matrix resolution	1280 × 720	1920 × 1080
Pixel size	1.4 × 1.4 μm^2	1.4 × 1.4 μm^2
Image format	24 bit, RGB	8 bit, Grayscale
Field of view per channel	35° × 40°	40° × 45°
Aperture	F/8	F/11
Effective focal length	1.8 mm	2.36 mm

The optical layouts of these systems are similar (figure 1). More information about the design of prism-based stereoscopic endoscopes may be found in [6] and [7].

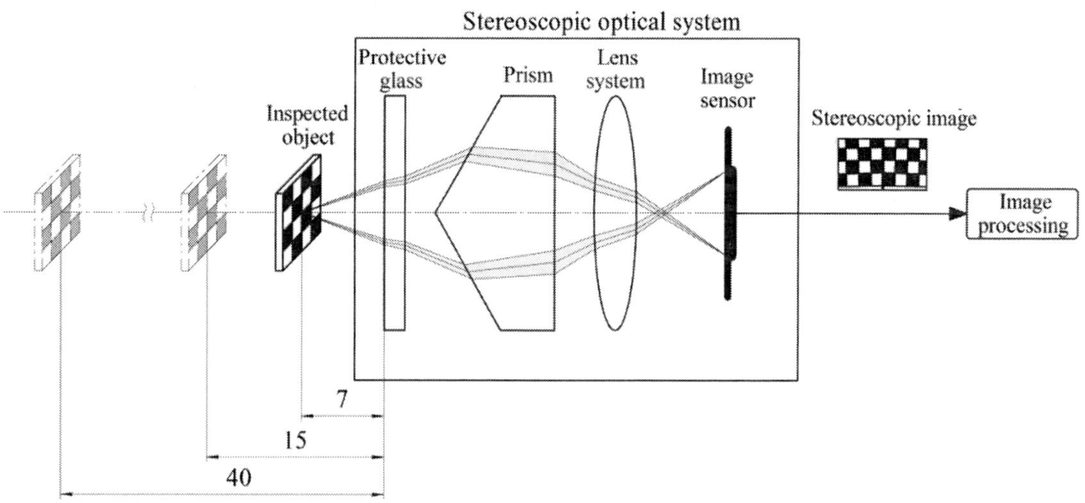

Figure 1. Scheme of MTF measurements for prism-based stereoscopic endoscopes.

The prism introduces significant aberrations, which can not be completely corrected by the lens system. These aberrations vary significantly across the field of view, so 9 field points in each channel were used for MTF assessment. The exact coordinates of these points are presented in table 2. MTF calculation for each image point was carried out in Zemax for three object distances (7, 15 and 40 mm from the protective glass, as shown in figure 1). The distances correspond to the boundaries and the middle of the working distance range of the system. The obtained data was exported from Zemax and then was processed in MATLAB.

Table 2. Points coordinates for MTF estimating.

System type	**OS1**				**OS2**			
Point number	**Image coordinate (px)**		**Zemax image coordinate (mm)**		**Image coordinate (px)**		**Zemax image coordinate (mm)**	
	x	*y*	*x*	*y*	*x*	*y*	*x*	*y*
1	80	80	-0.784	-0.392	80	80	-1.232	-0.644
2	80	360	-0.784	0	80	540	-1.232	0
3	80	640	-0.784	0.392	80	1000	-1.232	0.644
4	320	80	-0.448	-0.392	480	80	-0.672	-0.644
5	320	360	-0.448	0	480	540	-0.672	0
6	320	640	-0.448	0.392	480	1000	-0.672	0.644
7	560	80	-0.112	-0.392	880	80	-0.112	-0.644
8	560	360	-0.112	0	880	540	-0.112	0
9	560	640	-0.112	0.392	880	1000	-0.112	0.644
10	720	80	0.112	-0.392	1040	80	0.112	-0.644
11	720	360	0.112	0	1040	540	0.112	0
12	720	640	0.112	0.392	1040	1000	0.112	0.644
13	960	80	0.448	-0.392	1440	80	0.672	-0.644
14	960	360	0.448	0	1440	540	0.672	0
15	960	640	0.448	0.392	1440	1000	0.672	0.644
16	1200	80	0.784	-0.392	1840	80	1.232	-0.644
17	1200	360	0.784	0	1840	540	1.232	0
18	1200	640	0.784	0.392	1840	1000	1.232	0.644

The setup shown in figure 2 was used to obtain experimental data. The glass plate with chrome-etched chessboard pattern (2 mm chessboard cell size) was combined with white diffusing glass and illuminated from behind by a white-light source.

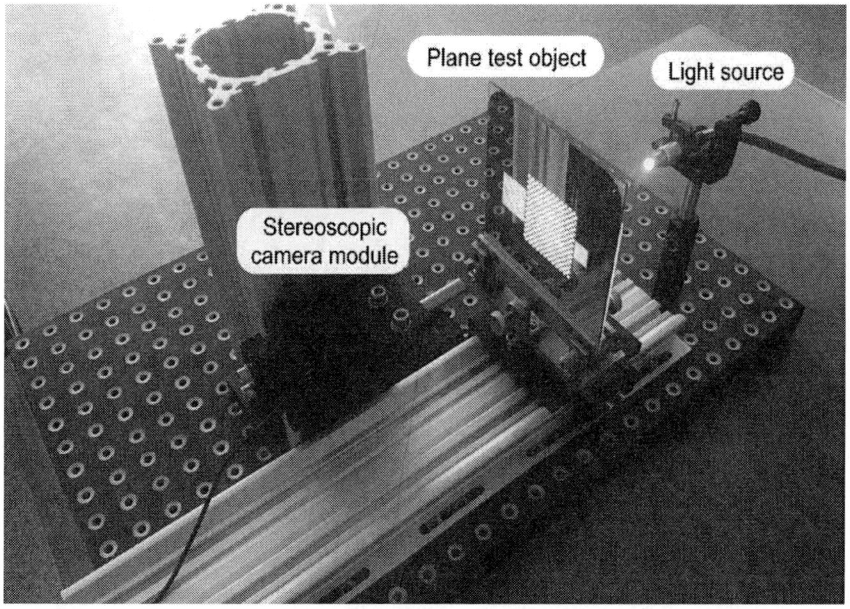

Figure 2. Setup for measuring MTF of stereoscopic optical systems.

Images of the test object were acquired at three distances of 7, 15 and 40 mm from the protective glass of the optical system for both endoscopes. MTF estimation was carried out in areas containing slanted high-contrast edges. The regions of estimation were chosen closely to the points used for MTF calculation in Zemax. Image processing and calculation of the MTF were performed according to the standard technique [5, 8]. For each region, a subpixel profile of the brightness difference on the boundary was constructed, then the Fourier transform of the derivative of the subpixel profile was calculated and normalized on the maximum value. An example of an image with highlighted points for MTF calculation in Zemax (green points) and areas for MTF estimation using image processing (red rectangles) is shown in figure 3 (here, we show the areas for measuring MTF in the vertical direction).

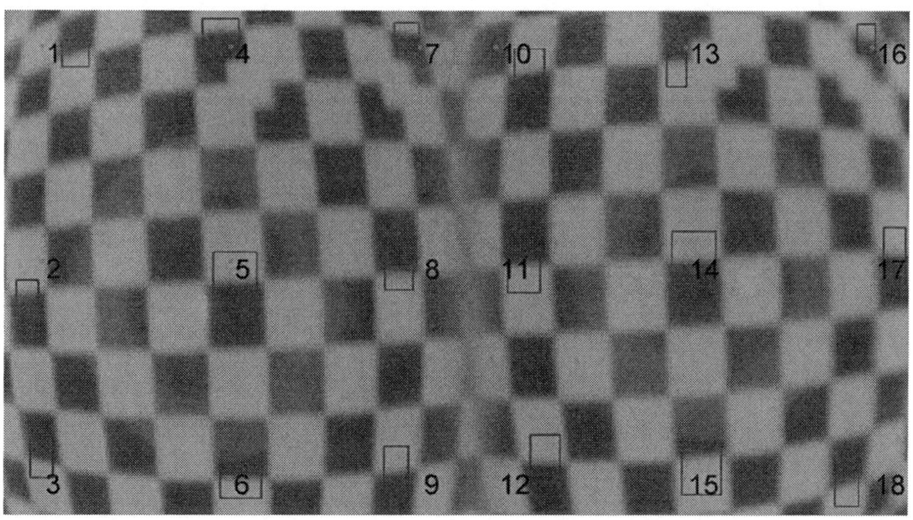

Figure 3. The image of the test object with highlighted points for MTF calculation in Zemax (green points) and areas for MTF estimation using image processing (red rectangles).

3. Results

We compared 2 calculated (in horizontal and vertical directions) and 2 experimental MTF curves as well as the values of spatial frequencies, at which the contrast values of 0.1 and 0.5 were achieved. The results are given in figures 5-7 and tables 4-6 for OS1 at three working distances. The results for OS2 are presented in figures 8-10 and tables 7-9.

The values in the tables are divided into the following subgroups:

- Left and Right channel - the channel of the stereo image in which the MTF is estimated.
- Level 0.5 and Level 0.1 - MTF levels necessary to define the key values of spatial frequencies.
- "Hor." and "Ver." - are horizontal and vertical directions respectively, used for MTF estimation.

Figure 4 presents the legends used for figures 5-10.

Description	Denotation
Estimated modulus of MTF in horizontal direction	
Zemax modulus of MTF in horizontal direction	
Estimated modulus of MTF in vertical direction	
Zemax modulus of MTF in vertical direction	
0.5 and 0.1 levels of MTF modulus	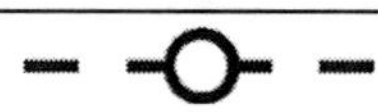

Figure 4. Legends for figures 5-10.

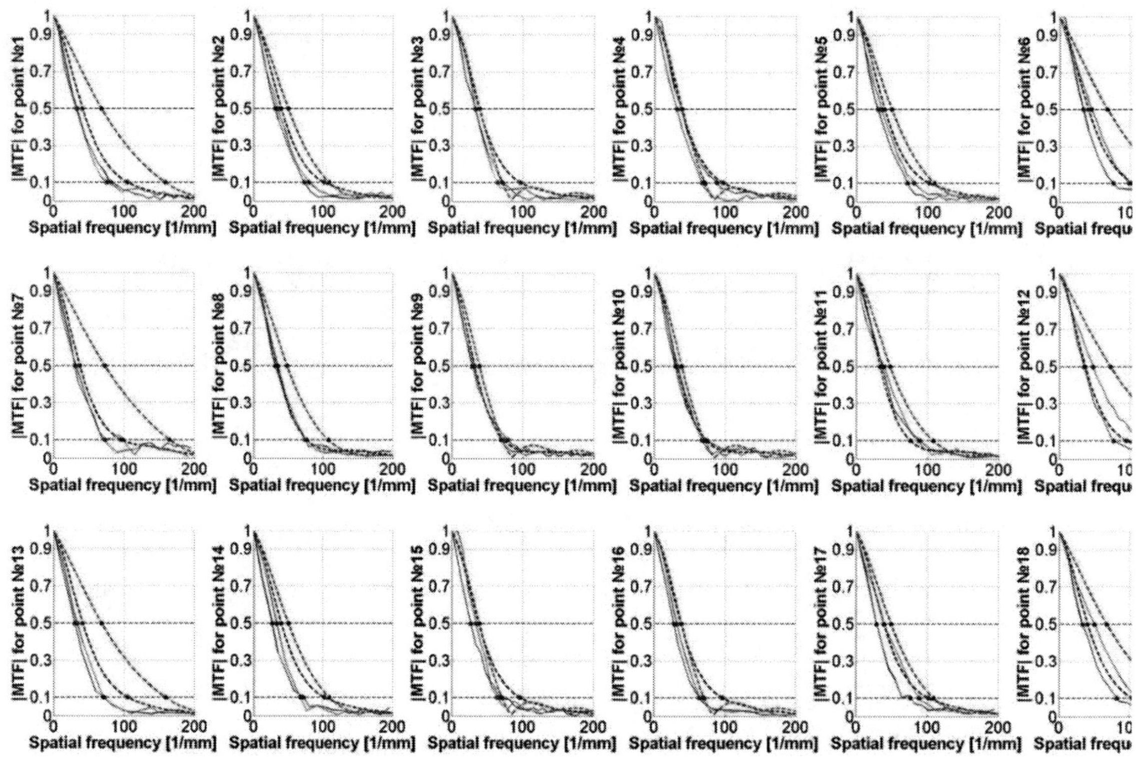

Figure 5. Estimated MTF at 18 points for OS1. Distance to the test object is 7 mm.

Table 3. The spatial frequencies at 0.5 and 0.1 levels of MTF for OS1 (in mm^{-1}). Distance to the test object is 7 mm.

			Point's number	*1*	*2*	*3*	*4*	*5*	*6*	*7*	*8*	*9*
Left channel	**Level 0.5**	**Hor.**	**Zemax**	41	39.2	38.2	38.2	39.2	41	37.2	33.3	32.5
			Estimated	33.3	31	33.1	38.1	29.7	34	30.3	31.4	29
		Ver.	**Zemax**	67.8	49.4	39.3	39.3	49.4	67.8	72.4	47.7	39
			Estimated	32.2	34.6	32.9	31.5	34.9	45.6	32	34.5	31.7
	Level 0.1	**Hor.**	**Zemax**	104.5	102	96.4	96.4	102	104.5	95.5	76.1	72.9
			Estimated	80.9	73	70.4	88.1	71.9	76.2	97.5	73.8	78.8
		Ver.	**Zemax**	158.5	109.3	70.8	70.8	109.3	158.5	164.4	108.1	71.2
			Estimated	75.5	79.5	64.5	67.6	80.8	99.1	72.4	75.6	69.2
			Point's number	*10*	*11*	*12*	*13*	*14*	*15*	*16*	*17*	*18*
Right channel	**Level 0.5**	**Hor.**	**Zemax**	32.5	33.3	37.2	41	39.2	38.2	38.2	39.2	41
			Estimated	30.1	36.4	35.6	30.1	26.4	27.3	28.1	29.1	33.9
		Ver.	**Zemax**	39	47.7	72.4	67.8	49.4	39.3	39.3	49.4	67.8
			Estimated	33.2	39.1	48.4	33.3	32.9	34.7	32.6	39.5	50.5
	Level 0.1	**Hor.**	**Zemax**	72.9	76.1	95.5	104.5	102	96.4	96.4	102	104.5
			Estimated	74.2	89.3	77	70.2	71	67.3	64.8	76.8	82.3
		Ver.	**Zemax**	71.2	108.1	164.4	158.5	109.3	70.8	70.8	109.3	158.5
			Estimated	67.5	88.3	116.4	71.2	66.9	64.6	65.7	89.5	111.7

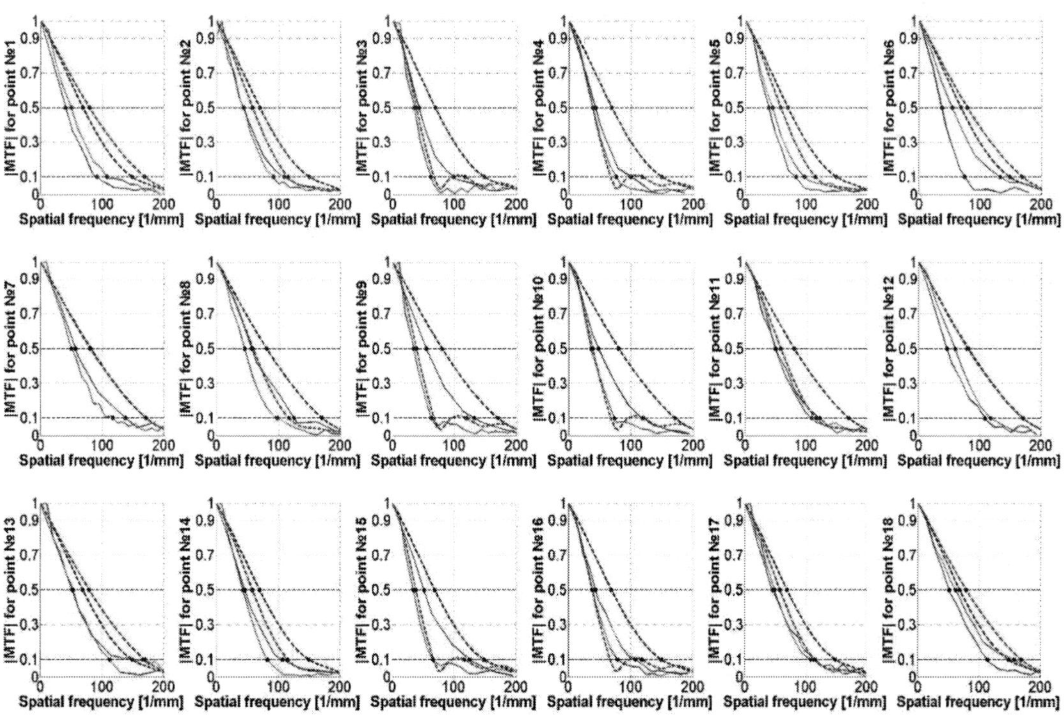

Figure 6. Estimated MTF at 18 points for OS1. Distance to the test object is 15 mm.

Journal of Physics: Conference Series **1421** (2019) 012044 doi:10.1088/1742-6596/1421/1/012044

Table 4. The spatial frequencies at 0.5 and 0.1 levels of MTF for OS1 (in mm^{-1}). Distance to the test object is 15 mm.

			Point's number	*1*	*2*	*3*	*4*	*5*	*6*	*7*	*8*	*9*
Left channel	**Level 0.5**	**Hor.**	**Zemax**	68.7	69.8	69.5	69.5	69.8	68.7	81.7	80.8	80.6
			Estimated	41	42.9	42.8	43.1	38.2	37.4	56.8	56.6	55.4
		Ver.	**Zemax**	79.2	58.1	38.6	38.6	58.1	79.2	80.3	58.8	39.1
			Estimated	49.8	43.8	35.2	38.8	44.2	54.5	50.6	45.4	34.5
	Level 0.1	**Hor.**	**Zemax**	148.4	149.6	149.1	149.1	149.6	148.4	170.8	170.4	170.4
			Estimated	90.7	109.2	99.4	99.9	83.3	75	137.7	125	133.1
		Ver.	**Zemax**	169	115.2	117.8	117.8	115.2	169	170.8	116.6	119.9
			Estimated	107.6	91.9	62.5	76.4	97.5	132.6	117	98.1	63.7
			Point's number	*10*	*11*	*12*	*13*	*14*	*15*	*16*	*17*	*18*
Right channel	**Level 0.5**	**Hor.**	**Zemax**	80.6	80.8	81.7	68.7	69.8	69.5	69.5	69.8	68.7
			Estimated	48.4	50.9	46.9	51.4	46	52	44.1	45.7	51.4
		Ver.	**Zemax**	39.1	58.8	80.3	79.2	58.1	38.6	38.6	58.1	79.2
			Estimated	36.3	49.3	59.9	53.2	42.9	34.5	39.6	49.3	62
	Level 0.1	**Hor.**	**Zemax**	170.4	170.4	170.8	148.4	149.6	149.1	149.1	149.6	148.4
			Estimated	118.4	123.4	108.5	111.8	106.9	126.5	107.9	116.6	113.8
		Ver.	**Zemax**	119.9	116.6	170.8	169	115.2	117.8	117.8	115.2	169
			Estimated	73.8	109.4	118.5	166.3	82.3	66.4	79.2	109.7	157.4

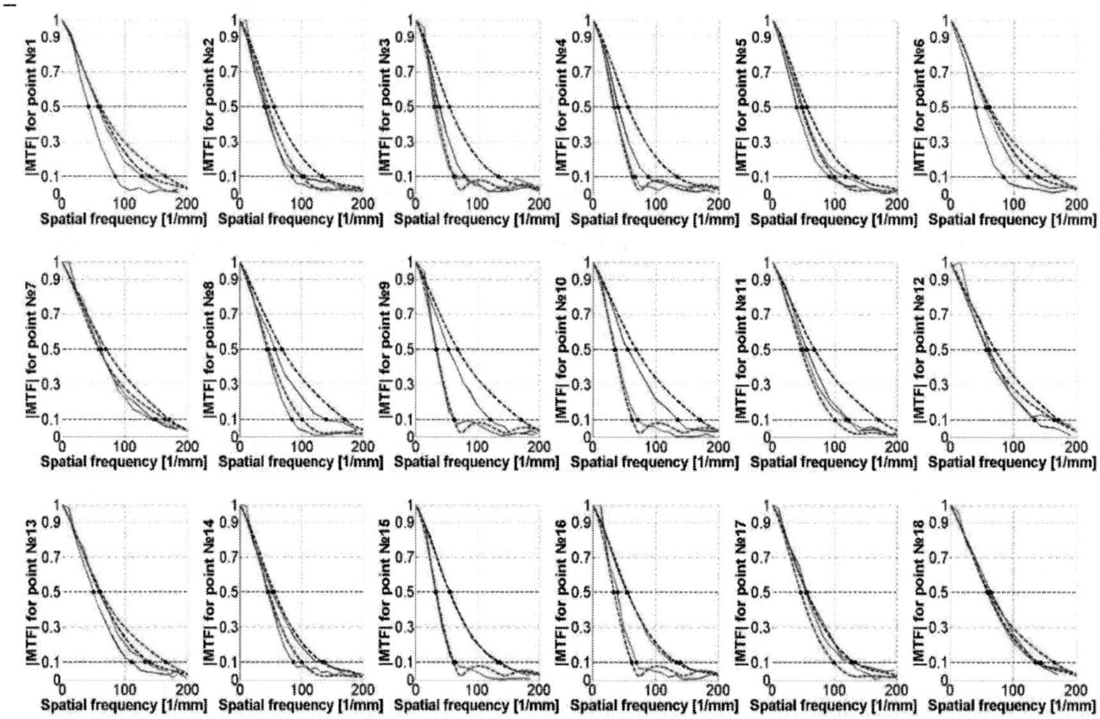

Figure 7. Estimated MTF at 18 points for OS1. Distance to the test object is 40 mm.

Table 5. The spatial frequencies at 0.5 and 0.1 levels of MTF for OS1 (in mm^{-1}). Distance to the test object is 40 mm.

			Point's number	*1*	*2*	*3*	*4*	*5*	*6*	*7*	*8*	*9*
Left channel	Level 0.5	Hor.	**Zemax**	59.7	55.4	54.9	54.9	55.4	59.7	70.6	67.9	68
			Estimated	43.2	39.7	38.8	39.5	38.8	39.4	59.7	56.7	53.5
		Ver.	**Zemax**	62	46	33.6	33.6	46	62	58.9	46.2	33.9
			Estimated	58.6	41.3	30.7	32.8	48.4	55.6	63	43.7	33.3
	Level 0.1	Hor.	**Zemax**	139.9	133.3	133.4	133.4	133.3	139.9	171.2	170.4	170.4
			Estimated	85.6	103.7	79.3	86.9	91.4	83.7	146.1	139.5	121
		Ver.	**Zemax**	165.2	99.6	61.6	61.6	99.6	165.2	164.3	100.4	61.9
			Estimated	127.9	84.7	64.4	66.2	117.4	122.1	149.4	83.6	63.1
			Point's number	*10*	*11*	*12*	*13*	*14*	*15*	*16*	*17*	*18*
Right channel	Level 0.5	Hor.	**Zemax**	68	67.9	70.6	59.7	55.4	54.9	54.9	55.4	59.7
			Estimated	55	55.8	57.2	50.3	52	56.3	53.7	56	65.3
		Ver.	**Zemax**	33.9	46.2	58.9	62	46	33.6	33.6	46	62
			Estimated	35.8	49.8	62.1	60.9	44.7	32.6	38.8	54.4	59
	Level 0.1	Hor.	**Zemax**	170.4	170.4	171.2	139.9	133.3	133.4	133.4	133.3	139.9
			Estimated	134.7	122	133.7	111.6	136.8	136.8	140.1	133.9	135.5
		Ver.	**Zemax**	61.9	100.4	164.3	165.2	99.6	61.6	61.6	99.6	165.2
			Estimated	70.8	118.2	165	131.8	85.9	64.6	69.4	127.5	143.8

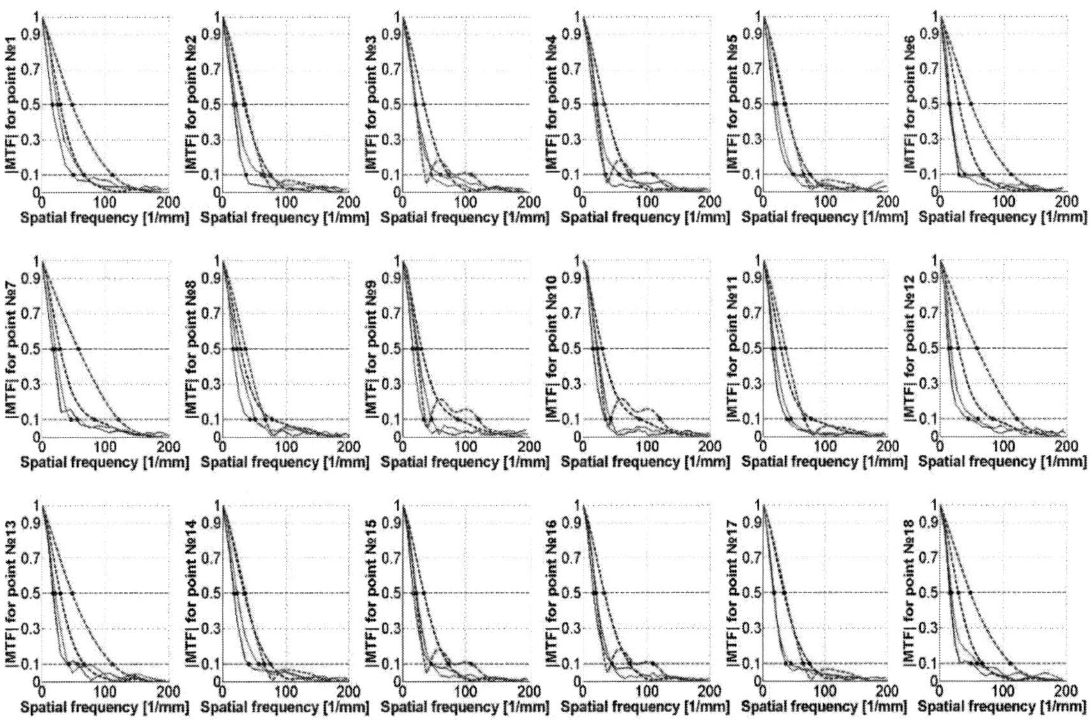

Figure 8. Estimated MTF at 18 points for OS2. Distance to the test object is 7 mm.

Table 6. The spatial frequencies at 0.5 and 0.1 levels of MTF for OS2 (in mm^{-1}). Distance to the test object is 7 mm.

			Point's number	*1*	*2*	*3*	*4*	*5*	*6*	*7*	*8*	*9*
Left channel	**Level 0.5**	**Hor.**	**Zemax**	29.1	33.7	33.6	33.6	33.7	29.1	28.9	28.2	29.8
			Estimated	16.7	17.7	20.9	17.2	17.1	13.5	17.3	16.2	15.8
		Ver.	**Zemax**	48.1	35.8	20.6	20.6	35.8	48.1	57.7	35.9	21.5
			Estimated	24.6	21.3	20	22	21.2	15.3	21.2	22.6	23.7
	Level 0.1	**Hor.**	**Zemax**	66.5	75.3	72.4	72.4	75.3	66.5	83.1	77.9	87.6
			Estimated	49.7	36.7	46.1	35.3	47.6	29.2	55.4	43.3	33.2
		Ver.	**Zemax**	111.6	65.8	109	109	65.8	111.6	120.8	66.9	119.7
			Estimated	66.4	61.3	60.4	56.5	48.9	38.7	46.2	51	46.1
			Point's number	*10*	*11*	*12*	*13*	*14*	*15*	*16*	*17*	*18*
Right channel	**Level 0.5**	**Hor.**	**Zemax**	29.8	28.2	28.9	29.1	33.7	33.6	33.6	33.7	29.1
			Estimated	15.9	14.9	13.1	17.9	18.2	17	16.7	17.9	15.9
		Ver.	**Zemax**	21.5	35.9	57.7	48.1	35.8	20.6	20.6	35.8	48.1
			Estimated	22.7	18.8	17	21.1	22.8	21.7	20.5	17.2	18.5
	Level 0.1	**Hor.**	**Zemax**	87.6	77.9	83.1	66.5	75.3	72.4	72.4	75.3	66.5
			Estimated	34.8	38	44.3	56.5	41.1	45.7	45.6	37.4	49.3
		Ver.	**Zemax**	119.7	66.9	120.8	111.6	65.8	109	109	65.8	111.6
			Estimated	43.1	44.9	59.3	42.3	56.8	45.4	44.8	45.2	58.8

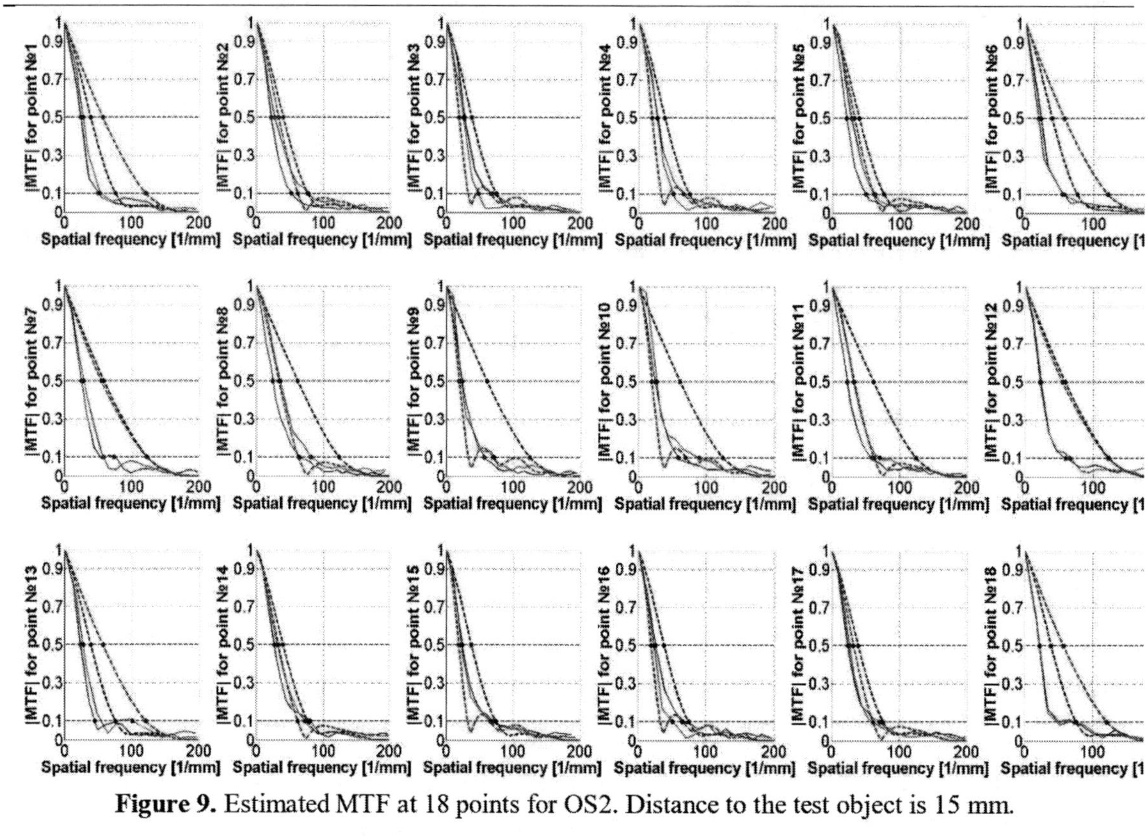

Figure 9. Estimated MTF at 18 points for OS2. Distance to the test object is 15 mm.

Journal of Physics: Conference Series **1421** (2019) 012044 doi:10.1088/1742-6596/1421/1/012044

Table 7. The spatial frequencies at 0.5 and 0.1 levels of MTF for OS2 (in mm^{-1}). Distance to the test object is 15 mm.

			Point's number	*1*	*2*	*3*	*4*	*5*	*6*	*7*	*8*	*9*
Left channel	**Level 0.5**	**Hor.**	**Zemax**	39	39.2	37.9	37.9	39.2	39	58.9	60.8	61.2
			Estimated	24.4	21.4	27.3	27.3	21.1	22.8	25.4	24.5	24
		Ver.	**Zemax**	57.2	32.4	18.9	18.9	32.4	57.2	54.8	32.8	19.3
			Estimated	27.8	26.7	25.4	26.6	28	19.2	29	35	22.9
	Level 0.1	**Hor.**	**Zemax**	76.1	75.9	74.9	74.9	75.9	76.1	123.1	124.3	124.7
			Estimated	51.1	51.7	68.7	65.7	49.3	56.5	74	63.7	57
		Ver.	**Zemax**	120.9	60.9	66.7	66.7	60.9	120.9	122.2	61.9	70.1
			Estimated	53.4	76.7	47.9	50.8	62.1	54.9	58.4	80.9	72
			Point's number	*10*	*11*	*12*	*13*	*14*	*15*	*16*	*17*	*18*
Right channel	**Level 0.5**	**Hor.**	**Zemax**	61.2	60.8	58.9	39	39.2	37.9	37.9	39.2	39
			Estimated	24.3	22.5	22.1	23.8	26.3	24.5	25.5	23.8	21.8
		Ver.	**Zemax**	19.3	32.8	54.8	57.2	32.4	18.9	18.9	32.4	57.2
			Estimated	26.7	31.5	23.5	27.7	30.9	23	23.2	26.6	21.9
	Level 0.1	**Hor.**	**Zemax**	124.7	124.3	123.1	76.1	75.9	74.9	74.9	75.9	76.1
			Estimated	59	59.1	59.6	45.3	79.3	70.2	64.7	71	74
		Ver.	**Zemax**	70.1	61.9	122.2	120.9	60.9	66.7	66.7	60.9	120.9
			Estimated	89.7	70.6	66.4	101.9	71.7	71	51.3	60.9	75.4

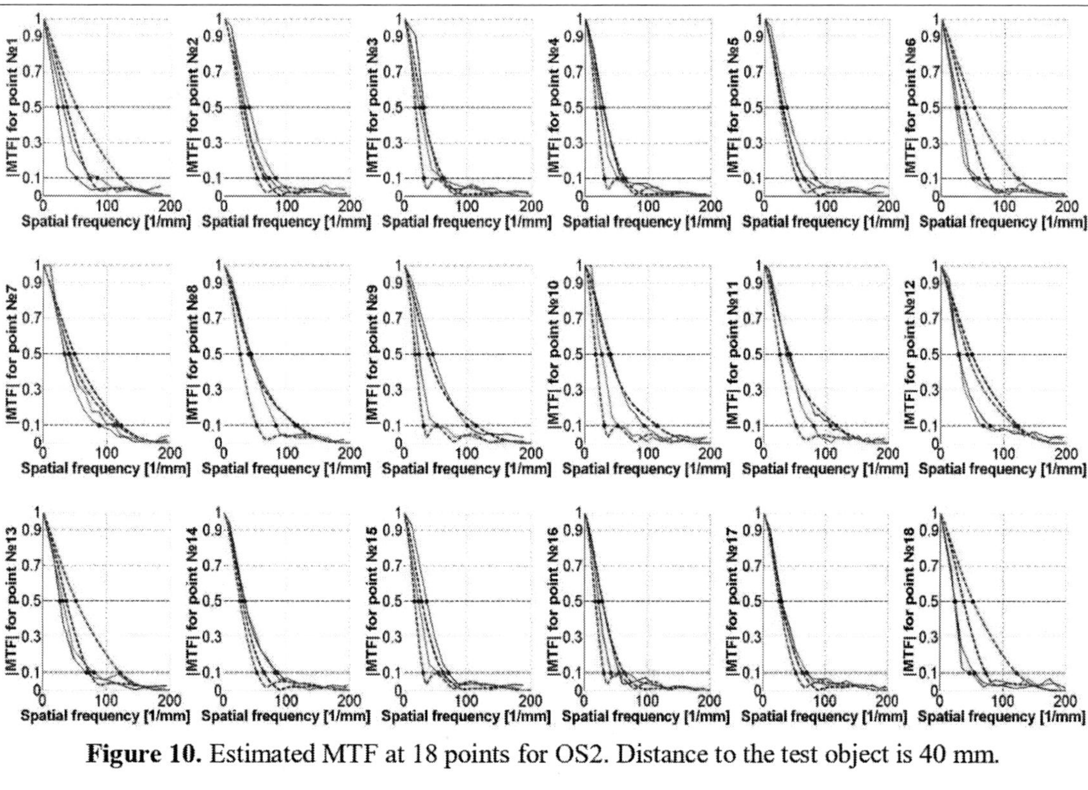

Figure 10. Estimated MTF at 18 points for OS2. Distance to the test object is 40 mm.

Table 8. The spatial frequencies at 0.5 and 0.1 levels of MTF for OS2 (in mm^{-1}). Distance to the test object is 40 mm.

Point's number				*1*	*2*	*3*	*4*	*5*	*6*	*7*	*8*	*9*
Left channel	**Level 0.5**	**Hor.**	**Zemax**	37.1	30.3	28.9	28.9	30.3	37.1	49.2	40.3	38.4
			Estimated	22.8	33.5	31.1	29.1	29.3	26.9	34	43.6	45.6
		Ver.	**Zemax**	52.3	26.5	17.6	17.6	26.5	52.3	42	26	17.7
			Estimated	31.5	39.8	24.1	24.4	36.6	22.9	40.2	42.4	23.8
	Level 0.1	**Hor.**	**Zemax**	72.7	64.7	63.4	63.4	64.7	72.7	122	116.6	114.3
			Estimated	52.1	71.7	59.6	58.7	65.7	58.7	88.1	113.5	100.8
		Ver.	**Zemax**	121	53.3	31.1	31.1	53.3	121	117	52.8	31.2
			Estimated	84.8	82.3	60.8	51.6	83.6	50.4	111.1	83.3	53
Point's number				*10*	*11*	*12*	*13*	*14*	*15*	*16*	*17*	*18*
Right channel	**Level 0.5**	**Hor.**	**Zemax**	38.4	40.3	49.2	37.1	30.3	28.9	28.9	30.3	37.1
			Estimated	40.7	42.5	28.6	27.3	34	36.8	28.4	30.8	23.4
		Ver.	**Zemax**	17.7	26	42	52.3	26.5	17.6	17.6	26.5	52.3
			Estimated	26.8	37.3	27.7	31.8	33.9	23.8	22.8	28.3	23
	Level 0.1	**Hor.**	**Zemax**	114.3	116.6	122	72.7	64.7	63.4	63.4	64.7	72.7
			Estimated	94	107.5	78.3	68.4	86.1	71.8	55.5	71.2	55.2
		Ver.	**Zemax**	31.2	52.8	117	121	53.3	31.1	31.1	53.3	121
			Estimated	52.1	80.9	67.4	80.2	82	55.5	42.2	71.3	46.8

In order to draw conclusions about the summary quality of the optical system, it is sufficient to compare the figures for OS1 and OS2 at one of the working distances, for example, at 15 mm (figures 6 and 9 for OS1 and OS2, respectively). Figure 9 shows that the experimentally measured MTF differs from the Zemax calculated one by 2-4 times at levels of 0.5 and 0.1. It indicates that OS2 has serious assembly faults. Figure 6 shows the estimated MTF for different areas of the image obtained using OS1. It indicates that the ratio of spatial frequencies for estimated MTFs to spatial frequencies for calculated MTFs at levels of 0.5 and 0.1 is less than 1.5 times. The main reason of that is possible noise distortions, which are not taken into account when estimating the MTF. Therefore this value is acceptable for our solution.

4. Conclusion

In this article, the possibility of using the slanted edge algorithm of MTF measurement for a stereoscopic endoscopic system was considered. It is shown that three images of plane test object with chessboard pattern on different working distances and 9 points of calculation per channel is enough for MTF estimation. A comparison of the estimated MTF with the Zemax calculated ones made it possible to identify assembly faults for the two self-developed prism-based stereoscopic endoscopy systems.

Further work will be directed towards automating the process of the MTF estimation from an image of chessboard pattern and developing user-friendly software for simultaneous spatial calibration and estimation of image quality. Usage of the detection algorithm of nodal points, the description of which can be found in [9], makes possible to select suitable areas for MTF measurements automatically. It allows us to build a map of the MTF values over the entire field of view. Capturing several images at the same position of the test object allows:

1. to evaluate the noise characteristics of the system and their impact on MTF measurements;

2. to reduce MTF measurement error by obtaining an average image over several frames.

Solution of the MTF estimating task allows us to create a universal user-friendly software that allows geometric calibration, measurement of distortion, MTF and noise characteristics. It will provide an opportunity to assess the quality of the operating system not only during assembly and adjustment, but also during the appraisal and working time of the device. Moreover, these procedures will make calibration and quality check available for users of the system without specialized personnel.

Acknowledgments
This research is supported by Russian Foundation for Basic Research (project 17-29-03469).

References

[1] Geng J and Xie J 2014 Review of 3-D endoscopic surface imaging techniques *IEEE Sensors Journal* **14** 945-60
[2] Bendall C A, Lia R, Salvati R and Chilek T A 2009 Stereo-measurement borescope with 3-D viewing, Patent US 7 564 626 B2
[3] Zhengyou Z, Flexible camera calibration by viewing a plane from unknown orientations, *Proc. of the Seventh IEEE Int. Con. on Com. Vis.* 666-73
[4] Gorevoy A V and Machikhin A S 2017 Optimal calibration of a prism-based videoendoscopic system for precise 3D measurements, *Computer Optics* **41** 536-45
[5] Photography–Electronic Still Picture Cameras–Resolution Measurements, ISO Standard 12233: 2000
[6] Batshev V, Machikhin A and Kachurin Y 2017 Stereoscopic tip for a video endoscope: problems in design *Proc. SPIE* 104664D (2017)004
[7] Machikhin A S , Batshev V I, Gorevoy A V, Khokhlov D D, Naumov A A and Kuznetsov A O 2019 Compact stereoscopic prism-based optical system with an improved accuracy of 3-D geometrical measurements *Optik* **185** 1172-81
[8] Kenichiro M, Takayuki Y, Yukihiro N and Masayuki S 2014 Modified slanted-edge method and multidirectional modulation transfer function estimation, *Optics Express* **22** 6040-46
[9] Yunsu B , Hyowon H and In So K 2016 Automated checkerboard detection and indexing using circular boundaries, *Pattern Recognition Letters* **71** 66-72

On computer modelling of operations, geometric characteristics and methods of pulsed optical tomography of nuclear fuel microobjects

E P Domoratskiy and T N Baybikova

School of Computer Engineering, National Research University Higher School of Economics, Myasnitskaya str., 20, Moscow, 101000, Russian Federation

E-mail: tbaibicova@hse.ru

Abstract. The paper refers to the area of morphological processing of projection images and its goal is to design some computer models of basic operations, geometric properties and methods of pulsed optical tomography which provide a high-speed production operational control and sorting of each microobjet (MO) of nuclear fuel in their flow according to their size and shape. Based on the proposed computer models, a precision laser method of industrial differential control and quality control of an MO flow of nuclear fuel was developed and experimentally tested. The method uses the statistical reconstruction of the size (*D*) and shape (*K*) of each microobject and takes into account the overall dimensions of the outlines of three mutually orthogonal two-dimensional pulsed discrete projection images of a microobject. The processing speed of this method is 100 MO/s in the diameter range of 400–1500 μm. The relative error of an MO diameter control is no more than 0.25% (at the reliability of P_D = 0.7 and K = 1.3 relative units), and the relative error of the non-sphericity coefficient control lies in the range of 2.3% (P_K = 0.7 and K = 1.3) to 0.6% (P_K = 0.96 and K = 1.05).

1. Introduction

One of the most promising and innovative ways of modern nuclear power development is the creation and application of High Temperature Gas-cooled Reactors (HTGR). The HTGR concept is being implemented in such countries as the USA (GA, ORNL), Japan (Fuji Electric), France (Framatom), Russia (National Research Center «Kurchatov Institute»), as well as in Germany, the UK, China, South Korea and other countries [1, 2]. There are two main HTGR reactor designs in the international development arena: pebble bed design and prismatic block design. The pebble-bed reactors use spherical fuel elements. The prismatic block reactors use cylindrical “fuel compacts”. The fuel compacts are massively loaded directly into the reactor active zone. The main part of fuel elements is granulated nuclear micro fuel particles which represent convex multi-layered three-dimensional microobjects (MO) of irregular spherical form with a size from 400 to 1500 μm, consisting of a solid kernel with a diameter of 500 μm and 4-5 layers of coatings with thickness from 50 to 90 μm which regulate the leakage of fission products. The technology of granulated nuclear materials (microobjects) manufacturing provides their mass (flow) production and consists of successive interrelated operations (steps): the formation of a dispersed jet of drops of original solvent based on UO_2 (Sol-gel process); the fabrication of solid kernels from drops of solvent; the sequential covering of kernels with one, two, three, four coatings; the fabrication of micro fuel particles (kernels with five coatings).

The most important quality characteristics of an MO of granulated nuclear fuel and fuel elements fabricated on their basis which determine their reliability, physical, chemical and consumer properties are the geometric properties of their size and shape. To ensure mass (flowing) production and quality control of nuclear fuel in the on-line mode, the geometric control method must meet a unique set of requirements and be simultaneously automatic, contactless, high-speed, dynamic, differential, multidimensional, precision and reliable, as well as provide classification and sorting of moving (flying) MOs according to their shape and size in real time.

2. Problem statement

The paper relates to the area of experimental physics and consists in the development of some computer models of operations (procedures), geometric properties and methods of pulsed optical tomography of moving three-dimensional objects which provide production operational geometric control as well as quality control of nuclear fuel MO flow [3].

3. Description of modelling methods

In general terms, the controlled geometric parameter (*CP*) of a three-dimensional object is characterized by an objective function (1) with a set of image properties necessary and sufficient for its adequate (with a given accuracy, reliability and processing speed) description [4]:

$$CP = f\left(Mo, Pi, Be, Md, Ai, Ni, Mr, T\right), \qquad (1)$$

where *Mo* is a method for obtaining an image; *Pi* are image parameters; *Be* are the basic elements (characteristics) of an image; *Md* is a method of determining the basic elements of an image; *Ai* is the viewing angle of an image (projection) relative to the object (Euler angles α, β, γ); *Ni* is the number of images (projections); *Mr* is the method of reconstruction (recovery) of an image phantom with the usage of image features; *T* is the total time (duration) of *CP* control and is determined as the sum $T = ti + tp + tr$, where *ti* is the time of image obtaining; *tp* is the time of image processing; *tr* is the reconstruction time.

In accordance with (1), the development of the method of geometric control of an MO involves the development of the following computer models of operations [4], which are understandable from the scheme (figure 1):

1. approximation (image) of a three-dimensional MO and spatial geometric properties of its size and shape;
2. spatial orientation of each MO in the flow;
3. determination of the optimal view, representative number and optimal spatial orientation (angles) of projection images of an MO;
4. method of producing (generating) the pulsed projection images of an MO;
5. determination of the numerical values of the optimal basic geometric properties of each projection image of an MO;
6. reconstruction methods of the size and shape of a three-dimensional MO on the basis of its discrete projection images.

Let us consider the operations of the modelling methods in more detail.

1. The ellipsoid of the General form is chosen as a mathematical model approximating the geometric properties of an MO (its three-dimensional image – phantom). The canonical equation of this ellipsoid is as follows:

$$\left(x'^2/A_1^2\right)+\left(y'^2/A_2^2\right)+\left(z'^2/A_3^2\right)=1, \qquad (2)$$

where *A1, A2, A3* are the semi – axes of the ellipsoid; *x', y', z'* belong to the original coordinate system of the ellipsoid.

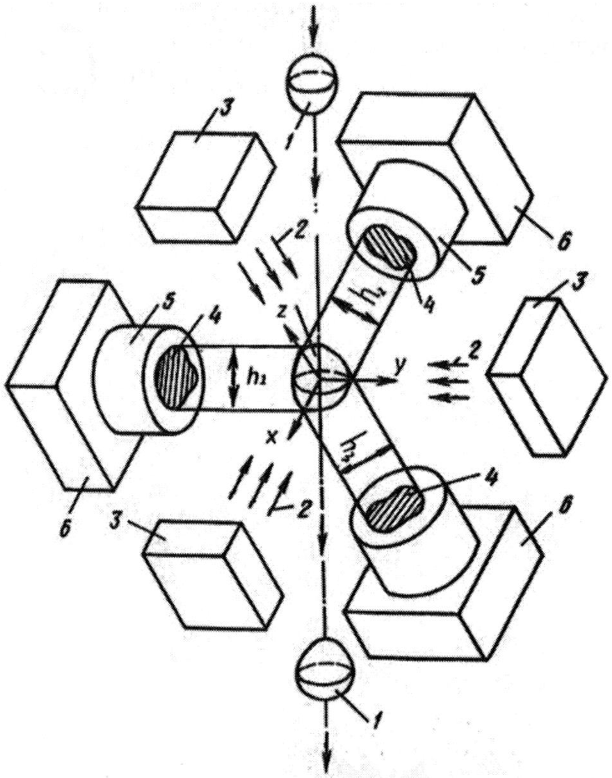

Figure 1. Method of geometric control of the size and shape of an MO flow. 1 – MO flow; 2 – parallel beam of optical radiation; 3 – pulsed light source; 4 – projection image of an object (s_i); 5 – depicting optics; 6 – position – sensitive video detector.

The geometric (overall) dimensions of an MO are determined by the numerical values of the mutually orthogonal axes of the ellipsoid *2A1, 2A2, 2A3*, its average diameter (*D*) and shape factor (*K*) are determined respectively from the relations (3) [5]:

$$D = \sum_{i=1}^{3} 2A_i/3\,;\; K = \max\{A_1, A_2, A_3\}/\min\{A_1, A_2, A_3\}. \tag{3}$$

2. The model of spatial orientation of an MO in the flux is determined by the rotation matrix of the ellipsoid *R* in a three-dimensional space to the Euler angles α, β, γ:

$$R = \begin{bmatrix} R_{11} & R_{12} & R_{13} \\ R_{21} & R_{22} & R_{23} \\ R_{31} & R_{32} & R_{33} \end{bmatrix} = \begin{bmatrix} \cos\alpha & \sin\alpha & 0 \\ -\sin\alpha & \cos\alpha & 0 \\ 0 & 0 & 1 \end{bmatrix} . \begin{bmatrix} \cos\beta & 0 & -\sin\beta \\ 0 & 1 & 0 \\ \sin\beta & 0 & \cos\beta \end{bmatrix} . \begin{bmatrix} \cos\gamma & \sin\gamma & 0 \\ -\sin\gamma & \cos\gamma & 0 \\ 0 & 0 & 1 \end{bmatrix} =$$

$$= \begin{bmatrix} \cos\alpha\cdot\cos\beta & \sin\alpha & -\cos\alpha\cdot\sin\beta \\ -\sin\alpha\cdot\cos\beta & \cos\alpha & \sin\alpha\cdot\sin\beta \\ \sin\beta & 0 & \cos\beta \end{bmatrix} \cdot \begin{bmatrix} \cos\gamma & \sin\gamma & 0 \\ -\sin\gamma & \cos\gamma & 0 \\ 0 & 0 & 1 \end{bmatrix} =$$

$$= \begin{bmatrix} \cos\alpha\cdot\cos\beta\cdot\cos\gamma-\sin\alpha\cdot\sin\gamma & \cos\alpha\cdot\cos\beta\cdot\sin\gamma+\sin\alpha\cdot\sin\gamma & -\cos\alpha\cdot\sin\beta \\ -\sin\alpha\cdot\cos\beta\cdot\cos\gamma-\cos\alpha\cdot\sin\gamma & -\sin\alpha\cdot\cos\beta\cdot\sin\gamma+\cos\alpha\cdot\cos\gamma & \sin\alpha\cdot\sin\beta \\ \sin\beta\cdot\cos\gamma & \sin\beta\cdot\sin\gamma & \cos\beta \end{bmatrix} \quad (4)$$

In this case, the size and shape of an MO model (ellipsoid) are determined deterministically, and its orientation (Euler angles) can be specified both deterministically and stochastically by even distribution.

3. The selection of the optimal type, number and angles of projection images of an MO, as well as their basic properties is determined on the basis of the optimal (best) combination of characteristics of accuracy, reliability and processing speed of the control method and is based on a numerical comparative analysis of their informativeness with the usage of simulation statistical modelling by the maximum entropy method. For this method, the optimal type of primary information is three discrete two-dimensional projection images. At the same time, their optimal spatial orientation is mutual orthogonality [6] (figure 1).

4. Pulsed projection images of an MO moving (flying) in the flow are formed by their exposure in parallel beams from three pulsed sources of optical radiation and their simultaneous synchronous registration by means of three corresponding optoelectronic sensors of discrete pulsed micro images [5]. The target exposure time of the video detector Te when obtaining the pulsed projection image of an MO is determined by the condition of the absence of the blurring of the image frame obtained from the moving MO in accordance with the equation:

$$Te \le R\cos\phi / (v + 0,5\omega D)K^*, \quad (5)$$

where R is the resolution of a video detector on an image; φ is the angle between the trajectory of the movement of the MO and the plane of the video detector target; v is the speed of the MO mass centre; ω is the maximum angular velocity of its rotation; D is the MO diameter; K^* is the image scale.

To reduce the information redundancy and improve the control speed, the spatial differentiation of pulsed electronic discrete images of an MO (that is, the creation of their outlines) is fulfilled. Thus, the computer model of each discrete outline is an ellipse (equation 6), the coordinates of the points of which are rounded to the nearest integers, which is adequate to the process of discretization of the primary information obtained from a real MO [5].

$$\sum_{i=1}^{3} A_i^2 R_{i2}^2 x_1^2 - 2\sum_{i=1}^{3} A_i^2 R_{i1}^2 x_1 y_1 + \sum_{i=1}^{3} A_i^2 R_{i1}^2 y_1^2 = A_1^2 A_2^2 A_3^2 \sum_{i=1}^{3}\left(R_{i3}^2 / A_i^2\right). \quad (6)$$

5. The optimal basic properties of projection images are chosen on the basis of numerical analysis of their informativeness by the method of maximum entropy. These properties include: the image length of the one-dimensional projection of an MO to the selected direction h_i; the area of the image of a two-dimensional projection s_i; the maximum $lmax_i$ and minimum $lmin_i$ linear dimensions of an image of the two-dimensional projection of an MO (i is a projection number). The numerical values of these basic properties are determined by the relations (7) [5]:

$$\begin{cases} h_i = 2\left[\sum_{j=1}^{3} A_j^2 R_{ji}^2\right]^{1/2} \\ s_i = \pi A_1 A_2 A_3 \left[\sum_{j=1}^{3} \left(R_{ji}^2 / A_j^2\right)\right]^{1/2}, \\ l\,maxi = 2\left\{\left\{I_i + \left[I_i^2 - 4A_1^2 A_2^2 A_3^2 \sum_{j=1}^{3} \left(R_{ji}^2 / A_j^2\right)\right]^{1/2}\right\} \Big/ 2\right\}^{1/2} \\ l\,mini = 2\left\{\left\{I_i - \left[I_i^2 - 4A_1^2 A_2^2 A_3^2 \sum_{j=1}^{3} \left(R_{ji}^2 / A_j^2\right)\right]^{1/2}\right\} \Big/ 2\right\}^{1/2} \end{cases} \quad (7)$$

where: $I_i = \sum_{j=1}^{3} A_j^2 \left(1 - R_{ji}^2\right)$; j, i are projection numbers and numbers of current orientation (angle);

6.1 The reconstruction method of a three-dimensional MO image based on the usage of basic characteristics h_i and s_i is defined by the simultaneous equations (8) [7]:

$$\begin{cases} A_1^2 + A_2^2 + A_3^2 = \sum_{i=1}^{3} h_i^2 / 4, \\ A_1^2 A_2^2 + A_1^2 A_3^2 + A_2^2 A_3^2 = \sum_{i=1}^{3} s_i^2 / \pi^2, \\ A_1^2 A_2^2 A_3^2 = \left[\sum_{i=1}^{3} h_i^2 s_i^2 / 4\pi^2\right] - \left(h_1^2 h_2^2 h_3^2 / 32\right)\left\{1 - \left\{\prod_{i=1}^{3}\left[1 - \left(16/\pi^2\right)\left(h_i^2 s_i^2 / h_1^2 h_2^2 h_3^2\right)\right]\right\}^{1/2}\right\}. \end{cases} \quad (8)$$

6.2 The reconstruction method of a three-dimensional MO image based on the usage of basic characteristics $lmax_i$ and $lmin_i$ is defined by the simultaneous equations (9) [8]:

$$\begin{cases} A_1^2 + A_2^2 + A_3^2 = \sum_{i=1}^{3} m_i, \\ A_2^2 A_3^2 + A_1^2 A_3^2 + A_1^2 A_2^2 = \sum_{i=1}^{3} c_i, \\ A_1^2 A_2^2 A_3^2 = \sum_{i=1}^{3} m_i c_i - 2 m_1 m_2 m_3 \left\{1 - \left[\prod_{i=1}^{3}\left[1 - \frac{m_i c_i}{m_1 m_2 m_3}\right]\right]^{1/2}\right\}. \end{cases} \quad (9)$$

where $m_i = \left[\sum_{j=1}^{3} \left(l_{1j}^2 + l_{2j}^2\right)/8\right] - \left[\left(l_{1i}^2 + l_{2i}^2\right)/4\right]$; $c_i = l_{1i}^2 l_{2i}^2 / 16$.

When solving the simultaneous equations (8) and (9) (for example, by the trigonometric method) with regard to the semi-axes of an ellipsoid *A1*, *A2*, *A3* and considering that only positive values of the roots make sense, we obtain the linear dimensions of the semi-axes of an ellipsoid which approximates an MO. In this case, the geometric properties of the size and shape of an MO are determined from the relations (3).

On the basis of the computer models proposed in this paper, a laser precision high-speed method of geometric optic-electronic differential contactless production and quality control is developed and experimentally tested. The method allows to realize these types of control in real-time of an MO flow of nuclear fuels and is based on dynamic spatial-temporal statistical few-viewed reconstruction of the size and shape of each MO on the basis of the maximum and minimum overall dimensions of the discrete outlines of three mutually orthogonal two-dimensional pulsed projection images of an MO [4] (figure 1).

When describing the size of each MO, the spatial geometric properties are its overall dimensions and the average projection diameter (*D*) of its approximating ellipsoid of the General form. When describing the shape of an MO, the spatial geometric property is the coefficient of non-sphericity (*K*), determined by the ratio of the maximum and minimum dimensions (axes) of the approximating ellipsoid.

The processing speed of this method is not less than 100 MO/s in the diameter range of 400–1500 μm for exposure time 0.8–1.0 μs. The relative error of an MO diameter control is no more than 0.25% (at the reliability of $P_D = 0.7$ and $K = 1.3$ relative units), and the relative error of the non-sphericity coefficient control lies in the range of 2.3% ($P_K = 0.7$ and $K = 1.3$ relative units) to 0.6% (with $P_K = 0.96$ and $K = 1.05$ relative units).

4. Conclusion

The direct and inverse problems of computational pulsed optical tomography are formulated and solved by the developing of some methods of computer simulation of operations (processes) of obtaining (registration), morphological analysis and space-time few-viewed dynamic reconstruction of a three-dimensional MO of nuclear fuel on the basis of the basic properties of pulsed discrete projection images.

The developed methods of dynamic reconstruction and methods of production geometric control on their basis have an optimal combination of high characteristics of accuracy, reliability and processing speed, fully meet the necessary requirements and minimize their hardware and software implementation and control time. Therefore, they can be used in the measurement and control of various spherical three-dimensional objects of irregular shape (ball fuel elements, elements of thermonuclear fuel, aerosols of diamond granules, etc.)

References

[1] IAEA 2010 *High Temperature Gas Cooled Reactor Fuels and Materials* IAEA-TECDOC-1645 Vienna (IAEA Publ.) 182

[2] Grebennik V I, Kukharkin N E and Ponomarev-Stepnoy N N 2008 *Vysokotemperaturnye gazookhlazhdaemye reaktory — innovatsionnoe napravlenie razvitiya atomnoy energetiki* Moscow (Energoatomizdat Publ.) 136

[3] Moder J J and Elmaghraby S E 1981 *Issledovanie operatsiy. Metodologicheskie osnovy i matematicheskie metody* Moscow(Mir Publ.) **1** 716

[4] Domoratskiy E P and Baybikova T N 2017 *Proc. XIV Int. Sci. Conf. on Optical methods of flow investigation (Electronic Materials)* (Moscow:MPEI) 275-81

[5] Domoratskiy E P 2016 Synthesis algorithm of the geometrical characteristics of the projection images of three-dimensional objects *J. Information Technologies* **22 8** 605–09

[6] Domoratskiy E P and Baybikova T N 2015 *Proc. II Int. Sci. Conf. on Regional issues of Earth remote sounding* (Krasnoyarsk: SibGAU) 139–42

[7] Domoratskiy E P 2017 Method of statistical few view reconstruction of geometrical characteristics of three-dimensional objects by their discrete projection images *J. Information Technologies* **23 3** 184–87

[8] Baybikova T N and Domoratskiy E P 2017 Technique for Dynamically Reconstructing Shape and Size of Three-Dimensional Objects from their Projected Dimensions *J. Herald of the Bauman Moscow State Tech. Univ., Nat. Sci.*. **5** 109–17

Hyperspectral camera application for remote gas analysis

S M Neverov

Scientific and Technological Center of Unique Instrumentation of the Russian Academy of Sciences, 117342, Moscow, Butlerov Street, 15

E-mail: snev85@yandex.ru

Abstract. The report reviewed current systems for monitoring gas leaks at industrial facilities. Since in industrial plants the emission of some gases can be technologically justified, and the presence of others indicates a violation of the technological process and is a sign of an emergency situation, a possible option is described not only for detection the fact of leakage, but also for the classification of detected gases. We considered existing hardware-software complexes including acousto-optical, and listed their shortcomings, such as insufficient sensitivity and low data processing speed. In general, we describe the novel technical implementation of imaging spectrometry. For simultaneous detection of spatial and spectral properties of objects, the infrared sensor is divided into a mosaic of the smallest optical filters. Each filter passes individual intervals of the spectrum, blocking the others. The uncooled detector 640×512 can be covered with 16 filters, which spectral characteristics should correlate with the absorption spectra of target gases.

1. Introduction

In recent years, hyperspectral cameras have become widespread in the food industry to assess food safety. The advent of cost-effective production allowed compact, lightweight and low-cost hyperspectral systems to become part of machine vision systems for food sorting and quality control in the pharmaceutical industry. Such applications require cameras to "see" targets on the conveyor and to ignore other background information. People are currently successfully solving this task, but the transition to automated systems eliminates the human factor and saves costs in the long term. The technology of machine vision using hyperspectral camera is more productive and faster than a human, and can see far beyond the visible spectral region, resolving damages inaccessible to the human eye.

Understanding of hyperspectral imaging principal and technical advantages and disadvantages of different source and detectors are important for obtaining reliable hyperspectral images with high quality. The choice of instrument components, their adjustment and calibration procedures require a good understanding of the hyperspectral imaging principles. The spectral imaging system creates a stack of images of the same object in different spectral wavelength ranges. There are three main classes (figure 1) in the field of spectral imaging: multiband, multispectral, and hyperspectral (ultraspectral). The main difference is the number of images inside the spectral cube and their bandwidth. Multiband devices provide detection in several rather wide bands (figure 1). Multispectral instruments area characterized by narrow channels usually of the same bandwidth. Hyperspectral systems detects the spectral range with no gaps. The last provide usually hundreds or thousands of images that make each pixel of a hyperspectral image have its own continuous spectrum. The result is a hyperspectral image, which is a three-dimensional (3D) Hyperspectral Cube. It consists of pixel vectors and contains the spectral information as well as two-dimensional spatial information [1].

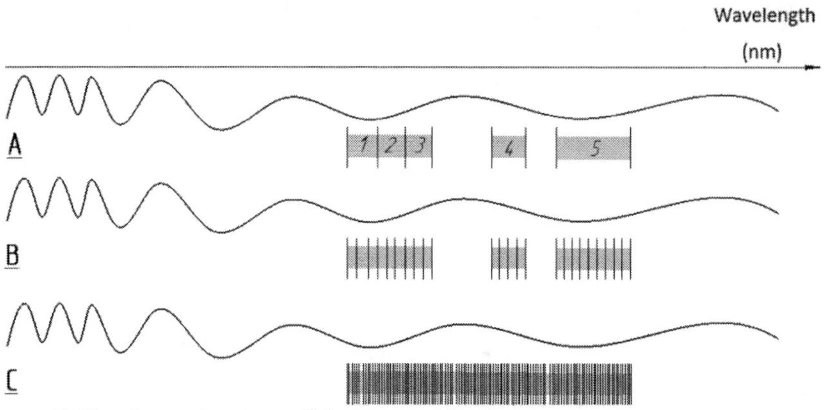

Figure 1. Bandpass structure of the spectral instruments:
(A) Multiband device; (B) Multispectral instrument; (C) Hyperspectral system.

As for the data acquisition modes, they are divided into four: spot scanning (or whiskbroom), line scanning (or pushbroom) scanning, spectral scanning and single shot (figure 2). While the whiskbroom mode gets all the bands, pixel by pixel, by moving the detector in the *x-y* space to store the data in array form pixel by pixel recording of spectral channels, this method called band-interleaved-by-pixel (BIP), a pushbroom mode works similarly, but instead of a pixel scanning a spatial line is formed, which, ultimately, is written to the array line by line recording simultaneously all the spectral bands, this type called band-interleaved-by-line (BIL) cube. Several other features of pushbroom mode include compact size, light weight, easier control and higher signal level due to use of small differential gratings as the spectral element. In flat scan mode, a per-channel recording array create spectral bands sequential (BSQ) cube, consisting of several images taken consequently, each of which contains spectral data relative to the entire *x-y* space. Finally, there is a mode that receives all spatial and spectral data at the same time, known as a single shot. It uses a multispectral detector. In addition, we indicate some of the problems of each mode. Whiskbroom is a slow-motion mode, and pushbroom should be used for a short time to avoid the risk of inconsistencies in the spectral band level (saturation or underexposure). Flat scanning is hardly suitable for moving environments, while single shot is a technology under development that still does not support high spatial resolution. [1]

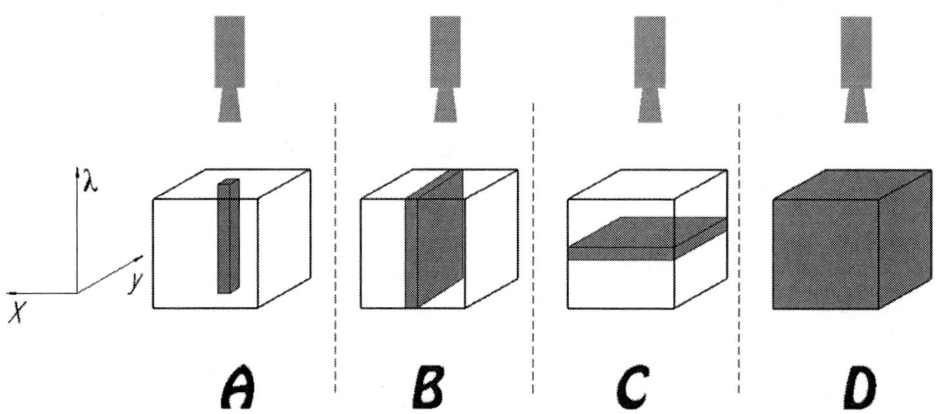

Figure 2. Hyperspectral data acquisition modes: (A) point scanning (whiskbroom mode); (B) line scanning (pushbroom mode); (C) spectral scanning (imaging mode); (D) single shot mode (adapted from [1]).

2. Element base

Below are analysis of the features of devices on various elemental base.

2.1. Sources

Light sources generate light to illumination of the target and are an internal part of optical control systems. Typical light sources used in hyperspectral imaging systems are: halogen lamps, light emitting diodes, lasers and tunable light sources.

- Halogen lamps. The main halogen lamps disadvantages are relatively short service life, high heat capacity, the thermal shift of the spectral peak due to the change of temperature, unstable power due to fluctuations of the operating voltage and sensitivity to vibration.
- Light emitting diodes (LEDs). Critical disadvantages of LEDs include sensitivity to large voltage fluctuations and transition temperature, low light intensity compared to halogen lamps and obtaining granular light when several LEDs are used together.
- Lasers. Light generated by lasers has a higher intensity and much narrower bandwidth than LEDs. However, in the spectral range from 8 μm to 14 μm lasers have big power consumption, dimensions and weight.
- Tunable light sources. This source is ineffective for point and line scanning, but is well-suited for BSQ mode. That is why, tunable light sources are practically unsuitable for conveyor systems.

For our application sources are not needed, as a detection distance must be close to 50 meters and system will be used like handheld device with small size and weight. For remote sensing, the natural radiation sources should be used like the Sun or thermal radiation of the environment (in MIR range).

2.2. Spectral elements and detection devices

Devices with wavelength dispersion are necessary for hyperspectral imaging systems using broadband light sources. They have the function of dispersing a broadband light into different wavelengths. Typical examples listed below: filter wheels, imaging spectrometers, acousto-optic tunable filters, liquid crystal tunable filters, Fourier-transform (FT) imaging spectrometers, and single-image devices.

- Filter wheels. Main disadvantages of filter wheels are mechanical vibration from moving parts, slow wavelength switching and unmatched image due to filter movement.
- Imaging spectrometers. They normally operate in linear scanning mode, have the ability to instantly disperse incident broadband light at different wavelengths and generate a spectrum for each point on the scanned line without the use of moving parts. The main disadvantages of reflective gratings are great difficulties with distortion correction.
- Tunable filters. Acousto-optic tunable filter (AOTF) and liquid crystal tunable filter (LCTF) are electronically tunable bandpass filters. On base of acousto-optic interactions in the crystal, AOTF selects light at one wavelength from a broadband source due to periodicity of applied acoustic field. The liquid crystal tunable filter (LCTF) has electronically controlled liquid crystal cells inserted between two parallel polarizers to transmit light with a certain wavelength, while the light outside the bandwidth is eliminated. Disadvantages of the tunable filter are small light acquisition angle and low efficiency of light collection, the need for linearly polarized incident light, which can cause 50% light loss, and a longer exposure time than that of spectrometers working with images in similar lighting conditions.
- Fourier-transform Image formation spectrometers. FT imaging spectrometers use an interferometer for the self-interference of broadband light, resulting in an interferogram that contains its spectral data. The generated interferogram is then computed using the inverse Fourier transform to resolve the composition of the optical frequencies (inverse wavelengths) of broadband light. However, spectrometer requires a time interval to shift the moving mirror, so it takes a long time to collect the interferogram to obtain accurate spectral resolution.
- Single-shot devices. Single-frame images can simultaneously collect multiplexed spatial and spectral data, providing a hypercube at a video frame rate. It should to say, single-frame scanners are still at an early stage.
- Zonal detectors. Zonal detectors have the function of quantifying the intensity of the light received by converting incident photons into electrons. Cameras used charge coupled devices (CCD) and metal-

Journal of Physics: Conference Series **1421** (2019) 012046 doi:10.1088/1742-6596/1421/1/012046

oxide-semiconductor structures (CMOS) are the two main types of solid-state detectors. Photodiodes made of light-sensitive materials are the main element of CCD and CMOS to convert radiation energy into an electrical signal. The main disadvantages of CCD are that both the photodetector and the reading amplifier at each pixel are not included in the CCD image sensor, as a result its need time to receive an image. The limitation of CMOS cameras is higher noise and dark current than in CCD, due to the built-in chips used to transmit and amplify signals, and as a result lower dynamic range and sensitivity than the CCD [2].

For the developed single-shot system we choose a microbolometer as a small and light detector. Infrared radiation with a wavelength of 7.5 to 14 microns falls on the detector material, heating it and thus changing its electrical resistance. This resistance change is measured and processed as temperatures that can be used to create the image. Main advantages of microbolometer camera: low power consumption relative to cooled detector thermal imagers, very long mean time between failures, less expensive compared to cameras based on cooled detectors. Also it provides real video output immediately after power on, and for applications that require relatively short distances, the physical dimensions of the camera are even smaller. It disadvantage is less sensitive than cooled thermal imagers and photon detectors (due to higher noise), but sensitivity of microbolometers rise every year.

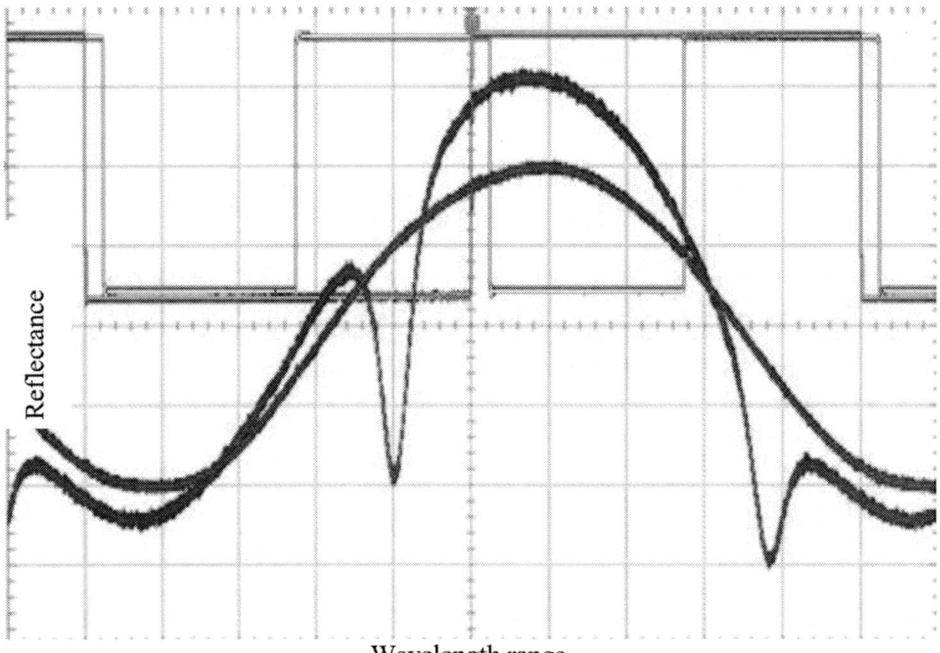

Figure 3. Hyperspectral image slice.

The hyperspectral camera creates a cube of data with a size of about 100 or more narrow spectral bands in each pixel, in the instantaneous field of view [4]. The number of pixels varies depending on the sensor tool. A typical GC system collects light (usually reflected from a topographic object) that passes through a slit, spectrally decomposed using a grid or prism and focuses on a detector or sensor. For each pixel or raster cell, the system receives and graphs the brightness value at each wavelength, thereby creating a continuous spectrum for the image cell. Scientists can compare these high-resolution spectra over the entire wavelength region with standard spectral libraries for identification. For example, concrete versus granite, healthy cultures compared with diseased cultures, healthy cells compared with cancers, and even determine the cause of human redness (from stress or not) (figure 3) [5].

3. The system developed

A hyperspectral camera measures adjacent spectral bands, in contrast to multispectral imaging, which measures diversity spectral bands. This difference favorably distinguishes multispectral cameras for our task, since no cumbersome computation and costly operating systems are needed for unfolding the spectrum study.

The advantage of a multispectral camera is that its infrared sensor is divided into a mosaic of the smallest optical filters. Each filter passes individual infrared regions of the spectrum, blocking the others. On the figure 5 below you can see graphs with absorption lines from 8 μm to 14 μm (spectral range that we choose for detection) of gases that is critical to indicate on the Oil & Gas facilities. We choose 8 gases: Propane C_3H_8, Butane C_4H_{10}, Methane CH_4, Ethan C_2H_6, Acetylene C_2H_2, Carbon monoxide CO, carbon dioxide CO_2, Ammonia NH_3, Hydrogen sulfide H_2S. We identified the strongest absorption lines, which are at the same time placed at a distance from each other (that is important item to avoid overlapping). Second critical point – all this lines are out of water absorption areas.

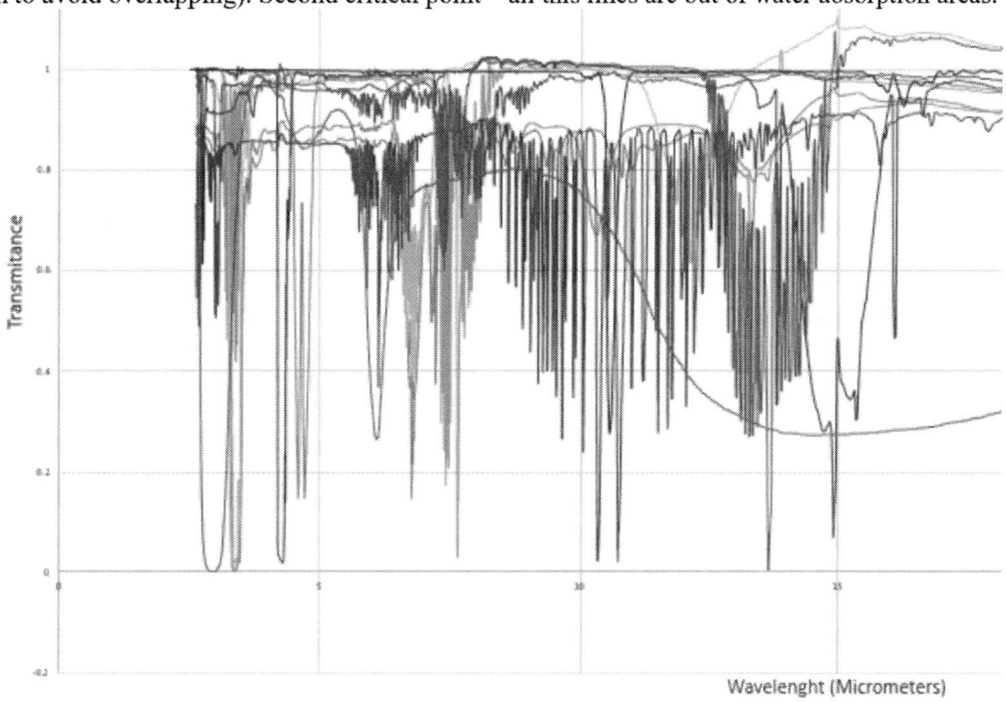

Figure 4. Critical gases absorption lines from 8 μm until 14 μm spectral range.

The uncooled detector 640 × 512 allows setting up to 16 filters, the spectral characteristics of which are selected after analyzing the absorption spectra of several gases. The presence of these gases at an industrial facility is associated with continuous monitoring of the concentration of these gases in the working area of the enterprise. To obtain the desired configuration was processed data array. In the future, based on the images obtained using the software included in the system, it will be possible to calculate the gas concentration and select the gas type in the image with the corresponding color in real time. Technically we placed very narrow optical filter on each channel (16 channels), we use duplication, so one gas – 2 channels. We calculate small lenses that must glue on the infrared detector together with filters. Usually in multispectral cameras use the turret of optical filters to receive on one detector a few spectral image, but huge disadvantages of this scheme is a time between images. For example to receive image in 10 spectral range takes around 10 seconds (based on Telops and Bertin multispectral cameras specification [6, 7]). This time is more than enough to lose position of the gas cloud – as a result each images takes in different environment condition. In our innovation scheme we receive image in wide spectral range between 8 μm and 14 μm less than one second.

The development of methods for the simultaneous analysis of a significant amount of gases can be simulated using acousto-optic tunable IR spectrometers [8], which allows one to tune in series to an

Journal of Physics: Conference Series **1421** (2019) 012046 doi:10.1088/1742-6596/1421/1/012046

arbitrary set of wavelengths. Finding optimal combinations of spectral lines and bands requires solving optimization problems [9] and can be modeled on a similar spectrometer that implements the function of controlling the width of the transmission window of an acousto-optic filter [10].

4. Conclusion

We have analyzed the optical and spectral characteristics of hyperspectral and multispectral cameras, revealed their distinctive features and principles of construction, and determined the trends of their design. As a result of the research, we identified an important problem: the absence of technical solutions that meet the full set of requirements for multispectral cameras and methods for industrial monitoring. We suggest a novel group of multispectral cameras for industrial use at hazardous production facilities. We plan to provide detection of multispectral image in wide spectral range 8 – 14 μm instantly (less than 1 second). Each of the 16 channels will provide information regarding critical gas components and its concentration.

References

[1] Wu D and Sun D -W 2013 Advanced applications of hyperspectral imaging technology for food quality and safety analysis and *assessment A review – Part II Fundamentals. Innov. Food Sci. Emerg. Technol.* **19** 1–14

[2] Sellar R G and Boreman G D 2005 Classification of imaging spectrometers for remote sensing applications *Optical Engineering* **44(1)** 013602

[3] Kastek M, Piątkowski T, Dulski R, Chamberland M, Lagueux P and Farley V 2012 Hyperspectral Imaging Infrared Sensor Used for Chemical Agent Detection and Identification *Sym. on Phot. and Opto.* SOPO 2012 art no 6270545

[4] Farley V, Chamberland M, Lagueux P, Vallières A, Villemaire A and Giroux J 2007 Chemical agent detection and identification with a hyperspectral imaging infrared sensor Proc. of SPIE 6661 66610L

[5] Kastek M, Piątkowski T, Trzaskawka P 2011 *Infrared imaging Fourier transform spectrometer as the stand–off gas detection systems, Metrology and Measurement Systems* **18** 4 607- 20

[6] Gagnon M-A, Tremblay P, Savary S, Farley V, Guyot É, Lagueux P, Morton V, Giroux J and Chamberland M 2017 Passive thermal infrared hyperspectral imaging for quantitative imaging of shale gas leaks Proc. SPIE *10428 Earth Resources and Environmental Remote Sensing/GIS Applications* **8** 1042818

[7] Doz S, Foucher P Y and Watremez X 2018 Methane leak near real time quantification with a hyperspectral infrared camera Proc. SPIE 10661 *Thermosense: Thermal Infrared Applications XL* 1066103

[8] Mazur M M and Pozhar V E 2015 *Optophysical measurements (Measurement Techniques)* **58** 982

[9] Fadeyev A V, Pozhar V E and Pustovoit V I 2013 Proc. SPIE 8890

[10] Mazur M M, Mazur L I, Suddenok Yu A and Chorin V N 2018 *Optics and Spectroscopy* **125** 272

Journal of Physics: Conference Series **1421** (2019) 012047 doi:10.1088/1742-6596/1421/1/012047

Study of convective structures and phase transition induced in a water layer by non-stationary boundary conditions

Yu N Dubnishchev[1,3], V A Arbuzov[1,3], E V Arbuzov[1,2,3], V S Berdnikov[1,3], O S Zolotukhina[1,3]and S A Kislitsin[1,3]

[1] Kutateladze Institute of Thermophysics Siberian Branch of the Russian Academy of Sciences, Prospekt Akademika Lavrent'eva 1, Novosibirsk, 630090, Russia
[2] B Sobolev Institute of Mathematics Siberian Branch of the Russian Academy of Sciences, Ac. Koptyuga ave. 4, Novosibirsk, 630090, Russia
[3] Novosibirsk State Technical University, Prospekt K. Marksa 20, Novosibirsk, 630073, Russia

E-mail: dubnistchev@itp.nsc.ru

Abstract. The problem of reconstructing interference and Hilbert structures from a numerical model of the evolution of the thermal field of convective flows in a water layer bounded by flat heat-exchanging surfaces under unsteady boundary conditions in the monotonic cooling mode and taking into account density inversion at a temperature of +4 °C was solved. The simulation of the thermal fields of convective flows in the form of the dynamic structure of isotherms was carried out taking into account the nonlinear dependence of thermal conductivity and water density on temperature. The field of the phase function, its Hilbert image and the interference field, which are compared with the results of the interference and Hilbert visualization of the fields of phase optical density obtained in the experiment, were reconstructed from the isotherms field supplemented by calculating of the velocity field and of the temperature gradients field. The presented films illustrate the qualitative adequacy of the coevolution of numerical models and real processes.

1. Introduction

Interest in the study of convective flows is associated with the special importance of convection in geodynamics, physics of the atmosphere and ocean, in hydrodynamic and thermophysical processes associated with the formation and growth of crystals [1, 2]. The relevance of this kind of research has recently increased due to the observed anomalies in the formation and melting of ice in the Arctic and Antarctic regions, as well as the intensification of the development of offshore oil and gas-bearing areas of the Arctic. Convective structures in a vertical water layer bounded by flat heat exchange surfaces under unsteady boundary conditions are visualized in [3] by methods of Hilbert optics and shear interferometry. Numerical simulation of temperature field as of field of isotherms in the mode of monotonic cooling of vertical walls is performed. The inverse problem of reconstruction of interferograms and hilbertograms from the numerical model of the temperature field is solved fragmentally.

In the development of these studies raises the problem of the reproduction of the co-evolution of the reconstructed interference fields and the Hilbert structures of the convective flows in comparison with the results obtained in the experiments. The novelty of the problem is due to the almost complete lack of data on the development of convective flows in the conditions of monotonic temperature changes on the vertical walls, taking into account the inversion of the water density in the vicinity of the isotherm (+4 °C) and consists in a comprehensive approach combining the results of a physical and numerical experiment in which the simulation of the isotherm field is supplemented by the calculation of the velocity field and temperature gradients.

2. Method

A sketch of a research complex containing an optical diagnostic system and an experimental stand is shown in figure 1 [3].

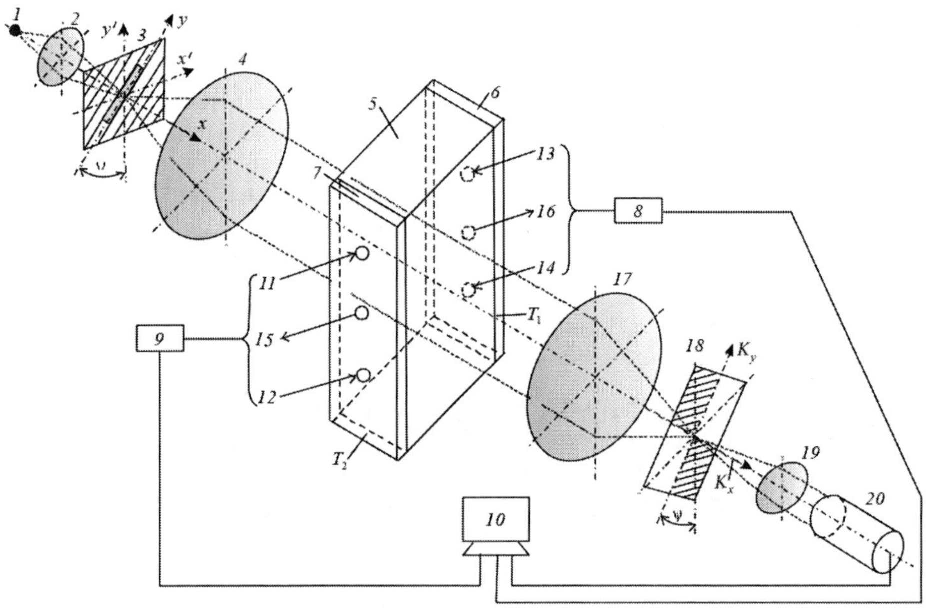

Figure 1. Sketch of the research complex.

The complex contains a lighting module consisting of a light source (1), a condenser lens (2) and a slit diaphragm (3) placed in the front focal plane of the lens (4), which forms a probing field in the test medium. Such an environment is water filling a rectangular cavity (5) with transparent glasses of optical quality.

The experimental stand is a rectangular cavity filled with water, with transparent optical-quality glasses. The internal dimensions of the cavity are 136×86×30 mm. Side walls are formed by parallel arranged cooled flat copper plates (6) and (7). On the outer side of the plates there are cavities into which coolant is supplied from the cryostat and thermostat (8) and (9). The temperature of one of the plates by pumping coolant from the cryostat can reach minus 28 °C. The temperature of the other plate is maintained in the range from room temperature to +8 °C. Thermostats are controlled by computer (10) according to a given program. The tubes supplying the refrigerant to the heat exchange side plates are designated as (11)–(14), the outflow tubes are (15), (16). The arrangement of the feed tubes and outflow tubes may vary according to the conditions of the experiment. The distance between the tubes is 80 mm. The lens (17) forms in the frequency plane the Fourier spectrum of the probing light field perturbed by the medium under study. The Hilbert filter (18) is placed in the Fourier plane of the lens (17). The frequency axis K_x of the Hilbert filter is orthogonal to the direction of the image of the slit diaphragm of the light source ($\psi = 45°$). Lens (19) performs the inverse Fourier transform of the filtered spatial-frequency spectrum. The Hilbert image of the phase structure of the light field perturbed by the medium under investigation is recorded by a digital video camera (20) connected to the computer (10). The optical diagnostics system was implemented on the basis of the serial IAB-463M shadow device [4], in which the following modules were modified: the light source module, the Fourier filtering module of the optical signal, the image recording module of the phase perturbations of the light field induced by the medium under study and module of shear interferometry. The probing field, formed by the lens (4), passes through the test medium (water), in which the temperature boundary conditions (T_1 and T_2, $T_1 < T_2$) are set according to a given program by lateral thermostatted surfaces (6) and (7).

The coherent transfer function of the spatial-frequency filter 18 that performs the one-dimensional Hilbert transform is described by the expression:

$$H(K_x, K_y) = [\exp(-i\varphi)\,\sigma(K_x) + \exp(i\varphi)\,\sigma(-K_x)]\sigma(K_y) + \\ +[\exp(i\varphi)\sigma(K_x) + \exp(-i\varphi)\sigma(-K_x)]\sigma(-K_y) = \\ = \cos\varphi - i\sin\varphi\,\mathrm{sgn}\,K_x, \quad (1)$$

where K_x and K_y are spatial frequencies; $\sigma(\pm K_x)$ and $\sigma(\pm K_y)$ are Heaviside functions; φ is the phase shift defined by the corresponding quadrant of the spatial-frequency filter. A filter with a coherent transfer function (1) performs a one-dimensional Foucault-Hilbert transform. In the Fourier plane $H(K_x, K_y)$ the spatial frequency axis K_x is orthogonal to the image of the slot light source. Fourier spectrum of the light field directly after the filter:

$$s(K_x, K_y)H(K_x, K_y) = s(K_x, K_y)[(\cos\varphi - i\sin\varphi\,\mathrm{sgn}\,K_x)] = \\ = s(K_x, K_y)\cos\varphi + \hat{s}_x(K_x, K_y)\sin\varphi. \quad (2)$$

Here $s(K_x, K_y)$ is the spatial-frequency Fourier spectrum of the light field perturbed by the medium under study; $\hat{s}_x(K_x, K_y) = -i\,\mathrm{sgn}(K_x)s(K_x, K_y)$ is the Fourier spectrum of the light field subjected to a one-dimensional Hilbert transform along the axis K_x. The phase shift φ is a function of the wave length λ of the probing light field: $\varphi = \varphi(\lambda)$. When the wavelength $\lambda = \lambda_0$, satisfies the condition $\varphi(\lambda) = \pi/2$, the Fourier filter (18) performs a one-dimensional Hilbert transform:

$$\hat{s}_x(K_x, K_y) = -i\,\mathrm{sgn}(K_x)s(K_x, K_y). \quad (3)$$

Convective structures appear in the space between the thermostated surfaces, which are manifested as perturbations of the optical phase density fields. These structures are induced by convection complicated by water density anomaly in the temperature range 0÷4 °C. The optical diagnostics system by the methods of Hilbert optics and shear interferometry performs visualization of gradients of perturbations of the phase optical density induced in the water layer by non-stationary temperature boundary conditions on vertical heat exchange surfaces limiting the medium under study.

Near the temperature-controlled surface, as the temperature gradient grows, water turns into a supercooled liquid, passing into a state of unstable equilibrium. In such a medium, the transition from liquid phase to solid-crystalline state occurs. This is a phase transition of the first kind. It becomes apparent through the appearance of a crystallization wave and is accompanied by an energy release. In its turn the release affects the dynamic distribution of the optical phase density gradient in supercooled water and induces phase perturbations in the probing light field, the Fourier spectrum of which forms in the frequency plane of the lens (17).

Figure 2 shows a frame from the video-film attached to the article, which illustrates the results of visualization of the phase density field by Hilbert-optics (a) and shear interferometry (b). As a verification of the results obtained in [4], numerical simulation of the temperature fields of convective flows in the form of a dynamic structure of isotherms is performed. From these structures, shear interferograms and hilbertograms were reconstructed, which were compared with the results obtained in the experiment.

Convective heat transfer in a liquid in a two-dimensional formulation was described by a dimensionless system of Navier-Stokes equations, energy and continuity in the approximation Boussinesq. Conductive heat transfer in massive horizontal walls of Plexiglas was modeled using the heat transfer equation.

Journal of Physics: Conference Series **1421** (2019) 012047 doi:10.1088/1742-6596/1421/1/012047

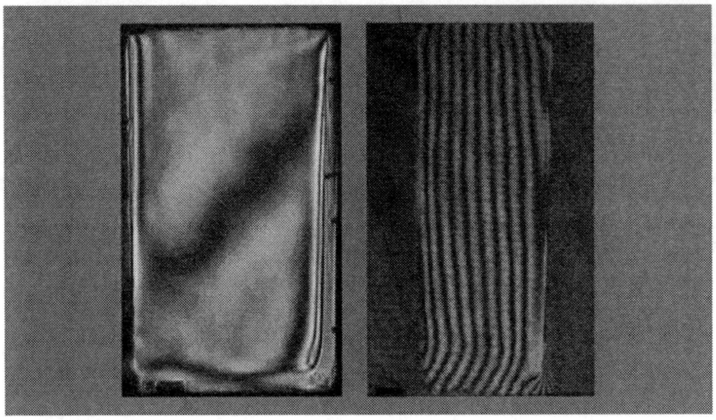

Figure 2. Visualization of the phase density field by Hilbert-optics (a) and shear interferometry (b).

The equations were solved numerically by the finite element method in the conjugate formulation. Dependences of density and coefficient of volumetric thermal expansion on temperature were taken into account. At the solid boundaries for the velocity (and, accordingly, for the current function), the conditions of adhesion and non-flow through were assigned. The boundary condition for the vortex is obtained from the field of the function of current using the method of conjugate results [3]. At the interface of the liquid with solid walls the condition of continuity of temperature and heat flow was set.

An uneven triangular grid, thickened bunched to varying degrees to all the boundaries of the computational domain, with the number of nodes ≈4×104 was used. Linear basis functions be given on the elements. A cellular step-by-step algorithm for the maximum angle was used to construct the triangulation [5].The program implements an iterative process, in which, if necessary, the necessary values of the variables from the previous steps and the calculated values of the coefficients on the parameters are substituted.

The problem is solved in a non-stationary situation: the initial temperature of the system is +20 °C, the left vertical wall is monotonically cooled to +8 °C for 2500 seconds. The right wall was cooled to 0 °C for 2500 seconds. The outer surfaces of the horizontal walls are adiabatic. In this paper, the simulation of the isotherms field is supplemented by the calculation of the velocity field and temperature gradients.

Temperature fields are reconstructed from the numerically models of isotherms (figure 2 (a)). The graphs are constructed in relative units, (in the figures the real size of the cuvette width equal to 86 mm is taken as a scale in the transition to dimensionless equations and boundary conditions).

Temperature boundary conditions on the left and right walls are indicated, respectively, T_1 and T_2. The visualized hilbertograms represent quasi-gradiented structure of the optical phase field $\Delta\varphi(x, y, T)$, induced in the under investigation environmental by the temperature boundary conditions:

$$\Delta\varphi(x, y, T) = k[n(x, y, T) - n_0]l, \tag{4}$$

where l is the thickness of the liquid layer, $k = 2\pi/\lambda$ is the wave number of the light field, $n(x, y, T)$ is the refractive index of the liquid as a function of the spatial coordinates x, y and temperature T.

This function is determined by the temperature dependence of the specific refraction of water. Temperature dependence of the refractive index of distilled water (figure 4) was calculated, according to [6, 7], on the density and specific refraction at a wavelength $\lambda = 589.3$ nm at normal atmospheric pressure.

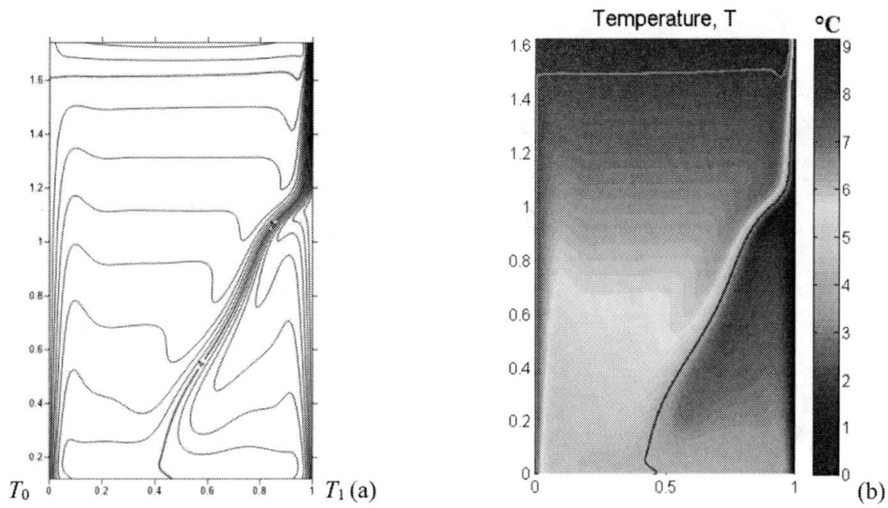

Figure 3. Numerical model: (a) – isotherm field (T_0 = 8 °C, T_1 = 0 °C);
(b) – numerical model of the temperature field from the initial field of isotherms
(3545 seconds from the moment the heat exchange surfaces are switched on, frame 901).

The optical phase field was calculated from the spatial distribution of the refractive index (figure 5). The Hilbert transformation of this field was performed and the shear interferograms was reconstructed.

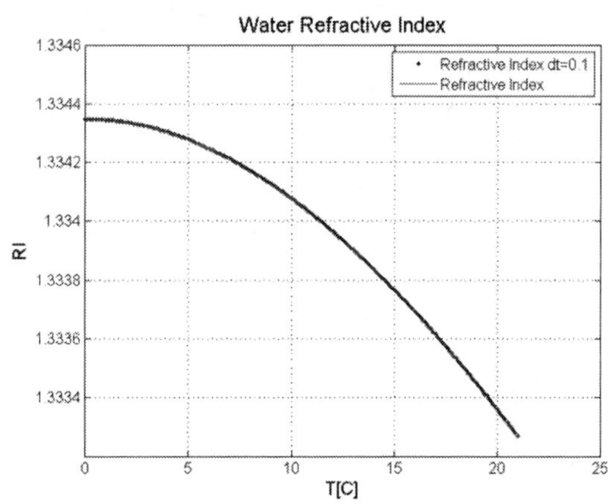

Figure 4. The dependence of the refractive index of distilled water on temperature from –2 °C to +21 °C [5,6].

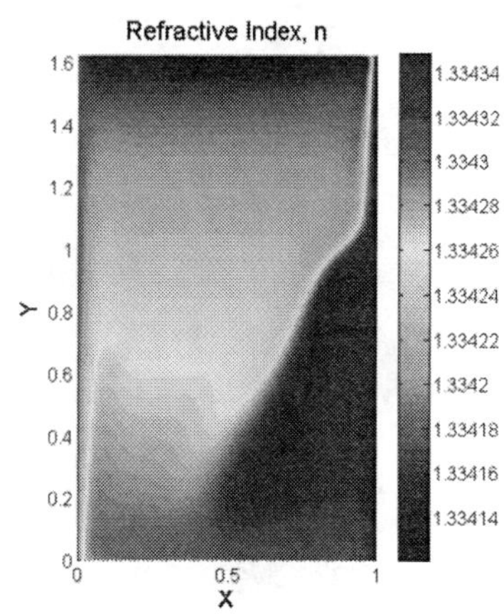

Figure 5. Field of refractive index of distilled water (*x*, *y* axis).

Figure 6 shows a frame from the video attached to the article, which shows the numerical fields of isotherms, temperature gradients and velocities (at the top left). At the bottom left is a Hilbert image of the phase density field (blue color) reconstructed from these numerically obtained isotherm fields. Isotherms corresponding to temperatures of +4 °C and +8 °C are highlighted in red and green.

On the right, the obtained Hilbert-visualized optical phase density field is shown, with which the evolution of the fields of isotherms and temperature gradients obtained numerically is synchronized. The film demonstrates the qualitative correspondence between the results of the experiment and simulation.

Journal of Physics: Conference Series **1421** (2019) 012047 doi:10.1088/1742-6596/1421/1/012047

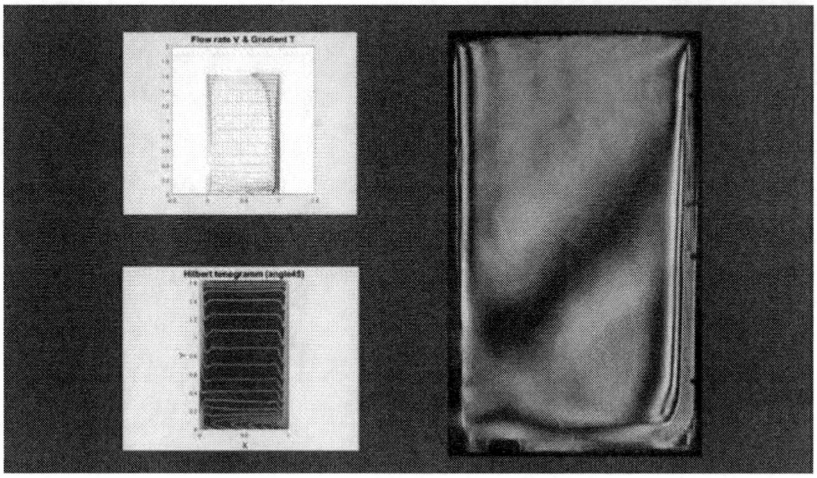

Figure 6. Reconstruction of Hilbert fields of phase optical density.

Figure 7 shows a frame from another video attached to the article.

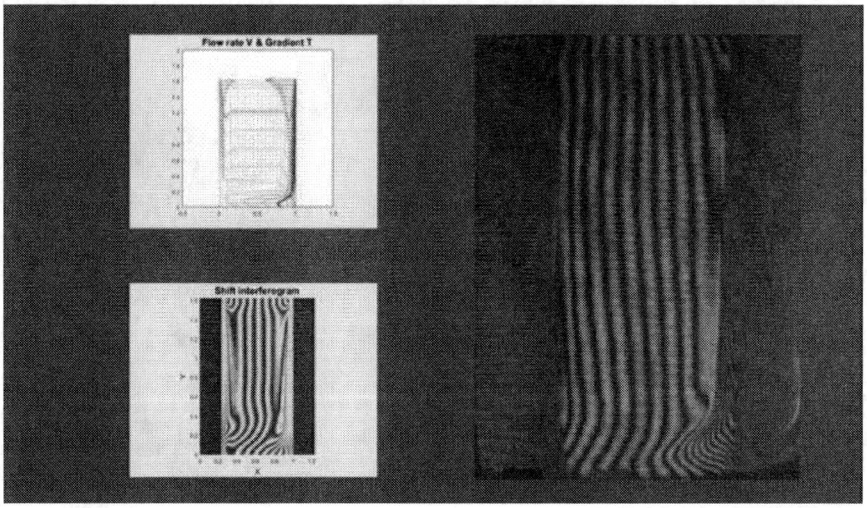

Figure 7. Reconstruction of shear interferograms of the optical density field.

At the top left a simulation of evolution of the fields of isotherms, temperature gradients and velocity fields is show, at the bottom left – shear interferograms reconstructed from the numerically obtained field of isotherms. On the right, the figure shows an experimentally obtained interference field that displays the structure of the convective flow induced by nonstationary boundary conditions. The film illustrates the qualitative agreement between the results of numerical simulation and experiment.

Journal of Physics: Conference Series **1421** (2019) 012047 doi:10.1088/1742-6596/1421/1/012047

(a)

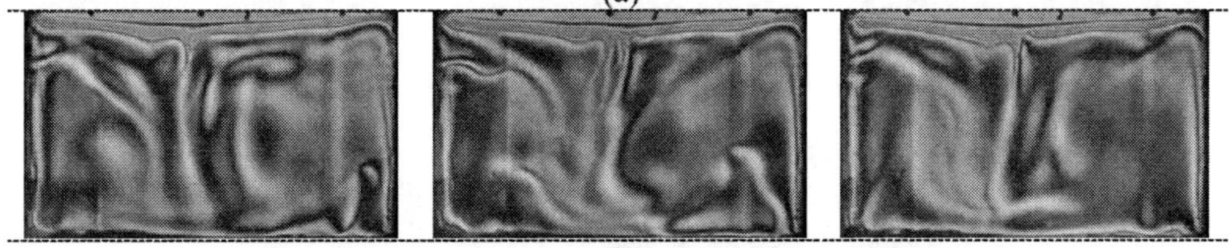

(b)

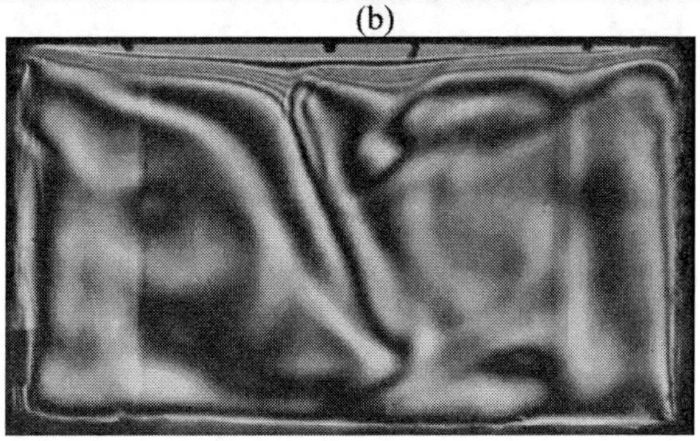

(c)

Figure 8. Convective structures induced by unsteady boundary conditions in the horizontal layer bounded by heat exchange surfaces: (a) – cooling; (b) – under the ice surface; ice formation; (c) – phase transition.

In development of the results obtained in [8], the convective structures and phase transition induced by unsteady boundary conditions in a horizontal water layer bounded by heat exchange surfaces are investigated. The fields of the phase optical density of convective structures are visualized by methods of the Hilbert optics. The dynamics of the crystallization wave front induced by a temperature gradient on the upper boundary of a horizontal layer of water bounded by thermostatted flat surfaces is studied. The dynamic profile of the isothermal surface bounding the spatial form of the solid phase on the water-ice interface is visualized. Video frames in figure 8 illustrate the evolution of convective structures in the cooling process (a), under the ice surface (b) and phase transition (c). The results obtained in the present study illustrate the efficiency of optical diagnostic methods based on the Hilbert optics. They are of interest for the development of physical models of phase transitions, studying ice formation processes, and various applications.

3. Conclusion

Methods the Hilbert-optics, shearing interferometry and numerical simulations are performed to study the evolution of convective structures induced by unsteady boundary conditions in the vertical layer of water bounded by flat heat exchange surfaces, in conditions of monotonic cooling. Solved the problem of the reconstruction of the Hilbert-images and shear interferograms fields of the phase optical density from numerically obtained fields of evolution of isotherms and velocity. The evolution of convective structures and phase transition induced by non-stationary boundary conditions in the horizontal water layer between the heat exchange surfaces is studied. Studies of convective flows induced by unsteady boundary conditions are relevant for solving various fundamental and applied problems, an example of which is the development of new technologies for crystal growth, understanding of convective processes in the water column in the study of the Arctic and Antarctic regions of the world ocean.

Acknowledgments
The work was performed within the state task of IT SB RAS (State. reg. AAA–AH17–117030310010–9 and AAA–A17–117022850021–3).

References
[1] Lappa M 2010 *Thermal Convection: Patterns, Evolution and Stability* (Chichester: John Wiley Sons)
[2] Smits A J and Lim N N 2010 *Flow Visualization* (Imperial College Press)
[3] Dubnishchev Yu N, Arbuzov V A, Arbuzov E V, Berdnikov V S, Kislitsyn S A and Melekhina O S 2018 *J. Scientific visualization* **10** 4 134–44
[4] Soloveychik Yu G, Royak M E and Persova M G 2007 *Finite Element Method for Solving Scalar and Vector Problems* (Novosibirsk: NSTU)
[5] Skvortsov A V 2002 *Delaunay Triangulation and its Application* (Tomsk: Tomsk state University)
[6] Ivanov A 1978 *Introduction to Oceanography* (Publishing House "Mir", Moscow) 1978 *Tables of standard reference data: Water. Density at atmospheric pressure and temperatures from 0°C to 1000°C* (Publishing standards. Moscow)
[7] Arbuzov V A, Arbuzov E V, Dubnishchev Yu N, Berdnikov V S and Melekhina O S 2017 *J. Optoelectronics, Instrumentations and Data Processing* **53** 2 39–44

Journal of Physics: Conference Series **1421** (2019) 012048 doi:10.1088/1742-6596/1421/1/012048

Research of the dynamics of two-phase flows by optical methods PIV/IPI

A I Guryanov, S V Veretennikov, O A Evdokimov, M M Guryanova and K L Kalinina

Department of general and technical physics, P.A. Soloviev Rybinsk State Aviation Technical University, Pushkina st. 173a, Rybinsk, 152934, Russian Federation

E-mail: marialex2004@mail.ru

Abstract. The paper describes conditions and experimental results of the dynamics of two-phase flows research by optical methods PIV/IPI. Studies were carried out for the finely dispersed flow generated by the centrifugal atomizer. Such atomizer is necessary for the implementation of certification of aircraft engine critical mode under rain conditions: height of 6100 m, the Mach number $M = 0.8$. The flow parameters were measured for 9 modes of liquid pressure in the nozzle from 0.2 to 1.0 MPa in increments of 0.1 MPa. Velocity vectors obtained for these pressure modes shown that maximum velocity of the droplets is achieved nearby the nozzle section and ranges from 20 to 30 m/s. Increasing the pressure reduces the difference between velocities at peripheral and axial regions of the flow.

1. Introduction

Two-phase mixtures are particular phenomena in engineering and the list of industrial facilities using two-phase flows is extremely wide. It contains steam boilers, evaporators, gas and oil pipelines, liquid atomization devices and much more [1, 2]. Better understanding of the processes occurring in two-phase mixtures requires both experimental studies and the development of computational models. Since a strict mathematical description of two-phase flows is extremely difficult due to the complexity of the definition of interfacial surfaces. Usually empirical or semi-empirical relationships obtained from the results of experimental studies is used in engineering practice [3, 4].

To obtain the most reliable research results, optical non-contact measurement methods, such as particle image velocimetry are used. They provide fixation of the instantaneous velocity values without additional flow turbulization and corresponding blunders. This work describes the optical investigation of the two-phase flow by methods of particle image velocimetry (PIV) and interferometric particle imaging (IPI) [5-7].

There was carried out a study of finely dispersed flow (droplets' size of about 500 μm) generated by the centrifugal atomizer. Such atomizer is necessary for simulation rain conditions and implementation of certification of aircraft engine critical mode at the height of 6100 m and value of Mach number $M = 0.8$. Certification in ground conditions, where the droplets are in undisturbed air (height of 0 m, Mach number $M = 0$), usually requires the formation of a flow of large droplets with a median diameter of 2.66 mm. The nozzle for these modes has already designed and described in paper [8]. The same nozzle does not allow testing in both modes. Therefore, it is necessary to develop and study a new nozzle designed to form finely dispersed air-water mixture.

Vectors of the velocity and sizes of droplets were obtained by using optical methods of PIV and IPI, respectively. Also, as a result of the studies, the ranges of the spray angle, the uniformity of the spray and the pressure/flow characteristic curves are determined.

2. Test Stand

To experimentally study the parameters of two-phase flow, a test stand has been designed and installed, the scheme of which is shown in figure 1.

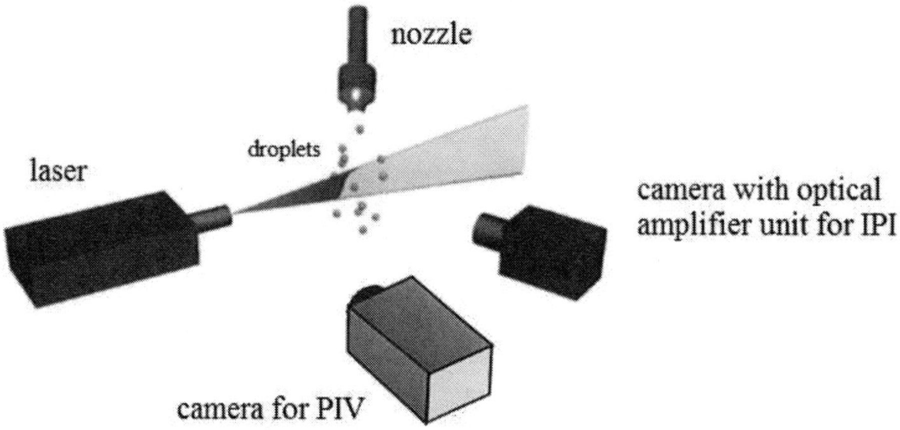

Figure 1. Test stand scheme.

The flow parameters were measured for 9 modes of liquid pressure in the nozzle from 0.2 MPa to 1.0 MPa in increments of 0.1 MPa.

For implementation of PIV and IPI methods, laser (532 nm), 2 high-speed cameras and optical amplifier were used. Processing of images was carried out in the program Actual Flow. Measurement of instantaneous velocity by PIV in the cross-section of the two-phase flow was based on the definition of the movement of an ensemble of particles during fixed time interval [9, 10]. IPI method was based on the interference of the reflection and refraction of glare points. The technique utilized the interferometry pattern created from a particle illuminated by a laser sheet. The particles must be transparent, homogeneous and spherical like water droplets in two-phase flow [11, 12].

3. Experimental results and their discussion

The photos taken with the high-speed camera (figure 2) show standard images of sprays from centrifugal nozzle: plumes in the form of a hollow cone. The droplets on the diameter of the plume are distributed nonuniformly – more drops fall on the periphery of the flow. Increasing pressure in the nozzle leads to fact that nonuniformity becomes more pronounced. As the droplets are removed from the nozzle section, the nonuniformity becomes less and the cone is filled with droplets.

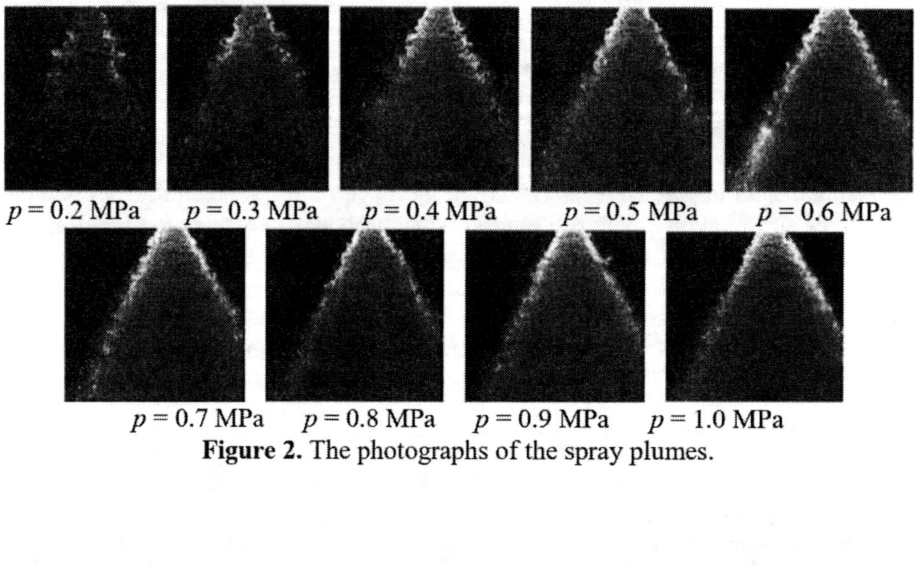

$p = 0.2$ MPa $p = 0.3$ MPa $p = 0.4$ MPa $p = 0.5$ MPa $p = 0.6$ MPa
$p = 0.7$ MPa $p = 0.8$ MPa $p = 0.9$ MPa $p = 1.0$ MPa

Figure 2. The photographs of the spray plumes.

Figure 3 shows vector distributions of flow velocity under different pressures in range from 0.2 MPa to 1.0 MPa. Figure 4 shows locations of three characteristic cross-sections at distances of 10 mm, 100 mm and 200 mm from the nozzle section. They both have the same palette shown in figure 4.

Figures 5, 6 and 7 show velocity dependences on dimensionless radius of the plume $\overline{R}$ which is equivalent to the coordinate x divided by nozzle diameter d_0, for three characteristic cross-sections (shown in figure 4). Values of pressure are equal to 0.4 MPa in Figure 5, 0.7 MPa in figure 6 and 1.0 MPa in figure 7.

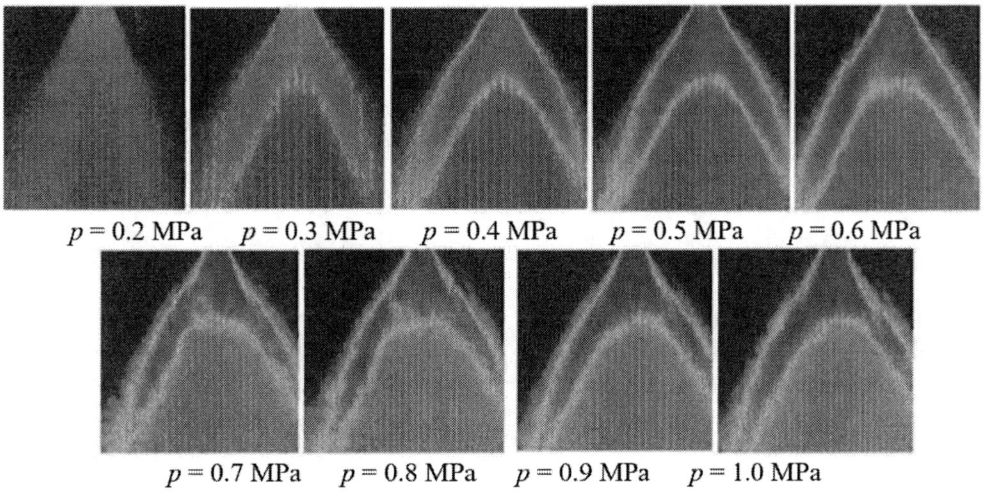

Figure 3. Vectors of flow velocity under different pressures obtained by PIV method.

These graphs show that the maximum velocity is achieved nearby the nozzle, i.e. at a distance of 3 nozzle's diameters (10 mm). In the middle part of the flow (100 mm cross-section), the velocity maxima on the periphery and the minimum on the axis of the flow are visible. This is typical situation for centrifugal nozzles' atomization. The section of 200 mm from the nozzle is characterized by a velocity field with less pronounced maxima. Increasing pressure leads to situation when the limiting velocities at the periphery in the middle and lower sections of the plume become more nonuniform.

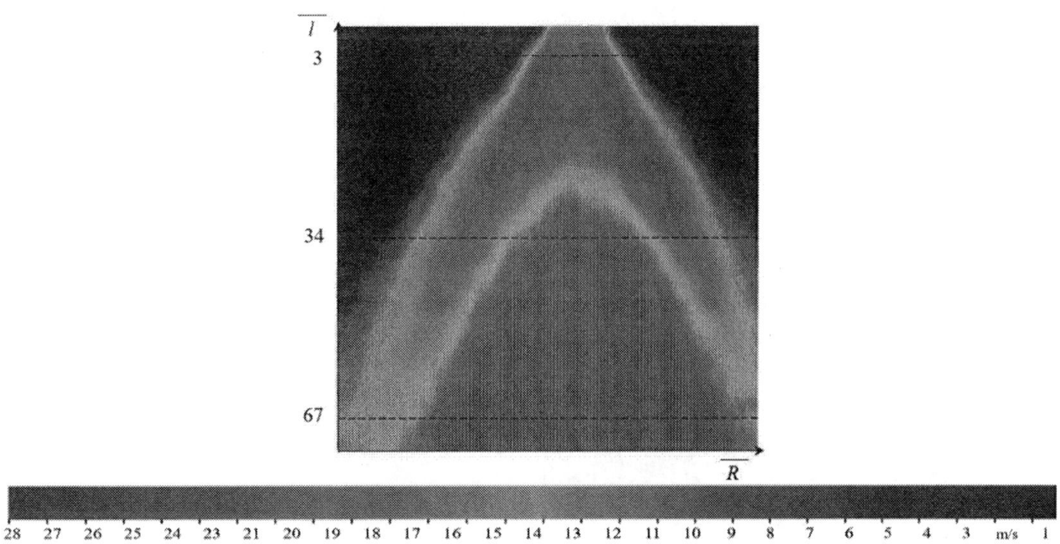

Figure 4. Vectors of flow velocity under pressure of 0.4 MPa.

Journal of Physics: Conference Series **1421** (2019) 012048 doi:10.1088/1742-6596/1421/1/012048

Figures 8-11 show the distribution of the mass fraction of the droplets by diameter *g* (figures 8 and 10) and the corresponding curves of the relative mass of the droplets *S* with different diameter formed by the centrifugal nozzle (figures 9 and 11). Since the flow pattern did not differ qualitatively with increasing pressure, so the dependencies only for two pressure modes of 0.2 MPa and 1.0 MPa are presented as examples.

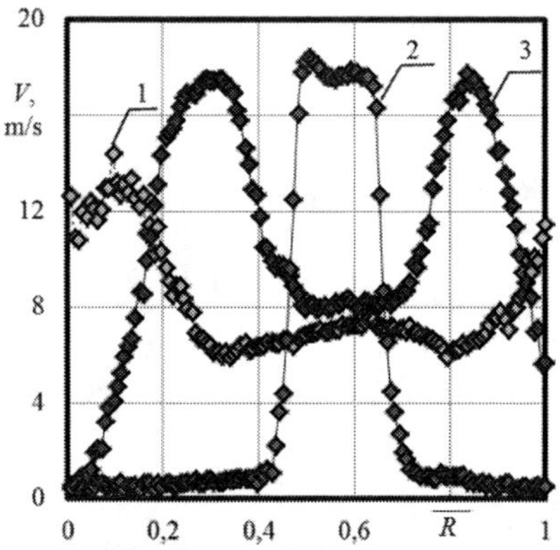

Figure 5. Velocity dependences on the dimensionless radius of the spray plume $\overline{R}$ at a pressure of 0.4 MPa: 1 – cross-section 200 mm; 2 – cross-section 10 mm; 3 – cross-section 100 mm.

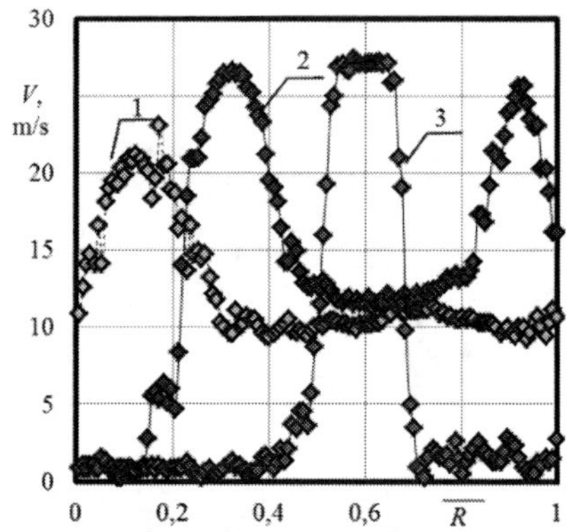

Figure 6. Velocity dependences on the dimensionless radius of the spray plume $\overline{R}$ at a pressure of 0.7 MPa: 1 – cross-section 200 mm; 2 – cross-section 10 mm; 3 – cross-section 100 mm.

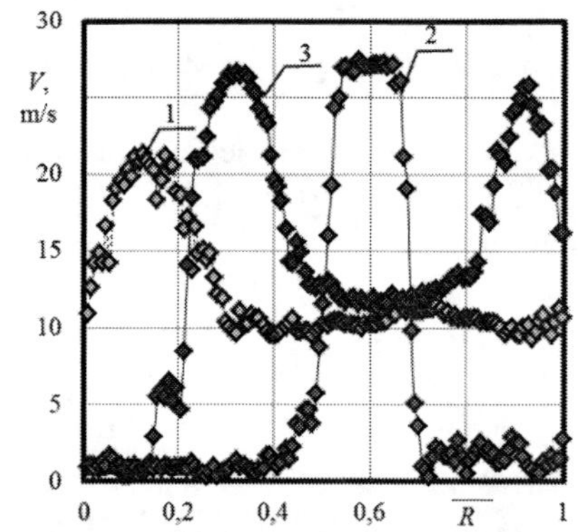

Figure 7. Velocity dependences on the dimensionless radius of the spray plume $\overline{R}$ at a pressure of 0.4 MPa: 1 – cross-section 200 mm; 2 – cross-section 10 mm; 3 – cross-section 100 mm.

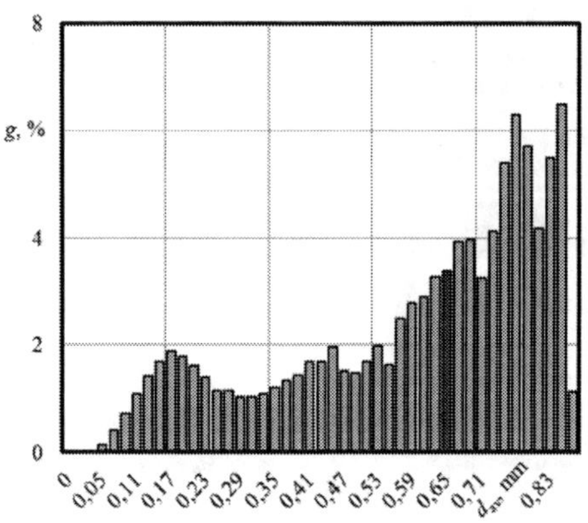

Figure 8. Droplets diameter distribution (pressure in the nozzle 0.2 MPa).

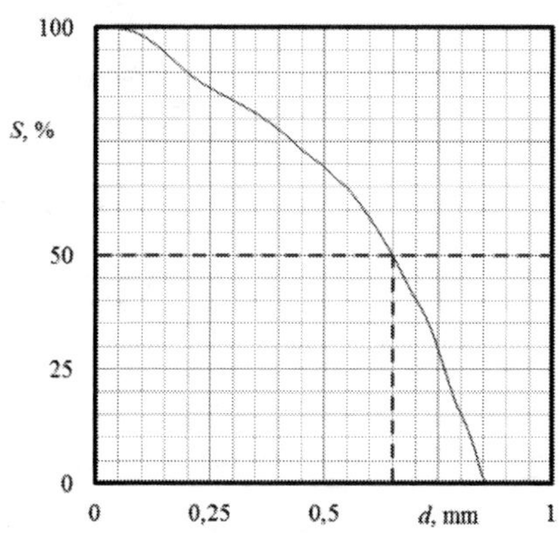

Figure 9. Distribution of the relative mass of droplets (pressure in the nozzle 0.2 MPa).

It can be seen that the formed flow at a pressure of 0.2 MPa contains the maximum number of droplets with the largest mass (figure 8). The bulk of the droplets are concentrated in the range of diameter from 0.5 mm to 0.8 mm. The median diameter obtained in this mode is d_m = 0.649 mm (figure 9).

In a percentage correlation (figure 10) there is a dominating number of droplets with a diameter of 0 to 0.2 mm. Therefore median diameter corresponds to the part of the graph with minimum number of droplets but their maximum mass.

The increase in pressure in the nozzle does not lead to a significant change in the droplet spectrum and median diameter ranges from 0.625 mm to 0.660 mm. Pressure mode of 1 MPa defines a median diameter d_m = 0.656 mm. In a percentage correlation (figure 11) there are more small droplets than under conditions of 0.2 MPa.

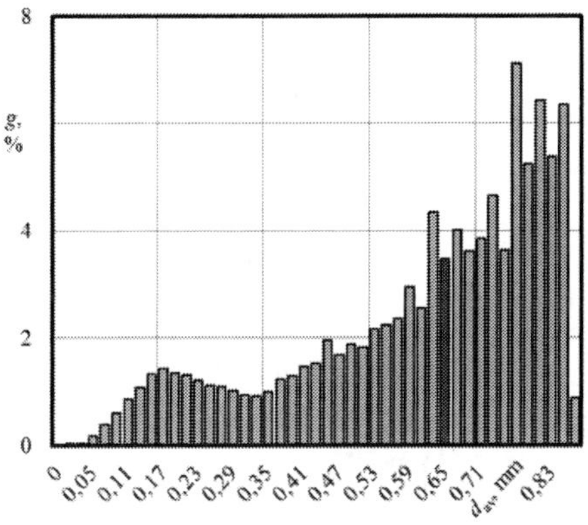

Figure 10. Droplets diameter distribution (pressure in the nozzle 1.0 MPa).

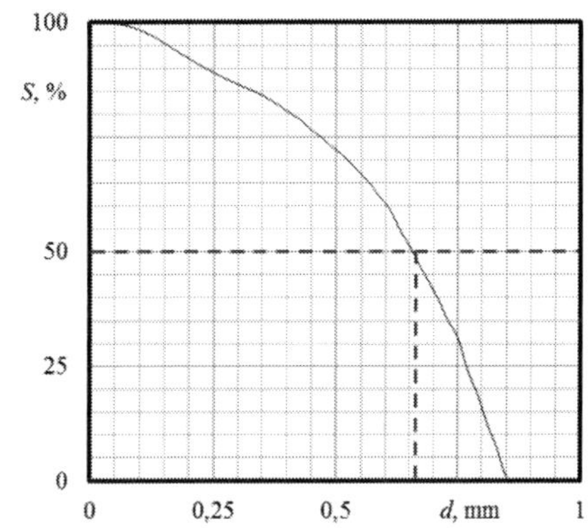

Figure 11. Distribution of the relative mass of droplets (pressure in the nozzle 1.0 MPa).

4. Conclusion

Using the methods of optical flow diagnostics (PIV/IPI), experimental studies of two-phase flow formed by a centrifugal nozzle for conditions H = 6100 m, M = 0.8 were performed. Studies have shown that the plume of droplets formed by the nozzle is unevenly distributed over the cross sections. Increase in pressure (more than 0.6 MPa) leads to the situation when circumferential and radial nonuniformities become more pronounced. However, the overall pattern of the flow from the centrifugal nozzle with a lack of droplets in the axial region and an excess at the periphery remains constant.

Distributions of velocity vectors obtained for different pressure modes shown that maximum velocity of the droplets is achieved nearby the nozzle section (10 mm below it and ranges from 20 to 30 m/s). Increasing the pressure reduces the speed difference between the periphery and the axis of the flow.

The formed spectrum contains droplets in the range of diameters from 0 to 0.9 mm. The bulk of the liquid is concentrated in large drops. Therefore, increase in the nozzle's pressure from 0.2 MPa to 0.7 MPa defines change in the median diameter from 0.625 mm to 0.660 mm.

Acknowledgments

This work was financially supported by RFBR (Russian Foundation for Basic Research) grant No. 18-31-00399 "Experimental and theoretical investigation of gas and two-phase swirling flows to increase the efficiency of vortex ejectors and to adapt it for perspective technical systems."

References

[1] Leonov A, Chudanov V and Aksenova A 2013 *Methods of direct numerical simulation in two-phase fluids* (Moscow: Nauka)
[2] Malyshev A, Mamchenko V and Kisser K 2016 *Heat transfer and hydrodynamics of two-phase flows of coolant* (St. Peterburg: ITMO)

[3] Barilovich V B 2009 *Fundamentals of therm gas dynamics of two-phase flows and their numerical soluition* (St. Peterburg: SPbPU)
[4] Barvinok V and Vashukov Yu 2012 *Methods of experimental research of technological processes in production of aircrafts* (Samara: SNACU)
[5] Akhmetbekov E, Bilsky A, Lozhkin Yu, Markovich D and Tokarev M 2006 Control system of experiment and processing data obtained by methods of particle image velocimetry (Actual Flow) *Computational methods and programming* **7** 79-85
[6] Zhdanova A, Zabelin M, Nyashina G and Strizhak P 2014 Particle image velocimetry of water sprays motion through high-temperature gases *Fundamental Research* **9(6)** 1225-29
[7] Lin J, Foucaut J-M, Laval J-P, Perenne N and Stanislas M 2008 Assessment of different SPIV processing methods for an application to near-wall turbulence *Topics in Applied Physics* **112** 191-221
[8] Guryanov A and Kalinina K 2018 Study of the nozzle for rain simulation at certification of aviation engines *MAI Journal* **25(1)** 18-27
[9] Wieneke B 2005 Stereo-PIV using self-calibration on particle images *Experiments in Fluids* **39** 267-80
[10] Lecordier B and Trinit'e M 2006 Accuracy assessment of image interpolation schemes for PIV from real images of particle *13th Int. Symp. on Applications of Laser Techniques to Fluid Mechanics* **10** 181-93
[11] Jasikova D, Kotek M, Tadeas L and Kopecky V 2012 The study of full cone spray using interferometric particle imaging method *EPJ Web of Conferences* **25** 01033
[12] Bilsky A, Lozhkin Yu and Markovich D 2011 Interferometric technique for measurement of droplet diameter *Thermophysics and Aeromechanics* **18(1)** 1-12

Infrared fiber based on AgCl-AgBr and AgBr-TlI crystals to transfer thermal radiation in pulsed

A S Shmygalev[1], A S Korsakov[1], B P Zhilkin[2] and A Turabi[2]

[1] Institute of Chemical Engineering, Ural Federal University, Yekaterinburg, 620002, Russia
[2] Ural Power Engineering Institute, Ural Federal University, Yekaterinburg, 620062, Russia

E-mail: a.s.shmygalev@urfu.ru

Abstract. The our research objective is to study IR energy transfer in pulsed and continuous mode through optical fibers produced from crystals of silver halide and a monovalent thallium solid solutions. It is assumed that the transfer of infrared radiation in the pulsed mode will significantly reduce losses. We designed the experimental setup, determined its operational parameters, and obtained the dependence of the detector's received radiation on the frequency of the pulses of thermal radiation. As a research subjects we used two different fibers compositions namely $AgCl_{0.25}Br_{0.75}$ and $Ag_{0.95}Tl_{0.05}Br_{0.95}I_{0.05}$.

1. Introduction

The temperature of a heated body can be measured using the properties of optical fibers based on solid solutions of silver halides and monovalent thallium (AgBr-TlI and AgCl-AgBr) [1] to receive and transmit radiation from heated bodies to sensors. Temperature determination by a noncontact method is based on this principle. Since in most applications infrared fiber receive radiation from bodies whose temperature is higher than the ambient temperature, in the establishment of thermal equilibrium, part of the transmitted wave energy (in the form of thermal energy) is lost. These losses [2] are especially critical in the case of IR optical fibers for noncontact temperature measurement, since they can cause significant errors due to parasitic radiation from a source, receiver noise and thermal properties of fiber material [3]. It should be noted that the radiation power of real heated bodies is very small. Thus, for example, energy luminosity of the body heated to 100 °C will be in the range 10^{-3} - 10^{-4} (W· m). Thus, the presence of even minor errors significantly complicates the signal processing. When using amplifiers of an optical signal (for example, lenses), there is also an increase in noise, which in turn substantially increases the errors.

One of the options for reducing errors and increasing the incoming signal can be the use of a pulsed mode of transferring thermal energy. This mode is used in laser technology, however, the laser beam has a much higher power and radiation density, therefore, the influence of the thermophysical properties of the material is small, in contrast to low-power radiation. In this connection, it becomes necessary to estimate the parameters of the pulsed mode at low input signal power from bodies emitting in the middle infrared wavelength range. In this paper, we are given a study to identify a

Journal of Physics: Conference Series **1421** (2019) 012049 doi:10.1088/1742-6596/1421/1/012049

transmission opportunity of thermal radiation in pulsed and continuous modes through the IR optical fiber based on AgBr-TlI and AgCl-AgBr crystal systems.

2. Experimental setup

To study transmission of the thermal radiation in the pulsed mode using an optical fiber, an optoelectronic system was designed and constructed, the scheme of which is shown in figure 1.

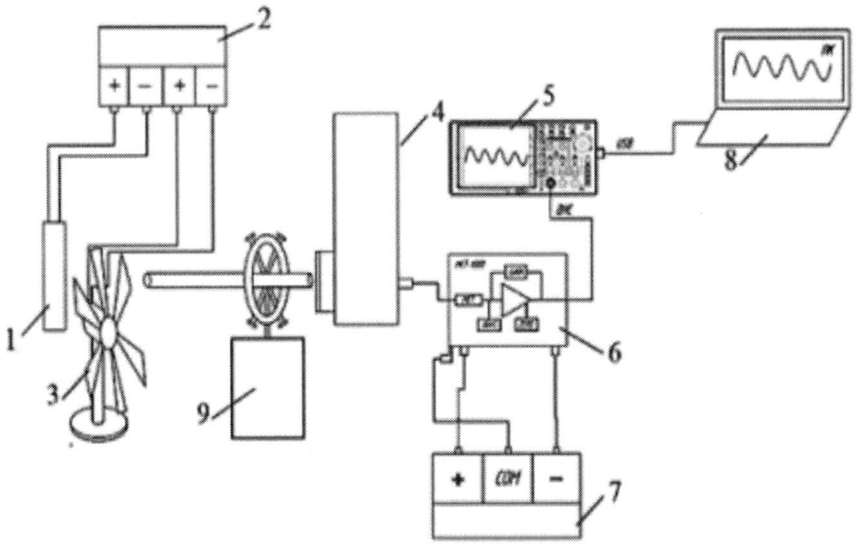

Figure 1. The scheme of optoelectronic system: 1 – Peltier's element, 2 – two-channel power supply unit, 3 – rotating shading obstruction impeller (modulator), 4 – MCT-detector, 5 – oscilloscope, 6 – operational amplifier, 7 – current source, 8 – PC, 9 – optical fiber in a special bracket.

The Peltier's element (1) was the source of thermal radiation and connected to the two-channel power supply unit Instek PPE-3323 (2) with a current strength of U = 3.3 V. Adjusting the voltage parameters on the power supply allowed to change the surface temperature of the hot side of the Peltier's element in the range from 290 to 400 K (measured by the thermal imager Testo 882).

The temperature range is comparable to a wavelength of about from 7 to 9 microns. Infrared radiation was transmitted directly to the cadmium-mercury-tellurium (MCT) detector (4) MCT-13.100 from InfraRed Inc., cooled by liquid nitrogen. MCT detector was working in the photoconductive mode with "dark" resistance of 35 ohms, the resistance detector operating range was in the range 20 - 100 ohms and the spectral region of measurement was 2 - 13 microns. The resistance of the detector in these experiments was 28-35 Ohm (with IR irradiation with a wavelength of 7-8 μm). The detector was connected to the operational amplifier (op-amp) (6) MST-1000 from InfraRed Inc., whose working in frequency range from 1.5 to 170 Hz at noise level $4\cdot10^{-10}$ (W · Hz $^{-1/2}$). The signal from the amplifier output was directed to the oscilloscope (5) Owon PDS 5022S, which displayed a graph of the change in the output voltage in time, and automatically measured the signal parameters. The graphs from the oscilloscope were transferred to the PC (8) and later processed using the software Oscilloscope PC suite (version 2.0.8.11).

To create a pulsed mode of the IR radiation transmission, a rotating shading obstruction impeller (modulator) was used (3). The impeller was attached to a low power drive with a starting voltage of 1 V and an operating voltage in the range from 0.6 V to 3.5 V, at which the speed range varied from 20 to 170 Hz. The rotational speed of the modulator was determined by the ST-5 stroboscope. Optical fiber (9) was placed between the impeller and the receiver to study the transfer of heat radiation in a pulsed mode.

3. Characterizations of the experimental setup

The influence of the pulsed operating mode on the perceiving power of the MCT detector is interesting from the viewpoint of the physical properties of the photocell itself, namely, the processes of generation and recombination of charge carriers. Generation-the production of free electrons-occurs under the action of photons whose energy exceeds the width of the forbidden band of the photocell material and is sufficient to form an electron-hole pair. Recombination is the reverse of the generation process and involves returning the electron to the valence band. In the process of recombination, energy is released equal to the difference between the initial and final energy states of the electron. The nature of energy release is one of the criteria for classifying recombination processes.

In the MCT detector, Auger recombination of carriers takes place, at which energy is transferred to the third particle when the main carrier returns to its valence band. So the recombination process has a longer duration and energy intensity than recombination of the "zone-zone" type. Pulsed mode of operation allows the MCT-element to take portions of the energy of the source, while the generation process lasts a longer time, as a result of which a deeper decrease in the detector's resistance occurs, which means that the voltage output from the op-amp is increased. There is a similar process with recombination, the inverse of generation, i.e. there is a gradual increase in resistance, accompanied by a voltage drop on the amplifier. The dependence of the generation and recombination processes on the increase in the pulse frequency is non-linear, and therefore the depense investigation and analysis should be performed for continuous and pulsed operation.

To estimate the operation characteristics of the MCT detector in the continuous and pulsed modes, the characteristics of the output voltage versus time were investigated without a special heat radiation source and with a Peltier's element as a source at a temperature of 350 K and a pulse frequency of 60 Hz. Sources of thermal radiation do not affect the characteristic of the output voltage in a continuous mode, and the amplitude of the oscillations is constant in time and has a noise character with a maximum value of 0.8 mV. Figure 2 shows the variation of the output voltage U_{out} on the op-amp in time τ under the pulsed mode of irradiation of the detector. In this case, the output voltage characteristic is in the form of a pseudo-sinusoid with a pulse frequency corresponding to the modulator frequency, and the output voltage correlates with the temperature of the heat source.

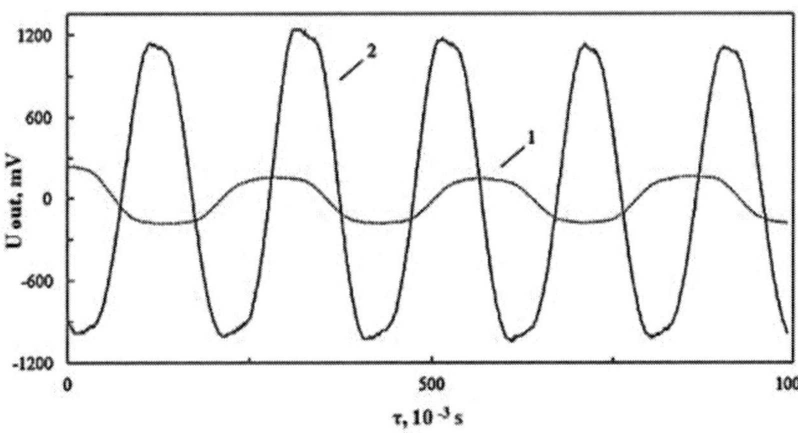

Figure 2. Graph of change in op-amp output voltage in time in the pulsed mode of irradiating the detector: 1 – without an irradiation source at a modulation frequency of 100 Hz, 2 – with an irradiation source at a modulation frequency of 60 Hz.

Journal of Physics: Conference Series **1421** (2019) 012049 doi:10.1088/1742-6596/1421/1/012049

A comparison of the MCT detector and the thermal radiation source operation modes shows that the highest sensing power was achieved in the pulsed mode: the maximum amplitude of the voltage U_{max} = 1200 mV, while in the continuous mode it is 1500 times smaller. Thus, the pulse mode of operation is most preferable, since high voltage values are achieved in this mode, which is especially important for studying the processes of mid-wavelength infrared transmission.

4. Results and discussion

As it was shown earlier, the radiation sensitivity of a CRT detector is significantly affected by the radiation transmission mode; therefore, in order to investigate the effect of a pulsed mode on the transmission of radiation through an optical fiber, it is necessary to compare the experimental results for two chemical compositions of infrared fibers at different pulse frequencies. Transmission of thermal energy by infrared fibers of various compositions in the pulsed mode was studied on $AgCl_{0.25}Br_{0.75}$ and $Ag_{0.95}Tl_{0.05}Br_{0.95}I_{0.05}$ fibers with a diameter of 1.12 mm, length of 150 mm. Initially, the spectral transmittance characteristics (figure 3) were obtained for the optical fibers under study using the IR -Fourier spectrometer IR-Prestige 21 Shimadzu.

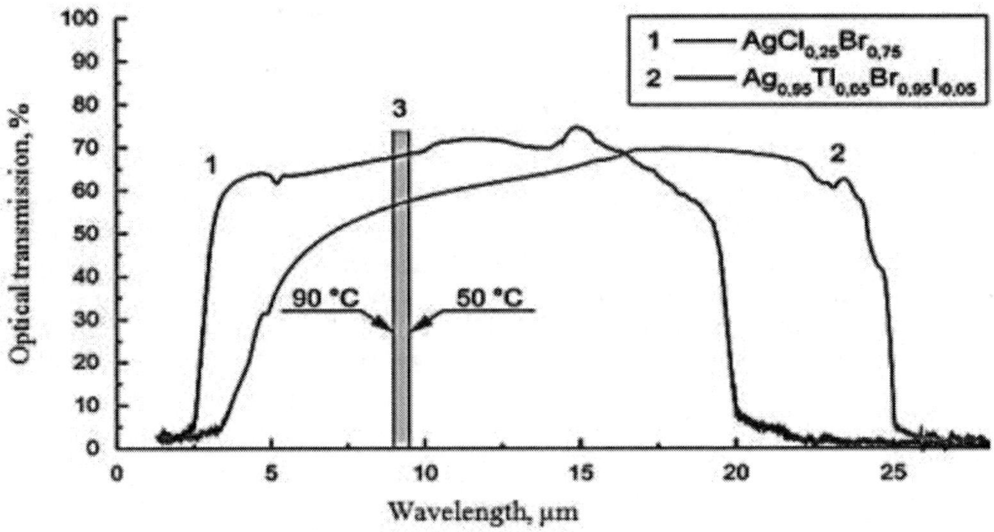

Figure 3. Spectral transmission characteristics of infrared fibers: 1 – $AgCl_{0.25}Br_{0.75}$, 2 – $Ag_{0.95}Tl_{0.05}Br_{0.95}I_{0.05}$, 3 – Peltier's element.

From the obtained spectral characteristics it can be seen that the infrared optical fibers have a high percentage transmittance over a broad spectral region including the analyzed mid -wavelength infrared range (7 - 9 microns). To study the transfer of heat radiation in a pulsed mode optical fiber placed between the modulator and receiver (see. figure 1), the radiation was transmitted from the source – the Peltier's element, and the modulation frequency was 60 Hz. The MCT detector operated in a linear mode depending on the oscillation frequency of the voltage exiting the op – amp from the rotational speed of the impeller. The results of the measurements are shown in figure 4.

Comparing maximum values of the pulsed voltage conduction modes infrared fibers of different compositions shows contrast by 15%. So, for the composition $AgCl_{0.25}Br_{0.75}$ U_{out} = 11.2 mV, and for the composition $Ag_{0.95}Tl_{0.05}Br_{0.95}$ $I_{0.05}$ U_{out} = 9.5 mV. Differences in the voltage values for the two optical fiber compositions are related to the spectral characteristics, as well as to their refractive indices and the Fresnel losses. For $AgCl_{0.25}Br_{0.75}$ fibers, the refractive index n = 2.138 at a wavelength of 8 µm, and, consequently, the Fresnel losses for the two interfaces are 26.3%.

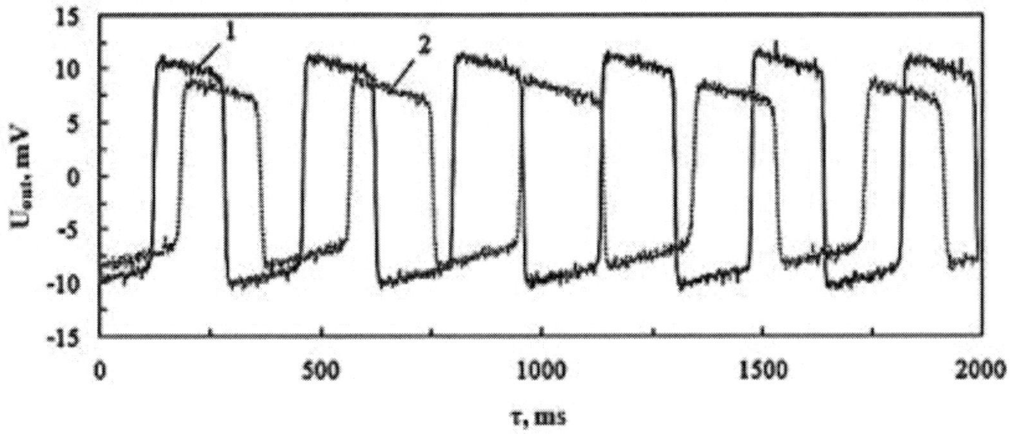

Figure 4. Spectral transmission characteristics of infrared fibers: 1 – $AgCl_{0.25}Br_{0.75}$, 2 – $Ag_{0.95}Tl_{0.05}Br_{0.95}I_{0.05}$, 3 – Peltier's element.

For the fiber of composition $Ag_{0.95}Tl_{0.05}Br_{0.95}I_{0.05}$, the refractive index is 2.239, Fresnel losses – 29.3% at a wavelength of 8 µm [4]. Thus, the use of an optical fiber with the chemical composition $AgCl_{0.25}Br_{0.75}$ will be the most effective in the pulsed mode of transmission of thermal radiation; in this mode, it is possible to obtain large values of the transmission of the IR wave in comparison with the $Ag_{0.95}Tl_{0.05}Br_{0.95}I_{0.05}$ fiber. However, the optical fiber of composition $Ag_{0.95}Tl_{0.05}Br_{0.95}I_{0.05}$ is advisable to apply in the wavelength range from 15 to 25 microns, since the transmission of the fiber of $AgCl_{0.25}Br_{0.75}$ composition is significantly reduced.

Acknowledgment
The research has been supported by the grants of President of the Russian Federation SP-2455.2018.1.

References
[1] Korsakov A, Zhukova L, Korsakova E and Zharikov V 2014 *J. Crys. Growth* **386** 94
[2] Lewi T and Katzir A 2012 *Opt. Lett.* **37** 2733
[3] Israeli S and Katzir A 2011 *Opt. Mat.* **33** 1579
[4] Korsakov A, Zhukova L and Vrublevsky D 2015 *Opt. Mat.* **50** 204

Study of effects of electromagnetic factors on the early development of biological organisms using comprehensive multispectral optical and scanning acoustic microscopy approach

A B Burlakov[1] D D Khokhlov[2] V L Domanskiy[2] and S A Titov[2]

[1]Department of Biology, Moscow State University, Moscow, 119991, Russia
[2]Scientific and Technological Center of Unique Instrumentation of Russian Academy of Sciences, Moscow, 117342, Russia

E-mail: sergetitov@mail.ru

Abstract. The effects of electromagnetic field on development of biological organisms were studied in this work by multispectral optical microscopy and scanning acoustic microscopy. The living embryos of loach *Misgurnus fossilis* at early stages of development were used as biological objects. The electric field applied to the living organism was generated by a specially designed unit that can operate independently or in combination with the acoustic microscope. It was shown that the sequentially recorded multispectral optical and acoustical images reveal dynamics of the development. The long lasting experiments confirm minor effect of the optical and acoustic radiation on the embryos. The scanning acoustic microscope was used for *in vivo* visualization of three dimensional structure of the egg, estimation of the acoustical properties of the yolk tissue and visualization of slow substance flows inside the object. The speed of motion depends on the location of the investigated area, status of the organism and its development stage. The estimated valued of the speed is in a range of 0.1 – 1 μm/s. It was shown that the application of the electric field causes deceleration of the embryos development.

1. Introduction

The study of the influence of various physical factors on the early stages of the development of biological organisms is of considerable interest [1, 2]. Particular attention is drawn to the influence of the weak endogenous and exogenous electric field (EF). Special EF generators with real-time adjustment of the amplitude and frequency parameters are required for the detailed study of the dynamics of spatial changes in biological structures under the external field. On the other hand the detection, visualization and monitoring of such changes in living biological objects require appropriate scientific instrumentation. Ordinary optical microscopy is an excellent method, but sometimes it suffers insufficient contrast of observed structural organs. Ultrasound is non invasive and posses deep penetration. However commonly used medical scanners demonstrate insufficient spatial resolution. In this work we use approach based on combined multispectral optical microscopy and scanning acoustic microscopy.

2. Materials and methods

There are two different methods used in this study for the *in vivo* monitoring of the developing small biological objects: multispectral optical microscopy and scanning acoustic microscopy. The electromagnetic field was generated with a specially designed electronic unit.

2.1. Preparation of biological objects

In this work, embryos of loach *Misgurnus fossilis* were selected as standard biological objects commonly used in developmental biology. Females were caught from nature and were kept in a refrigerator before the experiments. Chorionic gonadotropin (CG) hormonal stimulation was conducted to accelerate the maturation of females at room temperature. Production of sexual products and artificial insemination were carried out in accordance with standard methods described in detail elsewhere [3]. All procedures involving animals and their care conform to the institutional guidelines and with approval from the Government of Russian Federation.

2.2. Multispectral optical microscopy

The microscopic acousto-optic spectral imager based on a BW Optics MF-606 microscope with the acousto-optic module designed at the Scientific and Technological Center of Unique Instrumentation of RAS [4] was used in the experiments. The images shown in figure 1 as an example were obtained at wavelength of 660 nm. Images (a) and (b) were measured at different stages of the embryo development. Processing of the multispectral data reveals the relationship between the development stages and spectral data. More details about this technique can be found in another paper presented at the conference.

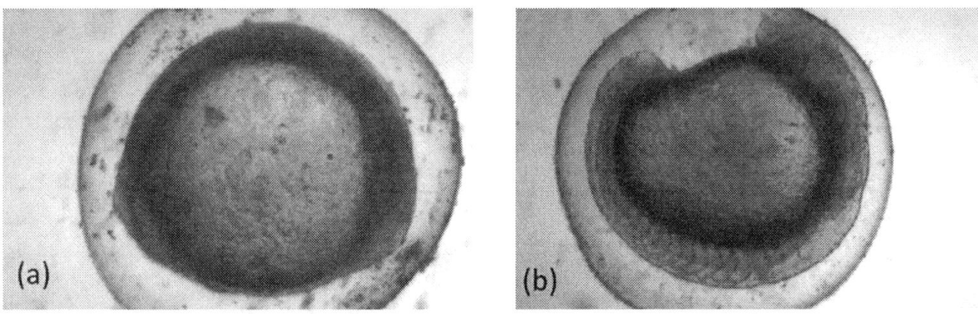

Figure 1. Acousto-optic spectral images of the embryo.

2.3. Scanning acoustic microscopy

The ultrasonic visualization of the biological objects in the course of the present study has been carried out by scanning acoustic microscopy (SAM). Ultrasound has a minor impact on the living organisms [5, 8], so the acoustic microscope can be considered as a suitable and effective tool for a long lasting observations of the structural changes in the evaluative bio object. In the acoustic microscope, high frequency, sharply focused ultrasonic waves are employed for the visualization. The microscope was designed in accordance with a traditional confocal scheme presented in figure 2.

The ultrasonic transducer 1 generates in immersion liquid (water) the wideband pulse wave focused inside the object 2. The waves scattered by the structural discontinuities propagate back and are converted by the same ultrasonic transducer to the electrical signal. The received signal is recorded as a function of the time of flight *t* which is proportional to the distance between the transducer and the reflector. To produce the three-dimensional data set the transducer is mechanically scanned in a raster manner over lateral coordinates x, y.

The central frequency and the bandwidth of the ultrasonic transducer used in the experiment were 50 MHz and 50 %, respectively. The aperture angle of the focused ultrasonic beam was $\theta_0 = 40°$. In accordance with the classical criteria [6], the lateral resolution in the focal plan of this confocal imaging system is about 30 μm. The spatial resolution along the vertical axis z is determined by the duration of the acoustic pulse, and it can be estimated at 20 μm.

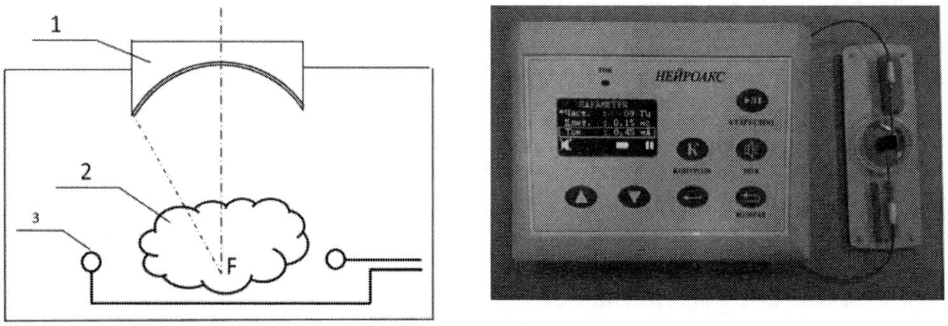

Figure 2. Scanning acoustic microscope and generator of electromagnetic field.

2.4. *Generator of electric field*

One of our tasks was the creation of a specialized generator of constant or alternative electric field with a flexible control of the field time parameters, its polarity, precise setting of the output pulse voltage and a programmable timer to implement a specified research protocol. Since the impedance of the "electrodes – biological object" system varies in a wide range of $10 - 10^5$ Ohms, the output amplifier of the designed device operates in the voltage source mode [9]. The controller of the device is build based on a microprocessor. It equipped with a display and autonomous power supply. The main specifications of the generator are as follows: output voltage – 0 ... 2500 mV with step of 50 mV, pulse duration – 0.1....10.0 ms with a step of 0.05 ms, pulse repetition rate – 0... 2000 Hz with a step of 1 Hz. Accuracy of all parameters setting is equal to 1%. The overall view of the generator is shown in figure 2 [10].

Electromagnetic stimulation of the organism was conducted *in vivo* by applying of the appropriate voltage to the electrodes 3 immersed in the water tank of the acoustic microscope (figure 2). Alternatively the biological object can be placed in a special separate electrode cell.

3. Experiment

In this work, loach eggs have been used as live convenient biological objects. The eggs were placed in the immersion liquid for the ultrasonic visualization. As an example, figure 3 shows ultrasonic images of the one live egg taken at different times. The images represents the cross sections of the object in the lateral dimensions using grayscale coding of the amplitude of the recorded echo The plane of visualization was set approximately in the middle of the yolk. The size of the field of view is 2 × 1.6 mm.

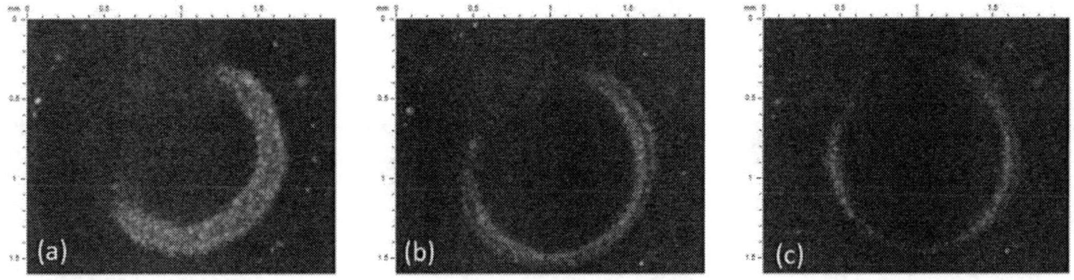

Figure 3. SAM images of the loach egg taken in the middle of the yolk at 2-hour intervals.

Continuous changes in the internal structure of the egg can be detected in the images. The external ring of the bright dots indicates the position of the skin. The dark circle in the middle of each picture (a) – (c) is produced by the yolk; and the bright noisy area around it is generated by the blastoderm. The observed contrast can be explained by the fact that there are many strong reflectors located inside the blastoderm.

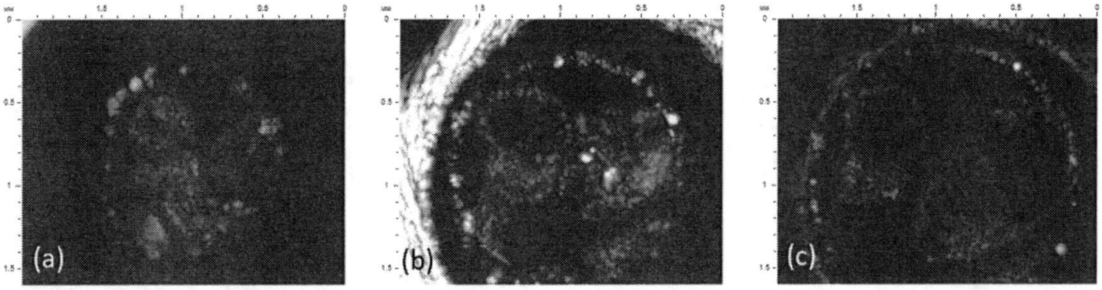

Figure 4. Ultrasonic images of the loach egg taken at various vertical positions of the focal plane with a step of 180 μm.

SAM is capable of visualization at various depths giving information about 3D structure of the object. Figure 4 (a) – (c) shows the cross sections of the well developed embryo as an example. Having a set of the scans of this kind, a volumetric reconstruction of the organs and calculation of their volumes and mutual positions are possible.

4. Results and discussion

To evaluate the influence of the electric field on the development of the embryo the ultrasonic images have been sequentially recorded. Figure 5 shows images of the blastoderm top of the loach egg at early stage of development taken at 10 min intervals. There was no electromagnetic field application between images (a) and (b) recorded with a delay of 30 min. Rapid changes in the blastoderm profile can be observed in the images. After image (b) was recorded, the alternative voltage 10 V with a frequency of 10 Hz was applied to the electrodes. After another 30 min interval since the voltage application image (c) was measured. Comparison of pictures (a) – (c) shows that the changes between (b) and (c) are much less pronounced then the changes between (a) and (b). Therefore it is possible to conclude that the electric field causes a slowdown in the development of the embryo.

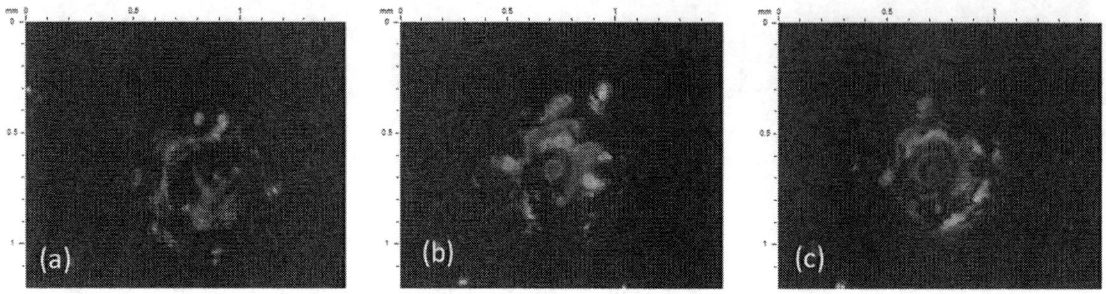

Figure 5. SAM images of the blastoderm top.

Scanning acoustic microscope is an attractive tool for detection and monitoring slow changes in elastic discontinuities inside living objects. To visualize these tiny variations the output signal of SAM is recorded as a function of the lateral coordinate x and the "slow" real time T. Often such data set $s(x,t)$ is called M–csan [6, 7]. Figure 6 presents the M–scans measured before (a) and after (b) electromagnetic field treatment. In this case a frequency of 100 Hz is used for the embryo stimulation.

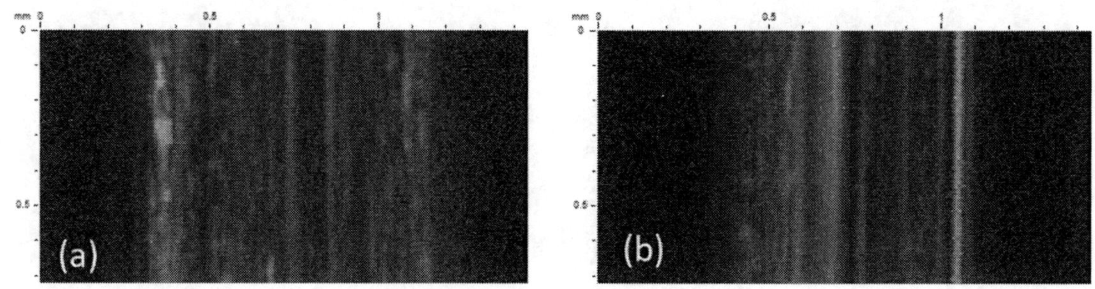

Figure 6. Ultrasonic M-scan data.

In these scans the horizontal and vertical axes represent the lateral coordinate x and time T, respectively. The length of x axis is equal to 1.5 mm, and the duration of T axis is 8 min at time step of 4 s. The changes in the image pattern in the vertical direction indicate the movement of the discontinuities inside the object. Generally it is possible to estimate the speed of the motion. Measurements conducted in a separated experiment shows that the reported speed is in a range of 0.1 – 1 µm/s. Comparison of data (a) and (b) presented in figure 6 leads to observation that the activity of the object is much lower after the electromagnetic stimulation.

5. Conclusion

The technique developed for study of development process in living biological objects is based on combination of the multispectral optical microscopy and scanning acoustic microscopy. In this work the influence of some electric fields on the development was successfully studied. It was shown that the complementary information provided by both instruments is valuable. Therefore it can be suggested that the proposed approach could be useful for much wider application area.

References

[1] Graves M S, Hassell T, Beier B L, Albors G O and Irazoqui P P 2011 *Ann. Biomed. Eng.* **39** 1759.
[2] Tang-Schomer M D 2018 *Brain Research* **1678** 288
[3] Burlakov A B, Machikhin A S, Khokhlov D D and Gadzaov A F 2019 *J. Biomedical Radioelectronics* **2** 63
[4] Polschikova O V, Machikhin A S, Ramazanova AG, Bratchenko I A, Pozhar V E, Danilycheva I V, Katunina O R and Danilychev M V 2018 *Optics and Spectroscopy* **125** 1074

[5] Tan G X Y, Jamil M, Tee N G Z, Zhong L and Yap C H 2015 *Annals Biomedical Engeneering* **43** 2780
[6] Foster F, Pavlin C, Harasiewicz K, Christopher D and Turnbull D 2000 *Ultrasound in Med. Biol.* **26** 1
[7] Sun L, Lein C. Xu X and Shung K. 2009 *Ultrasound in Med. Biol.* **34** 31
[8] Kneipp M, Estrada H, Lauri A, Turner J, Ntziachristos V, Westmeyer G and Razansky D 2015 *Mechanism of development* **138** 300
[9] Xiong G M, Do A T, Wang J K, Yeoh C L, Yeo K S and Choong C. 2015 *J. Biol. Eng.* **9** 1
[10] Domanskiy V, Koshelev S and Sobakin I. *Biomedical Engineering* 2017 **51** 6

Journal of Physics: Conference Series **1421** (2019) 012051 doi:10.1088/1742-6596/1421/1/012051

Application of the structured laser illumination to planar optical diagnostics of the flows with nonuniform density

D Sharaborin[1,2], R Tolstoguzov[1,2] and E Frolova[2]

[1]Kutateladze Institute of Thermophysics, 1 Ak. Lavrentyeva Avenue, Novosibirsk, 630090, Russia
[2]Novosibirsk State University, 2 Pirogova Str., Novosibirsk, 630090, Russia

E-mail: sharaborin.d@gmail.com

Abstract The aim of this work was to develop a method of panoramic optical diagnostics of the inhomogeneity of the density field in stratified flows based on the registration of the distribution of the Rayleigh scattering intensity using structured laser radiation. Several approaches have been implemented for organizing spatial modulation of laser illumination, intended for use with various values of the incident radiation energy density. To test the processing and reconstruction algorithms of the experimental data, a series of measurements was carried out using the method of planar laser-induced fluorescence. The distribution of local density in a jet of carbon dioxide flowing into atmospheric air was measured experimentally using a method based on the registration of Rayleigh scattering.

1. Introduction

Nowadays, numerical simulation is widely used for modeling of flows in the development and design of efficient and environmentally friendly combustion chambers and burners. The calculation of flows with combustion is a nontrivial problem, which requires the simulation of the hydrodynamic structure of the flow, taking into account local thermal emissions and non-uniform densities fields. In addition, this task is motivated by the fact that it is necessary to carry out calculations of tens and hundreds of intermediate chemical reactions occurring in the combustion process at each node of the computational grid. Such complications of the models lead to a significant increase in the required computing power and calculation time. Also, it should be noted that the introduction of new numerical models is impossible without their precise verification. Thus, the use of experimental research methods is a more effective approach (in terms of time and resources) in case of turbulent flows with combustion, rather than numerical simulation methods. Currently there are several approaches based on the use of laser radiation to measure temperature or density distribution in combustion systems: Rayleigh scattering [1], spontaneous Raman scattering [2], laser fluorescence [3]. The LIF method uses generally OH radicals [4, 5], also NO [6, 7] or the addition of labeled molecules/atoms [8] as temperature indicators. All these methods have their limitations for use in technical systems with combustion, in case of at least one of the following factors: spatial resolution, measurement area size, pressure, the presence of particles, toxicity, and often underestimated high price.

The structured illumination method of visualization was first introduced in 1997 in microscopy [9]. A distinctive feature of this method is ability to eliminate defocused light, which impairs the partitioning ability of optical devices within the line of sight. Over time, this technique has gained rapid development and widespread use in other areas [10-12]. Gustafsson illustrated that the method can be used to overcome the diffraction limit, reaching theoretically unlimited resolution [13, 14]. Most publications about structured illumination are concerned with a microscopic image, but the technique has also found popularity outside of this area. Cortizo et al. used a structured light source to study the topography of the sole of the foot during a walk [15], while Kristensson et. al. implemented a method for studying the dynamic flow of water droplets [16] and Rayleigh scattering thermometry [17].

The main aim of present paper is to develop a method of planar optical diagnostic of the nonuniformity of density and temperature fields in stratified flows based on the registration of the Rayleigh scattering intensity distribution.

2. Experimental setup

For the purpose of testing the reconstruction algorithms and processing of experimental data measurements were performed using the methods of planar laser-induced fluorescence (LIF) and Rayleigh scattering in combination with a structured laser illumination. In the first case, a drop of a fluorescent dye Rhodamine 6G dissolving in pure alcohol as a test object was investigated using the LIF method. In the second case, the method of Rayleigh scattering was used to study a round jet of carbon dioxide flowing into the atmospheric air at constant temperature.

Sketch of the experimental setup for LIF and Rayleigh measurements is shown in figure 1. The second harmonic of a pulsed Nd:YAG lasers Beamtech Vlite 200 and Quanta-Ray was used as excitation radiation. The energy of 6 ns laser pulses was monitored by an energy meter (Coherent LabMax-TOP). On average, the energy was 60 and 700 mJ for LIF and Rayleigh measurements, respectively, with 5% RMS fluctuations. A system of focusing and collimating optics was used to form a laser sheet 50 mm wide and less than 0.8 mm thick in the measurement area. For registration the LIF signal, a 16-bit ICCD camera (Princeton instruments PI-MAX-4, resolution: 2560×2160 pixels) equipped with a multi-notch holographic filter to block the emission at laser wavelength. The intensity of the Rayleigh scattering signal in the jet was recorded using an image intensifier with a multichannel amplifier (LaVision IRO) with a S20 photocathode (multialkali). After amplification, the phosphor coating of the intensifier illuminated a hybrid CCD-CMOS matrix (Imager sCMOS) of a 16-bit camera (resolution: 2560×2160 pixels, pixel size: 6.5×6.5 μm).

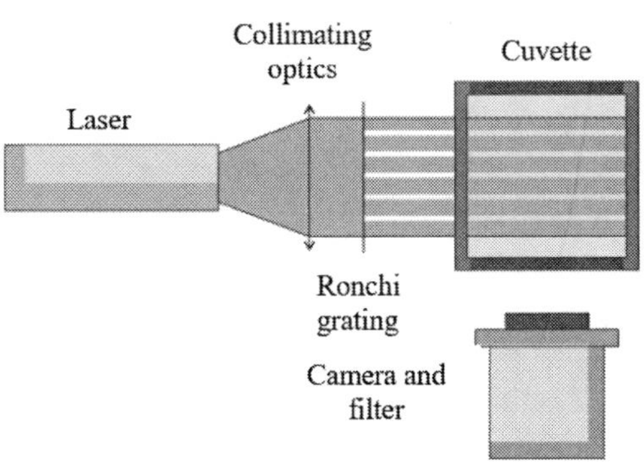

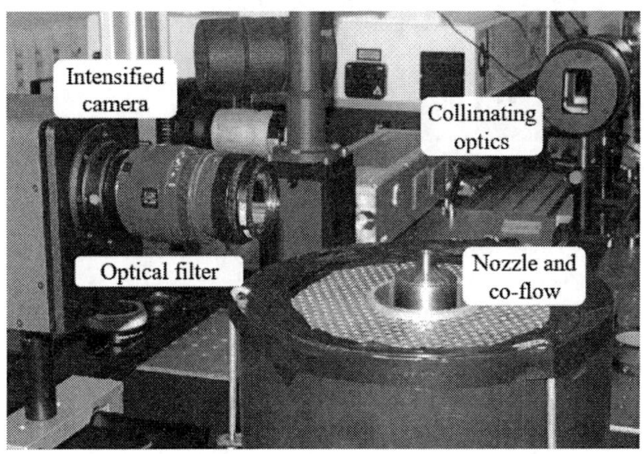

Figure 1. Sketch and photography of the experimental setup for LIF (left) and Rayleigh (right) measurements.

Two Ronchi grating with different line spacing were used to create a structured laser illumination in the first experimental study. The grating was placed close to the measurement area in the path of the laser sheet after collimating optics. Line spacing of 50 lines/inch and 5 lines/mm allowed to collect several data sets necessary for the development of the processing algorithms. Examples of LIF signal raw data in a drop of Rhodamine 6G excited by structured laser sheet are shown in figure 2.

A round jet of carbon dioxide flowing into the atmospheric air was organized using a contraction nozzle (with the exit diameter $d = 10$ mm). The Reynolds number Re was 5300 with volumetric flow rate. To protect the Rayleigh scattering signal recording system from the negative effect of the Mie scattering on dust particles around the carbon dioxide jet, an air co-flow was created. Spatial modulation of laser illumination using Ronchi gratings in the case of Rayleigh scattering measurements was impossible because the high density of laser energy in the sheet leads to the destruction of the grating (the grating bands evaporated from the substrate). Thus, to create a structured laser illumination, a diffraction grating placed between the focusing lens and the collimating optical system.

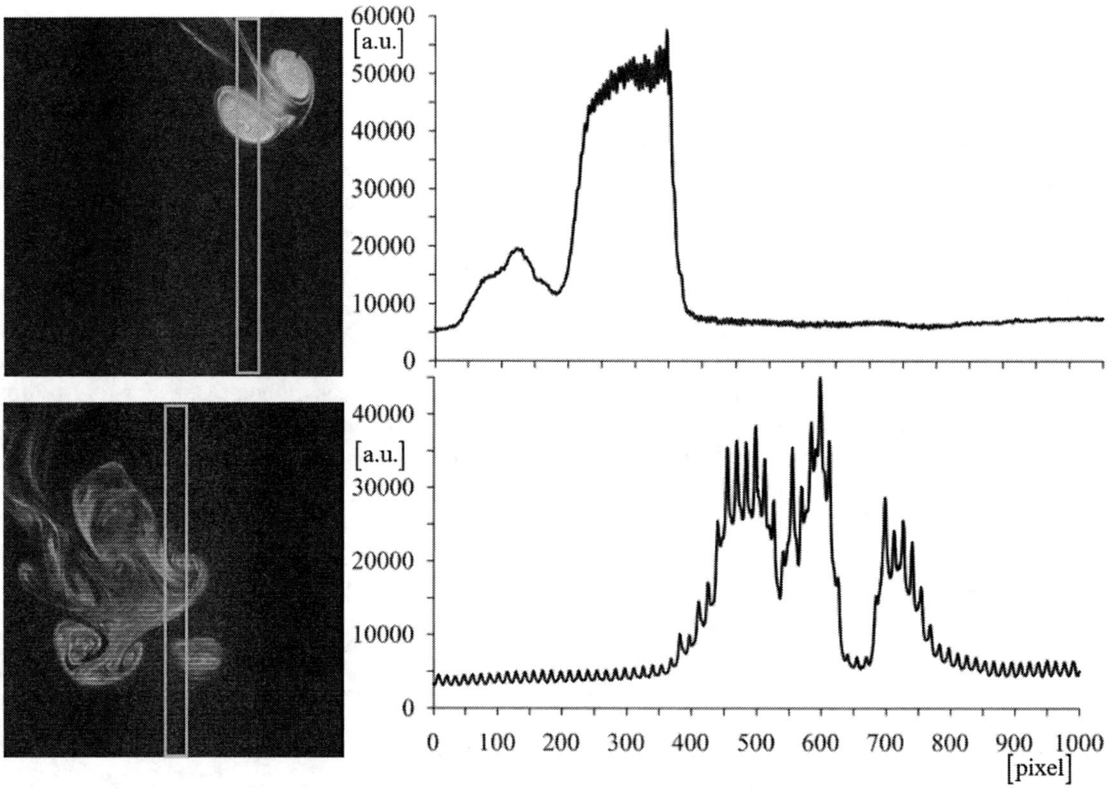

Figure 2. Raw data examples of LIF signal of a drop of Rhodamine 6G (left) and profiles in the selected image area with periodic signal (right).

3. Data processing and reconstruction approach

In most practical applications of Rayleigh scattering, for example, in engines, gas turbines or other combustion techniques, where a laser beam passes through optical windows. The main source of errors and the limiting factor for Rayleigh scattering is associated with the detection of scattered light – interference, including all the parasitic light obtained by the camera [18]. Rayleigh scattering is particularly sensitive to extraneous light noise for two reasons: (1) the Rayleigh signal is not wavelength shifted (this means that using optical filters to suppress, for example, laser light reflections is impossible) and (2) Rayleigh scattering is relatively weak compared to scattering from particles of dust and droplets (i.e., Mie scattering), which significantly exceed Rayleigh scattering intensity. In addition, the complexity of applying the LRS method in combustion processes caused by high temperature and, consequently, low numerical density, that lead to a low signal-to-noise ratio. Thus,

Journal of Physics: Conference Series **1421** (2019) 012051 doi:10.1088/1742-6596/1421/1/012051

all intensity contributions from other sources – reflections, flame emissions, Mie scattering on dust particles, etc. – have strong impact on the accuracy of measurements. The accuracy of the method depends on (1) the ability to accurately subtract the representative two dimensional field of background intensity, (2) preventing the creation of this unwanted light source. Unfortunately, none of the solutions are easy to use.

To provide a structured laser illumination, a laser sheet first passes through the Ronchi grating and then goes to the sample. The grating leaves a structural imprint on the profile laser sheet – sinusoidal modulation with a well-defined spatial frequency and phase. Photons that are scattered from the laser sheet directly to the camera (signal photons) retain this structure, while all background sources are displayed as offset (sharp or scattered) in the recorded image. The different characteristics of the two components of the image allow to remove unwanted offsets, which can be achieved in two ways. It is common practice to record three modulated images of a sheet of laser radiation, between which the modulation phase shifted by 120 degrees [9]. In another way, an image with a periodic signal component can be processed to eliminate background and noise using a discrete Fourier transform. The resulting spectrum shows peaks corresponding to the superimposed frequency of the structured laser illumination. For the presented data, the background, highlights, reflections and noise in the image is in the low-frequency range of the spectrum. Figure 3 shows an example of a sequence of image processing steps with a structured laser illumination. In the first step, using a discrete Fourier transform, we create a spectrum, figure 3 (left). Further, on the basis of the obtained spectrum, the frequency of the superimposed modulation is extracted. Sources of background are contained in the region of the spectrum between the peaks corresponding to the superimposed modulation, figure 3 (center). The spectral regions with a frequency equal to the frequency of the superimposed modulation and higher are shifted in such a way that the modulation peak takes the position of the zero mode, figure 3 (right). At the last stage of the data processing, the image is reconstructed using the inverse Fourier transform.

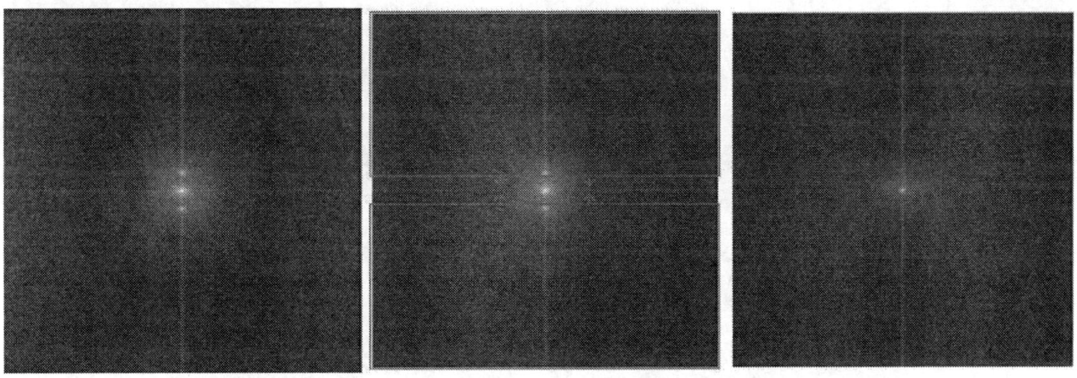

Figure 3. An example of processing stages of the Fourier spectrum. Original Fourier spectrum (left), low frequency region of the spectrum that contains the background signal (center) and Fourier spectrum after shifting superimposed modulation peaks into the position of the zero mode (right).

4. Results

Using the approach described before, the experimental data obtained by the LIF method of a drop of Rhodamine 6G in pure alcohol were processed and reconstructed. Raw data of LIF signal, processed Fourier spectrum and reconstructed image of a drop of Rhodamine 6G with spacing of Ronchi grating 5 lines/mm and 50 lines/inch are shown in figure 4 and figure 5 respectively. For both cases, a series of peaks are observed in the Fourier spectrum with frequency multiple to the frequency of the superimposed modulation. The presence of additional peaks in the spectrum leads to the formation of the same structure of the modulated signal as in the original data in the reconstructed image. In addition, the reconstructed image shows a significant decrease in signal intensity and the presence of large blurry spots (bright areas that are not in the original image). In order to eliminate the effect of additional peaks on the reconstructed image, a narrowband low-pass filter is imposed on the spectrum before applying the inverse Fourier transform. An example of a reconstructed image without and with filtering is shown in figure 6.

Figure 4. Raw data of LIF signal (left), shifted Fourier spectrum (center) and reconstructed image (right) of a drop of Rhodamine 6G with line spacing of Ronchi grating 5 lines/mm.

Figure 5. Raw data of LIF signal (left), shifted Fourier spectrum (center) and reconstructed image (right) of a drop of Rhodamine 6G with line spacing of Ronchi grating 50 lines/inch.

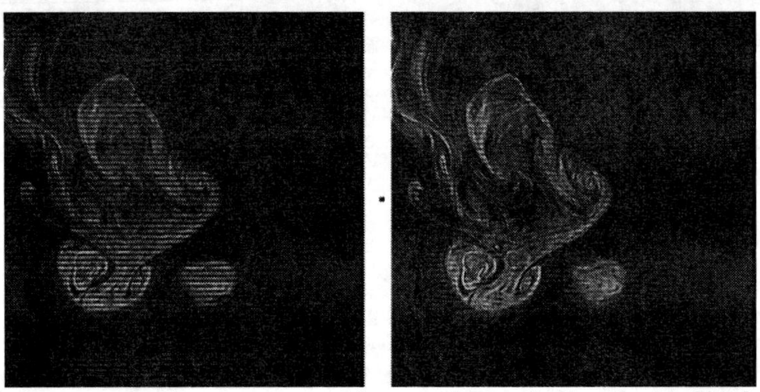

Figure 6. An example of a reconstructed image without (left) and with filtering (right).

Raw data of Rayleigh scattering signal and reconstructed image of a jet of carbon dioxide flowing into atmospheric air are shown in figure 7. Scales normalized to the same value of the signal intensity are shown below the images. In contrast to the results obtained for the LIF measurements, the reconstructed image for the Rayleigh scattering method shows an extreme decrease in signal intensity. The reason for this loss of signal intensity may be that the structured laser sheet is formed using diffraction grating. The interference pattern that is formed in the study area distributes the laser energy, and hence the signal intensity, not over a single modulation frequency, but over the set. Only one frequency is selected in processing and information about the others is lost.

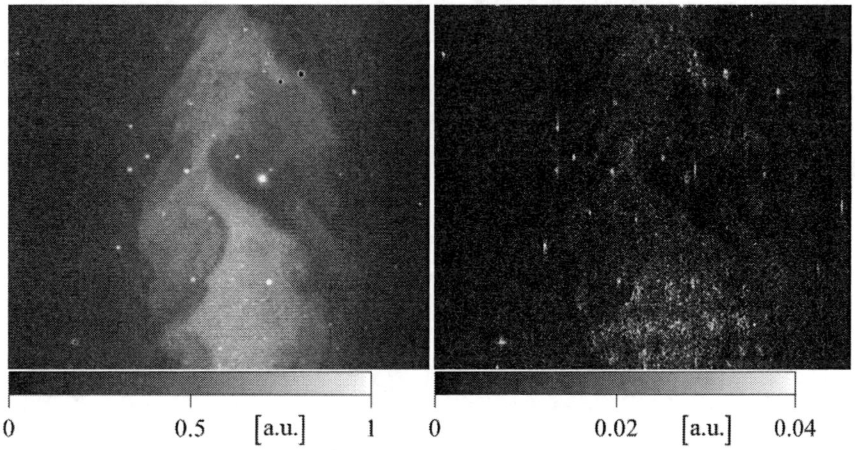

Figure 7. Raw data of Rayleigh scattering signal (left) and reconstructed image (right) of a jet of carbon dioxide.

5. Conclusion

In this work, several approaches have been implemented for organizing spatial modulation of laser illumination, intended for use with various values of the incident radiation energy density. To test the processing and reconstruction algorithms of the experimental data, a series of measurements was carried out using the method of planar laser-induced fluorescence. The distribution of local density in a jet of carbon dioxide flowing into atmospheric air was measured experimentally using a method based on the registration of Rayleigh scattering. These results were used to debug the measuring system and data processing algorithms under conditions of high energy density of laser radiation and to record a low-intensity signal. The results indicate the need for further development of the method, including approaches for the formation of structured laser illumination. In this study, the use of a diffraction grating led to an excessively low signal intensity, comparable to the noise level in the reconstructed image.

Acknowledgments

Experimental studies were carried out under state contract with IT SB RAS (AAAA-A19-119052190039-8), the development and testing of experimental data processing algorithms was financially supported by RFBR (Project No. 18-38-00924).

References

[1] Kampmann S, Seeger T and Leipertz A 1995 *Appl. Opt.* **34** 2780
[2] Wehr L, Meier W, Kutne P and Hassa C 2007 *Proc. Comb. Inst.* 3099
[3] Seitzman J and Hanson R K 1993 *Appl. Phys.* B **57** 385
[4] Giezendanner-Thoben R, Meier U, Meier W and Aigner M 2005 *Flow, Turbulence Combust.* 75 317
[5] Devillers R, Bruneaux G and Schulz C 2008 *Appl. Opt.* **47** 5871
[6] Vyrodov A O, Heinze J, Dillmann M, Meier U E and Stricker W 1995 *Appl. Phys.* B **61** 409
[7] Lee M P, McMillin B K and Hanson R K 1993 *Appl. Opt.* **32** 5379

Journal of Physics: Conference Series **1421** (2019) 012051 doi:10.1088/1742-6596/1421/1/012051

[8] Hult J, Burns I S and Kaminski C F 2005 Proc. *Comb. Inst.* **30** 1535
[9] Neil M A A, Juˇskaitis R and Wilson T, 1997 *Opt. Lett.* **22** 1905–7
[10] Krzewina L and Kim M 2006 *Opt. Lett.* **31** 477–79
[11] Neil M A A, Juˇskaitis R and Wilson T 1998 *Opt. Commun.* 153 1–4
[12] Cole M J, Siegel J, Webb S E D, Jones R, Dowling K, Dayel M J, Parsons-Karavassilis D, French P M W, Lever M J, Sucharov L O D and et al. 2000 *J. Microsc.* 203 246–57
[13] Gustafsson M G L 2000 J. *Microsc.* **198** 82–7
[14] Gustafsson M G L 2005 Proc. *Natl. Acad. Sci. U.S.A.* **102** 13081–86
[15] Cortizo E, Yeras A M, Lepore J and Garavaglia M 2003 *Opt. Laser Eng.* **40** 117–32
[16] Kristensson E, Richter M, Pettersson S-G, Ald´en M and Andersson-Engels S 2007 *Appl. Opt.* **47** 3927–31
[17] Kristensson E, Ehn A, Bood J and Aldén M 2015 Proc. *Comb. Inst.* **35 (3)** 3689-96.
[18] Elliott G S, Glumac N and Carter C D 2001 *Meas. Sci.Technol.* **12** 452–.66

Journal of Physics: Conference Series **1421** (2019) 012052 doi:10.1088/1742-6596/1421/1/012052

Experimental study of the multiphase flow in a pore doublet model

Yu S Zamula*, E S Batyrshin, R R Latypova, O A Abramova and Yu A Pityuk

Center for Micro- and Nanoscale Dynamics of Dispersed Systems, Bashkir State University, 32 Validi, Ufa, Russia

*E-mail: yuriyzamula@gmail.com

Abstract. The paper presents the results of an experimental study of the properties of single- and multiphase flows in a "pore doublet model" (PDM) using microfluidic and optical microscopy techniques. Polymer based microfluidic devices containing pore doublet were fabricated using soft lithography methods. The average cross-section of the microchannels was in the order of 100 μm, and the length reached several millimeters. A bubble generator in the form of a T-junction microchannel was fabricated to study the multiphase flows. High speed imaging and tracer visualization methods were used to study the flow patterns. The properties of multiphase flows were studied during liquid-gas displacement and bubbly liquid flow through pore doublet. The flow patterns were visualized using polymer micro particles suspended in flowing liquids.

1. Introduction

Microfluidics is new emerging multidisciplinary science of the 21st century, where fluid mechanics touches with many areas of physics, chemistry, and biology. Motivation for rapid development of these came from molecular analysis, molecular biology, microelectronics, and a number of prospective nanotechnology applications. Oil and gas, environmental, and other industries critical for economy require new technologies, which can be developed based on fundamental studies in micro- and nanofluidics. This is due to the fact that behavior of a macrosystems and macro-effects are often determined by the behavior of single microparticles as well as their aggregates.

Nowadays, micro models and microfluidic devices are widely used as effective equipment for quantitative and qualitative analyses of the features of fluid flow and other processes taking place in different porous materials at mili-, micro- and even nano-meter scales. Studying the processes in porous media is a difficult problem due to the large number of parameters affecting fluid flow in complex pore networks. The structure geometry is another factor that should be taken into account in micro-model simulations. Researchers have successfully fabricated various types of micronscale models representing the connected networks of channels designed to study a specific phenomenon, e.g. pore-doublet model (PDM) [1-3] and two-dimensional porous medium representations [4], in which flow of fluids, particles, and solutes are visualized and studied.

In this study, we are focused on the processes in a pore doublet model, which is represented as two connected flat channels of different width with common inlet and outlet. In this case, it can be a suitable simplification and can be considered as a building block of the complex porous domain. In PMD both pores are fulfilled with flow under the same inlet conditions, and the response of different pores may be studied. It allows one to figure out how the flow parameters of the one pore influence the other one. Such geometry is often utilized in past studies as an idealized model of the pore structure for the interpretation of the trapping of one phase by another immiscible phase during fluid-fluid displacements in permeable porous media [5]. Fluid-fluid and fluid-solid interactions mainly determine this phenomenon. The fluid-solid interaction is represented by a parameter known as solid's wettability, which indicates the ability of a fluid to spread itself on a solid surface, strongly influences displacements within a complex porous medium.

In the case of creeping flow, fluid-fluid displacement in a porous medium is controlled by the interplay between various surface and body forces, including capillary and viscous. The relative importance of these forces can be determined using dimensionless Capillary number *Ca*. Voids may form in the wider or narrow microchannel, depending on the velocity of movement of the impregnation fronts in the various branches of the PMD. Indeed, the voids occur in narrow part at high *Ca* because the viscous flow overcomes the capillary forces whereas the voids form in wider part at lower *Ca* because the capillary pressure dominates the viscous pressure [6].

Pore doublet model is widely used to study the processes of void formation during the fibrous preforms impregnation for Liquid Composite Molding (LCM) processes. [7-8]. Minimizing the void content in the composite material is a challenging task in the field of composites because of strong negative influence of voids on the mechanical properties of composite parts. The structure of the preform can be considered as a network of channels and a pore doublet can be used as an elementary building block of the network. Pore doublet model itself or as an element of a more extensive microchannel network [9] is used in the investigations of the dynamics of dispersed inclusions in the fluid flow. Microfluidic networks create very unique environments for droplets.

The origin of the periodic and chaotic droplet moving in a microfluidic loop device with microchannels of the same size was studied experimentally in [10]. It was found that the origin of chaotic behavior is through intermittency, which arises when drops enter and leave the junctions at the same time. Also the analytical expression to estimate the occurrence of these chaotic regions as a function of system parameters was derived. The similar device geometry representing the simplest microfluidic network (a loop) comprises two parallel channels with a common inlet and a common outlet, was utilized to study the droplet distribution of choice one of the loop branches [11]. In contrast to previous studies, the article provide the experiments and computer simulations to examine the impact of the changing of cross section of the channels along their length on the sequence of droplets entering the left (L) or right (R) arm of the loop. It was shown that even small variation of the cross section along channels completely shifts the dynamics either into the strong preference for highly grouped patterns (RRR. . . LLL) that generate system-size oscillations in flow or just the opposite—to patterns that distribute the droplets homogeneously between the arms of the loop (e.g., RLRLRL. . .). In [12] droplet sorting is studied in microfluidic channels forming an asymmetric loop with a long and a short arm. The loop is connected to an inlet and an outlet channel by two right angled T-junctions. The sorting diagram was represented depending on droplet size and distance between droplets in the inlet part of the channel, also the influence of loop asymmetry on the mobility ratio was taken into account.

Simple microfluidic loop [13] and more complicated network [14] used in biological applications as a model of microvascular network to conduct in vitro experiments, which can fill the gap between existing in vivo and in silico models as they provide better controllability than in vivo experiments without mathematical idealizations or simplifications inherent to in silico models. In [13] the mechanism non-uniform partitioning or phase separation of red blood cells (RBCs) is investigated in simple models of symmetric/asymmetric microvascular networks.

It should be noted that PDM was used to study not only multiphase systems, but also single-phase flows, as in [15] where PDM was used to study the transition of the flow from laminar to turbulent regime.

All the above results indicate that the pore doublet model is an attractive model system for researchers from various fields of science, and the research results have practical significance. At the same time, not all characteristics of flows, especially multiphase ones, are detailed studied. In the this paper, for the experimental study of the properties of flows in a pore doublet, we used the methods of microfluidics, optical microscopy, and visualization of flows using tracer particles.

2. Experimental Section

The pore doublet was made within a polymer material using standard methods for the manufacture of microfluidic devices [16]. The method of soft-lithography was first proposed in 1997 by the group of G.M. Whitesides and found wide application in the manufacture of prototypes of microfluidic chips for various studies. The dominant material among thermosetting elastomers for creating microfluidic devices is polydimethylsiloxane (PDMS). In this work we used a polymer Sylgard 184 (Dow Corning, USA), comprising of two parts: the base and the curing agent, which are traditionally mixed in a ratio of 10:1 by weight. The material has high optical transparency, good adhesion to glass. We applied the photolithography method using the negative photoresist SU-8 (MicroChem, USA). To increase the adhesion of the photoresist the glass slides with an ITO (indium tin oxide) coating were used. The microfluidic device was made in such a way that all the internal surfaces of the microchannels were made of PDMS. For experiments in this work, pore doublet models of PDMS with different ratio of channel widths were used, the height of microchannels in all cases was 38 μm (figure 1).

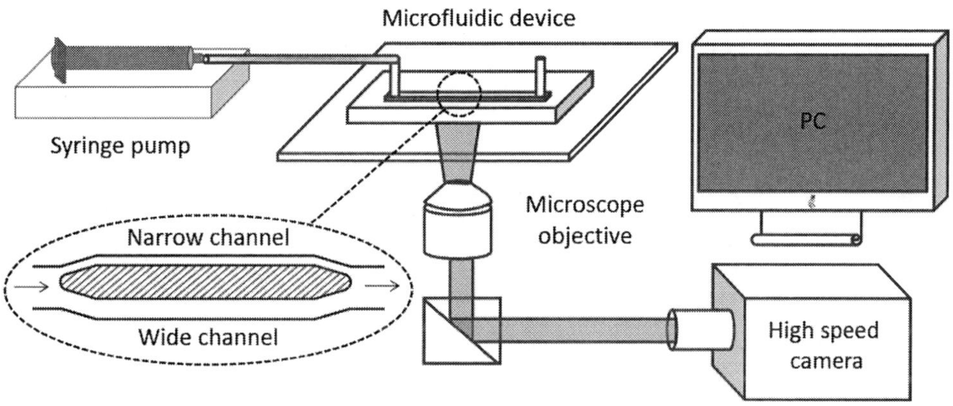

Figure 1. Experimental setup and scheme of pore doublet model.

To supply the liquid phase to the channels of the model with a constant flow rate, a syringe pump was used (figure 1). The visualization of the currents in the model was carried out using an Olympus IX-71 research microscope. Video recording of the processes was carried out using a high-speed camera Photron FASTCAM SA5, connected to a personal computer.

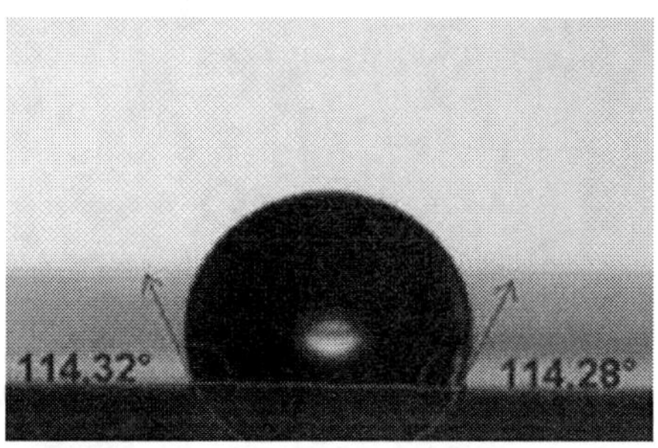

Figure 2. Contact angle of water on PDMS.

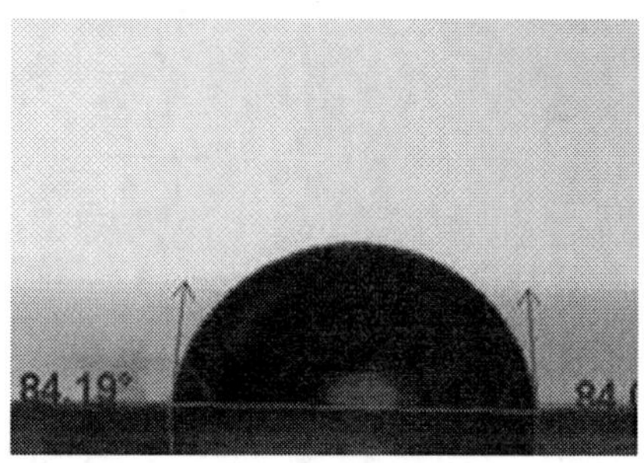

Figure 3. Contact angle of water-Tween 40 (1 wt%) on PDMS.

Deionized water from the Milli-Q and 1% solution of Tween 40 in water were used in experiments with fluid filling of the pore doublet. Tween 40 being a surfactant significantly changes the surface tension of water, and hence the contribution of capillary forces during the filling of the pore doublet. The contact angles of used liquids on the PDMS surface were determined using an optical tensiometer Attension Theta. The average contact angles of wetting on PDMS were: for water – 114.3° (figure 2), water-Tween 40 (1 wt%) in water – 84.1° (figure 3), from which follows the difference in their surface tension. For visualization of the flow structure in microchannels, polymer particles (microspheres) with a diameter of 6 μm (Fluoro-Max, Thermo Scientific) were used.

3. Results and Discussion

3.1. Bubble formation during filling of the pore doublet with liquid

To study the process of fluid flowing into the pore doublet model, a series of experiments was performed at various flow rates. The process of water flowing at a flow rate of 0.5 μl / min is shown in figure 4 (left column). As can be seen, the movement of fluid occurs only through the wide channel, and water does not enter the narrow channel, air remains in it. In the case of being a liquid in narrow channels or porous media, the influence of capillary forces becomes important and the movement of such an interfacial boundary is determined by the balance of viscous and capillary forces. The polymer surface is hydrophobic and the contact angle exceeds 90°, thus the curvature of the interface such that capillary pressure prevents the movement of the front of the liquid. Since the magnitude of capillary pressure depends on the radius of curvature of the surface, the contribution of capillary forces in the wide and narrow channels will be different, namely, in a narrow channel the counteraction to the flow will be greater. Capillary pressure at the interface in the microchannel [17]:

$$P_c = -2\gamma\cos\theta\left(\frac{1}{h} + \frac{1}{w}\right), \quad (1)$$

where P_c is the capillary pressure, γ is the surface tension of liquid in the microchannel, h, w are the channel height and width respectively, and θ are the contact angle of liquid with the microchannel wall. Knowing the geometrical dimensions of microchannels, one can estimate the ratio of capillary pressures $P_c^{narrow}/P_c^{wide} \approx 1.3$.

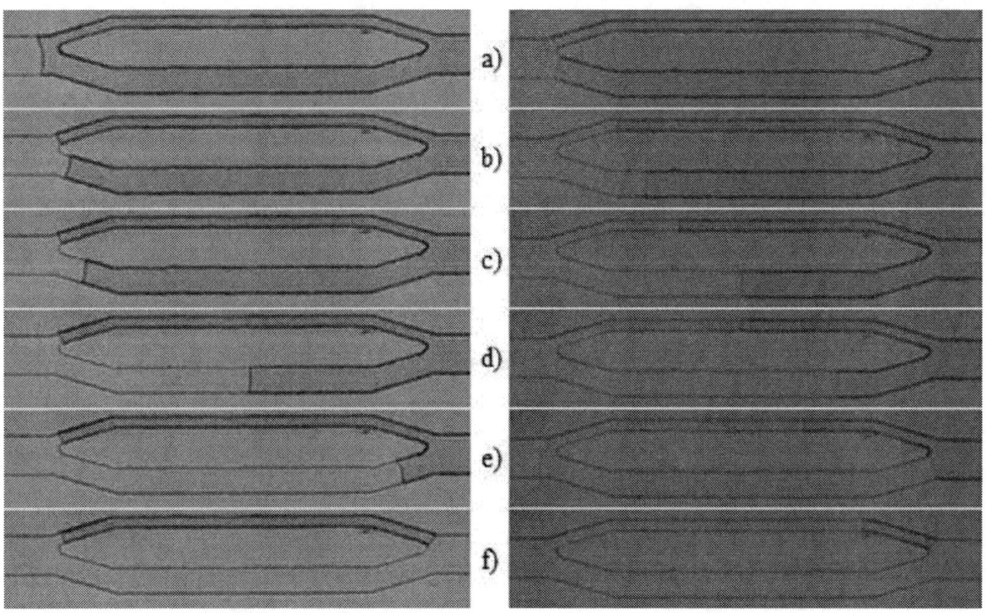

Figure 4. Process of water (left) and water-Tween 40 (1 wt%) (right) inflowing in pore doublet model after 15 s (a), 60 s (b), 105 s (c), 150 s (d), 195 s (e) from the beginning of flowing and in the end (f). Rate 0.5 μl/min. Channel width are 50 μm and 100 μm.

The results of the experiment for water-Tween 40 (1wt%) at a flow rate of 0.5 μl / min are shown in figure 4 (right column).

As can be seen from figure 4, the addition of surfactant Tween 40 at a concentration of 1 wt% changed the capillary pressure, as a result of which the air bubble in the narrow channel decreased significantly. Note, that the addition of Tween 40 to water leads to the change of the capillary pressure sign according to (1). Based on a series of experiments, the dependence of the volume of the air bubble remaining after the fluid flow through the pore doublet was determined from the volumetric flow rate of the fluid (figure 5).

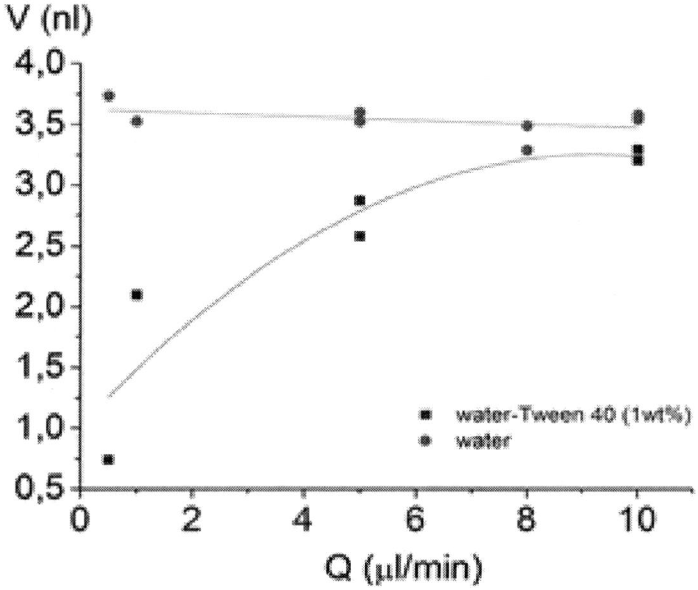

Figure 5. Volume of the remaining bubble after liquid inflowing in pore doublet model.

From the graph in figure 5 it can be seen that at the used flow rates the volume of the air bubble in the narrow channel with the water flow is almost unchanged and completely fills the entire narrow channel. When using a water-Tween 40 solution (1 wt%), the volume of the bubble decreases with decreasing flow rate. The results indicate the obvious contribution of capillary forces to the velocity of the front of the fluid in the microchannels of the pore doublet.

3.2. Flow visualization

The flow in the pore doublet was visualized using particles suspended in a viscous liquid (a solution of propylene glycol in water with a mass fraction of propylene glycol of 70%). The particles were polymer microspheres with an average size of 6 microns. The images were recorded in the dark field mode, which provided the high image contrast and intensity, sufficient for high-speed video recording. Image processing and the flow velocity mapping were performed using the PIVlab software package [18]. Figure 6 shows an image of a fluid flow with tracers in a pore doublet. In the dark field mode, the image intensity is proportional to the spatial gradient of the optical density of the sample; therefore, the bright areas in the image correspond to particles in the fluid and microchannel boundaries, i.e. areas where there is a change in the coefficient of optical refraction.

Journal of Physics: Conference Series **1421** (2019) 012052 doi:10.1088/1742-6596/1421/1/012052

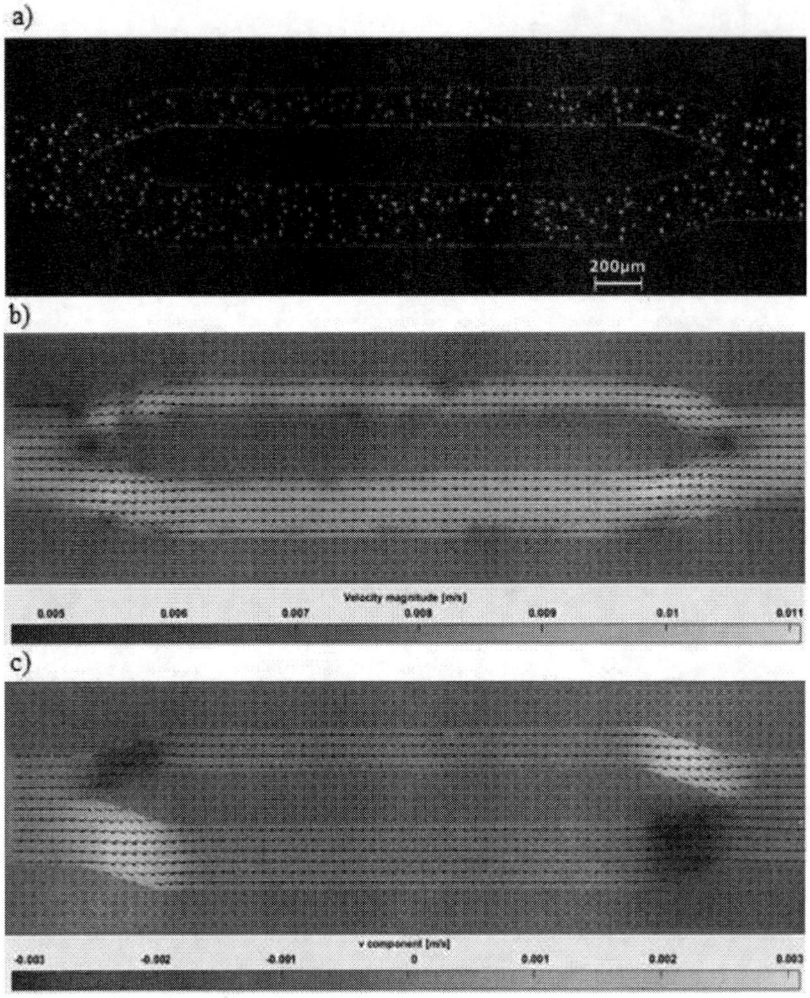

Figure 6. Velocity field obtained using PIV measurements: (a) – image of pore doublet containing liquid with suspended microspheres; (b) – velocity magnitude |(u,v)| map; (c) – velocity vertical component v map.

Figure 6 (b) and (c) show the fluid flow velocity field resulting from processing and averaging 100 frames. Vectors in the images indicate the direction of flow and the color corresponds to the magnitude of the flow velocity. Fig.6b shows the distribution of the velocity magnitude, and figure 6 (c) shows the vertical velocity component. The average value of the flow velocity in the wide (lower) microchannel was 9.0 m/s and was slightly higher than the average flow velocity in the narrow (upper) microchannel 8.1 m/s.

Figure 7 shows the distribution of flow around fixed bubbles attached to the walls of the channel.

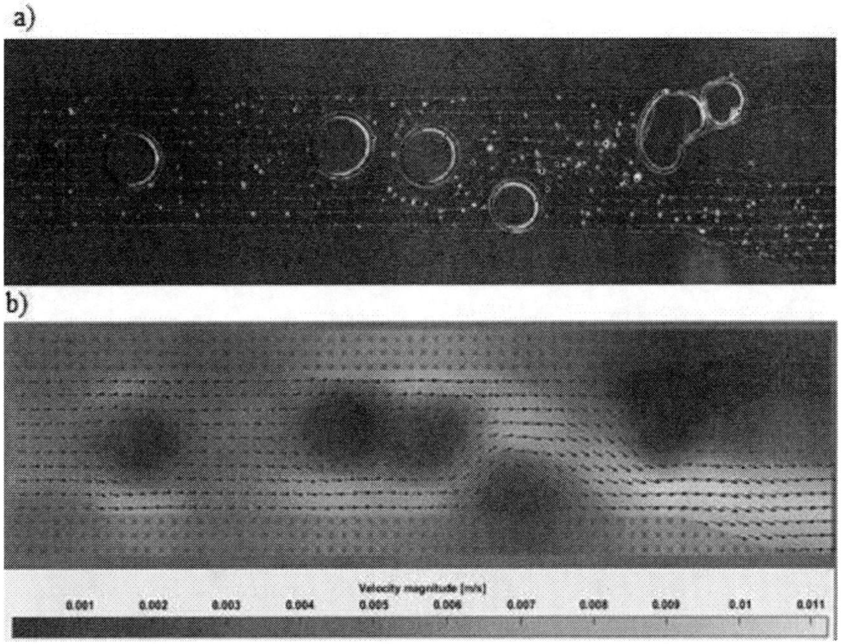

Figure 7. Velocity map near stationary bubbles: (a) – velocity vector map; (b) – velocity magnitude map.

It is interesting to analyze the flow distribution around a moving bubble. It can be seen that a zone arises in front of the bubble, where the particles move slower than behind the bubble (in the direction of flow). This is due to the generation of additional hydrodynamic flows behind the bubble, associated with its deformation.

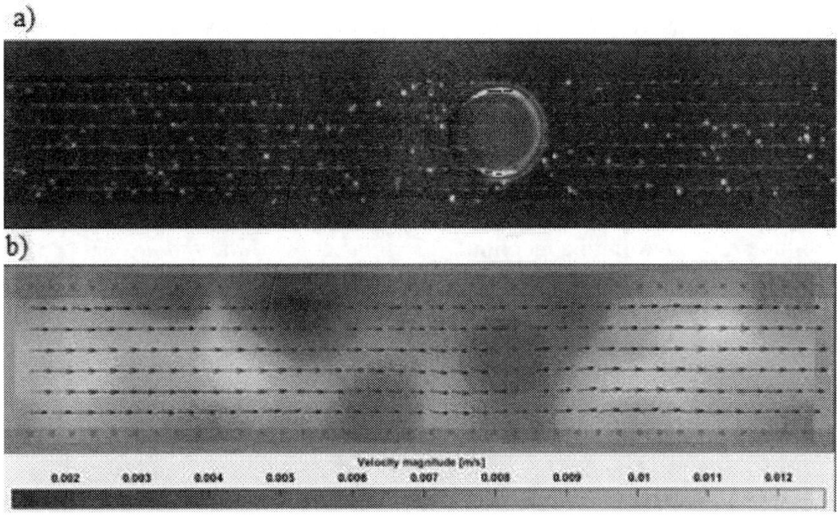

Figure 8. Velocity map near the moving bubble: (a) – velocity vector map; (b) – velocity magnitude map.

Thus, the presence of bubbles in the microchannel makes the structure of the fluid flow much more complicated than in single phase flow. As it is known, even in the case of symmetric channels both regular and chaotic dynamic modes can be realized [10]. So, in case of pore doublet with asymmetric channels the variety of complex dynamic regimes are expected.

4. Conclusions

In this paper we provide experimental study of the multiphase flow in a pore doublet model. The microfluidic device containing the microscale physical model of a pore doublet was fabricated using PDMS soft lithography technique. The formation of a bubble during the filling of the pore doublet with liquid was studied. The effect of wettability on bubble trapping was shown. The structure of the fluid flow in the pore doublet was visualized using tracer particles.

In our further studies we plan to use presented here experimental techniques to investigate the properties of the flow of liquid containing bubbles through the pore doublet during the variations of parameters: the size and concentration of bubbles, wettability conditions, the microchannel width, etc.

Acknowledgments

The study was funded by RFBR according to the research project No. 18-38-20102.

References

[1] Chatzis I and Dullien F A L 1983 Dynamic immiscible displacement mechanisms in pore doublets: theory versus experiment Journal of Colloid and Interface Science 91.1 199- 222

[2] Ehrlich R and Crane F E 1969 A model for two-phase flow in consolidated materials Society of Petroleum Engineers Journal 9(02) 221-31

[3] Al-Housseiny T T, Hernandez J and Stone H A 2014 Preferential flow penetration in a network of identical channels *Physics of Fluids* **26(4)**

[4] Oren P E, Billiotte J and W V Pinczewski 1992 Mobilization of waterflood residual oil by gas injection for water-wet conditions *SPE Formation Evaluation* **7.01** 70-8

[5] Nabizadeh A, Adibifard M, Hassanzadeh H, Fahimpour J and Moraveji M K 2019 Computational fluid dynamics to analyze the effects of initial wetting film and triple contact line on the efficiency of immiscible two-phase flow in a pore doublet model *Journal of Molecular Liquids* **273** 248-58

[6] Wielhorski Y, Abdelwahed M A B, Bizet L and Bréard J 2012 Theoretical and Experimental Modelling of Bubble Formation with Connected Capillaries in Liquid Composite Moulding Processes *arXiv preprint* arXiv:1207.4100

[7] Lundström T S, Gustavsson L H, Jēkabsons N and Jakovics A 2008 Wetting dynamics in multiscale porous media. Porous pore-doublet model, experiment and theory *AIChE Journal* **54(2)** 372-80

[8] Abdelwahed M A B, Wielhorski Y, Bizet L and Bréard J 2012 Void Prediction During Liquid Composite Molding Processes: Wetting and Capillary Phenomena *arXiv preprint* arXiv:1207.4184

[9] Zaremba D, Blonski S, Marijnissen M J and Korczyk P M 2019 Fixing the direction of droplets in a bifurcating microfluidic junction *Microfluidics and Nanofluidics* **23(4)** 55

[10] Maddala J, Vanapalli S A and Rengaswamy R 2014 Origin of periodic and chaotic dynamics due to drops moving in a microfluidic loop device *Physical Review E* **89(2)** 023015

[11] Cybulski O, Garstecki P and Grzybowski B A 2019 Oscillating droplet trains in microfluidic networks and their suppression in blood flow *Nature Physics* 1

[12] Kadivar E, Herminghaus S and Brinkmann M 2013 Droplet sorting in a loop of flat microfluidic channels *Journal of Physics: Condensed Matter* **25(28)** 285102

[13] Clavica F, Homsy A, Jeandupeux L and Obrist D 2016 Red blood cell phase separation in symmetric and asymmetric microchannel networks: effect of capillary dilation and inflow velocity *Scientific reports* **6** 36763

[14] Cybulski O, Jakiela S and Garstecki P 2015 Between giant oscillations and uniform distribution of droplets: The role of varying lumen of channels in microfluidic networks *Physical Review E* **92(6)** 063008

[15] Khayamyan S, Lundström T S and Gustavsson L H 2014 Experimental investigation of transitional flow in porous media with usage of a pore doublet model *Transport in porous media* **101(2)** 333-48

[16] Xia Y and Whitesides G M 1998 Soft lithography *Annual review of materials science* **28(1)** 153-84

[17] Olanrewaju A, Beaugrand M, Yafia M and Juncker D 2018 Capillary microfluidics in microchannels: from microfluidic networks to capillaric circuits *Lab on a Chip* **18(16)** 2323-47

[18] Thielicke W and Stamhuis E 2014 PIVlab–towards user-friendly, affordable and accurate digital particle image velocimetry in MATLAB *Journal of Open Research Software* **2(1)**

Detailed investigation of bag-breakup process in the context of spray of droplets generation within wind-wave interaction

A A Kandaurov[1], D A Sergeev[1] and Yu I Troitskaya[1]

[1] Institute of Applied Physics RAS, Nizhny Novgorod, Ulyanova 46, 603950, Russia

E-mail: daniil@ipfran.ru

Abstract. The main mechanism responsible for the generation of spume droplets in marine atmospheric boundary layer was investigated at the experiments carried out at the high-speed wind-wave flume of IAP RAS. Strong wind tears off water from the crest of the waves during bag breakup fragmentation of small-scale disturbances that arise at the air-water interface. Separate bag-breakup event forced to occur in a dried high-speed wind-wave flume was investigated qualitatively and quantitatively using multiperspective shadowgraph high-speed video recording.

1. Introduction

Recent studies of wind-wave interaction [1] have shown that bag-breakup fragmentation of small-scale disturbances that arise at the air-water interface under the strong wind the main mechanism responsible for the generation of spume droplets. This mechanism was first described in [2] where this mode of fragmentation of large drops in the lateral air flow, followed by the stage of inflating a large membrane, and its subsequent rupture into small drops was called "bag-breakup" mode. In [3], this mode was studied in more detail using high-speed video in a wide range of conditions, and in [4] a similar mode was observed when the jet fragmentized in the transverse air flow. The existence of such a mode in natural conditions in nature was first demonstrated in [5], by the interaction of an air flow with rough water surface within the framework of laboratory modeling of the interaction of the atmosphere and the ocean on a wind-wave flume.

Sea sprays are typical element of the marine atmospheric boundary layer and important environmental effect. But there are significant uncertainties in estimations of sea sprays influence on the marine atmospheric boundary layer [6, 7] due to difficulties of direct measurements in hurricane conditions and insufficient knowledge about the mechanisms of the spume droplet's formation that leads to insufficient knowledge on the sea spray generation function, which characterizes the size distribution of droplets injected from a unit of surface in a unit of time. There is a difference in estimations of empirical function of spray generation by six orders of magnitude based on different observation reports (see a review of experimental data in [8]). Therefore, constructing the spray generation function as the basis for calculating the contribution of spray to exchange processes at the air-sea interface is one of the most important problems, we are trying to solve.

In our previous work [1] high-speed recordings of the water surface at high winds in controlled laboratory environment allowed us to retrieve the statistics of number of bag-breakup events appearing on the surface at various wind and water conditions. But to build the spray generation function as convolution of the of single "bag" generation function and of the distribution function of droplets from one "bag-breakup", we need to know what droplets are formed during one bag-breakup fragmentation.

The second part of the article is devoted to the description of the methodology of experiments at high-speed wind-wave flume. We develop methods of creating a single "bag" and obtaining this distribution: due to spontaneous characterization of their occurrence on free water surface that complicated detailed studies the separate bag-breakup event was forced to occur in a dried high-speed wind-wave flume. The third part describes how high-speed video recordings of "bags" were captured using the shadowgraph technique. The fourth part is dedicated to processing by modified particle tracking techniques of high-speed image sequences of singles bag-breakup event to obtain a spray characteristic.

2. Experimental setup

The experiments were carried out in controlled laboratory environment on the High-speed wind-wave flume (HSWWF) of the Institute of Applied Physics (IAP RAS). The operating cross section of the airflow is the flume was 0.4×0.4 m², straight working part of 10 m. During the experiments axis air velocity in the flume varied from 10 to 17 m/s, that corresponds to 20-50 Hz frequency of centrifugal fan equipped with an electronic frequency converter. More details of the facility construction and parameters of air flow and surface waves are described in [9].

The principal scheme of the experimental setup is shown in figure 1. Bag-breakup event was forced to occur at a small reservoir installed at a distance of 7.5 m from the beginning of the working section. The flume was closed at its bottom with rigid plates at the water level used in previous experiments [1] to prevent spontaneous bag-breakup event appearing on free water surface for investigations of a separate bag-breakup event. The measured value of the surface tension was $\sigma = (7.0 \pm 0.15) \cdot 10^{-2}$ N/m.

Bag breakup developed from artificial initial disturbance that was created at the desired position using the underwater jet from a submerged nozzle placed in vertical orientation 1,5 cm under water surface. It is consisted of 5 nickel tubes 20×2 mm, lined up in a row in cross-wind direction to create elliptical leaf-like disturbance. In order to damp the disturbances on water surface around the nozzle a sheet of 20 mm thick foam rubber with a hole 30 mm in diameter was placed above it. The foam was additionally covered with a 2 mm cell size nylon net. Water supplied to the nozzle at a pressure of 3 bar through an electromagnetic valve, controlled by a microcontroller. This enabled us to synchronize the moment of disturbance creation with triggers of both top and size high-speed cameras. The dependence of the shape of the initial disturbance on time was investigated using image processing of frames from the side camera captured without wind. Image processing showed that this system allowed to inject up to 2.5 ml of water in 20 ms time interval with good repeatability: comparison of successive initial disturbances showed the stability of their shape within 10% for height and diameter.

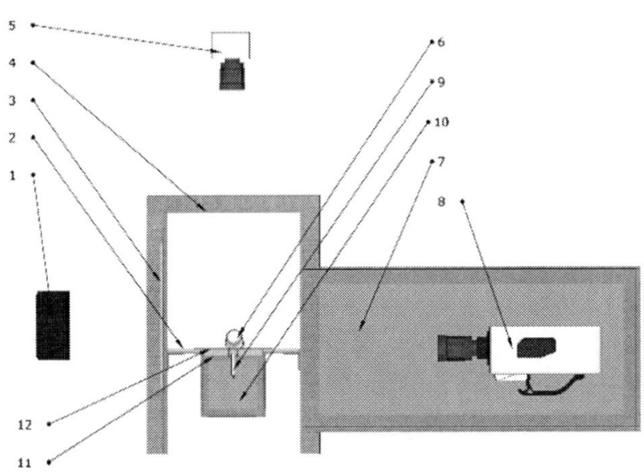

Figure 1. The cross-section of the high-speed wind-wave flume: 1 – LED lights, 2 – solid flat bottom, 3 – opaque screen, 4 – high-speed wind-wave flume, 5 – top high-speed camera, 6 – bag-breakup, 7 – semi-submerged box, 8 – high-speed camera, 9 – nozzle, 10 – reservoir with water, 11 – foam rubber in water, 12 – nylon mesh.

Experiments were performed for 5 fan rotation frequencies from 30 to 50 Hz with 5 Hz step, that corresponds to centerline wind speeds from 10 to 17 m/s. Wind velocity profiles were measured at the working section with reservoir by the profile method using the L-shaped Pitot tube on scanning device with the differential pressure transducer Baratron MKS 226A (the accuracy of 0.5% of full scale range, i.e., 3 cm/s). The profiles were measured in 1 to 20 cm height interval with continuously moving tube (1.1 mm/s vertical scanning speed). For each fixed wind parameters, at least five profiles were measured for subsequent averaging. Friction velocity u^* and equivalent speed U_{10} at a standard meteorological height of 10 m were obtained using profiling method with logarithmic approximation of bottom part of turbulent boundary layer. The scaling parameter for the air turbulent boundary layer flow above the water surface is the friction velocity, defined via vertical turbulent shear stress: $\tau = \rho_{air} \cdot u^{*2}$, where ρ_{air} is the air density. The friction velocity, u^*, varied between 0.46 and 0.73 m/s; the equivalent wind speeds U_{10} between 16 and 26 m/s, that corresponds to Beaufort number up to 10 in field conditions.

3. High-speed video recording setup with shadow illumination technique

High-speed video recording was used to investigate the details of the bag-breakup fragmentation both qualitatively and quantitatively. Records were made for all 5 different wind speeds (fan rotation frequencies of 30 to 50 Hz), 9 to 17 records of the bag-breakup were captured for each speed. Record was interrupted at the moment the last droplets escapes the imaging area. Shape and size of the initial disturbance were controlled before and after each series of record by recording the dynamics of the disturbance without wind.

High-speed recording was synchronously mad from top and from side. Side recording was made with high-resolution NAC Memrecam HX-3 high-speed digital video camera in shadowgraph configuration: vertical opaque screen was placed at the opposite to the camera wall of the flume and was illuminated from behind with two 300 W LED light source. The camera was placed in horizontal position at a distance of 60 cm from the center of the flume. Resolution of 63 µm/px at the frame size of 2560×960 px gave us total size of the imaging area of 161×60 mm that allowed to capture all evolution of bag-breakup. The recording was performed at 3990 fps with exposure time of 1/50000 s. There were up to 700 frames (175 ms) in each record. The side view was used to obtain detailed data on the dynamics of the resulting structures: the shape of the structure before and during bag-breakup, the speed and size of the droplets formed.

Top view was only used to assess shape of bag-breakup in horizontal plane. It was done with the Optronis CamRecord CR3000x2 high-speed camera. The camera was located at a distance of 112 cm from the surface (scale 102 µm/px). The recording was performed at 1250 fps with exposure time of 1/5000 s, the image size was 1696×720 px (173×74 mm). Since we used the same LED light source for side and top filming, only side view was made in shadowgraph setup.

4. Processing of high-speed images sequences

High-speed records of the bag-breakup process created using artificial disturbance (see figure 2) demonstrated that it developed though the same stages as bag-breakup at free water surface in laboratory conditions (compare with [1]). Analysis of high-speed records confirmed that there are two significantly different stages at which the spay in generated. First a rupture of the film takes place, which leads to the formation of small droplets, then the remaining after the rupture of the liquid film "rims" fragment into large drops.

Quantitative data on the characteristics and dynamics of droplets formed on both stages of bag-breakup fragmentation was obtained using specially developed software, that implements edge detection, segmentation, Particle Image Velocimetry (PIV) and Particle Tracking Velocimetry (PTV) methods. Image processing sequence consisted of two main steps: droplet detection and droplet tracking.

Position of all visible droplets on each frame of the record was detected using image processing sequence is shown in figure 3. for a fragment of one frame. We previously developed this method of

Journal of Physics: Conference Series **1421** (2019) 012053 doi:10.1088/1742-6596/1421/1/012053

image segmentations for water surface detection in laser measurements, it is described in details in [10]. The method consist of several steps: background calculation and subtraction, edge detection with sobel of canny operator, morphological operations to remove noise and fill closed areas. The main feature of this sequence is logical 'or' performed with binary initial image at this step: this preserves small droplets keeping the noise low. At the last step positions, sizes and shapes of all isolated regions are calculated. Initial disturbance and "rim" are detected the similar way with different thresholds and neglected. The result of algorithm is position and size of every droplet on every frame of side view records (figure 4). Tuning of the parameters of the algorithm was made manually for each series of records. The algorithm can only detect droplets larger than 100 μm for images captured with this side view setup, but same algorithm can be used for smaller droplets using video with higher resolution.

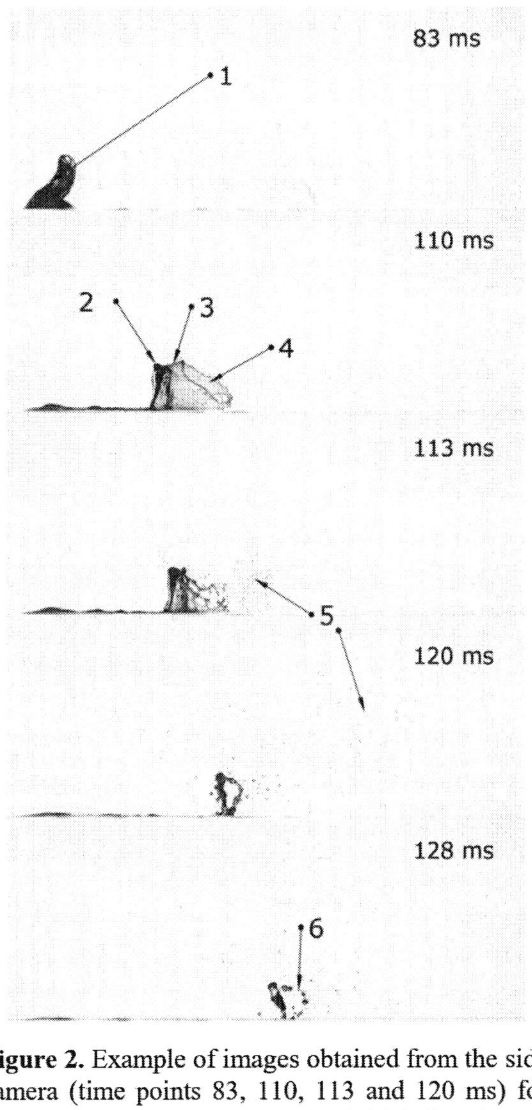

Figure 2. Example of images obtained from the side camera (time points 83, 110, 113 and 120 ms) for wind speed U_{10} = 21.8 m/s. 1 – initial disturbance, 2 – rim, 3 – film, 4 – film rupture, 5 – film drops, 6 – rim drops. Full image width 161 mm.

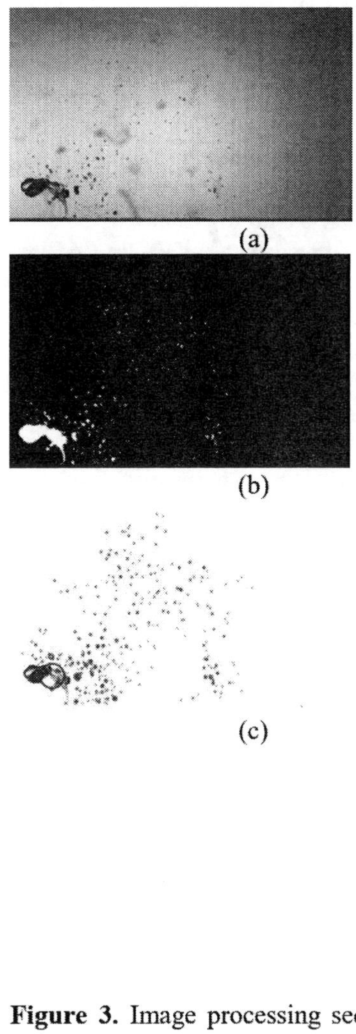

Figure 3. Image processing sequence for a frame at 113 ms after film puncture, U_{10} = 21.8 m/s, 81 mm downwind the initial disturbance, 33 mm above surface. (a) – initial image, (b) – binary image with marked droplet areas, (c) – detected droplets (green circles – detected droplets, red – neglected parts of "rim").

Straightforward particle tracking was not applicable for droplet tracking in present situation due to high visible density of particles at early moments after rapture. Velocities of droplets were roughly estimated by particle imaging velocimetry algorithm to simplify tracking problem. This was possible because most droplets in every particular area of image have similar velocities. We used custom PIV processing software that we developed for previous studies of air velocities in marine boundary layer [11]. Velocities were retrieved by splitting frames into small interrogation windows (128×128 px with 50% overlapping) in which overall displacement of particles was computed using cross-correlation. Only one iteration with tree-point interpolation for CCF maximum was used. Velocity fields were filtered using maximum and minimum velocity thresholds, 2-dimentional median filtering is space domain. PIV resulted in the velocity field on rectangular grid for every frame of the record (figure 5).

Custom tracking algorithm was developed in order to retrieve droplet trajectories. We run tracking backward in time as visible density of droplet decreases over time and it was easier to separate trajectories. Starting from the end of the record for every droplet on frame we tried to find appropriate droplet on previous frame in the vicinity of its assumed position. To define the search area position we used estimations of particle velocity and acceleration using two methods. PIV velocity field were interpolated to position of every droplet of the frame gave us estimate of velocity. Parabolic extrapolation for coordinated of four previously found points of current trajectory assuming a constant acceleration gave us velocity and acceleration. We used a combination of these data depending of its confidence: the first method is used for trajectories containing less than 4 known points, the average with weights (based on confidence) between two methods is used otherwise. Even if a drop has not been detected on one or two consecutive frames, the trajectory was constructed. The trajectory search stops when there are no droplets found in the vicinity of the expected position or the radius of the droplet found is more than 3 times different from the current one. If two trajectories find the same drop, the current drop position is assigned to the trajectory, the distance to the supposed point of which is smaller. The example of droplet tracking is shown on figure 6.

Postprocessing of trajectories consisted of filtering and combining split trajectories. Motionless trajectories with median displacement less than 0.001 px/frame and trajectories shorter than 10 frames are rejected as noise. The algorithm can lose the particle for several frames due to visible overlapping of the particles. To address this problem the standard deviation of the found positions and the polynomial extrapolation was calculated for each non-intersecting in time pair of trajectories. The trajectories were combined into one in case of low deviation. This combination is repeated until all split trajectories combined. This algorithm enables assigning more than 75% of droplets to trajectories for most part for the experimental condition and retrieving of more than 500 separate trajectories for some of the records. The example of retrieved trajectories is show on figure 7.

5. Conclusion

The dominant process of spray generation during wind-wave interaction – bag-breakup fragmentation can be investigates separately under controlled conditions by artificial creation of initial disturbance when separate bag-breakup event is forcefully generated in a dried high-speed wind-wave flume with water reservoir installed under the working section. Bag-breakup fragmentation from artificial disturbance developed through the same stages as bag-breakup on free water surface.

Characteristics of the generated spray can be obtained via high-speed shooting in shadowgraph configuration with special image processing sequence intended to identify the droplets on the images and track their trajectories based on the combination of PIV and reversed in time particle tracking. The data obtained by this method in combination with the statistics of the occurrence of "bags" can be used to construct the empirical function of spray generation.

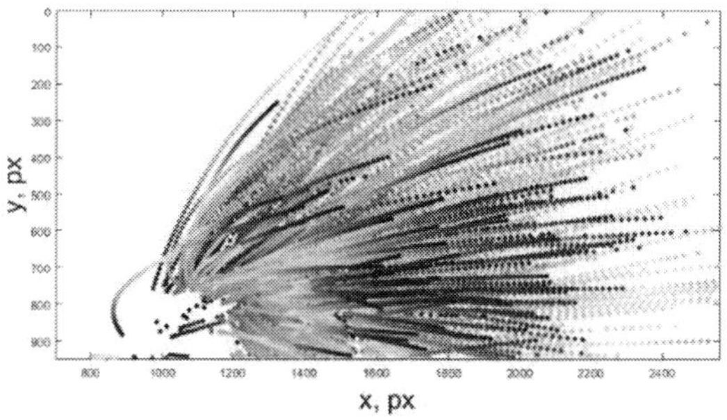

Figure 4. Position of all droplets found on all frames in one record, color represents frame number, full image width 117 mm, $U_{10} = 21.8$ m/s

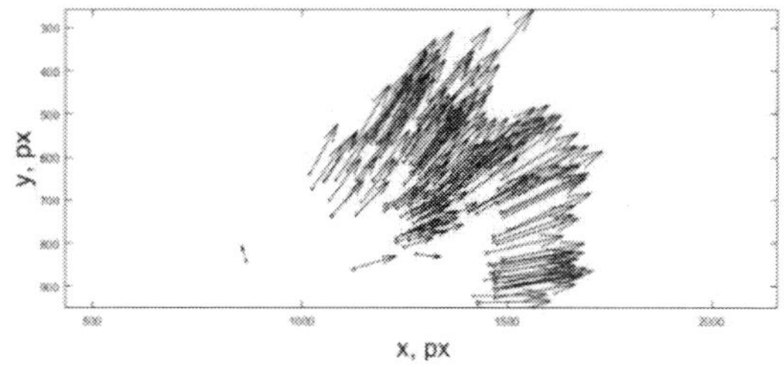

Figure 5. Approximate velocity field obtained by PIV interpolated to position of each droplet (shown in circles) for equivalent wind speed $U_{10} = 21.8$ m/s.

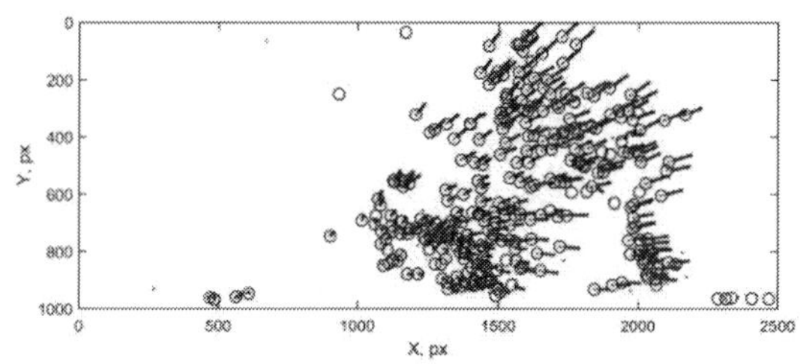

Figure 6. Example of a trajectory search for equivalent wind speed U_{10} = 21.8 m/s. Circles shows the search area near the assumed position. Blue lines are the previous 4 points of each trajectory. Red dots positions of all droplets on the frame.

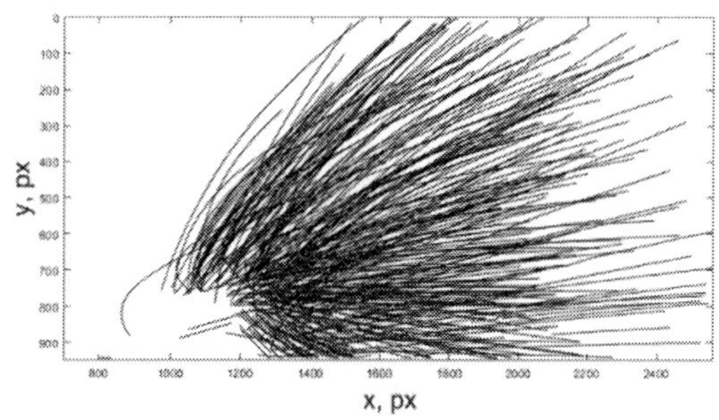

Figure 7. All trajectories found in one record (U_{10} = 21.8 m/s); example frame is 6.2 ms after film puncture. Full image width 117 mm.

Acknowledgements
Carrying out experiments themselves was supported by Russian Science Foundation (Agreement No. 19-17-00209); designing of methods of measurements including: optical and visualization scheme, methods of measurements of the air flow and wave field parameters were supported by Russian Science Foundation (No 18-77-00074); the development of the software for video processing was supported by the Grant of the President no. MK-3184.2019.5 and RFBR project No. 19-05-00249.

References

[1] Troitskaya Yu, Kandaurov A, Ermakova O, Kozlov D, Sergeev D and Zilitinkevich S 2017 *Sci. Rep.* **7(1)** 1614
[2] Gelfand B E 1996 *Energ. Combust. Sci.* **22** 201
[3] Kulkarni V and Sojka P 2014 *Phys. Fluids* **26** 072103
[4] Mashayek A and Ashgriz N 2011 *Handbook of Atomization and Sprays*, ed. Ashgriz N. (Springer: Boston, MA) pp 657–683
[5] Veron F, Hopkins C, Harrison E L and Mueller J A 2012 *Geophys. Res. Lett.* **39** L16602
[6] Andreas E L and Emanuel K A 2001 *J. Atmos. Sci.* **58** 374
[7] Bao J-W, Fairall C W, Michelson S A and Bianco L 2011 *Mon. Weather Rev.* **139** 3781
[8] Andreas E L 2002 *Atmosphere-Ocean Interactions* vol 1, ed. Perrie W (Southhampton: WIT Press) pp 1–46

[9] Troitskaya Yu I, Sergeev D A, Kandaurov A A, Baidakov G A, Vdovin M A, Kazakov V I 2012 *J. Geophys. Res. Ocean.* **117** C00J21
[10] Sergeev D A, Kandaurov A A, Vdovin M I, Troitskaya Y I 2015 *Science Visualization* **7(5)** 109
[11] Kandaurov A A, Troitskaya Y I, Sergeev D A, Vdovin M I, Baidakov G A 2014 *Izvestiya, Atmospheric and Oceanic Physics* **50**(4) 399

The method of compensation for local displacements of images of capillaries in the evaluation of capillary blood flow parameters

M V Volkov*, N B Margaryants, A V Potemkin and I P Gurov

Applied Optics Department, ITMO University, 49 Kronverksky Ave., Saint Petersburg, 197101, Russia

*E-mail: fosp@grv.ifmo.ru

Abstract. The paper discusses the features of applying the method of local compensation of video frame displacements in the study of microcirculation parameters by the methods of video capillaroscopy (VCS) and microphotoplethysmography (micro-PPG). The results of the assessment and compensation of local displacements of images of the capillary network in the region of the hand of a conditionally healthy volunteer are presented. The result of the computation of the synthesized image of capillaries with equalized contrast based on the processing of a locally-combined sequence of video frames is shown. An example of capillary visualization using the frequency analysis of locally aligned video frames is given. The effectiveness of the method of local image stabilization within the study of the capillary network of mucous membranes using a laparoscope is demonstrated.

1. Introduction

Studies of the parameters of capillary blood flow are important for the diagnosis of many socially significant diseases such as diabetes, Raynaud's disease, psoriasis, cardiovascular diseases. A number of diseases (for example, psoriasis, scleroderma) are characterized by local disorders of the morphology of the capillary network in an arbitrary area of arms [1-3]. Capillary network visualization and automation of capillary counting in the affected skin areas are the tasks to be solved in order to conduct diagnostics. This requires observing capillaries over an area of a few square millimeters. Preliminary studies have shown that it is not possible to solve these problems when analyzing a single frame due to the low image contrast and the local frame area defocusing. The task of visualization and counting of capillaries can be solved on the basis of analyzing the sequence of video frames, but the application of such methods requires compensation of local displacements of the capillary network in the process of recording video frames.

In contrast to the study of hand skin using a video capillaroscope, mobile devices such as a laparoscope are required for the study of the mucous membranes of the oral cavity or internal organs. At the same time, in the process of measurements, projective distortions of images are added to the local displacement of capillaries due to the displacement of the device relative to the study object. For this reason, processing and analyzing of video frames obtained with a laparoscope constitute a significant problem.

2. Methods and experimental results

In [4, 5], the usage of the method for full-frame stabilization and compensation for capillary blood flow video frames displacement in video capillaroscopy and photoplethysmography was considered. The implementation of the full-frame method has significant limitations, since it does not allow compensating for local distortions of the skin geometry in the studied area. The full-frame stabilization method was used in order to solve problems of restoring blood flow velocity in the capillaries of the nail bed. Note that in this area, the skin (and the capillary network) is well attached to the nail plate; therefore, in most experiments, the use of the full-frame stabilization method is justified. At the same time, in separate experiments during functional tests, local distortions of the geometry of the capillary network were observed. In such cases, it is necessary to apply the method of local displacement compensation in order to restore the blood flow velocity.

The skin areas outside the nail bed and the capillary network are more susceptible to local displacement than within the nail bed. Capillary displacement during the recording of video frames is a significant problem. Traditionally, such problems are solved involving the methods based on the selection of key points of the image (SURF [6], SIFT [7]). Such methods are ineffective to compensate for the displacements of biological objects due to the difficulty of tracking the key points.

The proposed method is based on the detection of displacements of local areas of images by the criterion of the minimum root mean square deviation. A feature of the proposed method of compensating for local displacements is the use of several key frames, which makes it possible to increase the accuracy. This allows applying the method of local displacement compensation for the considered tasks. The method of compensation for local displacements is as follows. Local points are set, in the neighborhood of which the estimation of displacements is performed. The obtained displacements are interpolated at each point of the frame and a continuous coordinate translation matrix is formed. After determining the local displacements for the constructed set of points for each frame and for each point of the frame, the interpolation of the displacement vector is performed. For each frame, a coordinate translation matrix is calculated, which makes it possible to compensate for local distortions of the geometry relative to the selected key frame. As a result of applying the coordinate translation matrix to the original video frames, compensation of local displacements is provided.

A study of the method of compensation for local displacements for capillary video frames was conducted in the hand area of a conditionally healthy volunteer. The investigated skin area has a size of about 2 square millimeters. A video sequence with 2000 video frames at the speed of 100 fps was recorded. In the course of the experiment, mechanical action was carried out on the skin (a shift in different directions) to ensure capillary displacements. Figure 1 shows video frames with local displacements of capillary images; for the first frame, 1 mm measure is displayed. Figure 2 shows the results of assessing local displacements for those video frames. It is seen that the detected displacements exceed typical size of the capillary. Figure 3 shows the locally aligned video frames, calculated using the coordinate translation matrix.

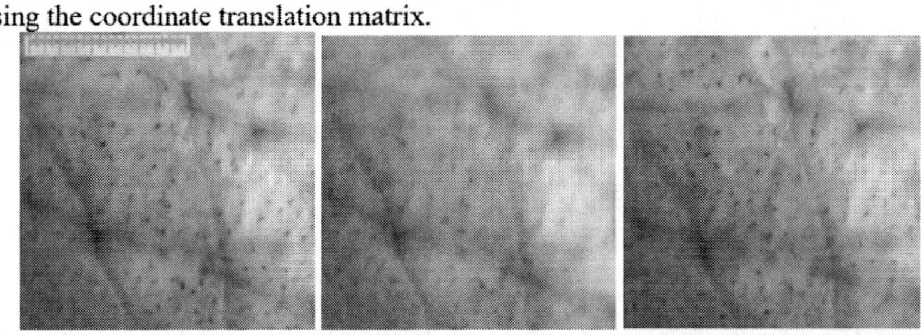

Figure 1. Video frames of the hand capillaries obtained at various displacements of the skin as a result of external exposure.

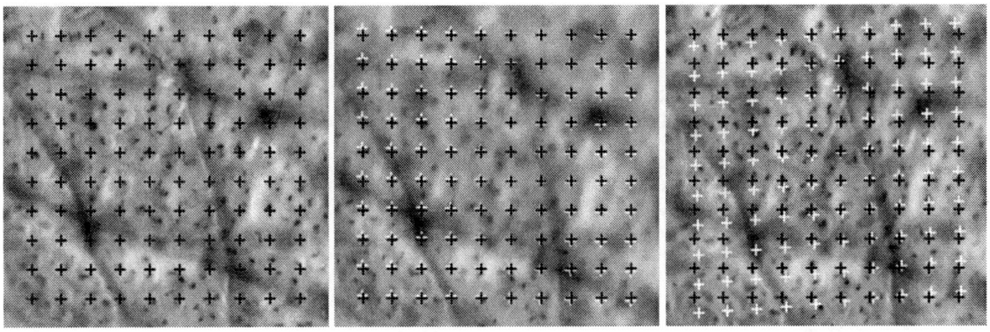

Figure 2. Local points (indicated by a black crosses) and points after detection of displacements (indicated by a white crosses) for some video frames.

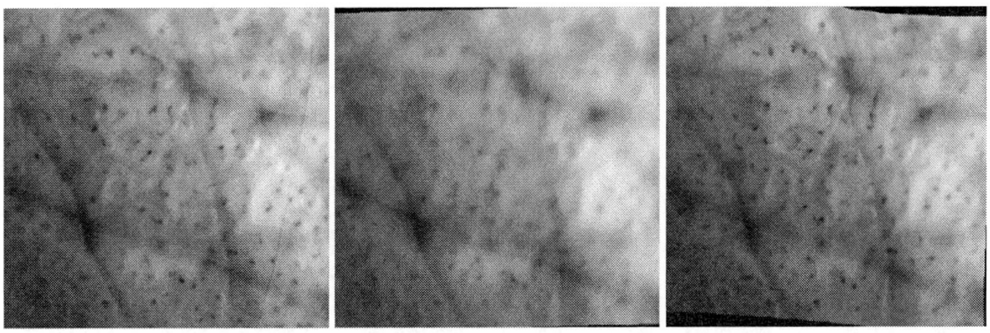

Figure 3. Video frames after local displacement compensation.

During the diagnosis of diseases associated with impaired microcirculation, the tasks of assessing the morphology and counting capillaries per skin area unit are relevant. Normally, capillary analysis is performed for a single image. When registering skin areas of several square millimeters in the hand area, it is extremely difficult to ensure the same focus and contrast of the capillaries for the entire frame. Figure 4 shows examples of video frames with various locally defocused areas. Processing of the locally-combined sequence of video frames allows estimating a synthesized image of capillaries that does not contain defocused areas, see figure 5 (a). The synthesized image does not contain the noise of the camera matrix and random capillary distortions due to the characteristics of the blood flow typical of individual video frames. The capillaries in figure 5 (a) are contrast. It is seen that the displacements of capillary images in the processed sequence of video frames are substantially smaller than the typical size of capillaries.

Capillaries can be visualized via analysis of the micro-PPG signal [8], which can be evaluated for a sequence of video frames. The result of capillary visualization based on Fourier analysis of the micro-PPG signal at each point of the frame is shown in figure 5 (b). With such a processing, one can select the capillaries and provide their visualization with suppression of other contrasting details of an image, such as skin folds, hair, pigment spots.

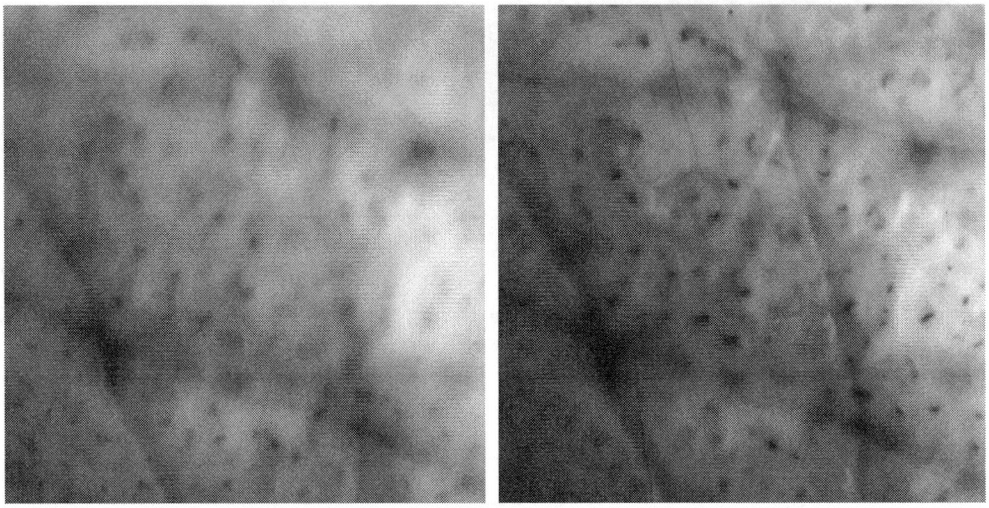

(a) (b)

Figure 4. Examples of video frames with a local decrease in the contrast of capillaries (a) in the upper right corner and (b) in the lower left corner.

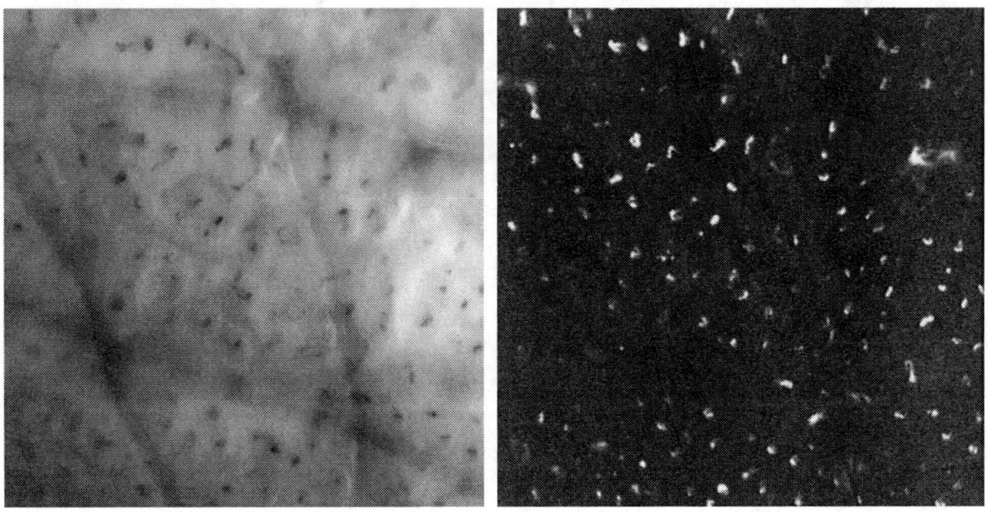

(a) (b)

Figure 5. The result of processing a sequence of aligned video frames: (a) a synthesized image with focus recovery over the entire field, and (b) the result of computation and visualization of capillaries by frequency methods.

The use of a laparoscope [9] allows exploring the parameters of microcirculation in the oral mucosa. The capillary network of the mucous membrane of the mouth of a conditionally healthy volunteer was registered on an area of approximately 3 square millimeters. Totally 1000 video frames were recorded at 66 fps.

Data processing is performed in two stages. At the first stage, the full-frame alignment method is used, which allows compensating for significant displacements of the laparoscope relative to the object of study, as shown in figures 6, 7.

Journal of Physics: Conference Series **1421** (2019) 012054 doi:10.1088/1742-6596/1421/1/012054

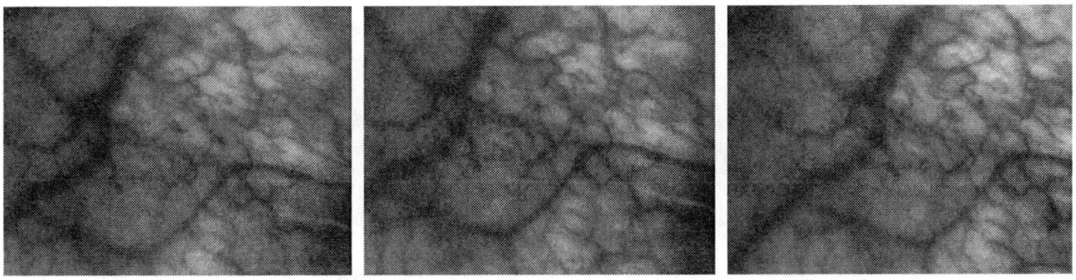

Figure 6. Original sequence of video frames.

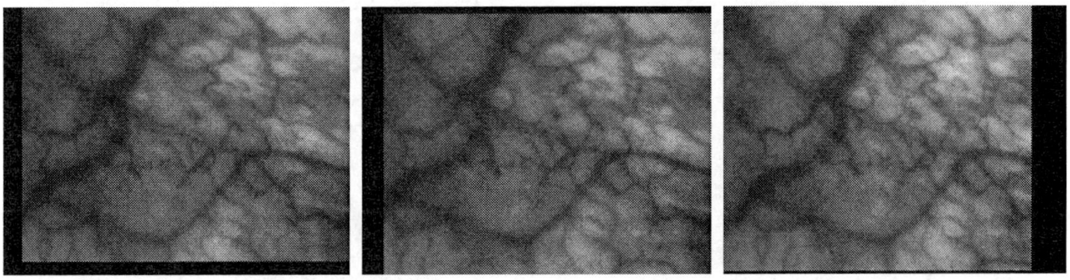

Figure 7. Sequence of video frames after full frame alignment.

At the second stage of processing, local displacements of images of the capillary network and projective distortions associated with displacements of the device are compensated. The detected local displacements and the results of their compensation are shown in figures 8 and 9.

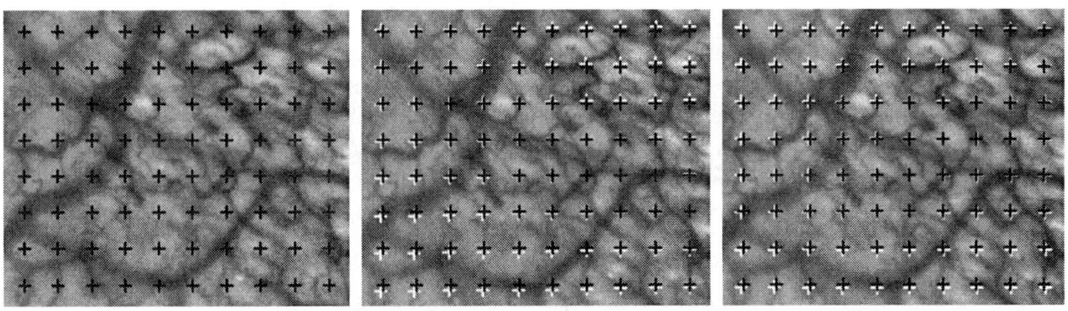

Figure 8. Sequence of video frames with calculated local displacements.

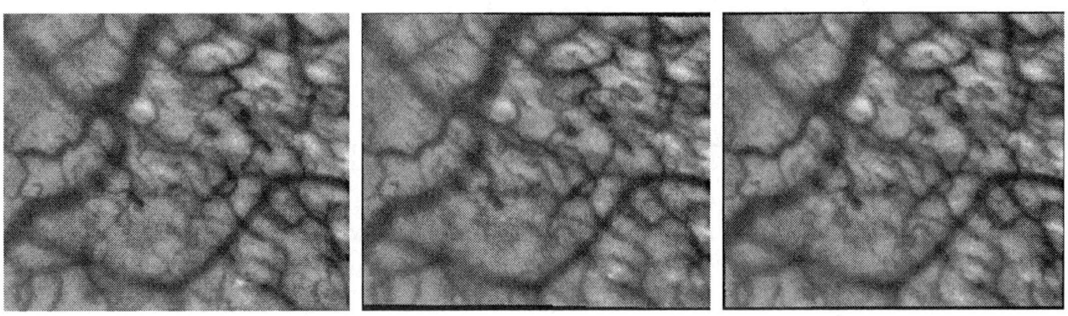

Figure 9. Video frames after local displacement compensation.

The efficiency of local alignment of images of capillaries can be estimated from the results of additional processing of video frames. An example of such processing based on calculation of the synthesized image with aligned capillary contrast is presented in figure 10. The synthesized image contains more details of the capillaries with complete suppression of the video camera matrix noise compared to the original video frames. Consequently, the local method for local alignment of video frames allows obtaining additional information about the capillary network under study and improving the quality of diagnostics.

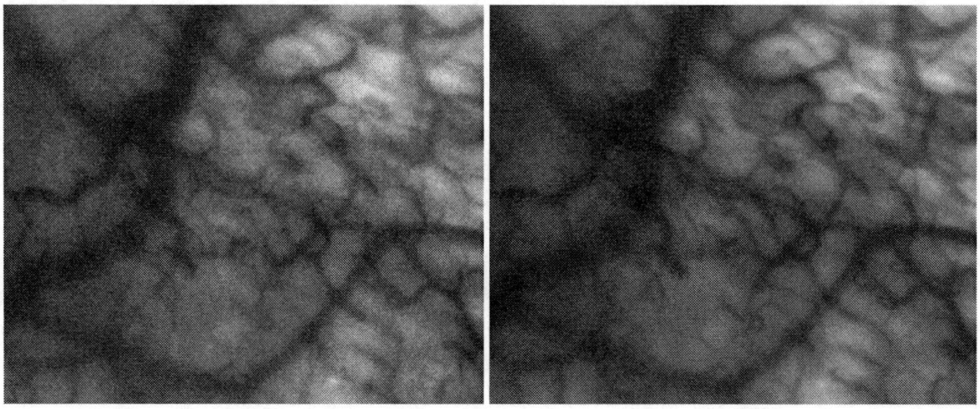

(a) (b)

Figure 10. (a) Video frames sample and (b) synthesized image estimated based on a pre-processed sequence of video frames.

3. Conclusion

The paper presents examples of applying the method of local compensation for local displacements of images of capillaries in the processing of video frames sequences recorded for hand and mucous membrane of the human mouth skin areas. It is shown that this method provides fairly accurate alignment of images of capillaries. The examples of processing locally-aligned sequences of video frames with the calculation of the synthesized image with aligned focus and maximum contrast are demonstrated. An example of capillary imaging based on frequency signal processing is presented. It can be concluded that the considered method for assessing the morphology of the capillary network is effective.

The considered method of local alignment of images of capillaries in video frame sequences is important for improving the quality of diagnostics of microcirculation disorders.

Acknowledgements

The work supported by the Ministry of Science and Higher Education of the Russian Federation, Project Number: 8.2501.2017/4.6.

References

[1] Hern S and Mortimer P S 2007 In vivo quantification of microvessels in clinically uninvolved psoriatic skin and in normal skin *Br. J. Dermatol.* **156** 1224–29

[2] Campanati A, Moroncini G, Ganzetti G, Pozniak K N, Goteri G, Giuliano A, Martina E, Liberati G, Ricotti F, Gabrielli A and Offidani A 2013 Adalimumab modulates angiogenesis in psoriatic skin *Eur. J. Inflamm.* **11** 489–98

[3] Grassi W, Core P, Carlino G, Blasetti P and Cervini M 1993 Labial capillary microscopy in systemic sclerosis *Ann. Rheum. Dis.* **52** 564–69

[4] Karimov K A and Volkov M V. 2015 The phase correlation algorithm for stabilization of capillary blood flow video frames *SPIE Videometrics, Range Imaging, and Applications XIII*

9528 952810
[5] Volkov M, Margaryants N, Pimenov A, Potemkin A and Gurov I 2018 High-speed video capillaroscopy method for imaging and evaluation of moving red blood cells *Opt. Lasers Eng.* **104** 244–51
[6] Bay H, Ess A, Tuytelaars T and Van Gool L 2008 Speeded-Up Robust Features (SURF) *Comput. Vis. Image Underst.* **110** 346–59
[7] Juan L 2010 A Comparison of SIFT , PCA-SIFT and SURF *Int. J. Image Process.* **3** 143–52
[8] Volkov M V., Margaryants N B, Potemkin A V., Volynsky M A, Gurov I P, Mamontov O V. and Kamshilin A A 2017 Video capillaroscopy clarifies mechanism of the photoplethysmographic waveform appearance *Sci. Rep.* **7** 1–8
[9] Machikhin A S, Khokhlov D D, Batshev V I and Pozhar V E 2018 An Acousto-Optic Endoscopic Module for Nondestructive Testing *Bull. Russ. Acad. Sci. Phys.* **82** 1403–05

Journal of Physics: Conference Series **1421** (2019) 012055 doi:10.1088/1742-6596/1421/1/012055

Multi-spectral phase imaging based on acousto-optic selection of light in common-path interferometer

L Burmak, L Zykova, A Machikhin and B Chichigin

Scientific and Technological Centre of Unique Instrumentation, Russian Academy of Sciences, Leninsky prospect 32A, Moscow, Russia

E-mail: burmakliu@yandex.ru

Abstract. Multispectral interference techniques are widely used in biomedicine and science for quantitative characterization of morphological, chemical and other properties. These techniques, for example, multispectral digital holography and spectral-domain optical coherence tomography, require wavelength tuning in the interferometer. In this paper, we discuss the effectiveness and potential applications of acousto-optic light filtration for this purpose. We show experimentally that acousto-optical tunable filters are effective for simultaneous spectral and phase data collection not only in conventional dual-path interferometers but also in a common-path interferometer. Scheme of a compact spectral interferometric module based on such interferometer is presented.

1. Introduction

Interference techniques for objects inspection are widely used in biomedicine, industry and science. They provide contactless operation, high precision and data acquisition speed. Along with the interference systems utilizing coherent monochromatic (laser) radiation, systems based on low-coherence radiation are in use, as they allow avoiding speckle noise and parasitic interference patterns. In the majority of such systems, spectral selection of light is necessary for implementing the interference measurements (for example, by optical coherence tomography or digital holography techniques) [1, 2]. For wavelength tuning in the interferometer, various spectral elements, such as diffraction gratings, liquid crystals and acousto-optic (AO) filters may be used. AO filters allow to acquire full two-dimensional (2D) image, do not require an additional scanning in one or two spatial coordinates [1] and provide narrow spectral bandwidth and high tuning speed [3]. The use of AO filters for filtration of interfering light beams in the output arm of the interferometer allows to create multimodal systems for spectral and interference analysis [4]. Spectral data can be used for contrast visualization of the object elements with particular physical and chemical properties as well as for spectral measurements, i.e. for calculating spectral dependences of object characteristics. On the one hand, special features of AO filters (absence of moving elements, fast electronic control, random spectral access, compactness, etc.) offer new opportunities for combination of spectral and low-coherence interference techniques. On the other hand, the feasibility of this approach is not evident since AO filtration may destroy the interference pattern as well as the angular spectrum of each of beams may be distorted and decrease the image quality. The possibility of applying AO filtration of interfering image-carrying light beams was demonstrated experimentally for low-coherence interferometry in classical versions of interferometers [4-6].

2. Conventional two-beam interferometers and common-path interferometers

Conventional low-coherence interferometry schemes are based on two-beam interferometers, such as Michelson, Mach-Zehnder and others (figure 1). The inspected object is located at one of the interferometer's arm, which is called "object arm". In this case, the object is located like "inside" the interferometer and the image of the object is formed by optics of the interferometer. These systems have relatively large dimensions, require time-consuming adjustment but they provide "flexibility", i.e. a possibility of beam convergence condition control for each specific task. AO filtration of interfering image-carrying light beams, in particular, in Michelson interferometer allows implementing measurements by spectral-domain optical coherence tomography technique and acquiring wideband and spectral image as well as phase profile without the inspected object displacement [4]. AO filtration in Mach-Zehnder interferometer allows implementing the multispectral holography mode and acquiring the phase maps and spectral images [5].

Lately, an alternative approach to the design of interference schemes, in which the inspected object is located "outside" the interferometer, is being actively developed. It is called "common-path interferometry". In this case, optical system (such as, for example, microscope, endoscope or telescope) forms the beam that carries image of the inspected object. This beam enters the common-path interferometer in which two interfering beams are formed from it in different ways. For example, it can be simple division into two for the autointerference (self-referencing) beams [7-11]. Alternatively, the reference beam can be formed by spatial filtration of the object beam [12-16]. The common-path interferometers are robust to the environmental vibrations (since the interfering beams propagate along approximately the same path), compact, relatively simple in terms of adjustment while they give less "freedom" for it. Based on common-path interferometers, compact modules for phase measurements using conventional imaging devices may be created. Applying of AO filtration in such modules could provide the device with a larger number of functions by adding a possibility of a spectral analysis for receiving more information about inspected object.

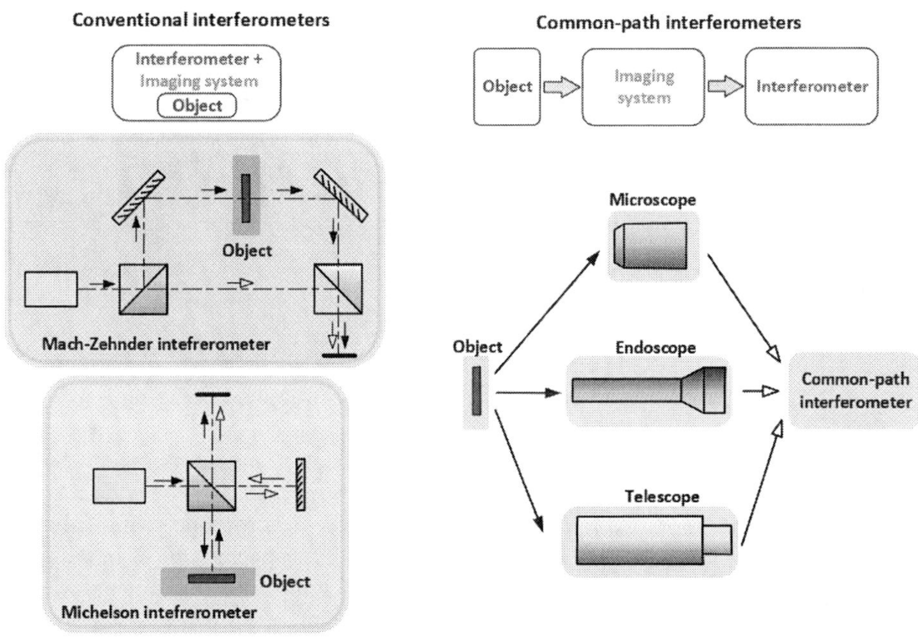

Figure 1. Approaches to interference schemes design: conventional two-beam interferometers (left) and common-path interferometers (right).

Journal of Physics: Conference Series **1421** (2019) 012055 doi:10.1088/1742-6596/1421/1/012055

3. Compact common-path interferometer based on acousto-optic filtration

To demonstrate the potentials of AO image filtration in common-path interferometers, we introduce spectral interferometric module based on off-axis τ-interferometer [16] with nearly common-path beams propagation (figure 2). The module consists of confocal lenses L1 and L2, beamsplitter BS, roof mirror M1, mirror M2 with pinhole P (point reflector), AO tunable imaging filter AOTF and digital camera. The wideband light beam enters the module from outer imaging system and then is divided into two. In figure 2, outer imaging system is a transmitted light microscope. One of these beams preserves the image of the inspected object while being reflected from roof mirror M1. The second beam is spatially filtered by point reflector (mirror M2 with pinhole P) to become a reference beam. Then both interfering beams are simultaneously spectrally filtered by AOTF. They form spectral holographic image on camera sensor. To calculate the phase map, acquired image is processed by digital holographic techniques, including double Fourier transform and phase unwrapping. With electronically controlled AO spectral tuning, several spectral interference images can be acquired. Multispectral holographic acquisition can be used to increase phase measurement precision [6] as well as to increase phase unwrapping algorithm accuracy [17]. Unlike the interferometers [12-14], it is possible to adjust the interference fringe appearance by tilting and moving roof mirror M1. The disadvantages of this arrangement include the difficulty of the pinhole alignment (to get a beam into the pinhole in the reflection mode is much harder than in transmission mode; this, however, can be solved as proposed in [15]) and significant energy losses (because of the pinhole, beam splitter and anisotropic diffraction mode implemented in AOTF).

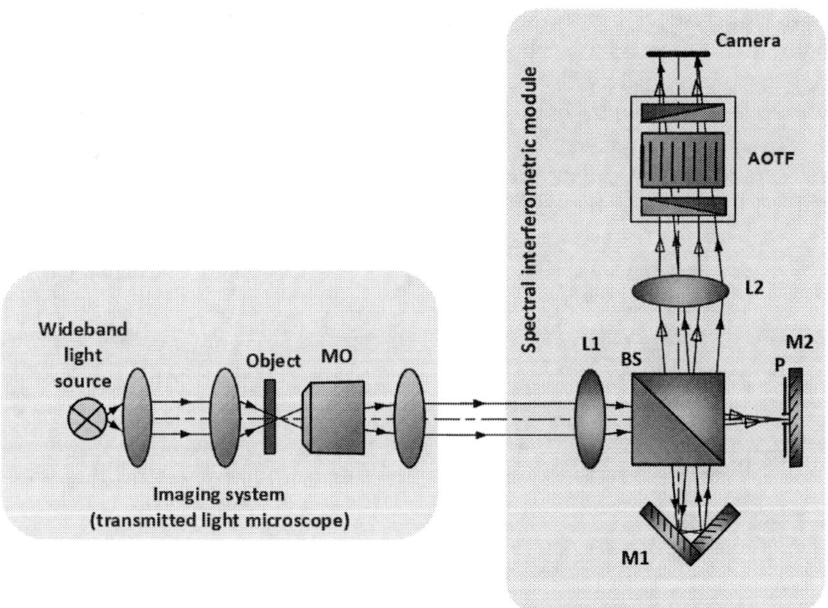

Figure 2. Proposed scheme of multi-spectral phase imaging.

For experimental research, we used the following components. Superluminescent diode Superlum S-785-B-I-15 ($\Delta\lambda$ = 770-810 nm) coupled into single-mode fiber was used as a light source. Single mode fiber is necessary to increase the spatial coherence. Microscope with infinity-corrected microscope objective MO 20x (NA 0.4) formed an image of the inspected object in the focal plane of the lens L1. Lenses L1 and L2 had focal lengths $f_1 = f_2 = 90$ mm. Pinhole diameter was 25 μm. AO filter was made of TeO2 cell placed between two crossed polarizers. In the filter, wide-angle non-collinear configuration of AO interaction was implemented for extraordinary polarized light. Its tuning range was $\Delta\lambda$ = 740-1000 nm with spectral bandwidth of approximately $\delta\lambda$ = 4 nm (at λ = 800 nm), entrance pupil 5 mm, angular aperture 2.5°× 2.5°. For image acquisition, we used monochrome CCD camera Prosilica GC1380.

Figure 3 shows the images of a Fresnel lens made of SU-8 photoresist. Upper row contains spectral holograms acquired with AO spectral tuning in range from 770 nm to 810 nm with 10 nm step. Bottom row contains numerically reconstructed phase maps. Phase maps are inverted due to "-1st" diffraction order selection during filtration in the spatial frequency domain, i.e. physically lens is thicker in regions with negative phase delay. Measured profile height is $h = \Delta\varphi \cdot \lambda / \left[2\pi(n-1)\right]$; $\Delta\varphi$ is measured phase delay, λ is a wavelength selected by AO filter, n is refractive index of the lens material at λ. An average measured height h = 460 nm. It coincides with the results acquired by atomic-force microscope.

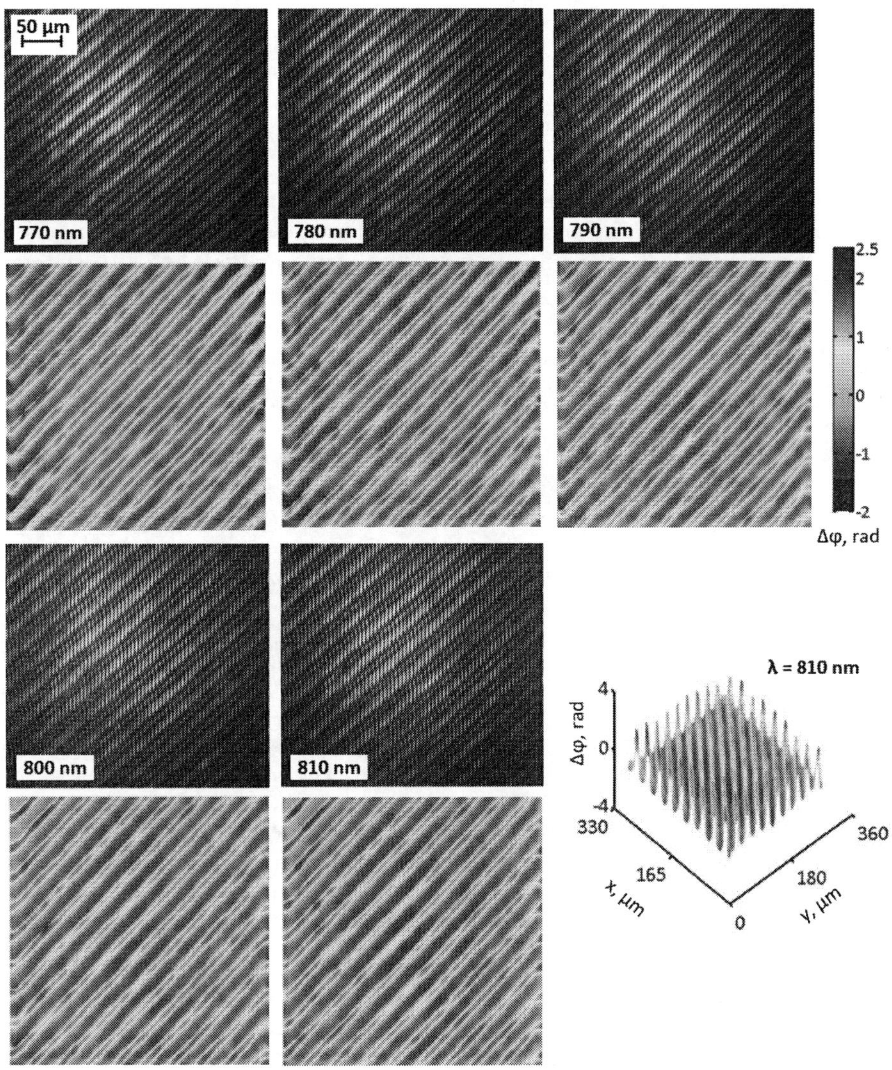

Figure 3. Experimental results: spectral holographic images acquired with AO tuning (λ = 770:10:810 nm) (upper row) and reconstructed phase maps (bottom row) of a Fresnel lens.

4. Conclusion

In this paper, we have shown that AO filtration of interfering light beams in common-path interferometer may be effectively used for multispectral low-coherence interferometry. It enables formation of two interfering narrow-band beams from one wideband beam and allows collecting both phase and spectral information about the inspected object. Based on this approach, the spectral interferometric modules for microscopes, endoscopes and other imaging systems may be produced.

Acknowledgments
This work was funded by Russian Foundation for Basic Research (project 18-38-00869).

References

[1] Drexler W and Fujimoto J G 2015 *Optical Coherence Tomography: Technology and Applications, 2nd edition* (Springer International Publishing Switzerland) p 2571

[2] Kim M K 2011 *Digital Holographic Microscopy: Principles, Techniques, and Applications* (New York: Springer) p 237

[3] Abdlaty R, Sahli S, Hayward J and Fang Q 2018 Hyperspectral imaging: comparison of acousto-optic and liquid crystal tunable filters *Medical Imaging 2018: Phys. of Medical Imaging* **10573** 105732P.

[4] Machikhin A S, Pozhar V E, Viskovatykh A V and Burmak L I 2015 Acousto-optical tunable filter for combined wideband, spectral and optical coherence microscopy *Appl.Opt.* **54**(25) 7508-13

[5] Machikhin A S, Burmak L I, Polschikova O V, Ramazanova A G, Pozhar V E and Boritko S V 2018 Multispectral phase imaging based on acoustooptic filtration of interfering light beams *Appl.Opt.* **57**(10) C64-C69

[6] Machikhin A S, Polschikova O V, Ramazanova A G and Pozhar V E 2017 Multi-spectral quantitative phase imaging based on filtration of light via ultrasonic wave *J. Opt.* **19** 075301

[7] Hillman T R, Lue N, Sung Y, Dasari R R and Yaqoob Z 2012 Near-common-path self-reference quantitative phase microscopy *IEEE Photon. Technol. Lett.* **24** 1812–14

[8] Singh A S G, Anand A, Leitgeb R A and Javidi B 2012 Lateral shearing digital holographic imaging of small biological specimens *Opt. Exp.* **20** 23617-22

[9] Roitshtain D, Turko N A, Javidi B and Shaked N T 2016 Flipping interferometry and its application for quantitative phase microscopy in a micro-channel *Opt. Lett.* **41** 2354-57

[10] Jang J, Bae C Yu, Park J K and Yel J C 2010 Self-reference quantitative phase microscopy for microfluidic devices *Opt. Lett.* **35** 514-16

[11] Latushko M I, Vishnyakov G N and Levin G G 2016 Shearing interference microscope with decoding of differential phase patterns of living cells using the phase stepping method *Meas.Tech.* **58** 1238-43

[12] Popescu G, Ikeda T, Dasari R R and Feld M S 2006 Diffraction phase microscopy for quantifying cell structure and dynamics *Opt. Lett.* **31** 775-77

[13] Polschikova O V, Machikhin A S, Ramazanova A G and Pozhar V E 2019 A compact system for multispectral quantitative phase imaging based on a lens-in-lens common-path interferometer *Proc.SPIE* **11030** 1103005

[14] Singh V, Tayal S and Mehta D S 2018 Highly stable wide-field common path digital holographic microscope based on a Fresnel biprism interferometer *OSA Continuum* **1** 48-55

[15] Nativ A and Shaked N T 2017 Compact interferometric module for full-field interferometric phase microscopy with low spatial coherence illumination *Opt. Lett.* **42** 1492-95

[16] Girshovitz P and Shaked N T 2013 Compact and portable low-coherence interferometer with off-axis geometry for quantitative phase microscopy and nanoscopy *Opt. Exp.* **21** 5701-14

[17] Gass J, Dakoff A and Kim M K 2003 Phase imaging without 2π ambiguity by multiwavelenght digital holography *Opt. Lett.* **28** 1141-43

Journal of Physics: Conference Series **1421** (2019) 012056 doi:10.1088/1742-6596/1421/1/012056

Spectral oil analyzer

A A Lashova, V S Korsakov, E A Korsakova, I A Kashuba and L V Zhukova

Ural Federal University named after the first President of Russia B.N. Yeltsin, 19 Mira St., Ekaterinburg, 620002, Russia

E-mail: viktor.korsakov@urfu.ru

Abstract. We worked out a spectral oil analyzer, implemented the method of Fourier transform infrared spectroscopy, which allow simultaneous detecting characteristic peaks of water absorption and sulfur-containing substances in oil within the wavelength range from 0.2 to 25.0 μm. The usage of fiber-optic delivering channels makes it possible to produce online industrial probes for the control of chemical composition and concentration, what is demanded at oil producing platforms and oil refinery plants. As identifiers, we used the most intense absorption peaks of water (λ – 1.45 μm, 1.95 μm, 3.00 μm, 6.00 μm), oil (λ – 1.72 μm, 2.31 μm, 3.41 μm) and sulfur-containing heterocycles (λ – 8.00 μm). The spectral analyzer consists of a broadband-light source, a flow-cell, a fiber-optic assembly, narrow-band optical filters and photodetectors. For delivering a broadband optical signal, a fiber-optic assembly was developed. It comprises silica fibers and polycrystalline fibers derived from solid solutions of silver and thallium (I) halides. It was calculated the effective area and diameter of each fiber type in the assembly, taking into account attenuation coefficients and overall optical losses of the system. Using eight channels for the probe at the same time, we were able to measure the content of water and sulfur-containing substances in oil in concentrations from 0.1 to 100.0 wt %.

1. Introduction

At present, the water-cut (water content, moisture) of oil products at wells reaches 75% on average, if considering wells in different countries. Both in Russia and abroad, the laboratory method remains the main method for determining the water content in oil at oil producing wells. It has not been changed for many years despite the fact that the error of the determination is unacceptably large. When the water content is high, the relative error is close to 90-100%. Moreover, considerable time is required for determining the water cut.

Among the online methods of oil moisture monitoring, there are two types which are separated according to the principle of measured physical characteristics: optical and electrical methods. The disadvantage of the electrical methods is low sensitivity in the case of considerable gas phase in the samples under study. Optical methods are based on the registration of absorption peaks which are inherent in test media, regardless of their state, and thus providing higher accuracy and rate of analysis.

One of commonly used methods for online moisture monitoring is a method based on measuring the electrical properties of oil-water mixtures using a capacitor probe. At high water-cut levels, such a capacitor probe doesn't give reliable results. Moreover, a change in the oil density leads to a change in the dielectric constant, what causes the increase of the measurement error. Paraffine depositions on the condenser plates, high influence of the gas factor are the main disadvantages of capacitor probes, what also limits their use [1]. Another principle of water-in-oil content measurement is based on the absorption of microwave radiation by the medium. The microwave radiation (at frequencies above 1 GHz) is intensively absorbed by polar water molecules and, to a much lesser extent, non-polar hydrocarbon molecules. The method allows to determine water in oil with high accuracy covering the entire range (0-100%), although at high water cut values, the accuracy slightly decreases.

The advantage of the method is low impact of the oil composition on the measurement accuracy. The disadvantages are the influence of the salt content in stratum water on the measurement, as well as its high cost [2]. Coriolis flowmeters can be used to determine the water composition. They simultaneously determine the mass flow rate, density, and calculates the percentage of water in oil based on the density of oil and stratum water. The main disadvantage of this method is a significant decrease in the measurement accuracy in the case of heavy oil, when the density of oil is just a little bit different from the density of water. The presence of free gas in the fluid flow also significantly impairs the measurement accuracy. It is important to note, that when we deal with the group of wells, the water and oil densities vary not only from well to well, but they also vary over time.

The second kind of impurities of oil are sulfur-containing substances. Its content can reach 14 vol.%. Sulfur in oil and oil products is presented partly by elemental sulfur, but mainly by various substances – hydrogen sulfide, mercaptans, sulfides, disulfides, thiophenes, and thiophans [3]. Methods for determining sulfur in oil can be divided into 3 groups: methods based on sulfur oxidation and subsequent determination of oxides [4]; methods based on sulfur reduction to H_2S: [5]; spectral methods (x-ray [6], atomic emission [7], optical [3]. All methods of the sulfur determination combine the general principle – sampling oil and its further processing. This approach, firstly, takes a lot of time, and secondly, it is associated with the statistical errors during the extrapolation of the experimental results on the entire volume of test medium.

Infrared spectroscopy is one of the main ways for analyzing oil and oil products. Its advantage is non-destructive technique, what allows using test samples after the analysis. The change in salinity and the presence of gas phases also do not have a significant effect on the measurement accuracy due to the fact that the method is based on detecting the intensity of characteristic absorption peaks of chemical bonds at certain wavelengths that do not change their positions. The use of fiber-optic delivering channels allows expanding the field of method application and implementing industrial online probes for monitoring chemical compositions and concentrations that are in demand at oil-producing and oil-refining stations.

2. Spectral oil analyzer

To quantify the moisture and sulfuric substances content in oil, the authors developed an immersion probe. It is intended for ATR Fourier transform infrared spectroscopy within the range from 1.2 to 6.5 μm. The specified range is justified by the absorption peaks of water at the wavelengths of 1.45 μm, 1.95 μm, 2.94 μm, and 6.02 μm, which are associated with stretching vibrations of OH groups and H–O–H deformation vibrations [8-10]. The main characteristic absorption bands of oil are 2920 cm^{-1} and 2860 cm^{-1} (the wavelengths of 3.42 μm and 3.49 μm, respectively), which are determined by the stretching vibrations of CH groups (namely, CH_2 and CH_3). In the near IR, there are also corresponding absorption bands at 1.725 μm and 2.31 μm [11]. Oil's sulfur-containing substances are represented by thiophenes and thiophans, which have the absorption peak around 8.0 μm, what corresponds to the sulfur vibrations in the molecular ring [3] (figure 1).

Water absorbs most intensively at a wavelength of 2.94 μm, which is 15 times higher than the absorption of radiation at $\lambda = 1.95$ μm, while the intensity of the absorption at a wavelength of 6.02 μm is almost not affected by temperature [8]. Thus, the use of the middle infrared range causes the significant increase of the sensitivity of the developed probe. The probe consists of a broadband IR source, a fiber probe, and optical signal detectors [12, 13]. Such fiber probes operate basing on the principles of attenuation total reflectance (ATR). As a sensitive element, we used a "cone" type element made of zinc selenide (the transmission range of 0.2–18 μm) (figure 2).

Fiber assemblies of the probe were made of optical fibers based on dielectric materials – SiO_2 fibers (the transmission range of 1.2–2.5 μm) and silver halide fibers (the transmission range of 2.5–20.0 μm) [14-16].

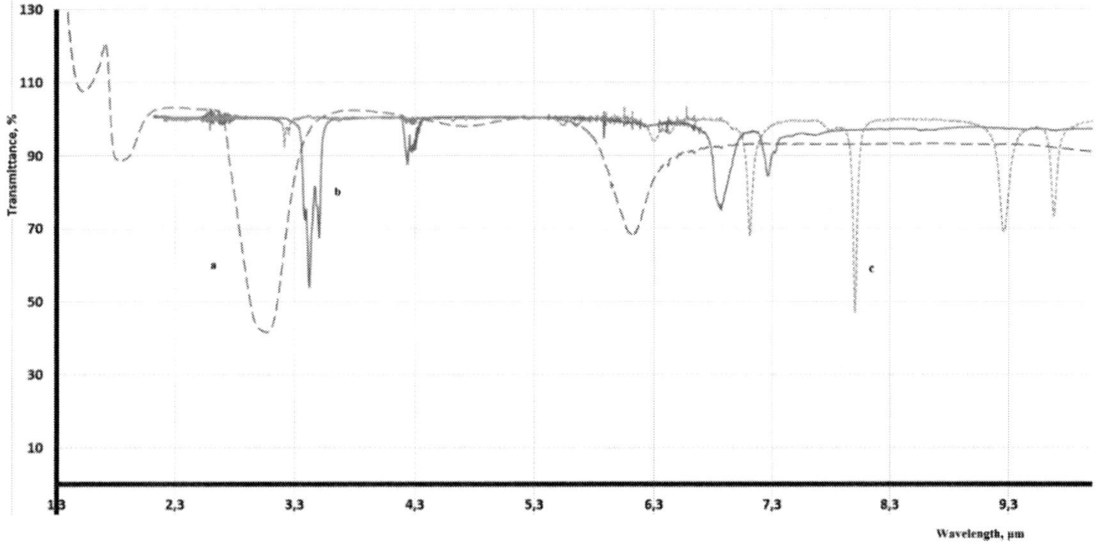

Figure 1. Infrared spectra of water (a), oil (b) and thiophene (c).

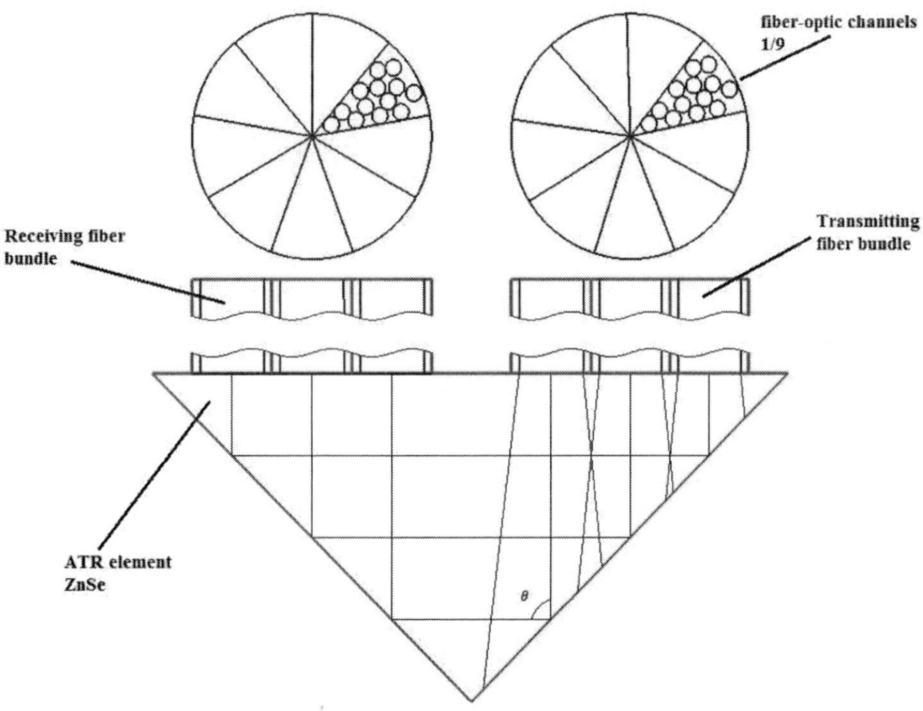

Figure 2. Model of radiation propagation from the transmitting assembly to the receiving assembly through the cone-type ATR-element.

The proposed fiber probe allows setting the units of IR-source and optical signal detector outside the strong field of EM-interference. One more advantage of the probe is that the immersible element does not react with test media. The total optical loss is one of the main functional parameters of sensor systems. It determines the requirements for electro-optical components of the sensor. In this paper, we estimated the total optical loss of a fiber probe depending on the refractive indices, the geometric configuration, and the detected wavelengths. The optical losses of the fiber probe depend on the power distribution efficiency over the receiving end, the optical losses in the fiber channels, and the efficiency of the ATR-element.

Each fiber assembly of the probe comprises nine hexagonally arranged fiber channels: four channels operate in the NIR range for the detection of water absorption peaks, three channels operate in the MIR range for the detection of oil absorption peaks, one channel also operates in the MIR range for detecting absorption peaks of sulfur-containing groups, and one channel operates at a wavelength of 5 μm for the reference signal. The optical fibers are arranged in the form of segments, thus achieving a uniform distribution of power (P_0) from a broadband-light source at the receiving end of the assembly. The ATR-element is a cone with 45^0 angle between the generator and the base. This form provides the multiple total reflection of optical signals from the cone surface and their transmission from each channel of the transmitting assembly to the corresponding channel of the receiving bundle (figure 2). The radius of the ATR cone base depends on the size of receiving and transmitting bundles and the value of the numerical aperture, and it is determined by the following relationship:

$$R_k \geq \frac{D_b}{1-2\sin(NA)}, \tag{1}$$

where R_k – the radius of the cone base, D_b – the diameter of the fiber probe, NA is the numerical aperture of the fibers. The parameter NA is calculated by the formula:

$$NA = \sqrt{n_{core}^2 - n_{clad}^2}, \tag{2}$$

where n_{core} – the refractive index of the fiber core, n_{clad} – the refractive index of the fiber cladding. In this case, the numerical aperture should not exceed 0.1.

The calculation of optical losses depending on the refractive index dispersion is presented in table 1. The refractive indices and intrinsic optical losses of AgBr-TlI fibers were determined experimentally by authors in [16]. The intrinsic losses of silica fibers were taken from [17]. The authors assumed that the intrinsic optical losses of ZnSe can be neglected because of the short optical path. Fresnel losses were defined as the coefficient of radiation reflection (r) from the media interface. They are calculated using the formula [18]:

$$r = \left(\frac{n_e - n_{core}}{n_e + n_{core}}\right)^2, \tag{3}$$

where n_e – the refractive index of the environment.

Total losses were defined as the sum of Fresnel losses (r) at (N) media interfaces and intrinsic losses (a) due to the attenuation of radiation inside each component of the probe:

$$\Delta P = Nr + a. \tag{4}$$

The power of transmitted radiation in a separate element of the optical scheme was defined as:

Journal of Physics: Conference Series **1421** (2019) 012056 doi:10.1088/1742-6596/1421/1/012056

$$P_{tr} = P_i \cdot \Delta P \,. \tag{5}$$

Table 1. Calculated optical losses of the fiber probe.

No. of channel	λ, µm	Material	P_i, %	n_{core}	a, dB/m	r, dB/m	ΔP, dB/m	P_{tr}, %
				Transmitting assembly				
1	1.45	Silica	100	1.443	$14\cdot10^{-6}$	0.0329	0.0658	93.4
2	1.725	Silica	100	1.442	$14\cdot10^{-6}$	0.0328	0.0655	93.4
3	1.95	Silica	100	1.439	$14\cdot10^{-6}$	0.0324	0.0648	93.5
4	2.31	Silica	100	1.433	$14\cdot10^{-6}$	0.0317	0.0633	93.7
5	2.94	AgBr-TlI	100	2.259	0.2	0.1492	0.4985	50.2
6	3.42	AgBr-TlI	100	2.258	0.2	0.1491	0.4982	50.2
7	5.0	AgBr-TlI	100	2.255	0.2	0.1487	0.4973	50.3
8	6.02	AgBr-TlI	100	2.254	0.2	0.1485	0.4970	50.3
9	8.0	AgBr-TlI	100	2.253	0.2	0.1484	0.4967	50.3
				"Cone" type prism				
Prism	1.45	ZnSe	93.4	2.46	-	0.1781	0.3561	60.2
	1.725	ZnSe	93.4	2.45	-	0.1766	0.3533	60.4
	1.95	ZnSe	93.5	2.446	-	0.1761	0.3522	60.6
	2.31	ZnSe	93.7	2.437	-	0.1761	0.3522	60.7
	2.94	ZnSe	50.2	2.436	-	0.1748	0.3496	32.6
	3.42	ZnSe	50.2	2.435	-	0.1747	0.3493	32.7
	5.0	ZnSe	50.3	2.429	-	0.1745	0.3490	32.7
	6.02	ZnSe	50.3	2.425	-	0.1737	0.3473	32.8
	8.0	ZnSe	50.3	2.416	-	0.1731	0.3462	32.9
				Receiving assembly				
1	1.45	Silica	60.2	1.443	$14\cdot10^{-6}$	0.0329	0.0658	56.2
2	1.725	Silica	60.4	1.442	$14\cdot10^{-6}$	0.0328	0.0655	56.5
3	1.95	Silica	60.6	1.439	$14\cdot10^{-6}$	0.0324	0.0648	56.7
4	2.31	Silica	60.7	1.433	$14\cdot10^{-6}$	0.0317	0.0633	56.8
5	2.94	AgBr-TlI	32.6	2.259	0.2	0.1492	0.4985	16.4
6	3.42	AgBr-TlI	32.7	2.258	0.2	0.1491	0.4982	16.4
7	5.0	AgBr-TlI	32.7	2.255	0.2	0.1487	0.4973	16.4
8	6.02	AgBr-TlI	32.8	2.254	0.2	0.1485	0.4970	16.5
9	8.0	AgBr-TlI	32.9	2.253	0.2	0.1484	0.4967	16.6

The Bouguer-Lambert-Beer law is used for the determination of the moisture content by means of IR-spectroscopy [19]:

$$I = I_0 e^{-\alpha x d} \quad , \tag{6}$$

where I – the intensity of transmitted radiation, I_0 – the intensity of incident radiation, α - the attenuation coefficient, x – the concentration of the substance, d – the optical path. According to this law, the optical density of radiation in the test medium is directly proportional to the concentration of the substance under investigation in this medium, while the attenuation coefficient and the optical path are known.

For ATR-elements, the optical path is calculated as a depth of the radiation penetration into the medium under study d_p [12]:

$$d_p = \frac{\lambda}{2\pi n_1 \sqrt{sin^2\theta - n_{21}^2}}, \tag{7}$$

where n_1 – the refractive index of an ATR-element, θ – the radiation incidence angle, n_{21} – the ratio of the ATR-element refractive index to the refractive index of the medium under study. The depth of the radiation penetration varies from 0.25 to 1.9 μm for the considered wavelengths.

The attenuation coefficient of water and sulfur-containing substances in oil was determined experimentally using the following facilities: IR Pestige 21 FTIR spectrometer (Shimadzu) within the wavelength range of 7800-1000 cm^{-1} (1.28-10.0 μm), MCT detector, KBr beam splitter with the resolution of 4 cm^{-1}, ATR-element. The spectrometer was placed in a sealed box for obtaining stable results, with the possibility of sweeping the box with dry N_2 (the volume fraction of water vapor in the nitrogen didn't exceed 0.0007%). The temperature in the laboratory room was maintained within the range of 22-25 °C by means of an air conditioner. We obtained calibration curves, which were used for plotting linear dependences of the optical density on the concentration of water and sulfur in oil at different wavelengths to determine attenuation coefficients (figure 3). The values of the coefficients for each wavelength are shown in table 2.

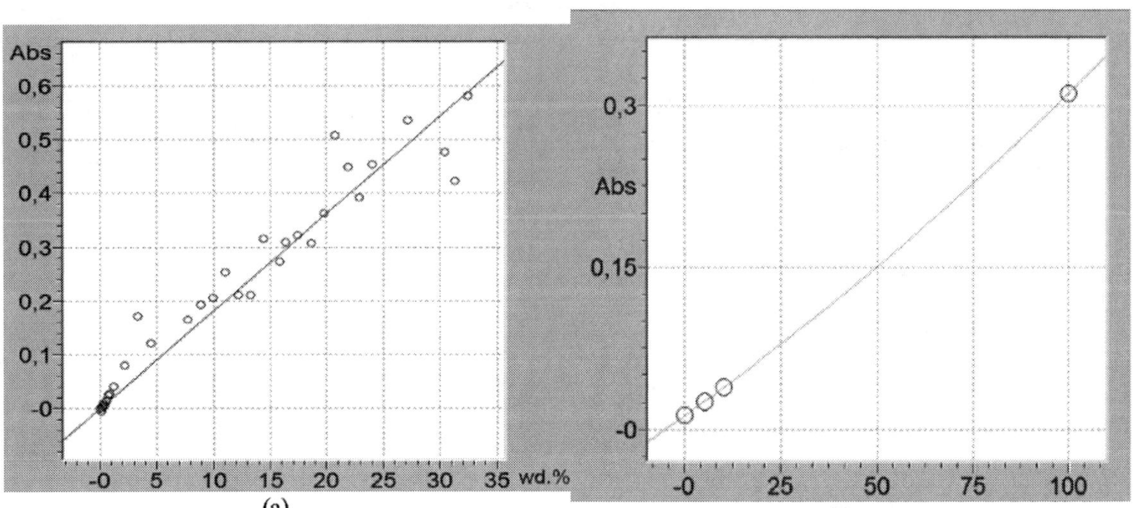

(a) (b)

Figure 3. The dependence of optical absorption at a wavelength of 2.94 μm and 8.0 μm on the composition of the water-oil system (a) and thiophene-oil system (b).

Table 2. Experimental values of extinction coefficients.

Wavelength λ, μm	Absorbing substance	Attenuation coefficient α, $*10^{-3}$
1.45	water	26.44
1.73	oil	21.87
1.95	water	71.37
2.94	water	535.13
3.42	oil	394.72
6.02	water	90.87
8.0	thiophene	197.36

The values of the attenuation coefficient and the depth of penetration, as well as the total optical loss at the corresponding wavelengths will be used to work out a mathematical model for processing optical signals of water and sulfur-containing substances in oil.

Thus, we developed a fiber-optic probe intended to the simultaneous determination of water and sulfur content in the oil. We modeled the optical scheme comprised of transmitting and receiving fiber assemblies, and an ATR-element. The total optical losses of the fiber probe were determined basing on the refractive indices of its components and the test medium. Using an ATR-element, we also calculated the depth of radiation penetration into the oil. Experimentally obtained values of attenuation coefficient will form the basis of a mathematical model for signal processing. All parameters were calculated for a fiber-optic probe with the length of 1 m, which allows deploying the analyzer on the main oil pipeline. If it is necessary to remove the most part of probe electronics, it is required to take into account the rise of intrinsic optical losses in the fiber-optic probe.

Acknowledgment
This work was supported by the Russian Science Foundation under grant No. 18-73-10063.

References

[1] Zhang N L and Xue Z M 2005 Measuring Technique for Water Content of Crude and its Development. *J. Technology Supervision in Petroleum Industry* **11** 25-28
[2] Bo O.L. and Nyfors E 2002 Application of Microwave Spectroscopy for the Detection of Water Fraction and Water Salinity in Water/Oil/Gas Pipe Flow *J. Non-Crystalline Solids* **305** 345–353.
[3] Han Y 2018 Molecular characterization of sulfur-containing compounds in petroleum *Fuel* **221** 144–158
[4] ASTM D129-18, Standard Test Method for Sulfur in Petroleum Products (General High Pressure Decomposition Device Method), ASTM International, West Conshohocken, PA, 2018, www.astm.org
[5] ASTM D4045-15, Standard Test Method for Sulfur in Petroleum Products by Hydrogenolysis and Rateometric Colorimetry, ASTM International, West Conshohocken, PA, 2015, www.astm.org
[6] ASTM D2622-16, Standard Test Method for Sulfur in Petroleum Products by Wavelength Dispersive X-ray Fluorescence Spectrometry, ASTM International, West Conshohocken, PA, 2016, www.astm.org
[7] ASTM D4951-14, Standard Test Method for Determination of Additive Elements in Lubricating Oils by Inductively Coupled Plasma Atomic Emission Spectrometry, ASTM International, West Conshohocken, PA, 2014, www.astm.org
[8] Yukhnevich G V 1973 *Infrared Spectroscopy of Water* (Moscow: Nauka Press)

[9] Zatsepina G N 1974 *Physical properties and structure of water* (Moscow: Moscow State University Press) p 168

[10] Karyakin A V,and Kriventsova G A 1973 *Sostoyanie vody vorganicheskikh i neorganicheskikh soedineniyakh* (Moscow: Nauka Press)

[11] Lynch. P F 1973 Identifying Source of Petroleum by Infrared Spectroscopy *J. Environ Sci. Technol.* **7** 1123-27

[12] Raichlin Y and Katzir A 2008 Fiber-Optic Evanescent Wave Spectroscopy in the Middle Infrared *J. Applied Spectroscopy* **62** 55A-72A

[13] Zhukova L 2013 Fiber probe for the spectral range of 2-45 μm for IR-Fourier spectrometer *Imaging and Applied Optics Congress*, OSA Technical Digest (online), Fourier Transform Spectroscopy (FTS). FTu3D.5

[14] Korsakov A S Zhukova L V Vrublevsky D S and Korsakova E A 2014 Investigating the properties of infrared PCFs based on AgCl-AgBr, AgBr-TlI, AgCl-AgBr-AgI(TlI) crystals theoretically and experimentally *J. Optics and Spectroscopy* **117** 960-63

[15] Korsakov A S 2014 Research of phase equilibriums and modelling of structure of AgBr-TlBr0.46I0.54 system *J. Tsvetnye Metally* **8** 50-54

[16] Zhukova L V 2008 AgClxBr1-x and AgClxBryI1-x-y crystals for IR engineering and optical fiber cables *J. Inorganic Materials* **44** 1372-77

[17] Blistanov A A 1982 *Acoustic crystals* (Moscow: Science Press) p 632

[18] Katsuyama T and Matsumura H 1992 *Infrared optical fibers* (Moscow: World Press) p. 272

[19] Beer A 1852 Determination of the absorption of red light in colored liquids *J. Annalen der Physik und Chemie* **86** 78–88

Journal of Physics: Conference Series **1421** (2019) 012057 doi:10.1088/1742-6596/1421/1/012057

Studies by PIV-methods of self oscillating hydrodynamic processes in the elements of pipeline

D A Sergeev[1,2], A A Kandaurov[1,2]

[1] Institute of Applied Physics RAS, Nizhny Novgorod, Ulyanova 46, 603950, Russia
[2] Nizhny Novgorod State Technical University n.a. R.E. Alekseev, Minina st. 24, Nizhny Novgorod, 603950 Russia

E-mail: daniil@ipfran.ru

Abstract. Investigation of the self-oscillation regimes in the flow chamber modelling elements of stop valves in the pipeline in the systems of heat exchange in the power plants were performed by using PIV methods. Dependency of parameters of self-oscillation regimes on the parameter of the gap between the inlet and outlet pipe line in the chamber was investigated. Simultaneous measurements of the hydrodynamic and acoustic fluctuations allows us to offer explanation of the observed regimes.

1. Introduction

Self-oscillation regimes are quite common in hydrodynamic processes. Their occurrence in various technical systems can lead to negative effects due to non-permanent (cyclic) and non-uniform (complicated spatial distribution) influence on structural elements, complicates diagnostics and control of flow patterns, etc. Also, these modes can be a source of narrow-band acoustic noise, which can also be a negative factor. In [1], a number of problems are discussed, which describe typical system configurations in which such regimes arise, and also illustrates the use of the Particle Image Velocimetry (PIV) method for obtaining the flow velocity field in them. At the same time, considerable attention is paid to the study of flows in a turbulent boundary layer, on a solid surface in the presence of various types of cavities and protrusions (see [2]). Such configurations can be very relevant from the point of view of applications associated with flows in pipelines of different configuration. Flow-through elements such as splitters (twins, tees, etc.), pipe joints of various transverse sizes and cross-section types, bends can be sources of self-oscillatory modes. These elements are typical for heat exchange systems in power plants including nuclear ones.

This work is devoted to the study of self-oscillation regimes in the simulation of hydrodynamics of flows in the pipeline elements of power plants. In particular, the case of the flow part (parallelepiped), the inlet and outlet pipes to which had a diameter of smaller transverse size of the flow-through area, was considered. This configuration simulated the area in which the valves of the pipeline system could be located. The dependence of the characteristics of the self-oscillatory flow regime on the gap between the pipes inside the flow area was investigated. An explanation of the physics of the observed process was proposed.

2. Experiment and measurements

2.1. Experimental setup

Experiments were carried out on a setup consisting of a model of the flow part, supply and discharge pipes, an exhaust fan, a absorption device to prevent pulsations caused by the fan operation (see figure 1).

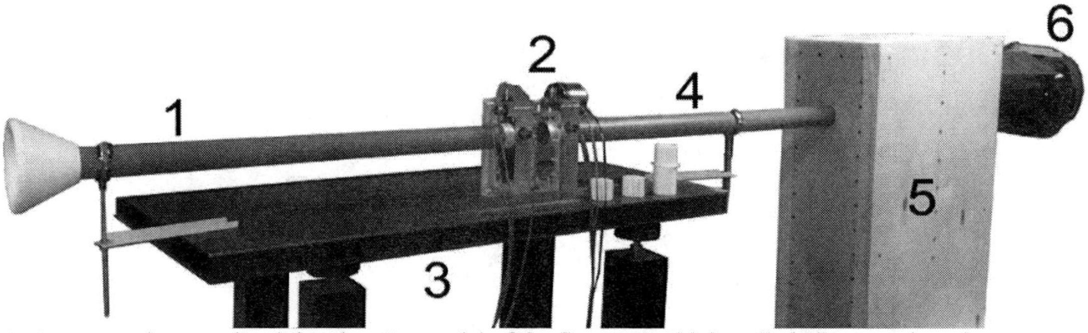

Figure 1. Aeroacoustic setup (1 – inlet pipe, 2 – model of the flow part with installed vibrator microphone, 3 – measuring table, 4 – outlet pipe, 5 – device for absorbing pressure pulsations, 6 – fan).

To exclude the influence of disturbances from the fan on the measurement processes in some experiments it was installed in an anechoic chamber and a device is installed in front of it to absorb pressure pulsations propagating along the air flow in the pipes.

The model of the flow part (chamber) was a parallelepiped 60×80 mm, and 140 mm high (all dimensions are internal) of plexiglas 8 mm fixed on the table. The table is mounted on special vibration dampers that prevent the transmission of parasitic vibrations. Air was supplied to the models and removed with the help of sets of polypropylene pipes 1.5 mm thick, of various length, but of the same outer diameter – 50 millimeters. The pipes were attached to the optical table, forming a single system together with the model flow part (figure 1). A conical nozzle was attached to the system inlet to reduce airflow resistance.

All measurements were performed for the constant fan speed.

A smoke spray generator seeding an aerosol with average droplet size of 5 microns was used for PIV-measurements. The generator was located in front of the entrance nozzle at a distance of about 1 m and operated in a pulsed mode, synchronously with the survey. The duration of smoke seeding was about one second.

The model of chamber was illuminated with a laser sheet from above in the middle along the direction the flow. A 6W (450 nm) blue LED continuous laser with a 9.7 mm cylindrical defocusing lens was used. The scheme is close to what was performed in [3]. Side view of the movement of smoke aerosol particles in the plane of the laser sheet was filmed with a high-speed light-sensitive camera NAC HX-3 Memrecam. The filming speed was 12220 frames/second, the scale is 0.0863 μm/pixel, the size in pixels is 768×960. The optical axis of the camera was perpendicular to the laser plane and horizontally centered. The arrangement of equipment for PIV measurements is shown in figure 2.

2.2 Image processing

It should be noted that in the process of carrying out PIV-experiments the seeding density varies significantly with time due to the irregular character of the smoke generator operation. Therefore, a dynamic background calculation was used: each pixel of the image synthesized for each frame is the 20th percentile for a set of pixels located in the same place in a sample of the 100 frames closest to a given moment. The background image was subtracted from the frame to reduce static noise and parasitic flare. After subtracting the background, a cross-correlated image processing in accordance to PIV-method. The velocity field was calculated based on the cross-correlation processing of successive frames on the rectangular grid. An adaptive two-pass scheme was used. At the first stage, the displacement field was roughly calculated with a comparison window of 128×128 px (overlapping

50%). The resulting field was filtered and interpolated to the grid in increments of 2 times less. On this grid, with a preliminary symmetric shift of the interrogation windows, the main calculation of the displacement were carried out: the window size was 64×64 px (overlap 50%). To clarify the fractional part of the bias we used the Gaussian approximation of the peak of the cross-correlation function at the three nearest points.

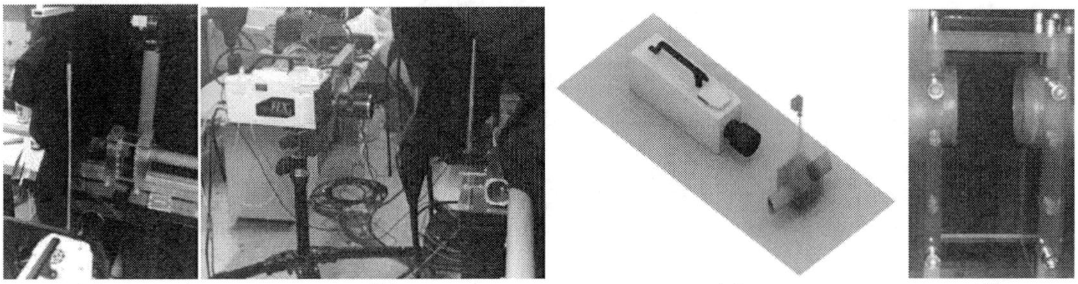

(a) (b) (c) (d)

Figure 2. (a) – view from above on the model of flow part and laser (on the top of the frame), (b) – side view он ерy high speed camera NAC-HX3, (c) – 3D scheme of experiment, (d) – gap between the pipes in the flow part.

The main filtering of obtained velocity values was carried out in time sequence for each point of the spatial grid. The values for which at least one of the found components by more than three standard deviations differs from the average value over the entire record (for a given point) were filtered out through two passes. As a result an average of 5% of the points were discarded.

Sound pressure levels were monitored using a microphone which could be placed in various places, both outside the layout and inside it, depending on the measurement mode, in addition to the main measurements using the PIV-method. The frequency of sound pressure was determined in the range of 50÷8000 Hz.

Initially the pipes were installed flush with the walls without protruding edges (L = 60 mm). In this configuration a sound was clearly heard which as shown by measurements from a microphone, had a frequency of 340 Hz. Also measurements recorded a weak noise at the second harmonic of 680 Hz. Then they began to push the inlet pipe into the flow area. Changing the size of L to 48 mm led to instability of the sound, a sharp decrease in the amplitude of the sound to the level of external interference. The absence of discrete noise was maintained in the interval L up to 38 mm. An even greater reduction in the gap between the pipes led to the resumption of the tonal sound of the system at a higher frequency of 680 Hz which was maintained up to a distance of 30 mm.

For performing PIV-measurements three different cases of gaps between the pipes: 55 mm, 42 mm, 35 mm were chosen, corresponding observed picture of acoustic noise. Figure 3 presents a comparison of averages, and figure 4 shows instantaneous velocity fields. The mean velocity fields practically do not differ, but the instantaneous ones demonstrate the presence of large-scale structures at the boundary of the jet in cases when acoustic noise is observed. Even more eloquent confirmation of the occurrence of the self-oscillatory regime in the flow is a comparison of the spectra of flow fluctuations shown in figure 5. At each point in the coordinate grid of the velocity field, there were spectra that were then averaged over the entire field. The fluctuation spectra of the vertical and horizontal velocity components confirm the presence of hydrodynamic perturbations of the corresponding frequencies.

3. Interpretation of the mechanism of self-oscillating regime

In order to find out how the distance L influence on the occurrence of the self-oscillation mode and the noise emission the results of preliminary numerical simulation of the flow in the model flow section were considered. Figure 6 shows one of the frames of the instantaneous implementation of the system of vortices formed in the flow part area.

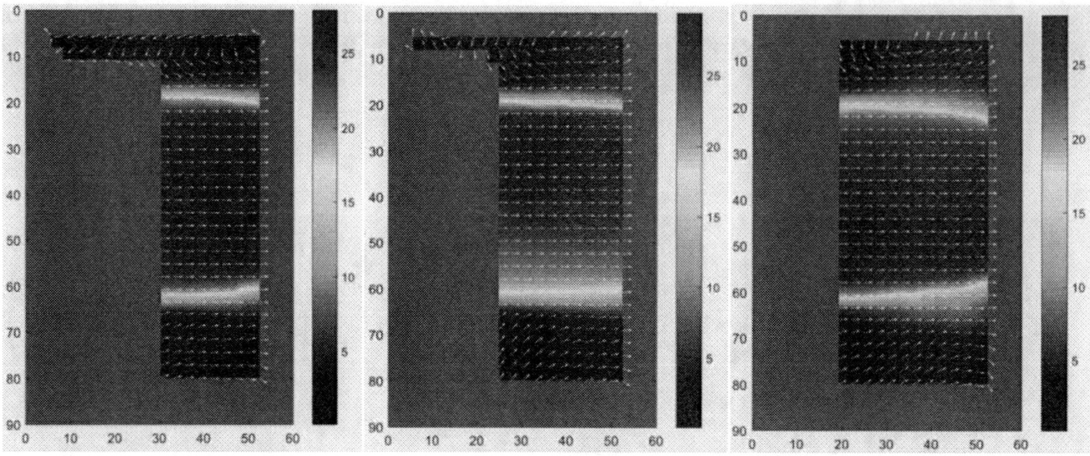

(a) (b) (c)

Figure 3. Mean velocity field (a) gap 35 mm, (b) 42 mm, (c) 55 mm.

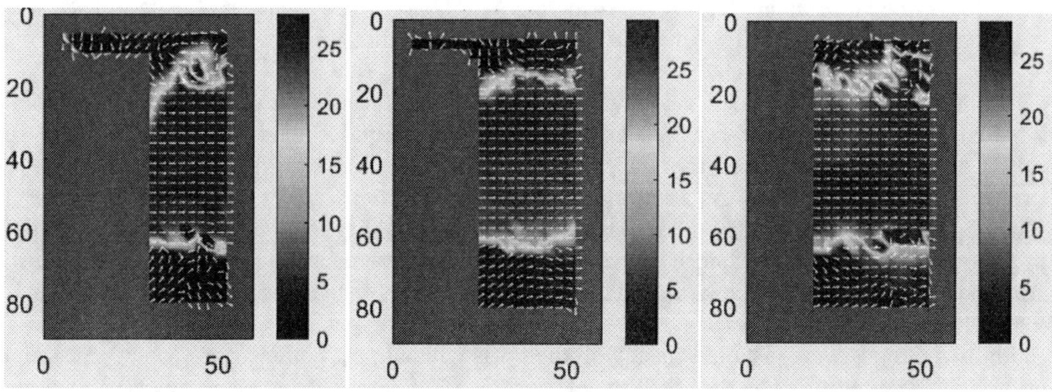

(a) (b) (b)

Figure 4. Instantaneous velocity field (a) gap 35 mm, (b) 42 mm, (c) 55 mm.

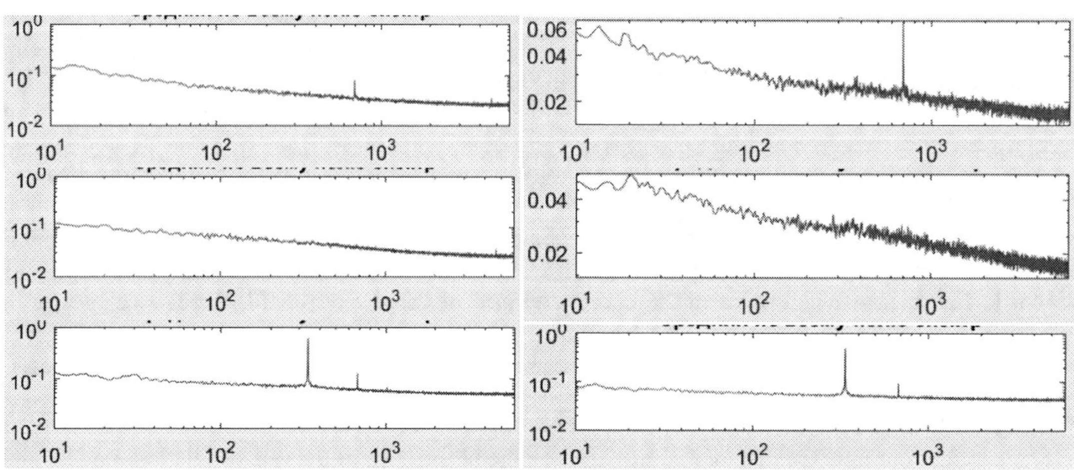

Figure 5. Spectrum of velocity pulsations. Left column corresponding horizontal component of velocity, right – vertical component of velocity. Top raw corresponding gap 35 mm, mean 42 mm, low 55 mm.

Journal of Physics: Conference Series **1421** (2019) 012057 doi:10.1088/1742-6596/1421/1/012057

As can be seen from figure 6 a vortex sheet (thin surface with nonzero vorticity) forms at the boundaries of the air jet penetrating from one pipe to another which separates the main flow air from the air in the stagnant zone of the chamber cavity. This vortex sheet represents a jump in the longitudinal component of velocity from the value in the main part of the flow to values close to zero in the stagnant zone of the chamber and extends the value of L from the edge of the incoming pipe to the edge of the outlet pipe.

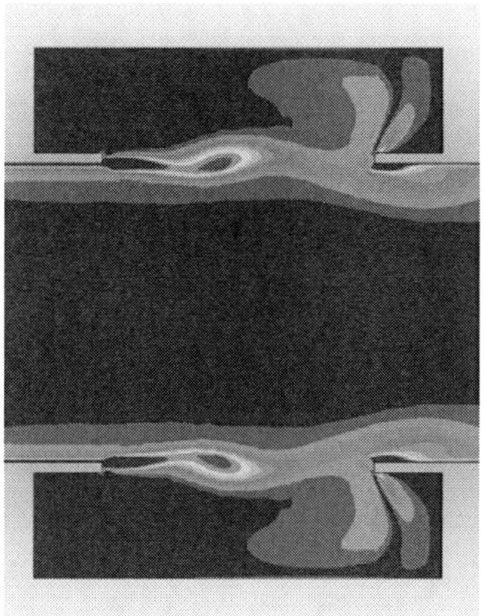

Figure 6. Vorticity distribution obtained by numerical simulations.

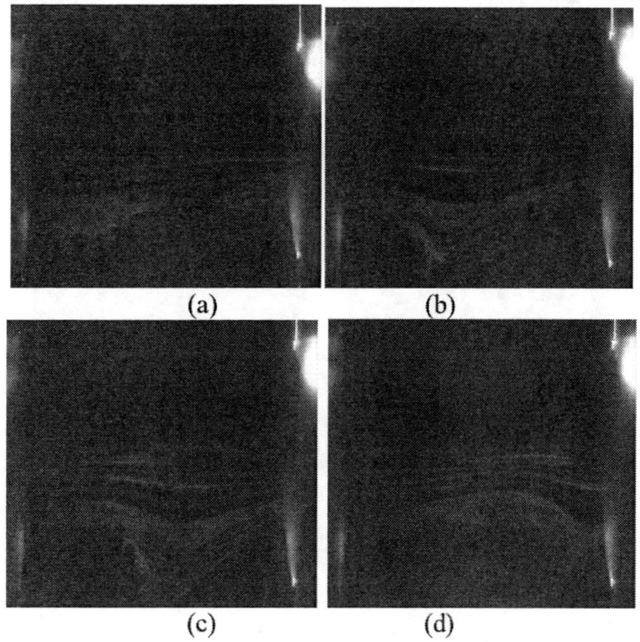

Figure 7. Evolution of the vortex flow in the chamber. Frames are obtained by a large exposure of 10 ms.

As is well known from hydrodynamics the vortex sheet is an unstable structure which with the slightest inhomogeneities tends to "collapse" into a set of large discrete vortices. The process of the occurrence of such a vortex can also be observed in figure 6.

This process is well confirmed by sequence of images obtained with a long (compared to PIV-measurements) exposure. Figure 6 shows the storyboard of a periodic process of the evolution of a vortex sheet in the gap between the pipes in the generation mode of the most intense vortices at a frequency of 340 Hz. This periodic process can be represented as four stages: the first stage is an increase in the intensity of vorticity near the edge of the inlet pipe (figure 7 (a)); the second stage is the loss of stability of the vortex sheet and its folding into a discrete ring vortex (figure 7 (b), right, above); the third stage – the movement of the vortex downstream (figure 7 (c), left, bottom); and the final fourth stage is the blow of the vortex against the edge of the outlet pipe and its resorption on its walls (figure 7 (d), right, bottom).

The most important thing in this process is that the indicated periodical motion of the vortices leads to a periodical change in the flow rate of air flowing through the inlet and outlet sections of the chamber. This vortex can be identified with a fictitious dipole source, oriented longitudinally, exciting acoustic waves.

The phenomenon described above explicitly reveals the reason for the influence of the distance L on the noise emission process, since L and the flow rate U explicitly determine the time during which a vortex forms and reaches the edge of the outlet pipe and hence the frequency of hydrodynamic instability. Obviously this frequency of hydrodynamic instability in some cases is synchronized with

Journal of Physics: Conference Series **1421** (2019) 012057 doi:10.1088/1742-6596/1421/1/012057

other resonances of the system which generates self-oscillations and the accompanying acoustic radiation.

Acknowledgements
Carrying out experiments themselves was supported by Russian Science Foundation project No. 18-19-00473; development of PIV-technique and analysis of results supported by Russian Foundation Basic Research project No 18-48-520023.

References

[1] Morris S C 2011 *Annu. Rev. Fluid Mech.* **43** 529
[2] Rockwell D, Lin J-C, Oshkai P, Reiss M and Pollack M 2003 *J. Fluid Struct.* **17** 381
[3] Sergeev D.A. 2009. *Instr. Exp. Tech.* **52**(3) 1

Technical implementation of acousto-optical instruments: basic types

V E Pozhar[1,2], M F Bulatov[1], A S Machikhin[1,3] and V A Shakhnov[1,2]

[1] Scientific and Technological Center of Unique Instrumentation of Russian Academy of Sciences; 15, Butlerova street, Moscow, 117342 Russia
[2] Bauman Moscow State Technical University; 5/1, 2nd Baumanskaya street, Moscow, 105005 Russia
[3] National Research University "MPEI" (Moscow Power Engineering Institute); Krasnokazarmennaya 14, Moscow, 111250 Russia

E-mail: vitold@ntcup.ru

Abstract. The comparative analysis of the basic characteristics of acousto-optical instruments of different levels of integration and complexity: functional elements, devices and systems. The basic characteristics and their interrelations are determined. It is shown that each class requires individual approach to design. We indicate a novel trend of development of special class of instruments – modules, which are easy-to-integrate into existant devices or systems extending their functionality. The principle requirements for this kind of instruments are their connectivity with the device in each channel, port, protocol, so that the module adding and removing need not the device retuning or adjustment. This tendency has universal nature. The presented analysis is convinient for prognosing prospects and basic directions of acousto-optical instruments development.

1. Introduction

The technology of dynamical Bragg grating based on light diffraction on acoustic waves, is widely used for creation of various functional units like modulators, deflectors, tunable optical filters, etc. [1]. These so-called acousto-optical (AO) units are used as components in devices and complicated stand-alone systems. In cases when they are key elements (like in optical spectrometers [2]), the total device or system is titled 'acousto-optical' and these units should meet specific requirements and usually should be developed together with the total complex.

Thus, AO equipment is developed at different levels of integration and complexity: element, device, and system. Taking AO tunable filters (AOTFs) as an example, we present a comparative analysis of the characteristic requirements and technical approaches to their design and development. We also show that there is a new essential class of AO instruments – add-on modules.

Schematically, the hierarchy of AO spectral instruments is presented in figure 1. Here, it displays 3 basic levels of complexity, simpler one being a component for more complex ones. The AO instruments stratification across the levels is following. An AO element is a functional unit, which performs the simplest operation: spectral filtration, deflection, modulation, etc. A device is intended for wide range of applications (for example, spectra detection). The device is a user-oriented

Journal of Physics: Conference Series **1421** (2019) 012058 doi:10.1088/1742-6596/1421/1/012058

instrument, so it contains all additional units necessary for performance, in particular power unit, control panel, display, information means.

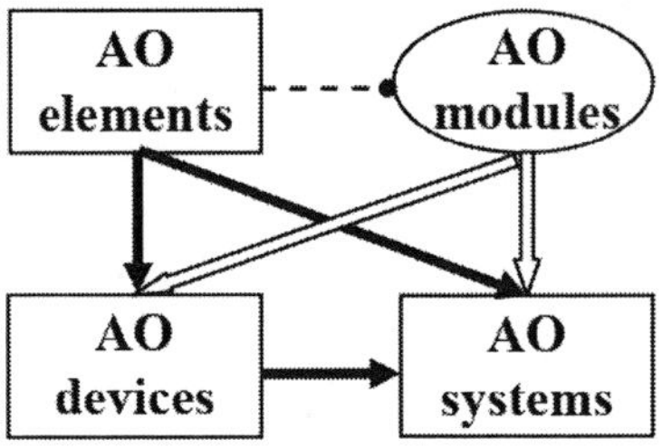

Figure 1. Relations between basic classes of AO instruments differing in complexity.

As a system, we regard the specialized instrumental complex intended for the solution of particular problem. Therefore, it comprises a set of units and devices as well as additional equipment, methods and software, which are sufficient for the problem solution. Usually, systems exists in one or several exemplars, each is in fact unique. And with every upgrade of the working technique, the system is usually modified by adding some new units. That is why, the characteristic feature of the systems is openness.

In addition to elements, devices and systems, there is another class of AO instruments – AO modules. In contrast to these three classes, the discriminating sign of modules is operational features: the module should be capable for prompt connection or disconnection to the instrument complex with no need to noticeable retuning, adjustment or calibration. In fact, the module is an extension of the element. It comprises the set of specialized units like device, but it is not an autonomous device as its goal is secondary or supplementary compared to the complex, which it is inserted into.

Consider the basic features of AO spectral instrument of different classes.

2. AO elements

Primary functional spectral AO element is AOTF. It performs the single operation – optical radiation selective transmission at wavelength λ_f specified by the user. It is characterized by the global transmission function $T(\nu,f,P;C)$, which depends on optical wavenumber $\nu=1/\lambda$, electric-control-signal frequency f and power P as well as on AO crystal cell temperature C detected by means of thermoelectrical sensor. This function originates a family of characteristic relationships (figure 2): spectral profile $T(\nu,\nu_f)$, power dependence $T(P;f)$ and tuning curve at different temperatures $\nu_f(f;C)$. The primary characteristics of AOTF are following: spectral range ν_{min} - ν_{max}, bandwidth $\Delta\nu$, diffraction efficiency dT/dP, temperature drift dν/dC, maximum achievable transmission value T_{max}, spectral contrast $K_{sp} = T_{peak}/T_{out}$, where T_{peak}, T_{out} – transmission coefficients inside the window and out of it (on the "tail"). Another important characteristic is relative background radiation level T_{bg}, which in particular depends on the contrast of the pair of cross-polarizers K_{pol}.

Theoretical form of the global transmission function is

$$T(\nu,f,P;C) = T_{peak}(P;f)\cdot \mathrm{sinc}^2[(\nu-\nu_f)/(\Delta\nu/2)], \quad (1)$$

where $\mathrm{sinc}(x) \equiv \sin(\pi x/2)/(\pi x/2)$. The real profile may significantly differ from it and be asymmetric, but always has the bandpass form. The window peak value is described by formula $T_{peak}(P;f) = \sin^2(f\cdot P^{1/2})$, though in practice, it does not reach 100%. Spectral tuning curve exhibits close to linear form $\nu_f(f;C) = \alpha(\nu_f;C)\cdot f$, but in the range of AO crystal dispersion, the factor α depends on λ, and, therefore, on a bandwidth location ν_f. As it depends on crystal temperature C, one needs to use thermostable crystal orientation or place a temperature sensor to AO crystal cell for tuning curve correction.

The most important additional characteristics are optical (form and area of the input pupil ΔS, field-of-view $\Delta\Omega$) and temporal (tuning time τ_t, delay time τ_{del}, warm-up time τ_{temp}). Also important are the weight (m), sizes ($L\times W\times H$), connecting dimensions, as well as permissible use conditions (temperature range, etc.).

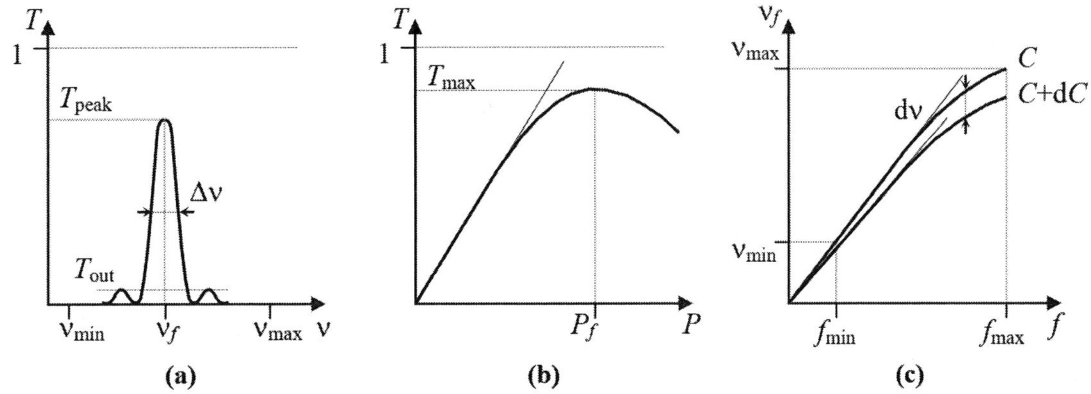

Figure 2. Characteristic relationships of AOTF: (a) spectral transmission function; (b) dependence of transmission coefficients on ultrasound power; (c) spectral tuning curve at different temperatures.

In principle, AOTF can be represented as functional element (figure 3 (a)) with optical input, optical output and control input, which is supplied with radiofrequency (RF) electrical signal generating ultrasound wave with use of piezotransducer and inducing in that way Bragg grating for wavelength selection. Except this simple AO element, there are more complicated configurations, which comprise a pair of AOTFs and some other additional elements [3]. Besides, for imaging AOTFs, the specific aberrational characteristics are of fundamental importance [4].

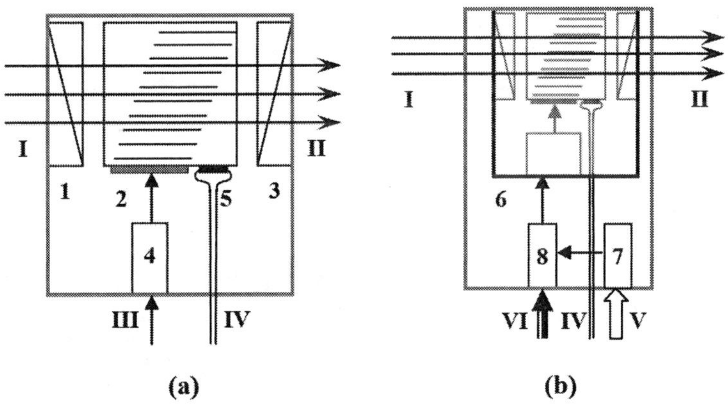

Figure 3. Diagrams for electronically-tuned AOTF (a) and program-controlled AO monochromator (b): 1, 3 – cross-polarizers, 2 – AO cell, 4 – electrical matching board, 5 – thermal sensor, 6 – AOTF, 7 – RF synthesizer, 8 – RF amplifier; I – optical input, II – optical output; III – control RF input, IV – temperature control output, V – digital port, VI – power supply input.

The simple AOTF driven by external RF generator can be converted into AO spectral module (figure 3 (b)), which contains computer-controlled built-in RF synthesizer. Such module must be equipped with electric power supply input to produce enough acoustic power.

3. AO devices

Depending on the problems to be solved, AOTF can be used in various ways. The most straightforward application is spectroscopy. AOTF can be a key element of the spectrometric devices, used for optical spectra detection $I(\nu)$. Fundamental characteristic of spectrometers is spectral sensitivity $U(\nu, \nu_f)$, which connects the photodetector signal S with the input radiation spectrum

$$S(\nu_f) = \int U(\nu, \nu_f)\, I(\nu)\, \mathrm{d}\nu . \tag{2}$$

For rather smooth spectra ($\mathrm{d}I/\mathrm{d}\nu << I/\Delta\nu$), the bandwidth is negligible $U(\nu, \nu_f) \approx A(\nu_f)\cdot\delta(\nu - \nu_f)$, so the input spectra can be calculated just by normalization $I_{calc}(\nu) \approx S(\nu)/A(\nu)$ with use of calibration spectral curve $A(\nu)$. In general case, we can reconstruct only smoothed spectral curve [5].

Spectral sensitivity is obviously connected with AOTF spectral transmission function $T_{\mathrm{AOTF}}(\nu, f, P)$, transmission of other optical elements $T_{\mathrm{opt}}(\nu)$ and photodetector sensitivity $\beta(\nu)$

$$U(\nu, \nu_f) = T_{\mathrm{AOTF}}(\nu, f, P)\cdot T_{\mathrm{opt}}(\nu)\cdot\beta(\nu). \tag{3}$$

This relationship, in fact, includes spectral tuning curve $\nu_f(f;C)$, and spectral profile $P(f)$ of the power supplied to the AOTF, which is adjusted in so way to equalize sensitivity across the spectral range.

The basic characteristics of the spectrometer are spectral range $\nu_{\min}$ - $\nu_{\max}$, the number of resolvable spectral lines $N_{\mathrm{sp}} = (\nu_{\min} - \nu_{\max})/\Delta\nu$, optical throughput $G = (\Delta S \times \Delta\Omega)\cdot T_{\mathrm{opt}}(\nu)\cdot T_{\mathrm{peak}}\cdot\Delta\nu$, sensitivity threshold $I_{\min}$, which is determined by the throughput, noises and leakage radiation connected with value T_{bg}.

Among the additional characteristics, the most important are temporal especially switching time τ_{sw} between random wavelengths. In fact, this time is greater than AOTF actuation time τ_{tr}, τ_{del}, being determined by electrical circuits. Also operation speed is influenced by the algorithm of spectral tuning: scanning or random spectral access. If the second mode is implemented, the spectrometer can be tuned over the informative spectral points so that the operation speed is comparable to the speed of spectrometers with parallel registration.

Besides, weight, dimensions, power consumption, power source and operation conditions do matter. Principle diagram of the spectrometer is depicted in figure 4.

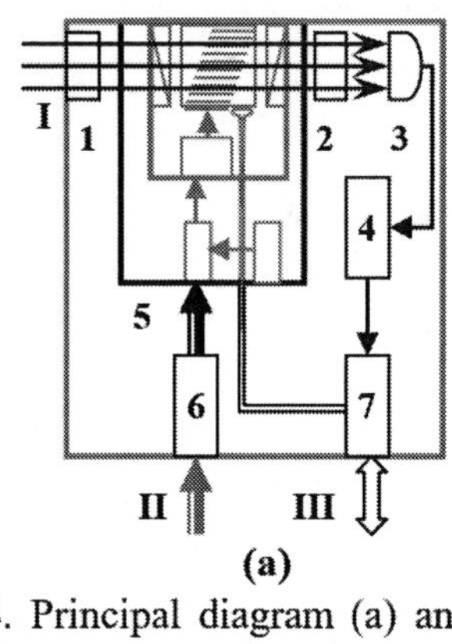

(a)

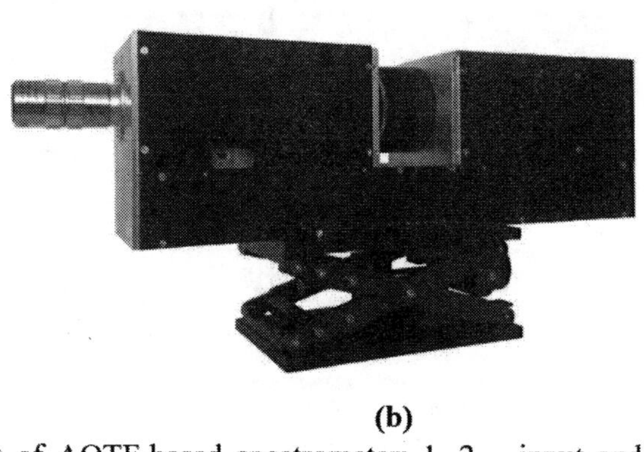

(b)

Figure 4. Principal diagram (a) and exterior (b) of AOTF-based spectrometer: 1, 2 – input and output lenses, 3 – photodetector, 4 – photocurrent amplifier, 5 – program-controlled AO monochromator, 6 – power supply; 7 – microcontroller; I – optical input, II – power supply input; III – digital data exchange port.

Except classical spectrometer based on AOTF, other devices are developed, which are capable of detection wider collection of data, such as hyper-spectrometers $I(\nu,x,y)$, imaging spectrometers $I(x,y;\nu)$, stereo-spectrometers $I(x,y,z;\nu)$ and similar devices with temporal resolution $I(\nu,t)$.

4. AO systems

Systems based on AO units possess the specific features like spectral selectivity, agility, configurability and variability. Such systems comprise a great complex of jointly operating units. AO component is a small but important part, which defines the principle characteristics of the complex. Many AO systems have been developed, in particular for gas analysis [6], biomedical research [7] and remote sensing [8] applications. In each AO system, AOTF performs the only, but the key function, and determines spectral and analytical capabilities. These features differ from those of spectral elements and devices and this is illustrated by UV AOTF-based spectral absorption gas analyzer (figure 5). This system detects concentration of contaminants in the air area between the optical unit and retroreflector with use of a spectral absorption database of pollutant gases.

Spectral characteristics (range, resolution) determine the list of substances, which can be detected, and resolvability of substances with overlapping spectra (for example, in [9] up to 10 substances can be detected in the range 0.25-0.45 μm). The sensitivity affects the detection limit, which must not exceed the allowable level for each substance. (In system "GAOS" [9], this requirements transfer into pathlength limitation ensuring substances detection at the threshold limit value for industrial zone: 15 m for aromatic hydrocarbons and 100 m for oxides SO_2 and NO_2). Temporal characteristics (measurement time) are determined by two factors: sensitivity and random-spectral access algorithm. (In "GAOS" system, for each substance the detection, it is specified of up to 3 spectral lines, each describing with triplet – so the measurement time turn out to be approximately 1 minute per substance [9].) The mass (3 kg) and sizes (6 dm^3) of AOTF spectrometer were not significant in the total weight and dimensions of the system (75 kg, 300 dm^3), while insensitivity of AOTF to vibrations and stability during temperature variations make it possible to use the system in out-of-lab environment as van-installed mobile apparatus.

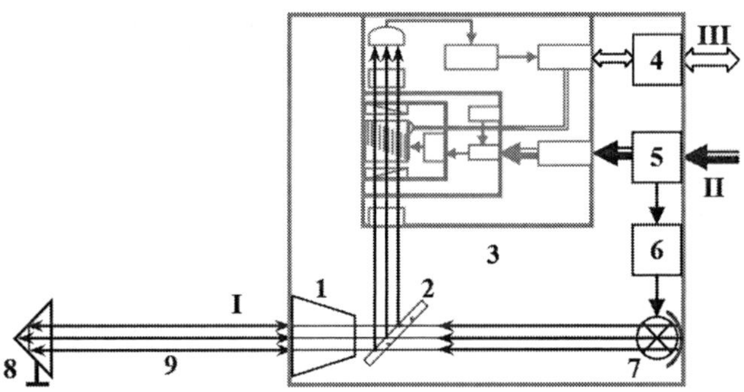

(a)

(b)

Figure 5. Principal diagram (a) and exterior (b) of AOTF-based for long-path gas analyzer "GAOS" 1 – input lenses, 2 – beam splitter, 3 – AOTF spectrometer, 4 – microcomputer, 5 – power supply, 6 – ignition unit for discharge lamp; 7 – light source, 8 – standalone corner retroreflector, 9 – monitored air mass; I – optical input, II – power supply input; III – digital data exchange port.

Operation of AOTF gas analyzer is described by multi-stage algorithm, which presumes determination of target pollutant substances with use of measured absorption data at the characteristic wavelengths of the substances [10]. The system inherently includes a program unit and a database referring the substances spectral specific absorption characteristics. In result, the source data for AO unit development for its integration into a complex system are the characteristics of the system: the number of target substances the detection threshold, detection time, the exploration requirements. Therefore, a descending iterative development scenario is necessary.

5. AO modules

An extremely important sphere of AO units instrumentation is development of modules. They can be inserted into the device or system without any affect on that. Their development principles are quite similar to other units, but they should only be "insertable", that is must be matched with all the parameters of the target device. Thus, modules must be unified in each channel (conjugation): optical path, control contour, data exchange protocols, software. One of the examples is endoscopic add-on spectral AO module (figure 6) [11].

The importance of AO module indication as a separate instrument class is caused by the modern trends in optical instrumentation. Rapid development of new measurement procedures and analytical techniques demand intensive modification of instrument base. That is why, many classical devices are now based on the modular concept acquiring features of open systems. It gives opportunity for modules exchange or an upgrade for acquiring new properties. For this purpose, a lot of standards are developed: from mechanical connectors, plugs and sockets to analog and digital signal characteristics. The modules being developed must correspond to standards and must be insertable into the existent devices. The same principles are also applied to control software.

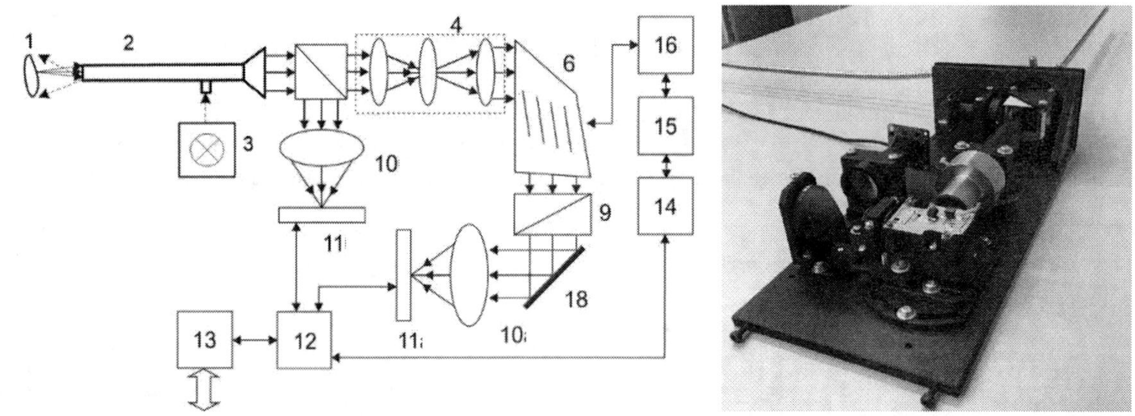

(a) (b)

Figure 6. Principal diagram (a) and interior (b) of AO endoscopic module [12]. 1 – object; 2 – conventional optical endoscope; 3 – light source; 4 – optical coupling system; 6 – AO crystal cell; 9 – polarizer; 10– objective lenses; 11 – cameras; 12 – frame grabber; 13 – processing unit; 14 – controller board; 15 – RF generator; 16 – RF amplifier; 18 – flat mirror

6. Conclusions

Thus, each class of AO spectral instruments is characterized by specific technical parameters as well as individual additional features. It means that their development processes follow different algorithms, in particular, in design of AO crystal cell.

It should be noted that some earlier developed AO elements [3, 12, 13, 14] have module features (see figure 3 (b)). Surely, the collection of spectral modules will be permanently expanded and in near future they will become the typical units of optical systems such as light sources, lenses, photodetectors. The significant problem on this way is that classical ray-tracing methods of optical system design do not suit for AO instruments, operating on the wave optics principles. This technique [15, 16] should use the special designed program units (subroutines) for ray-tracing optical system design. These units describe light beams transmission through AOTF and are based on the theory of image transmission [4].

For convenient comparing AO instruments we joint their basic characteristics in table 1.

Table 1. Characteristics of Acousto-optical instruments.

Characteristics	**AO element**	**AO device**	**AO system**
Spectral	Range Bandwidth	Range Resolvable lines number	Detectable substances list
Optical	Input aperture Field-of-View	Throughput	Range distance
Energetic	Diffraction efficiency Transmission contrast	Transmission coefficient Sensitivity	Detection threshold
Temporal	Switching time Delay time	Single-point detection time Random-spectral access	Detection time
Constructional	Sizes	Dimensions and Weight Power consumption	Weight and Dimensions
Operational	Temperature stabilization	Portability Power supply	Mobility Working environment
Software and methodical	Spectral tuning curve Power transmission profile	User software Calibration procedure	Measurement procedure Analytical technique

It is obvious that the classification does not cover all the types of AO spectral units. For instance, there are multi-band AOTFs [17], variable-bandwidth AOTFs [18], bichannel AOTFs [19]. However, in principle, it displays the characteristic features of AO instruments on different integration levels.

References

[1] Goutzulis A and Rape D 1994 *Design and fabrication of acousto-optic devices* (New-York: Dekker)

[2] Pozhar V E, Balashov A A and Bulatov M F 2018 *Nauchnoe Priborostroenie* **28**(4) 49

[3] Pustovoit V I, Pozhar V E, Mazur M M, Shorin V N, Kutuza I B and Perchik A V 2005 *Proc. SPIE* **5953** 200

[4] Pozhar V and Machihin A 2012 *Applied optics* **51**(20) 4513

[5] Ananiev E, Pozhar V and Pustovoit V 1987 *Optics and Spectroscopy* **62** 159

[6] Mazur M M, Pozhar V E and Shorin V N 1998 *Journ. Analytical Chemistry* **53** 876

[7] Mazur M and Pozhar V 2015 *Optophysical measurements (Measurement Techniques)* **58** 982

[8] Pustovoit V I and Pozhar V E 1999 *Proc.SPIE* **3750** 243

[9] Mazur M M, Shorin V N, Pustovoit V I, Pozhar V E and Fadeyev A V 2011 *Instruments and Experimental Techniques* **54** 276

[10] Fadeyev A V and Pozhar V E 2013 *Journal of Optical Technology* **80** 444

[11] Machikhin A, Pozhar V and Batshev V 2013 *Journal of Optical Technology* **80**(7) 439

[12] Machikhin A S, Khokhlov D D, Batshev V I and Pozhar V E 2018 *Bulletin of Russian Academy of Sciences: Physics* **82** 1403

[13] Polschikova O V, Machikhin A, Batshev V, Ramazanova A, Belov A and Pozhar V 2017 *Proc. SPIE* **10592** 105920H

[14] Machikhin A, Burmak L, Polschikova O, Ramazanova A, Pozhar V and Boritko S 2018 *Applied Optics* **57**(10) C64.

[15] Batshev V I, Machikhin A S and Pozhar V E 2017 *Technical Physics Letters* **43** 216

[16] Machikhin A, Batshev V and Pozhar V 2017 *J. Optical Society America* **A34** 1109
[17] Proklov V V, Byshevski-Konopko O A and Filatov A L 2015 *Technical Physics Letters* **41**(10) 987
[18] Mazur M M, Mazur L I, Suddenok Yu A and Shorin V N 2018 *Optics and Spectroscopy* **125**(4) 594
[19] Machikhin A, Batshev V, Pozhar V, Naumov A and Gorevoy A 2018 *Optics Letters* **43**(5) 1087

Installation for investigation of optical and mechanical system of fiber optical sensors of parameters of liquid or air flows

E Badeeva, E Shachneva, T Murashkina and A Kukushkin

Department of Instrument Engineering, Penza State University, Krasnaya st. 40, Penza, 440026, Russia

Abstract A special measuring unit has been developed for experimental studies of micro-optomechanical systems of fiber-optic micro-angle transducers based on a bellows and micrometer, which are the basic elements of fiber-optic sensors of aerodynamic angles, imitating the physical effect of the flux on the sensing element.

1. Introduction

In solving many problems of measuring the parameters of air or liquid flows, at present, either mechanical (potentiometric, pneumatic) or ultrasonic transducers and sensors based on them are used [1-6]. For example, in wind vane sensors of aerodynamic angles, the measuring circuit contains a mechanical conversion system, including gearing, bearing connections, etc. [11]. This factor leads to a decrease in the reliability of the sensor when exposed to various mechanical effects, the emergence of resonance phenomena in the vibrations of the aircraft, significant dynamic errors, large mass-dimensional characteristics. Therefore, these measuring instruments cannot guarantee the high accuracy of measuring the parameters of fluxes under severe mechanical, climatic, and electromagnetic environmental conditions [7].

A good alternative to "electrical" measuring instruments are fiber-optic sensors of flow parameters [8, 9].

2. Methods

It is proposed to use the fiber-optical conversion principle for measuring the parameters of fluid and air streams, which allows to completely move away from the mechanical conversion system with moving elements [10].

In papers [11, 12, 13] considered fiber-optic sensors (FOS) aerodynamic angles (FOSAA) and parameters of fluid flows of bellows type, the basic element of which are fiber-optic micro-angle transducers (FOMAT) with an elastic element in the form of a bellows (figure 1).

For many years, bellows have been the subject of close attention when considering the problem of compensating for the stretching and contraction of products subjected to movement under the influence of temperature, pressure and external sources.

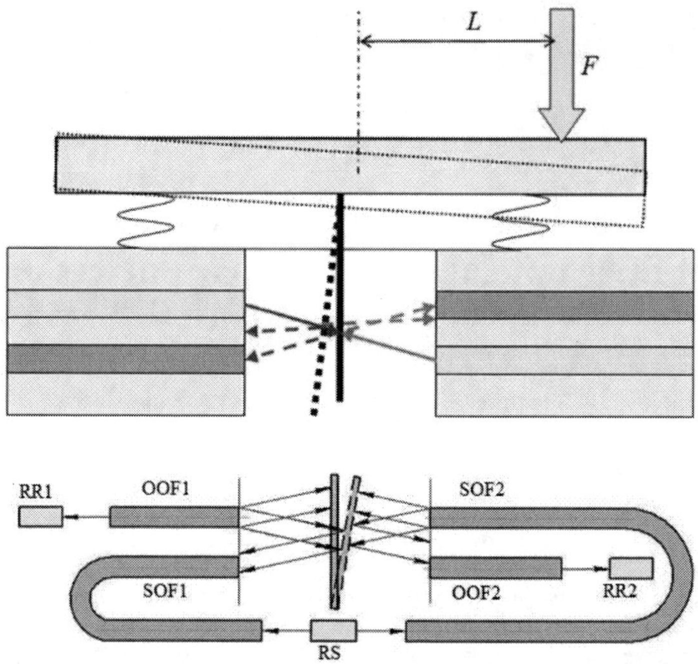

Figure 1. Simplified design diagram, reflecting the processes occurring in the ULF bellows type.

The bellows basically performed the function of compensating or balancing the influence of various factors on the operation of a system or mechanism. Their use as compensators is characterized by reliable and effective protection of products against statistical and dynamic loads arising during deformations. Due to the use in the manufacture of high-quality materials, the bellows are able to work in the toughest conditions with working temperatures ranging from « absolute zero » before 1000 °C and perceive working pressures from vacuum to 100 atmospheres depending on the design and working conditions [14].

The listed advantages of the bellows make it possible to consider the possibility of its use as part of FOSAA a bellows type to convert the force effect of an air or liquid flow into the angular displacement of its axis (figure 2) [15, 16].

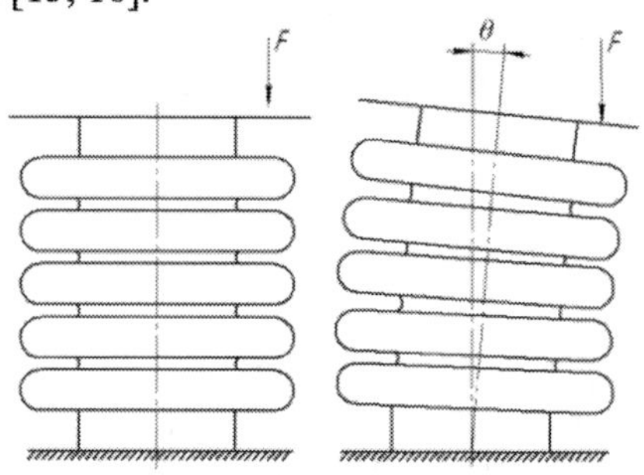

Figure 2. Sketch of the bend of the bellows under the influence of the load.

The elastic characteristic of any bellows is more or less nonlinear and manifests itself in the relationship between the movement of the bellows and the load value applied to it [14]. When designing the FOSAA, which is based on the FOMAT of the bellows type, we adhere to the statement that the bellows performs angular microdisplacements ($\theta = \pm 3 .. 5°$) – thus, the linearity of the elastic characteristic of the bellows is ensured in a given range of microdisplacements (see figure 2). Studies conducted by the authors of the article determined the restrictions on the angular micromovement of a reflecting element in the range ± 3°.

2.1 Objective

In order to achieve improved metrological characteristics of fiber optic sensors with FOSAA bellows type, it is necessary to determine the type of conversion function $I = f(\langle)$,angle range $\langle$, design parameters of the node alignment of the supply and output optical fibers, the optical modulating element: the distance between the common end of the optical fibers and the optical modulating element, the center-to-center distance between the optical fibers This requires specialized metrological support.

3. Results

To conduct experimental studies of micro-optomechanical systems, a special laboratory measuring facility has been developed that simulates the physical effect of air or liquid flows on a perceiving element (figure 3).

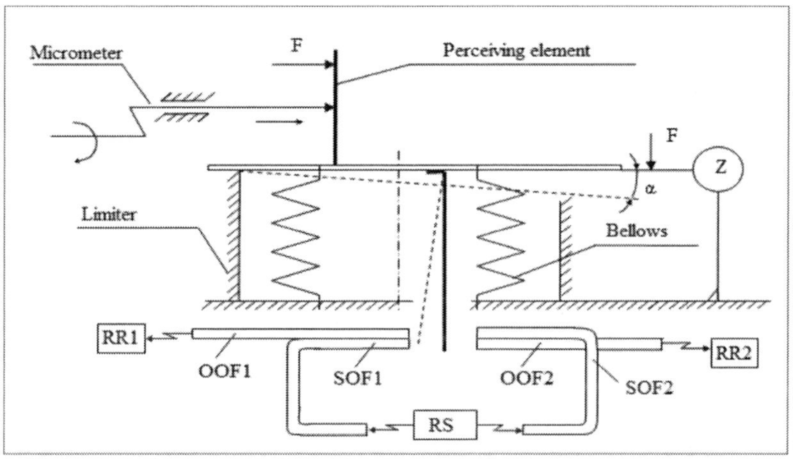

Figure 3. Installation simulating the effect of air flow on the sensing element.

Figure 4 shows the 3-D model of the mechanical part of this installation.

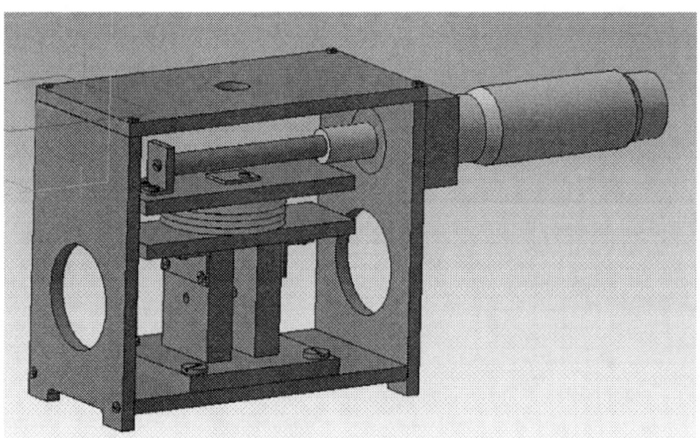

Figure 4. 3D-model of the mechanical system of the installation, simulating the processes occurring in the fiber-optic sensor aerodynamic angles.

The input and output optical fibers in FOMAT located relative to the reflective surfaces of the modulating element at equal distances and coaxially relative to each other. The element perceiving the measured physical quantity is made in the form of a metal plate having mirror outer surfaces. On both sides of the plate are optical fibers of two measuring channels, and the supplying optical fibers SOF1 the first measuring channel are located coaxially with the output optical fibers OOF2 the second measuring channel and, conversely, the supply optical fibers SOF2 of the second measuring channel are aligned with the outgoing optical fibers OOF1 of the first measuring channel. The width and length of the plate are chosen in such a way that the dimensions of the light spot in the measurement range do not exceed its dimensions. If the optical fibers have a core diameter $d_c = 0{,}2$ mm, external diameter $d_{OF} = 0{,}5$ mm, then the calculated dimensions of the plate: width – 3 mm, thickness – 0,2 mm, length – 20 mm, distances between optical axes of optical fibers $D = 0{,}7$ mm, and the initial distance from the fiber end to the plate $x_0 = 1{,}5$ mm.

The sensing element is rigidly fixed on the lid of the bellows with a certain displacement relative to the center of the bellows. In the inner part of the bellows, a metallic platysna is also rigidly fixed with an offset from the center of the bellows, but in the opposite direction. The bellows is fixed at the base of the stopper. The limiter does not allow the bellows to shrink more than 5 degrees.

FOMAT optical scheme has design parameters that meet the following requirements:

- in the absence of an impact of the measured physical quantity on the sensing element, the light radiation emerging from the ends of SOF1 and SOF2 after reflection from the mirror surfaces of the plate should be distributed so that the overlap area of the light spot and the receiving ends of the OOF1 and OOF2 is half of;
- under the influence of a physical quantity on the sensing element, an angular displacement of the plate through an angle must occur ⟨ and, accordingly, the change in the area of overlap of the light spot and the receiving ends OOF1 and OOF2;
- plate deflection at maximum angle 3° to the right leads to a complete overlap of the area of the receiving end OOF1 the light flux reflected from the plate, and the light flux will not come to the receiving end of the OOF2.

Similar processes occur when the plate is deflected to a maximum angle 3° to the left.

Installation for experimental studies consists of an optical tester, fiber-optic cable, installation to simulate the parameters of air or liquid flow (figure 5).

FOMAT is designed for contactless conversion of changes in the angle of inclination of the plate in the change in the amplitude of the DC.

Installation to simulate the parameters of air or liquid flow is designed to set the angle of inclination (offset from the vertical position) of the plate in the range 0…5°.

The angle is set using the angle setting system based on the microscrew. Since the micrometer screw is used, it is necessary to compare the displacements in microns (H) with a deviation from the vertical in degrees (α) (table 1).

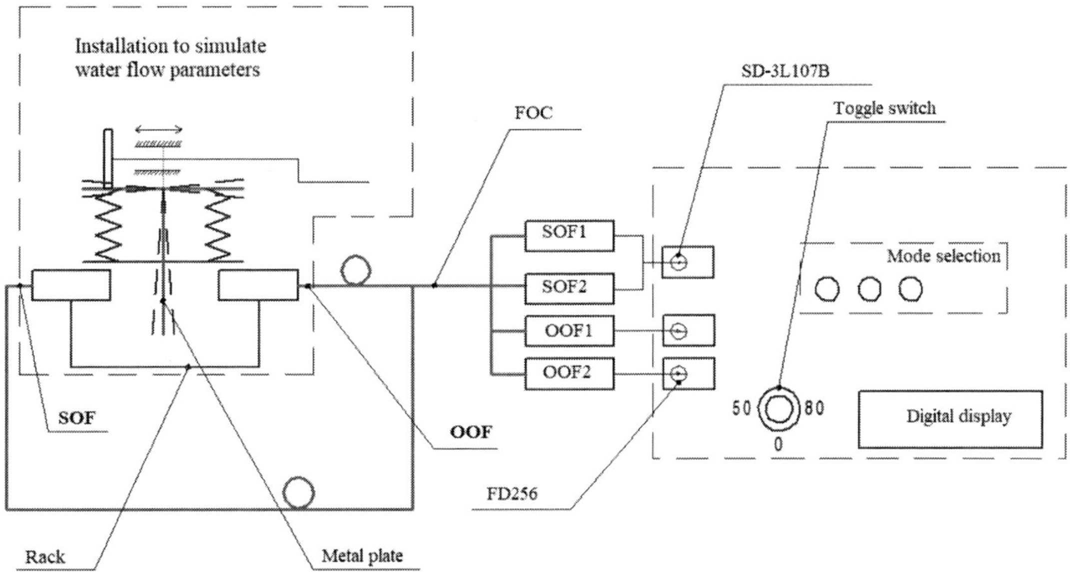

Figure 5. Block diagram of the laboratory layout:
FOC – fiber optic cable; P – photodiode; LED – Light-emitting diode; SOF – the supply optical fibers; OOF – the outgoing optical fibers.

Table 1. Correspondence of movement to angular movement.

H, мкм	0	0,261	0,525	0,768	1,047	1,308	1,569	1,83	2,094	2,355	2,616
α, град	0	0,5	1	1,5	2	2,5	3	3,5	4	4,5	5

When mounting the plate in the lid of the bellows, it is necessary to ensure strict perpendicularity of their planes and parallelism of the reflecting planes of the plate and the working ends of optical fibers. To provide an initial distance, for example x_0 = 1,5 mm, it is necessary to install on both sides of the plate unambiguous measures of the length of 1.5 mm and to bring the tips with optical fibers close to them. At the same time, the output signals of photodiodes, docked to the tapping optical fibers should be minimal (the numerical value is determined experimentally for a specific sample FOMAT). If the values of the output signals are minimal, then the initial distance x_0 = 1,5 mm between the plate and the ends of the optical fibers corresponds to the calculated value. This position of the plate determines the point of reference in the measurements.

The installation works as follows. With the help of the micrometer screw, a force is applied to the perceiving element fixed in the upper part of the installation.

In the zero position with no inclination angle of the plate, the rays of light from the SOF1 and SOF2 under the aperture angle $\cup_{NA}$ to the optical axis of the fiber are in the forward direction x_0 to the reflecting plane of the plate and the path x_0 in the opposite direction to OOF1 and OOF2. In this case, in the plane of the receiving ends of OOV1 and OOV2, there is an illuminated annular zone with a width $h = r_C$, where r_C – the radius of the core of the optical fiber. The annular zone is converted into an ellipsoid zone, which is displaced relative to OOF1 and OOF2 in the direction perpendicular to the common fiber end [10]. In this case, the areas of the receiving ends of OOF1 and OOF2, illuminated by light streams reflected from the mirrors, change. Then there is a further conversion of the output optical signal into a current signal.

In the optical tester circuit, two radiation detectors (photodiode) are provided to provide a differential circuit. The optical tester contains a radiation source (an infrared LED of type 3L107B) with adjustable radiation power, two photoreceiving channels (photodiodes with amplifiers), an information processing unit (programmable microcontroller), a digital indicator and a power supply. Three buttons are located on the front panel of the tester: the first button allows you to compensate for the initial bias voltage of the photoreceiving channels ("Const. 0"); the second button is the choice of information processing mode; the third button turns off the display of the number of the selected information processing mode.

The current passing through the LED can take three values -0; 50; 80 nA, selected with the help of a toggle switch: "Const. 0"- to compensate for the initial bias voltage of the photodetector channels (zero setting), «50 nA» и «80 nA» – to measure. Used type photodiodes FD256 work in the photovoltaic mode, which ensures a low level of intrinsic noise. They are connected to a "current-to-voltage" converter, providing near-zero bias voltage (equally U_{SM} operational amplifier) and low (close to zero) load impedance of the photodiode. This allows to obtain a high linearity of the conversion function of the photodetector channels in the range of variation of the intensity of the received optical signal from 0 before 10^6 nA.

Maximum output of photovoltaic cells – 5 V, is determined by the ADC embedded in the microcontroller type PIC16F873/SP. In accordance with the program, the microcontroller provides three modes of information processing:

- in the first mode, the digital display shows the digital values of the signals from both channels (I_1 and I_2);
- in the second mode, the ratio of these signals is displayed (I_1/ I_2);
- in the third mode, the result of the calculation by the formula $(I_1 - I_2)/(I_1 + I_2)$.

The microcontroller, if necessary, allows you to compensate for the initial bias voltage of the photoreceiver channels. When the radiation source is off, the button is pressed "Const. 0". The controller remembers the voltages present at the moment at the outputs of the photoreceiver channels, and automatically takes them into account in further work.

3.1 The results of experimental studies

Experimental studies have been carried out that made it possible to obtain a graphical representation of the dependence of the photocurrent on the movement of the microscrew (figure 6).

Table 2 and figure 7 present the experimentally obtained results of the dependence of the photocurrent value on the angular displacement.

Conclusion

The obtained experimental data confirmed the theoretical positions of the feasibility of the optical signal conversion circuit using FOMAT with bellows, allowed to determine the optimal design and technological parameters of the FOMAT optical system: $\alpha = \pm 3$ °, $x_0 = 1{,}5$ mm, $D = 0{,}7$ mm, required for maximum sensitivity of optical signal conversion.

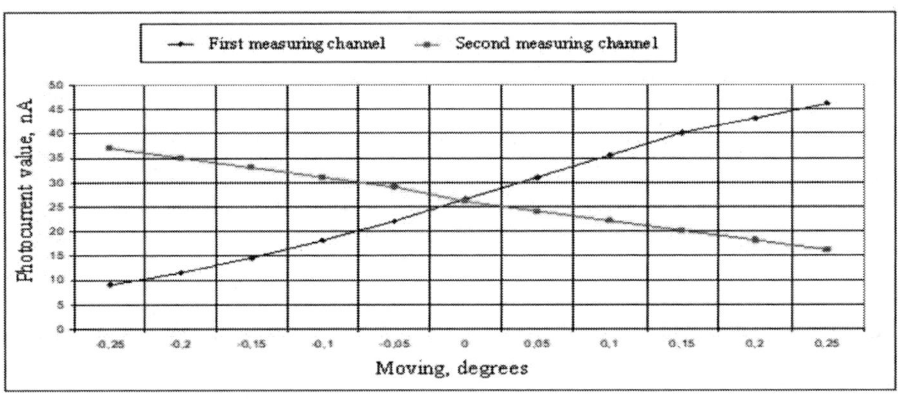

Figure 6. Graphic dependences of the photocurrent on the movement of the microscrew.

Table 2. Experimental Results.

h	0	0,261	0,525	0,768	1,047	1,308	1,569	1,83	2,094	2,355	2,616
α	0	0,5	1	1,5	2	2,5	3	3,5	4	4,5	5
clockwise											
I_1 first channel, nA	0,21	0,18	0,16	0,14	0,12	0,11	0,1	0,09	0,08	0,07	0,06
I_2 second channel, nA	0,22	0,24	0,25	0,32	0,35	0,38	0,37	0,27	0,13	0,06	0,03
counterclock-wise											
I_1 first channel, nA	0,21	0,24	0,27	0,3	0,34	0,37	0,4	0,4	0,3	0,19	0,1
I_2 second channel, nA	0,22	0,2	0,17	0,15	0,13	0,11	0,1	0,09	0,08	0,07	0,06

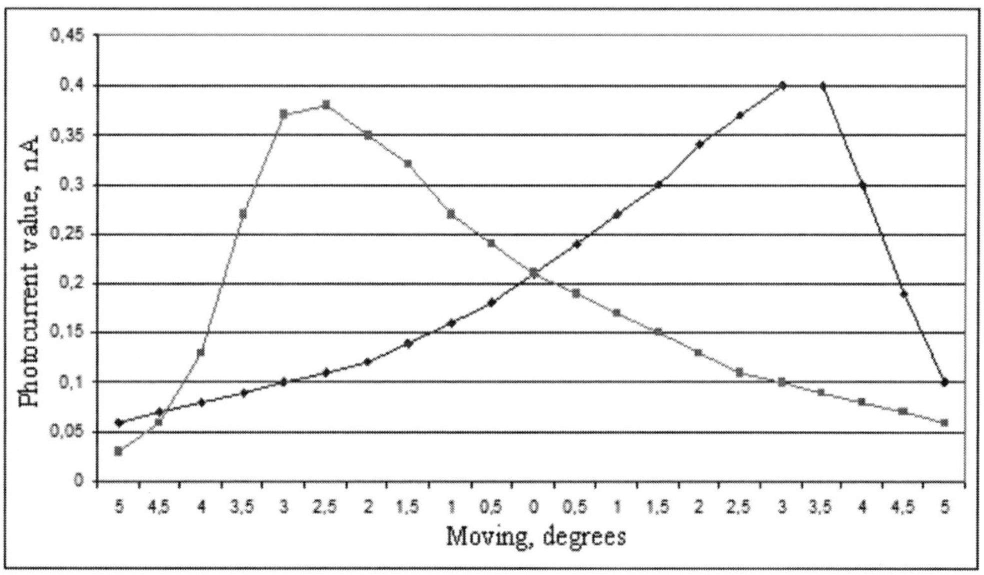

Figure 7. Graphic dependences of the photocurrent on the movement of the microscrew (0° < α < 5 ° both sides of the vertical).

Acknowledgments

The scientific studies were carried out as part of the State Order 8.11785.2018 / 11.12., with financial support from the Russian Foundation for Basic Research in the form of grant No. 18-38-20045.

Journal of Physics: Conference Series **1421** (2019) 012059 doi:10.1088/1742-6596/1421/1/012059

References

[1] Idelchik I E 1975 Reference book on hydraulic resistance (Moscow: Mechanical Engineering) 560
[2] Bashta T M 1971 Mechanical engineering hydraulics: Reference book (Moscow: Mechanical Engineering) 672
[3] Vasilchenko V A 1983 Hydraulic equipment of mobile machines: Reference book (Moscow: Mechanical Engineering) 301
[4] Weinberg V B 1977 Optics of light guides (Moscow: Mechanical Engineering) 320
[5] Alekseeva T V, Galdin N S and Sherman E B 1994 *Hydraulic machines and hydraulic drive of mobile machines* (Novosibirsk: NSU) 212
[6] Girgidov A D Kozhevnikova E N, Laksberg A I and Loktionova E A 2007 *Mechanics of fluid and gas (hydraulics)* (Saint-Petersburg: Publishing house of the Polytechnic University) 72

Journal of Physics: Conference Series **1421** (2019) 012060 doi:10.1088/1742-6596/1421/1/012060

Analysis of the statistical errors of the emissivity measurement of a laser heated surface by acousto-optical tunable filter

Yu V Mantrova, P V Zinin, A A Bykov and K M Bulatov

Scientific-Technological Center of Unique Instrumentation of the Russian Academy of Sciences, Butlerova 15, Moscow 117342, Russia

E-mail: mantrovayv@gmail.com

Abstract. The aim of this report is to conduct a rigorous data analysis of the emissivity and temperature measurements on the surface of a solid specimen during laser heating (LH). Measurements of the emissivity and temperature is based on the use of a tundem acousto-optic filter (tandem acousto-optical tunable filter or TAOTF) connected with a high-resolution video camera. The set of spectral images obtained with the help of TOTF in the range of 750-800 nm makes it possible to calculate the dependence of the radiation intensity of the heated surface on the wavelength at each point of the surface. The distribution of the temperature and the emissivity of the heated body surface is obtained by fitting the experimental spectral dependence of the radiation intensity at each point of the heated object to the Planck distribution using the least squares method (LSM). A development of the mathematical analysis of the statistical errors of the emissivity determination by the LSM is described.

1. Introduction

The emissivity of a material describes the ability of a material's surface to emit radiation at a specific wavelength relative to a black body radiation. Measurement of emissivity at high temperatures is necessary for understanding many physical phenomena, including heat exchange in the center of the Earth, processes in a diffusion flame, obtaining solar cells with low emissivity, insulation and energy saving of buildings, etc. The laser-heating is the only experimental tool able to create extreme temperatures and is widely used in research on materials science and geophysics [1]. Non-contact radiometric methods are based on comparison surface radiation of the studied materials with a calibration source (absolutely black body) under the same temperature, geometric and spectral conditions [2]. Since the measurement of the emissivity is closely related to the measurement of temperature, the measurement methods differ in the methods of heating the samples. Measurement of emissivity is a fundamental constant, and also plays an important role in high-pressure physics, when it is a task to establish heat balance, for example, in the Earth's crust. Previously, the measurement of the emissivity was difficult because it was impossible to obtain the temperature distribution. The possibility of using acousto-optic filters made it possible to measure the temperature distribution in the laser beam [3]. Unfortunately, distribution of the emissivity during laser heating was studied. In this paper, we will conduct (a) a rigorous data analysis of the emissivity and temperature measurements of laser heated areas of the tungsten plate and (b) will provide an estimate of measurement errors of the emissivity.

2. Experimental Method

Description of the laser heating system combined with TAOTF can be found elsewhere [3]. Briefly, the system consists of four components: (1) 100 W fiber laser LH (1064 nm, IPG Photonics), which is designed to allow precise control of the total power in the range from 2 to 100 W by changing the diode current, for heating samples; (2) TAOTF; (3) motorized sample stage; and (4) high magnification imaging system based on long working distance infinity-corrected objective. This system is unique and allows: (a) measurement of 2-D temperature distribution (TD) with a resolution less than 2 μm in a sample under high temperature using Planck's law; and (b) measurement of 2-D intensity distribution of an IR laser beam on the surface of specimen [4].

An experimental LH-TAOTF setup operating in the reflection configuration is shown in figure 1. The fiber laser beam is guided to the sample position with mirrors (M5, M6), and a narrow line mirror (M4). A pi-shaper is located between mirrors M5 and M4 and designed to allow control of the IR laser spot shape (e.g., gauss, flat-top, donut) and size (8-100 μm) [5]). The M4 separates the laser beams and visible radiation. The thermal radiation, coming from the sample during LH, is directed toward the TAOTF by mirrors M1 and M2. The image of the heated specimen at a selected wavelength is taken by a camera after a long focal distance lens.

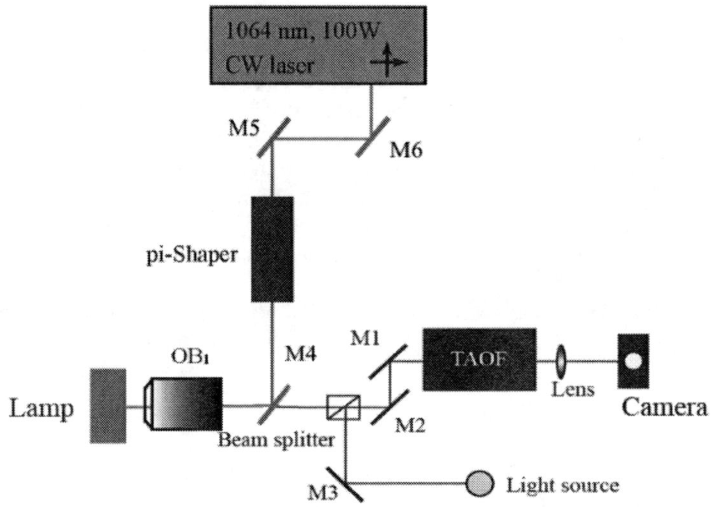

Figure 1. The sketch of the LH-TAOTF system. OB1: geoHEAT, 40NIR, AldOptica, GmbH; TAOTF: tunable acousto-optic filter, A-Optic, Russia; Beam splitter; lens, f = 400 mm, AC254-400-B Thorlabs; M: mirrors on kinematic mounts; M4: narrow line mirror (1064 nm); pi-Shaper; Camera: monochrome camera, Allied Vision Mako G-030B, which runs 309 frames per second at resolution 644×484 pixels; CW laser: continuous wave laser (IPG Photonics, YLR-100-AC-Y11).

The calibration of the field of view was done using a glass plate with a 50 μm grid. The camera was used to image 0.36 μm at the object plane per 7.4 μm pixel, indicating that magnification is around 20.6x. The software written for collecting TAOTF images compensates for this shift by taking into account the linear dependence of the shift on the wavelength.

Temperature and emissivity can be calculated from the radiation emitted by a specimen using Planck's equation (for details see [6, 7]):

$$I(\lambda,\varepsilon,T)=\varepsilon g(\lambda,T), \quad g(\lambda,T)=\frac{c_1}{\left\{\lambda^5\left[\exp(\frac{c_2}{\lambda T})-1\right]\right\}} \tag{1}$$

where $I(\lambda,\varepsilon,T)$ is the spectral intensity, ε is the sample emissivity, λ is the wavelength, T is the temperature, and c_1 and c_2 are physical constants ($c_1 = 2\pi c^2$, $c_2 = h{\cdot}c/k$ =0.01432 $m{\cdot}K$). To calibrate the optical system, a lamp with a known temperature (1700 K) is used, which is placed in the measurement area. In our experimental set-up calibration is based on the imaging of a certified lamp at a temperature of 1700 K [8]. A radiometric power supply is used to provide a very precise and reproducible lamp current, giving a constant radiance. The lamp is placed at the position normally occupied by the sample place so that the spectral intensity $I(\lambda)_{optics}$ of the lamp is acquired through the optical pathways. To determine the spectral radiance of the heated specimen, $I(\lambda)_{corrected}$, the following equation is used

$$I(\lambda)_{corrected} = I(\lambda)_{measured} \cdot I(\lambda)_{standard}/I(\lambda)_{optics} \tag{2}$$

where $I(\lambda)_{measured}$ is the spectral intensity of the specimen during heating [3].

3. Least squares method

The standard method of the temperature and emissivity determination from the thermal radiation data, $I_{corrected}(\lambda_k)$, is to find such values of T_o and ε_o at which the sum of squares or S function,

$$S(\varepsilon,T)=\sum_{i=1}^{N}\left[I(\lambda_i)-\varepsilon g(\lambda_i,T)\right]^2, \tag{3}$$

has a minimum (2D non-linear least square fitting) [9-11]. To reduce the problem of the 2D non-linear least square fitting (3) to the one dimensional (1D) least-squares fitting we used a method proposed in [12]. It is based on the fact that emissivity ε in (1) is a linear parameter. It allows us to obtain an analytical expression for ε_o using the following procedure. If the function $S(T,\varepsilon)$ has a minimum at T_0 and ε_0, then the following conditions should be satisfied: 1) $\partial S/\partial\varepsilon\,|_{\varepsilon=\varepsilon_0,\,T=T_0} = 0$; 2) $\partial S/\partial T\,|_{\varepsilon=\varepsilon_0,\,T=T_0} = 0$. The first equation gives the expression for the ε_0:

$$\varepsilon_o=\frac{\sum_{i=1}^{N}\left[I(\lambda_i)g(\lambda_i,T)\right]}{\sum_{i=1}^{N}\left[g^2(\lambda_i,T)\right]}. \tag{4}$$

The second equation can be written as

$$\frac{\partial S}{\partial T}=-\frac{2c_2\varepsilon_o}{T^2}\sum_{i=1}^{N}\frac{1}{\lambda_i}\left[I(\lambda_i)g(\lambda_i,T)-\varepsilon_o\right]=0. \tag{5}$$

The 1D least square fitting strategy is following. The temperature is varied and the (4) is used to determine ε for a given T. Then S function (3) is calculated and values of T_0 and ε_0, at which the S function has a minimum, are determined.

Because the emissivity ε is a linear parameter in (1) it is also possible to obtain an analytical expression for the standard deviation for ε_o following the method described elsewhere [13]. To use this method we rewrite (4) in the form

$$\varepsilon_o=\sum_{i=1}^{N}\left[I(\lambda_i)G(\lambda_i,T_o)\right], \text{ where } G(\lambda_i,T_o)=\frac{g(\lambda_i,T_o)}{\sum_{i=1}^{N}\left[g^2(\lambda_i,T_o)\right]}. \tag{6}$$

If we assume that all the measurement of $I(\lambda_i)$ are independent then the standard deviation of the emissivity can be written as

$$\sigma_\varepsilon^2=\sum_{i=1}^{N}\left[\frac{\partial\varepsilon(\lambda_i)}{\partial I}\sigma_I\right]^2, \text{ where } \sigma_I=\sqrt{\frac{\sum_{i=1}^{N}\left[I(\lambda_i)-\varepsilon_o g(\lambda_i,T_o)\right]^2}{N-2}}. \tag{7}$$

The derivative in (6) can be taken: Then the expression for the standard deviation of the emissivity ε_0 can be written

$$\sigma_\varepsilon = \frac{\sigma_I}{\sum_{i=1}^{N}\left[g^2(\lambda_i, T_o)\right]}. \quad (8)$$

We will use this expression to estimate the error of the emissivity measurement.

4. Results

To conduct the laser heating experiments we used a tungsten lamp, which is a standard test object in the pyrometry. The tungsten ribbon of the lamp was heated by the IR laser with a fixed power of 8 W. Figure 2 shows the temperature (figure 2 (a)) and emissivity (figure 2 (b)) distributions simulated by the least square method, (3). Evaluations of the statistical errors of temperature and emissivity were conducted at three points with different temperatures: in the center of the heating spot (area I), where the heating temperature was highest, 2640 K (figure 2 (a)); and in two areas (area II and area III) along its edge, where the heating temperature was 2451 K (area II, figure 4 (a)); and 2135 K (area III, figure 6 (a)). At these points, emissivity was also measured (figures 2 (b), 4 (b) and 6 (b)). For all the points, the thermal radiation measurements were made in the wavelength range from 750 to 800 nm.

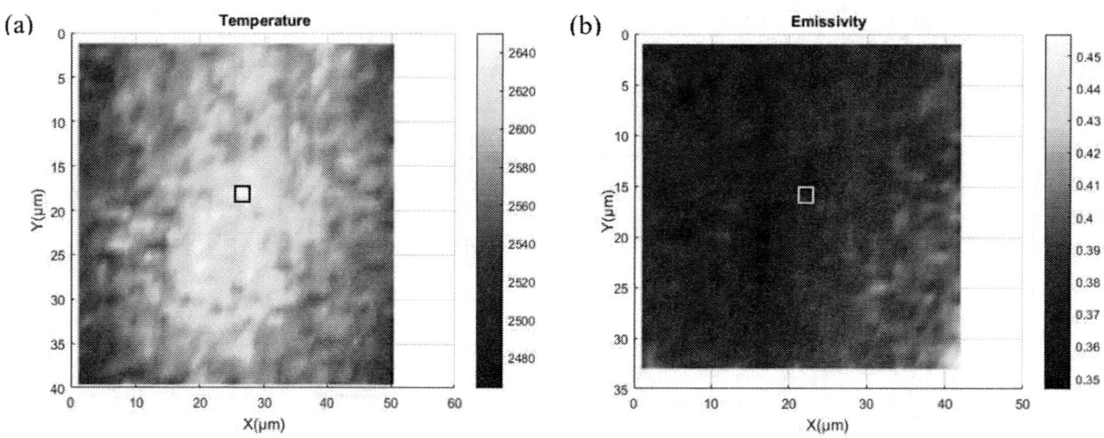

Figure 2. Temperature and emissivity distributions of the laser heated area I: (a) temperature distribution; (b) emissivity distribution. W. Exposure time 1/4 s. For the center of the heating spot.

The confidence interval of the T_o and ε_o was evaluated using the F-test developed for the non-linear regression technique [14], where the confidence level is derived from the expression,

$$S(T,"(") = S(T_{1}o, (_{1}o)[1 + P/(N-P)\, F(N,P,\alpha)]. \quad (9)$$

where N is the number of measurements, P is the number of variables, $F(P,N,1\text{-}\alpha)$, is the Fisher coefficient, and the coefficient (1-α) is the level of confidence. For the current experiment $N = 51$, $P = 2$. Expression (9) provides the confidence interval for temperature and emissivity measurements. For the level of confidence of 0.95, it is shown in figures 3-5 as a red line.

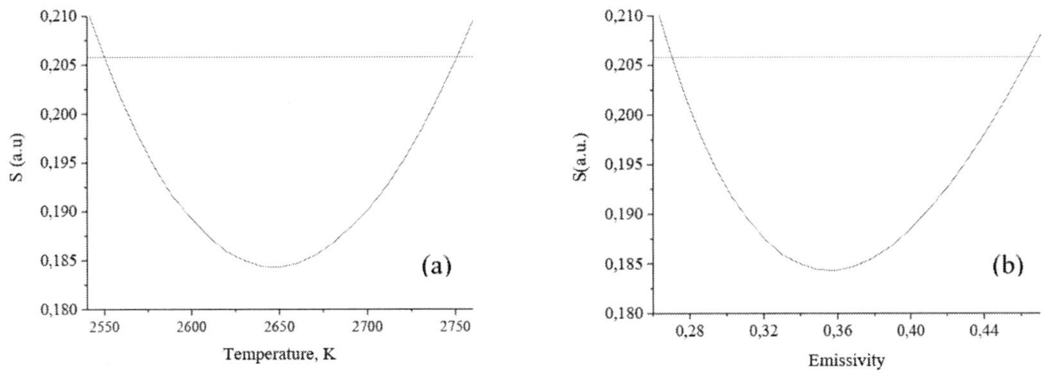

Figure 3. Behavior of S as a function of (a) of the temperature, and (b) of the emissivity simulated at the point in the center of the heating spot, area I (figure 2 (a) and (b)).

Figure 3 shows that the S function has a minimum, $S_0 = 0.1843$, at $\varepsilon_0 = 0.3548$ and $T_0 = 2640$ K. The confidence interval for the emissivity is 0.084 indicating that the relative error is pretty high 23%. Similarly, we calculate the confidence interval for the temperature. It is found to be 89 K indicating that the relative error is around 3.4%. Now, we can compare the confidence interval obtained with the F-test with that obtained with (8). The value of the σ_ε was measured to be 0.268. The confidence interval can be obtained using the formula,

$$\Delta_\varepsilon = \frac{\sigma_\varepsilon \cdot t}{\sqrt{N}}, \tag{10}$$

where t is the Student coefficient equal to 2.006. Then the confidence interval should be 0.075 for the emissivity (relative error is around 21%) which pretty close to the value obtained by the F-test.

Figure 4 shows the temperature distribution (figure 4 (a)) and emissivity (figure 4 (b)) for the area II. Figure 5 shows the behavior of the S function as a function of the temperature (figure 5 (a)) and emissivity (figure 5 (b)).

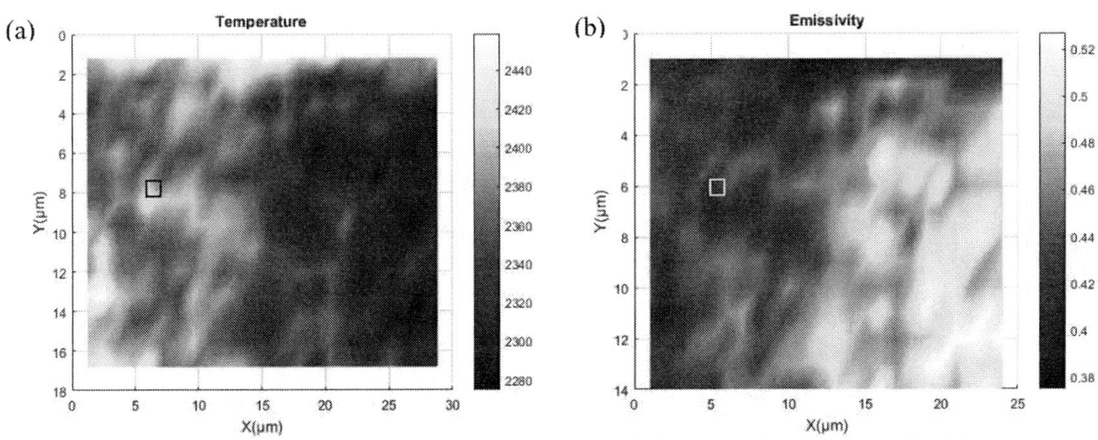

Figure 4. Temperature and emissivity distributions of the laser heated area II: (a) temperature distribution; (b) emissivity distribution. Exposure time 1/4 s.

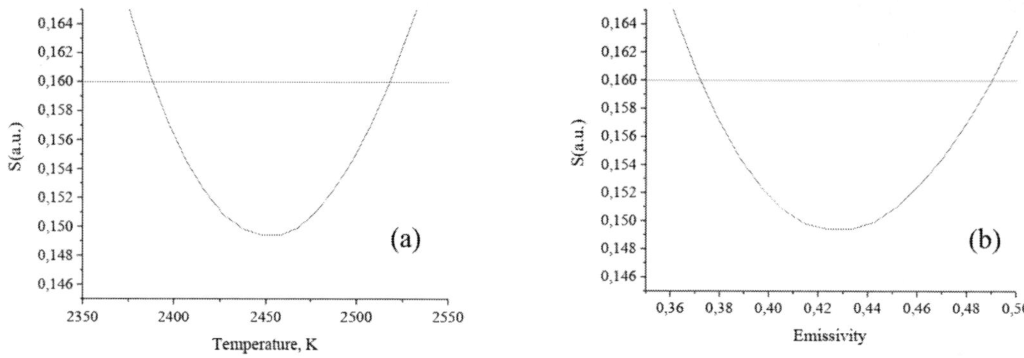

Figure 5. Behavior of the S as function (a) of the temperature, and (b) of the emissivity simulated at the point of the area II (Figure 4a and 4b).

The value of the S function at the minimum, $\varepsilon_0 = 0.4274$ and $T_0 = 2451$ K, is $S_0 = 0.14934$. The F-test confidence interval for the emissivity is 0.06 (the relative error is 14%) and is 66 K for temperature (the relative error is 2.7%). The confidence interval using (10) is 0.0066 for the emissivity indicating that the relative error of the emissivity measurement is around 15%. It is in a good agreement with the value obtained by the F-test.

Simulations of the temperature and temperature distributions at the third point in the area III are shown in figure 6. Figure 7 shows the behavior of the S function for temperature (figure 7 (a)) and emissivity (figure 7 (b)). The S function has its minimum ($S_0 = 0.08123$) at $\varepsilon_0 = 0.4478$ and $T_0 = 2135$ K. The F-test confidence interval for emissivity is 0.035 (the relative error is 7.9%) and for the temperature is 43 K (the relative error is 2.0%). The confidence interval for emissivity obtained using (8) is 0.041 (the relative error is 9%), and is close to that obtained by the F-test.

Our results demonstrate show the higher the temperature the lower the value of emissivity in the wavelength range from 750-800 nm which is in a good agreement with published experimental data [15]. We note that emissivity values measured by TAOTF system are also in a good agreement (within experimental errors) with those given in a handbook [15]: (a) $\varepsilon_0 = 0.355 \pm 0.084$ measured at $T_0 = 2640$ K is close to the value given in [15], $\varepsilon = 0.394$ obtained at $T = 2600$ K; (b) $\varepsilon_0 = 0.427 \pm 0.066$ measured at $T_0 = 2451$ K is within experimental error with the value $\varepsilon = 0.398$ obtained at $T = 2400$ K; (c) $\varepsilon_0 = 0.448 \pm 0.041$ measured at $T_0 = 2135$ K is close to that $\varepsilon = 0.403$ obtained at $T = 2200$ K.

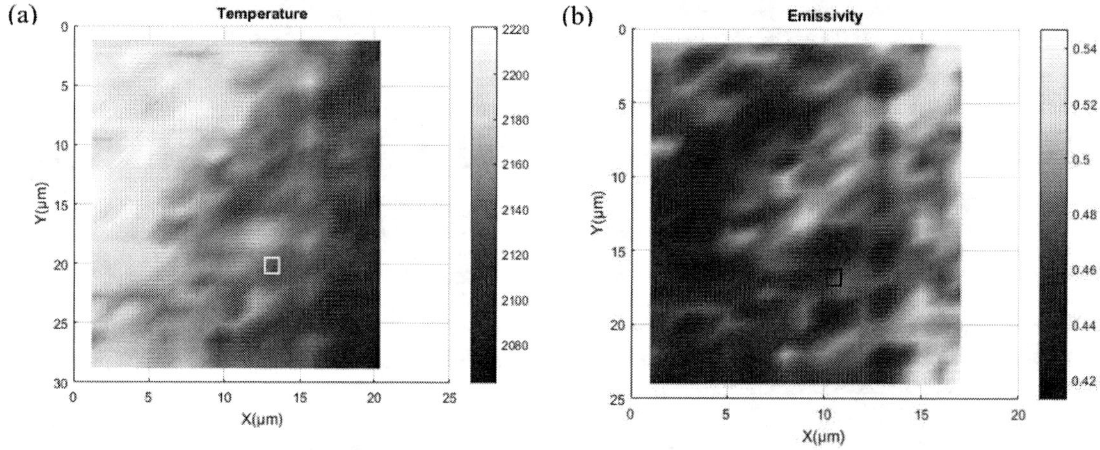

Figure 6. Temperature and emissivity of the laser heated area III: (a) temperature distribution; (b) emissivity distribution. Exposure time 1/4 s.

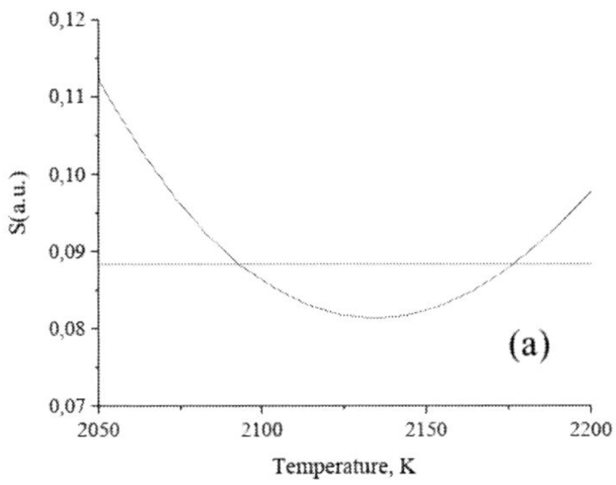

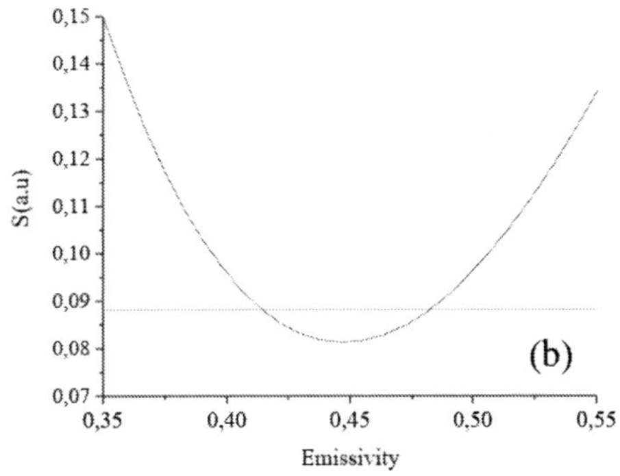

Figure 7. Behavior of the S as function (a) of the temperature, and (b) of the emissivity simulated at the point of the area III (figure 6 (a) and 6 (b)).

5. Conclusion

Analysis of the experimental data demonstrates that the application of the TAOFT allows us to measure emissivity as well as temperature with relative errors of 23% and 4% respectively. Relative statistical errors of the emissivity appeared to be nearly seven times higher as that for the temperature. We also found the statistical error of the emissivity increased with temperature. The possible reason of this phenomenon might be related to the strong dependence of the emissivity value on the wavelength at high temperature: $\varepsilon_0(\lambda)$. In future work, we plan to determined dependence of the emissivity on the wavelength from thermal radiation measurements in the wavelength range from 600 to 850 nm.

Acknowledgment This work was performed using the Unique Scientific Instrument "Laser heating at diamond anvil cell" [http://ckprf.ru/usu/507563/].

References

[1] Bassett W A 2001 *Review of Scientific Instruments* **72** 1270
[2] Magunov A N 2009 *Instrum. Exp. Tech.* **52** 451
[3] Zinin P V, Bykov A A, Machikhin A S, Troyan I A, Bulatov K M, Mantrova Y V, Batshev V I, Gaponov M I, Kutuza I B, Rashchenko S V, Prakapenka V B and Sharma S K 2019 *High Pressure Research* **39** 131
[4] Bulatov K M, Zinin P V, Mantrova Y V, Bykov A A, Gaponov M I, Machikhin A S, Troyan I A and Kutuza I B 2018 *Comptes rendus Geoscience* doi.org/10.1016/j.crte.2018.06.011
[5] Prakapenka V B, Kubo A, Kuznetsov A, Laskin A, Shkurikhin O, Dera P, Rivers M L and Sutton S R 2008 *High Press. Res.* **28** 225
[6] Heinz D L, Sweeney J S and Miller P 1991 *Review of Scientific Instruments* **62** 1568
[7] Ribaud G 1931 *Traite De Pyrometrie Optique.* (Paris: Ed. De La Revue)
[8] Machihin A S and Batshev V 2014 *Instrum. Exp. Tech.* **57** 736
[9] Plotnichenko V G, Pyrkov Y N and Svet D Y 1999 *High Temp.* **37** 444
[10] Zhang Z M, Tsai B K and Machin G 2010 *Radiometric Temperature Measurements I. Fundamentals* vol. 42 Elsevier
[11] Rusin S P 2018 *Thermophys. Aeromechanics* **25** 539
[12] Bulatov K M, Mantrova Y V, Bykov A A, Gaponov M I, Zinin P V, Machikhin A S, Troyan I A, Batshev V I and Kutuza I B 2017 *Computer Optics* **41** 864
[13] Taylor J R 1982. *An introduction to error analysis : the study of uncertainties in physical measurements.* Mill Valley (Calif.: University Science Books)
[14] Draper N R and Smith H 1966 *Applied regression analysis* (New York: Wiley)
[15] Latyev L N, Petrov V A, Chekhovskoi V Y and Shestakov E N 1974 *Radiative Propertiesof Solid, Handbook* (Moscow: Energiya Press)

The specifics of single-mode laser modules application in optical-electronic complexes based on light scattering

V Parshin, V Bliznyuk and O Evtikhieva

National Research University «Moscow Power Engineering Institute», Krasnokazarmenaya ulitsa 14, Moscow, Russia

E-mail: ParshinVasA@gmail.com

Abstract. The polarization state of edge-emitting single-mode laser modules radiation and their conditions of using in laser measuring systems of scattered radiation diagnosis are considered. For laser modules from two batches, both the polarization degree of collimated radiation and polarization degree spatial dependence in two orthogonal planes when there is no collimation are measured. It is established the angular dependence of non-collimated radiation polarization degree at different planes is various. We also considered the impact laser diodes degradation on these dependencies. It is shown that while exploitation of single-mode laser diodes in laser systems for scattered radiation characteristics researching it is needed to make control the polarization state of their radiation while service.

1. Introduction

Radiation scattering phenomena is widely used at such fields of science as ecology, medicine, astronomy, as well as for study mixture properties containing nanoparticles [1-4]. The scattered radiation properties analyzing need to do taking into account the properties of the specifically analyzed medium. For correct analysis of experimental results in the scattering study it is important to know the probing radiation properties of source itself: power and spectrum, radiation pattern and polarization state. It's known that the last property significant influences on scattered radiation structure – at scattering observation of source radiation p- and s- components all else being equal conditions the pattern of scattered radiation is different principally. This difference makes it possible to obtain more complete information about the characteristics of the test substance and often is taken into account when studying the properties of biological tissues, as well as the characteristics of the atmosphere, various solutions and crystals [5-12].

Depending on the task, LEDs and various types of lasers are used as radiation sources. The coherence and high polarization of the radiation of the latter allows them to occupy their niche in almost all scientific, technical and applied problems associated with the study of the properties of scattered radiation. Of particular interest are single-mode laser modules, which advantages most of all is associated with their compactness, energy efficiency and low cost relatively of gas, solid-state and fiber lasers. Besides, coherence of single-mode laser diodes (LD) radiation as their improvement starts to approach the level of solid-state and fiber lasers. It gives a reason to predict of wider laser diodes use in future at optical-electronic complexes for scattered radiation properties studying.

The free-space radiation of edge-emitting single-mode LD has a large divergence. This feature allows exploring an extensive space near LD. This feature allows you to explore the region of space near the LD within a solid angle of 0.015 sr. But at the moment polarization properties of such radiation are not studied in detail therefore in this work attention is paid to them. It is shown that the degree of linear polarization in different places of the LD radiation pattern may differ. It is also known that as the degradation occurs, the current-voltage and watt-ampere characteristics of the LD, as well as the spectral and polarization properties of its radiation are changed [13-15]. These parameters allow us to estimate the parameters and the possibility of using a separately selected model of a laser module in problems of studying the properties of scattered radiation.

2. Measurement technique and installation scheme

Usually the radiation at the output of laser modules is collimated and transformed into slightly divergent beam with a contrast close to 1. Of particular interest is the radiation into free space, the contrast of which has an angular dependence. In this case, the formation of various normalized indicatrices of scattering from different parts of the medium being probed is occurred. In this work, we studied the contrast of collimated radiation of single-mode LDs and angular dependence of the contrast of their radiation into free space at two planes. However, special attention was paid to the effect of LD degradation on its characteristics.

The polarization state of the radiation at the output of the LD cavity is determined by the anisotropy of the waveguide, the internal stresses of the heterostructure, and the reflectivity of the resonator mirrors for TE and TM modes [16]. The free-space radiation polarization state investigation of considered LD type is difficult because of large radiation divergence which is caused by diffraction on output LD mirror. As will be shown later the polarization degree in various areas of radiation pattern is different. Taking into account the astigmatism of the laser beam, it is analyzed along the wave front in the vertical *XZ* and horizontal *YZ* planes to simplify the measurements of the radiation pattern (figure 1).

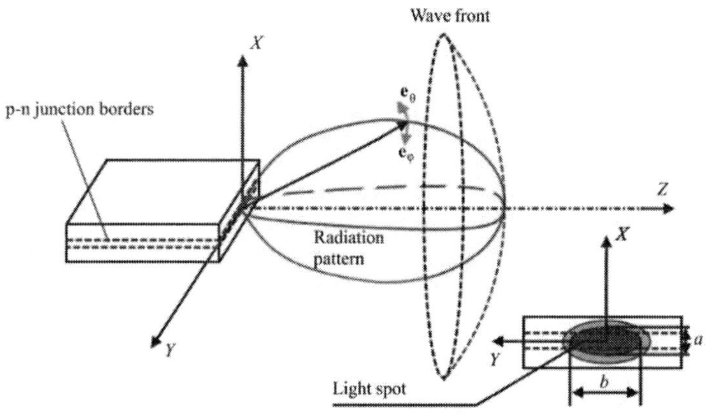

Figure 1. General scheme of single-mode LD and its radiation into free-space.

It is determined experimentally that at the scanning along wave front in horizontal the directions of vector $\boldsymbol{E}$ oscillations with maximum and minimum amplitude are oriented along $\boldsymbol{e}_\theta$ and $\boldsymbol{e}_\varphi$ basis vectors of spherical coordinate system, respectively. And when scanning in the vertical plane – on the contrary, along the unit vectors $\boldsymbol{e}_\varphi$ and $\boldsymbol{e}_\theta$, respectively. Thus there are only two analyzing polarizer positions when the maximum and minimum LD radiation transmission in exploring areas of radiation pattern is provided. Strictly speaking the radiation of explored LD types is not monochromatic and the wave front concept using is not so correct here. It means the scanning along surface the point of which are at equal distance from light spot, and normal radiation incidence on polarizer face and receiving area of the photoreceiver is provided. If the LD radiation were completely monochromatic, then the wave front would coincide with this surface in a vertical and horizontal plane.

The full width at half maximum (FWHM) of the spectrum of the edge-emitting single-mode LD is 1.0÷1.5 nm, which corresponds to the radiation coherence length many times smaller than 1 mm. It means a low degree of mutual coherence between two transversal components of radiation linear polarization at Fraunhofer field. Therefore we may neglect the energy of circular polarization component of radiation. It allows counting that the polarization degree is equal to contrast (1):

$$K = \frac{P_{max} - P_{min}}{P_{max} + P_{min}}, \quad (1)$$

where P_{max} and P_{min} is respectively maximum and minimum power value of radiation transmitting across linear polarizer.

Here it should be noted separately that the concept of "wave front" is usually applicable to monochromatic radiation. The edge-emitting single-mode LD radiation cannot be considered monochromatic, therefore, it means a surface that would coincide with the radiation wave front of the same laser, if it were completely monochromatic. When scanning the radiation pattern along this surface in the considered planes, both the normal incidence of radiation on the receiving optical system and the equidistant location of this system from the light spot are provided.

Measurements were made using a measuring complex, which scheme is shown on a figure 2.

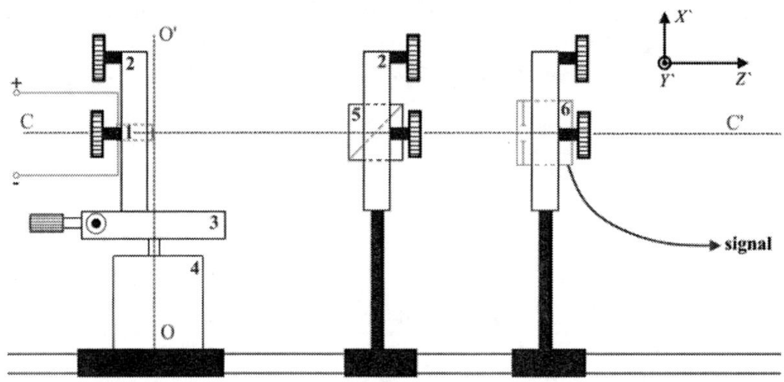

Figure 2. Measuring complex scheme:1 – explored laser module, 2 – optical holders with adjusting screws, 3 – two-axis linear positioner, 4 – stepper motor, 5 – Thompson-Glan polarization prism, 6 – receiver unit with diaphragm, 7 – optical rail, OO' – rotation axis of a stepper motor, CC' – laser beam axis.

The laser module (1) is placed in one of the holders (2), the adjustment screws of which allow you to adjust its inclination in a vertical and horizontal direction. It's necessary the holder also provide LDs rotation around beam axis CC' what gives possibility for measuring angular dependence of contrast at radiation pattern various planes. The position adjustment of LD in $Y'Z'$ plane is realizing using two-axis linear positioner (3), on which the holder with LD is mounted. Whole system is mounted on stepper motor (4) (57HM76-3004) which rotation rack axis is denoted as OO'. Using two-axis linear positioner is achieving that axis OO' led across center of light spot (figure 1). Thus when rotating LD its radiation always normal incidences both onto polarization prism (5) front surface and on receiving area of the photoreceiver (6) that provides radiation scanning along imaginary wave front. The receiving unit is an aperture diaphragm and a photodiode (FD-24K), the signal from which is read by the Arduino microcontroller board and displayed on the PC screen (is not shown in the figure 2). Whole system is rigidly fixed on optical rail (7).

Journal of Physics: Conference Series **1421** (2019) 012061 doi:10.1088/1742-6596/1421/1/012061

Insofar as stepper motor has a step discretisation of rotation angle what makes additional limits in measuring system possibilities, it's also possible using the goniometer turntable as device providing rotation around OO' axis at less step.

3. Measurement of LD radiation contrast and discussion

3.1 The angular dependence study of the radiation contrast

A radiation pattern into free space is characterised by various divergence angles in a vertical and horizontal plane. In a vertical plane the whole divergence angle on half maximum of intensity is 27° and on horizontal is about 8°. The advantage of such LD service mode is possibility of simultaneously probing the mixture large volume. But in this case it is necessary to take into account not only the angular intensity distribution but also the angular radiation contrast distribution.

On figures 3 and 4 the results of the angular dependence measurements of the radiation contrast in a vertical and horizontal plane at various operating time for single-mode laser module LD1 of KLM-E650-5-3 model are presented. The LD1 heterostructure has no quantum well and besides we removed the collimating optics for radiation into free-space analysis. Since the vertical plane is perpendicular to the border of the p-n-junction, and the horizontal plane is parallel to this border, the corresponding contrast distribution distributions for these planes are denoted by $K_{\perp}(\theta)$ for the vertical plane; $K_{\parallel}(\theta)$) – for the horizontal plane. Dots show experimental results and solid lines are the approximation by smooth functions.

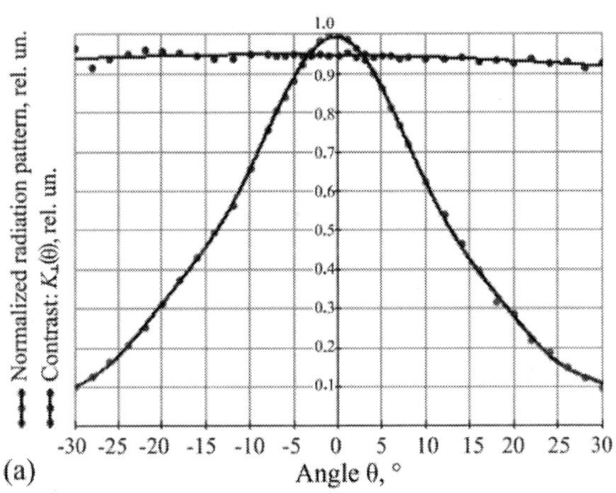

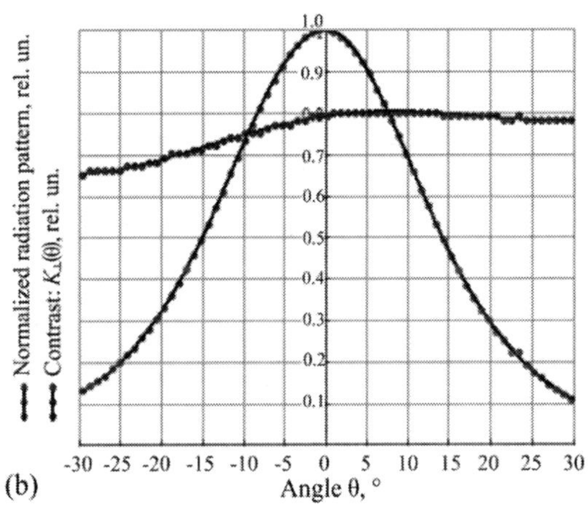

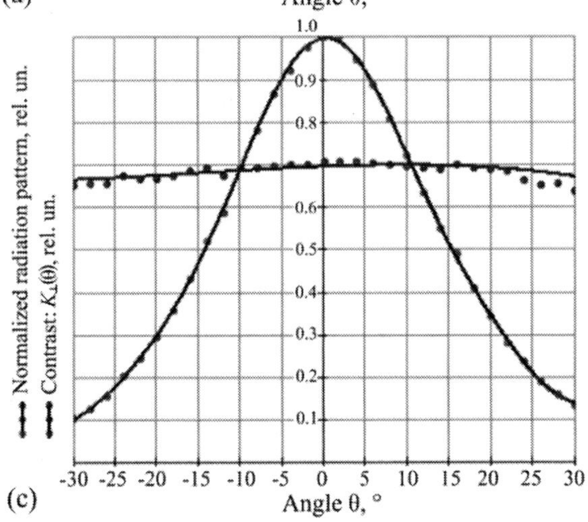

Figure 3. Normalized measured radiation pattern of LD1 in a vertical plane and the angular dependence of contrast $K_{\perp}(\theta)$: (a) at the beginning of operation, (b) after 50 hours of operation, (c) after 100 hours of operation.

Journal of Physics: Conference Series **1421** (2019) 012061 doi:10.1088/1742-6596/1421/1/012061

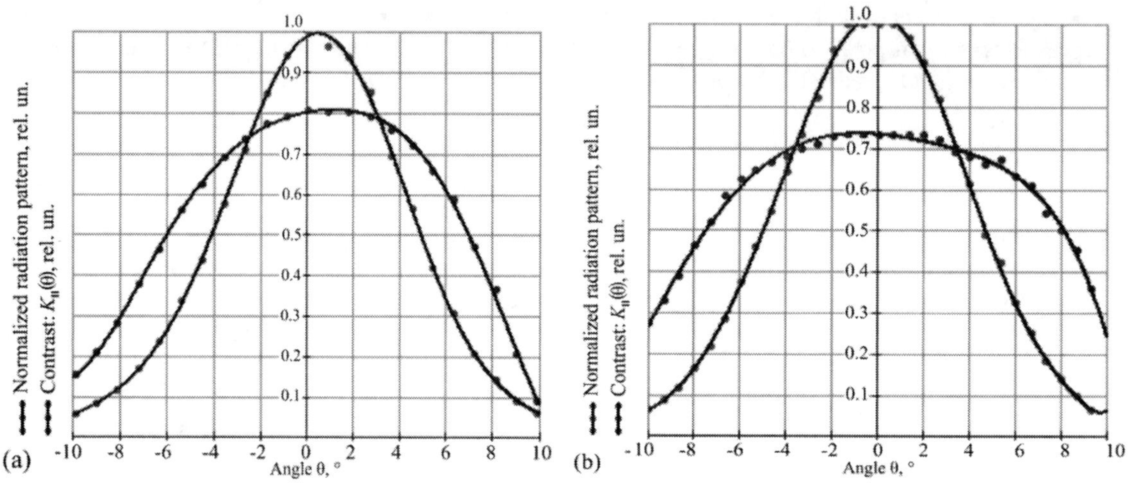

Figure 4. Normalized measured radiation pattern of LD1 in a horizontal plane and the angular dependence of contrast $K_{\|}(\theta)$: (a) after 50 hours of operation, (b) after 100 hours of operation.

Let us now consider the polarization properties the radiation of a single-mode laser module of KLM-D650-5-5 model (LD2) with a nanoheterostructure. Here, just as in the case of LD1, radiation into free space was investigated, without using collimating optics. The results of the measurements in a vertical plane at various operation times are presented in figure 5.

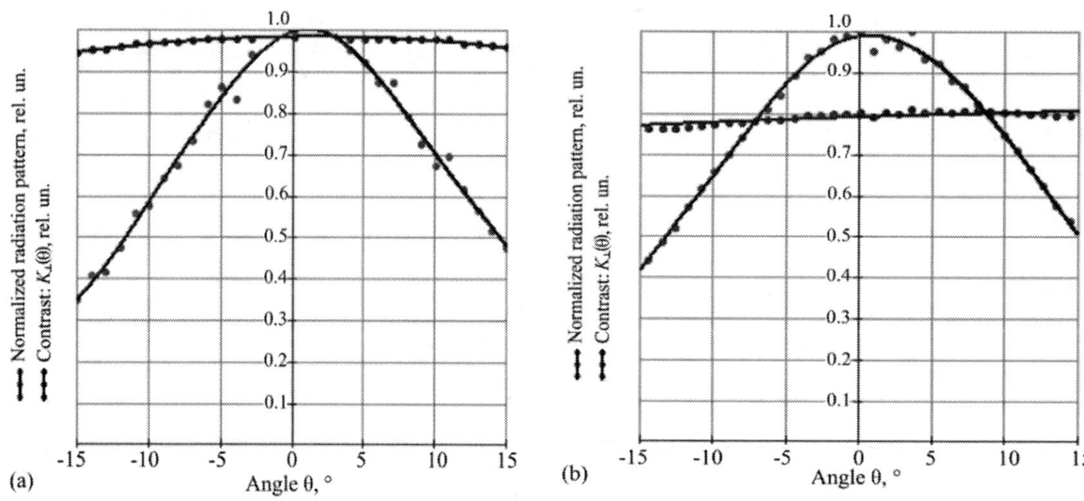

Figure 5. Normalized measured radiation pattern of LD2 in a vertical plane and the angular dependence of contrast $K_{\perp}(\theta)$: (a) at the beginning of operation, (b) after 50 hours of operation.

In figure 6 the measurements in a horizontal plane at 50 hours of operation time are also shown.

Analysis of the measurement results when scanning a beam along a wave front in a vertical plane showed that as the laser operating time increases, there is a general tendency for the contrast to decrease both on the axis and on the periphery of the beam. In the initial period of operation, when scanning the beam in the indicated plane for LD1 and LD2, one can note a value of contrast close to 1 in the entire scanning area. However, in the case of LD1, over time, the contrast at the periphery of the beam decreases somewhat faster than on the axis, and then, as shown in figure 3 (c), the graph of the contrast function $K_{\perp}(\theta)$ is smoothed again. At the same time, its value in the entire scanning area in just 100 hours of operating time decreases by 25%.

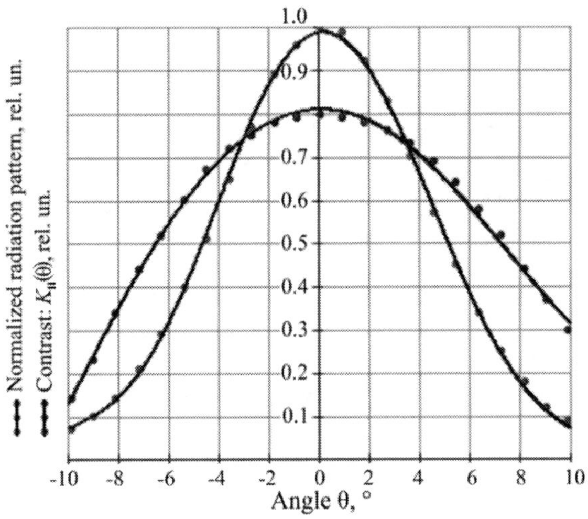

Figure 6. Normalized measured radiation pattern of LD2 in a horizontal plane and the angular dependence of contrast $K_{\|}(\theta)$ after 50 hours of operation.

In the horizontal plane *YZ* a pronounced angular dependence of the contrast $K_{\|}(\theta)$ is observed for both LDs. In figures 4 and 6, it is shown that its value rapidly decreases with distance from the beam axis at any operating time. From this it follows that the radiation pattern of orthogonal linearly polarized components in this plane is different – the normalized radiation pattern, due to the diffraction of the TM mode on the output mirror, has a flatter appearance in the peripheral region than the normalized radiation pattern, due to the diffraction of the TE mode. This is due primarily to the different amplitude-phase of the TE and TM modes energy distribution on the output mirror in the direction under consideration. In addition, here the cylindrical shape of the wave surface also affects [16]. Because of this, in the *YZ* plane there are differences between the diffraction conditions for the TE and TM field components at the laser output mirror, which also affects the shape of the angular dependence $K_{\|}(\theta)$.

Separately, we note the effect of increasing the operating time on the radiation polarization state of this LD type. Such a change is associated with the redistribution of energy between the TE and TM modes due to changes in the properties of the heterostructure. As degradation occurs, additional stresses arise in it due to the uneven temperature distribution associated with a non-uniform pump current density in a semiconductor. The reason for this is the formation of defects in the process of the heterostructure creation during its production. As a result, the energy of the TE mode goes into the TM mode, while the total radiation energy at the first stages of the degradation onset does not significantly decrease and the distribution of the total intensity on the LD light spot does not change. Therefore, the radiation pattern of the entire radiation almost completely retains its spatial-energy characteristics. However, the distribution of LD radiation contrast undergoes significant changes. Although we considered only two planes of scanning the radiation pattern, it is also worth assuming that the angular distribution of contrast changes in other areas of the radiation pattern as the operating time increases.

In laser-measuring complexes for diagnosing scattered radiation, the stability of the polarization characteristics of the probing radiation is often important. Since the state of polarization of LD radiation into free space changes rather quickly, their use in such systems must be accompanied by preliminary measurements of the contrast angular distribution within the beam region used for probing the medium. At the same time, such monitoring should be carried out before each series of measurements of the scattered radiation indicatrices.

3.2 Measurements of collimated radiation contrast

For laser modules of the same models, the total contrast K_Σ of collimated radiation was measured at the beginning of operation, and then after 50 and 100 hours of operating time. Measurements have shown that during the specified time the LD degradation does not affect K_Σ which remains close to 1. However, with longer operating times, from 500 hours, the contrast of even collimated radiation decreases with degradation along with changes in the spectrum and the onset of a fall in the total output power of the laser radiation. Thus, a decrease in the contrast of K_Σ signals a change in the other parameters and characteristics of the radiation. This means that with long-term use in measuring systems with LD with collimating optics, it is also worthwhile to periodically control the state of polarization of its radiation.

A detailed analysis of the distribution of contrast in the cross section of a collimated beam of the investigated LDs wasn't made in this paper. However, it is known that the use of optics changes the state of polarization of radiation transmitted through it. Therefore, it should be expected that the distribution of contrast in the beam cross section during collimation won't be equivalent to its angular dependence in the case of freely propagating radiation. To study the collimated beam in the scheme in figure 2, the receiving unit 6 should be replaced with a calibrated CCD array, which provides a one-time measurement of the intensity distribution for each polarization component of the laser beam. But the taking into account the effect of the nonuniform distribution of the collimated beam contrast on the scattering indicatrix is a rather laborious task. In the diagnosis of scattering, this is usually neglected, considering that at each point of the cross section the contrast is equal to its total value $K_\Sigma \approx 1$; the probe radiation of the LD in this case is considered as linearly polarized with the plane of polarization lying in the horizontal plane *YZ*.

Also, one of the main methods of ensuring a high degree of linear polarization of probe radiation is its transmission through a linear polarizer which the transmission axis is oriented along the direction of the electromagnetic field oscillations maximum amplitude. For other types of polarization the phase plates are additionally used. However, as the contrast of the probe radiation of the LD decreases as it passes through the polarizer, the energy loss will be more significant. In this connection, even when polarization optical elements are used into the measurement complexes for the scattering diagnostics a preliminary monitoring of the LD radiation polarization state is also required.

4. Conclusion

The choice of radiation sources in laser measuring systems for diagnosing the properties of various media is based on the requirements for such radiation characteristics as spectrum, power density and polarization state. It is especially important to ensure the stability of these characteristics over a long period of the source operation. In some cases, the properties of the scattered radiation substantially depend on the polarization of the probing radiation [1-12].

We have considered the specifics of the radiation polarization structure of single-mode edge-emitting LD from two batches for two cases: when the radiation is collimated by an optical system and when it propagates into free space. It is shown that the integral contrast K_Σ of their collimated radiation is almost equal to 1 retaining its value for about 500 hours of operation. At the same time, the radiation in free space is characterized by a complex angular distribution of contrast along the radiation pattern. In addition, for two different types of end LDs, the same angular dependence of contrast is observed in the vertical and horizontal planes. Separately, it should be noted that in the central region of the beam, within a solid angle of 0.004 sr, the contrast retains approximately the same value in these planes. Therefore, in the absence of collimation it is recommended that this area of the beam be used to probe the medium. This eliminates that part of the measurement error of the normalized scattering indicatrices, which is associated with an uneven contrast distribution of within the laser beam.

The influence of LD degradation on the contrast of its radiation which is sensitive to the heterostructure degradation, especially in the absence of collimation, is considered separately. Therefore, in the measuring systems of the scattering radiation diagnosis where edge-emitting single-mode LD is used as a source its operation should be accompanied by preliminary measurements and subsequent monitoring of the state of polarization of its probing radiation. Also, in order to slow down the rate of degradation it is necessary to optimize its operation which includes both control over the stability of the power supply and ensuring uniform cooling of the LD crystal. Fulfillment of these conditions ensures the most efficient use of the considered types of LD in the composition of such measuring systems.

Acknowledgments
This work was funded by Russian Foundation for Basic Research (project 19-07-00921A).

References

[1] Deirmendjian D 1969 *Electromagnetic scattering on spherical polydispersions* (NY: American Elsevier Publishing Company Inc.)
[2] H C van de Hulst 1981 *Light Scattering by Small Particles* (NY: Dover Publications Inc.)
[3] Levin A D, Shmytkova E A and Khlebtsov B N 2017 Multipolarization Dynamic Light Scattering of Nonspherical Nanoparticles in Solution *J. Phys. Chem. C* **121**(5) 3070-7
[4] Zaman M, Ang S, and Singh S 2016 Characterizing Nanoparticle Size by Dynamic Light Scattering *Journal of the Arkansas Academy of Science* **70** 255-9
[5] Tuchin V V 2016 Polarized light interaction with tissues *Journal of Biomedical Optics* **21**(7) 071114
[6] Gorodnichev E E 2018 Decay of correlations between cross-polarized electromagnetic waves in a two-dimensional random medium *Journal of the Optical Society of America A* **35**(4) 681-9
[7] Kunnen B, MacDonald S, Doronin A, Jacques S, Eccles M and Meglinski I 2015 Application of circularly polarized light for non invasive diagnosis of cancerous tissues and turbid tissue like scattering media *Journal of Biophotonics* **8**(4) 317-23
[8] Huang Q, Cheng M, Guo L, Li J, Yan X and Liu S 2017 Scattering of aerosol particles by a Hermite–Gaussian beam in marine atmosphere *Applied Optics* **56**(19) 5329-35
[9] Clays K and Persoons A 2000 Application of hyper-rayleigh scattering for the experimental determination of molecular structure in monomers and polymers *Journal of molecular structure* **521**(1-3) 303-13
[10] Němec P, Pásztor F, Brajer M and Němec I 2017 Spectrally- and polarization-resolved hyper-rayleigh scattering measurements with polarization-insensitive detection *Optics communications* (Elsevier Science Publishing Company, Inc.) **388** 21-8
[11] Vidal X, Fernandez-Corbaton I, Barbara A F and Molina-Terriza G 2015 Polarization properties of light scattered off solutions of chiral molecules in non-forward direction *Appl. Phys. Lett.* **107** 211107
[12] Nair R V 2016 Polarization-selective branching of stop gaps in three-dimensional photonic crystals *Phys. Rev. A* **93** 063850
[13] Bliznyuk V V, Berezovskaya N V, Parshin V A, Gadaev I S, Koval O I, Rzhanov A G and Solovyev G A 2015 Spectral, Space-Energy and Polarization Characteristics of Laser Diodes with Lasing Wavelengths of 530 nm *Bulletin of the Russian Academy of Sciences: Physics* (Allerton Press Inc, United States) **79**(12) 1453-7
[14] Bliznyuk V V, Berezovskaya N V, Brit M A, Koval O I, Parshin V A and Rzhanov A G 2017 Experimental study of the characteristics of high-power laser diode radiation in the above-threshold mode for degradation analysis *Physics of Wave Phenomena* **25**(2) 95-100
[15] Epperlein P W 2013 *Semiconductor Laser Engineering, Reliability and Diagnostics: A Practical Approach to High Power and Single Mode Devices* (John Wiley & Sons Limited)
[16] Casey H C and Panish M B 1978 *Heterostructure lasers: Part B* (New York: Academic Press)

Medical control of a pulse wave fiber optic sensors

M Gerashchenko, E Badeeva, T Murashkina and S Gerashchenko

Penza state university, Krasnaya str., 40, Penza, 440026, Russia

E-mail: gsm@pnzgu.ru, badeeva_elena@mail.ru, timurashkina@mail.ru, mpo@list.ru

Abstract. The modern level of development of systems of measurement, processing and data transmission promote creation of optical multiple parameter medical diagnostic systems. One of intensively developed directions is creation of fiber-optical diagnostic systems on the basis of fiber optic sensors. Fiber optic sensors are characterized by high sensitivity to mechanical oscillations of a surface of a body of the patient under power influence of a blood flow. This property creates premises for design miniature, steady against noises, safe sensors of micromovements. Their fixing on a body of the patient allows to reveal a number of the vital parameters characterizing normal and critical condition of the patient. Results of a research of fiber optic sensors of pressure on the basis of standard single-mode optical SMF-28 fiber are presented in article. The principle of operation of sensors consists in use of physical properties of the fiber - the impact which had in some pressure point on throughput characteristic. The purpose of the real work is the research of a possibility of use of fiber optic sensors of the tunnel, flexural and microflexural types on Bragg's grid for control of a pulse wave providing more exact determination of parameters of a pulse wave. The offered monitoring systems of a status of patients on the basis of fiber optic sensors in medicine are very perspective.

1. Introduction

Weak energy of pulse fluctuations creates difficulties for reproduction of vibration amplitude and conversion of their values to units of arterial blood pressure. An important point is transfers of an undistorted form and phase frequency characteristics of the pulse wave representing energy of distribution of a flow of blood up to an artery.

As practice shows, the existing pressure sensors expected measurement range from zero to 300 mm Hg. have low sensitivity of conversion, essential nonlinearity of conversion function, are subject to influence of x-ray and electromagnetic radiations and also big mass and dimensional characteristics. There is no possibility of positioning of the sensor on the surface of a body of the person, at the same time there is a need of leading of dimensional hoses that significantly complicates their construction. Besides, sensors are sources of electromagnetic impact on the patient.

Accuracy of representation of the haemo dynamic parameters based on processing of amplitude time characteristics depends on reliability of a form of reproduction of a pulse wave. First of all it is connected with assessment of systolic and diastolic components of pressure, the parameters characterizing elasticity and rigidity of arteries, productivity of work of heart [1-4].

Semiconductor sensors now in use possess the big errors leading to significant distortion of true values of the evaluated haemo dynamic parameters and also are not capable to perform trebuyemy functions when performing the magnetic resonance imaging (MRI), the nuclear magnetic resonance (NMR), positron emission tomography (PET).

From the analysis of the available arsenal of the sensor equipment functioning with the low pressure range at impact of the x-ray and electromagnetic influences which were widely adopted not

Published under licence by IOP Publishing Ltd

only in the equipment, but also in medicine, fiber optic sensors which are convenient in operation in wearable systems with the minimum inconveniences in the systems of continuous monitoring.

The developments of fiber optic sensors connected with determination of heart rate, breath, body temperature are known [5, 6]. Advanced developments of the similar sensors which are built in clothes elements, the bed or a medical chair which are not demanding additional actions for training of the patient for monitoring [7, 8].

Prospects of creation of monitoring systems of a cardiovascular system using pressure fiber optic sensors on Bragg's grid, tunnel, flexural and microflexural types are offered [9-24].

The purpose of the real work is the research of a possibility of use of fiber-optical sensors of parameters of a pulse wave on Bragg's lattices, the tunnel, flexural and microflexural types providing more exact determination of parameters of a pulse wave.

2. Methods of researches

Distinctive feature of the offered technique of assessment of haemo dynamic parameters is the combination of a liquid cuff to the fiber-optical sensors (FOS) for perception of micromovements of a surface of tissues of forearm, a wrist, an anklebone or a phalanx of a finger when passing the pulse wave formed by warm reductions.

The layout of sensors and main units of a system is shown in figure 1.

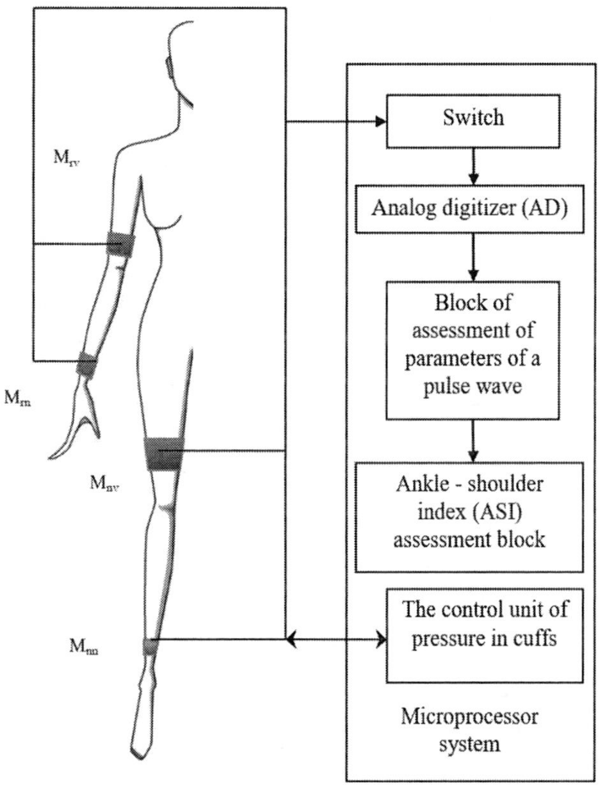

Figure 1. Layout scheme of sensors and structure of main units.

On the patient's body on the lower and top parts of hands and legs cuffs M_{nn}, M_{nv}, Mrn and M_{rv} with pressure sensors located on them are fixed. The provided layout of hydrocuffs allows to determine on the basis of the analysis of a form of a pulse wave of value of systolic, diastolic pressure on a hand and a leg, and taking into account time of its distribution and the measured distances between sensors as well the speed of a pulse wave υ. It is in addition possible to define the ankle-shoulder index (ASI) - one of key parameters of assessment of a condition of a cardiovascular system [27].

Journal of Physics: Conference Series **1421** (2019) 012062 doi:10.1088/1742-6596/1421/1/012062

For perception of micromovements of a surface of tissues of forearm, a wrist or a phalanx of a finger when passing the pulse wave created by warm reductions it is offered to use the pressure fiber optic sensors (FOSP) which will be located near arteries.

3. Results of researches

In tunnel FOSP modulation of power of a flow of optical radiation happens owing to rapprochement (under the influence of pressure) two environments in one of which the flow of optical radiation, extends to the distance commensurable with the wavelength (about 1 μm). One of possible constructions of tunnel PFOS is given in figure 2.

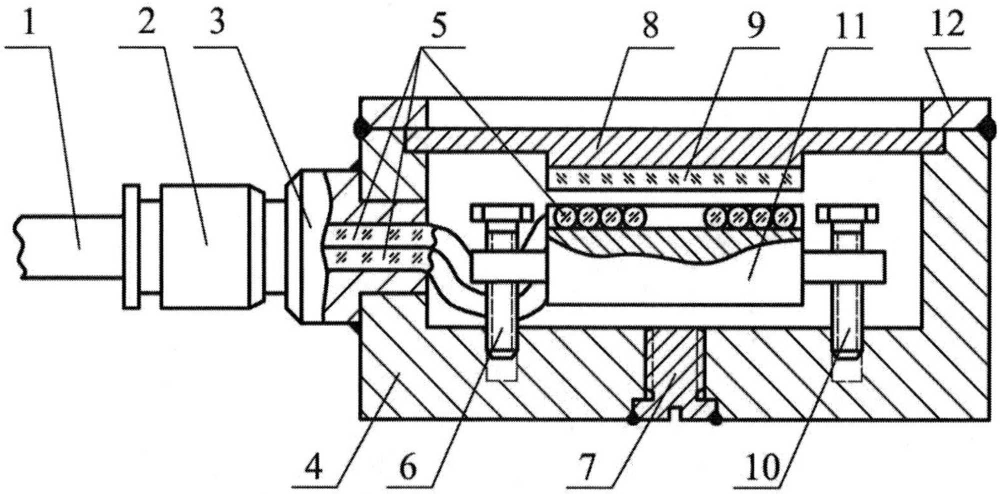

Figure 2. Construction of the tunnel PFOS. 1 – optical fiber cable; 2 – fixer; 3 – tip; 4 – case; 5 – optical fiber which part is contracted into a flat spiral; 6, 10 – adjustment screws; 7 – an opening with a stub for pressure control in a submembrane cavity; 8 – membrane; 9 – flat wafer; 11 – hard reference drive; 12 – cover.

The sensor consists of optical fiber (5) and the flat wafer (9) attached to a membrane (8) [14]. The reflecting cover is removed from a certain section of optical fiber (5). This section of optical fiber is contracted into a flat spiral and located under a plate (9). The last is made of material which has an index of refraction more, than the index of refraction of material of a core of optical fiber. At rapprochement of a plate (9) and optical fiber (5) under the influence of pressure a part of a flow of optical radiation «flows» from optical fiber (5) in a plate (9).

In the PFOS on the basis of curved optical fibers modulation of power of a flow of optical radiation is caused by change of bend radius of optical fiber which depends on applied pressure [14, 17].

The PFOS on the basis of microcurved optical fibers (in some works they are called «microflexural PFOS») are characterized by the fact that in them modulation of power of a flow of optical radiation happens at its distribution on optical fiber which owing to influence of pressure is exposed to several microbends. At the same time the radius of each of microbends is commensurable with a diameter of the used optical fiber [18, 19].

For increase in sensitivity of conversion of the fiber-optical transformer of microflexural type it is offered to use an intermediate element between fiber and the element perceiving pressure in the form of two metal rods (figure 3) [20].

Pressure from the perceiving element (1) is transmitted through metal cores (2) with a diameter of 1 mm, under optical fiber (5) there is an elastic rubber lining (3). When rendering influence there is a considerable falling of level of an optical signal that it is connected with violation of a condition of full internal reflection and effluence from fiber of extra aperture beams.

By way of illustration in figure 4 the fragment of a design of FOSP in which for creation of microbends of optical fiber (5) a set of cylindrical rollers (2) and (4) is used is shown.

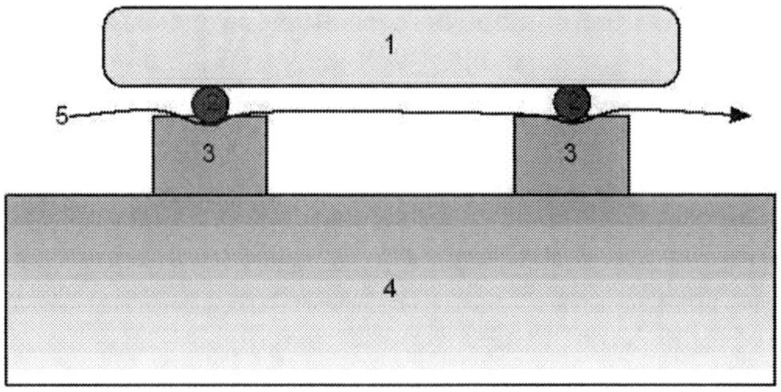

Figure 3. The scheme of the measuring transducer of the fiber optic sensor on the basis of microcurved optical fiber. 1 – detector; 2 – metal rods; 3 – elastic rubber lining; 4 – basis; 5 – microcurved optical fiber.

The last are squeezed by the deformer (1) made in the form of two gear plates. One of deformer is not mobile, another – moves at a deflection of the perceiving element (3). Optical fiber (5) is surrounded with elastic substance (6) which index of refraction is more, than the index of refraction of a cover of optical fiber.

Sensors are exposed to the deformation caused by concentration of tension which value is correlated with pressure attached to the bearing element.

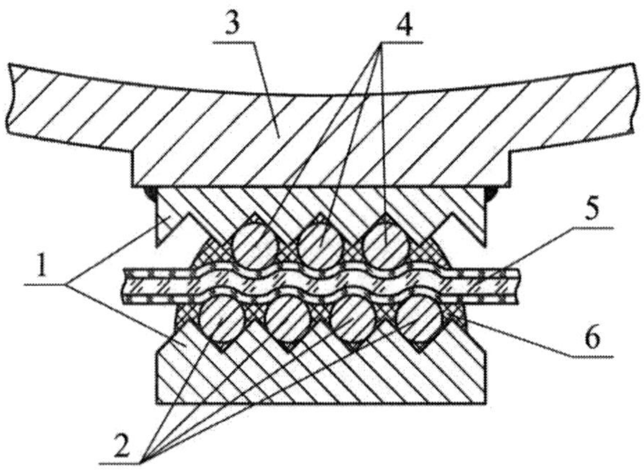

Figure 4. A sensor design fragment on the basis of microcurved optical fiber. 1 – deformer; 2, 4 – cylindrical rollers; 3 – a membrane with the rigid center; 5 – microcurved optical fiber; 6 – elastic substance.

The main lack of the above-considered fiber optic sensors of pressure is a low sensitivity of conversion of an optical signal in a measurement zone, insufficient for measurement of parameters of arterial blood pressure. For increase in sensitivity of conversion the quantity of rounds of optical fiber in a measurement zone increases. Fiber optic sensors on the basis of FBG (fiber Bragg grating) are considered as the most perspective [21]. They will usually transform the measured value to shift of the central wavelength of the spectral response. Their application in the equipment is limited a number of factors:

- according to conversion function their signal output bears information on two physical quantities at once: «deformation temperature», «a bend temperature» therefore additional measures are taken for

division of the different measured values, for example, is added to the scheme the second bregovsky grid used as reference (compensatory), respectively significantly increases system complexity [22];
- for secondary information processing interrogator by means of which bregovsky grids are polled are used. They are under construction on the basis of optical analyzers of a range or optical filters: the scanning filter of Fabri-Perot, the combined space filter on the basis of diffraction grating and the CCD-camera, or the optical narrowband filter with the spectral characteristic of a slope depending on wavelength [23]. Interrogators are difficult in production, also expensive devices are bulky.

These shortcomings are less essential to medicine. So, if blood-groove parameters are measured, then it is possible to consider that temperature has the constant value equal to the body temperature of the patient, and dimensions and cost of interrogators are comparable with dimensions of other diagnostic equipment.

Let's stop on classical FBG.

Profile of a range of reflection of FBG depending on length of FBG L and a detune δ [24]:

$$R = \frac{\sinh^2\left[kL\sqrt{1-(\delta/k)^2}\right]}{\cosh^2\left[kL\sqrt{1-(\delta/k)^2}\right]-(\delta/k)^2}, \tag{1}$$

k – coefficient of communication of the falling and reflected wave, δ/k – a relative detune who for a bregovsky lattice with the period Λ makes:

$$\frac{\delta}{k} = (2\pi n_{\text{EFF}}/\lambda) - \frac{\pi}{\lambda}, \tag{2}$$

λ - length of an optical wave.

For demonstration of changes of a range of a fiber bregovsky grid in work [24] the dependence of reflection coefficient R for its different lengths is presented to L. The effective index of index of refraction for a fundamental mode of n_{EFF} =1.5. The grid period Λ was selected so that the central wavelength of reflection of FBG was 1500 nanometers. During computational modeling values of length of a grid of L in the range (2 … 5) mm, kL=0.38 were set. On the basis of data retrieveds the conclusion is drawn that to reduction of length of a grid of Bragg there is a decrease in reflection coefficient and broadening of a profile. Taking into account these results for the developed sensor of a pulse wave the grid 5 mm long of Bragg was selected.

The simplified design of the measuring converter of pressure of a blood-groove on the basis of FBG differing in hypersensibility of transformation of parameters of a blood-groove to change of parameters of an output signal is given in figure 5.

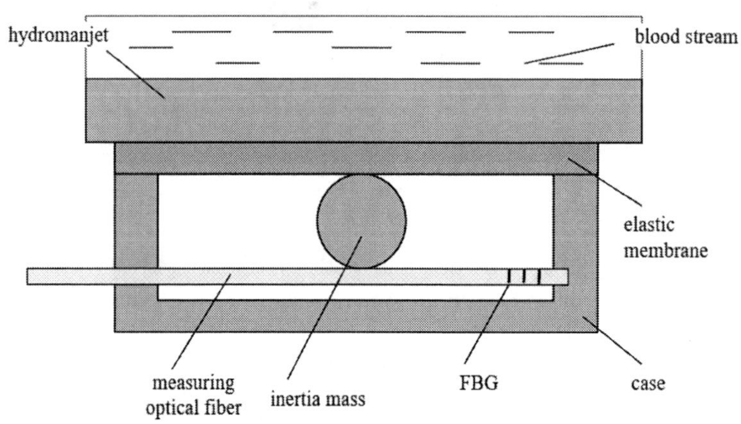

Figure 5. The simplified design of the measuring converter of a blood-groove The simplified design of the measuring converter of a blood-groove on the basis of FBG.

The fiber-optical measuring transducer (FOMT) contains optical fiber with Bragg's grid as a detector. The detector is executed in the form of an elastic membrane on which internal surface in the center by means of the gluing structure inertial weight is located (for example, in the form of a glass sphere or a parallelepiped). Over inertial weight optical fiber with FBG is fixed. Section of optical fiber with FBG is located between the inertial weight and the sensor housing in hover.

The power influence of a pulse wave changing with acceleration of an is transferred to the cup fixed on the patient's body along a y axis, further to the inertial mass of m. $F_y = ma_y$, last under the influence of force, will fluctuate concerning the neutral situation along a y axis (figure 6) [24].

Fluctuations of inertial weight will cause deformation of optical fiber. At the same time section of a grid of FBG stretches (is extended). Deformation of optical fiber leads to proportional change of the reflected light wavelength in Bragg's grid from power influence, that is bears information on parameters of a pulse wave.

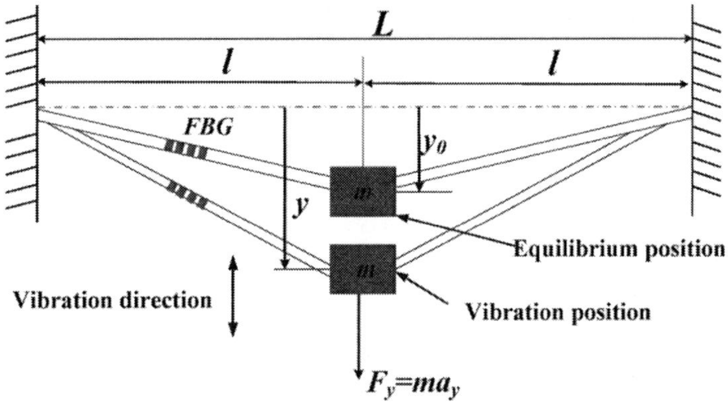

Figure 6. Geometrical model of vertical fluctuation of optical fiber with Bragg's lattice and inertial weight.

The flowchart of the developed monitoring system using fiber optic sensors is submitted in figure 7.

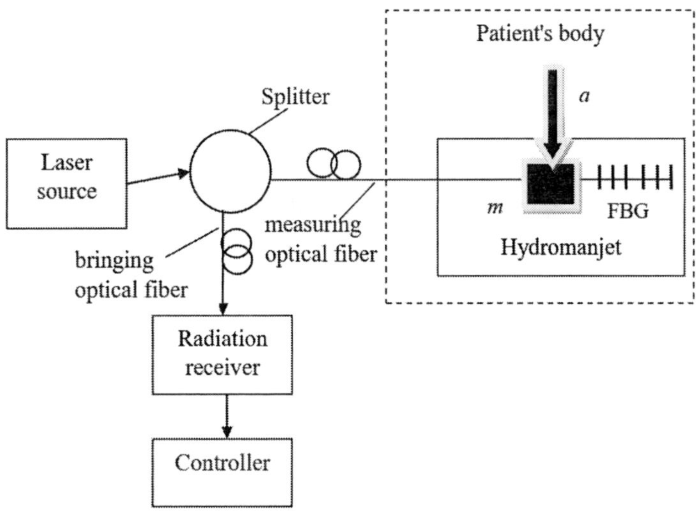

Figure 7. Flowchart of a fiber optical system of condition monitoring of a cardiovascular system of the patient.

For measurement of parameters of a blood flow the lasing source generates the continuous radiation which via the beam splitter arrives on sensing optical fiber and through it on the FOMT on an FBG.

The response from the FOMT through the sensing optical fiber, the first and second output of the beam splitter bringing optical fiber arrives on a detector in which its amplitude is registered. The acquired information comes to the controller in which pressure of a blood flow proportional to pressure created by a pulse wave in an artery under a cup (figure 8) is determined by the received values of amplitude.

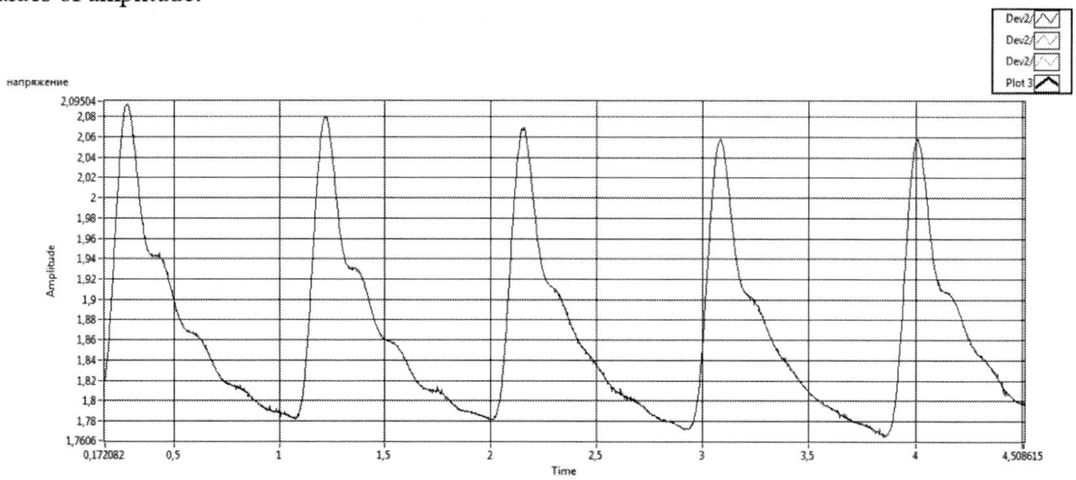

Figure 8. The diagram of a pulse wave received from the output of the sensor on the basis of the fiber-optical vibrotransformer.

4. Conclusion

Fiber optic sensors are characterized by high sensitivity to mechanical oscillations of a surface of a body of the patient under power influence of a blood flow. This property creates premises for design miniature, steady against noises, safe sensors of micromovements.

The offered monitoring systems of a status of patients on the basis of fiber optic sensors in medicine are very perspective.

Fixing of pulse waves on extremities of the patient allows to determine the haemo dynamic parameters characterizing a status of heart and vessels with a high accuracy.

Acknowledgments
Work is performed within the State task 8.11785.2018/11.12.

References

[1] Istomina T, Badeeva E and Murashkina T 2016 Development of sensors and systems for remote multidiagnostics and rehabilitation for medicine 5th *Mediterranean Conference on Embedded Computing MECO-2016* (Montenegro Bar, 12–16 June) 432–5

[2] Boutouyne P, Laurent S, Girerd X, Benetos A, Lacolley P, Abergel E. and Safar M 1995 Common carotid artery stiffness and patterns of left ventricular hypertrophy in hypertensive patients *Hypertension* **25 (4 Pt 1)** 651-9

[3] Gerashchenko M, Gerashchenko S I and Gerashchenko S M 2016 Development of the gidromanzhetny device for monitoring of the haemo dynamic Measurement parameters. *Monitoring. Management. Control* **4 (18)** 112–7

[4] London G 2003 Cardiovascular disease in chronic renal failure: pathophysiologic aspects *Semin Dial* **16 (2)** 85-94

[5] Taffoni F, Formica D, Saccomandi P, Di Pino G and Schena E 2013 Optical Fiber-Based MR-Compatible Sensors for Medical Applications: An Overview *Sensors* **13** 14105-20

[6] Ding Y, Chen N, Chen Z, Pang F, Zeng X and Wang T 2010 Dynamic temperature monitoring and control with fully distributed fiber Bragg grating sensor *Proceedings of SPIE, Optics in Health Care and Biomedical Optics IV* **7845**

[7] Presti D, Massaroni C, Formica D, Saccomandi P, Giurazza F, Caponero M and Schena E 2017

Smart textile based on 12 fiber bragg gratings array for vital signs monitoring *IEEE Sensors Journal* **17** 6037 – 43

[8] Yui K, Hiroaki I, Katsuragawa Y and Ishizawa H 2015 Non-invasive blood pressure measurement by pulse wave analysis using FBG sensor *IEEE International Instrumentation and Measurement Technology Conference (I2MTC) Proceedings*, Pisa, Italy, **15291530** 511-5

[9] Mishra V, Singh N, Tiwari U and Kapur P 2011 Fiber grating sensors in medicine: current and emerging applications *Sensors and Actuators A:Physical* **167(2)** 279-290

[10] Ceyssens F, Driesen M, Wouters K and Puers R 2007 A low-cost and highly integrated fiber optical pressure sensor system *Sensors and Actuators A:Physical* **145-146** 81-6

[11] Chavko M, Koller W, Prusaczyk W and McCarron R 2007 Measurement of blast wave by a miniature fiber optic pressure transducer in the rat brain *Journal of Neuroscience Methods* **159(2)** 277-281

[12] Grattan K and Sun T 2000 Fiber optic sensor technology: an overview *Sensors and actuators a : Physical* **82(1-3)** 40-61

[13] Hamel C and Pinet E 2006 Temperature and pressure fiber-optic sensors applied to minimally invasive diagnostics and therapies *Progress in Biomedical Optics and Imaging* **7(6)**

[14] Redko V and Kondratov V 2009 Classification of Fiber-Optic Pressure Sensors with Amplitude Modulation of Optical Signal *Sensors & Transducers* **100(1)** 146-169

[15] Jeong Y, Baek S and Lee B 2002 Macrobend sensor via the use of hollow-core splice fiber : theory and experiments *Optical Engineering* **41(8)** 1815-20

[16] Istomina T, Chulkov V, Stepanov M, Istomin V and Murashkina T 2016 Improvement of Information Support of the Complex for Remote Multidiagnostics and Rehabilitation *International Journal of Applied Engineering Research* **11(20)** 10326-31

[17] Brostilov S and Murashkina T 2010 Pressure fiber optic sensor on the basis of a tunnel effect. Description of construction. Operation principle *News of higher educational institutions. Volga region region. Technical science* **4** 62-6

[18] Wlodarczyk M, Krage M and Vickers D 1990 Fiber optic pressure sensor using pressure sensitive fiber different from input and output fibers *Patent* 4918305 US International Patent Classification (IPC) H01 J 5/16 pp 1-6

[19] Cuadrado-Laborde C 2013 Current trends in short and long-period fiber gratings IntechOpen

[20] Patrick H, Williams G, Kersey A, Pedrazzani J and Vengsarkar A 1996 Hybrid fiber Bragg grating/long period fiber grating sensor for strain and/temperature discrimination *IEEE Photonics Technology Letters* **8(9)** 1223–25

[21] Kashyap R 2009 *Fiber Bragg Gratings* (London : Academic Press) p 632

[22] Srimannarayana K, Shankar M, Prasad R, and Mohan K 2008 Fiber Bragg grating and long period grating sensor for simultaneous measurements and discrimination of strain and temperature effects *Optica Applicata* **XXXVIII 3** 601-9

[23] Morozov O G, Nureev I I, Sakhabutdinov A Zh, Kuznetsov A A and Faskhutdinov L M 2015 Fiber optic sensors of wear and temperature of the rubbing surfaces *Foton-ekspress* **6 (126)** 210-1

[24] Tianliang L,Yuegang T and Zude Z 2016 A Fiber bragg grating sensing-based micro-vibration sensor and its application *Sensors* **16 (4)** 547

Journal of Physics: Conference Series **1421** (2019) 012063 doi:10.1088/1742-6596/1421/1/012063

Optical methods for droplets evaporation on graphene nanocomposite surfaces

D D Babenko and A S Dmitriev

National Research University «Moscow Power Engineering Institute»
Russia, 111250, Moscow, Krasnokazarmennay 14

E-mail: asdmitriev@mail.ru

Abstract. We study the evaporation of water droplets on the surface of graphene nanocomposites of different compositions. Methods for manufacturing samples of nanocomposites, optical methods for studying contact angles and evaporation dynamics are described. New experimental data are presented on the behavior of contact angles and evaporation rates at various temperatures for samples of nanocomposites in both the constant contact line mode and the constant contact angle mode. The features of evaporation for nanocomposites with inclusions of boron nitride and aluminum nitride nanoparticles inside graphene samples are noted.

1. Introduction

The paper presents research in the field of thermal physics of surfaces of functional thermal interface materials (TIM) based on graphene nanocomposites [1-4]. Contact angles and hysteresis angles were measured for various nanocomposite materials, including nanocomposites based on graphene flakes, boron oxide nanoparticles, nanodiamonds, aluminum nitride, as well as hybrid nanocomposites with the inclusion of polymers and microscopic metal microspheres. A study was carried out by optical methods, including optical observations of the nonstationary dynamics of the evaporation of droplets of various liquids on the surfaces of graphene nanocomposites. The processing of optical images made it possible to reveal the features of droplet evaporation on various surfaces in the modes of constant contact angle and constant contact line. These measurements are important for optimizing drip evaporative cooling processes in electronic and optoelectronic devices. New experimental data on the study and analysis of hydrodynamic and thermophysical properties of various thermal interface materials interacting with droplet flows, including the Leidenfrost effect, have also been obtained. One of the main objectives of the work is the study of the properties of thermal interface materials for the intensification of heat exchange and increasing the efficiency of cooling systems in microelectronics and high-current electronics. The physicochemical and thermohydrodynamic properties of such surfaces are important for a more efficient heat release in cooling systems using the droplet cooling method. Droplet evaporation is one of the most important phenomena and has various applications in spin-coating processes, cooling electronics, inkjet printing, Leidenfrost effect and other [5, 6]. For these reasons, the study of the evaporation of droplets on nanocomposite substrates at different temperatures is an important and promising task. Since graphene-based nanocomposites are being used increasingly actively in recent years, there are very few publications on this subject, our research seems to be relevant.

2. Experimental setup, materials and methods

Generation, observation, measurements of droplet evaporation time were carried out on an Kruss EasyDrop goniometer (the schematic of the current experimental setup is shown in figure 1). Kruss EasyDrop measuring system was used with special heater and temperature control system. This goniometer is equipped with a system for automatically dispensing drops with high accuracy. The volume of drops in all experiments was 5 μl. To change the temperature of the samples, the goniometer was equipped with a heater mounted on a standard holder. Heater power was controlled by changing the voltage on it. The surface temperature was measured with an Extech EA10 thermometer. The thermocouples of the instrument were located directly on the surface of the samples. The production of tablets was carried out with the help of an automatic machine for hot pressing the Buehler SimpliMet 1000 (figure 2). Samples of three types of graphene nanocomposites were manufactured: sample 1 (S1): graphene nanoflakes pressed into a tablet (p = 400 bar), sample 2 (S2): graphene nanoflakes (96%) + BN nanoparticles (4%) pressed into a tablet (p = 400 bar) and sample 3 (S3): graphene nanoflakes (96%) + AlN nanoparticles (4%) pressed into a tablet (p = 400 bar). In detail, methods for producing tablets have been described in [8-10]. Electron microscopy of the surface of nanocomposites is shown in figure 3 and figure 4: in figure 3 - graphene nanoflakes sample, in figure 4 - graphene nanoflakes (96%) + BN nanoparticles (4%) sample.

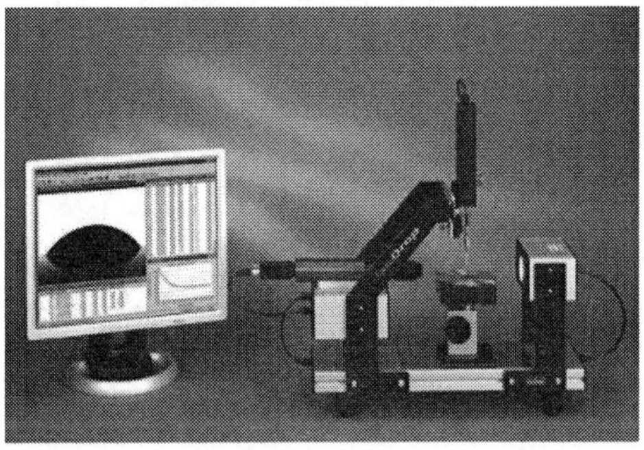

Figure 1. Kruss EasyDrop goniometer.

Figure 2. Automatic machine for hot pressing Buehler SimpliMet 1000.

Optical images of samples and optical microscopy are shown in figures 5-8. It is clearly seen that the surface of the nanocomposites is rather rough, therefore, it can be expected that there is a significant pinning of the contact line on such surfaces. This fact is indeed confirmed by experiments, the results of which are presented below.

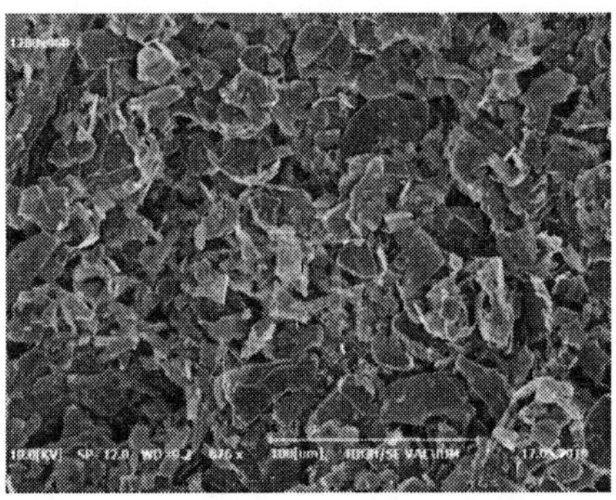

Figure 3. Graphene nanoflakes sample.

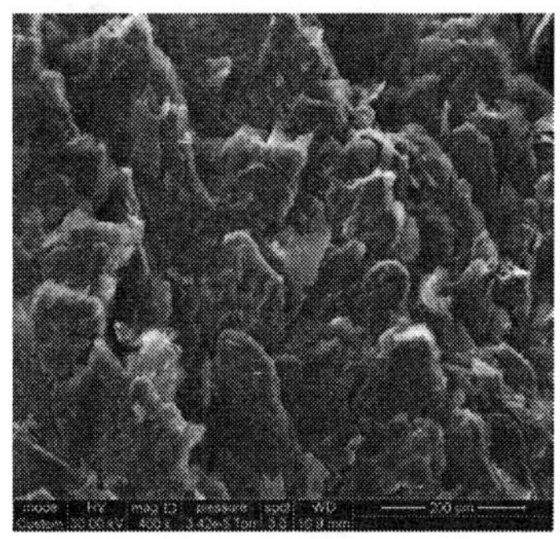

Figure 4. Graphene nanoflakes (96%) + BN nanoparticles (4%) sample.

Figure 5. Sample 1 (S1): GNF pressed into a tablet (p = 400 bar).

Figure 6. Optical microscopy of the sample surface GNF (S1).

Figure 7. Sample 2 (S2): graphene nanoflakes (96%)+BN nanoparticles (4%) pressed into a tablet (p = 400 bar).

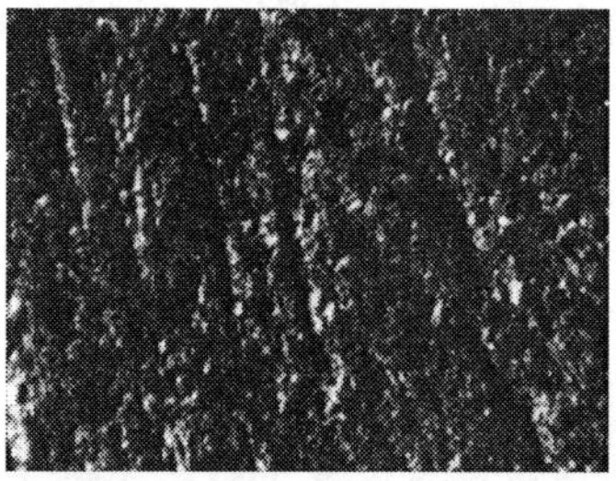

Figure 8. Optical microscopy of the sample surface (S2).

3. Drop Evaporation Experiments

The influence of surface properties on sessile drop evaporation has only recently been taken into account. The thermal conductivity of the substrate appears to be a key parameter in the evaporation process with respect to the internal flow and dynamics of evaporation. Previously proposed two modes of evaporation: constant contact angle (CCA) and constant contact radius (CCR) modes [5-7]. In the CCA mode, the contact area of the droplet on the surface decreases, whereas in the CCR mode, the contact angle shrinks. The evaporation rates of various samples of graphene nanocomposites were investigated. For this purpose, droplets of distilled water with a volume of 5 mkl were placed on nanocomposite surfaces from the top a dispenser. The evaporation rates for various nanocomposites are shown in figure 9. It is clearly seen that, qualitatively, the rate of evaporation from the surface of various nanocomposites is about the same.

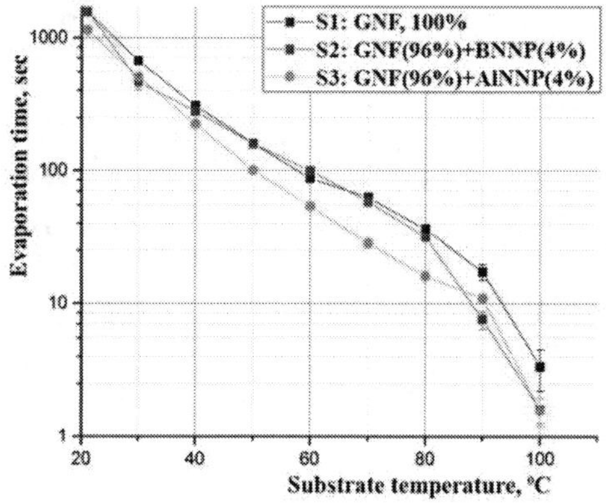

Figure 9. Evaporation rates of various samples of graphene nanocomposites.

At the same time, a higher evaporation rate is observed for graphene nanoflakes (96%) + AlN nanoparticles (4%) nanocomposite. Apparently this is due to smaller contact wetting angles on a similar surface. This leads to the fact that near the contact line the mass evaporation flow is higher for smaller contact angles. At the beginning of the process, as can be seen from the figure 10 and figure 11, evaporation proceeds at an almost constant contact angle with an increase in the contact line. Upon reaching a certain critical value, the contact line remains constant, and the wetting angle decreases.

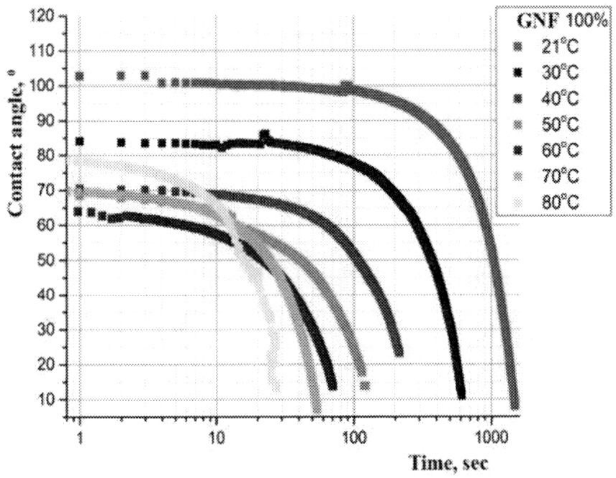

Figure 10. Contact angle for the droplets evaporation on the surface GNF sample (S1) at different temperatures.

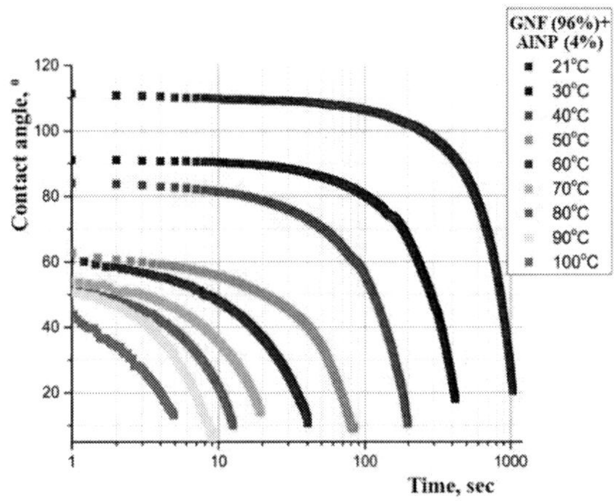

Figure 11. Contact angle for the droplets evaporation on the nanocomposite surface GNF (96%) + AlNNP (4%) at different temperatures.

It should be noted that the experimental data for a graphene sample (figure 10) at a temperatures 70 °C and 80 °C clearly fall out of the overall picture of the dynamics of evaporation. It is possible that at such temperatures, bubbles of dissolved gas (air, for example) appear inside the drop, which affect the rate of evaporation. At the same time, this effect is not observed for nanocomposites GNF (96%) + AlNNP (4%) and GNF (96%) + BNNP (4%) the dependences of the contact angles on temperature have a clear character - as the temperature rises, the contact angles decrease.

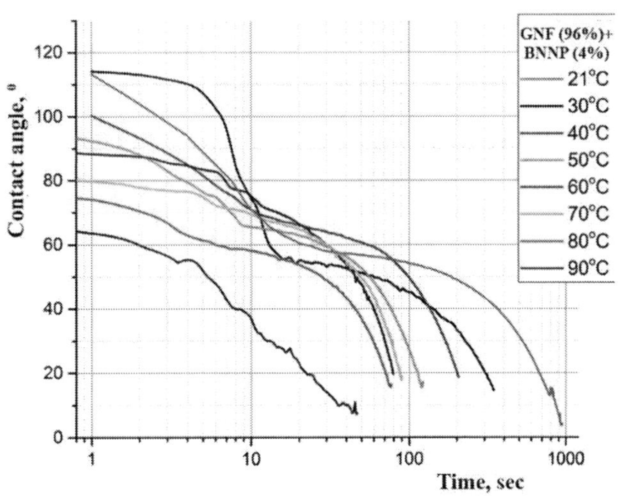

Figure 12. Contact angle for the droplets evaporation on the surface nanocomposite GNF (96%) + BNNP (4%) at different temperatures.

Journal of Physics: Conference Series **1421** (2019) 012063 doi:10.1088/1742-6596/1421/1/012063

However, at a temperature 21 °C and 30 °C for nanocomposite GNF (96%) + BNNP (4%), an anomalous behavior of the contact angle is observed as a function of time. This effect has a high repeatability and we could not find an explanation for this phenomenon.

The study of the contact angle dynamics for a nanocomposite GNF (96%) + BNNP (4%) (T = 50 °C) is shown in the figure 13. It can be seen that in the first period of droplet evaporation, the contact line moves. At the next stages of evaporation, the contact line undergoes pinning and the contact angle decreases.

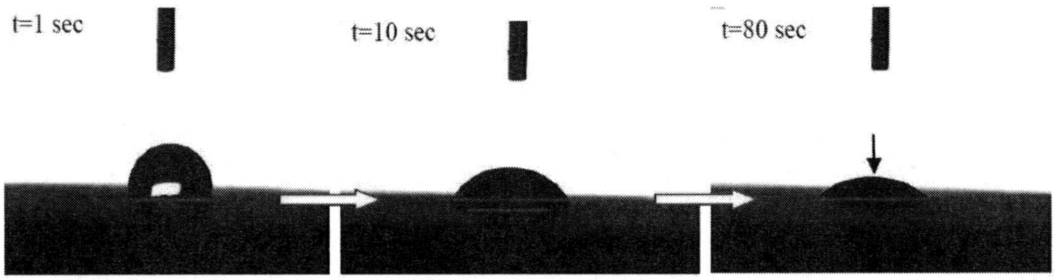

Figure 13. Contact angle when droplets evaporate on the surface of nanocomposite GNF (96%) + BNNP (4%) (T = 50 °C).

We also experimentally investigated the mass of vapor that is generated when the droplets evaporate on different surfaces of nanocomposites (figures 14 and 15). It is not difficult to see that the character of steam generation is monotonic with increasing temperature. It can also be noted that the evaporation rate is weakly dependent on the type of nanocomposite. In our opinion, the reason is the weak difference in the thermal properties of the nanocomposite surfaces GNF and GNF (96%) + BNNP (4%).

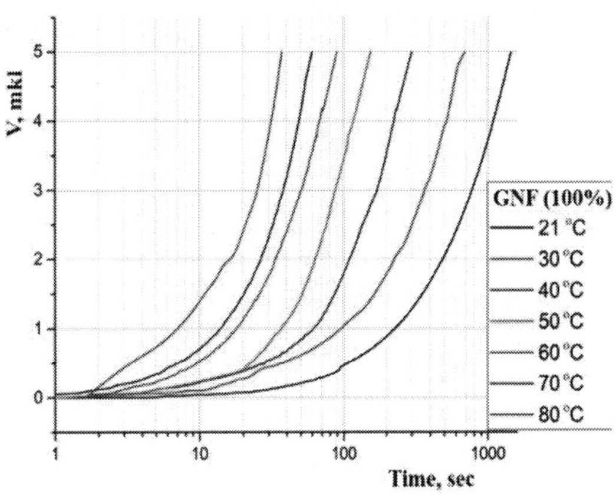

Figure 14. The mass volume of steam during evaporation of a droplet on the surface GNF.

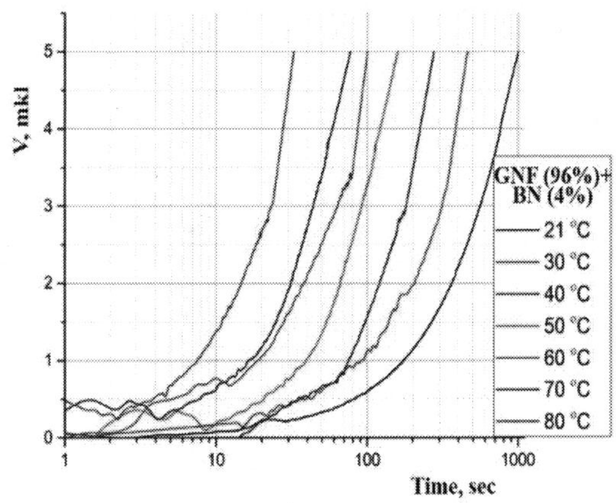

Figure 15. The mass volume of steam during evaporation of a droplet on the surface GNF (96%) + BNNP (4%).

4. Results and discussion

The influence of the surface properties of various substrates on the evaporation of sessile drops is studied. Three types of nanocomposite substrates were used to study the evaporation of distilled water droplets: sample 1 (S1): graphene nanoflakes pressed into a tablet (p = 400 bar), sample 2 (S2): graphene nanoflakes (96%) + BN nanoparticles (4%) pressed into a tablet (p = 400 bar) and sample 3 (S3): graphene nanoflakes (96%) + AlN nanoparticles (4%) pressed into a tablet (p = 400 bar). By varying the roughness and surface energy of the substrate surfaces using various dynamics of the triple line and a large range of wettability (contact angle from 10° to 110°) are obtained. The influence of these parameters on the dynamics of evaporation is studied in order to highlight the separated and coupled roles of the triple line and the wettability of the drops on the sessile drop evaporation process. Strong dependences of the contact angles and evaporation rates were found for various temperatures for all types of nanocomposites. These data clearly show the importance and the effect of surface properties on sessile drop evaporation. They can be used for new functional materials, including thermal interface materials, as well as materials for thermal stabilization and thermal management in electronics, optoelectronics, and power engineering [9, 10].

Acknowledgments
This work was supported by the Russian Science Foundation (project No. 17-19-01757).

References

[1] Goyal V and Balandin A 2012 Thermal properties of the hybrid graphene–metal nano-microcomposites: applications in thermal interface materials *Appl. Phys. Lett.* **100** 073113–5
[2] Khan M and Balandin A 2012 Graphene-multilayer graphene nanocomposites as highly efficient thermal interface materials *Nano Lett.* **12** (2) 861–7
[3] Semenov S, Trybala A, Rubio R, Kovalchuk N, Starov V and Velarde M 2014 Simultaneous spreading and evaporation: Recent developments *Advances in Colloid and Interface Science* **206** 382–98
[4] Günay A, Sett S, Oh J and Miljkovic N 2017 Steady method for the analysis of evaporation dynamics *Langmuir* **33** 12007−15
[5] Pan Z, Weibel J and Garimella S 2014 Influence of surface wettability on transport mechanisms governing water droplet evaporation *Langmuir* **30** (32) 9726−30
[6] Khan S, Atieh M and Koç M 2018 Micro-nano scale surface coating for nucleate boiling heat transfer: a critical review *Energies* **11** 3189
[7] Brutin D and Starov V 2018 Recent advances in droplet wetting and evaporation *Chem Soc Rev.* **47**(2) 558-85
[8] Alekseev S, Dmitriev A S, Dmitriev A A, Makarov P and Mikhailova I 2017 Functional energy nanocomposites surfaces based on mesoscopic microspheres, polymers and graphene flakes *IOP Conf. Series: Journal of Physics: Conf. Series* **891** 012361
[9] Dmitriev A S, Dmitriev A A, Makarov P and Mikhailova I 2018 New nanocomposite surfaces and thermal interface materials based on mesoscopic microspheres, polymers and graphene flakes *AIP Conference Proceedings* **1957** 020003
[10] Babenko D, Dmitriev A S, Mikhailov V and Mikhailova I 2018 Physical features of Leidenfrost effect on the surface of a grapheme nanocomposite for the problems of thermal and nuclear energy *IOP Conf. Series: Materials Science and Engineering* **447** 012044

Journal of Physics: Conference Series **1421** (2019) 012064 doi:10.1088/1742-6596/1421/1/012064

The combination of PIV and heat flux measurement in study of flow and heat transfer near a circular finned cylinder

S Z Sapozhnikov, V Yu Mityakov, V V Seroshtanov and A A Gusakov

Peter the Great St.Petersburg Polytechnic University (SPbPU), 195251, Russia, St.Petersburg, Polytechnicheskaya, 29

E-mail: vladvik1992@gmail.com

Abstract. The flow and heat transfer of the finned cylinder are investigated by the methods of PIV, gradient heat flux measurement and thermometry. On the hollow model of a circular cylinder heated by saturated water vapor at atmospheric pressure, five annular fins of 20 mm high and of 10 mm thick are mounted. For the first model, the "acting" fin was hollow, for the second, it was solid and made of titanium alloy VT22. The remaining four fins simulated flow blockage. To visualize the flow in the intercostal space, the fin simulators were made of Perspex. Battery gradient heat flux sensors (GHFSs) with volt-watt sensitivity of 10 mV/W, were installed on the "acting" fin at different distances from the surface of the carrier cylinder. The cylinder rotated around its axis at an angle of $\varphi = 0...180°$, which made it possible to obtain the distribution of heat flux over the surface of the fin. By combining PIV diagnostics and gradient heat flux measurement, it was possible to obtain a complex 3D structure of the flow in the intercostal space and the distribution of heat flux on the surface of the fin. The values of the local heat transfer coefficients (HTCs) were also obtained. The dependence of HTC on flow mode and intercostal space is revealed. Comparison of the obtained characteristics for models with hollow and solid fins allowed us to determine in experiment fin efficiency for different intercostal spaces and free-stream velocities. The combination of all three technologies opens up new possibilities in study of flow and convective heat transfer.

1. Introduction

A common elements of most recuperative heat exchangers are transversely streamlined bundles of round tubes. To increase the heat transfer coefficient (HTC) finned tubes with circular fins are often used. Structure of the flow near the finned tubes directly affects heat transfer between the fins and streaming fluid. In real heat exchangers, influence of the height of the fins, the non-cross wrap, blockage, etc. on the heat transfer is important.

The study of flow and heat transfer in transversely finned tubes is a multifactorial problem. So, in the works [1, 2] 3D flow at the base of the fins is described. Due to the difference in velocities near the fin and in the core of the flow, a pressure drop occurs along the cylinder during flow past the pipe. Due to the differential, there is a movement of fluid from the centre of the intercostal space to the root of the fin. The authors of work [1] note that with a low fin height ($0.2...0.35 \times h/d$, where h is the fin height, m, d is diameter of the carrier cylinder, m) the influence of the cylinder as a barrier to the main flow increases [1]. The flow visualization in this work was carried out with the help of a soot-paraffin suspension. In studies on this topic, the method of recalculating of the velocity field from measurements of static pressure field [3] and hot-wire method are widely used [4].However, both of these methods are invasive, and to estimate the distortion of the flow pattern in the intercostal gap with its small value ($u/\delta = 0.4....1$, where u is the intercostal gap, m, δ is the fin thickness, m) seems complicated. With the expansion of the scope of application of lasers, it became possible to apply the method of PIV (Particle Image Velocimetry), which is based on measuring of particle velocity. Since PIV diagnostics does not imply introduction of measuring probes into the flow, the prospects for the development of this scientific direction are obvious.

In the study of heat transfer, the main task is to determine the HTC. Works devoted to the measurement of this parameter are based on the use of various methods for different forms of the fins and flow regimes. Thus, the method based on the analogy between the processes of heat and mass transfer [3], allows you to determine the coefficients of mass transfer by photometric method. There are paintings in which the colour intensity of different parts of the fin is proportional to the value of the local HTC. However, this method was not widely used due to the complexity of processing the data and the methodical lack of the rigor.

More common methods are based on the use of sensors – heat flux meters. In [1], measurements were performed on fully heated models of finned tubes. The sensors were installed on the edge and on the surface of the carrier cylinder, and their position relative to the direction of the oncoming flow was varied by turning the cylinder around the axis. Studies were conducted on single pipes [2] and pipe matrix [3].

Note that the results of experiments differ both qualitatively and quantitatively. Some authors note [1] that the distribution of HTC over the surface of the fin is uneven, and at the top of the fin the value of α is higher than at the base, where a thicker boundary layer is formed. It is also noted that the maximum values of α fall on the azimuth angle of rotation $\varphi = 70...90°$. On the other hand, in studies of [2] under similar conditions, the local HTC maxima were found near the root of the fin. In addition, the results of the work [3] indicate bursts of HTC value in the region of $\varphi = 100...130°$.

In practice, the more important parameter is the average HTC over the surface. For its definition, use is made of semi-empirical dependences that require clarification, or numerical methods that require verification by physical experiment.

Summarizing the above, it can be concluded that a joint study of heat transfer and edge flow around fins using gradient heat flux measurement, PIV and thermal imaging diagnostics can yield significantly new and useful results and will allow the refinement of existing calculation methods and models.

2. Experimental setup

The experimental model was a hollow cylinder with a diameter of 66 mm and a length of 600 mm, made of steel sheet with a thickness of 0.1 mm. The experiment was carried out in several stages:

- On the cylinder is mounted a single hollow fin heated by steam with a diameter of 106 mm (1).
- A single "acting" hollow fin and two simulator fins made of Perspex are mounted on the cylinder (2).
- On the cylinder one "acting" hollow fin and four simulator fins are mounted (3).
- Repetition of the series of experiments (1) – (3) in the same hydrodynamic regimes with the only difference being that the fin made of VT22 alloy was used as an "acting" one (4).

Figure 1 shows photographs of experimental models. Simulator fins are made of optically transparent Perspex, which made it possible to visualize the flow inside the intercostal space.

The finned cylinder was installed in the working part of the subsonic wind tunnel of the Department of Thermophysics of Power Units of Peter the Great St. Petersburg Polytechnic University (figure 2).

Figure 1. Photographs of finned cylinder models.

It should be noted that the wind tunnel is equipped with a heat exchanger (air cooler) connected to the cold water supply system, which made it possible to conduct long-term experiments keeping the free-flow temperature almost constant (0.1...0.2 °C). The operating characteristics of the wind tunnel are also shown in the figure.

Figure 2. The appearance of the wind tunnel.

3. Experimental techniques

Within the framework of this study, several modern methods were used at once, therefore each of them requires a detailed description.

3.1. PIV diagnostics

The wind tunnel described in section 2 is equipped with the POLIS PIV system [4]. In the PIV diagnostics (figure 3), the air flow is seeded with tracers with the diameter of 2…3 μm. A laser knife obtained by transforming a laser beam using a system of cylindrical lenses illuminates each tracer with double flashes at a known time interval.

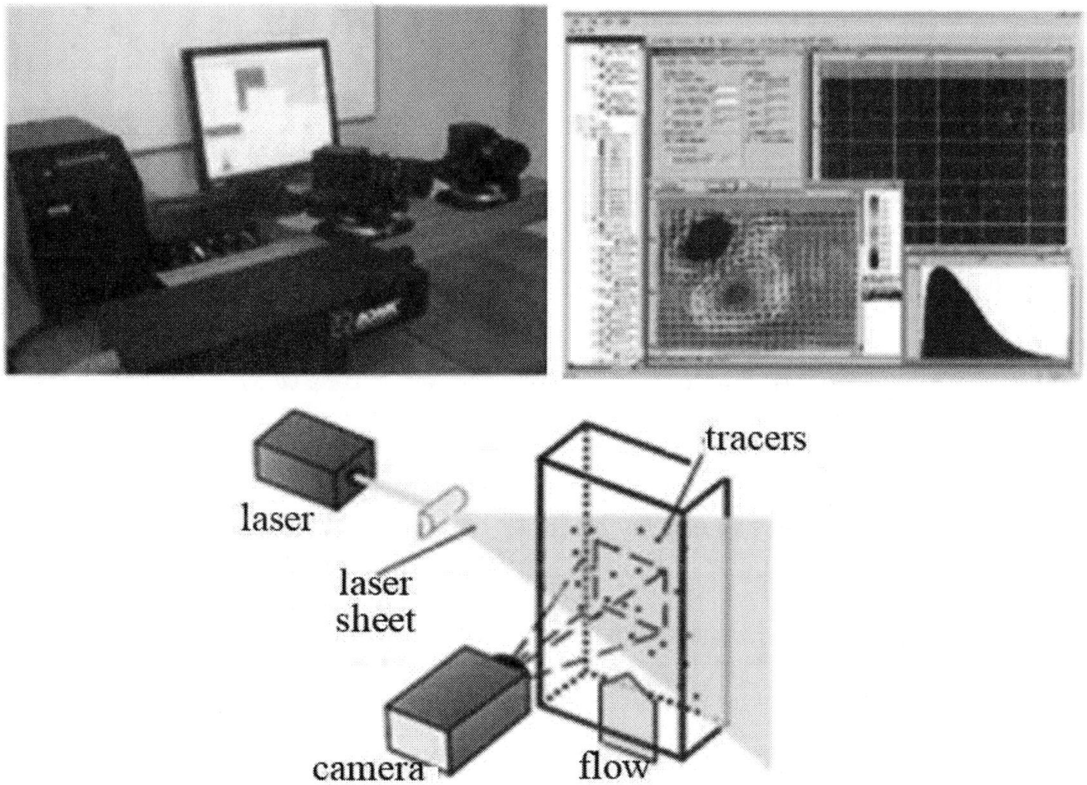

Figure 3. POLIS PIV system [4].

Digital cameras (their number may vary depending on the PIV configuration) during flashes captures the image of the tracers. The synchronization unit provides simultaneous operation of the laser and cameras. The system allows you to adjust the flow of tracers, laser power and frequency of photographs in accordance with the flow mode [5].

3.2. Gradient heat flux measurement

Since 1996, gradient heat flux sensors [6] have been actively used by the scientific group of the Department of Thermophysics of Power Units of Peter the Great St. Petersburg Polytechnic University [7]. To date, the GHFSs are successfully used in studies of heat transfer during condensation and boiling, measurement of heat transfer parameters in ICE [8, 9] and also in industrial experiments [10]. A unique feature of the GHFSs is record low value of their time constant (up to $10^{-8}…10^{-9}$ s [6]), which makes them practically an instantaneous measuring tool in most heat transfer tasks. The principle of the GHFS operation is based on the transverse Seebeck effect: when a heat flux

passes through a plate with anisotropy of thermal and electrical properties, thermopower occurs in it, normal to the heat flux vector direction and proportional to its module [6, 11].
In our study, three battery GHFSs based on monocrystalline bismuth were used (figure 4).

Figure 4. GHFS and its location on the surface of the fin.

The size of the GHFSs in plan was of 2×2 mm, and the thickness was of 0.2 mm. The volt-watt sensitivity of the GHFSs was at the level of 10 mV/W and was determined by the method of absolute graduation according to the Joule heat flux [11].

3.3. Thermal Imaging Diagnostics

To calculate the HTC, in addition to the heat flux per unit area, it is necessary to know the difference between temperature on the surface of the fin at the locations of the GHFS and free-flow temperature. For a single annular fin model, a FLIR P640 thermal imager (Forward-Looking InfraRed) was used to measure the fin surface temperature. The principle of operation of a thermal imager is as follows: bodies with a temperature other than absolute zero, emit electromagnetic IR waves. The spectral power density of this radiation has a maximum, the wavelength of which depends on temperature. Position of the maximum with increasing temperature shifts towards shorter wavelengths. Bodies heated to temperatures of 40…100 °C are characterized by a radiation maximum in the mid-IR range (figure 5).

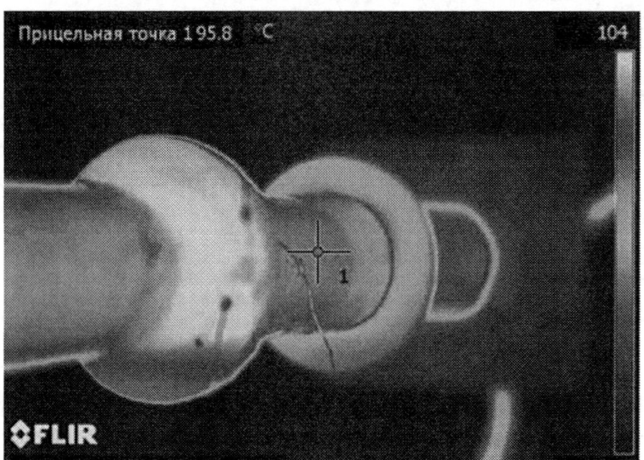

Figure 5. IR pattern of the model made by FLIR.

The use of thermal imaging diagnostics allowed non-invasive measurement of temperature on the surface of the fin, and the software of the FLIR camera made it possible to simultaneously measure temperature at several points on these surfaces with accuracy of 1 K.

4. Results

To compare the results obtained for hollow and solid fins, and then to estimate the effect of blocking, the experiments were carried out in the same ranges of Reynolds numbers $Re = (0.4 \dots 5)\times10^4$, calculated for the diameter of the carrier cylinder d.

The flow visualization around a single hollow annular fin is shown in figure 6. The flow patterns are averaged over 1000 frames and agree well with the flow pattern described in [1].

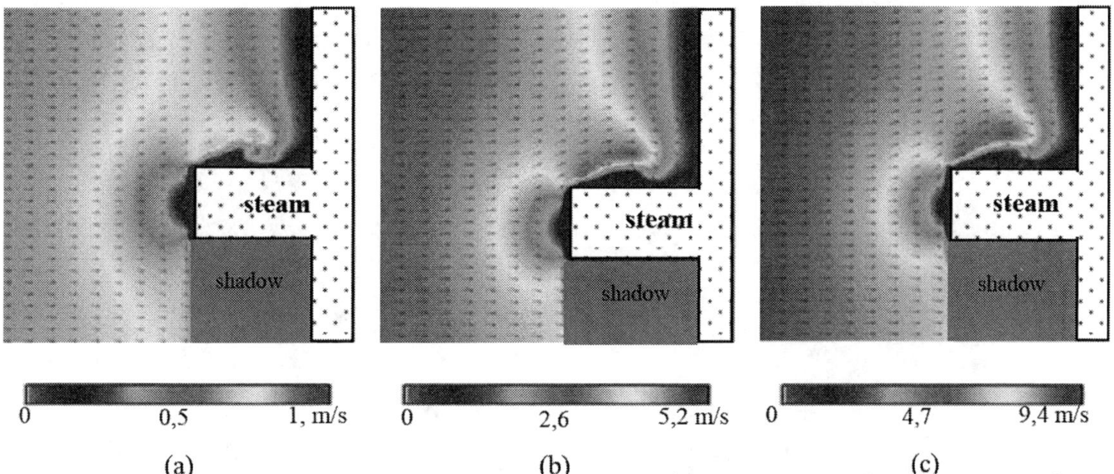

Figure 6. Velocity field near a single fin: (a) $Re = 0.9\times10^4$; (b) $Re = 2.1\times10^4$; (c) $Re = 4.2\times10^4$.

Due to the low height of the fin and presence of a sharp edge in the pictures, air flow separation is seen, and its further adjacency with a small spin of flow, which, however, does not have time to form a full-fledged vortex over the fin. At a sufficient distance from the fin (near its calibre), one can see the flow structure inherent in the flow around a smooth circular cylinder.

Figure 7 shows 3D field of HTC for hollow and solid fins, corresponding to the Reynolds number $Re = 4.2\times10^4$.

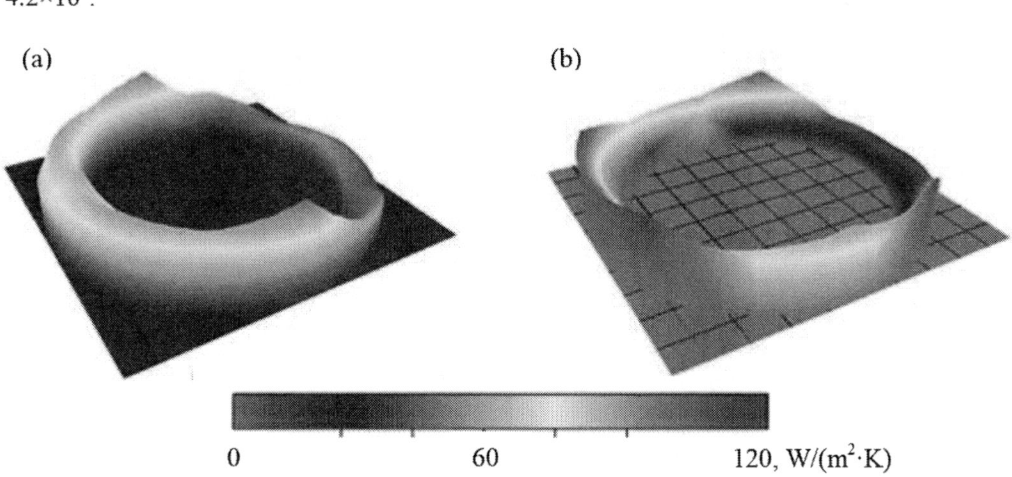

Figure 7. 3D field of HTC for hollow (a) and solid (b) fins.

It can be seen that the value of the local HTC h_φ along the height of the isothermal (hollow) fin varies much less than the same one for the fin made of titanium alloy. For a solid fin, the value of h_φ near the root is lower for all values of the angle φ than for a hollow fin. Also interesting is the reduction of the local HTC in the aft of the carrier cylinder. It is assumed that minimum of HTC is caused by the presence of a stagnant zone beyond the mid-section of the carrier cylinder.

In the next series of experiments, the flow in the intercostal space is visualized in the case when one "acting" and two simulator fins are installed on the cylinder. Results for different intercostal spaces are presented in figure 8.

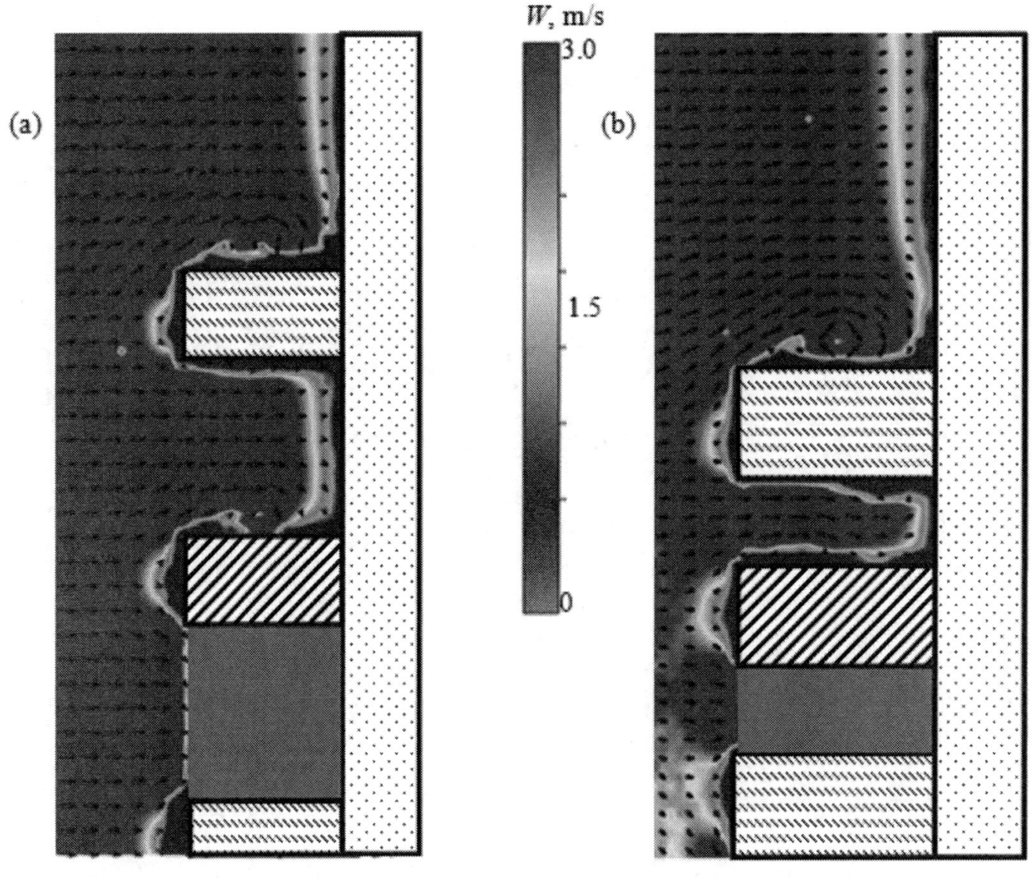

Figure 8. Velocity field near the finned cylinder with $Re = 1.3\times10^4$ at the intercostal space of (a) $\delta = 20$ mm; (b) $\delta = 10$ mm.

Above the first fin one can see a vortex typical for the flow around a single fin. However, with a smaller distance between the fins, the flow in the space is essentially asymmetrical. This character affects the values of the local HTCs. In this regard, in the next series of experiments, the flow and heat transfer in the intercostal space were investigated in the case when there were five annual fins mounted on the cylinder (one "acting" and four simulator ones). Figure 9 shows the flow patterns in the intercostal spaces for different modes at intercostal distance of $u = 15$ mm. In this case, a vortex is also observed over the last fin, which loses intensity with increasing velocity. The flow in the space between the first and second fins is more symmetrical than for a cylinder having three fins. In the space between the second and third fins, one can see reverse currents.

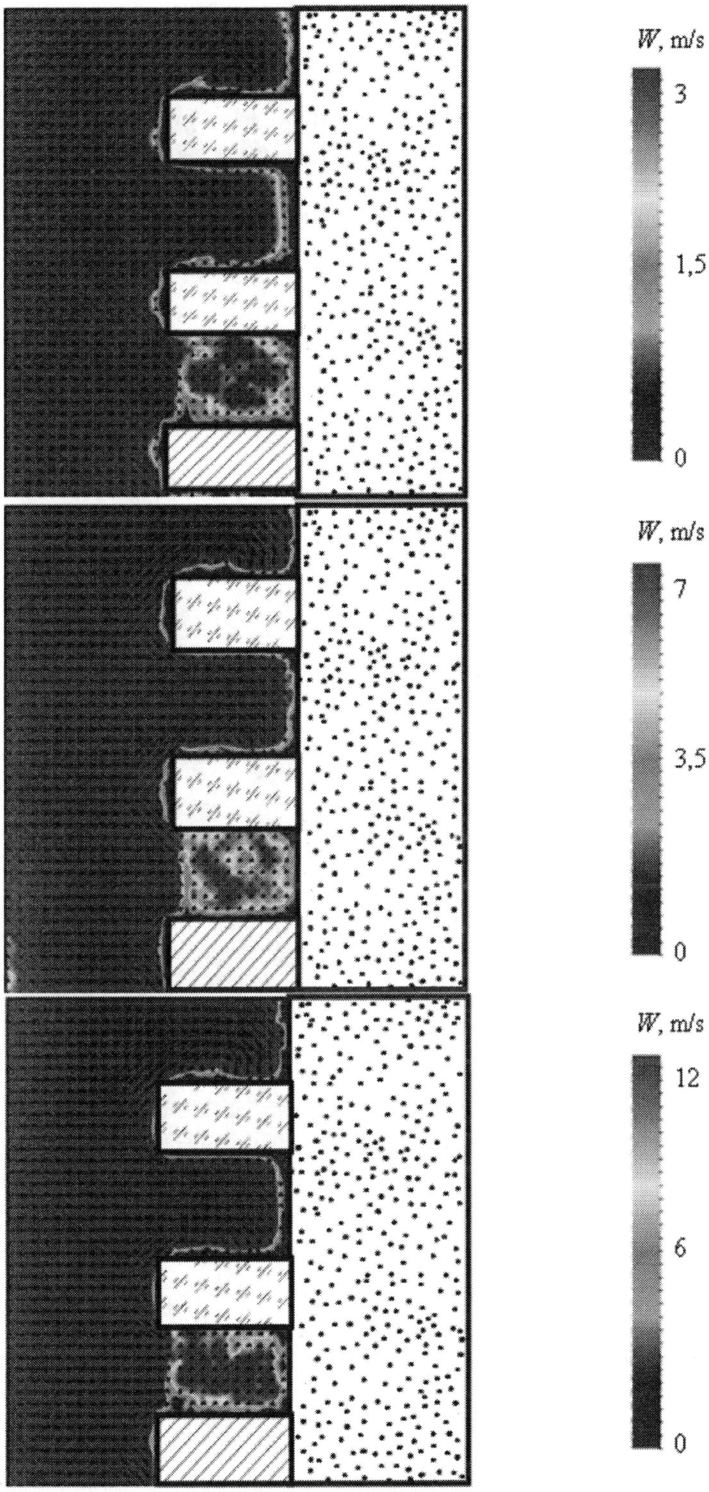

Figure 9. Velocity field near the finned cylinder. The intercostal space is of 15 mm.

Using, together with velocity fields, the data of heat flux measurement and thermometry also succeeded in experimentally determining annual fin efficiency *E*. Since the experiments were conducted in similar modes and forms, the ratio of heat fluxes measured on the surface of solid and hollow (ideal) fins is fin efficiency. Figure 10 shows the analytic curve [12] and experimental values of efficiency of the fin of 20 mm high and of 10 mm thick. Here *m* is product of fins height and expression $\beta = (2\times h / (k\times\delta))^{0.5}$.

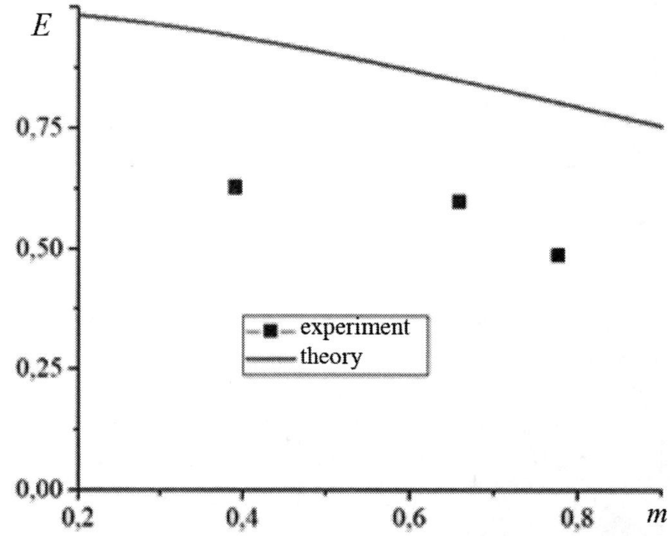

Figure 10. Analytic curve and experimental points for efficiency of the annual fin of 20 mm high.

It can be seen that the experimental points are significantly different from the analytic curve. This may be due to a sufficiently large thickness of the ribs and the presence of a sharp edge.

5. Conclusion

The paper shows the combined use of PIV diagnostics, gradient heat flux measurement and thermometry (invasive and non-invasive) for a comprehensive experimental study of flow and heat transfer in a flow around a single annular fin on a carrier cylinder and finned cylinder with "acting" and simulating fins. The values of local HTCs are revealed over the surface of the fins on a finned cylinder model. The results obtained on a solid (non-isothermal) fin are compared with those obtained for a hollow (isothermal) one. Fin efficiency are experimentally obtained in different modes.

References

[1] Pismenny Ye N 2004 *Heat Transfer and Aerodynamics of Packages of Cross-Finned Tubes* (Kiev: Alterpres)
[2] Mon, M S and Gross U 2004 Numerical study of fin-spacing effects in annular-finned tube heat exchangers *International Journal of Heat and Mass Transfer* **47** 1953-64
[3] Sung H J, Yang J S and Park T S 1996 Local Convective mass transfer on circular cylinder with transverse annular fins in a cross-flow *Int J. Heat Mass Transfer* **39** 1093-1101
[4] «POLIS» velocity profile measuring instrument ([website]/ Institute of Thermophysics SB RAS, Novosibirsk – access mode: http://www.polis-instruments.ru.
[5] Gusakov A A, Kosolapov A S, Markovich D M, Mityakov A V, Mityakov V Y, Mozhayskiy S A and Sapozhnikov S Z 2014 Simultaneous PIV and gradient heat flux measurement of a circular cylinder in cross-flow *Appl. Mech. Mater.* **629** 444–49

[6] Sapozhnikov S Z , Mityakov V Y and Mityakov A V 2012 *Gradient Heat Flux Measurement Basic Concepts* (Russian: Saint Petersburg St. Petersburg State Polytechnical University) p 203

[7] Sapozhnikov S Z, Mityakov V Y, Mityakov A V and Gusakov A A 2017 *J. Phys.: Conf. Ser.* **891** 012371

[8] Sapozhnikov S, Mityakov V, Babich A and Zainullina E 2018 Study of condensation at the surfaces of tube with gradient heat flux measurement. *MATEC Web of Conferences*. **245** 06010

[9] Sapozhnikov S Z, Mityakov V Y, Mityakov A V and Subbotina V V 2019 *Tech. Phys. Lett.* **45** 253

[10] Mityakov V, Mityakov A, Vintsarevich A and Gerasimov, D 2018 Gradient heat flux measurement as monitoring tool for the diesel engine *MATEC Web of Conferences* **245** 14001

[11] Sapozhnikov S Z, Mityakov V Y, Mityakov A V, Pokhodun A I, Sokolov N A and Matveev M S 2012 *Meas Tech* **54** 1155

[12] Zhukauskas A A 1982 *Convective Transfer in Heat* p 472

Design features of the glossmeter system

A S Beliaeva[1,2] and G E Romanova[1]

[1] National Research University of Information Technologies, Mechanics and Optics, 49, Kronverksky pr., St. Petersburg, 197101 Russia
[2] "K-M" LLC (CONSTANTA Group of Companies), 29, Marshal Govorov street, St. Petersburg, 198097 Russia

E-mail: belyaevalina@inbox.ru, romanova_g_e@mail.ru

Abstract. The paper discusses the stages of development of the glossmeter device. Possible options are analysed and the choice of the optical scheme is justified, as well as the choice and matching of the source and detector. The issues of ensuring the linearity of the device characteristics are considered. The results of the device testing are given.

1. Introduction

This device is necessary in industry, both in the fabrication and in the inspection of coatings, for example, in cars painting. The geometry of the optical system of the glossmeter varies; the most common angles for inspecting the gloss are 20/20, 45/45, 60/60, 85/85, and a gloss meter with the geometry of 75/75 is sometimes used to inspect paper materials. The choice of angle depends on the degree of glossiness of the coating, for high-gloss coatings an angle of 20/20 is used, for medium-gloss ones – 60/60, for matte ones – 85/85. This device can be used not only to inspect the glossiness of coatings, but also to determine the brightness of coatings and products in there is a special geometry in the device 45/0 [1].

The glossiness of coatings is measured in gloss units. The reflection of black velvet is taken as 0 gloss units, the gloss of a black glass plate with a refractive index of 1.567 is taken as 100 gloss units; the height of irregularities should not exceed 2 interference fringes.

The device has been designed and tested at the base of K-M LLC. The device has been developed in accordance with the following regulatory documents: Russian State Standard 31975-2017 [1], Russian State Standard 896-69 [2], Russian State Standard 16143 [3], ISO 7668 [4], ASTM C346 [5], ASTM D523 [6], ASTM D2457 [7], ASTM D4039 [8], ASTM E430 [9], DIN EN 14086 [10].

In accordance with the considered documents, the beam incidence on the inspected surface shall be parallel, the surface irradiance shall be uniform, the source shall have a spectrum corresponding to a source of A or C type [11]. International standards offer the Keller system as a standard optical system (figure 1 (a)). Russian State Standards do not offer a specific optical system, and as a principle of operation they offer a schematic diagram of the device (figure 1 (b)).

2. Optical system of the device

In the first version of the device "Constanta BF", a single-lens optical system has been used as a lighting system, the lens material is an optical polymer, the light source is the LED KPT-1608YC from Kingbright, and the photodetector is the Photodiode BPW34S from Vishay. Calibration in VNIIOFI has shown that the device has errors of up to 19 units. The calibration results are printed in figure 2.

Published under licence by IOP Publishing Ltd

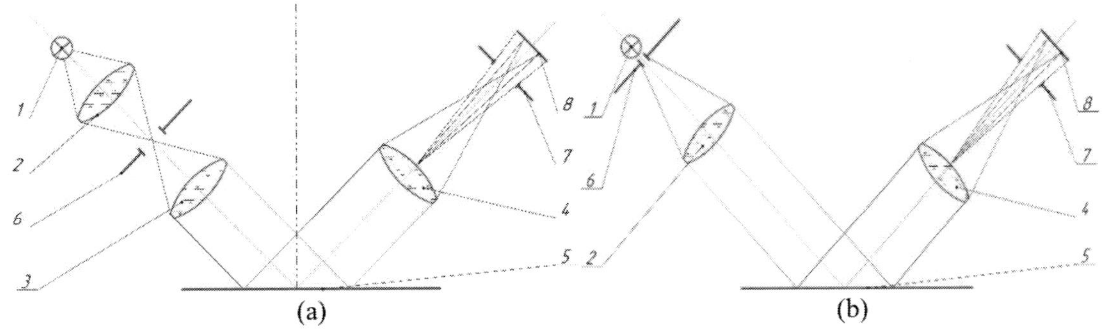

Figure 1. Schematic optical diagrams in the gloss measurement device:
(a) according to ASTM E 430, (b) according to Russian State Standards 31975-2017.
1 – light source; 2, 3 – lighting channel lens; 4 – collector lens; 5 – inspected surface; 6 – field aperture; 7 – field diaphragm of the detector; 8 – detector.

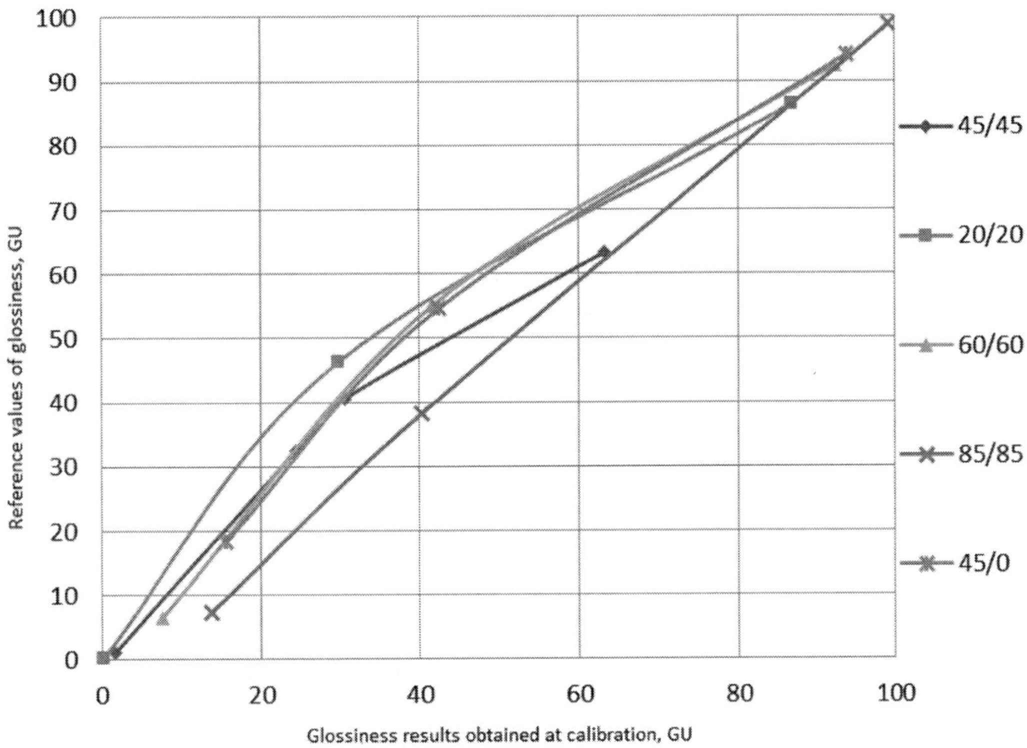

Figure 2. The results of calibration of the first version of the gloss meter "Constant".

The errors can be caused by the following reasons:

- characteristic of the lighting channel (non-parallel light beam, non-uniformity of the energy distribution over the surface).
- inconsistency between source and detector, and the detector is sensitive to the IR radiation.

The LED 16-led of XD series from CREE has been chosen as a new light source. The advantages of this LED are that it has small dimensions (1.6x1.6 mm^2) and the luminous flux is about 150 Lm at 350 mA (the maximum operating current of the LED is 2 A), the LED in the device will operate at the current of 100 mA (the luminous flux is 43 Lm). The color temperature of the chosen LED is 5000 K. The detector channel uses the photodiode from Blue Enhanced Photodiodes of the series VTB9413BH with the sensitive area of 1.6 mm^2, the spectral sensitivity range is within 330-720 nm, and the maximum sensitivity corresponds to 580 nm. Since the sensitivity curve of the photodiode lies in the visible range, changes in the temperature of the surface under test does not affect the instrument readings.

The technical assignment has restricted the dimensions of the optical system. The lighting and detecting parts should be less than 50 mm in length. Taking into account the overall dimensions of the housing, the diameter of the optical system should not be greater than 10 mm. The diameter of the optical system for the geometry of 85/85 should not exceed 6 mm.

The following options for the optical lighting system have been considered:

- single lens system without aspherical surfaces;
- single lens optical system with aspherical surfaces;
- keller system;
- optical system of two plano-convex lenses.

Various options of a single-lens optical system have been designed. In case of using a system without aspherical surfaces, either a high energy concentration in the center of the illuminated area, or a large energy concentration in the ring has been observed (figure 3 (a)) [12].

This effect can be excluded by adding the asphericity coefficient, or by adding the optical elements. As can be seen from figure 3 (b), the use of the aspherical lens gives more uniform energy distribution, but the stock aspherical lenses cannot be used to achieve necessary characteristics, and manufacturing a small set of lenses of this type is expensive, therefore, it has been decided not to use such an optical system in the model.

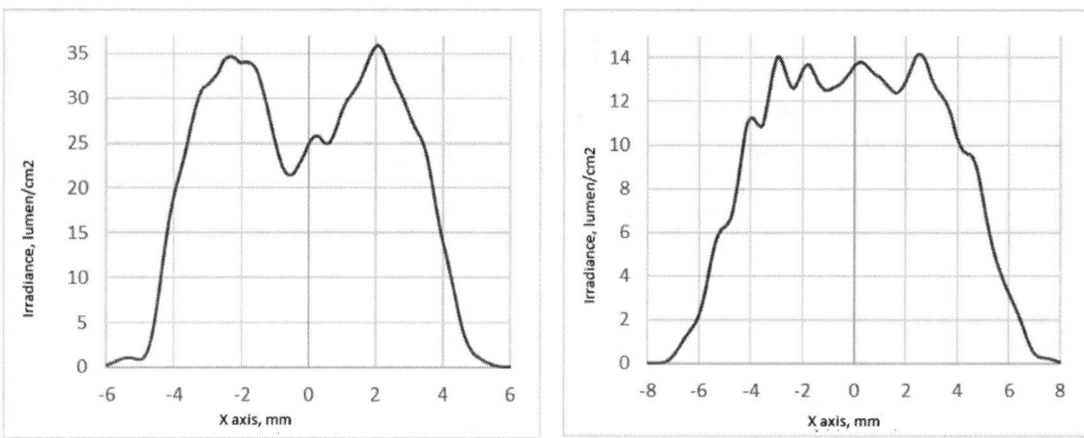

(a) (b)

Figure 3. Surface irradiation (a) in a system with a single lens with spherical surfaces, (b) in a system with an aspherical lens.

As an alternative, the system proposed by international standards has been calculated using the Keller system (figure 4 (a)). As can be seen from figure 4 (b), the system gives a uniform distribution of energy and collimates the beam. The first lens in a system is flat-convex with a focal length of 24 mm, the lens diameter is 18 mm, the material is TF3; the second lens is also flat-convex with the focal length of 40 mm, the diameter is 24 mm, the material is K8; the overall length of the system is 159 mm.

However, the overall dimensions of this system exceed dimension of the ones, specified in the technical assignment, and it is impossible to reduce the dimensions of the system without complicating the components, therefore, this system cannot be implemented in the considered case [6].

The adopted practical version of the system is a system consisting of two lenses (figure 5 (a)). It has suitable dimensions, provides uniform distribution of energy (figure 5 (b)). Due to the use in the system of flat convex lenses with spherical surfaces made of K8 glass, the cost of manufacturing of this system is optimum compared to others.

Journal of Physics: Conference Series **1421** (2019) 012065 doi:10.1088/1742-6596/1421/1/012065

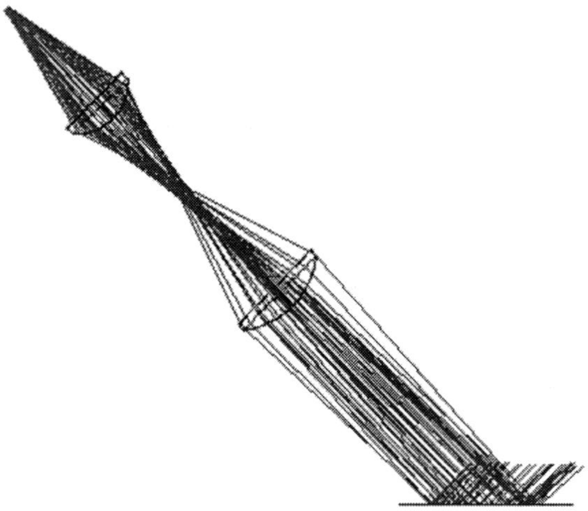

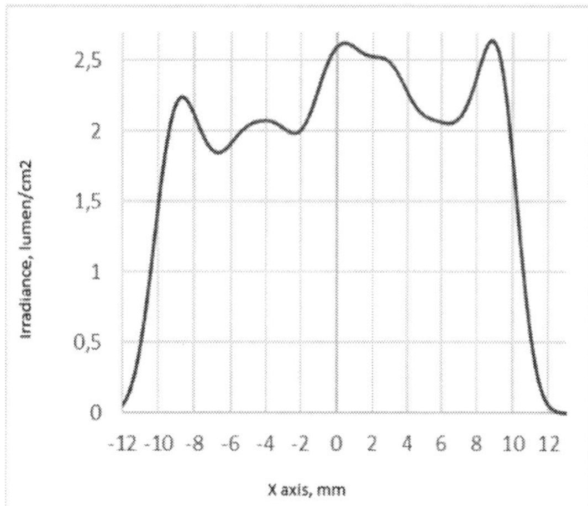

(a) (b)

Figure 4. Keller system (a) general view of the system, (b) surface irradiance.

In the lighting channel with the geometry of 85/85, the same optical system is used, but with a smaller diameter. The detecting channel with the geometry of 85/85 uses a system of a biconvex lens made of K8 material with the focal length of 7.4 mm. The other detecting channels use a flat-convex lens with the focal length of 18.9 mm.

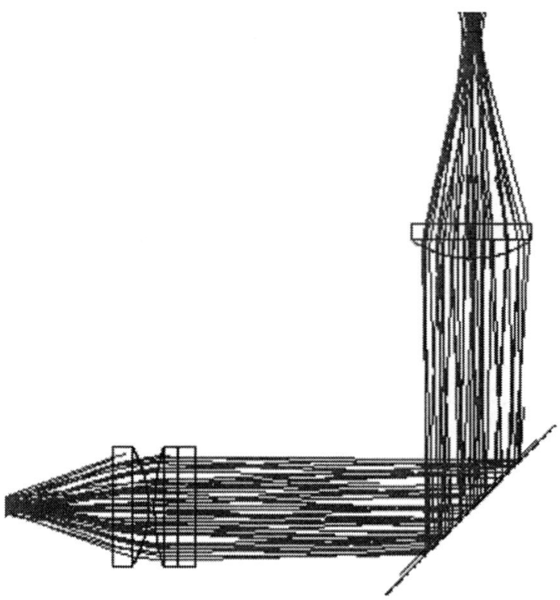

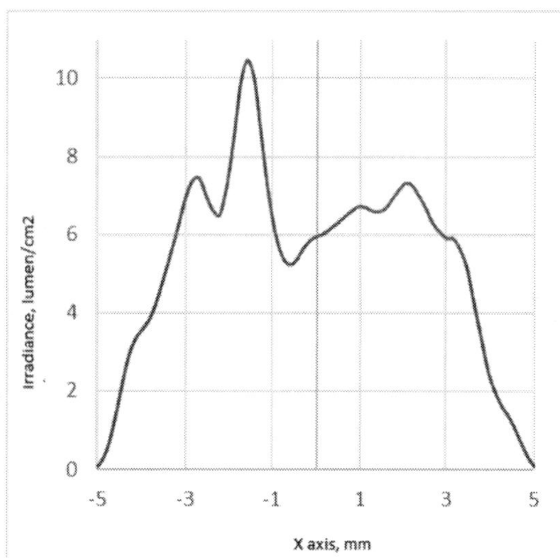

(a) (b)

Figure 5. System in use (a) general view of the system, (b) surface irradiance.

3. Color Temperature Correction

Since all regulatory documents require that the source have the color temperature of A type (2856 K) or C type (6774 K), in addition the system should include a light filter (a plane-parallel plate that does not affect the path of rays), the purpose of which is to correct the color temperature of the chosen LED

Journal of Physics: Conference Series **1421** (2019) 012065 doi:10.1088/1742-6596/1421/1/012065

in accordance with the requirements. To determine the parameters of the light filter, a series of measurements of spectral characteristics of the radiation in the optical system have been performed in order to take into account not only the spectral composition of the source, but also the spectral transmittances of the glasses, after that the values have been averaged and the light filter has been made according to them. The spectral transmission of the filter is illustrated in figure 6 [11].

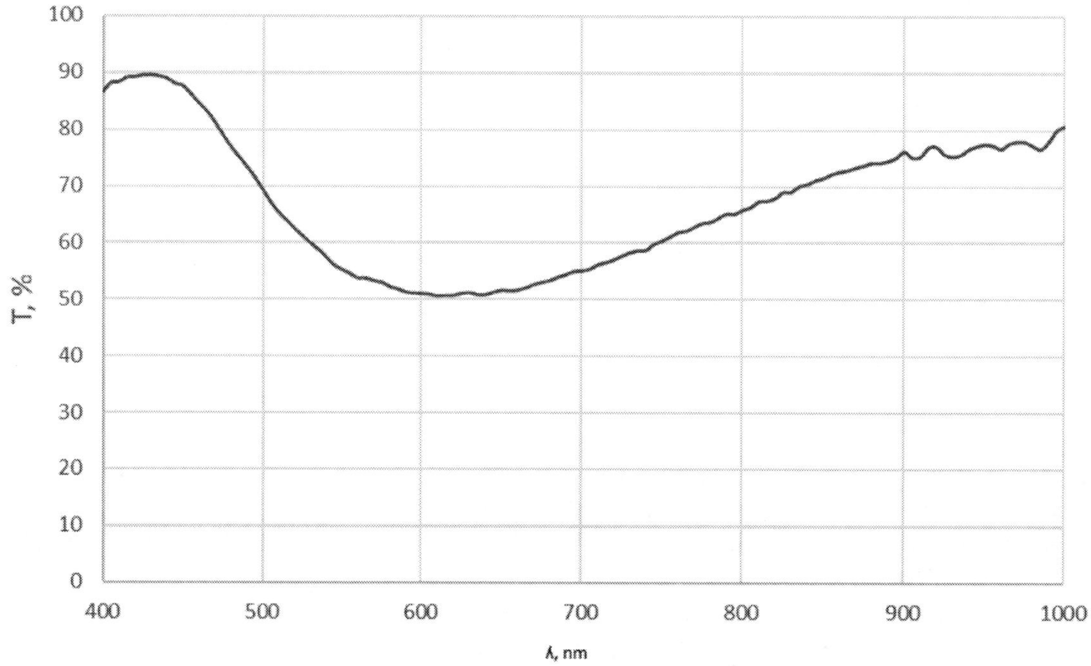

Figure 6. Light filter transmittance to implement the spectral radiation of C type.

To measure the color temperature, a series of spectrum measurements have been also performed in the system with the light filter. According to the results of measurement of the color temperature, and in particular of X and Y coordinates, the LEDs have been excluded, the total error margin of which, ΔX and ΔY, exceed ± 0.015.

4. Principle of operation of the electrical circuit of the device

The device operates under the control of a microcontroller that uses embedded software. The microcontroller in the glossmeter controls the following processes:

- LED driver, LEDs operate in pulsed mode;
- analog-to-digital converter (ADC) in the receiving channel;
- SD card;
- temperature sensor;
- battery device;
- speakers;
- USB;
- display.

5. Testing the components for physical characteristics

To ensure the linearity of the device characteristics it is necessary to ensure the photodiode operation in the photodiode mode. In this circuit, a variable resistor of 10-100 KOhm is used.

The device has been tested for the following:

- to determine the photodiode's signal dependence on gloss function;

- to define the photodiode signal shift as a temperature function. Finding of the deviation coefficient.

Figure 7 illustrates the photodiode signal vs the brightness curve, using a modified optical system. The results are the code from the analog-to-digital converter (ADC).

As it can be seen, the results have become more linear, however, the absolute measurement error satisfies the requirements only at the angles of 85/85 and 45/0.

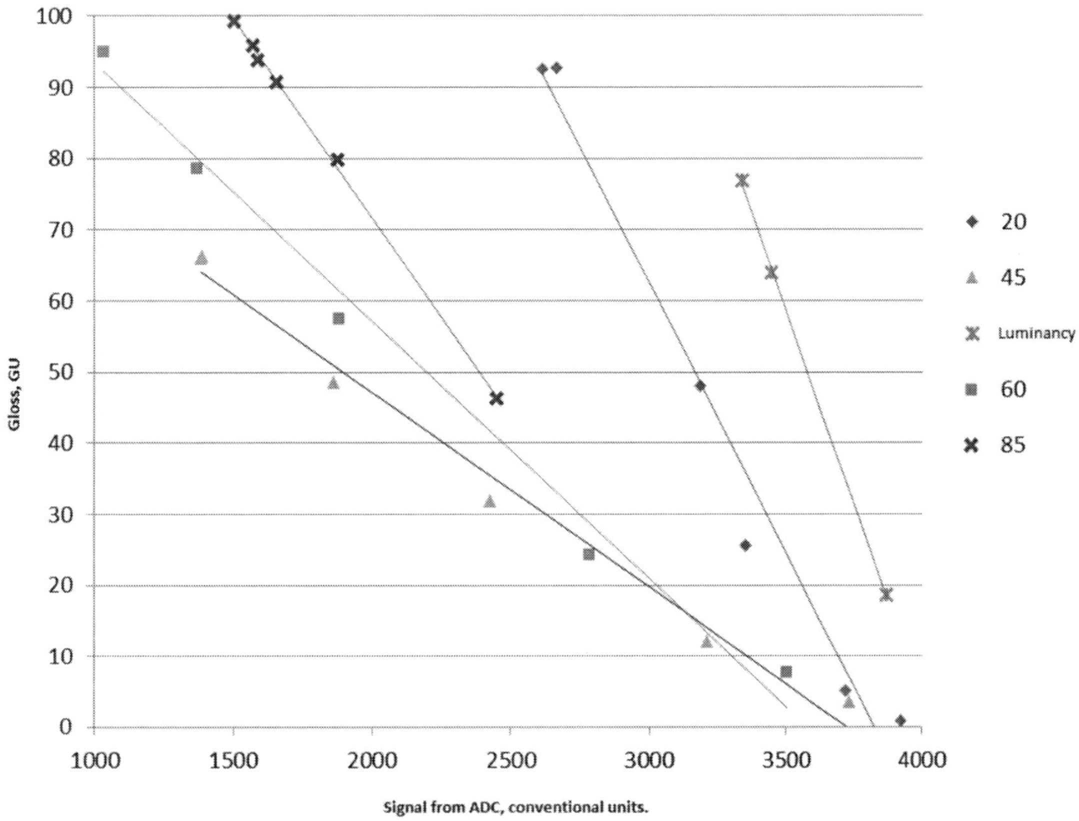

Figure 7. Results of the device testing using reference samples without diaphragm.

In figure 7 the points mark the measurement results, also the trend lines are shown, which allow to evaluate the deviations of the obtained results from linearity.

To improve the linearity of results, it is necessary to bring the size of the source to a point one, that is, to add a diaphragm. Measurements have been performed with the following diaphragm diameters: 1 mm, 0.75 mm, 0.5 mm.

Figures 8, 9 show the results of tests with the diaphragms of 1 mm and 0.5 mm. As it can be seen, the smaller the diaphragm, the more linear the curve, but the diaphragm size is most critical at small angles: this is most noticeable if compare the angles of 20/20 and 60/60 (85/85). For the angle 20/20, the size of the source is most critical. For unification, it has been decided to add the diaphragm of 0.5 mm to the optical systems 20/20, 60/60, 45/0/45. In case of a system with the geometry of 85/85, no modifications have been made.

The tests for temperature deviation have been conducted in a thermobox with the temperature settings from 15 to 35 °C. The signal vs temperature function in the photodiode in the experiment has been found to be linear; therefore the ambient temperature will not introduce nonlinear distortions into the measurement results. If there is a temperature sensor, the appropriate correction can be introduced, therefore there is no need to wait until the device heats up to ambient temperature for starting measurements.

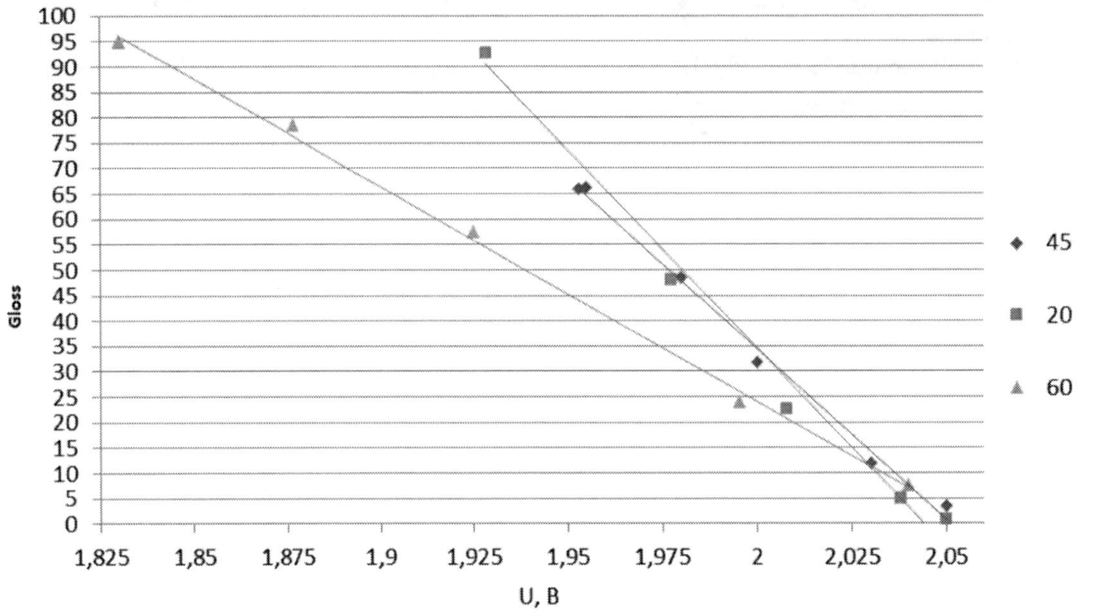

Figure 8. Results of the device testing using the diaphragm with the diameter of 1 mm.

6. General view of the device

A general view of the device with three angles in the section is shown in figure 10. This setup is the latest version at the moment.

Similarly, the device with the geometry of 45/0/45 (figure 11) has been simulated and set up. In the channel, the photodiode FDUK-2P is fixed at 0 degrees, no optical elements in this geometry are used.

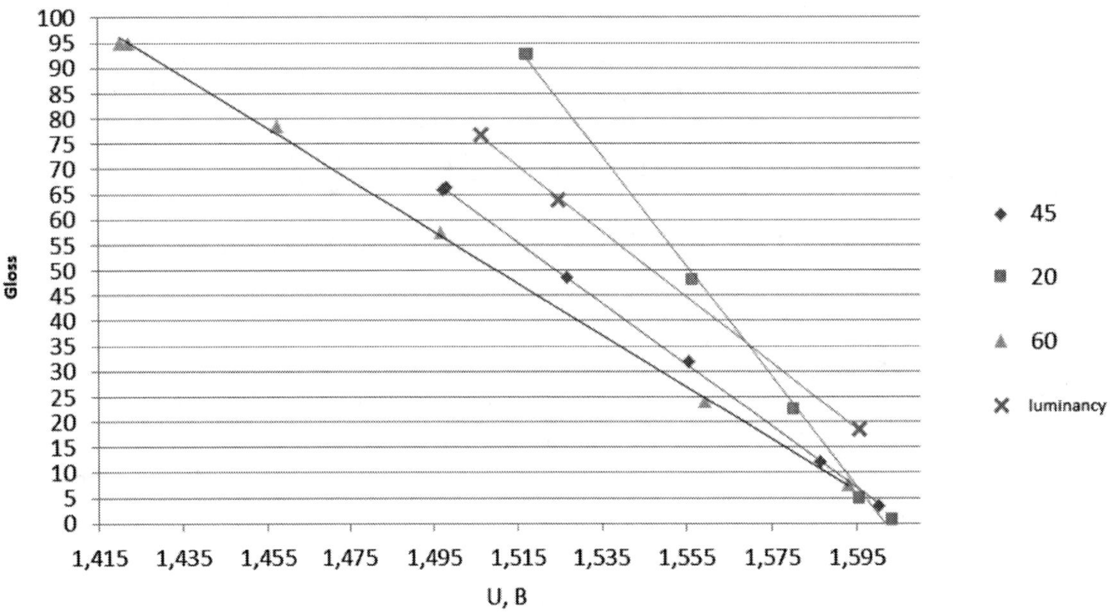

Figure 9. Measurement results using the diaphragm with the diameter of 0.5 mm.

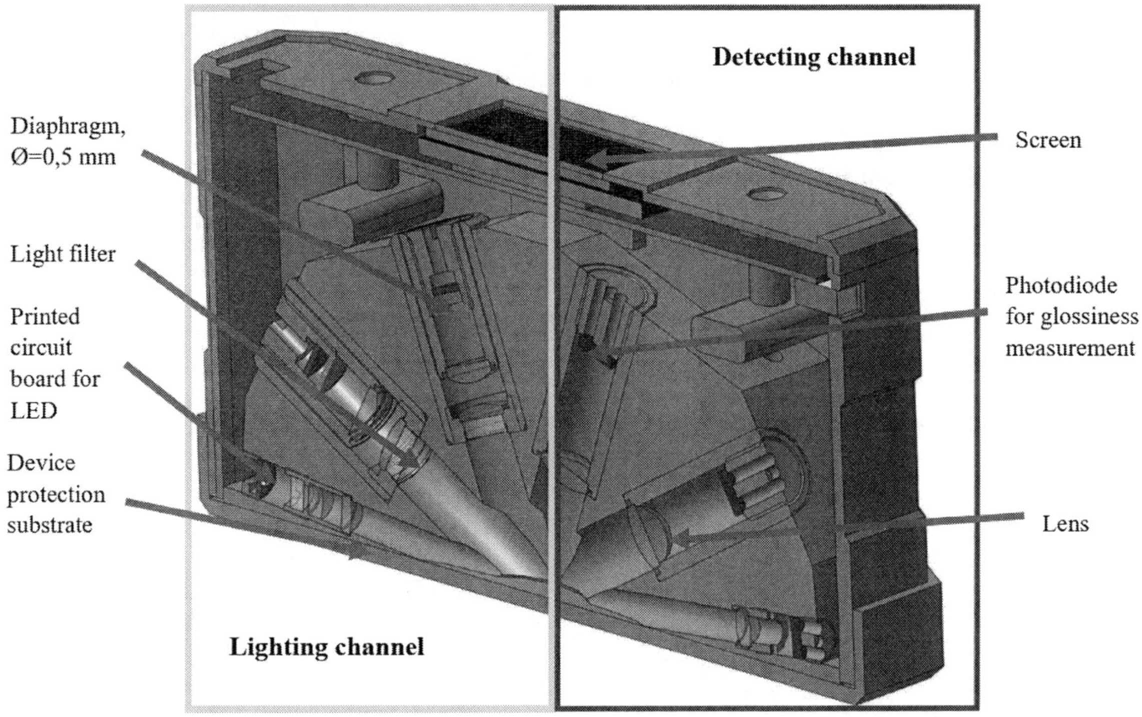

Figure 10. 3D model of the device with three angles in the section.

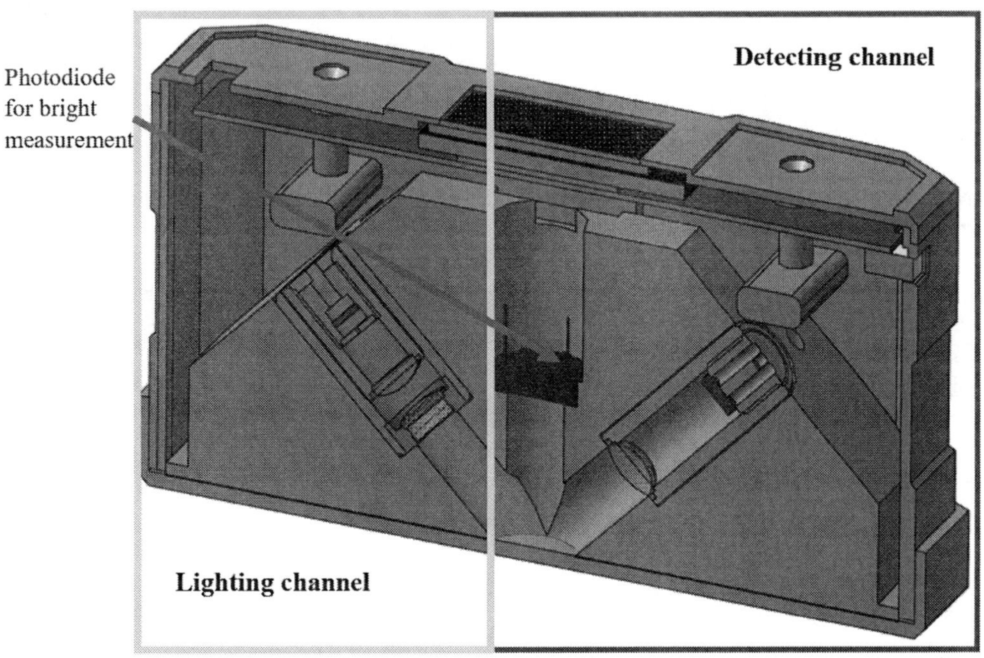

Figure 11. 3D model of the device 45/045 in the section.

7. Conclusion

The paper describes the developed device for measuring the gloss of surfaces. The developed device has passed through all stages of development – from the choice of the scheme to the design and assembly. The model of the device with the geometry 20/20, 45/45, 60/60, 85/85, 45/0 has been tested using the reference samples. The test results have demonstrated a decrease of the measurement error, the causes of deviations from the linearity of the results have been identified and taken into account; the best diaphragm size for the source has been selected based on the measurement results and their comparisons. In the future, it is planned to assemble the device and calibrate it with the aim of launching into mass production and for sale in the company "K-M" LLC, "Constant" Group of Companies.

References

[1] Russian State Standards 31975-2013 2014 *Method for determining the gloss of non-metallic paint coatings at the angles of 20°, 60° and 85°.* (Moscow: Standartinform Publ.) (In Russian)

[2] Russian State Standards 896-1969 2003 *Paintwork materials. Photoelectrical method for determination of coatings gloss.* (Moscow: Standartinform Publ.) (In Russian)

[3] Russian State Standards 16143-1981. *Parts and products of wood and wooden materials. Method for determination of transporent varnish coating gloss.* Moscow, Standartinform Publ., 1999 (In Russian)

[4] ISO 7668:2018 2018 *Anodizing of aluminium and its alloys – Measurement of specular reflectance and specular gloss of anodic oxidation coating at angles of 20 degrees, 45 degrees, 60 degrees or 85 degrees* Geneva, Switzerland: International Organization for Standardization (ISO)

[5] ASTM C 346-87 2018 *Standard test method for 45-deg specular gloss of ceramic materials.* (West Conshohocken, PA, USA: ASTM International)

[6] ASTM D523-14 2018 *Standard test method for specular gloss.* (West Conshohocken, PA, USA: ASTM International)

[7] ASTM D2457-13 2013 *Standard Test Method for Specular Gloss of Plastic Films and Solid Plastics.* (West Conshohocken, PA, USA: ASTM International)

[8] ASTM D4039-09 2015 *Standard test method for reflection haze of high-gloss surfaces.* (West Conshohocken, PA, USA: ASTM International)

[9] ASTM E430-11 2015 *Standard test methods for measurement of gloss of high-gloss surfaces by abridged gonophotometry.* (West Conshohocken, PA, USA: ASTM International)

[10] DIN EN 14086:2003-04. *Paper and board – Measurement of specular gloss – 45° gloss with a parallel beam, DIN method.* (In German)

[11] Russian State Standards 7721-89 1989 *Illuminants for colour measurements. Types. Technical requirements.* Moscow, Standartinform Publ. (In Russian)

[12] ZEMAXOpticStudio *User Manual* ZEMAX LLC p 2545

Immersion laser-induced ultrasound imaging of solids with complex macroscopic geometry of surface

V Zarubin[1,2], A Bychkov[1,2], V Simonova[3], E Cherepetskaya[1] and A Karabutov[1,3,4]

[1] The National University of Science and Technology "MISiS", 4 Leninskiy prospekt, 119049, Moscow, Russia
[2] The Faculty of Physics, Lomonosov Moscow State University, 1 Leninskie Gory, 119991, Moscow, Russia
[3] The Institute on Laser and Information Technologies of the Russian Academy of Sciences, 1 Svyatozerskaya st., 140700, Shatura, Moscow, Russia
[4] The International Laser Center of Lomonosov Moscow State University, 1 Leninskiye gory, 119991, Moscow, Russia

E-mail: zarubin.vasily@gmail.com

Abstract. In this paper a real-time experimental system and algorithms for immersion two-dimensional laser ultrasonic imaging of solids with complex geometry of surface are presented. Wideband probe ultrasonic pulse is generated photoacoustically in an opaque plate, which absorbs pulsed laser radiation. The back-scattered acoustic field is recorded by a multi-element piezoelectric array. The signals are used to reconstruct refraction-corrected ultrasonic images, which are post-processed to determine positions of the external and internal boundaries of the object. The numerical simulations are carried out to verify the developed algorithms. The proposed approach allowed measuring the positions of the boundaries of the polymethyl methacrylate (PMMA) sample with accuracy of ~0.1 mm in the direction of the probe beam propagation, and ~0.2-0.3 mm in the perpendicular direction.

1. Introduction

The common approach to ultrasonic imaging of solids is to use contact phased arrays [1]. In this case synthetic aperture focusing techniques based on the ray tracing are used. However, contact ultrasonic imaging can encounter difficulties when non-flat surfaces are inspected. Flexible phased arrays [2] are widely used in practice to inspect such surfaces, but they need a tight contact with surface, and cannot be used for inspection of non-smooth surfaces.

Another approach is immersing both the phased array and the object in a body of liquid. However, accurate determination of the position and the shape of the object's external boundary is important for correct reconstruction of the scatterers inside the object. The accuracy of the surface measurement accuracy strongly affects the accuracy of localization of internal scatterers. Malkin et. al. [3] studied resolution achieved by phased arrays in the immersion imaging of objects with strongly curved surfaces. The results show, that surface imaging provides rather poor resolution due to complex waveform, side-lobes of the probe pulse and surface curvature.

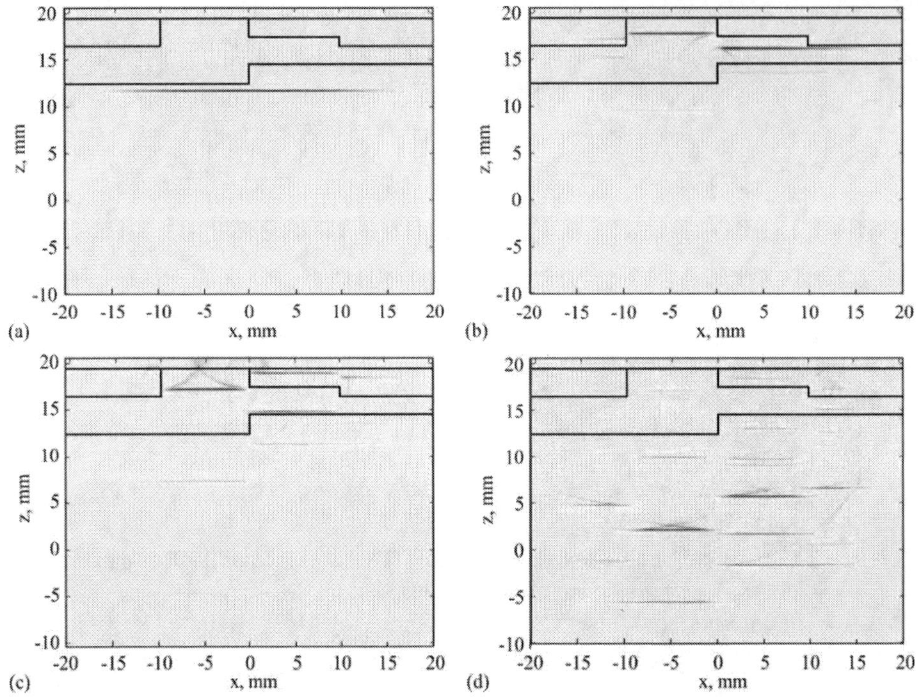

Figure 1. Numerically simulated propagation and scattering of the probe laser ultrasonic beam through the immersion liquid and the PMMA sample on a steel rod. Time elapsed since the laser pulse absorption at $z = z_0 = -20$ mm: (a) 22 μs, (b) 25.7 μs, (c) 27.5 μs, (d) 38 μs. Black lines outline the boundaries of the specimen (figure 3 (a)): the PMMA sample, the steel rod, and a liquid between them.

In this study we develop the laser ultrasonic imaging approach to accurate inspection of solids which decouples excitation and detection of ultrasound, allowing their separate optimization. The advantages of laser excitation of probe ultrasonic pulse include short duration, aperiodic bipolar waveform, and smooth directivity pattern devoid of side-lobes [4]. All this significantly improves the accuracy and quality of the reconstructed ultrasonic images in comparison with the traditional approach. The presented system and algorithms are suitable for samples with piecewise-linear internal and external profile cross-sections, but could be generalized for more complex objects.

2. Method

The experimental setup used for imaging of solid bodies is based on the combined photoacoustic and laser ultrasonic real-time imaging system [5]. A short laser pulse for excitation of the ultrasonic beam is provided by a Nd:YAG laser (pulse duration – 8 ns, pulse energy – 1 mJ, pulse repetition rate – 20 Hz). The probe ultrasonic pulse has a duration of ~100 ns, and frequency band of ~0.1-12 MHz. The beam is focused by a cylindrical PMMA acoustic lens. The sample is immersed in water. The external surface of the sample and the internal inhomogeneities reflect and scatter the probe pulse backwards, and the scattered waves are recorded by an array of 16 wideband polyvinylidene fluoride sensors. The signals are digitized and averaged on a FPGA, filtered and used for image reconstruction with its further post-processing. The acoustic lens allows focusing the probe beam and reconstructing the cross-section of the object by the image plane. The array and the sample are mounted in a 4-axis positioning system (3 translational axes and rotational axis), which is used to align the sample and image plane. The image reconstruction algorithms are accelerated using NVIDIA CUDA on a graphical processors to provide real-time operation (frame rate – 10 Hz).

Numerical simulations of the 2D probe pulse propagation through the immersion liquid and the sample (modelled as a liquid to speed up the simulations) were carried out using k-Wave Toolbox [6].

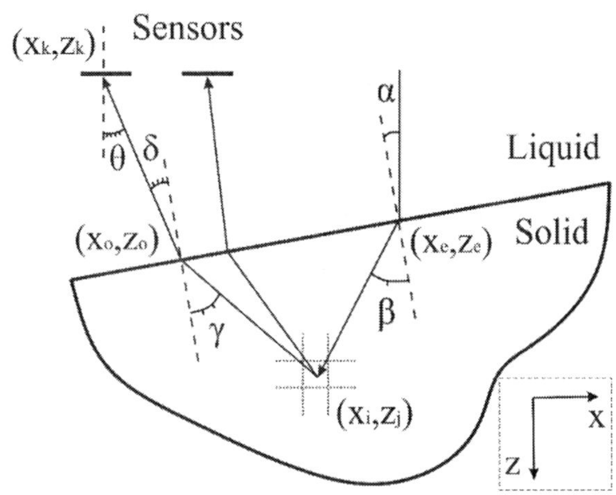

Figure 2. The schematic of the proposed algorithm and variable notation.

A CAD model of the sample (PMMA cylinder with complex internal and external generatrices fixed on a steel rod) used for simulations and experiments is shown in Figure 3 (a). Figure 1 shows the normalized acoustic pressure field $p(x, z, t)$ at different time instants. Red and blue colors correspond to positive and negative pressures respectively. Figure 1 (a) shows the positive probe pulse before it is refracted at the external boundary. In figure 1 (b), the pulse is reflected from an acoustically hard sample surface, remains positive and propagates backwards. Inside the sample, the pulse widens due to the difference between the speed of sound in water and PMMA. In figure 1(c) and (d), the pulses reflected from the acoustically hard wall of the steel rod are positive, and the pulses reflected from the water-PMMA boundary are negative.

The signals are recorded by 16 point-like sensors located at $z = z_0 = -20$ mm, where the Gaussian probe pulse is generated.

The image reconstruction algorithm is based on geometrical acoustic ray tracing (figure 2). First, the imaging region is assumed to be filled entirely with acoustically homogeneous immersion liquid (water), and a filtered back-projection algorithm is used to reconstruct the primary laser-ultrasonic image [7]. The image is a distribution of scatterers' strength $\epsilon(\mathbf{r})$ calculated using [8]:

$$\epsilon(\mathbf{r}) = \sum_{k=1}^{N} p\left(\mathbf{d}_k, t_k^{of}(\mathbf{r})\right) \cdot |\mathbf{d}_k - \mathbf{r}| \cdot D(\mathbf{d}_k,\mathbf{r}), \quad (1)$$

where $t_k^{of}(\mathbf{r}) = |\mathbf{d}_k - \mathbf{r}|/c_0$ is time of flight of the acoustic pulse from the optical absorber to the scatterer at $\mathbf{r} = (x_i, z_j)$ and from the scatterer to the k-th sensor at $\mathbf{d}_k = (x_k, z_k)$, N – number of sensors, c_0 – speed of sound in water, $p(\mathbf{d}_k, t)$ – pressure waveform, recorded by the k-th sensor, and $D(\mathbf{d}_k,\mathbf{r})$ is its directivity pattern. The primary distribution of scatterers $\epsilon(\mathbf{r})$ can be processed to segment the external boundary as the set of local maxima closest to the sensor array [8].

Second, the image pixels located inside the object are re-calculated accounting for refraction at the calculated external boundary. The distribution $\epsilon'(\mathbf{r})$ within the sample is calculated similarly to (1):

$$\epsilon'(\mathbf{r}_{ij}) = \sum_{k=1}^{N} p(\mathbf{d}_k, t_k(\mathbf{r})) \cdot D(\mathbf{d}_k,\mathbf{r}) \cdot W_{ijm}^{el} W_{ijm}^{ol}, \quad (2)$$

where W_{ijm}^{el} is coefficient of transformation of the longitudinal wave in water to the longitudinal wave in the sample on the forward path, and W_{ijm}^{ol} is the corresponding coefficient on the backward path. However, the travel time of probe pulse from the absorber to the scatterer and to the k-th sensor $t_k(\mathbf{r})$ is the sum of forward and backward travel times, which are calculated from the Snell's law and the

Journal of Physics: Conference Series **1421** (2019) 012066 doi:10.1088/1742-6596/1421/1/012066

Fermat's principle [4] using the known positions of the sample boundaries. The internal boundaries are determined as local maxima or minima of $\epsilon'(\mathbf{r})$ depending on the sign of the reflected pulse.

3. Results

Figure 3 shows the images reconstructed using the numerically simulated data. Figure 3 (b) is a primary image used to segment the external profile (lines 1-2). The piecewise linear profile segmented from primary image is shown in figure 3 (c) and used to recalculate the pixels inside the sample. Figure 3 (d) is a refraction-corrected image calculated with equation (2). Figure 3 (b) appears to be a ~2 times "squeezed" version of figure 3 (d) due to speed of sound mismatch. The dimensions of the sample calculated using this image are in agreement with the model within 5 μm due to the finite grid step. This result confirms the correctness of the proposed approach and developed algorithms.

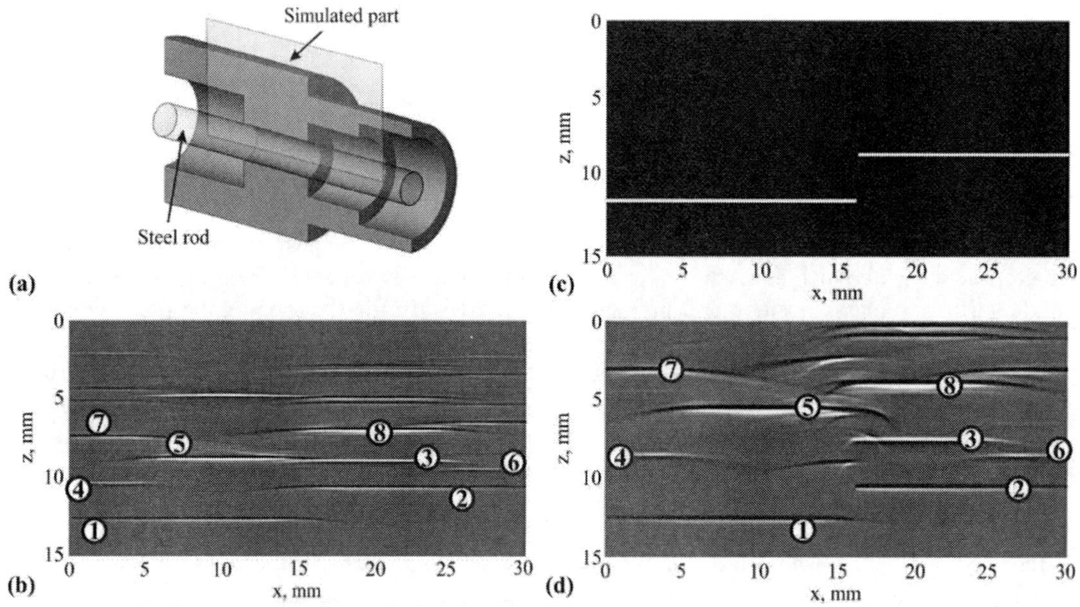

Figure 3. (a) The cross-section of the sample model used for numerical simulation and for experiments. (b) The primary laser ultrasonic image reconstructed under the assumption of a homogeneous medium. (c) The sample profile segmented from figure 3 (b). (d) The final image reconstructed accounting for refraction of waves at the sample boundary. Black and white stripes show positions of the boundaries and sign of the reflected pulse. 1, 2 – external boundary of the sample; 3, 4, 5, 6 – internal boundaries; 7, 5, 8 – the wall of the steel rod. The unnumbered stripes are all artifacts corresponding to the reverberations or the spurious signals.

Figure 4 shows the refraction-corrected images reconstructed from experimental data (sensors are located at $z \approx 40$ mm). Although the images are in a good agreement with figure 3, the lines in figure 4 are wider and the number of artifacts is increased due to the finite detection bandwidth, finite size of the piezoelectric transducers and reverberations within the acoustic lens. The dimensions of the sample measured by the laser ultrasonic imaging and standard gauge instrument are in agreement within 0.1 mm along z-axis, and 0.2-0.3 mm along x-axis. The sample internal surface was then damaged by a file. Lines 3 and 4 in figure 4 (b) are inclined and widened compared to those in figure 4 (a). The results suggest that laser ultrasonic imaging allows quantitative assessment of wear of solid parts.

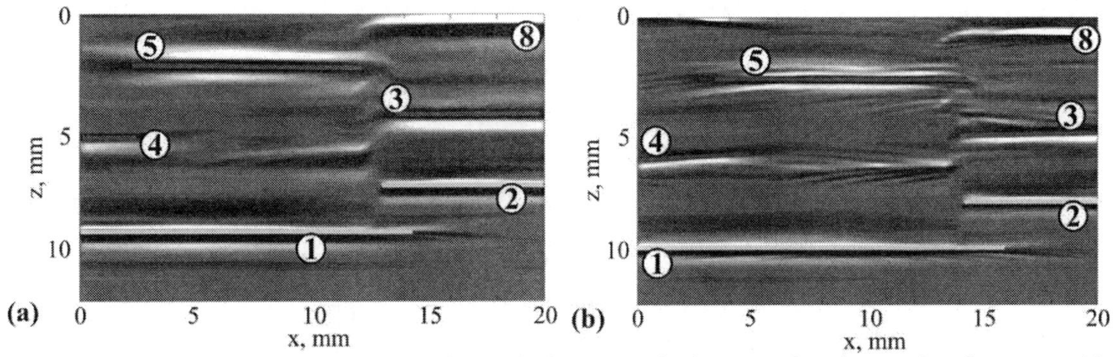

Figure 4. Refraction-corrected experimental images of the sample shown in figure 3 (a). (a) Undamaged sample. (b) Damaged sample. The numbering is the same as in figure 3. Yellow lines show the segmented external sample profile, blue lines show the segmented internal sample profile.

4. Conclusions

The paper describes the results of application of the immersion laser ultrasonic imaging for measurement of dimensions of solid parts with complex geometry of internal and external surfaces. The refraction-corrected algorithm for reconstruction of 2D images of internal structure and measurement of their dimensions is presented. The algorithms were tested using the numerical simulations carried out in k-Wave Toolbox. The approach and the experimental system were tested on the PMMA sample in damaged and intact states, which suggests the possibility of assessing the wear of parts. The experimental accuracy is ~0.1 mm in the direction of the probe pulse propagation, and ~0.2-0.3 mm in the perpendicular direction. The proposed technique is promising for fast industrial control of the dimensions of solid objects, underwater and immersion non-destructive inspection.

Acknowledgements

This work was supported by the Ministry of Education and Science of the Russian Federation grant No. K2-2019-004 (algorithms, simulation) and as part of the work on the State order of the Research Centre "Crystallography and Photonics" of Russian Academy of Sciences (experimental array).

References

[1] Olympus NDT 2007 Advances in phased array ultrasonic technology applications

[2] Nakahata K, Tokumasu S, Sakai A, Iwata Y, Ohira K and Ogura Y 2016 *NDT & E International* **82** 13

[3] Malkin R E, Franklin A C, Bevan R L, Kikura H and Drinkwater B W 2018 *NDT & E International* **96** 26

[4] Zarubin V, Bychkov A, Zhigarkov V, Karabutov A, Simonova V and Cherepetskaya E 2018 *Applied Physics Letters* **112**(21) 214102

[5] Bychkov A S, Zarubin V P, Karabutov A A, Simonova V A and Cherepetskaya E B 2017 *Photoacoustics* **5** 10

[6] Treeby B E and Cox B T 2010 *Journal of Biomedical Optics* **15**(2) 021314

[7] Wurzinger G, Nuster R, Schmitner N, Gratt S, Meyer D and Paltauf G 2013 *Applied Physics Letters* **4**(8) 1380

[8] Zarubin V, Bychkov A, Karabutov A, Simonova V and Cherepetskaya E 2018 *Applied Optics* **57**(10) C70

Acoustic properties of LNM crystal

D Yu Velikovskii[1], A I Kokshaiskii[2], M G Milkov[2] and V A Sukharev[3]

[1] Scientific and Technological Center of Unique Instrumentation (STC UI) RAS, Butlerova str., 15, Moscow, 117342, Russia
[2] M.V.Lomonosov Moscow State University, Faculty of Physics, Leninskie Gory, Moscow, 119991, Russia
[3] Dmitry Mendeleev University of Chemical Technology of Russia, Miusskaya sq., 9, Moscow, 125047, Russia

E-mail: velikovskii@ntcup.ru

Abstract. We report linear acoustic and acousto-optical properties of $LiNa_5Mo_9O_{30}$ (LNM) single crystal. This relatively new crystal, known for its complex crystallographic structure, is optically biaxial with high birefringence of 0.2545 at 450.2 nm, and crystals that consist of molybdates are widely used in different applications, from cintillators, double β-decay detectors, "dark matter" sensors to electro optics. The lowest measured propagation velocity of shear wave was 1200 m/s, and for the longitudinal mode it was 4100 m/s, which are a relatively low value for crystalline media. The maximum attenuation of the longitudinal wave at a frequency of 180 MHz was moderate and equal 20 dB/cm·GHz^2. The values of the acousto-optic figure of merit for light diffraction on a longitudinal acoustic wave were also determined, the maximum value of the coefficient was 13 c.u., which exceeds the efficiency of the one used in AO devices material α-SiO_2 . Thus, the set of measured acoustic and elastic-optical properties shows a new promising optically biaxial material for acousto-optics, which needs for further researching.

1. About the $LiNa_5Mo_9O_{30}$ crystal: properties and growth

We report linear acoustic and acousto-optical properties of $LiNa_5Mo_9O_{30}$ (LNM) single crystal. This relatively new crystal is known for its complex crystallographic structure, it is optically biaxial with high birefringence of 0.2545 at 450.2 nm, and crystals that consist of molybdates are widely used in different applications, from cintillators, double β-decay detectors, "dark matter" sensors to electro optics. The LNM single crystal is characterized by the lowest mm^2 orthorhombic symmetry, with unit cell parameters $a = 7.2265$ A, $b = 37.1575$ A, $c = 17.9398$ A. The dielectric axis X fits with the crystallographic axis a, the Y axis with the c axis, and the Z axis fits with the b axis. The crystal is not water soluble and has a Mohs hardness of 5. For the first time $LiNa_5Mo_9O_{30}$ phase was obtained in [1], the LNM single crystal was firstly grown in [2], the crystal density is 4.04 g/sm^3 [2]. The scope of complex molybdates crystals is very wide. They are used as scintillators, double β-decay detectors, "dark matter" sensors, as well as acousto-optic modulators of laser radiation, etc. [3, 4].

The crystalline sample provided for the research has dimensions of 4.6x11.5x8.7 mm along the a, b and c crystallographic axes, respectively, and were synthesized at the Dmitry Mendeleev University of Chemical Technology of Russia in Moscow. For the growth of a single crystal, ultrapure reagents,

Li_2CO_3, Na_2CO_3 with purity4N, manufactured by Fox-Chemocals GmbH, and especially pure molybdenum oxide with purity6N, manufactured by ARMOLED, were used. The quality control of reagents was carried out using ICP-MS. The single crystal of the composition $LiNa_5Mo_9O_{30}$ has a congruent character of melting and was grown by the low gradient Czochralski method with a temperature gradient in the melt of 1 K/cm. The draw rate was 1 mm/hour, the rotation speed of the crystal ranged from 8 to 25 rpm. The crystal was grown from a stoichiometric composition on the seed, oriented in the direction of the main axis. The control of the stoichiometry of the grown crystal was carried out by the SIMS method, along the entire length of the boule. It was found that 75% of the boules have a stoichiometric composition.

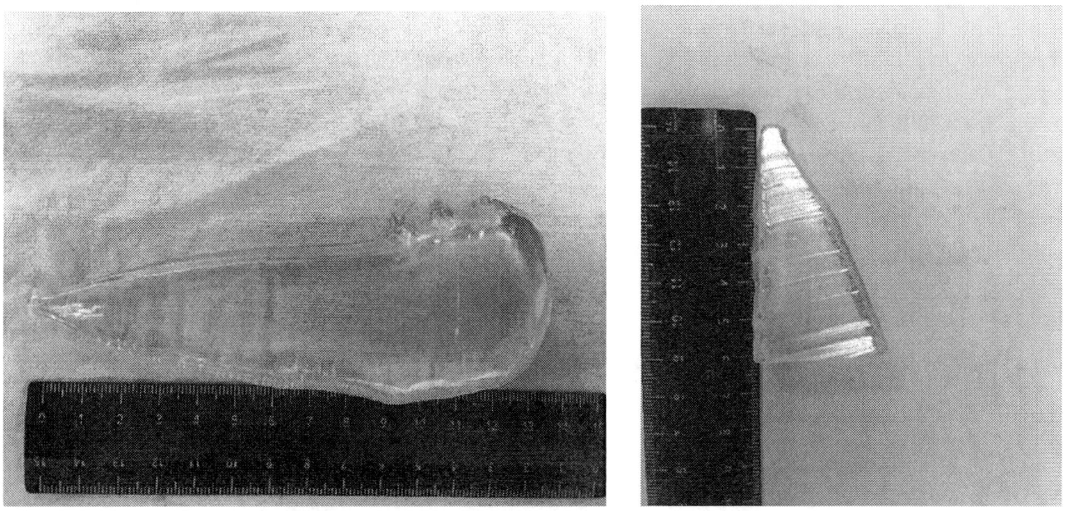

Figure 1. Figure with short caption (caption centred).

To determine the optimal conditions for growing a single crystal, a numerical simulation of the thermal node was performed. The calculated configuration of the heat node allowed growing a high-quality single crystal with a mass of more than 250 grams.

The material is transparent in the visible range of optical radiation, has a large transmission window (0.357-5.26 μm), is optically biaxial, and has relatively large values of the refractive index, with noticeable anisotropy (1.8266, 2.0123 and 2.0414 at a wavelength of 0.6365 μm) [2].

2. Acoustic properties

2.1. Measurement of the bulk speed and attenuation of ultrasonic longitudinal waves by acousto-optic methods

The propagation velocities of longitudinal acoustic waves are determined by the acousto-optic (AO) method by comparing the angle between +1 and -1 diffraction orders arising from the diffraction of light by a sound wave propagating in a reference (SiO_2) and probe (LNM) materials. The diffraction pattern was taken on the screen, equally distant from the acoustic column in both materials. Thus, to restore the speed of sound in the material under study, it is sufficient to compare the ratio of the distances between the diffraction orders and the speed of ultrasonic in fused quartz. The measured propagation velocities of the longitudinal mode in the LNM crystal were $V_X = 4100$ m/s, $V_Z = 5235$ m/s and $V_Y = 4780$ m/s.

Measurements of the attenuation of ultrasound in a single crystal LNM were carried out by an acousto-optic method. It is known that the intensity of diffracted light with an acousto-optic effect depends on many factors [5]. As for the dependence of the intensity of the diffracted light on the intensity of the ultrasonic wave in the region of acousto-optic interaction, with a low diffraction efficiency, it is almost linear. This circumstance allows us to measure the attenuation of ultrasound [5]. The ultrasonic wave is launched into the crystal in a pulsed mode. The pulse duration must be sufficiently short, and the pulse repetition period must be much longer than the pulse duration. The diameter of the laser beam passing through the medium must be substantially smaller than the dimensions of the sample under study. An ultrasound pulse, launched into the test medium through the entrance face, propagates in it along the test direction, passes through the laser beam, and, after being reflected from the face opposite to the input edge, goes back and passes through the laser beam again.

Due to the attenuation of the ultrasonic wave, the intensity of the diffracted light I_2 on the pulse going back will be less than the intensity I_1. The diffracted beams fall on the photodetector and are recorded by an oscilloscope, which is synchronized with the master pulse generator. The recording of the intensity of each pulse is carried out after the angle of incidence of the light on the ultrasound corresponds to the fulfillment of the Bragg phase matching condition [6]. The distance L, that the acoustic wave passed in the medium between these two scattering events, can easily be defined as the product of the ultrasonic wave propagation velocity along the selected direction in the medium υ and the time delay between the first and second diffracted light pulse $L = v \cdot \tau$. The ratio $a = I_2/I_1$ gives us the absolute value of the attenuation of ultrasound in intensity in the case of an ultrasonic wave with frequency f passing a distance L in the medium. The value

$$\Gamma = \frac{10 \cdot \lg(I_2 / I_1)}{(10^{-9} \cdot f)^2 \cdot 10^{-2} \cdot L}$$

gives the attenuation in units of dB/sm·GHz^2.

This approach to measuring the attenuation of ultrasound avoids the need to pass a laser beam through different areas of the sample under study and, as a result, does not require normalization of the intensity of the diffracted light to the intensity of the transmitted beam I_0. Such a normalization introduces an additional error if the optical faces of the crystal are not processed well enough and have a profile that is different from the planar one. It should also be noted that attenuation measurements should be carried out at such high frequencies at which it begins to manifest itself.

This method has some of the following limitations: the implementation of acousto-optical scattering and registration of the two pulses mentioned above is important in the medium under study, the divergence of ultrasound should be small, and the ultrasonic wave should propagate along those directions of the medium where there is no significant acoustic demolition.

Single crystal LNM, belongs to the orthorhombic system. The calculated crystal density is 4.0 g/cm^3 [1]. Crystalline sample was oriented its faces orthogonal to the crystallographic axes. The sample has the following dimensions: 4.6 mm along the crystallographic axis a, 11.5 mm along the axis b and 8.7 mm along the axis c. The velocities of longitudinal ultrasonic waves along these axes, measured in operation, are $V_a = 4100$ m/s, $V_b = 5235$ m/s, and $V_c = 4780$ m/s (table 1).

180 MHz longitudinal ultrasonic wave was generated in an acousto-optic cell (buffer) based on fused silica and launched into the LNM sample alternately along each of the three axes — a, b, and c. Acoustic contact of the ultrasound input face of the sample and the output face of the buffer was carried out through a layer of glycerin. The sample face opposite to the input face (reflecting face) did not has acoustic contact with any solid or liquid body so that the reflection coefficient of the ultrasonic pulse from this face could be considered equal to unity. The crystal was illuminated by a laser beam from a He-Ne laser with a wavelength of 633 nm and aperture 0.8 mm. Ultrasound attenuation measurements were performed as described above, namely by measuring the intensity ratio I_2 of diffracted light on an acoustic pulse after it was reflected from the corresponding face with its intensity I_1 as a result of diffraction on the same pulse going in the direction of the reflecting face. The beam of light passed through the sample at two different distances from the reflecting face, which gave 2 different values of the absolute attenuation a (a <1) for two different distances (wavelengths), passed by ultrasound. The attenuation values Γ were also calculated in units of dB/sm·GHz^2. The uncertainty in determining the speed of ultrasound is 0.5%, the uncertainty in determining the time interval τ, and,

consequently, the path length L is about 5%. The uncertainty in determining the intensity of an ultrasonic wave (the intensity of diffracted light) is about 10%. The attenuation of ultrasound in all three directions is not significant.

2.2. Measurement of propagation velocity of shear ultrasonic waves by pulse method

The harmonic wave velocity may be found by measuring the wave phase. The phase of a plane wave that propagates in the direction of increase of x is written as $\varphi(x, t) = \omega t - kx + \varphi$. In the source at $x = 0$, the wave phase equals to $\varphi(0, t) = \omega t + \varphi$. As is evident, the difference of phases Φ between the signals of source and receiver is not dependent on time and appears to be a rather simple distance and frequency function:

$$\Phi(x, f) = \varphi(0, t) - \varphi(x, t) = kx = \frac{2\pi}{c} \cdot f \cdot x$$

According to the equation, the difference of phases Φ is in linear dependence on both distance x and frequency f, with phase increase rate inversely proportional to wave velocity c. Thus since we know the distance x between the source and receiver it is possible to measure the ratio of phases Φ displacement to frequency f and from the resulting line slope ($\partial\Phi/\partial f = 2\pi\, x/c$) finally find the velocity c value.

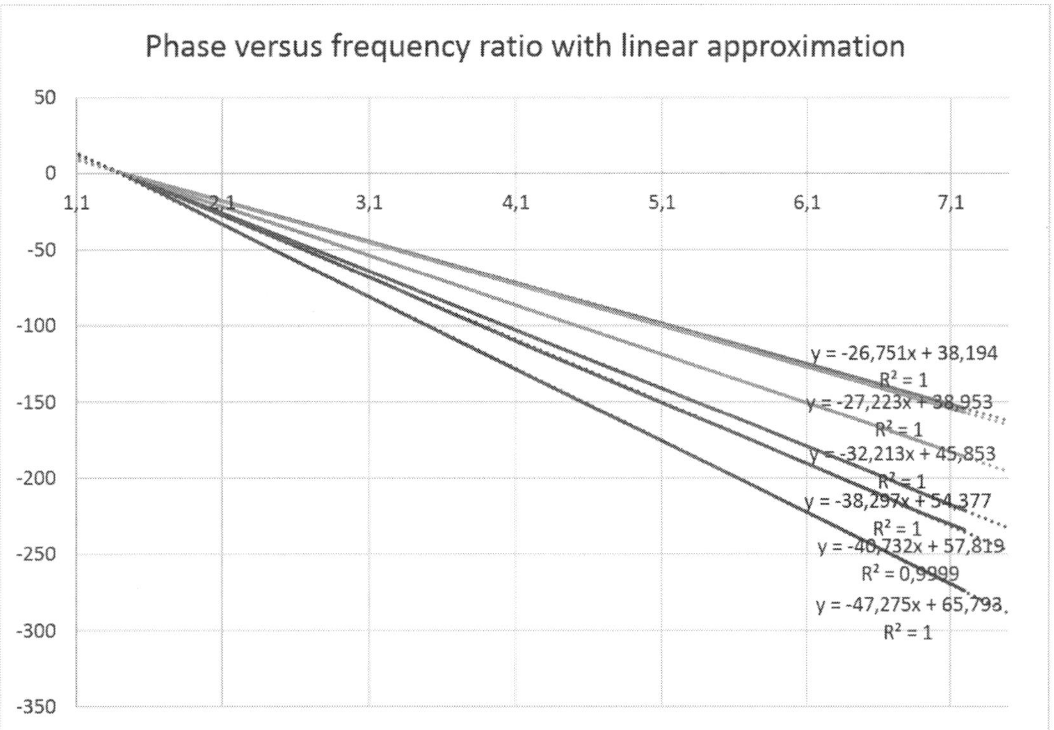

Figure 2. Phase versus frequency ratio with linear approximation.

The measurement of shear acoustic waves velocity in the orthorhombic crystal sample was carried out by phase method. On the opposing sides of the sample, two transducers were attached. Then one of the transducers excited a shear wave pulse in the sample, and the other one detected it. In the course of measurement for each direction and orientation of shear waves approximately 30 000 ultrasonic pulses in the series of 50 were send. Within each series, the properties of the pulses were uniform therefore they were averaged to provide more accurate results. Various pulse series had different frequency, as it changed linearly from 1.4 MHz to 7.2 MHz. The changes in the phase of the signal were also linear.

Table 1. Propagation velocity of longitudinal and transverse ultrasonic modes along axial directions of LNM crystal, and attenuation of longitudinal ultrasonic waves on a frequency of 180 MHz.

Polarization	$C_{\perp}$, **[m/sec]**	**L [mm]**		
ab XZ	1279±3	4,64±0,1	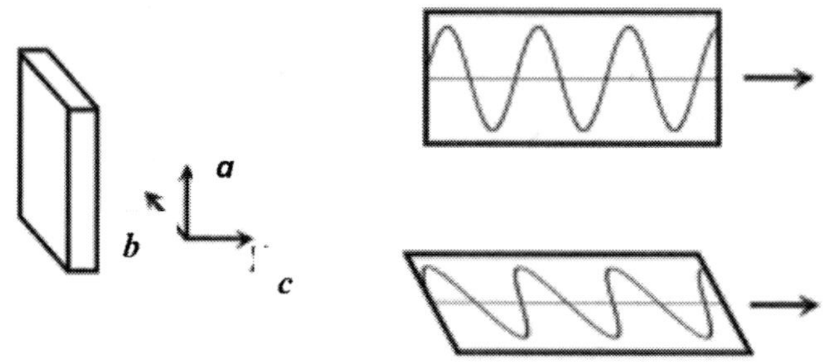	
ac XY	1187±3			
aa XX	4100±10			Attenuation $a=I_2/I_1 = 0{,}94$; $\Gamma = 20\pm2$ dB/sm·GHz^2
ba ZX	1283±2	8,74±0,1	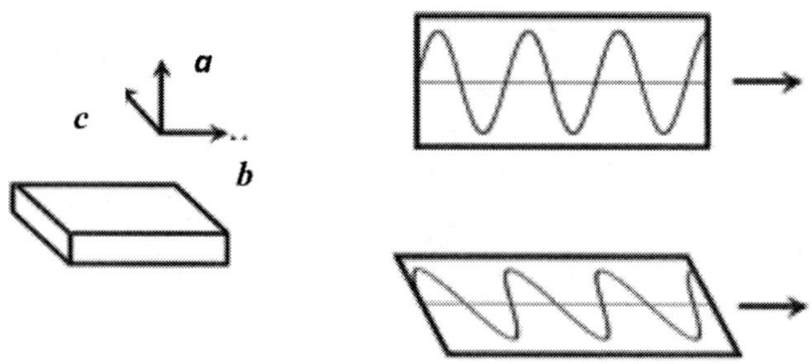	
bc ZY	1796±2			
bb ZZ	5235±10			Attenuation $a=I_2/I_1 = 0{,}94$; $\Gamma = 9\pm1$ dB/sm·GHz^2
ca YX	1192±2	11,53±0,1	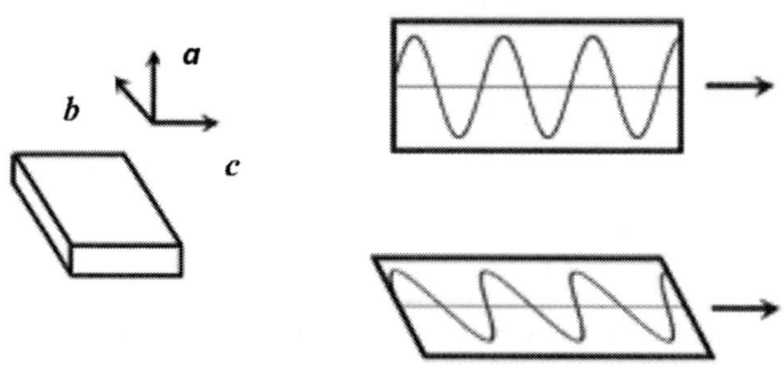	
cb YZ	1801±2			
cc YY	4780±10			Attenuation $a=I_2/I_1 = 0{,}92$; $\Gamma = 11\pm1$ dB/sm·GHz^2

If we know a simple equation for wave phase $\varphi(x, t) = \omega t - kx + \varphi_0$, it is possible to combine the changes in signal frequency with changes if phases difference in the source and receiver $\frac{\partial\varphi}{\partial f} = \frac{2\pi l}{c}$. Consequently, with the help of values of the sample length along the propagation direction of the wave L and the angle of the graph of phase versus frequency ratio, it is possible to estimate wave velocity c.

2.3. *The coefficient of AO figure of merit of crystal LNM*

The values of the acousto-optics figure of merit M_2 for LNM crystal were measured by the Dixon-Cohen method [7]. These measurements taken on a longitudinal ultrasonic wave with a frequency of 80 MHz. Liquid glycerol gluing was used for acoustic contact of the sample and reference AO cell. The reference AO cell (buffer) based on fused silica also used. Research was carried out for three different directions of propagation of longitudinal ultrasound wave (along the axes a, b, and c). Each direction of the ultrasound corresponds to two directions of propagation of light with a wavelength of $\lambda = 633$ nm and each of the beams had two polarizations corresponds to the intrinsic optical modes of the sample. The results of measuring the M_2 listed at the table 2. The absolute accuracy of the method is approximately 20%.

Table 2. The values of the coefficients of AO figure of merit M_2 for LNM crystal under isotropic diffraction on a longitudinal wave.

Direction of propagation of the ultrasound wave	Direction of propagation of the light	Polarization (**E** vector along axis)	Value of the Acousto-optic figure of merit M_2, ($x10^{-18}$ s^3/g)
a (X)	b (Z), c (Y)	a (X)	10,2±2,0
a (X)	b (Z)	c (Y)	**13,1±2.6**
a (X)	c (Y)	b (Z)	10,9±2.2
b (Z)	a (X), c (Y)	b (Z)	1,9±0,4
b (Z)	a (X)	c (Y)	0,9±0,2
b (Z)	c (Y)	a (X)	1.1±0.2
c (Y)	a (X), b (Z)	c (Y)	5.6±1.2
c (Y)	a (X)	b (Z)	8.1±1.6
c (Y)	b (Z)	a (X)	5.5±1.1

Measurements show that the elastic-optical effect in the LNM crystal is relatively good, but the AO figure of merit M_2 of the LNM crystal in case of diffraction of light at longitudinal ultrasound waves has relatively small values. The maximum value of the measured figure of merit was 15 a.u.. So the maximum value of M_2 is many time more than M_2 of the crystalline α-SiO_2 and as well as the KGW crystal [8].

It would be better to combine measurements of AO figure of merit on a new specimen larger than in present measurements. It is necessary to use another AO buffer to excite shear acoustic wave to make a research of the AO properties of LNM on shear waves..

3. Conclusion

A new optically biaxial crystal $LiNa_5Mo_9O_{30}$ was under research. In such a media, it is possible to create new configurations of AO interaction, which provides new types of electro-optical devices [9, 10]. The maximum received coefficient of AO figure of merit was $15x10^{-18}$ s^3/g. The minimum propagation velocity of ultrasound was 1200 m/s for the shear mode and 4100 m/s for the longitudinal mode. The attenuation of ultrasound in all three directions at a frequency of 180 MHz is not significant, attenuates less or substantially less than 3 dB/cm (<< 2 times per 1 cm of the wave propagation). All this speaks of the great promise of the LNM crystal for the purposes and tasks of acousto-optic.

Acknowledgments
This work was carried out through a Russian Science Foundation grant (project RSF 17-72-10300)

References

[1] Hamza H and et al 2012 $LiNa_5Mo_9O_{30}$: *ActaCrystallographica Section E Structure Reports Online* **68(11)** 80-1

[2] Zhang W and et al 2016 $LiNa_5Mo_9O_{30}$:Crystal growth, linear, and nonlinear optical properties *Chemistry of Materials* **28(12)** 4483-91

[3] Shlegelh V N and et al 2016 *Astroparticle Physics* **72** 38-45

[4] Martynyuk-Lototska I and et al 2015 Acoustic anisotropy for acoustooptic $NaBi(MoO_4)_2$ crystals *Ukrainian journal of physical optics* **16(2)** pp 69-76

[5] Balakshii V I, Parygin V N and Chirkov L E 1985 *Physical Fundamentals of Acoustooptics* (Russian: Moscow Radio i svyaz)

[6] Voloshinov V B and Lemyaskina E A 1996 Acousto-optic measurements of ultrasound attenuation in tellurium dioxide crystal *Acta Physica Slovaca* **46(6)** 733-38

[7] Dixon R W and Cohen M G 1966 A new technique for measuring magnitudes of photoelastic tensors and its application to lithium niobate *Appl. Phys. Lett.* **8(8)** pp 205-06

[8] Mazur M M , Velikovskii D Yu, Kuznetsov F A, Mazur L I, Pavlyuk A A, Pozhar V E and Pustovoit V I 2012 Elastic and Photoelastic Properties of $KGd(WO_4)_2$ Single Crystals *Acoustical Physics* **58(6)** 658–65

[9] Lee Hyuk 1987 Acoustooptic light modulation with large bandwidth and angular aperture *IEEE Transactions on Ultrasonics, Ferroelectrics and Frequency Control* **34(4)** 485-6

[10] Velikovskii D Yu, Pozhar V E and Mazur M M 2015 Acousto-optic devices based on potassium rear-earth tungstates laser crystals *Acta physica polonica a* **127** 75-7

Application of a laser-Doppler anemometry for the study of unsteady flow structure in a model micro-hydro turbine

S I Shtork[1], I V Litvinov[1,2], E Yu Gorelikov[1,2], D G Mukhin[1], S V Dremov[1,2] and D A Suslov[1,2]

[1] S S Kutateladze Institute of Thermophysics SB RAS, 630090, Russia, Novosibirsk, Lavrentyev Avenue, 1
[2] Novosibirsk State University, 630090, Russia, Novosibirsk, Pirogova Street, 1

E-mail: shtork@itp.nsc.ru

Abstract. Current work deals with an LDA characterization of the velocity distributions and pressure pulsation maps in an axisymmetric draft tube for the flow conditions corresponding to wide range of the operating regimes of the micro-hydro turbine unit. For the detailed analysis of the velocity profiles, three characteristic regimes corresponding to part-load regime with precessing vortex core formation, best efficiency regime and overload regime of hydro turbine operation were selected. To capture spatial characteristics of the unsteady flow for part-load regime a phase averaging procedure for the LDA data has been employed.

1. Introduction

Hydropower is the oldest and currently most developed clean energy area employing renewable energy sources [1, 2]. Hydropower has high efficiency and flexibility in load control. Currently, the share of hydropower in the global balance of power generation is about 16%. In many countries, it represents the major source of energy production, as, for example, in Norway (99%), Brazil (86%), Switzerland (76%), Sweden (50%). At present, along with large hydropower plants, small hydropower industry occupies a rather prominent place in the world. The great interest in small hydropower is due to the fact that small hydropower plants can operate both autonomously and are included in the power grid. Low-power hydro stations can be installed on small rivers or streams. As energy sources for small hydropower plants, natural elevation changes on various spillways, irrigation systems [3], pipelines and industrial spillways [4] can be used. Small hydropower plants can be installed at low cost on existing reservoirs. Small hydropower is probably one of the most clean technology to produce electricity, because small hydropower plants do not pollute the atmosphere with harmful emissions, do not block the migration ways of fish, and do not need creation of reservoirs for their work that flood large areas of land. Small HPPs do not pose a serious danger in the event of emergency situations on the plant and are characterized by short commissioning times and low operating costs. Modern equipment for micro hydro power plants can be made quite compact with automatic control system, which allows the station to operate in an automatic mode, without the involvement of staff on duty.

To develop recommendations on the organization of optimal geometric boundary conditions and operating modes of a micro-hydro turbine in terms of flow stability, minimal energy losses in a wide range of load control of a hydro-turbine apparatus, detailed studies of flow characteristics in the micro-turbine elements are necessary [5]. In that context, the current work deals with an LDA characterization of the velocity distributions and pressure pulsation map in an axisymmetric draft tube model for the flow conditions corresponding to wide range of the operating regimes of the micro-hydro turbine. In particular, the study focus was on determining regimes with generation of powerful flow pulsations in form of precessing vortex core (PVC). To capture spatial characteristics of the unsteady flow an ingenious phase averaging procedure for the LDA data has been employed.

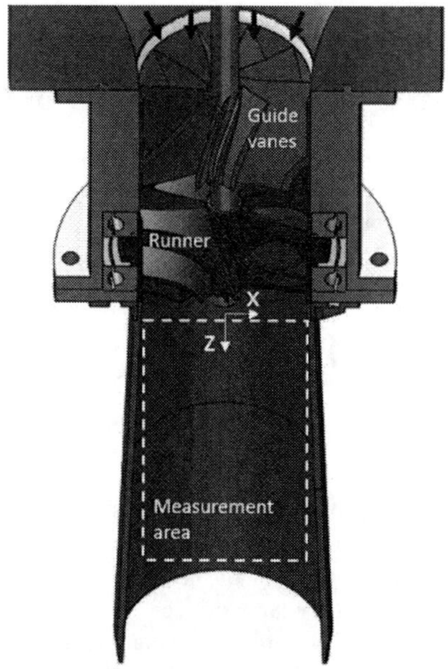

Figure 1. Scheme of experimental setup (measuring cross-section shifted 4 mm down).

2. Experimental setup

A concept based on a propeller-type turbine has been chosen as a prototype microhydroturbine device [6-10]. This turbine consist of sequentially installed stationary and rotating blade elements (figure 1). As a reference case, the geometry of the axial swirlers simulating velocity distribution at the exit of a real hydroturbine [11] has been used. In particular, the geometry used was calculated for the turbine operation at the best efficiency point (BEP) for the flow rate $Q_c = 48.5$ l/s and the runner rotation frequency $n_c = 40.5$ Hz of the laboratory scale model. All the parts of the model test section were manufactured making use of a rapid prototyping technique. By applying a sophisticated software, it was possible to maintain the specified flow regime parameters within the required time with an uncertainty of 1.5 and 0.5% for setting a discharge Q and the rotation frequency of the runner n, respectively. The inlet diameter of the cone draft tube was $D = 100$ mm. The length of the cone was 280 mm; the cone angle was 4°.

To characterize the various regimes of the hydro turbine operation, pressure pulsations were measured using two acoustic probes with the same procedure as described in [12]. Acoustic probes included measuring microphones Behringer ECM8000, which were connected to a pressure wave transmitter in the form of thin capillary. Acoustic measurements were performed on diametrically opposed sides of the draft tube cone. The acoustic probe ends were installed in the holes drilled in the cone wall at a distance of 50 mm from the cone inlet.

The averaged and phase-averaged velocity distributions of the flow were measured using a laser Doppler anemometer "LAD-06i" in the selected central cross-section as showed in figure 1. As tracers we used particles of paraffin oil aerosol, produced by Laskin atomizer, which allows obtaining droplets with a characteristic size of 1-3 μm [13].

3. Results

3.1. Characteristics of the flow

Figure 2 (a) shows the maximum amplitudes in the spectra of the differential signal of two sensors as a function of the flow rate and rotation speed of the impeller. This procedure allows retrieving the fluctuating signal component corresponding to the PVC and eliminate the in-phase component of pulsations [12]. It can be noticed that a shift away from the BEP, for example by reducing the flow rate below Q_c at a constant rotor frequency n_c (partial load) results in a rise of the pressure pulsation intensity as can be seen in the upper left area of the plot.

The figure 2 (b) shows frequencies of the maximum amplitudes in the spectra of differential signals as a function of the volumetric flow rate and the impeller rotation frequency. The diagram evidences a linear growth of the frequency with increasing of the parameters Q and n.

To survey the velocity profiles over the full space of investigated operating parameters we have chosen the flow rates corresponding to $0.51{\cdot}Q_c$, Q_c, and $1.51{\cdot}Q_c$ at a constant rotation speed of the impeller equal to n_c. These three distinctive regimes correspond to part-load regime with PVC formation, BEP regime and overload regime, respectively. In figure 3, the time-averaged velocity field in the central cross-section of the draft tube cone is presented. It can be noted inspecting the velocity field plots that main features of the flow structure are very similar to those observed for the draft tube model with an elbow part [12]. For the optimal regime (figure 3 (a)) attributed to a subtle angular-to-axial momentum ratio, the axial velocity is invariably directed along the z-axis throughout the overall surveyed region except for the insignificant backward flow area under the cowl, which ends at a distance $z < 10$ mm. The tangential velocity distribution at the optimal conditions with flowrate Q_c shows a flow pattern consisting of two concentric counter-rotating swirling flows (see two white dot lines at figure 3 (b)). The inner part of swirling flow with a transversal size of about 20 mm, slightly shifted from the cone central axes, rotates in the counterclockwise direction following the runner exit angle, but opposite to the impeller rotation direction. The external rode type flow area is unwraped in a clockwise direction superimposed by guide vanes of the stationary swirler.

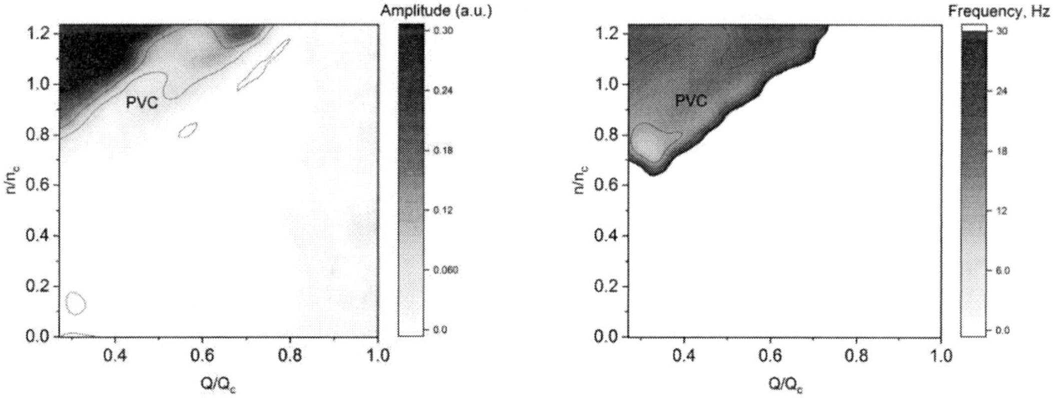

a b

Figure 2. Pressure pulsations map with pressure pulsations.

In the case of over load at flowrate $1.51{\cdot}Q_c$ (figure 4), the axial velocity distribution shows that the structure of the flow significantly changed (figure 4 (a)). In the bottom part of the cone, it can be seen that there is a reducing of the flow velocity in the center of the cone. This region is associated with the formation of a PVC at the cone exit. As can be seen from the tangential velocity distribution, an axisymmetric concentrated vortex core appears in the centerline of the cone (size of vortex $\approx$10 mm), and the flow on the periphery of the cone occupied by free vortex (figure 4 (b)). In the bottom part of the cone, PVC forms without observation of strong pressure pulsations.

Journal of Physics: Conference Series **1421** (2019) 012068 doi:10.1088/1742-6596/1421/1/012068

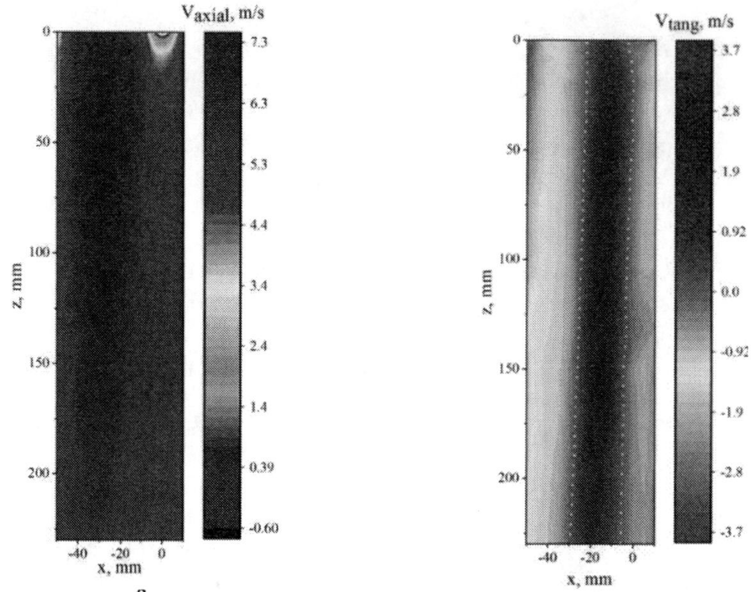

a b

Figure 3. The time-averaged velocity field in the central cross-section of the draft tube cone: axial (a) and tangential velocity (b) at flowrate Q_c.

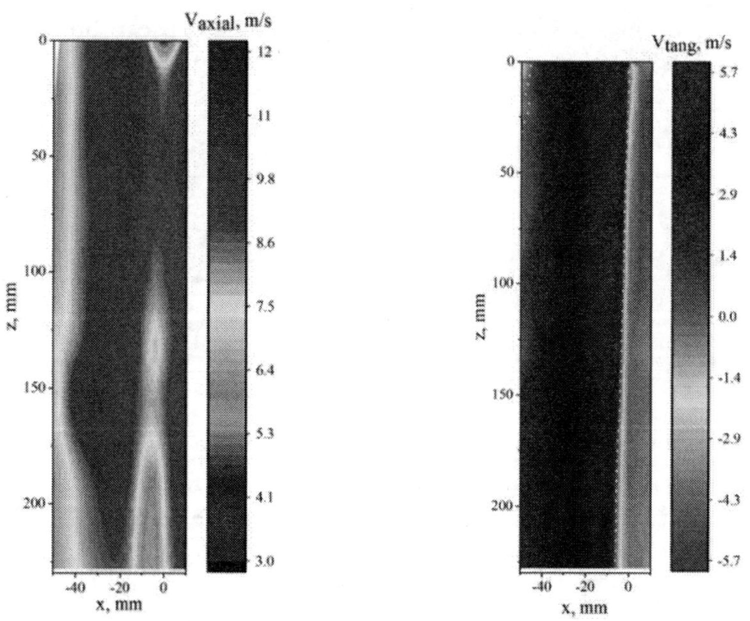

a b

Figure 4. The time-averaged velocity field in the central cross-section of the draft tube cone: axial (a) and tangential velocity (b) at flowrate $1.51{\cdot}Q_c$.

In the case of part load conditions at flowrate $0.51{\cdot}Q_c$ (figure 5), the axial momentum is much weaker than the angular one and the flow is forced to squeeze along the cone wall where the axial velocity attains its maximum (figure 5 (a)). The much wider central area, progressively increasing in transversal size when moving downstream, is near a stagnant flow. The tangential velocity shows that the vortex deviates from the central position in the cone, i.e. strong PVC motion observed in the cone (figure 5 (b)).

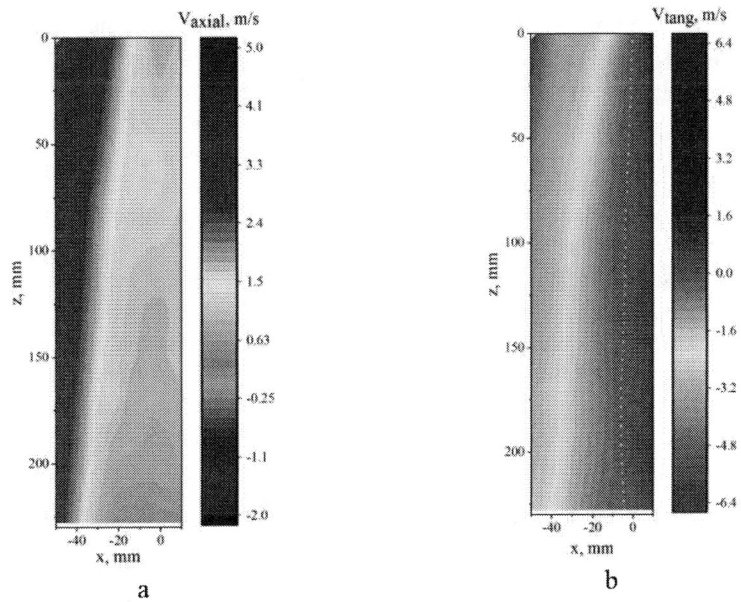

Figure 5. The time-averaged velocity field in the central cross-section of the draft tube cone: axial (a) and tangential velocity (b) at $0.51 \cdot Q_c$.

To understand the three-dimensional structure of the PVC, a phase averaging of the velocity field was constructed with applying the reference signal from the acoustic probe following to procedure outlined in [14]. Figure 6 presents a sample result on the phase-averaged of three components of velocity over an *x-y* cross-section. It can be noticed that the recirculation zone (area with negative axial velocity) and the PVC are shifted apart from the geometric centre of the draft tube cone.

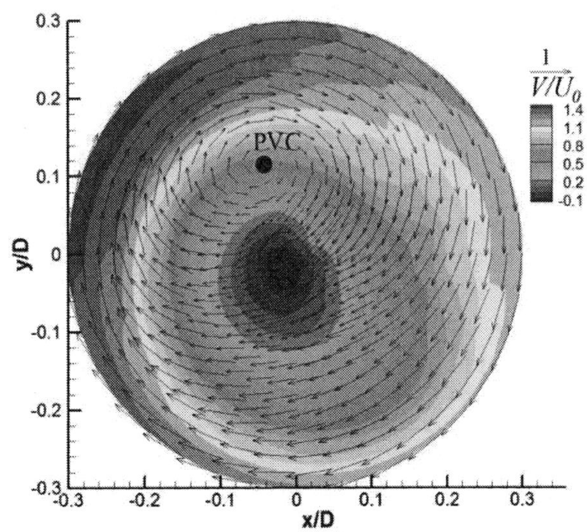

Figure 6. Phase-averaging velocity distribution at $z = 4$ mm cross-section.

4. Summary

Current work deals with an LDA characterization of the velocity distributions and pressure pulsation map in the axisymmetric draft tube of a model micro-hydro turbine for the flow conditions corresponding to wide range of the operating regimes of the hydro turbine unit. To survey the velocity profiles over the full space of investigated operating parameters the three characteristic regimes correspond to part-load regime with PVC formation, BEP regime and overload regime of hydro turbine operation were selected. To capture spatial characteristics of PVC for part-load regime a phase averaging procedure for the LDA data has been employed. In particular, it was shown that a shift away from the BEP, for example, by reducing the flow rate below Q_c at a constant rotor frequency n_c (partial load) results in a rise of the intensity of pressure pulsations superimposed by the PVC. At these conditions, the axial momentum is much weaker than the angular one and the flow is drastically pushed towards the cone periphery where the axial velocity reaches rather high maximum. While the overly broad central region is occupied by a stagnant flow. Therefore, the overall flow is forced to squeeze through the narrow annulus area along the cone wall. The revealed heavy unsteadiness and nonuniformity of the flow at overload conditions would naturally lead to energy losses and reducing the hydroturbine efficiency. Thus, the problem of extending range of the stable and efficient operation of the microhydroturbine device demands for discovering ways to suppress the PVC and to perform levelling of the axial flow distribution.

Acknowledgments

The study of pressure pulsation map was carried out under state contract with IT SB RAS (AAAA-A17- 117030910025-7), the measurements of time-mean and phase-averaged velocity fields were funded by Russian Foundation for Basic Research and Government of the Novosibirsk Region of the Russian Federation according to the research project No. 18-48-540033.

References

[1] Sobolin G V, Satunkin I V and Korovin Yu I 2004 Problems of using small rivers and canals of irrigation systems for the development of small hydropower *News of the Orenburg State Agrarian University* **2** (2-1) 32-35 [In Russian]

[2] Jain S V and Patel R N 2014 Investigations on pump running in turbine mode: A review of the state-of-the-art *Renewable and Sustainable Energy Reviews* **30** 841–68

[3] Shchedrin V N, Baklanova D V, Bondarenko V L and Lobanov G L 2017 Assessment of the prospects for the use of small hydropower on irrigation systems to meet the internal needs of electricity *Scientific Journal of the Russian Research Institute of the Land Reclamation Problems* **3** (27) 160–78 [In Russian]

[4] Popelyukh I A and Zhdanovich A A 2013 Analysis of the installation sites of micro-hydropower plants in industrial effluents of Novosibirsk *In: Science, Technology, and Innovation - Materials of the All-Russian Scientific Conference of Young Scientists: in 10 parts - Novosibirsk State Technical University* 191-94 [In Russian]

[5] Zhang Y, Liu K, Xian H and Du X 2018 A review of methods for vortex identification in hydroturbines *Renewable and Sustainable Energy Reviews* **81** 1269–85

[6] Ivanov V M, Ivanova T Yu, Zhdanov E P, Blinov A A and Pchelintsev S G 2010 Hydropower plant with axial hydraulic turbine of a new original design and hydraulic stand for complex modeling of flow parts of hydroturbines *Polzunovsky Vestnik* **4** (2) 54-60. [In Russian]

[7] Yassi Y and Hashemloo S 2010 Improvement of the efficiency of the Agnew micro hydro turbine at part loads due to installing guide vanes mechanism *Energy Conversion and Management* **51** 1970–75

[8] [8] Shojaeefard M H, Mirzaei A and Babaei A 2014 Shape optimization of draft tubes for Agnew microhydro turbines *Energy Conversion and Management* **79** 681–89

[9] [9] Nishi Ya, Kobayashi Yu, Inagaki T and Kikuchi N 2016 The design method of axial flow runners focusing on axial flow velocity uniformization and its application to an ultra-small

axial flow hydraulic turbine *International Journal of Rotating Machinery* **2016** 5390360
[10] Kaunda Ch S, Kimambo C Z and Nielsen T K 2014 A technical discussion on microhydropower technology and its turbines *Renewable and Sustainable Energy Reviews* **35** 445–59
[11] Sonin V, Ustimenko A, Kuibin P, Litvinov I and Shtork S 2016 Study of the velocity distribution influence upon the pressure pulsations in draft tube model of hydro-turbine *IOP Conference Series: Earth and Environmental Science* vol 49 (IOP Publishing) p 82020
[12] Litvinov I, Shtork S, Gorelikov E, Mitryakov A and Hanjalic K 2018 Unsteady regimes and pressure pulsations in draft tube of a model hydro turbine in a range of off-design conditions *Experimental Thermal and Fluid Science* **91** 410–22
[13] Echols W H and Young J A 1963 Studies of portable air-operated aerosol generators *Naval research laboratory report* 16–16
[14] Litvinov I V, Sharaborin D K and Shtork S I 2015 Finding of parameters of helical symmetry for unsteady vortex flow based on phase-averaged PIV measurement data *Thermophysics and Aeromechanics* **22** 647–50

Journal of Physics: Conference Series **1421** (2019) 012069 doi:10.1088/1742-6596/1421/1/012069

Selective scheme of gas concentration measurement based on acousto-optic spectral filters with multi-frequency control signal

Yu V Pisarevsky[1,2], S A Kolesnikov, V A Lomonov[1] and A V Vinogradov[1]

[1] FSRC «Crystallography and Photonics» RAS, Leninskiy Prospekt 59, Moscow 119333 Russia
[2] National Research Centre "Kurchatov Institute", Akademika Kurchatova pl., 1, Moscow 123182 Russia

E-mail: yupisarev@yandex.ru

Abstract. A scheme for measuring the concentration of components of gas mixtures improving the selectivity of conventional optical absorption spectroscopy is proposed. The scheme is based on the use of acousto-optic spectral filter as a dispersion element which is fed simultaneously several control signals with frequencies corresponding to the maxima and minima of absorption in the optical spectrum of the measured gas. The scheme is implemented on the basis of the acousto-optic filter based on paratellurite TeO_2 crystal, providing the high resolution achieved by reducing the absorption of ultrasound due to the special technology of the growth of high quality crystals and the use of absorption anisotropy of ultrasound in paratellurite. Experimental verification of this approach with measuring the concentration of nitrogen dioxide has confirmed the effectiveness of the proposed scheme.

1. Introduction

Optical spectroscopy is widely used to detect and determine the composition of gas mixtures. Absorption spectroscopy provides the possibility to measure in a wide range of gas concentration; however, it usually meets serious problems with selectivity in the allocation of individual component in the gas mixture [1]. Selectivity can be significantly improved by using a correlation optical scheme based on the use of optical polarization filters [2-4]. However, such schemes operate upon a certain gas and changeover from one gas to another requires the replacement of the filter, which does not allow measuring several gas components using just one device.

Solving this problem may be possible by using of tunable acousto-optic filters [5-13]. This paper describes one of the solutions using several frequencies and modulation of ultrasound in the control of an acousto-optic filter based on paratellurite applied to the detection of such an atmospheric pollutant as nitrogen dioxide.

2. Experimental part

The experimentally proposed scheme was tested on a sealed cuvette with nitrogen dioxide having an optical absorption band of 2-3 eV. An acousto-optic spectral filter based on paratellurite was manufactured for this band and a control scheme based on several RF signal generators was created. The diagram of the measuring unit is shown in figure 1.

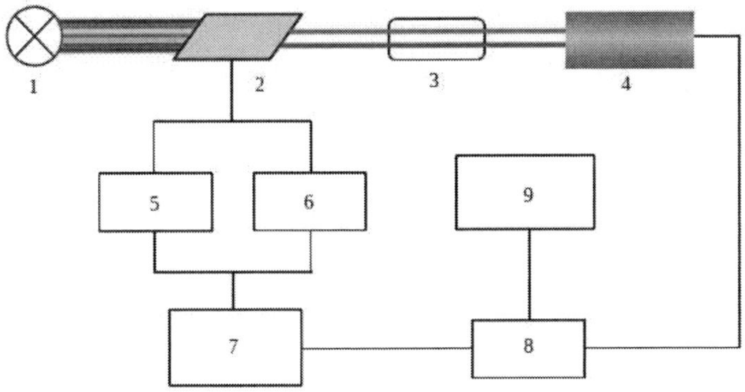

Figure 1. 1 – light source; 2 – paratellurite acousto-optic spectral filter; 3 - NO_2 filled cuvette; 4 - photomultiplier; 5 and 6 - radio frequency generators; 7 - kHz frequency generator; 8 - lock-in amplifier; 9 - data recorder.

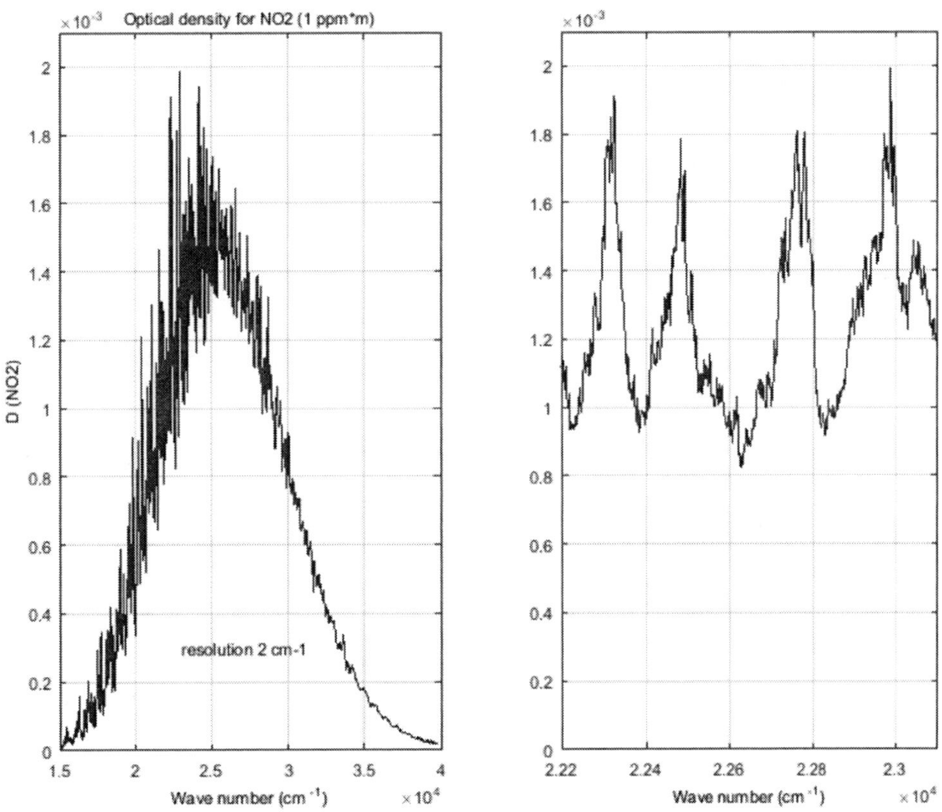

Figure 2. Optical density (the relative absorption) D of NO_2 with concentration 1 ppm m on the right hand side of the picture there is a section of the most intensive absorption which was used for present measurements.

The light from light source (1) was collected in a collimated beam and passed through an acousto-optic filter (2) and a cuvette with the test gas (nitrogen dioxide) (3). The acousto-optic filter received signals from two high-frequency generators (5) and (6) simultaneously. Frequency control signals were 169.7 MHz and 172.1 MHz. The control signals were modulated in the antiphase from the sound generator, which simultaneously gave a reference signal to the synchronous detector. The supply of signals of these frequencies from two generators to the acousto-optic filter led to the transmission of light by the acousto-optic filter in two narrow bands and corresponding to the maximum and minimum absorption of nitrogen dioxide in the 2-3 eV band. The left part of figure 2 shows a general view of the absorption band of nitrogen dioxide in the blue region of the spectrum. The right part of figure 2 shows a large-scale portion of the spectrum of the maxima and minima which were used to measure in this paper. Optical signals passed through a cell with nitrogen dioxide were detected by a photomultiplier and than fed into synchronous detector.

The described features of the used measurement scheme required high selectivity of the acousto-optic filter for control signals in the range of 150-200 MHz. This imposed a special requirement on the design of the filter and the quality of the crystal by the features of its own sound absorption determined by the viscosity tensor. Growth of paratellurite crystals was carried out by the Czochralski method. Of paramount importance is the provision of low attenuation of ultrasound on the working light aperture at relatively high frequencies for paratellurite. The attenuation of ultrasound is determined by losses due to scattering on defects crystals were grown from raw materials of high purity and with optical homogeneity better than 0.1 wavelength. The anisotropy calculations and measurements given in figure 3 were carried out during the development of the filter design.

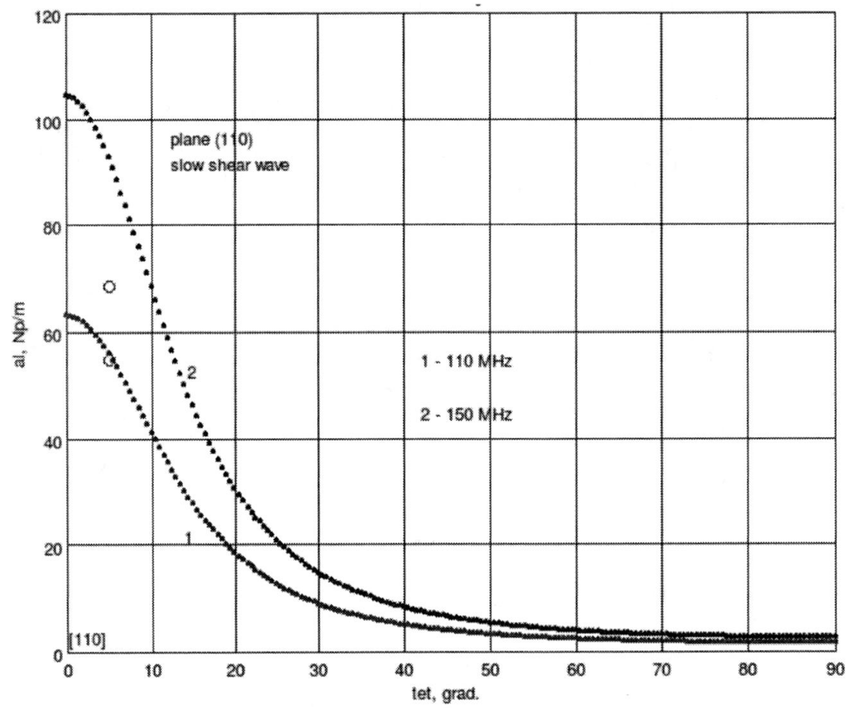

Figure 3. The coefficient of attenuation with the angle; $F_0 = 110$ MHz, 150 MHz.

It can be seen from figure 3 that the change in the propagation direction allows to obtain a significant decrease in attenuation and, as a consequence, the possibility of increasing the operating aperture and increasing selectivity. On this basis, an acousto-optic filter providing the necessary spectral resolution was manufactured and tested with a sealed cuvette filled with nitrogen dioxide at a pressure of 7.5 mm Hg. The measurements were carried out for a long time and showed a high signal level with the introduced cuvette and a low noise level when measuring without the cuvette. In this case, the use of one absorption frequency causes a decrease in the recorded signal by two orders of magnitude.

3. Discussion of results and conclusions

The proposed scheme can be used to detect and measure the concentrations of gas mixtures in particular for monitoring atmospheric pollution, as it combines the reliability and large dynamic range of the optical absorption method with the selectivity determined by adjusting to the fine structure of the spectrum of the measured gas. The range of measured gases is determined by the operating range of the acousto-optic filter, within which the tuning from the spectrum of one gas to another is made easily and quickly by the tuning of the set of frequencies of the control generators.

Acknowledgements

This work was supported by the Ministry of Science and Higher Education within the State assignment FSRC «Crystallography and Photonics» RAS and RFBR grant 15-12-00030 and RFBR grant 16-29-14053 ofi_m.

References

[1] Zaidel' A N, Ostrovskaya G V and Ostrovskii Yu I 1976 *Technique and Practice of Spectroscopy* (Moscow: Nauka)

[2] Kolesnikov S A, Kolesnikova E S, Pisarevsky Yu V and Turutin Yu A 2011 Application of interference polarization filters for the optical absorption method of gas analysis *Crystallography Reports* **56** 1 52–6

[3] Pisarevsky Yu V, Kolesnikov S A, Kolesnikova E S, Turutin Yu A, Konopelko L A and Shor N B 2016 Sensitivity of the optical absorption method based on an instrumental selection out of Fourier components of the absorption spectrum *Doklady Physics* **61** 4 160-3

[4] Pisarevsky Yu V, Kolesnikov S A, Kolesnikova E S, Turutin Yu A, Konopelko L A and Shor N B 2016 Selectivity of the optical absorption method based on an instrumental pick out of Fourier components in the absorption spectrum *Doklady Physics* **61** 6 253–6

[5] Pustovoit V I and Pozhar V E Controlling the characteristics of a collinear acousto-optic filter by modeling ultrasound *J. Commun. Technol. Electron.* **43** 2 121-7

[6] Proklov V V, Byshevski-Konopko O A and Grigorievski V I 2013 *Commun. Technol. Electron.* **58** 9 905

[7] Mazur M M, Pozhar V E and Shorin V N 1998 *Journ. Analytical Chemistry* **53** 9 876-8

[8] Vainer A V, Proklov V V, Rezvov Y G and Sivkova O D 2017 *J. Commun. Technol. Electron.* **62** 1152-5

[9] Fadeyev A V, Pozhar V E and Pustovoit V I 2013 The Principle of fragmentary spectrum registration for acousto-opical spectrometers based on differential optical absorption spectroscopy *SPIE Proceedings* **8890**

[10] Proklov V V and Rezvov Y G 2018 Features of formation of multiband transmission function based on multi-frequency acousto-optic diffraction *Opt. Spectrosc.* **124** 121

[11] Byshevski-Konopko O A and Grigorievski V I 2013 On the possibility of developing incoherent fibre-optic data transmission systems based on signal spectral coding with matched acousto-optical filters *Quantum electron* **43** 542–5

[12] Proklov V V, Byshevski-Konopko O A and Filatov A L 2015 On generation of spectral-coded signals in incoherent optical communication systems based on multiband acousto-optic filters. *Tech. Phys. Lett.* **41** 987

[13] Proklov V V, Byshevski-Konopko O A, Filatov A L, Lugovskoi A V and Pisarevsky Yu V 2016 Experimental study of the use of multiband acousto-optic filters for spectral encoding/decoding the optical signals *Journal of Physics: Conference Series* **737** 012060

Journal of Physics: Conference Series **1421** (2019) 012070 doi:10.1088/1742-6596/1421/1/012070

Lasing in slurry-like active medium on LiF:F_2 microparticles

O A Burdukova[1], V A Konyushkin[2], V A Petukhov[1], Yu V Senatsky[1] and P G Zverev[2*]

[1] P.N. Lebedev Physics Institute of RAS, 53 Leninskiy Prospekt, Moscow, 119991, Russia

[2] A.M. Prokhorov General Physics Institute of RAS, 38 Vavilov str., Moscow, 119991, Russia

[*]E-mail: zverev@lst.gpi.ru

Abstract. Active media composed of the particles of solid-state laser-active material immersed in the liquid are the subject of research since this approach can provide new solutions to the problem of heat removal from the laser medium. Investigation of the active medium on microparticles of LiF crystal with F_2 color centers is presented. The technique of the preparation and characterization of the index-matched slurry-like mixture was developed. Powders of LiF crystal particles with F_2 color centers with size from 50 to 1000 μm immersed in various liquids and their mixtures were tested. Laser radiation of a compact cavity containing two flat dichroic mirrors and a glass cuvette with the LiF:F_2 slurry with grain size of 500 ÷ 1000 μm and the mixture of heptane and vaseline oil was obtained under Coumarin 120 dye laser pumping. The pulses of the laser radiation with the beam divergence of 2 ÷ 5 mrad at the wavelength of about 680 nm, duration of 10 ns and energy of 83 μJ were obtained under pulsed pumping with the energy of 1.34 mJ. The real and slope efficiencies of the LiF:F_2 slurry laser reached 6.2% and 12.9% correspondingly.

1. Introduction

Active media based on particles of solid-state laser materials distributed in liquid or gas media have been the object of research, since such an approach can provide a new solution to the problem of the heat removal during the laser operation [1-6]. Suspensions of glass or crystal nanoparticles doped with Nd or Yb ions were considered as laser active media as the scattering losses from particles with the size of about 10 nm are negligible [4-6]. However nanoparticles may have some properties (emission lifetime, quantum yield) different from their bulk counterparts and so for laser action a surface modification of Nd_2O_3 nanoparticles was required [4, 5]. Particles of solid-state laser materials with dimensions over 1 μm are easier to manufacture, they do not need a complicated modification, as their properties are identical to those in the original single crystal or bulk glass samples. For such particles index-matching liquids may be used to suppress Fresnel reflection and scattering. A "slurry laser" with an active medium in the form of a mixture of LiF microparticles with F_2 color centers and an immersion liquid was demonstrated in [7]. The “slurry laser” concept may be attractive due to the possibility of receiving potentially higher laser output energy from a dense pulp, circulating through the laser active element. This report considers the method of fabrication of an active medium in the form of an index-matched slurry and testing of compact LiF:F_2 slurry laser.

2. Fabrication of a slurry-like active medium

Slurry-like laser is possible only with an active medium with high gain, because it is necessary to overcome the internal parasitic losses in the mixture of solid particles and liquid. Incomplete immersion of particles as well as various scattering centers at the surface and inside the particles can be the source of parasitic losses. LiF crystal with F_2 color centers (CC) is one of the most promising materials for preparing an active medium for the slurry laser. LiF:F_2 CC crystal have broad absorption (0.42-0.47 μm) and emission (0.56-0.84 μm) bands in the visible spectral region. F_2 CCs have high emission and absorption cross sections $\sigma \approx 7 \cdot 10^{-17}$ cm^2. Their intrinsic feature is very high concentration in LiF crystal which results in high gain up to tens of cm^{-1} [8-11]. LiF crystal has high thermal conductivity (4 Wm^{-1}K^{-1}) and can be used in cw and pulse-periodic lasers. Its low refractive index $n_d \approx 1.39$ simplifies search and selection the immersion among various liquids.

F_2 CCs with the concentration of 10^{18} cm^{-3} were produced in the nominally pure oxygen-free samples of LiF crystal by irradiation at room temperature with γ-quanta from Co^{60} source with the doze of $2 \cdot 10^7$ Rad. The grinded LiF:F_2 powder was sieved serially through the sieves with different mesh size for separation of the fractions from 200 μm to 1000 μm and over 1000 μm. Several pieces of transparent non-irradiated LiF crystal samples were subjected to grinding and fractionation by the similar procedure for comparison. Microphotographs of non-irradiated LiF powder fractions obtained using mortar with pestle and then sieved through 200-500 μm and > 1000 μm cells are shown in figure 1. Fractions of crushed material were placed and poured with the immersion liquid into rectangular glass cuvettes with dimensions of 24 × 40 × 8 mm^3 and 3 mm gap between the walls (figure 2).

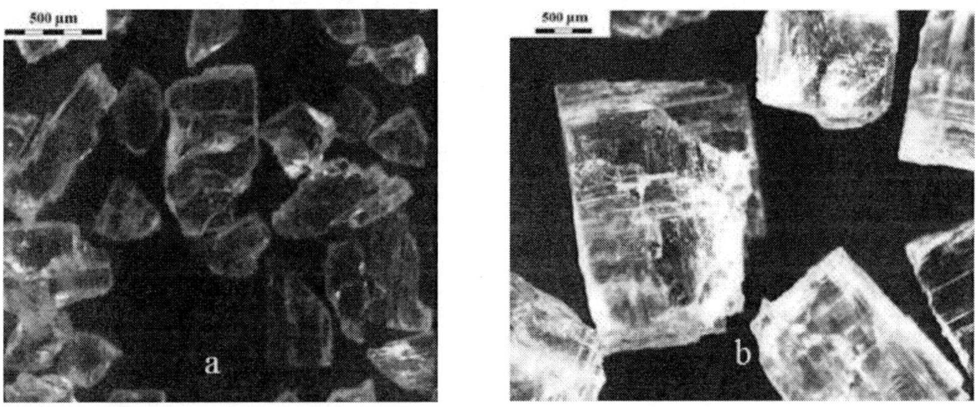

Figure 1. Microphotographs of fractions of non-irradiated LiF crystal powder sieved through 200-500 μm (a) and > 1000 μm (b) cells.

Figure 2. Photo of the glass cuvette containing the 3 mm thick slurry of 500 - 1000 μm LiF:F_2 granules in the immersion liquid heptane/vaseline oil.

The values of the refractive index of irradiated and non-irradiated LiF crystal are practically the same and equal to 1.3909 at 656.3 nm. The refractive indexes of common liquid chemicals are presented in table 1. All these liquids are transparent in the visible and near IR spectral range (400 - 1000 nm). One can see that there is no liquid with refractive index that equals to that in LiF. However, it is possible to use the mixture of two liquids with higher and lower refractive indexes to fit it. In this case the value of the particle refractive index n_p equals to

$$n_p = x \cdot n_1 + (1 - x) \cdot n_2 \tag{1}$$

where n_1 and n_2 are refractive indexes of the first and second liquids, x is the mixing fraction. We had the following requirements for the immersion liquid: complete coincidence of its refractive index and LiF crystal, high hydrophilicity with LiF, insolubility of LiF in the liquid, high solubility of one component of immersion in another component, safety and non-toxic.

The index-matching liquid was prepared by mixing two components: the first one with higher refractive index was anise oil or vaseline oil and the second one with the lower index was acetone or heptane. Following equation (1) the mixture anise oil/heptane, vaseline oil/heptane and vaseline oil/acetone should be prepared as 16:1, 6:1 and 6:2.5, correspondingly. It is necessary to note that refractive index depends on the operating wavelength, temperature, on the manufacturer. That's why the preparation of the index-matched slurry was carried out using the special laser set-up shown in figure 3. It used a diode and a He–Ne lasers as reference light sources operating at the wavelengths of 650 nm and 632 nm, that was close to the expected operating wavelength of LiF:F_2 CC laser. Their beam diameters were about 2 mm. To measure the power transmitted through a cuvette with slurry, a power meter was placed behind an iris aperture near the focus of the lens with a focal length of 85 mm. Liquids were added to the layer of powder (≈ 1 cm in height) in a cuvette using a syringe, drop-wise with stirring, so that a buffer layer of liquid was formed above the slurry.

Table 1. Refractive indexes of LiF and $YLiF_4$ crystals and investigated liquids.

Material	Refractive index (656.3 nm)
Methanol	1.3241
Acetone	1.3567
Ethanol	1.3606
Heptane	1.3878
LiF	1.3909
Isobutyl alcohol	1.3959
Ethylene glycol	1.4283
Vaseline oil	1.45-1.46
$YLiF_4$	1.4523 (o), 1.4748 (e)
Castor oil	1.47-1.48
Glycerine	1.4717
Benzene	1.4966
Benzyl alcohol	1.5396
Anise oil	1.54-1.57

The optimum composition of the immersion liquid was achieved by recording the maximum of the reference beam transmission through the slurry and by the observations of the quality of transmitted laser beam spot at a remote screen. While adding the immersion liquid to the powder, a noticeable decrease of light scattering of the reference beam, the formation of a thin light channel in the slurry and a compact beam spot on the screen were observed. At the optimal composition, the transmission of reference beam through the slurry of a fixed thickness depended on the size of granules. Reducing the thickness of the layer of particles in the cuvette, as well as the use of particles of larger size (500 ÷ 1000 μm and above) contribute to the reduction of the scattering in the medium. For 500 ÷ 1000 μm LiF or $LiF:F_2$ grains the fraction of the transmitted laser radiation in the angle of 0.01 rad reached 70%. For the particles of smaller sizes (200 ÷ 500 μm and < 200 μm), the transmission reduced to 30% ÷ 60%. All these observations correspond to the characteristics of the Christiansen filter – a well-known device in the form of a cuvette with a powder of passive (non-activated) particles of glass or crystal in immersion [12, 13].

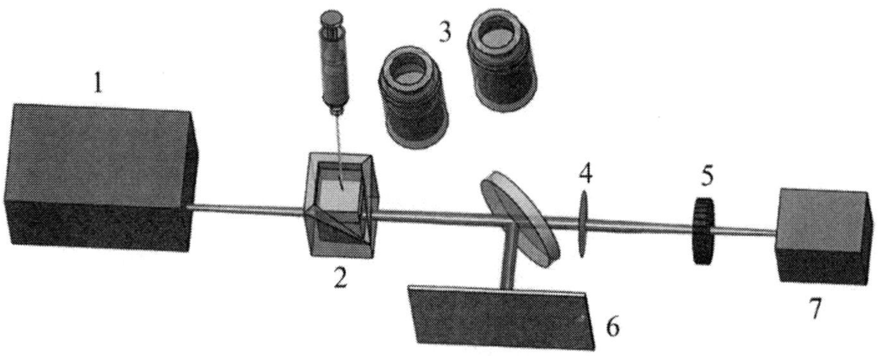

Figure 3. Scheme of the laser set-up for the preparation and testing of the index-matched $LiF:F_2$ slurry, where 1 – laser sources; 2 – cuvette with the slurry; 3 – components for preparing immersion liquids; 4 – lens; 5 – diaphragm; 6 – screen;7 – power meter.

Note that reduction in scattering losses in a cuvette is possible for particles of isotropic materials (glass, cubic crystals etc.). This circumstance is illustrated in figure 4, which shows the pictures of the spot of the He-Ne laser passed through the cuvette of 3 mm thickness containing a powder of the isotropic non-irradiated LiF crystal in an immersion mixture of heptane and vaseline oil and a powder from the anisotropic Nd:YLF crystal in an immersion mixture of isobutyl and benzyl alcohols. The composition of the mixture was chosen for the highest power transmission of the slurry. Comparison of beam profiles on the screen indicates extremely high scattering losses for the cuvette with Nd:YLF powder. The reason of these losses is the fundamental impossibility to fit the refractive index of the liquid to values of ordinary and extraordinary refraction indexes simultaneously.

The transmission spectra of $LiF:F_2$ slurries with 500 ÷ 1000 μm and 200 ÷ 500 μm granules are similar (in the region 300 ÷ 900 nm) to the spectrum of $LiF:F_2$ bulk plate (figure 5).The slurry spectra also reveal the absorption peak of the immersion liquid near 1200 nm. Comparing these spectra with the spectrum of the immersion liquid it was possible to estimate the particles' filling factor in the slurry which was found to be about 0.5.

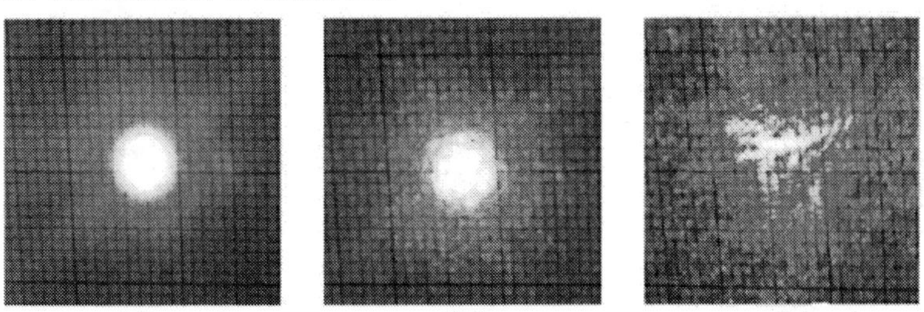

(a) (b) (c)

Figure 4. Images of the He-Ne laser beam on the screen at a distance of 1 m from the stage with cuvettes: without the cuvette (a); cuvette with powder of isotropic LiF crystal (b); cuvette with powder of anisotropic Nd:YLF crystal (c). The size of the grains was 500 ÷ 1000 μm in both cases.

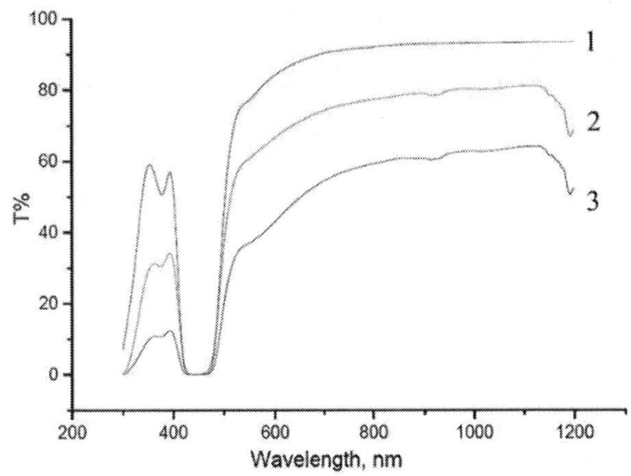

Figure 5. Transmission spectra of $LiF:F_2$ bulk crystal 2.5 mm thick (1) and cuvettes 3 mm thick containing the slurry of $LiF:F_2$ crystal granules with dimensions of 500 ÷ 1000 µm (2) and 200 ÷ 500 µm (3) in the immersion liquid of heptane/vaseline oil.

3. $LiF:F_2$ slurry-like laser

The scheme of the experiment on $LiF:F_2$ slurry laser is shown in figure 6. The transversely pumped Coumarin 120 (C120) dye laser operating at the wavelength of about 0.45 µm was used as a pump source. The windows of the cuvette with the dye solution in the ethanol formed a resonator with the length of 30 mm. Windows were composed of two flat mirrors: one with high reflectivity $R \approx 99\%$ and an output coupler with $R \approx 30\%$ at $\lambda_p \approx 0.45$ µm. The C120 dye was pumped by the 3rd harmonic (354 nm) of a pulsed Nd:YAG laser LOTIS TII. The energy in a single 12 ns pulse of the 3rd harmonic was 8.7 mJ. Linear polarized 3rd harmonic radiation was focused through the side wall of the cuvette into the dye solution by a cylindrical lens, $f = 35$ mm. The length of the dye solution along the resonator axis was 8 mm. The C120 dye laser emitted linearly polarized pulses with the duration of 11 ns and energy up to 1.6 mJ at the wavelength of $\lambda_p = 446$ nm. The diameter of the output beam was 2 mm. This radiation was focused on the samples ($LiF:F_2$ plate or a cuvette with the $LiF:F_2$ slurry) by the lens $f = 172$ mm. The $LiF:F_2$ sample was placed into the resonator ($L \approx 40$ mm) formed by two flat mirrors. The diameter of the pump spot at the entrance of the active media was 0.5 mm, and the maximum pump intensity reached 60 $MW{\cdot}cm^{-2}$. The dye laser radiation was focused along the resonator axis through the input mirror with 90% transmission at λ_p and 99% reflection in the red spectral region. The pump radiation was absorbed almost completely (> 98%) in the $LiF:F_2$ active media. Mirrors with transmission from $T = 1\%$ to $T = 23\%$ at the wavelength near 0.68 µm were used as an output coupler (OC). The pulses of red collimated radiation were produced at the output of the resonator containing $LiF:F_2$ plates or cuvettes with $LiF:F_2$ granules. It is known that the parameters of $LiF:F_2$ laser under intensive pumping may degrade due to ionization of the F_2 active color centers in a two photon excitation process, $F_2 \rightarrow F_2^+$ [8].

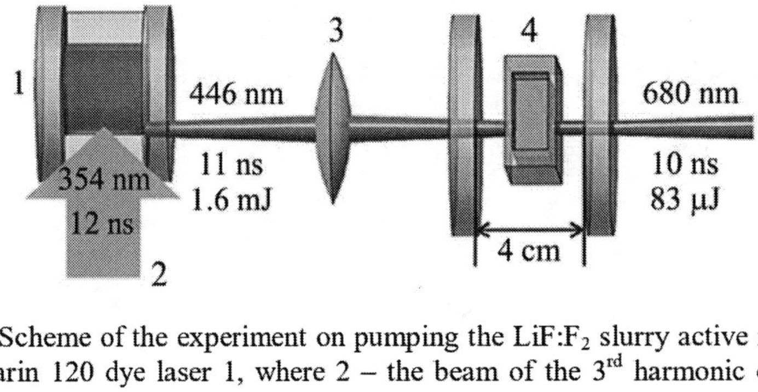

Figure 6. Scheme of the experiment on pumping the $LiF:F_2$ slurry active medium by the Coumarin 120 dye laser 1, where 2 – the beam of the 3rd harmonic of a pulsed Nd:YAG laser; 3 – focusing lens; 4 – cuvette with the $LiF:F_2$ slurry.

This process was observed at high pump power levels in our experiments as well. At 1 Hz pulse repetition rate and pump energy close to the generation threshold of 0.5-0.6 mJ per pulse, the LiF:F_2 laser output energy reduced by a factor of 2 after 3-4 shots for single crystal sample and after 6-7 shots for the slurry without stirring. At the maximum pump pulse energy of 1.4 mJ the generation degrades just after the first shot. Because of this, the measurements of LiF:F_2 laser generation parameters were carried out in a single shot regime. In the study of LiF:F_2 slurry, after each shot the medium in the cuvette was stirred to ensure the presence of fresh active granules in the pumped region. In the experiments with bulk plates, each new measurement was recorded at a neighbor place by shifting the active element.

Measurements of the pulse energy from all types of LiF:F_2 active elements were performed using a calibrated silicon photodiode FD24 connected to a digital oscilloscope. Figure 7 compares the dependences of the LiF:F_2 slurry laser output energy (for 2 compositions) and that of the LiF:F_2 single crystal laser on the pump. All measurements were carried out in the same resonator configuration with the OC with the transmission of 13%. From the graphs in figure 7 one can see that the efficiency of the slurry laser is only twice as less than that of single crystal laser. When comparing these data, one should take into account 0.5 filling factor of granules in the slurry and, also, the fact that the pump radiation at λ_p was noticeably scattered in the slurry, since the immersion in the cuvettes was balanced for the laser oscillation wavelength λ_e. The maximum generation energy of slurry laser was about 83 μJ at the pump energy of 1.34 mJ, the slope efficiency was 12.9%. It was found that the lasing threshold is smaller and the laser output energy is higher for 500 ÷ 1000 μm granules as compared with 200 ÷ 500 μm granules. For the cuvettes with the slurry of 2 mm thickness, the laser output energy was weaker compared to the 3 mm thick cuvettes. For particles < 200 μm the lasing threshold was not achieved, even for the OC with the reflectivity of 99%.

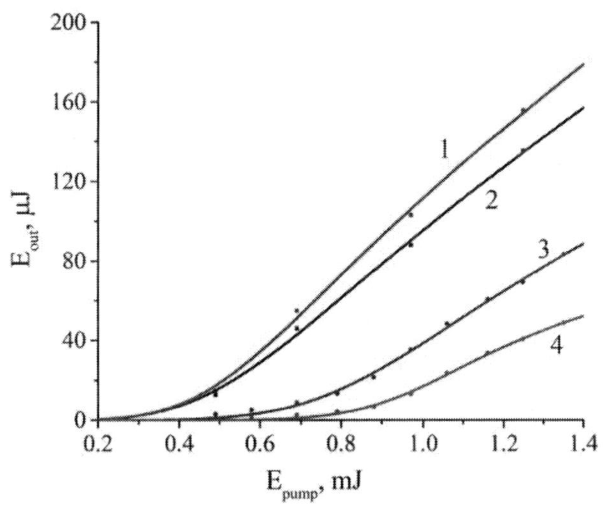

Figure 7. The dependences of the LiF:F_2 laser output pulse energies on the pump pulse energy: LiF:F_2 bulk plates of 3.8 mm (1) or 2.5 mm (2) thickness in the resonator; LiF:F_2 slurry with 500 ÷ 1000 μm (3) or 200 ÷ 500 μm (4) granules in the resonator.

Time profiles of LiF:F_2 slurry laser pulses were recorded with a GW Instek GDS-2104 digital oscilloscope (100 MHz band, 1 $GS \cdot s^{-1}$) using a fast MRD500 photodiode. Pulse duration of LiF:F_2 slurry-like laser was measured to be 10 ns. A diffraction spectrograph with a dispersion of 2.4 $nm \cdot mm^{-1}$ was used to measure the spectrum of the LiF:F_2 slurry laser. Due to nonselective cavity, the central wavelength of the generation spectrum may jump from shot to shot in the region 675 - 695 nm. The spectral width of the slurry laser output may extend for about 60 nm.

The intensity distribution in the LiF:F_2 slurry output beam was usually not reproduced from shot to shot. This can be apparently explained by the disordered internal structure of the active medium based on particles of various dimensions and shapes, which somehow manifest themselves due to unavoidable particles and liquid small index-mismatch. Figure 8 illustrates the beam spot in the near field at the 6 cm distance from the OC and in the far field (in the focus of the lens with the focal length of 1 m) and intensity profiles. The most symmetrical and homogeneous beam profiles were obtained when a 'fresh' place of the slurry was pumped near the generation threshold. Beam divergence of LiF:F_2 slurry laser near the generation threshold was estimated to be about 2 mrad and increased up to 5 mrad at higher pump energies. The degree of linear polarization of the LiF:F_2 slurry radiation was about 75%.

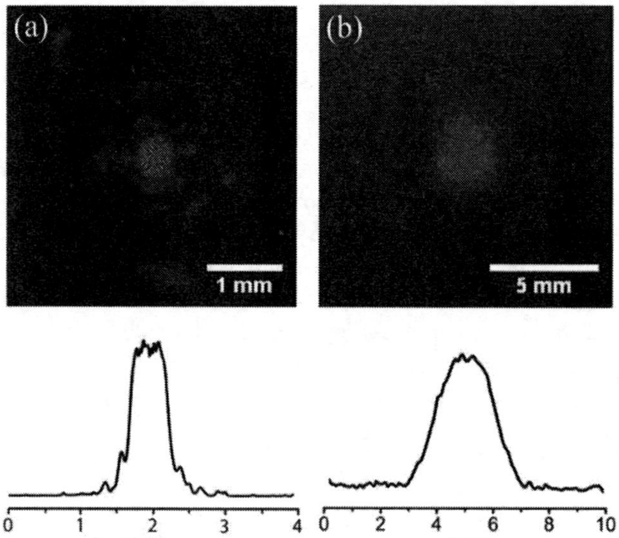

Figure 8. Pictures of LiF:F_2 slurry laser beam spots (up) and their intensity profiles (below): (a) in the near field at the distance of 6 cm from the OC; (b) in the far field at the focus of the lens with the focal length of 1 m.

4. Conclusion

The observed collimated beams of LiF:F_2 slurry emission, the dependences of the beam energy on pump energy and other registered parameters of LiF:F_2 emission give grounds for the conclusion that the lasing action from an active medium based on microparticles from a solid-state laser material in an immersion liquid (a slurry laser) was demonstrated in our experiments. Further research is needed to improve the slurry laser characteristics. One of the important research objectives of such an active medium is the search for the solution to the problem of heat removal from the laser active medium by using a flow of dispersed laser material through the pumped region. In addition to the perspective in solving the problem of heat removal, the use of an active medium on the particles has a number of other advantages, such as an easy replacement of the active medium in case of damage, a possibility of combining several activators in one laser element, new capabilities for inversion profiling, new pumping schemes and some others. The continuation of the study into the generation and amplification of cw and pulsed laser radiation in active media of this type at LD and laser pumping seems to be reasonable. The technique of preparation and characterization of index-matched slurry containing solid-state particles and an immersion liquid was tested not only for fluoride crystals, but for laser glasses as well. It is of interest to conduct experiments with the slurry active media based on standardized particles, for example, those on glass beads highly doped by Nd or Yb ions.

Acknowledgment
This study is supported by RFBR under the grant No. 18-02-00285.

References

[1] Stefanov V J 1989 Laser with two-phase active medium *Infrared Phys.* **29** 381–4
[2] Perry M 2003 *US Patent* № 0161364 A1
[3] Morgan R D, Keto J W and Ditmire T 2011 Colloidal nanoparticles: a promising gain medium for high average power lasers *J. Opt. Soc. Am.* B **28** 2726
[4] Tzuk Y, Goren C, Raanan D and Strum G 2012 Nanoparticle dispersion laser *Optics Lett.* **37** 939
[5] Tzuk Y, Goren C, Strum G, Greenblatt J and Raanan D 2015 Flashlamp-pumped nanoparticle dispersion laser *Appl. Opt.* **54** 1157
[6] Yang X, Wang H, Yang Z and Xu X 2017 Diode pumped nanoparticle gas laser physics: a preliminary modelling study *Opt. Express* **25** 10574
[7] Burdukova O A, Konyshkin V A, Petukhov V A, Senatsky Yu V and Zverev P G 2018 Laser with the slurry active medium *Laser Phys. Lett.* **15** 095805.
[8] Basiev T T, Mirov S B and Osiko V V 1988 Room-temperature color center lasers *IEEE J. Quantum. Electron.* **24** 1052
[9] Mollenauer L F 1987 *Tunable Lasers* ed L F Mollenauer et al (New York: Springer) p 225
[10] Basiev T T, Zverev P G and Mirov S B 2004 Color center lasers *Handbook of Laser Technology and Applications* (*Laser Design and Laser Systems* vol 2) ed C Webb *et al* (Bristol: Institute of Physics Publishing) ch B1.8, pp 499–522
[11] Baldacchini G, De Nicola E, Montereali R M, Scacco A and Kalinov V 2000 Optical bands of F2 and F3+ centers in LiF *J. Phys. Chem. Solids* **61** 21–6
[12] Hoffmann H-J and Steinhart R 1998 Spectral transmittance of Christiansen filters-Experimental observations *Glastech. Ber. Glass Sci. Technol.* **71** 319
[13] Exarhos G J, Ferris K F, Windisch C F Jr, Bozlee B J, Risser S M and Swam S L 2001 Index-mismatch scattering approach to optical limiting *Proc. SPIE* **4347** 432

Estimation of the image quality correction efficiency in optical measurement systems under Scheimpflug condition

O V Pechinskaya

National Research University "Moscow Power Engineering Institute"
Krasnokazarmennaya 14, Moscow, 111250 Russia

E-mail: zhukovaov@mpei.ru

Abstract. A criterion for evaluating the image quality correction efficiency in optical systems under Scheimpflug conditions is proposed. The limits of applicability of the geometric optics approximation are clarified for the sensor tilt angle calculating. The results of a simulation optical stereoscopic system arm for three cases: normal, oblique system not under the Scheimpflug condition and oblique system under the Scheimpflug condition are presented. The quantitative estimation of the image quality correction efficiency is performed. The ranges of object distances and viewing angles, providing the maximum image quality correction without loss of accuracy due to perspective distortion are defined.

1. Introduction

Optical research methods are widely used in scientific and production aims. These are flow visualization systems; measuring systems, for example, stereo-PIV [1, 2], close-range photogrammetry [3], automated surface shape control production systems and many others. The design features of stereoscopic, tomographic and some photogrammetric measurement systems do not allow placing the image recording equipment normally to the optical axis. In such cases tilt and shift lens or Scheimpflug adapters are used to aligning the image plane and the sensor plane in accordance with the Scheimpflug principle [4]. Despite the fact that perspective distortions caused by oblique view are increased due to Scheimpflug condition, these distortions are successfully corrected by the different calibration methods [5-10]. Although the image quality is improved, it will not be the same as in the case of the classical optical system, when the object plane and the image plane are parallel to each other and perpendicular to the optical axis. Typically the measurements error is estimated like the standard deviation between the measured results and model predictions [1, 2, 11]. This estimation does not allow identifying individual error components. The reconstruction quality factor (the quality factor) dependences upon the particle image diameter, particle image number density, number of independent cameras are presented in [1] and there is no information about the Scheimpflug correction influence to the quality factor. In the present article, the quantitative estimation criterion of the Scheimpflug correction effectiveness in optical measurement systems based on the information about the deviation of the size of the circle of confusion of nominal is proposed.

In the classical optical system, when the object plane and the sensor plane are perpendicular to the optical axis, the circle of confusion size is usually minimal for the center of the field of view and

Journal of Physics: Conference Series **1421** (2019) 012071 doi:10.1088/1742-6596/1421/1/012071

increases with increasing field of view. The circle of confusion size distribution in the field of view is symmetrical with respect to the optical axis. In the case of images registration at a certain viewing angle to the optical axis, the distribution ceases to be axisymmetric. For the part of the object plane that is removed from the lens when the optical system is oblique, the size of the scattering spots will be larger than it was for classical images and larger than for the part of the object plane that approaches the lens when the optical system is oblique. The rotation of the sensor plane at a certain viewing angle in accordance with the Scheimpflug principle allows combining image plane and sensor one, and the size of the scattering spots will be smaller than for images captured at a viewing angle not under the Scheimpflug condition. The dispersion of the scattering spot radius on the field of view will also decrease, thus the asymmetry of the scattering circles distribution across the field of view is compensated. Figure 1 shows the spot diagrams for three cases: the classical optical system (a), the oblique system without correction (b) and under the Scheimpflug condition (c).

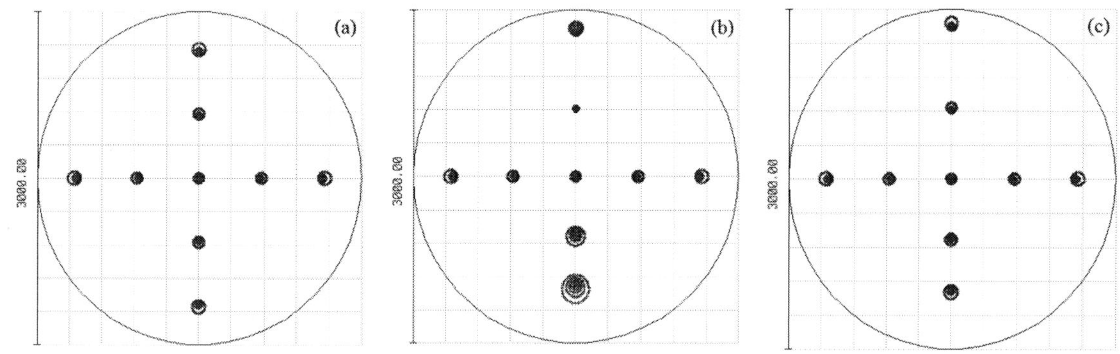

Figure 1. Examples of scattering spots distribution across the field of view: (a) the classical optical system; (b) the oblique system without correction; (c) the oblique system under the Scheimpflug condition.

Figure 2 schematically shows the dependences of the scattering circle size across the field of view. The dashed line corresponds to the images obtained by the optical system normally located to the object plane without tilting the sensor plane (normal images). The red line corresponds to the images obtained by the optical system at a certain viewing angle to the object plane, without tilting the sensor plane (uncorrected images). The green line corresponds to the images obtained by the optical system under the Scheimpflug condition (corrected images). The value $\Delta r(y)$ is the difference between the scattering spots radii of the corrected and normal images for the linear field of view y. The value $\Delta r_0(y)$ is the difference between the scattering spots radii of uncorrected and normal images for the linear field of view y.

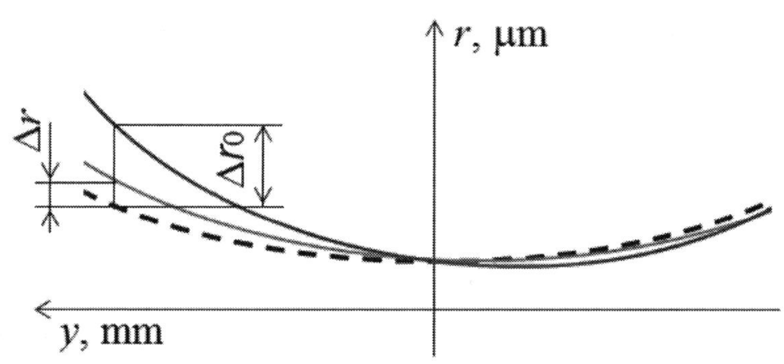

Figure 2. Distributions of scattering spot radii across the field of view for normal, uncorrected and corrected images.

To estimate the image quality correction effectiveness in optical measurement systems under Scheimpflug condition, a criterion can be introduced that contains information on the spot size RMS distribution deviation for uncorrected and corrected images compared to the normal image. Perform the averaging over the pupil

$$\sigma R = \left(\frac{1}{n}\sum_n \Delta r(y)^2\right)^{1/2}, \quad \sigma R_0 = \left(\frac{1}{n}\sum_n \Delta r_0(y)^2\right)^{1/2}, \tag{1}$$

n is number of points, y is linear field of view, σR is the RMS across the field of view of the spot radii difference for corrected and normal images, μm; σR_0 is the RMS across the field of view of the spot radii difference for uncorrected and normal images, μm. The criterion for evaluating the image quality correction effectiveness is determined as

$$\eta = \frac{\sigma R_0 - \sigma R}{\sigma R_0}. \tag{2}$$

If no correction is made $\sigma R = \sigma R_0$ and $\eta = 0$, in case of full correction $\sigma R = 0$ and $\eta = 1$. To determine the dependence of the image correction efficiency on such parameters of the optical measuring system as the relative object distance and the viewing angle, the optical system with oblique image plane was simulated.

2. Simulation of the optical stereoscopic system arm with image plane tilt

A photographic lens with a focal length of 35 mm was chosen for modeling. The aperture diaphragm located between the second and third lenses, the modelling was carried out for the aperture diaphragm size $f'/22$. The simulation was performed for three wavelengths: 450 nm, 550 nm and 650 nm; primary wavelength is 550 nm. Figure 3 shows the optical system layout with tilted image plane. Object distance $A = 350$ mm, it is corresponds to 10 focal lengths.

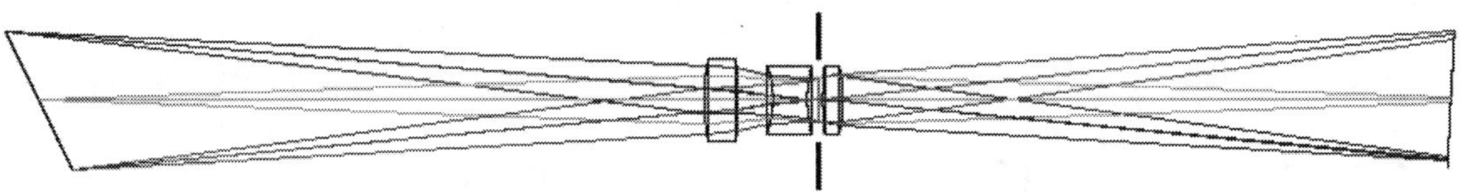

Figure 3. Rays layout for the following optical system parameters: $\theta = 40°$, $\alpha = 5.2°$.

In the course of the preliminary work the limits of applicability of the geometric optics approximation were clarified, and it was shown that the equation [12]

$$\mathrm{tg}\alpha = \frac{f'}{A}\mathrm{tg}\theta, \tag{3}$$

gives a result that is in good agreement with the results of numerical simulation for the values of the relative object distance $A \geq 10 f'$. In the equation (3) are indicates: θ is the viewing angle, α is the angle of the image plane inclination. Figure 4 shows graphs of the image plane inclination on the viewing angle for the values of the object distance $A/f' = 2, 4, 10, 20, 30$.

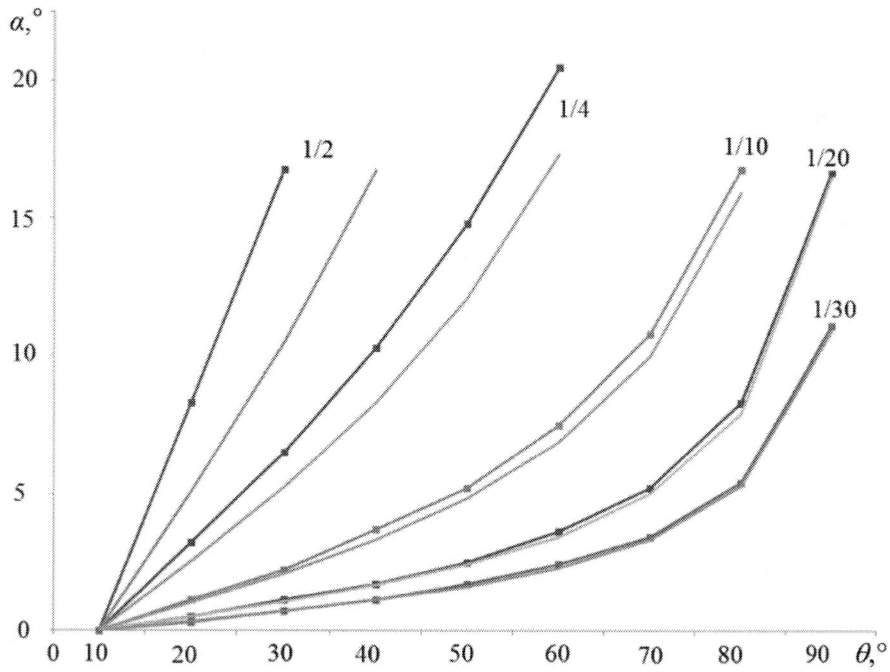

Figure 4. Functions $\alpha(\theta)$ obtained by MTF analysis (curves with markers) and analytically using the equation (3) (curves without markers).

The modulation transfer function of the optical system with tilted image plane analysis showed that each viewing angle corresponds to the image plane tilting angle in which the MTF of the top and bottom pupil zones are the same. Pairs of values θ and α, determined by the coincidence of MTF for two symmetrical relative to the object rotation axis points, are marked in figure 3. It should be noted that for the viewing angles from 0° to 40°, the functions $\alpha(\theta)$ are almost linear. For the object distance over 10 focal ones requires precision of sensor positioning not worse than 1° for every 10° the object plane tilt. For each pair of the viewing angle and the object distance by the equation (3) the corresponding image plane tilting angle was calculated, which provides the maximum image correction for this configuration. The calculation results are shown in table 1.

Table 1. Image plane tilting angle α,°.

Viewing angle θ		10°	20°	30°	40°	50°	60°
Relative object distance	1/30	0.3	0.7	1.1	1.6	2.3	3.3
	1/25	0.4	0.8	1.3	1.9	2.7	3.7
	1/20	0.5	1.0	1.7	2.4	3.4	4.9
	1/15	0.7	1.4	2.2	3.2	4.4	6.6
	1/10	1.0	2.1	3.3	4.8	6.8	9.8

Thus, three series of images were obtained:

1) the object plane and the image plane are perpendicular to the optical axis (normal images);
2) the tilt of object plane is not corrected (uncorrected images);
3) the tilt of object plane is corrected according to Scheimpflug principle (corrected images).

The first and second images series are reference and allow estimating the corrected images deviation from the normal and uncorrected ones at the same object distance. It is also possible to estimate the deviation of the corrected image from the normal, i.e. the image quality reduction. Spot diagrams are shown on the figure 5. In the bottom field displays the RMS scattering spot radii for each field of view (inside the red outline).

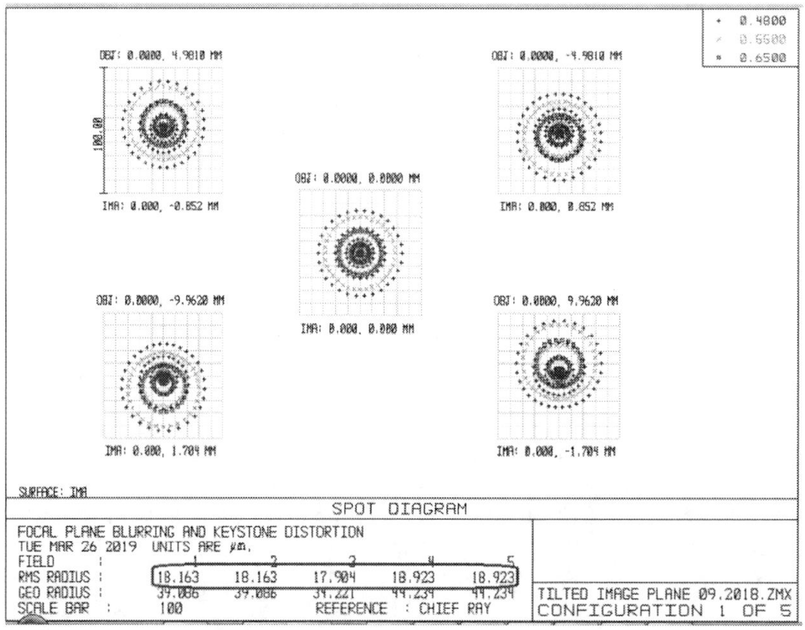

Figure 5. Spot diagrams analysis window.

3. Simulation results

To analyze the dependence of the standard deviation of the scattering spot size of the corrected and normal images on the object distance and the viewing angle by the equation (1) the values σR were calculated and functions graphs were plotted (figures 6 - 7).

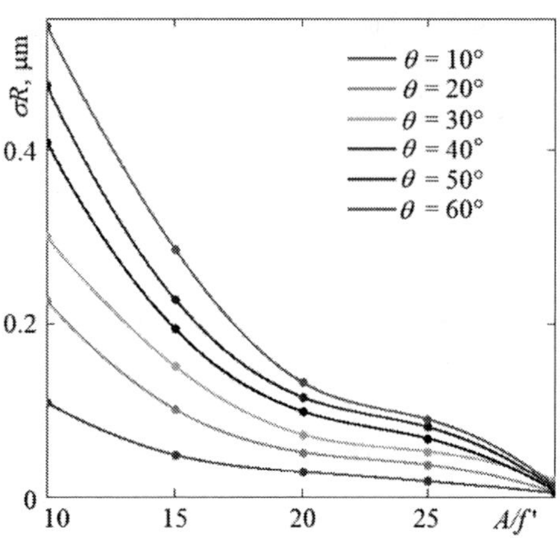

Figure 6. Dependency $\sigma R(A/f')$ graphs for different viewing angles.

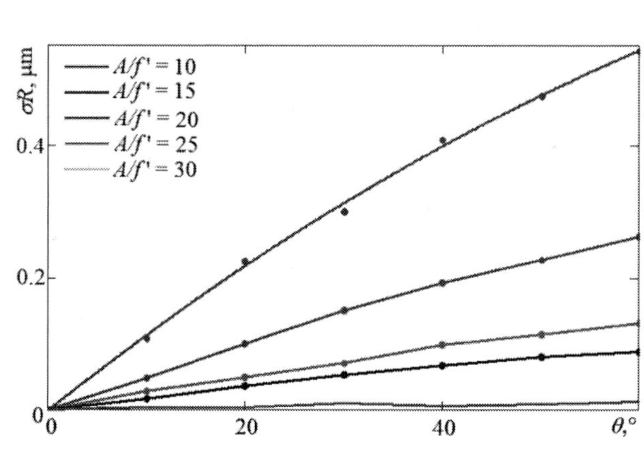

Figure 7. Dependency $\sigma R(\theta)$ graphs for different relative object distance.

It is possible to distinguish the range of values A/f' from 20 to 25, within which the value σR does not change significantly in the entire range of the investigated viewing angles. As the viewing angle increases, the value σR increases for all distance values. Moreover, the smaller object distances, the greater gradient of the function $\sigma R(\theta)$ and its nonlinearity. Using the obtained results, in accordance with the proposed criterion for estimating the image quality correction effectiveness in optical measurement systems under the Scheimpflug condition, the values of the parameter η were calculated and graphs η as a function of the object distances and viewing angles were constructed (figure 8).

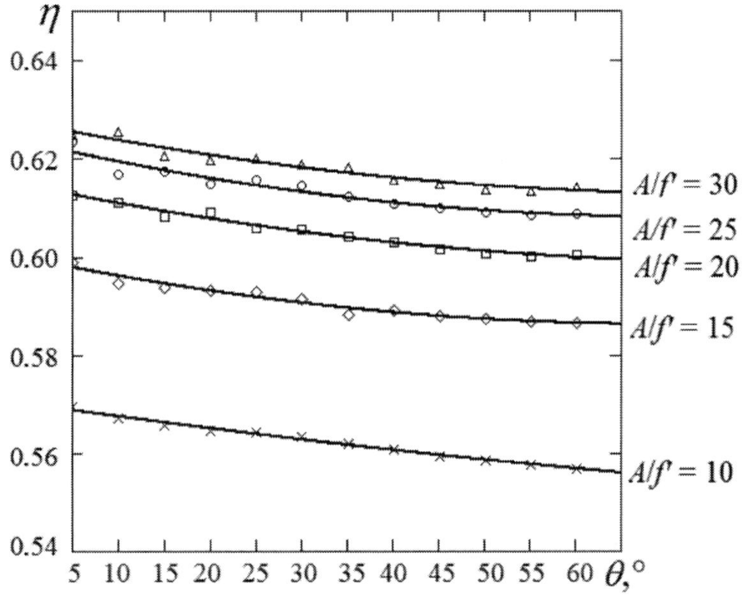

Figure 8. Image correction efficiency as a function of viewing angle.

The correction effectiveness increases nonlinearly with increasing object distance and decreases slightly with increasing viewing angle (not more than 1%). In the manual [12] it is shown that while the viewing angle increases to 45°, the error of the normal velocity component determining by the stereo-PIV method decreases nonlinearly; the viewing angle values more than 45° is also undesirable because of the considerable calibration coefficients inhomogeneity. The accuracy of measurements by optical systems under the Scheimpflug condition is determined by many factors, including the lens and camera parameters. With increasing object distance and viewing angle the resultant error may increase, despite the high correction of optical image quality, by increasing contributions to the overall error of constituents due to other mechanisms.

4. Conclusion

The criterion for the estimating image quality correction in optical measurement systems under the Scheimpflug condition based on the analysis of scattering spot size dynamics is proposed. It is shown that the results of calculating the image plane angle in the approximation of geometric optics and obtained by ray tracing coincide for the object distances greater than 10 focal ones. Using the optical system model with the image tilt is obtained that the Scheimpflug correction efficiency increases from 55% to 63% for the relative distances range from 10 to 30 focal lengths in the range of viewing angles from 10° to 60°. The high correction in which there is no loss of measurement accuracy due to perspective distortions is 61-62% and is achieved at a relative object distance of 20-25 focus ones for the viewing angles is in the range of 45°.

The results are obtained for a computed model in which the inclination angles of the object plane and the image one are precisely matched in accordance with the Scheimpflug principle. In real measurements systems, taking into account the fact that the required accuracy of the image tilt is about 0.1° at the object distances of more than 10 focal lengths, we can expect an image quality correction efficiency decrease due to inaccurate positioning of the sensor plane.

Acknowledgments
The work was supported by the Ministry of higher education and science of the Russian Federation (project 3.8009.2017/BCh).

References

[1] Raffel M, Willert C E, Scarano F, Kähler C, Wereley S T and Kompenhans J 2018 *Particle Image Velocimetry. A Practical Guide* (Springer International Publishing AG, part of Springer Nature) 669
[2] Scarano F 2013 *Meas. Sci. Technol.* **24(1)** 012001
[3] Nocerino E, Menna F, Remondino F, Beraldin J A, Cournoyer L and Reain G 2016 *The Int. Archives of the Photogrammetry, Remote Sensing and Spatial Information Sciences* **XLI-B5** 99–105
[4] Merklinger H M 2010 *Focusing the view camera: A scientific way to focus the view camera and estimate depth of field* Internet edition 1.6.1 (Canada) 146
[5] Fournel T, Louhichi H, Barat C and Menudet J F 2006 *In The 12th Int. Symp. on Flow Visualization (Göttingen Germany)* 1–10
[6] Louhichi H, Fournel T, Lavest J M and Aissia H B 2007 *Meas. Sci. Technol.* **18(8)** 2616– 22
[7] Albers O, Poesch A and Reithmeier E 2015 *Optics Express* **23(23)** 29592
[8] Fasogbon P, Duvieubourg L, Lacaze P A and Macaire L 2015 *Proc. SPIE 9534, Twelfth International Conference on Quality Control by Artificial Vision (Le Creusot France)* 953416
[9] Fasogbon P, Duvieubourg L and Macaire L 2016 *Proc. of Int. Conf. on Computer Vision and Image Processing, Advances in Intelligent Systems and Computing* **1** 159–169
[10] Cong S, Haibo L, Mengna J and Shengyi C 2018 *J. of Sensors* **2018** 3901431
[11] Sutton M A, Orteu J-J and Schreier H W 2009 *Image Correlation for Shape, Motion and Deformation Measurements: Basic Concepts, Theory and Applications* (New York: Springer Science+Business Media) 321
[12] User manual software ActualFlow 2004-2016 LTD SigmaPRO

Spherical surface control by Background Oriented Schlieren method

M M Nyein

V.A. Fabrikant Physics Department, National Research University "Moscow Power Engineering Institute", Krasnokazarmennaya 14, Moscow, 111250 Russia

E-mail: myomyintnyein49@gmail.com

Abstract. The technique of automated quality spherical surfaces control using the background oriented schlieren method was presented. The results of theoretical refractive index gradient modelling for spherical lens were presented. The experimental results of spherical lenses quality diagnostics were shown.

1. Introduction

The quality of the manufactured optical components is subject to strict control [1-3]. Various purity classes are introduced for each type of optical details. Accordingly, a control system is needed either during the production process or at the acceptance stage. The dimensions and number of permissible defects on the surface and in the material of any optical component are usually known. Currently proposed methods are time consuming or require sophisticated special equipment.

The principle of the background oriented schlieren method was proposed in 1999. It allows visualizing the gradient of the investigated object refractive index [4-6]. Any defect on the surface or inside the optical part will be appearing by a change in the optical path of the radiation passing through it. Disturbance of the medium optical homogeneity is expressed in the difference of refractive index n in its volume. Often optical inhomogeneities in media are defects in the structure of a substance or the incorporation of one substance into another, which can be expressed by the presence of a refractive index gradient. Another common defect in optical details is scratches and chips, in the presence of which the optical path of the propagating light changes and a shift in the points of the background screen is also observed. For spherical surfaces and lenses, there may be disturbances in sphericity and a displacement of the center from a given one.

Accordingly, the background oriented schlieren method allows remotely diagnose the presence of defects in optical components at the stage of their production or acceptance. The use of the developed complex will significantly reduce the number of manufactured optical parts with various defects.

For the correct interpretation of experimentally obtained pictures, it is necessary to simulate the pictures of the background oriented schlieren method with all defects kinds and create a database of possible images from them [7, 8].

2. Modeling the effect of lens-like heterogenity

A plane-convex lens with know geometric parameters (curvature radius and thickness) and refractive index n is considered as a model of a constant refractive index gradient.

The beam passage through a flat-convex lens is shown in figure 1. The beam falls parallel to the optical axis at a distance h from the lens center and enters the lens spherical surface. The angle of the beam incidence can be calculated by the following formula:

$$\alpha = 2 \cdot arctg\left(\frac{-\frac{r}{h} \pm \sqrt{\frac{r^2}{h^2}+3}}{3}\right).$$

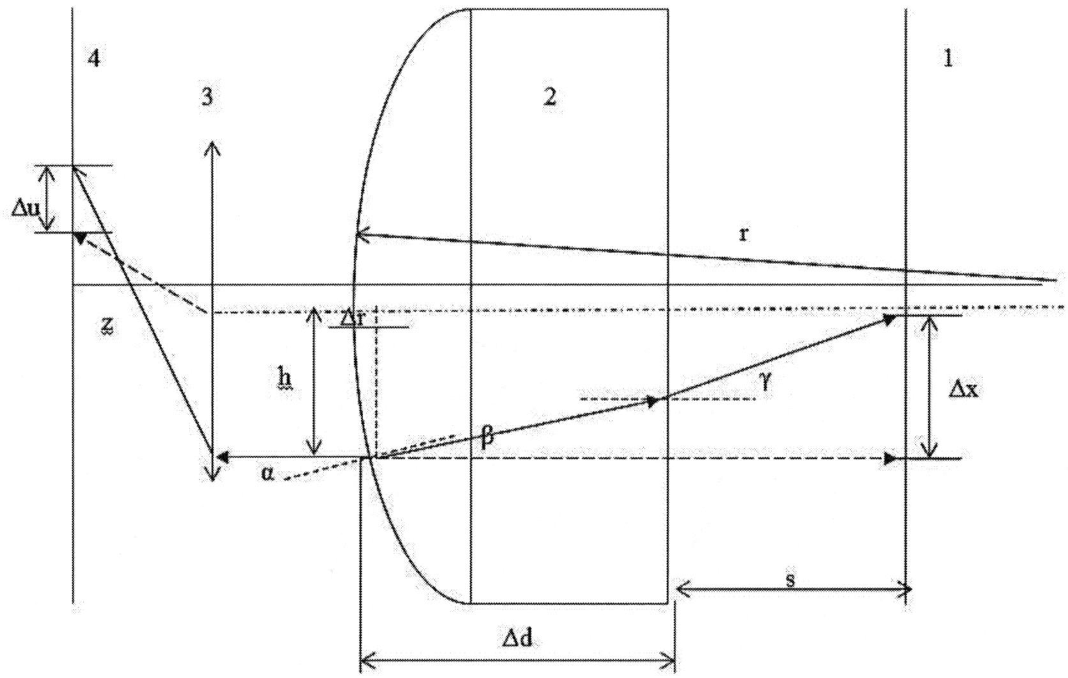

1 – background screen, 2 – plano-convex lens, 3 – lens, 4 – CCD
Figure 1. Screen imaging through a plano-convex lens.

In terms of geometric optics, it is possible to calculate the image displacement of the background pattern points located at a distance s from the lens caused by this lens:

$$\Delta u = M\Delta dtg\left(2arctg\left(\frac{-\frac{r}{h} \pm \sqrt{\frac{r^2}{h^2}+3}}{3}\right) - \arcsin\left(\frac{1}{n}\sin\left(2arctg\left(\frac{-\frac{r}{h} \pm \sqrt{\frac{r^2}{h^2}+3}}{3}\right)\right)\right)\right) + stg\gamma.$$

The results of modeling a lens-like inhomogeneity are presented in figure 2. Using the obtained vector field of displacements, it is possible to clearly determine the center of a spherical lens. Ideal displacement patterns can be obtained for components with any parameters using the developed simulation program. Accordingly, any deviations from these patterns of more than the error of the measuring installation will indicate the presence of defects in the optical details.

Journal of Physics: Conference Series **1421** (2019) 012072 doi:10.1088/1742-6596/1421/1/012072

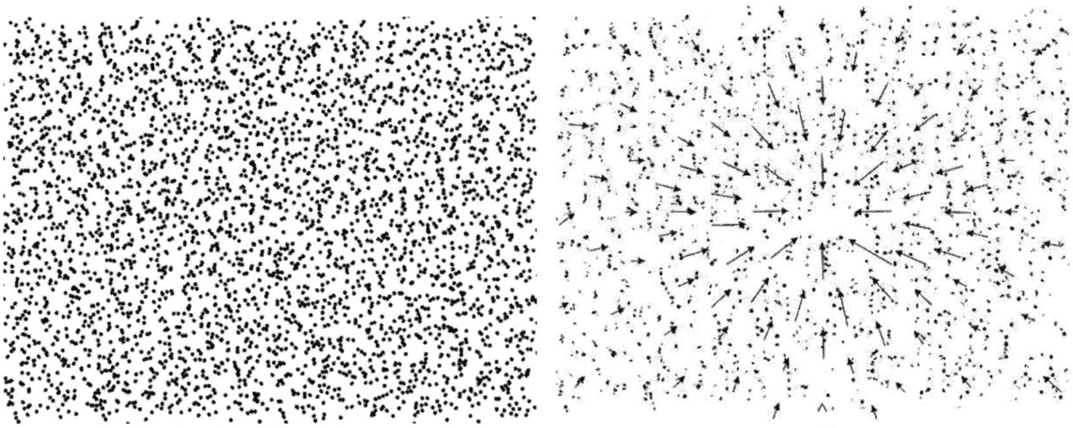

Simulated image Processed image

Figure 2. Background pattern with a lens-like heterogeneity model superimposed on it and a vector field obtained as a result of processing/

3. The results of experimental studies

Experimental studies of spherical lenses set with various defects were carried out to further verify the operability of the proposed method for diagnosing defects in optical details. Image registration is carried out at different distances from the camcorder to the background screen using optical and digital zoom. After that, a couple of images with the best quality were selected.

Figures 3-5 show the results of registering a background screen image with positive lenses and their processing [9]. The vectors show the direction of the refractive index gradient. Their absolute length is proportional to its value. It can be seen from the processing result that figure 3 shows a study of a good quality lens. Figure 4 shows a shift in the center of sphericity from the lens geometric center. Figure 5 visualizes the stresses of the lens material near the frame, however, the central part of the lens has no defects.

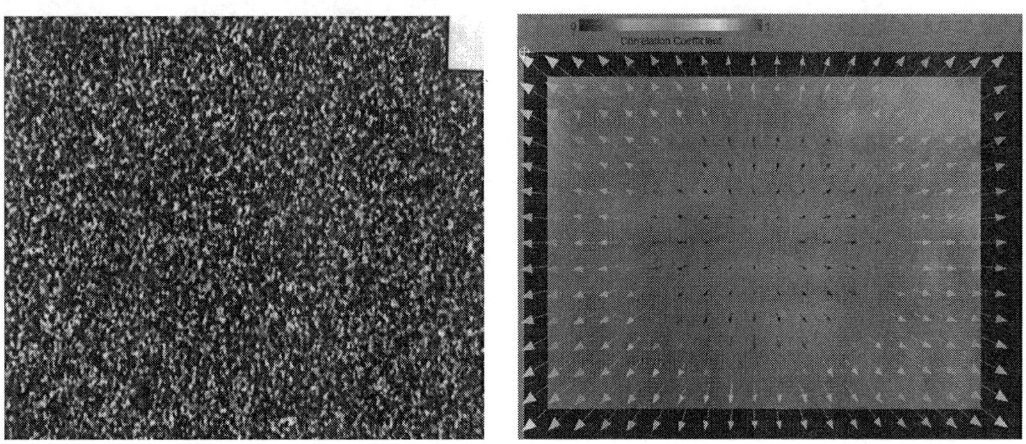

Experimental image Processed image

Figure 3. Experimental image of lens without defects and a vector field obtained as a processing result.

Journal of Physics: Conference Series **1421** (2019) 012072 doi:10.1088/1742-6596/1421/1/012072

Experimental image Processed image

Figure 4. Experimental image and a vector field obtained as a processing result (displacement of the center).

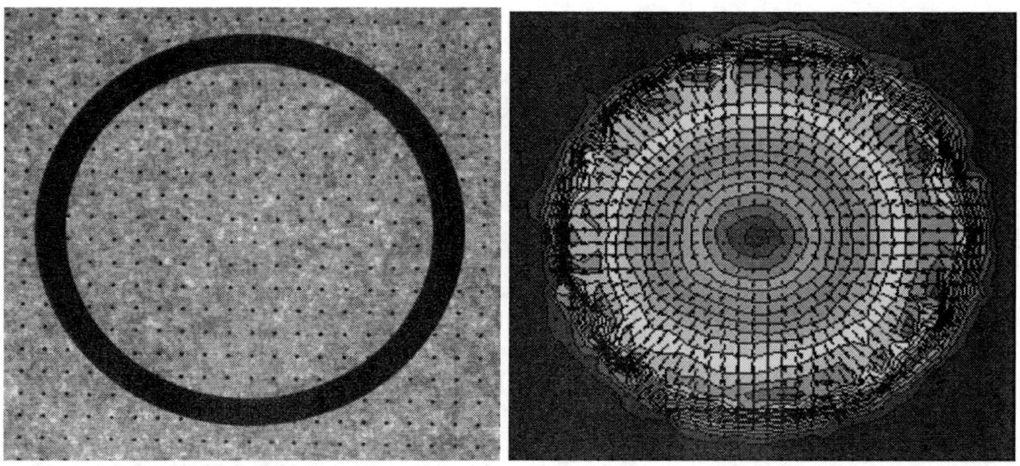

Experimental image Processed image

Figure 5. Experimental image and a vector field obtained as a processing result.

4. Conclusion

The possibility of defects visualization and quantitative characteristics of optical spherical elements quality using the background oriented schlieren method has been shown experimentally and using computer modeling. Also, during the research it was revealed that the detection of various types of defects is affected by the experimental setup parameters and window cross-correlation processing. Using the presented method of contactless diagnostics, it is possible to implement an automated complex for controlling the optical parts quality.

References

[1] Russian State Standards 28869–90 *Optical materials. Methods of measuring the refractive index.* (Moscow: Standartinform Publ.) (In Russian)

[2] Russian State Standards 11200–75 *Objectives and tubes of microscopes. The connecting sizes.* (Moscow: Standartinform Publ.) (In Russian)

[3] Russian State Standards 11141–84 *Optical details. Surface cleanliness classes. Control methods.* (Moscow: Standartinform Publ.) (In Russian)
[4] Meier G E A 2002 Computerized background-oriented schlieren *Experiments in Fluids* **33** 181-7
[5] Skornyakova N M 2009 Application of the shadow background method *Optical methods for the study of flows: proceedings of X Int. scientific and technical conferences* (Moscow: Publishing House of MPEI)
[6] Venkatakrishnan L and Meier G E A 2004 Density measurements using the background oriented schlieren technique *Exp. Fluids* **37** 237-47
[7] Sharma A, Kumar D V and Ghatak A K 1982 Tracing rays through graded–index media: a new method *Appl. Opt.* **21** 984-7
[8] Doric S 1990 Ray tracing through gradient-index media: recent improvements *Appl. Opt.* **29** 4026-9
[9] Raffel M, Willert C, Wereley S and Kompenhans Ju 2007 Particle Image Velocimetry, 2nd edn. (Berlin: Springer)

AUTHOR INDEX

Abdurakipov, S 27, 118
Abramova, O A 347
Afshar, E 266
Anikeev, A V 21, 65, 266
Arbuzov, E V 315
Arbuzov, V A 315
Babenko, D D 428
Badeeva, E 397, 420
Barinov, V Yu 72
Batshev, V I 93, 225, 239, 290
Batyrshin, E S 347
Baybikova, T N 302
Bednyakov, S A 243
Beliaeva, A S 444
Belov, A S 239
Belov, S Yu 180
Belyi, V N 127
Berdnikov, V S 315
Bilsky, A V 175
Bliznyuk, V 412
Blokhin, A A 239
Bogomolov, A B 261
Bokuchava, N I 229
Boukharov, A V 275
Bulatov, K M 208, 405
Bulatov, M F 261, 389
Burdukova, O A 476
Burlakov, A B 10, 334
Burmak, L 370
Butakov, E 27, 118
Bychkov, A 453
Bykhanov, A N 103
Bykov, A A 208, 405
Chelebyan, O G 51, 58
Cherepetskaya, E 453
Chichigin, B 370
Dmitriev, A S 428
Dobroselsky, K G 97
Dolnikov, G G 243
Domanskiy, V L 334
Domoratskiy, E P 302
Dremov, S V 465
Dubnishchev, Yu N 315
Dulin, V 118
Dzlieva, E S 202
Evdokimov, O A 283, 323
Evtikhieva, O 412
Evtikhieva, O A 21
Frolova, E 340
Gadzaov, A F 10
Gavlina, A E 93, 143
Gerashchenko, M 420
Gerashchenko, S 420
Gerasimov, V V 82
Gorelikov, E Yu 465
Gorevoy, A V 225, 290
Gurov, I P 363
Guryanov, A I 323
Guryanova, M M 323
Gusakov, A A 434
Ilin, D V 134
Ilina, N S 219
Inochkin, F M 229
Ivanova, N A 257
Kalinina, K L 323
Kandaurov, A A 356, 383
Karabutov, A 453
Karasev, V Yu 202
Kashuba, I A 375
Kazak, N S 157
Khant, P T 72
Khasanov, I Sh 82
Khitrov, O V 82
Khokhlov, D D 10, 208, 225, 334
Kinzhagulov, I 147
Kislitsin, S A 315
Kokshaiskii, A I 458
Kolesnikov, S A 472
Konyushkin, V A 476
Korotkikh, I I 43
Korsakov, A S 329
Korsakov, V S 375
Korsakova, E A 375
Koshelev, D A 168
Kovalev, A V 175
Krokh, G V 127
Krylov, I R 202
Kuchmenko, A V 43, 168
Kuibin, P A 195
Kukanov, S I 275
Kukushkin, A 397
Kulak, G V 127, 157
Kulakov, S A 261
Kutuza, I B 208, 261
Kutvitsky, V A 261
Kuvshinnikov, A E 16
Kuzmin, V I 10
Kuznetsov, I A 243
Lapitskaya, I A 219
Lapitskiy, K M 219
Lashova, A A 375
Latypova, R R 347
Litvinov, I V 465
Liu, Shuhong 195
Loginova, A A 51, 58
Lomonov, V A 472
Lyash, A N 243
Machikhin, A 370

AUTHOR INDEX

Machikhin, A S 225, 290, 389
Malakhov, Yu I 43
Mantrova, Yu V 405
Margaryants, N B 363
Mashek, I Ch 202
Matveeva, A G 157
Milkov, M G 458
Mineev, B I 51
Mityakov, V Yu 434
Mukhin, D G 465
Murashkina, T 397, 420
Naumov, A A 290
Naumov, I V 188
Neverov, S M 309
Nikilaenko, T V 157
Nikitin, A K 82
Nikitin, P 212
Nikulin, V V 38
Nosov, P A 249
Novikov, D A 143
Novikov, L A 202
Nyein, M M 491
Panenko, R A 38
Parshin, V 412
Parshin, V A 65
Pavlov, I N 76
Pavlov, S I 202
Pechinskaya, O V 484
Petukhov, V A 476
Pisarevsky, Yu V 472
Piskunov, D E 249
Pityuk, Yu A 347
Poroykov, A Yu 219, 243
Potemkin, A V 363
Poyatsyka, D A 219
Pozha, V E 290
Pozhar, V E 389
Proskuryakov, K N 266
Romanova, G E 444
Ropot, P I 127
Ryzhova, T A 82
Sapozhnikov, S Z 434
Sapronov, M V 109
Senatsky, Yu V 476
Sergeev, D A 356, 383
Sergeeva, M V 143
Seroshtanov, V V 434
Shachneva, E 397
Shakhnov, V A 389
Shakin, O V 127, 157
Sharaborin, D 340
Shashkova, I A 243
Shcherbakov, P P 168
Shirokov, S V 10
Shmygalev, A S 329
Shtork, S I 465
Simonova, V 453
Sitnikov, D S 1
Skornyakova, N M 21, 65, 109, 168
Skripkin, S G 195
Stepanova, K 147
Sukharev, V A 458
Suslov, D A 465
Sviridenkov, A A 58
Sychev, D G 161
Thu, Win 134
Tigaev, V O 249
Titov, S A 334
Tokarev, M 118
Tolstoguzov, R 340
Troitskaya, Yu I 356
Tsoy, M A 188, 195
Turabi, A 329
Usmanova, S S 21
Vasilyev, A Y 58
Vasilyev, A Yu 51
Vedyashkina, A V 76
Velikovskii, D Yu 458
Veretennikov, S V 283, 323
Vinogradov, A V 472
Vinogradov, M A 249
Volkov, M V 363
Yablokova, A A 249
Yagodnitsyna, A A 175
Yakovlev, Y 147
Yanina, G M 76
Zaharov, A V 243
Zamula, Yu S 347
Zarubin, V 453
Zhilkin, B P 329
Zhukova, L V 375
Zinin, P V 208, 261, 405
Zolotukhina, O S 315
Zuo, Zhigang 195
Zverev, P G 476
Zykov, A Yu 257
Zykova, L 370